Fluid Mechanics

F. Durst

Fluid Mechanics

An Introduction to the Theory
of Fluid Flows

With 347 Figures and 13 Tables

 Springer

Prof. Dr. Dr. h.c. Franz Durst
FMP Technology GmbH
Am Weichselgarten 34
91058 Erlangen
Germany
f.durst@fmp-technology.com

English Translation:
Ingeborg Arnold
Fliederstrasse 40
66119 Saarbrücken
Germany
arnoldwi@t-online.de

ISBN: 978-3-540-71342-5   e-ISBN: 978-3-540-71343-2

Library of Congress Control Number: 2007937409

© 2008 Springer-Verlag Berlin Heidelberg

This work is subject to copyright. All rights are reserved, whether the whole or part of the material is concerned, specifically the rights of translation, reprinting, reuse of illustrations, recitation, broadcasting, reproduction on microfilm or in any other way, and storage in data banks. Duplication of this publication or parts thereof is permitted only under the provisions of the German Copyright Law of September 9, 1965, in its current version, and permission for use must always be obtained from Springer. Violations are liable to prosecution under the German Copyright Law.

The use of general descriptive names, registered names, trademarks, etc. in this publication does not imply, even in the absence of a specific statement, that such names are exempt from the relevant protective laws and regulations and therefore free for general use.

Cover design: eStudio Calamar S.L., F. Steinen-Broo, Pau/Girona, Spain

Printed on acid-free paper

9 8 7 6 5 4 3 2 1

springer.com

This book is dedicated to my wife Heidi and my sons Bodo André and Heiko Brian and their families

Preface of German Edition

Some readers familiar with fluid mechanics who come across this book may ask themselves why another textbook on the basics of fluid mechanics has been written, in view of the fact that the market in this field seems to be more than saturated. The author is quite conscious of this situation, but he thinks all the same that this book is justified because it covers areas of fluid mechanics which have not yet been discussed in existing texts, or only to some extent, in the way treated here.

When looking at the textbooks available on the market that give an introduction into fluid mechanics, one realizes that there is hardly a text among them that makes use of the entire mathematical knowledge of students and that specifically shows the relationship between the knowledge obtained in lectures on the basics of engineering mechanics or physics and modern fluid mechanics. There has been no effort either to activate this knowledge for educational purposes in fluid mechanics. This book therefore attempts to show specifically the existing relationships between the above fields, and moreover to explain them in a way that is understandable to everybody and making it clear that the motions of fluid elements can be described by the same laws as the movements of solid bodies in engineering mechanics or physics. The tensor representation is used for describing the basic equations, showing the advantages that this offers.

The present book on fluid mechanics makes an attempt to give an introductory structured representation of this special subject, which goes far beyond the potential-theory considerations and the employment of the Bernoulli equation, that often overburden the representations in fluid mechanics textbooks. The time when potential theory and energy considerations, based on the Bernoulli equation, had to be the center of the fluid mechanical education of students is gone. The development of modern measuring and computation techniques, that took place in the last quarter of the 20th century, up to the application level, makes detailed fluid-flow investigations possible nowadays, and for this aim students have to be educated.

Using the basic education obtained in mathematics and physics, the present book strives at an introduction into fluid mechanics in such a way that each chapter is suited to provide the material for a one-week or two-week lectures, depending on the educational and knowledge level of the students. The structure of the book helps students, who want to familiarize themselves with fluid mechanics, to recognize the material which they should study in addition to the lectures to become acquainted, chapter by chapter, with the entire field of fluid mechanics. Moreover, the present text is also suited to study fluid mechanics on one's own. Each chapter is an introduction into a subfield of fluid mechanics. Having acquired the substance of one chapter, it is easier to read more profound books on the same subfield, or to pursue advanced education by reading conference and journal publications.

In the description of the basic and most important fluid characteristic for fluid mechanics, the viscosity, much emphasis is given so that its physical cause is understood clearly. The molecular-caused momentum transport, leading to the τ_{ij}-terms in the basic fluid mechanical equations, is dealt with analogously to the molecular-dependent heat conduction and mass diffusion in fluids. Explaining viscosity by internal "fluid friction" is physically wrong and is therefore not dealt with in this form in the book. This text is meant to contribute so that readers familiarizing themselves with fluid mechanics gain quick access to this special subject through physically correctly presented fluid flows.

The present book is based on the lectures given by the author at the University of Erlangen-Nürnberg as an introduction into fluid mechanics. Many students have contributed greatly to the compilation of this book by referring to unclarified points in the lecture manuscripts. I should like to express my thanks for that. I am also very grateful to the staff of the Fluid Mechanics Chair who supported me in the compilation and final proof-reading of the book and without whom the finalization of the book would not have been possible. My sincere thanks go to Dr.-Ing. C. Bartels, Dipl.-Ing. A. Schneider, Dipl.-Ing. M. Glück for their intense reading of the book. I owe special thanks to Mrs. I.V. Paulus, as without her help the final form of the book would not have come about.

Erlangen, *Franz Durst*
February 2006

Preface of English Edition

Fluid mechanics is a still growing subject, due to its wide application in engineering, science and medicine. This wide interest makes it necessary to have a book available that provides an overall introduction into the subject and covers, at the same time, many of the phenomena that fluid flows show for different boundary conditions. The present book has been written with this aim in mind. It gives an overview of fluid flows that occur in our natural and technical environment. The mathematical and physical background is provided as a sound basis to treat fluid flows. Tensor notation is used, and it is explained as being the best way to express the basic laws that govern fluid motions, i.e. the continuity, the momentum and the energy equations. These equations are derived in the book in a generally applicable manner, taking basic kinematics knowledge of fluid motion into account. Particular attention is given to the derivations of the molecular transport terms for momentum and heat. In this way, the generally formulated momentum equations are turned into the well-known Navier–Stokes equations. These equations are then applied, in a relatively systematic manner, to provide introductions into fields such as hydro- and aerostatics, the theory of similarity and the treatment of engineering flow problems, using the integral form of the basic equations. Potential flows are treated in an introductory way and so are wave motions that occur in fluid flows. The fundamentals of gas dynamics are covered, and the treatment of steady and unsteady viscous flows is described. Low and high Reynolds number flows are treated when they are laminar, but their transition to turbulence is also covered. Particular attention is given to flows that are turbulent, due to their importance in many technical applications. Their statistical treatment receives particular attention, and an introduction into the basics of turbulence modeling is provided. Together with the treatment of numerical methods, the present book provides the reader with a good foundation to understand the wide field of modern fluid mechanics. In the final sections, the treatment of flows with heat transfer is touched upon, and an introduction into fluid-flow measuring techniques is given.

On the above basis, the present book provides, in a systematic manner, introductions to important "subfields of fluid mechanics", such as wave motions, gas dynamics, viscous laminar flows, turbulence, heat transfer, etc. After readers have familiarized themselves with these subjects, they will find it easy to read more advanced and specialized books on each of the treated specialized fields. They will also be prepared to read the vast number of publications available in the literature, documenting the high activity in fluid-flow research that is still taking place these days. Hence the present book is a good introduction into fluid mechanics as a whole, rather than into one of its many subfields.

The present book is a translation of a German edition entitled "Grundlagen der Strömungsmechanik: Eine Einführung in die Theorie der Strömungen von Fluiden". The translation was carried out with the support of Ms. Inge Arnold of Saarbrücken, Germany. Her efforts to publish this book are greatly appreciated. The final proof-reading was carried out by Mr. Phil Weston of Folkestone in England. The author is grateful to Mr. Nishanth Dongari and Mr. Dominik Haspel for all their efforts in finalizing the book. Very supportive help was received in proof-reading different chapters of the book. Especially, the author would like to thank Dr.-Ing. Michael Breuer, Dr. Stefan Becker and Prof. Ashutosh Sharma for reading particular chapters. The finalization of the book was supported by Susanne Braun and Johanna Grasser. Many students at the University of Erlangen-Nürnberg made useful suggestions for corrections and improvements and contributed in this way to the completion of the English version of this book. Last but not least, many thanks need to be given to Ms. Isolina Paulus and Mr. Franz Kaschak. Without their support, the present book would have not been finalized. The author hopes that all these efforts were worthwhile, yielding a book that will find its way into teaching advanced fluid mechanics in engineering and natural science courses at universities.

March 2008 *Franz Durst*

Contents

1 Introduction, Importance and Development of Fluid Mechanics 1
   1.1 Fluid Flows and their Significance 1
   1.2 Sub-Domains of Fluid Mechanics 4
   1.3 Historical Developments 9
   References 14

2 Mathematical Basics 15
   2.1 Introduction and Definitions 15
   2.2 Tensors of Zero Order (Scalars) 16
   2.3 Tensors of First Order (Vectors) 17
   2.4 Tensors of Second Order 21
   2.5 Field Variables and Mathematical Operations 23
   2.6 Substantial Quantities and Substantial Derivative 26
   2.7 Gradient, Divergence, Rotation and Laplace Operators 27
   2.8 Line, Surface and Volume Integrals 29
   2.9 Integral Laws of Stokes and Gauss 31
   2.10 Differential Operators in Curvilinear Orthogonal Coordinates 32
   2.11 Complex Numbers 36
      2.11.1 Axiomatic Introduction to Complex Numbers 37
      2.11.2 Graphical Representation of Complex Numbers 38
      2.11.3 The Gauss Complex Number Plane 39
      2.11.4 Trigonometric Representation 39
      2.11.5 Stereographic Projection 41
      2.11.6 Elementary Function 42
   References 47

3 Physical Basics 49
   3.1 Solids and Fluids 49
   3.2 Molecular Properties and Quantities of Continuum Mechanics 51

	3.3	Transport Processes in Newtonian Fluids	55
		3.3.1 General Considerations	55
		3.3.2 Pressure in Gases	58
		3.3.3 Molecular-Dependent Momentum Transport	62
		3.3.4 Molecular Transport of Heat and Mass in Gases	65
	3.4	Viscosity of Fluids	69
	3.5	Balance Considerations and Conservation Laws	73
	3.6	Thermodynamic Considerations	76
	References	81
4	**Basics of Fluid Kinematics**	83
	4.1	General Considerations	83
	4.2	Substantial Derivatives	84
	4.3	Motion of Fluid Elements	85
		4.3.1 Path Lines of Fluid Elements	86
		4.3.2 Streak Lines of Locally Injected Tracers	90
	4.4	Kinematic Quantities of Flow Fields	94
		4.4.1 Stream Lines of a Velocity Field	94
		4.4.2 Stream Function and Stream Lines of Two-Dimensional Flow Fields	98
		4.4.3 Divergence of a Flow Field	101
	4.5	Translation, Deformation and Rotation of Fluid Elements	104
	4.6	Relative Motions	108
	References	112
5	**Basic Equations of Fluid Mechanics**	113
	5.1	General Considerations	113
	5.2	Mass Conservation (Continuity Equation)	115
	5.3	Newton's Second Law (Momentum Equation)	119
	5.4	The Navier–Stokes Equations	123
	5.5	Mechanical Energy Equation	128
	5.6	Thermal Energy Equation	130
	5.7	Basic Equations in Different Coordinate Systems	135
		5.7.1 Continuity Equation	135
		5.7.2 Navier–Stokes Equations	136
	5.8	Special Forms of the Basic Equations	142
		5.8.1 Transport Equation for Vorticity	143
		5.8.2 The Bernoulli Equation	144
		5.8.3 Crocco Equation	146
		5.8.4 Further Forms of the Energy Equation	147
	5.9	Transport Equation for Chemical Species	150
	References	151

Contents

6 Hydrostatics and Aerostatics 153
  6.1 Hydrostatics 153
  6.2 Connected Containers and Pressure-Measuring
       Instruments 163
       6.2.1 Communicating Containers 163
       6.2.2 Pressure-Measuring Instruments 166
  6.3 Free Fluid Surfaces 168
       6.3.1 Surface Tension 168
       6.3.2 Water Columns in Tubes and Between Plates 172
       6.3.3 Bubble Formation on Nozzles 175
  6.4 Aerostatics 183
       6.4.1 Pressure in the Atmosphere 183
       6.4.2 Rotating Containers 187
       6.4.3 Aerostatic Buoyancy 188
       6.4.4 Conditions for Aerostatics: Stability of Layers 191
  References 192

7 Similarity Theory 193
  7.1 Introduction 193
  7.2 Dimensionless Form of the Differential Equations 197
       7.2.1 General Remarks 197
       7.2.2 Dimensionless Form of the Differential Equations ... 199
       7.2.3 Considerations in the Presence of Geometric
              and Kinematic Similarities 204
       7.2.4 Importance of Viscous Velocity,
              Time and Length Scales 207
  7.3 Dimensional Analysis and π-Theorem 212
  References 219

8 Integral Forms of the Basic Equations 221
  8.1 Integral Form of the Continuity Equation 221
  8.2 Integral Form of the Momentum Equation 224
  8.3 Integral Form of the Mechanical Energy Equation 225
  8.4 Integral Form of the Thermal Energy Equation 228
  8.5 Applications of the Integral Form of the Basic Equations ... 230
       8.5.1 Outflow from Containers 230
       8.5.2 Exit Velocity of a Nozzle 231
       8.5.3 Momentum on a Plane Vertical Plate 232
       8.5.4 Momentum on an Inclined Plane Plate 234
       8.5.5 Jet Deflection by an Edge 236
       8.5.6 Mixing Process in a Pipe
              of Constant Cross-Section 237
       8.5.7 Force on a Turbine Blade in a Viscosity-Free Fluid .. 239
       8.5.8 Force on a Periodical Blade Grid 240

		8.5.9 Euler's Turbine Equation 242
		8.5.10 Power of Flow Machines 245
	References 247

9 Stream Tube Theory 249
    9.1   General Considerations 249
    9.2   Derivations of the Basic Equations 251
           9.2.1 Continuity Equation 251
           9.2.2 Momentum Equation 253
           9.2.3 Bernoulli Equation 254
           9.2.4 The Total Energy Equation 256
    9.3   Incompressible Flows 257
           9.3.1 Hydro-Mechanical Nozzle Flows 257
           9.3.2 Sudden Cross-Sectional Area Extension 258
    9.4   Compressible Flows 260
           9.4.1 Influences of Area Changes on Flows 260
           9.4.2 Pressure-Driven Flows Through
                  Converging Nozzles 263
    References 273

10 Potential Flows 275
    10.1 Potential and Stream Functions 275
    10.2 Potential and Complex Functions 280
    10.3 Uniform Flow 283
    10.4 Corner and Sector Flows 284
    10.5 Source or Sink Flows and Potential Vortex Flow 288
    10.6 Dipole-Generated Flow 291
    10.7 Potential Flow Around a Cylinder....................... 293
    10.8 Flow Around a Cylinder with Circulation 296
    10.9 Summary of Important Potential Flows 299
    10.10 Flow Forces on Bodies 302
    References 307

11 Wave Motions in Non-Viscous Fluids 309
    11.1 General Considerations 309
    11.2 Longitudinal Waves: Sound Waves in Gases 313
    11.3 Transversal Waves: Surface Waves....................... 318
           11.3.1 General Solution Approach..................... 318
    11.4 Plane Standing Waves 323
    11.5 Plane Progressing Waves............................... 325
    11.6 References to Further Wave Motions 329
    References 330

12 Introduction to Gas Dynamics ... 331
- 12.1 Introductory Considerations ... 331
- 12.2 Mach Lines and Mach Cone ... 335
- 12.3 Non-Linear Wave Propagation, Formation of Shock Waves . 338
- 12.4 Alternative Forms of the Bernoulli Equation ... 341
- 12.5 Flow with Heat Transfer (Pipe Flow) ... 344
    - 12.5.1 Subsonic Flow ... 347
    - 12.5.2 Supersonic Flow ... 347
- 12.6 Rayleigh and Fanno Relations ... 351
- 12.7 Normal Compression Shock (Rankine–Hugoniot Equation) . 355
- References ... 360

13 Stationary, One-Dimensional Fluid Flows of Incompressible, Viscous Fluids ... 361
- 13.1 General Considerations ... 361
    - 13.1.1 Plane Fluid Flows ... 362
    - 13.1.2 Cylindrical Fluid Flows ... 363
- 13.2 Derivations of the Basic Equations for Fully Developed Fluid Flows ... 364
    - 13.2.1 Plane Fluid Flows ... 364
    - 13.2.2 Cylindrical Fluid Flows ... 366
- 13.3 Plane Couette Flow ... 366
- 13.4 Plane Fluid Flow Between Plates ... 369
- 13.5 Plane Film Flow on an Inclined Plate ... 372
- 13.6 Axi-Symmetric Film Flow ... 376
- 13.7 Pipe Flow (Hagen–Poiseuille Flow) ... 379
- 13.8 Axial Flow Between Two Cylinders ... 383
- 13.9 Film Flows with Two Layers ... 386
- 13.10 Two-Phase Plane Channel Flow ... 388
- References ... 391

14 Time-Dependent, One-Dimensional Flows of Viscous Fluids ... 393
- 14.1 General Considerations ... 393
- 14.2 Accelerated and Decelerated Fluid Flows ... 397
    - 14.2.1 Stokes First Problem ... 397
    - 14.2.2 Diffusion of a Vortex Layer ... 399
    - 14.2.3 Channel Flow Induced by Movements of Plates ... 402
    - 14.2.4 Pipe Flow Induced by the Pipe Wall Motion ... 407
- 14.3 Oscillating Fluid Flows ... 414
    - 14.3.1 Stokes Second Problem ... 414
- 14.4 Pressure Gradient-Driven Fluid Flows ... 417
    - 14.4.1 Starting Flow in a Channel ... 417
    - 14.4.2 Starting Pipe Flow ... 422
- References ... 427

15 Fluid Flows of Small Reynolds Numbers ... 429
- 15.1 General Considerations ... 429
- 15.2 Creeping Fluid Flows Between Two Plates ... 431
- 15.3 Plane Lubrication Films ... 433
- 15.4 Theory of Lubrication in Roller Bearings ... 438
- 15.5 The Slow Rotation of a Sphere ... 443
- 15.6 The Slow Translatory Motion of a Sphere ... 445
- 15.7 The Slow Rotational Motion of a Cylinder ... 451
- 15.8 The Slow Translatory Motion of a Cylinder ... 453
- 15.9 Diffusion and Convection Influences on Flow Fields ... 459
- References ... 461

16 Flows of Large Reynolds Numbers Boundary-Layer Flows ... 463
- 16.1 General Considerations and Derivations ... 463
- 16.2 Solutions of the Boundary-Layer Equations ... 468
- 16.3 Flat Plate Boundary Layer (Blasius Solution) ... 470
- 16.4 Integral Properties of Wall Boundary Layers ... 474
- 16.5 The Laminar, Plane, Two-Dimensional Free Shear Layer ... 480
- 16.6 The Plane, Two-Dimensional, Laminar Free Jet ... 481
- 16.7 Plane, Two-Dimensional Wake Flow ... 486
- 16.8 Converging Channel Flow ... 489
- References ... 492

17 Unstable Flows and Laminar-Turbulent Transition ... 495
- 17.1 General Considerations ... 495
- 17.2 Causes of Flow Instabilities ... 501
  - 17.2.1 Stability of Atmospheric Temperature Layers ... 502
  - 17.2.2 Gravitationally Caused Instabilities ... 505
  - 17.2.3 Instabilities in Annular Clearances Caused by Rotation ... 507
- 17.3 Generalized Instability Considerations (Orr–Sommerfeld Equation) ... 512
- 17.4 Classifications of Instabilities ... 517
- 17.5 Transitional Boundary-Layer Flows ... 519
- References ... 522

18 Turbulent Flows ... 523
- 18.1 General Considerations ... 523
- 18.2 Statistical Description of Turbulent Flows ... 527
- 18.3 Basics of Statistical Considerations of Turbulent Flows ... 528
  - 18.3.1 Fundamental Rules of Time Averaging ... 528
  - 18.3.2 Fundamental Rules for Probability Density ... 530
  - 18.3.3 Characteristic Function ... 537
- 18.4 Correlations, Spectra and Time-Scales of Turbulence ... 538

	18.5	Time-Averaged Basic Equations of Turbulent Flows 542
		18.5.1 The Continuity Equation 543
		18.5.2 The Reynolds Equation 544
		18.5.3 Mechanical Energy Equation for the Mean Flow Field 546
		18.5.4 Equation for the Kinetic Energy of Turbulence 550
	18.6	Characteristic Scales of Length, Velocity and Time of Turbulent Flows 553
	18.7	Turbulence Models 557
		18.7.1 General Considerations 557
		18.7.2 General Considerations Concerning Eddy Viscosity Models 560
		18.7.3 Zero-Equation Eddy Viscosity Models 565
		18.7.4 One-Equation Eddy Viscosity Models............. 573
		18.7.5 Two-Equation Eddy Viscosity Models 576
	18.8	Turbulent Wall Boundary Layers....................... 578
	References 585

19 Numerical Solutions of the Basic Equations 587
 19.1 General Considerations 587
 19.2 General Transport Equation and Discretization of the Solution Region 591
 19.3 Discretization by Finite Differences..................... 595
 19.4 Finite-Volume Discretization 598
   19.4.1 General Considerations 598
   19.4.2 Discretization in Space 600
   19.4.3 Discretization with Respect to Time................ 611
   19.4.4 Treatments of the Source Terms 613
 19.5 Computation of Laminar Flows 614
   19.5.1 Wall Boundary Conditions 615
   19.5.2 Symmetry Planes 615
   19.5.3 Inflow Planes................................... 615
   19.5.4 Outflow Planes 615
 19.6 Computations of Turbulent Flows 616
   19.6.1 Flow Equations to be Solved 616
   19.6.2 Boundary Conditions for Turbulent Flows.......... 620
 References 626

20 Fluid Flows with Heat Transfer 627
 20.1 General Considerations 627
 20.2 Stationary, Fully Developed Flow in Channels 630
 20.3 Natural Convection Flow Between Vertical Plane Plates ... 633
 20.4 Non-Stationary Free Convection Flow Near a Plane Vertical Plate 637

	20.5	Plane-Plate Boundary Layer with Plate Heating at Small Prandtl Numbers 641
	20.6	Similarity Solution for a Plate Boundary Layer with Wall Heating and Dissipative Warming 644
	20.7	Vertical Plate Boundary-Layer Flows Caused by Natural Convection 647
	20.8	Similarity Considerations for Flows with Heat Transfer 649
	References 651

21 Introduction to Fluid-Flow Measurement 653
    21.1 Introductory Considerations 653
    21.2 Measurements of Static Pressures 656
    21.3 Measurements of Dynamic Pressures 660
    21.4 Applications of Stagnation-Pressure Probes 662
    21.5 Basics of Hot-Wire Anemometry 664
        21.5.1 Measuring Principle and Physical Principles 664
        21.5.2 Properties of Hot-Wires and Problems of Application 667
        21.5.3 Hot-Wire Probes and Supports 672
        21.5.4 Cooling Laws for Hot-Wire Probes 676
        21.5.5 Static Calibration of Hot-Wire Probes 680
    21.6 Turbulence Measurements with Hot-Wire Anemometers.... 685
    21.7 Laser Doppler Anemometry 694
        21.7.1 Theory of Laser Doppler Anemometry 694
        21.7.2 Optical Systems for Laser Doppler Measurements ... 701
        21.7.3 Electronic Systems for Laser Doppler Measurements 705
        21.7.4 Execution of LDA-Measurements: One-Dimensional LDA Systems 715
    References 717

Index 719

Chapter 1
Introduction, Importance and Development of Fluid Mechanics

1.1 Fluid Flows and their Significance

Flows occur in all fields of our natural and technical environment and anyone perceiving their surroundings with open eyes and assessing their significance for themselves and their fellow beings can convince themselves of the far-reaching effects of fluid flows. Without fluid flows life, as we know it, would not be possible on Earth, nor could technological processes run in the form known to us and lead to the multitude of products which determine the high standard of living that we nowadays take for granted. Without flows our natural and technical world would be different, and might not even exist at all. Flows are therefore vital.

Flows are everywhere and there are flow-dependent transport processes that supply our body with the oxygen that is essential to life. In the blood vessels of the human body, essential nutrients are transported by mass flows and are thus carried to the cells, where they contribute, by complex chemical reactions, to the build-up of our body and to its energy supply. Similarly to the significance of fluid flows for the human body, the multitude of flows in the entire fauna and flora are equally important (see Fig. 1.1). Without these flows, there would be no growth in nature and human beings would be deprived of their "natural food". Life in Nature is thus dependent on flow processes and understanding them is an essential part of the general education of humans.

As further vital processes in our natural environment, flows in rivers, lakes and seas have to be mentioned, and also atmospheric flow processes, whose influences on the weather and thus on the climate of entire geographical regions is well known (see Fig. 1.2). Wind fields are often responsible for the transport of clouds and, taking topographic conditions into account, are often the cause of rainfall. Observations show, for example, that rainfall occurs more often in areas in front of mountain ranges than behind them. Fluid flows in the atmosphere thus determine whether certain regions can be used for agriculture, if they are sufficiently supplied with rain, or whether entire areas turn

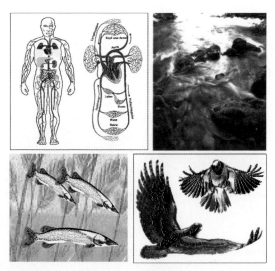

Fig. 1.1 Flow processes occur in many ways in our natural environment

Fig. 1.2 Effects of flows on the climate of entire geographical regions

arid because there is not sufficient rainfall for agriculture. In extreme cases, desert areas are sometimes of considerable dimensions, where agricultural use of the land is possible only with artificial irrigation.

Other negative effects on our natural environment are the devastations that hurricanes and cyclones can cause. When rivers, lakes or seas leave their natural beds and rims, flow processes can arise whose destructive forces are known to us from many inundation catastrophes. This makes it clear that humans not only depend on fluid flows in the positive sense, but also have to learn to live with the effects of such fluid flows that can destroy or damage the entire environment.

1.1 Fluid Flows and their Significance

Leaving the natural environment of humans and turning to the technical environment, one finds here also a multitude of flow processes, that occur in aggregates, instruments, machines and plants in order to transfer energy, generate lift forces, run combustion processes or take on control functions. There are, for example, fluid flows coupled with chemical reactions that enable the combustion in piston engines to proceed in the desired way and thus supply the power that is used in cars, trucks, ships and aeroplanes. A large part of the energy generated in a combustion engine of a car is used, especially when the vehicles run at high speed, to overcome the energy loss resulting from the flow resistance which the vehicle experiences owing to the momentum loss and the flow separations. In view of the decrease in our natural energy resources and the high fuel costs related to it, great significance is attached to the reduction of this resistance by fluid mechanical optimization of the car body. Excellent work has been done in this area of fluid mechanics (see Fig. 1.3), e.g. in aerodynamics, where new aeroplane wing profiles and wing geometries as well as wing body connections were developed which show minimal losses due to friction and collision while maintaining the high lift forces necessary in aeroplane aerodynamics. The knowledge gained within the context of aerodynamic investigations is being used today also in many fields of the consumer goods industry. The optimization of products from the point of view of fluid mechanics has led to new markets, for example the production of ventilators for air exchange in rooms and the optimization of hair driers.

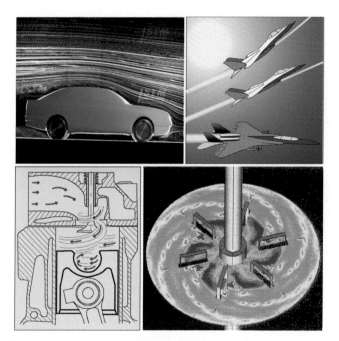

Fig. 1.3 Fluid flows are applied in many ways in our technical environment

We also want to draw the attention of the reader to the importance of fluid mechanics in the field of chemical engineering, where many areas such as heat and mass transfer processes and chemical reactions are influenced strongly or rendered possible only by flow processes. In this field of engineering, it becomes particularly clear that much of the knowledge gained in the natural sciences can be used technically only because it is possible to let processes run in a steady and controlled way. In many areas of chemical engineering, fluid flows are being used to make steady-state processes possible and to guarantee the controllability of plants, i.e. flows are being employed in many places in process engineering.

Often it is necessary to use flow media whose properties deviate strongly from those of Newtonian fluids, in order to optimize processes, i.e. the use of non-Newtonian fluids or multi-phase fluids is necessary. The selection of more complex properties of the flowing fluids in technical plants generally leads to more complex flow processes, whose efficient employment is not possible without detailed knowledge in the field of the flow mechanics of simple fluids, i.e. fluids with Newtonian properties. In a few descriptions in the present introduction to fluid mechanics, the properties of non-Newtonian media are mentioned and interesting aspects of the flows of these fluids are shown. The main emphasis of this book lies, however, in the field of the flows of Newtonian media. As these are of great importance in many applications, their special treatment in this book is justified.

1.2 Sub-Domains of Fluid Mechanics

Fluid mechanics is a science that makes use of the basic laws of mechanics and thermodynamics to describe the motion of fluids. Here fluids are understood to be all the media that cannot be assigned clearly to solids, no matter whether their properties can be described by simple or complicated material laws. Gases, liquids and many plastic materials are fluids whose movements are covered by fluid mechanics. Fluids in a state of rest are dealt with as a special case of flowing media, i.e. the laws for motionless fluids are deduced in such a way that the velocity in the basic equations of fluid mechanics is set equal to zero.

In fluid mechanics, however, one is not content with the formulation of the laws by which fluid movements are described, but makes an effort beyond that to find solutions for flow problems, i.e. for given initial and boundary conditions. To this end, three methods are used in fluid mechanics to solve flow problems:

(a) Analytical solution methods (analytical fluid mechanics):
    Analytical methods of applied mathematics are used in this field to solve the basic flow equations, taking into account the boundary conditions describing the actual flow problem.

1.2 Sub-Domains of Fluid Mechanics

(b) Numerical solution methods (numerical fluid mechanics):
Numerical methods of applied mathematics are employed for fluid flow simulations on computers to yield solutions of the basic equations of fluid mechanics.
(c) Experimental solution methods (experimental fluid mechanics):
This sub-domain of fluid mechanics uses similarity laws for the transferability of fluid mechanics knowledge from model flow investigations. The knowledge gained in model flows by measurements is transferred by means of the constancy of known characteristic quantities of a flow field to the flow field of actual interest.

The above-mentioned methods have until now, in spite of considerable developments in the last 50 years, only partly reached the state of development which is necessary to be able to describe adequately or solve fluid mechanics problems, especially for many practical flow problems. Hence, nowadays, known analytical methods are often only applicable to flow problems with simple boundary conditions. It is true that the use of numerical processes makes the description of complicated flows possible; however, feasible solutions to practical flow problems without model hypotheses, especially in the case of turbulent flows at high Reynold numbers, can only be achieved in a limited way. The limitations of numerical methods are due to the limited storage capacity and computing speed of the computers available today. These limitations will continue to exist for a long time, so that a number of practically relevant flows can only be investigated reliably by experimental methods. However, also for experimental investigations not all quantities of interest, from a fluid mechanics point of view, can always be determined, in spite of the refined experimental methods available today. Suitable measuring techniques for obtaining all important flow quantities are lacking, as for example the measuring techniques to investigate the thin fluid films shown in Fig. 1.4. Experience shows that efficient solutions of practical flow problems therefore require the combined use of the above-presented analytical, numerical and experimental methods of fluid mechanics. The different sub-domains of fluid mechanics cited are thus of equal importance and mastering the different methods of fluid mechanics is often indispensable in practice.

When analytical solutions are possible for flow problems, they are preferable to the often extensive numerical and experimental investigations. Unfortunately, it is known from experience that the basic equations of fluid mechanics, available in the form of a system of nonlinear and partial differential equations, allow analytical solutions only when, with regard to the equations and the initial and boundary conditions, considerable simplifications are made in actually determining solutions to flow problems. The validity of these simplifications has to be proved for each flow problem to be solved by comparing the analytically achieved final results with the corresponding experimental data. Only when such comparisons lead to acceptably small differences between the analytically determined and experimentally investigated velocity field can the hypotheses, introduced into the analytical

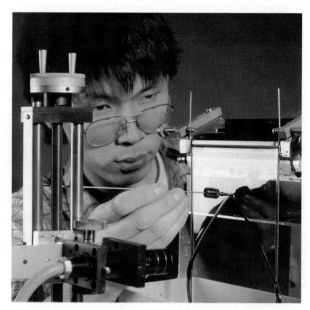

Fig. 1.4 Experimental investigation of fluid films

solution of the flow problem, be regarded as justified. In cases where such a comparison with experimental data is unsatisfactory, it is advisable to justify theoretically the simplifications by order of magnitude considerations, so as to prove that the terms neglected, for example in the solution of the basic equations, are small in comparison with the terms that are considered for the solution.

One has to proceed similarly concerning the numerical solution of flow problems. The validity of the solution has to be proved by comparing the results achieved by finite volume methods and finite element methods with corresponding experimental data. When such data do not exist, which may be the case for flow problems as shown in Figs. 1.5 and 1.6, statements on the accuracy of the solutions achieved can be made by the comparison of three numerical solutions calculated on various fine grids that differ from one another by their grid spacing. With this knowledge of precision, flow information can then can be obtained from numerical computations that are relevant to practical applications. Numerical solutions without knowledge of the numerically achieved precision of the solution are unsuitable for obtaining reliable information on fluid flow processes.

When experimental data are taken into account to verify analytical or numerical results, it is very important that only such experimental data that can be classified as having sufficient precision for reliable comparisons are used. A prerequisite is that the measuring data are obtained with techniques that allow precise flow measurements and also permit one to determine fluid flows

1.2 Sub-Domains of Fluid Mechanics

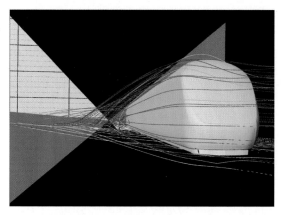

Fig. 1.5 Numerical calculation of the flow around a train in crosswinds

Fig. 1.6 Flow investigation with the aid of a laser Doppler anemometer

by measurement in a non-destructive way. Optical measurement techniques fulfill, in general, the requirements concerning precision and permit measurements without disturbance, so that optical measuring techniques are nowadays increasingly applied in experimental fluid mechanics (see Fig. 1.6). In this context, laser Doppler anemometry is of particular importance. It has developed into a reliable and easily applicable measuring tool in fluid mechanics that is capable of measuring the required local velocity information in laminar and turbulent flows.

Although the equal importance of the different sub-domains of fluid mechanics presented above, according to the applied methodology, has been outlined in the preceding paragraphs, priority in this book will be given to analytical fluid mechanics for an introductory presentation of the methods for solving flow problems. Experience shows that it is better to include analytical solutions of fluid mechanical problems in order to create or deepen with their help students' understanding of flow physics. As a rule, analytical methods applied to the solution of fluid flow problems, are known to students from lectures in applied mathematics. Hence students of fluid mechanics bring along the tools for the analytical solutions of flow problems. This circumstance does not necessarily exist for numerical or experimental methods. This is the reason why in this introductory book special significance is attached to the methods of analytical fluid mechanics. In parts of this book numerical solutions are treated in an introductory way in addition to presenting results of experimental investigations and the corresponding measuring techniques. It is thus intended to convey to the student, in this introduction to the subject, the significance of numerical and experimental fluid mechanics.

The contents of this book put the main emphasis on solutions of fluid flow problems that are described by simplified forms of the basic equations of fluid mechanics. This application of simplified equations to the solution of fluid problems represents a highly developed system. The comprehensible introduction of students to the general procedures for solving flow problems by means of simplified flow equations is achieved by the basic equations being derived and formulated as partial differential equations for Newtonian fluids (e.g. air or water). From these general equations, the simplified forms of the fluid flow laws can be derived in a generally comprehensible way, e.g. by the introduction of the hypothesis that fluids are free from viscosity. Fluids of this kind are described as "ideal" from a fluid mechanics point of view. The basic equations of these ideal fluids, derived from the general set of equations, represent an essential simplification by which the analytical solutions of flow problems become possible.

Further simplifications can be obtained by the hypothesis of incompressibility of the considered fluid, which leads to the classical equations of hydrodynamics. When, however, gas flows at high velocities are considered, the hypothesis of incompressibility of the flow medium is no longer justified. For compressible flow investigations, the basic equations valid for gas dynamic flows must then be used. In order to derive these, the hypothesis is introduced that gases in flow fields undergo thermodynamic changes of state, as they are known for ideal gases. The solution of the gas dynamic basic equations is successful in a number of one-dimensional flow processes. These are appropriately dealt with in this book. They give an insight into the strong interactions that may exist between the kinetic energy of a fluid element and the internal energy of a compressible fluid. The resulting flow phenomena are suited for achieving the physical understanding of one-dimensional gas dynamic fluid flows and applying it to two-dimensional flows. Some two-dimensional flow problems are therefore also mentioned in this book. Particular significance in these considerations is given to the physical understanding of the fluid flows that occur. Importance is also given, however, to representing the basics of the applied analytical methods in a way that makes them clear and comprehensible for the student.

1.3 Historical Developments

In this section, the historical development of fluid mechanics is roughly sketched out, based on the most important contributions of a number of scientists and engineers. The presentation does not claim to give a complete picture of the historical developments: this is impossible owing to the constraints on allowable space in this section. The aim is rather to depict the development over centuries in a generally comprehensible way. In summary, it can be said that already at the beginning of the nineteenth century the basic equations with which fluid flows can be described reliably were known. Solutions of these equations were not possible owing to the lack of suitable solution methods for engineering problems and therefore technical hydraulics developed alongside the field of theoretical fluid mechanics. In the latter area, use was made of the known contexts for the flow of ideal fluids and the influence of friction effects was taken into consideration via loss coefficients, determined empirically. For geometrically complicated problems, methods based on similarity laws were used to generalize experimentally achieved flow results. Analytical methods only allowed the solution of academic problems that had no relevance for practical applications. It was not until the second half of the twentieth century that the development of suitable methods led to the numerical techniques that we have today which allow us to solve the basic equations of fluid mechanics for practically relevant flow problems. Parallel to the development of the numerical methods, the development of experimental techniques was also pushed ahead, so that nowadays measurement techniques are available which allow us to obtain experimentally fluid mechanics data that are interesting for practical flow problems.

Some technical developments were and still are today closely connected with the solution of fluid flows or with the advantageous exploitation of flow processes. In this context, attention is drawn to the development of navigation with wind-driven ships as early as in ancient Egyptian times. Further developments up to the present time have led to transport systems of great economic and socio-political significance. In recent times, navigation has been surpassed by breathtaking developments in aviation and motor construction. These again use flow processes to guarantee the safety and comfort which we take for granted nowadays with all of the available transport systems. It was fluid mechanics developments which alone made this safety and comfort possible.

The continuous scientific development of fluid mechanics started with Leonardo da Vinci (1452–1519). Through his ingenious work, methods were devised that were suitable for fluid mechanics investigations of all kinds. Earlier efforts of Archimedes (287–212 B.C.) to understand fluid motions led to the understanding of the hydromechanical buoyancy and the stability of floating bodies. His discoveries remained, however, without further impact on the development of fluid mechanics in the following centuries. Something similar holds true for the work of Sextus Julius Frontinus (40–103), who provided the basic understanding for the methods that were applied in the Roman Empire for measuring the volume flows in the Roman water supply system. The work of Sextus Julius Frontinus also remained an individual achievement. For more than a millennium no essential fluid mechanics insights followed and there were no contributions to the understanding of flow processes.

Fluid mechanics as a field of science developed only after the work of Leonardo da Vinci. His insight laid the basis for the continuum principle for fluid mechanics considerations and he contributed through many sketches of flow processes to the development of the methodology to gain fluid mechanics insights into flows by means of visualization. His ingenious engineering art allowed him to devise the first installations that were driven fluid mechanically and to provide sketches of technical problem solutions on the basis of fluid flows. The work of Leonardo da Vinci was followed by that of Galileo Galilei (1564–1642) and Evangelista Torricelli (1608–1647). Whereas Galileo Galilei produced important ideas for experimental hydraulics and revised the concept of vacuum introduced by Aristoteles, Evangelista Torricelli realized the relationship between the weight of the atmosphere and the barometric pressure. He developed the form of a horizontally ejected fluid jet in connection with the laws of free fall. Torricelli's work was therefore an important contribution to the laws of fluids flowing out of containers under the influence of gravity. Blaise Pascal (1623–1662) also dedicated himself to hydrostatics and was the first to formulate the theorem of universal pressure distribution.

Isaac Newton (1642–1727) laid the basis for the theoretical description of fluid flows. He was the first to realize that molecule-dependent momentum transport, which he introduced as flow friction, is proportional to the velocity gradient and perpendicular to the flow direction. He also made some additional contributions to the detection and evaluation of the flow resistance. Concerning the jet contraction arising with fluids flowing out of containers, he engaged in extensive deliberations, although his ideas were not correct in all respects. Henri de Pitot (1665–1771) made important contributions to the understanding of stagnation pressure, which builds up in a flow at stagnation points. He was the first to endeavor to make possible flow velocities by differential pressure measurements following the construction of double-walled measuring devices. Daniel Bernoulli (1700–1782) laid the foundation of hydromechanics by establishing a connection between pressure and velocity, on the basis of simple energy principles. He made essential contributions to pressure measurements, manometer technology and hydromechanical drives.

Leonhard Euler (1707–1783) formulated the basics of the flow equations of an ideal fluid. He derived, from the conservation equation of momentum, the Bernoulli theorem that had, however, already been derived by Johann Bernoulli (1667–1748) from energy principles. He emphasized the significance of the pressure for the entire field of fluid mechanics and explained among other things the appearance of cavitations in installations. The basic principle of turbo engines was discovered and described by him. Euler's work on the formulation of the basic equations was supplemented by Jean le Rond

1.3 Historical Developments

d'Alembert (1717–1783). He derived the continuity equation in differential form and introduced the use of complex numbers into the potential theory. In addition, he derived the acceleration component of a fluid element in field variables and expressed the hypothesis, named after him and proved before by Euler, that a body circulating in an ideal fluid has no flow resistance. This fact, known as d'Alembert's paradox, led to long discussions concerning the validity of the equations of fluid mechanics, as the results derived from them did not agree with the results of experimental investigations.

The basic equations of fluid mechanics were dealt with further by Joseph de Lagrange (1736–1813), Louis Marie Henri Navier (1785–1836) and Barre de Saint Venant (1797–1886). As solutions of the equations were not successful for practical problems, however, practical hydraulics developed parallel to the development of the theory of the basic equations of fluid mechanics. Antoine Chezy (1718–1798) formulated similarity parameters, in order to transfer the results of flow investigations in one flow channel to a second channel. Based on similarity laws, extensive experimental investigations were carried out by Giovanni Battista Venturi (1746–1822), and also experimental investigations were made on pressure loss measurements in flows by Gotthilf Ludwig Hagen (1797–1884) and on hydrodynamic resistances by Jean-Louis Poiseuille (1799–1869). This was followed by the work of Henri Philibert Gaspard Darcy (1803–1858) on filtration, i.e. for the determination of pressure losses in pore bodies. In the field of civil engineering, Julius Weissbach (1806–1871) introduced the basis of hydraulics into engineers' considerations and determined, by systematic experiments, dimensionless flow coefficients with which engineering installations could be designed. The work of William Froude (1810–1879) on the development of towing tank techniques led to model investigations on ships and Robert Manning (1816–1897) worked out many equations for resistance laws of bodies in open water channels. Similar developments were introduced by Ernst Mach (1838–1916) for compressible aerodynamics. He is seen as the pioneer of supersonic aerodynamics, providing essential insights into the application of the knowledge on flows in which changes of the density of a fluid are of importance.

In addition to practical hydromechanics, analytical fluid mechanics developed in the nineteenth century, in order to solve analytically manageable problems. George Gabriel Stokes (1816–1903) made analytical contributions to the fluid mechanics of viscous media, especially to wave mechanics and to the viscous resistance of bodies, and formulated Stokes' law for spheres falling in fluids. John William Stratt, Lord Rayleigh (1842–1919) carried out numerous investigations on dynamic similarity and hydrodynamic instability. Derivations of the basis for wave motions, instabilities of bubbles and drops and fluid jets, etc., followed, with clear indications as to how linear instability considerations in fluid mechanics are to be carried out. Vincenz Strouhal (1850–1922) worked out the basics of vibrations and oscillations in bodies through separating vortices. Many other scientists, who showed that applied mathematics can make important contributions to the analytical solution of flow problems, could be named here. After the pioneering work of Ludwig Prandtl (1875–1953), who introduced the boundary layer concept into fluid mechanics, analytical solutions to the basic equations followed, e.g. solutions of the boundary layer equations by Paul Richard Heinrich Blasius (1883–1970).

With Osborne Reynolds (1832–1912), a new chapter in fluid mechanics was opened. He carried out pioneering experiments in many areas of fluid mechanics, especially basic investigations on different turbulent flows. He demonstrated that it is possible to formulate the Navier–Stokes equations in a time-averaged form, in order to describe turbulent transport processes in this way. Essential work in this area by Ludwig Prandtl (1875–1953) followed, providing fundamental insights into flows in the field of the boundary layer theory. Theodor von Karman (1881–1993) made contributions to many sub-domains of fluid mechanics and was followed by numerous scientists who engaged in problem solutions in fluid mechanics. One should mention here, without claiming that the list is complete, Pei-Yuan Chou (1902–1993) and Andrei Nikolaevich Kolmogorov (1903–1987) for their contributions to turbulence theory and Herrmann Schlichting (1907–1982) for his work in the field of laminar–turbulent transition, and for uniting the fluid-mechanical knowledge of his time and converting it into practical solutions of flow problems.

The chronological sequence of the contributions to the development of fluid mechanics outlined in the above paragraphs can be rendered well in a diagram as shown in Fig. 1.7. This information is taken from history books on fluid

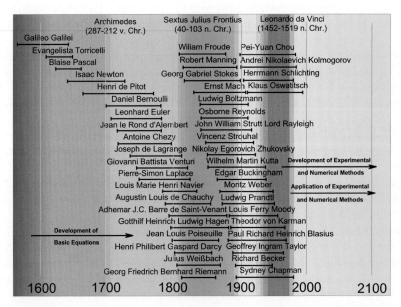

Fig. 1.7 Diagram listing the epochs and scientists contributing to the development of fluid mechanics

1.3 Historical Developments

mechanics as given in refs [1.1] to [1.6]. On closer examination one sees that the sixteenth and seventeenth centuries were marked by the development of the understanding of important basics of fluid mechanics. In the course of the development of mechanics, the basic equations for fluid mechanics were derived and fully formulated in the eighteenth century. These equations comprised all forces acting on fluid elements and were formulated for substantial quantities (Lagrange's approach) and for field quantities (Euler's approach). Because suitable solution methods were lacking, the theoretical solutions of the basic equations of fluid mechanics, strived for in the nineteenth century and at the beginning of the twentieth century, were limited to analytical results for simple boundary conditions. Practical flow problems escaped theoretical solution and thus "engineering hydromechanics" developed that looked for fluid mechanics problem solutions by experimentally gained insights. At that time, one aimed at investigations on geometrically similar flow models, while conserving fluid mechanics similarity requirements, to permit the transfer of the experimentally gained insights by similarity laws to large constructions. Only the development of numerical methods for the solution of the basic equations of fluid mechanics, starting from the middle of the twentieth century, created the methods and techniques that led to numerical solutions for practical flow problems. Metrological developments that ran in parallel led to complementary experimental and numerical solutions of practical flow problems. Hence it is true to say that the second half of the twentieth century brought to fluid mechanics the measuring and computational methods that are required for the solution of practical flow problems. The combined application of the experimental and numerical methods, available today, will in the twenty-first century permit fluid mechanics investigations that were not previously possible because of the lack of suitable investigation methods.

The experimental methods that contributed particularly to the rapid advancement of experimental fluid mechanics in the second half of the twentieth century were the hot-wire and laser-Doppler anemometry. These methods have now reached a state of development which allows their use in local velocity measurements in laminar and turbulent flows. In general, one applies hot wire anemometry in gas flows that are low in impurities, so that the required calibration of the hot wire employed can be conserved over a long measuring time. Reliable measurements are possible up to 10% turbulence intensity. Flows with turbulence intensities above that require the application of laser Doppler anemometry. This measuring method is also suitable for measurements in impure gas and liquid flows.

Finally, the rapid progress that has been achieved in the last few decades in the field of numerical fluid mechanics should also be mentioned. Considerable developments in applied mathematics took place to solve partial differential equations numerically. In parallel, great improvements in the computational performance of modern high-speed computers occurred and computer programs became available that allow one to solve practical flow

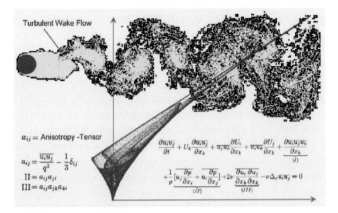

Fig. 1.8 Diagram of the turbulence anisotropy due to the invariants of the anisotropy tensor

problems numerically. Numerical fluid mechanics has therefore also become an important sub-domain of the entire field of fluid mechanics. Its significance will increase further in the future.

One can expect in particular new ansätze in the development of turbulence models which will use invariants of the tensors $\overline{u_i u_j}$, ϵ_{ij}, etc., so that the limitations of modelling turbulent properties of flows can be taken into consideration. This is indicated in Fig. 1.8. Information of this kind can be used for advanced turbulence modeling.

References

1.1. Bell ET (1936) Men of Mathematics. Simon & Schuster, New York
1.2. Rouse H (1952) Present day trends in hydraulics. Applied Mechanics Reviews 5:2
1.3. Bateman H, Dryden HL, Murnaghan FP (1956) Hydrodynamics. Dover, New York
1.4. Van Dyke M (1964) Perturbation Methods in Fluid Mechanics. Academic press, New York
1.5. Rouse H, Ince S (1980) History of Hydraulics. The University of Iowa, Institute of Hydraulic Research, Ames, IA
1.6. Sžabo I (1987) Geschichte der mechanischen Prinzipien und ihrer wichtigsten Anwendungen. Birkhaeuser, Basel

Chapter 2
Mathematical Basics

2.1 Introduction and Definitions

Fluid mechanics deals with transport processes, especially with the flow- and molecule-dependent momentum transports in fluids. Their thermodynamic properties of state such as pressure, density, temperature and internal energy enter into fluid mechanics considerations. The thermodynamic properties of state of a fluid are scalars and as such can be introduced into the equations for the mathematical description of fluid flows. However, in addition to scalars, other kinds of quantities are also required for the description of fluid flows. In the following sections it will be shown that fluid mechanics considerations result in conservation equations for mass, momentum, energy and chemical species which comprise scalar, vector and other tensor quantities. Often fundamental differentiations are made between such quantities, without considering that the quantities can all be described as tensors of different orders. Hence one can write:

Scalar quantities $\quad=$ tensors of zero order $\quad\rightsquigarrow \{a\} \quad\rightarrow a$
Vectorial quantities $=$ tensors of first order $\quad\rightsquigarrow \{a_i\} \quad\rightarrow a_i$
Tensorial quantities $=$ tensors of second order $\rightsquigarrow \{a_{ij}\} \rightarrow a_{ij}$

where the number of the chosen indices i, j, k, l, m, n of the tensor presentation designates the order and 'a' can be any quantity under consideration. The introduction of tensorial quantities, as indicated above, permits extensions of the description of fluid flows by means of still more complex quantities, such as tensors of third or even higher order, if this becomes necessary for the description of fluid mechanics phenomena. This possibility of extension and the above-mentioned standard descriptions led us to choose the indicated tensor notation of physical quantities in this book, the number of the indices i, j, k, l, m, n deciding the order of a considered tensor.

Tensors of arbitrary order are mathematical quantities, describing physical properties of fluids, with which "mathematical operations" such as addition, subtraction, multiplication and division can be carried out. These may be well known to many readers of this book, but are presented again below as a summary. Where the brevity of the description does not make it possible for readers, not accustomed to tensor descriptions, to familiarize themselves with the matter, reference is made to the corresponding mathematical literature; see Sect. 2.12. Many of the following deductions and descriptions can, however, be considered as simple and basic knowledge of mathematics and it is not necessary that the details of the complete tensor calculus are known. In the present book, only the tensor notation is used, along with simple parts of the tensor calculus. This will become clear from the following explanations. There are a number of books available that deal with the matter in the sections to come in a mathematical way, e.g. see refs. [2.1] to [2.7].

2.2 Tensors of Zero Order (Scalars)

Scalars are employed for the description of the thermodynamic state variables of fluids such as pressure, density, temperature and internal energy, or they describe other physical properties that can be given clearly by stating an amount of the quantity and a dimensional unit. The following examples explain this:

$$P = \underbrace{7.53 \times 10^6}_{\text{Amount}} \underbrace{\left[\frac{\text{N}}{\text{m}^2}\right]}_{\text{Unit}}, \quad T = \underbrace{893.2}_{\text{Amount}} \underbrace{[\text{K}]}_{\text{Unit}}, \quad \rho = \underbrace{1.5 \times 10^3}_{\text{Amount}} \underbrace{\left[\frac{\text{kg}}{\text{m}^3}\right]}_{\text{Unit}} \quad (2.1)$$

Physical quantities that have the same dimension can be added and subtracted, the amounts being included in the adding and subtracting operations, with the common dimension being maintained:

$$\sum_{\alpha=1}^{N} a_\alpha = \sum_{\alpha=1}^{N} \underbrace{|a_\alpha|}_{\text{Amount}} \underbrace{[a]}_{\text{Unit}} \quad a \pm b = \underbrace{(|a| \pm |b|)}_{\text{Amount}} \underbrace{[a \text{ or } b]}_{\text{Unit}}, \text{ with } [a] = [b] \quad (2.2)$$

Quantities with differing dimensions cannot be added or subtracted.

The mathematical laws below can be applied to the permitted additions and subtractions of scalars, see for details [2.5] and [2.6].

The amount of 'a' is a real number, i.e. $|a|$ is a real number if $a \in \mathbb{R}$. It is defined by $|a| := +a$, if $a \geq 0$ and $|a| := -a$, if $a < 0$.

The following mathematical rules can be deducted directly from this definition:

$$-|a| \leq a \leq |a|, \quad |-a| = |a|, \quad |ab| = |a||b|, \quad \left|\frac{a}{b}\right| = \frac{|a|}{|b|} \text{ (if } b \neq 0\text{)}$$

$$|a| \leq b \Leftrightarrow -b \leq a \leq b$$

2.3 Tensors of First Order (Vectors)

From $-|a| \leq a \leq |a|$ and $-|b| \leq b \leq |b|$, it follows that $-(|a|+|b|) \leq a+b \leq (|a|+|b|)$. Thus for all $a, b \in \mathbb{R}$:

$$|a+b| \leq |a| + |b| \qquad \text{(triangular inequality)}$$

The commutative and associative laws of addition and multiplication of scalar quantities are generally known and need not be dealt with here any further. If one carries out multiplications or divisions with scalar physical quantities, new physical quantities are created. These are again scalars, with amounts that result from the multiplication or division of the corresponding amounts of the initial quantities. The dimension of the new scalar physical quantities results from the multiplication or division of the basic units of the scalar quantities:

$$a \cdot b = \underbrace{(|a| \cdot |b|)}_{\text{Amount}} \underbrace{\big[[a] \cdot [b]\big]}_{\text{Unit}} \quad \text{and} \quad \frac{a}{b} = \underbrace{\frac{|a|}{|b|}}_{\text{Amount}} \cdot \underbrace{\left[\frac{[a]}{[b]}\right]}_{\text{Unit}} \qquad (2.3)$$

It can be seen from the example of the product of the pressure P and the volume V how a new physical quantity results:

$$P \cdot V = |P| \cdot |V| \underbrace{\left[\frac{N}{m^2} \cdot m^3\right]}_{[J = N\,m]} = |P| \cdot |V| \big[N\,m\big] \qquad (2.4)$$

The new physical quantity has the unit $J = $ joule, i.e. the unit of energy. When a pressure loss ΔP is multiplied with the volumetric flow rate, a power loss results:

$$\Delta P \cdot \dot{V} = |\Delta P||\dot{V}| \left[\frac{N}{m^2} \cdot \frac{m^3}{s}\right] = |\Delta P||\dot{V}| \underbrace{\left[\frac{N\,m}{s}\right]}_{[W = \frac{N\,m}{s}]} \qquad (2.5)$$

The power loss has the unit $W = watt = joule/s$.

2.3 Tensors of First Order (Vectors)

The complete presentation of a vectorial quantity requires the amount of the quantity to be given, in addition to its direction and its unit. Force, velocity, momentum, angular momentum, etc., are examples for vectorial quantities. Graphically, vectors are represented by arrows, whose length indicates the amount and the position of the arrow origin and the arrowhead indicates the direction. The derivable analytical description of vectorial quantities makes use of the indication of a vector component projected on to the axis of a coordinate system, and the indication of the direction is shown by the signs of the resulting vector components.

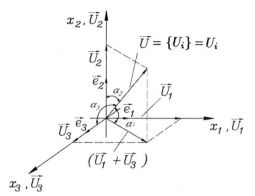

Fig. 2.1 Representation of velocity vector U_i in a Cartesian coordinate system

To represent the velocity vector $\{U_i\}$, for example, in a Cartesian coordinate system, the components $U_i (i = 1, 2, 3)$ can be expressed as follows:

$$\boldsymbol{U} = \{U_i\} = \begin{Bmatrix} U_1 \\ U_2 \\ U_3 \end{Bmatrix} = |U| \begin{Bmatrix} \cos\alpha_1 \\ \cos\alpha_2 \\ \cos\alpha_3 \end{Bmatrix} \left[\frac{\text{m}}{\text{s}}\right] \rightsquigarrow U_i = \underbrace{\pm}_{\text{Direction}} \underbrace{|U| \cdot |\cos\alpha_i|}_{\text{Amount}} \underbrace{\left[\frac{\text{m}}{\text{s}}\right]}_{\text{Unit}} \tag{2.6}$$

Looking at Fig. 2.1, one can see that the following holds:

$$\boldsymbol{U}_1 = U_1 \cdot \boldsymbol{e}_1, \quad \boldsymbol{U}_2 = U_2 \cdot \boldsymbol{e}_2, \quad \boldsymbol{U}_3 = U_3 \cdot \boldsymbol{e}_3 \tag{2.7}$$

where the unit vectors $\boldsymbol{e}_1, \boldsymbol{e}_2, \boldsymbol{e}_3$ in the coordinate directions x_1, x_2 and x_3 are employed. This is shown in Fig. 2.1. α_i designates the angle between $\boldsymbol{U}$ and the unit vector $\boldsymbol{e}_i$. Vectors can also be represented in other coordinate systems; through this, the vector does not change in itself but its mathematical representation changes. In this book, Cartesian coordinates are preferred for presenting vector quantities.

Vector quantities which have the same unit can be added or subtracted vectorially. Laws are applied here that result in addition or subtraction of the components on the axes of a Cartesian coordinate system:

$$\boldsymbol{a} \pm \boldsymbol{b} = \{a_i\} \pm \{b_i\} = \{(a_i \pm b_i)\} = \{(a_1 \pm b_1), (a_2 \pm b_2), (a_3 \pm b_3)\}^T$$

Vectorial quantities with different units cannot be added or subtracted vectorially. For the addition and subtraction of vectorial constants (having the same units), the following rules of addition hold:

$\boldsymbol{a} + \boldsymbol{0} = \{a_i\} + \{0\} = \boldsymbol{a}$ (zero vector or neutral element $\boldsymbol{0}$)
$\boldsymbol{a} + (-\boldsymbol{a}) = \{a_i\} + \{-a_i\} = \boldsymbol{0}$ ($\boldsymbol{a}$ element inverse to $-\boldsymbol{a}$)
$\boldsymbol{a} + \boldsymbol{b} = \boldsymbol{b} + \boldsymbol{a}$, d.h. $\{a_i\} + \{b_i\} = \{b_i\} + \{a_i\}$
    $= \{(a_i + b_i)\}$ (commutative law)
$\boldsymbol{a} + (\boldsymbol{b} + \boldsymbol{c}) = (\boldsymbol{a} + \boldsymbol{b}) + \boldsymbol{c}$, d.h. $\{a_i\} + \{(b_i + c_i)\}$
    $= \{(a_i + b_i)\} + \{c_i\}$ (associative law)

2.3 Tensors of First Order (Vectors)

With $(\alpha \cdot \boldsymbol{a})$ a multiple of $\boldsymbol{a}$ results, if $\alpha > 0$. α has no unit of its own, i.e. $(\alpha \cdot \boldsymbol{a})$ designates the vector that has the same direction as $\boldsymbol{a}$ but has α times the amount. In the case $\alpha < 0$, one puts $(\alpha \cdot \boldsymbol{a}) := -(|\alpha| \cdot \boldsymbol{a})$. For $\alpha = 0$ the zero vector results: $0 \cdot \boldsymbol{a} = \boldsymbol{0}$.

When multiplying two vectors two possibilities should be distinguished yielding different results.

The **scalar product** $\boldsymbol{a} \cdot \boldsymbol{b}$ of the vectors $\boldsymbol{a}$ and $\boldsymbol{b}$ is defined as

$$\boldsymbol{a} \cdot \boldsymbol{b} := \begin{cases} |\boldsymbol{a}| \cdot |\boldsymbol{b}| \cdot \cos(\boldsymbol{a},\boldsymbol{b}), & \text{if } \boldsymbol{a} \neq \boldsymbol{0} \text{ and } \boldsymbol{b} \neq \boldsymbol{0} \\ 0, & \text{if } \boldsymbol{a} = \boldsymbol{0} \text{ or } \boldsymbol{b} = \boldsymbol{0} \end{cases} \qquad (2.8)$$

where the following mathematical rules hold:

$$\begin{array}{ll} \boldsymbol{a} \cdot \boldsymbol{b} = \boldsymbol{b} \cdot \boldsymbol{a} & \boldsymbol{a} \cdot \boldsymbol{b} = \vec{0} \Leftrightarrow \text{ if } \boldsymbol{a} \text{ orthogonal to } \boldsymbol{b} \\ (\alpha \boldsymbol{a}) \cdot \boldsymbol{b} = \boldsymbol{a} \cdot (\alpha \boldsymbol{b}) = \alpha(\boldsymbol{a} \cdot \boldsymbol{b}) & |\boldsymbol{a}| \stackrel{\text{def}}{=} \sqrt{\boldsymbol{a} \cdot \boldsymbol{a}} \\ (\boldsymbol{a} + \boldsymbol{b}) \cdot \boldsymbol{c} = \boldsymbol{a} \cdot \boldsymbol{c} + \boldsymbol{b} \cdot \boldsymbol{c} & \end{array}$$

(2.9)

When the vectors $\boldsymbol{a}$ and $\boldsymbol{b}$ are represented in a Cartesian coordinate system, the following simple rules arise for the scalar product $(\boldsymbol{a} \cdot \boldsymbol{b})$ and for $\cos(\boldsymbol{a},\boldsymbol{b})$:

$$\boldsymbol{a} \cdot \boldsymbol{b} = a_1 b_1 + a_2 b_2 + a_3 b_3, \qquad |\boldsymbol{a}| = \sqrt{a_1^2 + a_2^2 + a_3^2} \qquad (2.10)$$

$$\cos(\boldsymbol{a},\boldsymbol{b}) = \frac{\boldsymbol{a} \cdot \boldsymbol{b}}{|\boldsymbol{a}||\boldsymbol{b}|} = \frac{a_1 b_1 + a_2 b_2 + a_3 b_3}{\sqrt{a_1^2 + a_2^2 + a_3^2}\sqrt{b_1^2 + b_2^2 + b_3^2}} \qquad (2.11)$$

The above equations hold for $\boldsymbol{a},\boldsymbol{b} \neq \vec{0}$. Especially the directional cosines in a Cartesian coordinate system are calculated as

$$\cos(\boldsymbol{a},\boldsymbol{e}_i) = \frac{|a_i|}{\sqrt{a_1^2 + a_2^2 + a_3^2}} \qquad i = 1,2,3 \qquad (2.12)$$

i.e. a_i represents the angles between the vector $\boldsymbol{a}$ and the base vectors

$$\boldsymbol{e}_1 = \begin{Bmatrix} 1 \\ 0 \\ 0 \end{Bmatrix}, \ \boldsymbol{e}_2 = \begin{Bmatrix} 0 \\ 1 \\ 0 \end{Bmatrix}, \ \boldsymbol{e}_3 = \begin{Bmatrix} 0 \\ 0 \\ 1 \end{Bmatrix} \qquad (2.13)$$

The **vector product** $\boldsymbol{a} \times \boldsymbol{b}$ of the vectors $\boldsymbol{a}$ and $\boldsymbol{b}$ has the following properties:

$\boldsymbol{a} \times \boldsymbol{b}$ is a vector $\neq \boldsymbol{0}$, if $\boldsymbol{a} \neq \boldsymbol{0}$ and $\boldsymbol{b} \neq \boldsymbol{0}$ and $\boldsymbol{a}$ is not parallel to $\boldsymbol{b}$;

$|\boldsymbol{a} \times \boldsymbol{b}| = |\boldsymbol{a}| \cdot |\boldsymbol{b}| \sin(\boldsymbol{a},\boldsymbol{b})$ (area of the parallelogram set up by $\boldsymbol{a}$ and $\boldsymbol{b}$);

$\boldsymbol{a} \times \boldsymbol{b}$ is a vector standing perpendicular to $\boldsymbol{a}$ and $\boldsymbol{b}$ and can be represented with $(\boldsymbol{a},\boldsymbol{b},\boldsymbol{a} \times \boldsymbol{b})$, a right-handed system.

It can easily be seen that $\boldsymbol{a} \times \boldsymbol{b} = \boldsymbol{0}$, if $\boldsymbol{a} = \boldsymbol{0}$ or $\boldsymbol{b} = \boldsymbol{0}$ or $\boldsymbol{a}$ is parallel to $\boldsymbol{b}$. One should take into consideration that for the vector product the associative law does not hold in general:

$$\boldsymbol{a} \times (\boldsymbol{b} \times \boldsymbol{c}) \neq (\boldsymbol{a} \times \boldsymbol{b}) \times \boldsymbol{c}$$

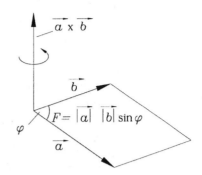

Fig. 2.2 Graphical representation of a vector product $a \times b$

The following computation rules can be stated (Fig. 2.2):

$$a \times a = 0, \quad a \times b = -(b \times a),$$
$$\alpha(a \times b) = (\alpha a) \times b = a \times (\alpha b) \quad \text{(for } \alpha \in R\text{)}$$
$$a \times (b + c) = a \times b + a \times c$$
$$(a + b) \times c = a \times c + b \times c \quad \text{(distributive laws)}$$
$$a \times b = 0 \quad \Leftrightarrow \quad a = 0 \text{ or } b = 0 \text{ or } a, b \text{ parallel} \quad \text{(parallelism test)}$$
$$|a \times b|^2 = |a|^2 \cdot |b|^2 - (a \cdot b)^2$$

If one represents the vectors a and b in a Cartesian coordinate system with e_i, the following computation rule results:

$$\begin{Bmatrix} a_1 \\ a_2 \\ a_3 \end{Bmatrix} \times \begin{Bmatrix} b_1 \\ b_2 \\ b_3 \end{Bmatrix} = \begin{vmatrix} e_1 & a_1 & b_1 \\ e_2 & a_2 & b_2 \\ e_3 & a_3 & b_3 \end{vmatrix} = \begin{Bmatrix} a_2 b_3 - a_3 b_2 \\ a_3 b_1 - a_1 b_3 \\ a_1 b_2 - a_2 b_1 \end{Bmatrix} \quad (2.14)$$

The tensor of third order $\epsilon_{ijk} := e_i \cdot (e_j \times e_k)$ that will be introduced in Sect. 2.5 permits, moreover, a computation of a vector product according to

$$\{a_i\} \times \{b_j\} := \epsilon_{ijk}\, a_i\, b_j \quad (2.15)$$

A combination of the scalar product and the vector product leads to the scalar triple product (STP) formed of three vectors:

$$\left[a, b, c\right] = a \cdot (b \times c) \quad (2.16)$$

The properties of this product from three vectors can be seen from Fig. 2.3. The STP of the vectors a, b, c leads to six times the volume of the parallelopiped (ppd), V_{ppd}, defined by the vectors: a, b and c.

The "parallelopiped product" of the three vectors a, b, c is calculated from the value of a triple-row determinant:

$$[a, b, c] = \begin{vmatrix} a_1 & b_1 & c_1 \\ a_2 & b_2 & c_2 \\ a_3 & b_3 & c_3 \end{vmatrix} \quad (2.17)$$

2.4 Tensors of Second Order

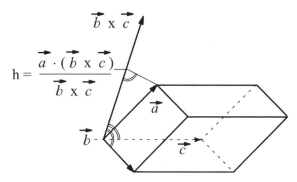

Fig. 2.3 Graphical representation of scalar triple product by three vectors

$$V_{\text{ppd}} = \frac{1}{6} V_{\text{STP}} = \frac{1}{6} [a, b, c] \qquad (2.18)$$

It is easy to show that for the STP

$$\begin{aligned} a \cdot (b \times c) = b \cdot (a \times c) &= c \cdot (a \times b) \\ &- b \cdot (a \times c) \end{aligned} \qquad (2.19)$$

For the vector triple product $a \times b \times c$, the following relation holds:

$$a \times (b \times c) = (a \cdot c)b - (a \cdot b)c \qquad (2.20)$$

Further important references are given in books on vector analysis; see also Sect. 2.12.

2.4 Tensors of Second Order

In the preceding two sections, tensors of zero order (scalar quantities) and tensors of first order (vectorial quantities) were introduced. In this section, a summary concerning tensors of second order is given, which can be formulated as matrices with nine elements:

$$\{a_{ij}\} = \begin{Bmatrix} a_{11} \ a_{12} \ a_{13} \\ a_{21} \ a_{22} \ a_{23} \\ a_{31} \ a_{32} \ a_{33} \end{Bmatrix} = a_{ij} \qquad (2.21)$$

In the matrix element a_{ij}, the index i represents the number of the row and j represents the number of the column, and the elements designated with $i = j$ are referred to as the diagonal elements of the matrix. A tensor of second order is called symmetrical when $a_{ij} = a_{ji}$ holds. The unit second-order tensor is expressed by the Kronecker delta:

$$\delta_{ij} = \begin{Bmatrix} 1 \ 0 \ 0 \\ 0 \ 1 \ 0 \\ 0 \ 0 \ 1 \end{Bmatrix}, \text{ i.e. } \delta_{ij} = \begin{cases} +1 & \text{if } i = j \\ 0 & \text{if } i \neq j \end{cases} \qquad (2.22)$$

The transposed tensor of $\{a_{ij}\}$ is formed by exchanging the rows and columns of the tensor: $\{a_{ij}\}^T = \{a_{ji}\}$. When doing so, it is apparent that the transposed unit tensor of second order is again the unit tensor, i.e. $\delta_{ij}^T = \delta_{ij}$.

The sum or difference of two tensors of second order is defined as a tensor of second order whose elements are formed from the sum or difference of the corresponding ij elements of the initial tensors:

$$\{a_{ij} \pm b_{ij}\} = \left\{ \begin{array}{ccc} a_{11} \pm b_{11} & a_{12} \pm b_{12} & a_{13} \pm b_{13} \\ a_{21} \pm b_{21} & a_{22} \pm b_{22} & a_{23} \pm b_{23} \\ a_{31} \pm b_{31} & a_{32} \pm b_{32} & a_{33} \pm b_{33} \end{array} \right\} \tag{2.23}$$

In the case of the following presentation of tensor products, often the so-called Einstein's summation convention is applied. By this one understands the summation over the same indices in a product.

When forming a product from tensors, one distinguishes the outer product and the inner product. The outer product is again a tensor, where each element of the first tensor multiplied with each element of the second tensor results in an element of the new tensor. Thus the product of a scalar and a tensor of second order forms a tensor of second order, where each element results from the initial tensor of second order by scalar multiplication:

$$\alpha \cdot \{a_{ij}\} = \{\alpha \cdot a_{ij}\} = \left\{ \begin{array}{ccc} \alpha \cdot a_{11} & \alpha \cdot a_{12} & \alpha \cdot a_{13} \\ \alpha \cdot a_{21} & \alpha \cdot a_{22} & \alpha \cdot a_{23} \\ \alpha \cdot a_{31} & \alpha \cdot a_{32} & \alpha \cdot a_{33} \end{array} \right\} \tag{2.24}$$

The outer product of a vector (tensor of first order) and a tensor of second order results in a tensor of third order with altogether 27 elements. The inner product of tensors, however, can result in a contraction of the order. As examples are cited the products $a_{ij} \cdot b_j$:

$$\{a_{ij}\} \cdot \{b_j\} = \left\{ \begin{array}{ccc} a_{11} & a_{12} & a_{13} \\ a_{21} & a_{22} & a_{23} \\ a_{31} & a_{32} & a_{33} \end{array} \right\} \left\{ \begin{array}{c} b_1 \\ b_2 \\ b_3 \end{array} \right\} = \left\{ \begin{array}{c} a_{11}b_1 + a_{12}b_2 + a_{13}b_3 \\ a_{21}b_1 + a_{22}b_2 + a_{23}b_3 \\ a_{31}b_1 + a_{32}b_2 + a_{33}b_3 \end{array} \right\} \tag{2.25}$$

and

$$\{b_i\}^T \cdot \{a_{ij}\} = \{b_1, b_2, b_3\} \cdot \left\{ \begin{array}{ccc} a_{11} & a_{12} & a_{13} \\ a_{21} & a_{22} & a_{23} \\ a_{31} & a_{32} & a_{33} \end{array} \right\} = \left\{ \begin{array}{c} b_1 a_{11} + b_2 a_{21} + b_3 a_{31} \\ b_1 a_{12} + b_2 a_{22} + b_3 a_{32} \\ b_1 a_{13} + b_2 a_{23} + b_3 a_{33} \end{array} \right\}^T \tag{2.26}$$

In summary, this can be written as

$$\{a_{ij}\} \cdot \{b_j\} = \{(a_{ij} b_j)\} = \{(ab)_i\} \tag{2.27}$$

and

$$\{b_i\} \cdot \{a_{ij}\} = \{(b_i a_{ij})\} = \{(ab)_j\} \tag{2.28}$$

If one takes into account the above product laws:

$$\{\delta_{ij}\} \cdot \{b_j\} = \{b_i\} \quad \text{and} \quad \{b_i\}^T \cdot \{\delta_{ij}\} = \{b_j\}^T \tag{2.29}$$

The multiplication of a tensor of second order by the unit tensor of second order, i.e. the "Kronecker delta", yields the initial tensor of second order:

$$\{\delta_{ij}\} \cdot \{a_{ij}\} = \begin{Bmatrix} 1 & 0 & 0 \\ 0 & 1 & 0 \\ 0 & 0 & 1 \end{Bmatrix} \cdot \begin{Bmatrix} a_{11} & a_{12} & a_{13} \\ a_{21} & a_{22} & a_{23} \\ a_{31} & a_{32} & a_{33} \end{Bmatrix} = \{a_{ij}\} \tag{2.30}$$

Further products can be formulated, as for example cross products between vectors and tensors of second order:

$$\{a_i\} \cdot \{b_{jk}\} = \epsilon_{ikj} \cdot a_i \cdot b_{jk} \tag{2.31}$$

but these are not of special importance for the derivations of the basic laws in fluid mechanics.

2.5 Field Variables and Mathematical Operations

In fluid mechanics, it is usual to present thermodynamic state quantities of fluids, such as density ρ, pressure P, temperature T and internal energy e, as a function of space and time, a Cartesian coordinate system being applied here generally. To each point $\mathcal{P}(x_1, x_2, x_3) = \mathcal{P}(x_i)$ a value $\rho(x_i, t), P(x_i, t), T(x_i, t), e(x_i, t)$, etc., is assigned, i.e. the entire fluid properties are presented as field variables and are thus functions of space and time Fig. 2.4. It is assumed that in each point in space the thermodynamic connections between the state quantities hold, as for example the state equations that can be formulated for thermodynamically ideal fluids as follows:

$\rho = \text{constant}$ \quad (state equation of the thermodynamically ideal liquids)

$P/\rho = RT$ \quad (state equation of the thermodynamically ideal gases)

Entirely analogous to this, the properties of the flows can be described by introducing the velocity vector, i.e. its components, as functions of space and time, i.e. as vector field Fig. 2.5. Furthermore, the local rotation of the flow field can be included as a field quantity, as well as the mass forces and mass acceleration acting locally on the fluid. Thus the velocity $\boldsymbol{U}_j = U_j(x_i, t)$, the rotation $\omega_j = \omega_j(x_i, t)$, the force $\boldsymbol{K}_j = K_j(x_i, t)$ and the acceleration $\boldsymbol{g}_j(x_i, t)$ can be stated as field quantities and can be employed as such quantities in the following considerations.

In an analogous manner, tensors of second and higher order can also be introduced as field variables, for example, $\tau_{ij}(x_i, t)$, which is the molecule-dependent momentum transport existing at a point in space, i.e. at the point

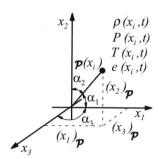

Fig. 2.4 Scalar fields assign a scalar to each point in the space and as a function of time

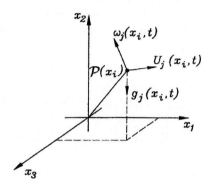

Fig. 2.5 Vector fields assign vectors to each point in the space as functions of time

$\mathcal{P}(x_i)$ at time t. It represents the j-momentum transport acting in the x_i direction. Further, $\epsilon_{ij}(x_i, t)$ represents the fluid element deformation depending on the gradients of the velocity field at the location $\mathcal{P}(x_i)$ at time t.

The properties introduced as field variables into the above considerations represented tensors of zero order (scalars), tensors of first order (vectors) and tensors of second order. They are employed in fluid mechanics to describe fluid flows and the corresponding fluid description is usually attributed to Euler (1707–1783). In this description, all quantities considered in the representations of fluid mechanics are dealt with as functions of space and time. Mathematical operations such as addition, subtraction, division, multiplication, differentiation and integration, that are applied to these quantities, are subject to the known laws of mathematics.

The differentiation of a scalar field, for example the density $\rho(x_i, t)$, gives

$$\begin{aligned}\frac{d\rho}{dt} &= \frac{\partial \rho}{\partial t} + \frac{\partial \rho}{\partial x_1}\left(\frac{dx_1}{dt}\right) + \frac{\partial \rho}{\partial x_2}\left(\frac{dx_2}{dt}\right) + \frac{\partial \rho}{\partial x_3}\left(\frac{dx_3}{dt}\right) \\ &= \frac{\partial \rho}{\partial t} + \sum_{i=1}^{3}\left(\frac{\partial \rho}{\partial x_i}\right)\left(\frac{dx_i}{dt}\right) = \frac{\partial \rho}{\partial t} + \left(\frac{\partial \rho}{\partial x_i}\right)\left(\frac{dx_i}{dt}\right)\end{aligned} \quad (2.32)$$

In the last term, the summation symbol $\sum_{i=1}^{3}$ was omitted and the "Einstein's summation convention" was employed, according to which the double index

2.5 Field Variables and Mathematical Operations

i in $\left(\frac{\partial \rho}{\partial x_i}\right)\left(\frac{dx_i}{dt}\right)$ prescribes a summation over three terms $i = 1, 2, 3$, i.e.:

$$\sum_{i=1}^{3}\left(\frac{\partial \rho}{\partial x_i}\right)\left(\frac{dx_i}{dt}\right) = \left(\frac{\partial \rho}{\partial x_i}\right)\left(\frac{dx_i}{dt}\right) \qquad (2.33)$$

The differentiation of vectors is given by the following expressions:

$$\frac{d\mathbf{U}}{dt} = \left\{\frac{dU_1}{dt}, \frac{dU_2}{dt}, \frac{dU_3}{dt}\right\}^T \Rightarrow \frac{dU_i}{dt}, \qquad i = 1, 2, 3 \qquad (2.34)$$

i.e. each component of the vector is included in the differentiation. As the considered velocity vector depends on the space location x_i and the time t, the following differentiation law holds:

$$\frac{dU_j}{dt} = \frac{\partial U_j}{\partial t} + \frac{\partial U_j}{\partial x_i}\left(\frac{dx_i}{dt}\right) \qquad (2.35)$$

When one applies the Nabla or Del operator:

$$\nabla = \left\{\frac{\partial}{\partial x_1}, \frac{\partial}{\partial x_2}, \frac{\partial}{\partial x_3}\right\}^T = \left\{\frac{\partial}{\partial x_i}\right\}, \qquad i = 1, 2, 3 \qquad (2.36)$$

on a scalar field quantity, a vector results:

$$\nabla a = \left\{\frac{\partial a}{\partial x_1}, \frac{\partial a}{\partial x_2}, \frac{\partial a}{\partial x_3}\right\}^T = \text{grad } a = \left\{\frac{\partial a}{\partial x_i}\right\}, \qquad i = 1, 2, 3 \quad (2.37)$$

This shows that the Nabla or Del operator ∇ results in a vector field deduced from the gradient field. The different components of the resulting vector are formed from the prevailing partial differentiations of the scalar field in the directions x_i.

The scalar product of the ∇ operator with a vector yields a scalar quantity, i.e. when $(\nabla \cdot)$ applied to a vector quantity results in:

$$\nabla \cdot \mathbf{a} = \frac{\partial a_1}{\partial x_1} + \frac{\partial a_2}{\partial x_2} + \frac{\partial a_3}{\partial x_3} = \text{div } \mathbf{a} = \frac{\partial a_i}{\partial x_i} \qquad (2.38)$$

Here, in $\partial a_i/\partial x_i$ the subscript i again indicates summation over all three terms, i.e.

$$\sum_{i=1}^{3} \frac{\partial a_i}{\partial x_i} \Longrightarrow \frac{\partial a_i}{\partial x_i} \qquad \text{(Einstein's summation convention)} \qquad (2.39)$$

The vector product of the ∇ operator with the vector $\mathbf{a}$ yields correspondingly

$$\nabla \times \mathbf{a} = \begin{vmatrix} \mathbf{e}_1 & \partial/\partial x_1 & a_1 \\ \mathbf{e}_2 & \partial/\partial x_2 & a_2 \\ \mathbf{e}_3 & \partial/\partial x_3 & a_3 \end{vmatrix} = \left\{\begin{array}{c} \partial a_3/\partial x_2 - \partial a_2/\partial x_3 \\ \partial a_1/\partial x_3 - \partial a_3/\partial x_1 \\ \partial a_2/\partial x_1 - \partial a_1/\partial x_2 \end{array}\right\} = \text{rot } \mathbf{a} \qquad (2.40)$$

or

$$\nabla \times \boldsymbol{a} = \text{rot } \boldsymbol{a} = -\epsilon_{ijk}\frac{\partial a_i}{\partial x_j} = \epsilon_{ijk}\frac{\partial a_j}{\partial x_i} \quad (2.41)$$

The Levi–Civita symbol ϵ_{ijk} is also called the alternating unit tensor and is defined as follows:

$$\epsilon_{ijk} = \begin{cases} 0 : \text{if two of the three indices are equal} \\ +1 : \text{if} \quad ijk = 123,\ 231 \text{ or } 312 \\ -1 : \text{if} \quad ijk = 132,\ 213 \text{ or } 321 \end{cases} \quad (2.42)$$

Concerning the above-mentioned products of the ∇ operator, the distributive law holds, but not the commutative and associative laws.

If one applies the ∇ operator to the gradient field of a scalar function, the Laplace operator ∇^2 (alternative notation Δ) results. When applied to a, the result can be written as follows:

$$\nabla^2 a = (\nabla \cdot \nabla)a = \frac{\partial^2 a}{\partial x_1^2} + \frac{\partial^2 a}{\partial x_2^2} + \frac{\partial^2 a}{\partial x_3^2} = \frac{\partial^2 a}{\partial x_i \partial x_i} \quad (2.43)$$

The Laplace operator can also be applied to vector fields, i.e. to the components of the vector:

$$\nabla^2 \boldsymbol{U} = \begin{pmatrix} \nabla^2 U_1 \\ \nabla^2 U_2 \\ \nabla^2 U_3 \end{pmatrix} = \begin{pmatrix} (\partial^2 U_1/\partial x_1^2) + (\partial^2 U_1/\partial x_2^2) + (\partial^2 U_1/\partial x_3^2) \\ (\partial^2 U_2/\partial x_1^2) + (\partial^2 U_2/\partial x_2^2) + (\partial^2 U_2/\partial x_3^2) \\ (\partial^2 U_3/\partial x_1^2) + (\partial^2 U_3/\partial x_2^2) + (\partial^2 U_3/\partial x_3^2) \end{pmatrix} \quad (2.44)$$

2.6 Substantial Quantities and Substantial Derivative

A further approach to describing fluid mechanics processes is to derive the basic equations in terms of substantial quantities. This approach is generally named after Lagrange (1736–1813) and is based on considerations of properties of fluid elements. The state quantities of a fluid element $\Re$ such as the density $\rho_\Re$, the pressure $P_\Re$, the temperature $T_\Re$ and the energy $e_\Re$ are employed for the derivation of the laws of fluid motion. If one wants to measure or describe these properties of a fluid element in a field, one has to move with the element, i.e. one has to follow the path of the element:

$$(x_i)_{\Re,T} = (x_i)_{\Re,0} + \int_0^T (U_i)_\Re \, dt \quad (2.45)$$

As the path of a fluid element is only a function of time t and an initial space coordinate $(x_i)_{\Re,0}$, the substantial quantities, i.e. the thermodynamic state quantities of a fluid element can also only be functions of time.

2.7 Gradient, Divergence, Rotation and Laplace Operators

Thus the total differentials of all substantial quantities can be formulated as follows, with reference to (2.35):

$$\frac{da_\Re}{dt} = \frac{\partial a}{\partial t} + \frac{\partial a}{\partial x_i}\left(\frac{dx_i}{dt}\right)_\Re, \quad \text{where} \quad \left(\frac{dx_i}{dt}\right)_\Re = (U_i)_\Re \qquad (2.46)$$

The quantity specified with $(dx_i/dt)_\Re$ indicates the change of position of a fluid element $\Re$ with time, i.e. the substantial velocity of a fluid element.

If a fluid element is positioned at time t at location x_i, then $(U_i)_\Re = U_i$ results and from this arises the final equation of the substantial derivative of a field variable Da/Dt:

$$\frac{da_\Re}{dt} = \frac{Da}{Dt} = \frac{\partial a}{\partial t} + U_i \frac{\partial a}{\partial x_i} \qquad (2.47)$$

This equation results also from the identity relationship. This states, for a representation of a fluid flow in two different ways, without contradiction, that the following equality of Euler and Lagrange variables holds:

$$a_\Re(t) = a(x_i, t) \text{ if } (x_i)_\Re = x_i \text{ at time } t$$

From this one can derive

$$\frac{da_\Re}{dt} = \frac{\partial a}{\partial t} + (U_i)_\Re \frac{\partial a}{\partial x_i} = \frac{\partial a}{\partial t} + U_i \frac{\partial a}{\partial x_i} = \frac{Da}{Dt} \qquad (2.48)$$

From (2.47), it can be seen that the operator $\frac{D}{Dt}$ (= substantial derivative) can be written as follows:

$$\frac{D}{Dt} = \frac{\partial}{\partial t} + U_i \cdot \frac{\partial}{\partial x_i} = \frac{\partial}{\partial t} + (\boldsymbol{U} \cdot \nabla) \qquad (2.49)$$

This operator can be applied to field variables and is very important for the subsequent derivations of the basic equations of fluid mechanics, as it permits the formulation of the basic equations in Lagrange variables and in a second step the subsequent transformation of all terms in this equation into Euler variables. In this final form, i.e. expressed in Euler variables, the equations are suited for the solution of practical flow problems.

2.7 Gradient, Divergence, Rotation and Laplace Operators

$a(x_i, t)$ represents a scalar field, i.e. it is defined or given as a function of space and time. The gradient field of the scalar 'a' can be assigned the following components at each point in a space:

$$a(x_i,t) \Longrightarrow \left\{\frac{\partial a}{\partial x_i}\right\} = \operatorname{grad}(a) = \left\{\begin{array}{c} \partial a/\partial x_1 \\ \partial a/\partial x_2 \\ \partial a/\partial x_3 \end{array}\right\}; \quad \operatorname{grad}(a) = f(x_i,t) \quad (2.50)$$

Thus the operator grad( ) is defined as follows:

$$\operatorname{grad}() = \left\{\frac{\partial()}{\partial x_i}\right\} = \left\{\frac{\partial()}{\partial x_1}, \frac{\partial()}{\partial x_2}, \frac{\partial()}{\partial x_3}\right\} \quad (2.51)$$

i.e. $\operatorname{grad}(a)$ is a vector field, whose components are marked by the index i.

The $\operatorname{grad}(a)$ vectors exhibit directions which are perpendicular to the lines of $a = $ constant of the considered scalar field, i.e. perpendicular to $a(x_i,t) = $ constant.

Furthermore, the Laplace operator can be assigned to each scalar field $a(x_i,t) \Rightarrow \Delta a(x_i,t)$ (J.S. Laplace (1749–1827)). Here, $\Delta a(x_i,t)$ is a scalar field, e.g. to each space point the quantity $\Delta(a)$ is assigned

$$\Delta a(x_i,t) = \frac{\partial^2 a}{\partial x_i \partial x_i} = \frac{\partial^2 a}{\partial x_1^2} + \frac{\partial^2 a}{\partial x_2^2} + \frac{\partial^2 a}{\partial x_3^2} \quad (2.52)$$

Employing the previously defined divergence operator, the following equation results:

$$\Delta a(x_i,t) = \operatorname{div}(\operatorname{grad} a) = \operatorname{div}\left(\frac{\partial a}{\partial x_i}\right) = \frac{\partial^2 a}{\partial x_i \partial x_i} = \frac{\partial^2 a}{\partial x_i^2} \quad (2.53)$$

For the mathematical treatment of flow problems, there are other mathematical operators of importance, in addition to the operators div( ) and rot( ) = curl( ) that are applicable to vector fields such as $\boldsymbol{U}(x_i,t)$ and are defined as follows:

$$\operatorname{div}(\boldsymbol{U}(x_j,t)) = \frac{\partial U_i}{\partial x_i} = \frac{\partial U_1}{\partial x_1} + \frac{\partial U_2}{\partial x_2} + \frac{\partial U_3}{\partial x_3} \quad (2.54)$$

and

$$\operatorname{rot} \boldsymbol{U}(x_j,t) = \epsilon_{ijk} \cdot \frac{\partial U_j}{\partial x_i} = \left\{\begin{array}{c} \partial U_3/\partial x_2 - \partial U_2/\partial x_3 \\ \partial U_1/\partial x_3 - \partial U_3/\partial x_1 \\ \partial U_2/\partial x_1 - \partial U_1/\partial x_2 \end{array}\right\} \quad (2.55)$$

Here, div $\boldsymbol{U}$ is a scalar field and rot or curl $\boldsymbol{U}$ are vector fields. When $\boldsymbol{U}$ is a velocity field, the value of div $\boldsymbol{U}$ describes the temporal change of the volume $\delta V_\Re$ of a fluid element with constant mass $\delta m_\Re$, i.e.

$$\operatorname{div}(\boldsymbol{U}) = \frac{\partial U_i}{\partial x_i} = \frac{1}{\delta V_\Re} \frac{d(\delta V_\Re)}{dt} \quad (2.56)$$

If the density $\rho = $ constant is included, $\operatorname{div}(\rho \boldsymbol{U})$ implies a mass density source at the point x_i at time t. Correspondingly, rot $(\boldsymbol{U})$ or curl $(\boldsymbol{U})$ represent the vortex density of the velocity field at the point x_i at time t. If curl $(\boldsymbol{U}) = 0$, a fluid element at the point x_i, and at time t, experiences no contribution to its rotation by the velocity field. For rot $(\boldsymbol{U}) \neq 0$ at time t and at point x_i, a fluid element consequently experiences, at the corresponding point, a contribution to its rotational motion.

In summary, the operators described above can be formulated as follows:

$$\operatorname{grad}(a) = \left(\frac{\partial a}{\partial x_i}\right) = \nabla a \tag{2.57}$$

$$\operatorname{div}(\boldsymbol{U}) = \nabla \cdot U_i = \frac{\partial U_i}{\partial x_i} \tag{2.58}$$

$$\Delta a = \nabla^2 a = \nabla \cdot \nabla a = \frac{\partial}{\partial x_i}\left(\frac{\partial a}{\partial x_i}\right) = \frac{\partial^2 a}{\partial x_i \partial x_i} \tag{2.59}$$

$$\operatorname{rot}(\boldsymbol{U}) = \nabla \times \boldsymbol{U} = \epsilon_{ijk}\frac{\partial U_j}{\partial x_i} \tag{2.60}$$

These operators will be employed for the derivation of the basic equations of fluid mechanics and also when dealing with flow problems.

2.8 Line, Surface and Volume Integrals

The line integral of a scalar function $a(x_i, t)$ along a line S is defined as follows:

$$I_s(t) = \int_S a \, \mathrm{d}S = \int_S a(x_i, t)\mathrm{d}S \quad \text{with } x_i \in S \tag{2.61}$$

Line integrals of this kind are required in fluid mechanics to define the position of the "center of gravity" of a line. Their computation is carried out in three steps as follows for $t = $ constant:

1. The viewed curve is parameterized:

$$S : s(\gamma) = \{s_i(\gamma)\}^T, \quad \alpha \leq \gamma \leq \beta \tag{2.62}$$

2. The arched element $\mathrm{d}s$ is defined by differentiation:

$$\mathrm{d}s = \left|\frac{\mathrm{d}s_i(\gamma)}{\mathrm{d}\gamma}\right|\mathrm{d}\gamma = \sqrt{\left(\frac{\mathrm{d}s_i(\gamma)}{\mathrm{d}\gamma}\right)^2}\mathrm{d}\gamma \tag{2.63}$$

3. The computation of the defined integral from $\gamma = \alpha$ to $\gamma = \beta$:

$$I_s = \int_\alpha^\beta a(s_i(\gamma))\sqrt{\left(\frac{\mathrm{d}s_i}{\mathrm{d}\gamma}\right)^2}\mathrm{d}\gamma \quad \rightsquigarrow \quad I_s \tag{2.64}$$

The application of the above steps for the computation of the defined integral leads for $a = 1$ to the length of the considered curve $s(\gamma)$ between $\gamma = \alpha$ and $\gamma = \beta$.

Analogous to the above considerations, the integration of a vector field along a curve can be carried out in the following way:

$$I_{s_i}(t) = \int_S a_i \cdot \mathrm{d}s_i = \int_\alpha^\beta a_i(s_j(\gamma)) \frac{\mathrm{d}s_i(\gamma)}{\mathrm{d}\gamma} \mathrm{d}\gamma \quad \text{at time } t \tag{2.65}$$

Computations of the work done in the fields of forces, the circulation of mass and momentum flow in the case of two-dimensional flow fields are effected via such defined integrations of vector fields along space lines.

Analogous to the integrals along lines or along line segments, space integrals for scalar and vectors can also be defined and computed according to the following computation rules:

$$I_F(t) = \iint_F a \, \mathrm{d}F \quad \text{at time } t \tag{2.66}$$

If F is the surface area of a considered fluid element and $a(x_i, t)$ a scalar field, that is continuous on the surface, the above integral as the surface integral is named from a to F. The surface-averaged value of a is computed as follows:

$$\tilde{a} = \frac{1}{F_0} \iint_{F_0} a \, \mathrm{d}F_0 \quad \text{(surface mean value)} \tag{2.67}$$

For the surface integral of a vector field holds that

$$I_F(t) = \iint_F a_i \, \mathrm{d}F_i = \iint_F a_i \, n_i \, \mathrm{d}F \quad \text{at time } t \tag{2.68}$$

For the case $a_i = U_i$, i.e. the execution of a surface integration over the velocity field, an integral value is obtained that corresponds to the instantaneous volume flow through the surface F:

$$\dot{Q}(t) = \iint_F U_i \, \mathrm{d}F_i \quad \text{(volume flow through } F \text{ at time } t\text{)} \tag{2.69}$$

Analogously, the mass flow through F is computed by

$$\dot{M}(t) = \iint_F \rho U_i \, \mathrm{d}F_i \quad \text{(mass flow through } F \text{ at time } t\text{)} \tag{2.70}$$

The mean mass flow density is given by

$$\tilde{\dot{m}}(t) = \frac{1}{F} \iint_F \rho U_i \, \mathrm{d}F_i \tag{2.71}$$

The above integrations can be extended to volume integrals, which again can be applied to scalar and vector fields. If V designates the volume of a regular field and $a(x_i, t)$ a steady scalar field (occupation function) given in this space, the total occupancy of the space is computed as follows:

$$I_V(t) = \iiint_V a\, dV \quad \text{(total occupancy of } V \text{ through } a \text{ at time } t) \quad (2.72)$$

The mass of a regular space with the density distribution $\rho(x_i, t)$ results in the following triple integral of the density distribution $\rho(x_i, t)$:

$$\widehat{M(t)} = \iiint_V \rho(x_i, t) dx_1 dx_2 dx_3 \quad (2.73)$$

Here, variables with $\sim$ and $\wedge$ symbols indicate surface- and volume-averaged quantities in this chapter of the book.

For the practical implementation of surface and volume integrations, it is often advantageous to employ the laws of Guldin:

1. Law of Guldin (1577–1643): The surface area of a body with rotational symmetry is given by $\int_s 2\pi \cdot r(s)\, ds$, with d$s$ denoting an arc element of the plane curve s generating the body and $r(s)$ denoting the distance of ds from the axis of rotation.
2. Law of Guldin (1577–1643): The volume of a body with rotational symmetry is given by $\int_F 2\pi \cdot r_s(F) \cdot dF$, with d$F$ denoting an area element of the area enclosed by the plane curve s generating the body and $r_s(F)$ denoting the distance of dF from the axis of rotation.

2.9 Integral Laws of Stokes and Gauss

The integral law named after Stokes (1819–1903) reads

$$\oint_s \boldsymbol{a} \cdot d\boldsymbol{s} = \iint_{O_s} \text{rot } \boldsymbol{a} \cdot d\boldsymbol{F} \quad (2.74)$$

which means that the line integral of a vector $\boldsymbol{a}$ over the entire edge line of a surface is equal to the surface integral of the corresponding rotation of the vector quantity over the surface. Thus the integral law of Stokes represents a generalization of Green's law (1793–1841), which was formulated for plane surfaces, i.e. for "spatial areas". If one stretches two different surfaces over a boundary of surface S, Stokes' law gives

$$\iint_{O_{S_1}} \text{rot } \boldsymbol{a} \cdot d\boldsymbol{F} = \iint_{O_{S_2}} \text{rot } \boldsymbol{a} \cdot d\boldsymbol{F} \quad (2.75)$$

where S is equal to the stretching quantities of O_{S_1} and O_{S_2}. If one introduces by $\varGamma = \oint_S \boldsymbol{U} \mathrm{d}\boldsymbol{s}$ the term of circulation of a vector field $\boldsymbol{U}$ along a boundary S employing the mean-value law of the integral calculus from the rot integral of velocity field U_i over a surface O_S with normal $\boldsymbol{n}$, surface area F and boundary curve S, the surface integral of the vector field $\boldsymbol{U}$ when $F \to 0$ in the borderline case results:

$$\varGamma = \boldsymbol{n} \cdot \operatorname{rot} \boldsymbol{U} = \lim_{F \to 0} \frac{1}{F} \oint_S \boldsymbol{U} \cdot \mathrm{d}\boldsymbol{S}$$

This relation makes it clear that the rotation effect of a fluid element is at its maximum when the surface normal $\boldsymbol{n}$ is in the direction of the rot $\boldsymbol{U}$ vector.

The integral law named after Gauss can be formulated as follows:

$$\iiint_V \operatorname{div} \boldsymbol{a} \cdot \mathrm{d}V = \iiint_V \frac{\partial a_i}{\partial x_i} \mathrm{d}V = \oiint_{O_V} \boldsymbol{a} \cdot \mathrm{d}\boldsymbol{F} = \oiint_{O_V} a_i \cdot n_i \, \mathrm{d}F \qquad (2.76)$$

Thus the flow of the vector field $\boldsymbol{a}(x_i, t)$ through the surface of a regular space, i.e. the flow "from the inside to the outside", is equal to the volume integral of the divergence over the space. The mean-value theorem of the integral calculus for $V \to 0$ and consideration of a velocity field U_i give

$$\operatorname{div} \boldsymbol{U} = \frac{\partial U_i}{\partial x_i} = \lim_{V \to 0} \frac{1}{V} \iint_{O_V} \boldsymbol{U} \cdot \mathrm{d}\boldsymbol{F} \qquad (2.77)$$

The divergence of a velocity field thus measures the flow emerging from the volume unit, i.e. it is the source density of U_i in the point x_i at time t.

2.10 Differential Operators in Curvilinear Orthogonal Coordinates

The compilation of important equations and definitions of vector analysis to date is based on the Cartesian coordinate system. A great number of problems can, however, be treated more easily in a curvilinear coordinate system usually adapted to a considered special geometry. As examples are cited the creeping flow around a sphere or the flow through a tube with a circular cross-section which can be described appropriately in spherical and cylindrical coordinates, respectively. In addition, the solution of a flow problem can often be simplified considerably by exploiting the symmetry properties of the problem in a curvilinear coordinate system adapted to the geometry.

In this section, only some frequently used relationships for the differential operators in curvilinear orthogonal coordinate systems will be recalled without strict derivations. More detailed and mathematically precise presentation

2.10 Differential Operators in Curvilinear Orthogonal Coordinates

can be found in the corresponding literature, as for example [2.3], [2.4] and [2.5], on whose presentation this section is oriented.

General curvilinear coordinates (x'_1, x'_3, x'_3) can be computed from Cartesian coordinates (x, y, z) by (local) unequivocally reversible relations:

$$\begin{aligned} x'_1 &= x'_1(x, y, z) \\ x'_2 &= x'_2(x, y, z) \\ x'_3 &= x'_3(x, y, z) \end{aligned} \tag{2.78}$$

Conversely the Cartesian coordinates depend on the curvilinear ones:

$$\begin{aligned} x &= x(x'_1, x'_2, x'_3) \\ y &= y(x'_1, x'_2, x'_3) \\ z &= z(x'_1, x'_2, x'_3). \end{aligned} \tag{2.79}$$

If one holds on to two coordinates, one obtains, with the third coordinate as a free parameter, a space curve, the so-called coordinate line, for example:

$$\boldsymbol{r} = \boldsymbol{r}(x'_1, x'_2 = a, x'_3 = b) \tag{2.80}$$

Concerning the respective tangential vectors

$$\boldsymbol{t}_i = \frac{\partial \boldsymbol{r}}{\partial x'_i}, \quad i = 1, 2, 3 \tag{2.81}$$

of the coordinate lines in point $P(x'_1, x'_2, x'_3)$ with the definition of the so-called metric coefficients

$$h_i := \left| \frac{\partial \boldsymbol{r}}{\partial x'_i} \right|, \quad i = 1, 2, 3 \tag{2.82}$$

the unit vectors

$$\boldsymbol{e}_i = \frac{1}{h_i} \boldsymbol{t}_i, \quad i = 1, 2, 3 \tag{2.83}$$

can be defined that form the basic vectors for a local reference system in point P. To be emphasized here is the local character of this reference system for general curvilinear coordinates, as the basic vectors themselves can depend on coordinates, contrary to coordinate-independent basic vectors $(\boldsymbol{e}_x, \boldsymbol{e}_y, \boldsymbol{e}_z)$ of the Cartesian coordinate system. If the coordinate lines at each point stand vertically on one another in pairs, i.e. the following relationship

$$\boldsymbol{e}_i \cdot \boldsymbol{e}_j = \delta_{ij} \tag{2.84}$$

holds, one designates the coordinate system curvilinear orthogonal coordinate system. Curvilinear orthogonal coordinate systems are the subject of this section. The reader interested in general curvilinear coordinate systems is referred, for example, to the book by R. Aris [2.3].

If one considers two infinitesimal closely neighbouring points $P_1(x_1', x_2', x_3')$ and $P_2(x_1' + dx_1', x_2' + dx_2', x_3' + dx_3')$, for the difference of their position vectors $\boldsymbol{r}_1 = \boldsymbol{r}(x_1', x_2', x_3')$ and $\boldsymbol{r}_2 = \boldsymbol{r}(x_1' + dx_1', x_2' + dx_2', x_3' + dx_3')$ at the lowest order (Taylor expansion)

$$d\boldsymbol{r} = \boldsymbol{r}_2 - \boldsymbol{r}_1 = \sum_{i=1}^{3} \frac{\partial \boldsymbol{r}}{\partial x_i'} dx_i' \tag{2.85}$$

holds. The length of the distance vector $d\boldsymbol{r}$, the so-called line element ds, when employing the definition for the metric coefficients, is given by

$$ds^2 = d\boldsymbol{r}^2 = \sum_{i=1}^{3} h_i^2 dx_i^2 \tag{2.86}$$

A vector field $\boldsymbol{f}$ is represented by its components in curvilinear coordinate systems:

$$\boldsymbol{f} = f_1 \boldsymbol{e}_1 + f_2 \boldsymbol{e}_2 + f_3 \boldsymbol{e}_3 \tag{2.87}$$

Without derivation, the following relationships for differential operators are stated in curvilinear orthogonal coordinates:

- Surface elements:

$$d\boldsymbol{S} = \left(\frac{\partial \boldsymbol{r}}{\partial x_i'} \times \frac{\partial \boldsymbol{r}}{\partial x_j'} \right) dx_i' \, dx_j', \quad i \neq j = 1, 2, 3 \tag{2.88}$$

- Volume elements:

$$dV = h_1 h_2 h_3 \, dx_1' dx_2' dx_3' \tag{2.89}$$

- Gradient of a scalar field Φ:

$$\text{grad}\, \Phi = \nabla \Phi = \sum_{i=1}^{3} \frac{1}{h_i} \frac{\partial \Phi}{\partial x_i'} \boldsymbol{e}_i \tag{2.90}$$

- Divergence:

$$\text{div}\, \boldsymbol{f} = \nabla \cdot \boldsymbol{f} = \frac{1}{h_1 h_2 h_3} \left(\frac{\partial h_2 h_3 f_1}{\partial x_1'} + \frac{\partial h_1 h_3 f_2}{\partial x_2'} + \frac{\partial h_1 h_2 f_3}{\partial x_3'} \right) \tag{2.91}$$

- Rotation:

$$\text{rot}\, \boldsymbol{f} = \nabla \times \boldsymbol{f} = \frac{1}{h_1 h_2 h_3} \begin{vmatrix} h_1 \boldsymbol{e}_1 & h_2 \boldsymbol{e}_2 & h_3 \boldsymbol{e}_3 \\ \frac{\partial}{\partial x_1'} & \frac{\partial}{\partial x_2'} & \frac{\partial}{\partial x_3'} \\ h_1 f_1 & h_2 f_2 & h_3 f_3 \end{vmatrix} \tag{2.92}$$

- Laplace operator:

$$\Delta \Phi = \nabla \cdot \nabla \Phi = \text{div}\,\text{grad}\, \Phi =$$
$$\frac{1}{h_1 h_2 h_3} \left[\frac{\partial}{\partial x_1'} \left(\frac{h_2 h_3}{h_1} \frac{\partial \Phi}{\partial x_1'} \right) + \frac{\partial}{\partial x_2'} \left(\frac{h_1 h_3}{h_2} \frac{\partial \Phi}{\partial x_2'} \right) + \frac{\partial}{\partial x_3'} \left(\frac{h_1 h_2}{h_3} \frac{\partial \Phi}{\partial x_3'} \right) \right] \tag{2.93}$$

2.10 Differential Operators in Curvilinear Orthogonal Coordinates

When employing differential operators in curvilinear coordinates, the dependence of the (local) unit vectors and metric coefficients of the coordinates is also to be taken into account at least in principle.

Example 1: Cylindrical Coordinates (r, φ, z) (Fig. 2.6)

- Conversion in Cartesian coordinates:

$$\begin{aligned} x &= r \cos \varphi \\ y &= r \sin \varphi \\ z &= z \end{aligned} \qquad (2.94)$$

$(0 \leq r < \infty, \quad 0 \leq \varphi \leq 2\pi, \quad -\infty < z < \infty)$

- Position vector:

$$\begin{aligned} \boldsymbol{r} &= x(r, \varphi, z)\boldsymbol{e}_x + y(r, \varphi, z)\boldsymbol{e}_y + z(r, \varphi, z)\boldsymbol{e}_z \\ &= r\boldsymbol{r}_\rho(\varphi) + z\boldsymbol{e}_z \end{aligned} \qquad (2.95)$$

- Local unit vectors:

$$\begin{aligned} \boldsymbol{e}_r &= \cos \varphi \, \boldsymbol{e}_x + \sin \varphi \, \boldsymbol{e}_y \\ \boldsymbol{e}_\varphi &= -\sin \varphi \, \boldsymbol{e}_x + \cos \varphi \, \boldsymbol{e}_y \\ \boldsymbol{e}_z &= \boldsymbol{e}_z \end{aligned} \qquad (2.96)$$

- Metric coefficients or scaling factors:

$$h_r = 1, \quad h_\varphi = r, \quad h_z = 1 \qquad (2.97)$$

- Gradient:

$$\operatorname{grad} \Phi = \frac{\partial \Phi}{\partial r}\boldsymbol{e}_r + \frac{1}{r}\frac{\partial \Phi}{\partial \varphi}\boldsymbol{e}_\varphi + \frac{\partial \Phi}{\partial z}\boldsymbol{e}_z \qquad (2.98)$$

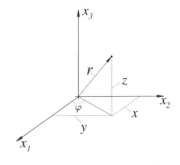

Fig. 2.6 Cylindrical coordinates

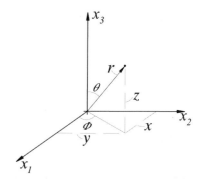

Fig. 2.7 Spherical coordinates

Example 2: Spherical Coordinates (r, θ, ϕ) (Fig. 2.7)

- Conversion in Cartesian coordinates:

$$\begin{aligned} x &= r \sin \theta \cos \phi \\ y &= r \sin \theta \sin \phi \\ z &= r \cos \theta \end{aligned} \quad (2.99)$$

$(0 \leq r < \infty, \quad 0 \leq \theta \leq \pi, \quad 0 \leq \phi \leq 2\pi)$

- Metric coefficients or scaling factors:

$$h_r = 1, \quad h_\theta = r, \quad h_\phi = r \sin \theta \quad (2.100)$$

2.11 Complex Numbers

The introduction of complex numbers permits the generalization of basic mathematical operations, as for example the square rooting of numbers, so that the extended grouping of numbers can be stated as follows:

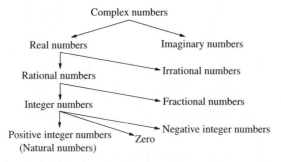

By extending to complex functions, mathematically interesting descriptions of technical problems become possible, for example the entire field of potential

2.11 Complex Numbers

flows; see Chap. 10. Complex numbers and complex functions therefore have an important role in the field of fluid mechanics. As will be shown, potential flows can be dealt with very easily through functions of complex numbers. It is therefore important to provide here an introduction to the theory of complex numbers in a summarized way.

2.11.1 Axiomatic Introduction to Complex Numbers

A complex number can formally be introduced as an arranged pair of real numbers (a, b) where the equality of two complex numbers $z_1 = (a, b)$ and $z_2 = (c, d)$ is defined as follows:

Equality: $z_1 = (a, b) = (c, d) = z_2$ holds exactly when $a = c$ and $b = d$ holds, where $a, b, c, d \in \mathbb{R}$.

The first component of a pair (a, b) is named the real part and the second component the imaginary part.

For $b = 0$, $z = (a, 0)$ is obtained, with the real number a, so that all the real numbers are a sub-set of the complex numbers. When determining basic arithmetics operations, one has to keep in mind that operations with complex numbers lead to the same results as in the case of arithmetics of real numbers, provided that the operations are restricted to real numbers in the above sense, i.e. $z = (a, 0)$.

Additions and multiplications of complex numbers are introduced by the following relationships:

$$\begin{aligned}\text{Addition:} \quad & (a, b) + (c, d) = (a + c, b + d) \\ \text{Multiplication:} \quad & (a, b) \cdot (c, d) = (ac - bd, ad + bc)\end{aligned} \quad (2.101)$$

Then,

$$\begin{aligned}(a, 0) + (c, 0) &= (a + c, 0) = a + c \\ (a, 0) \cdot (c, 0) &= (ac, 0) = ac\end{aligned} \quad (2.102)$$

i.e. no contradictions to the computational rules with real numbers arise. The quantity of the complex numbers (denoted $\mathbb{C}$ in the following) is complete as far as addition and multiplication are concerned, i.e. with $z_1, z_2 \in \mathbb{C}$ follows:

$$\begin{aligned}z_3 &= z_1 + z_2 \in \mathbb{C} \\ z_3 &= z_1 \cdot z_2 \in \mathbb{C}\end{aligned} \quad (2.103)$$

Furthermore, it can be shown that the above operations of addition and multiplication satisfy the following laws:

Commutative concerning addition: $\quad z_1 + z_2 = z_2 + z_1$
Commutative concerning multiplication: $\quad z_1 z_2 = z_2 z_1$

Associative concerning additon: $(z_1 + z_2) + z_3 = z_1 + (z_2 + z_3)$
Associative concerning multiplicaton: $(z_1 z_2) z_3 = z_1 (z_2 z_3)$

Distributive properties: $(z_1 + z_2) z_3 = z_1 z_3 + z_2 z_3$

Analogous to the case of the real number $z(a, 0)$, a so-called purely imaginary number can also be introduced: $z = 0, b$.

A complex number $z = (a, b)$ is called imaginary if $a = 0$ and $b \neq 0$. Moreover, one puts $i = (0, 1)$ and calls i an imaginary unit.

According to the multiplication rules, introduced for complex numbers, this complex number i, i.e. the number pair $(0, 1)$, has a special role, namely,

$$i^2 = (0, 1) \cdot (0, 1) = (-1, 0) = -1 \qquad (2.104)$$

i.e. the multiplication of the imaginary unit number by itself yields the real number -1. Based on equation (2.103), it can be represented as

$$i = \sqrt{-1} \qquad (2.105)$$

where the unambiguity of the root relationship for i requires some special considerations. Because

$$z = (a, b) = (a, 0) + (0, b) = (a, 0) + (0, 1) \cdot (b, 0) = a + ib \qquad (2.106)$$

each complex number $z = (a, b)$ can also be written as the sum of a real number a and an imaginary number ib.

Subtraction and division can be achieved by inversion of the addition and multiplication, i.e., $(z_1 - z_2)$ is equal to the complex number z_3, for which

$$z_2 + z_3 = z_1 \qquad (2.107)$$

holds. Following the above notation with $z_1 = (a, b)$, $z_2 = (c, d)$ results in

$$z_1 - z_2 = (a - c, b - d) \qquad (2.108)$$

$$\frac{z_1}{z_2} = \left(\frac{ac + bd}{(c^2 + d^2)}, \frac{bc - ad}{(c^2 + d^2)} \right) \qquad (2.109)$$

In the above presentations, elementary mathematical operations based on the quantity $\mathbb{C}$ of the complex numbers were introduced. All other properties of the complex numbers are followed in the implementation of these definitions.

2.11.2 Graphical Representation of Complex Numbers

In order to explain the above properties of complex numbers, they are often shown graphically in ways summarized below. Several kinds of presentations are chosen in the literature for a better understanding.

2.11 Complex Numbers

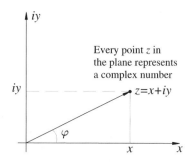

Fig. 2.8 Diagram of a complex number in the Gauss number plane

2.11.3 The Gauss Complex Number Plane

As the complex number $z = x + iy$ represents an arranged pair of numbers, a rectangular coordinate system is recommended for the graphical representation of complex numbers, in which a real axis for x and an "imaginary axis" for iy is defined. The complex number $z = x + iy$ is then defined as a point in this plane, or as a vector z from the origin of the coordinate system to the point Z with the coordinates (x, iy). This is illustrated in Fig. 2.9, where the addition and subtraction of complex numbers are stated graphically.

2.11.4 Trigonometric Representation

If one considers the graphical representation in Fig. 2.8, the following trigonometric relations can be given for complex numbers:

$$x = r\cos\varphi \quad \text{and} \quad y = r\sin\varphi \quad \text{with} \quad r = |z| \tag{2.110}$$

A complex number can therefore be written as follows:

$$z = r\cos\varphi + i(r\sin\varphi) = r(\cos\varphi + i\sin\varphi) \tag{2.111}$$

or

$$z = re^{i\varphi} \tag{2.112}$$

The connection between the exponential function and the trigonometric functions follows immediately by a series expansion of the exponential function and rearrangement of the series, i.e.

$$e^{i\varphi} = \sum_{k=0}^{\infty} \frac{(i\varphi)^k}{k!} = \sum_{k=0}^{\infty} (-1)^k \frac{\varphi^{2k}}{(2k)!} + i\sum_{k=0}^{\infty} (-1)^k \frac{\varphi^{2k+1}}{(2k+1)!} = \cos\varphi + i\sin\varphi \tag{2.113}$$

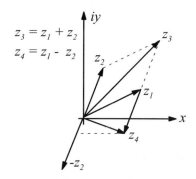

Fig. 2.9 Diagram of the addition and subtraction of complex numbers in the Gauss number plane

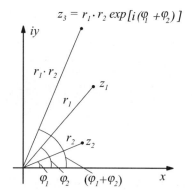

Fig. 2.10 Graphical representation of multiplication of complex numbers

With the following relationship, the multiplication and division of complex numbers can be carried out:

$$z_1 z_2 = r_1 r_2 e^{i(\varphi_1+\varphi_2)} = r_1 r_2 (\cos(\varphi_1+\varphi_2) + i \sin(\varphi_1+\varphi_2)) \quad (2.114)$$

$$\frac{z_1}{z_2} = \frac{r_1}{r_2} e^{i(\varphi_1-\varphi_2)} = \frac{r_1}{r_2}(\cos(\varphi_1-\varphi_2) + i \sin(\varphi_1-\varphi_2)) \quad (2.115)$$

These multiplications and divisions of complex numbers can be represented graphically as shown in Figs. 2.9 and 2.10.

At this point, it is advisable to discuss the treatment of the roots of complex numbers. It is explained below, how the mathematical operator $\sqrt[n]{(\,)}$ is to be applied to a complex number.

It is agreed that $\sqrt[n]{z}$ ($n \in \mathbb{N}$ = (natural number)) is the set of all those numbers raised to the $1/n$th power of the number z. Therefore, if one puts

$$z = r(\cos\varphi + i\sin\varphi) \quad (2.116)$$

Then

$$\sqrt[n]{z} = \sqrt[n]{r}\left(\cos\frac{\varphi+2k\pi}{n} + i\sin\frac{\varphi+2k\pi}{n}\right)$$
$$= \sqrt[n]{r}\, e^{i\left(\frac{\varphi}{n}+\frac{2k\pi}{n}\right)} \qquad k = 0, 1, 2, \ldots, n-1 \quad (2.117)$$

2.11 Complex Numbers

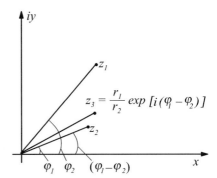

Fig. 2.11 Graphical representation of division of complex numbers

i.e. $\sqrt[n]{z}$ is a set of complex numbers consisting of n numbers of values that can be interpreted geometrically in the complex plane as corner points of a polynomial, which is inscribed in a circle with radius $\sqrt[n]{r}$ around the zero point.

Specifically for $k = 0$, for example:

$$\sqrt[n]{z} = \sqrt[n]{r}\left(\cos\frac{\varphi}{n} + i\sin\frac{\varphi}{n}\right) = \sqrt[n]{r}e^{i\frac{\varphi}{n}} \tag{2.118}$$

2.11.5 Stereographic Projection

The above representations of complex numbers were described by using the plane employed in the field of analytical geometry and well known trigonometric relationships were used. For many purposes it proves more favorable to understand the points in the $x - iy$ plane as projections of points lying on a unit sphere, whose poles lie on the axis perpendicular to the complex plane. One of the poles of the sphere lies at the zero point, whereas the other takes the position coordinates $(0, 0, 1)$. Stereographic projections are carried out from the latter pole as indicated in Fig. 2.12. Thus each point of the plane corresponds precisely to a point of the sphere which is different from N and vice versa, i.e. the spherical surface is, apart from the starting point of the projection, projected reversibly in an unequivocal manner on to the complex plane. The figure is circle-allied and angle-preserving.

- The property of the circle-allied figure indicates that each circle on the sphere is projected as a circle or a straight line on the plane (and vice versa).
- The angle-preserving figure signifies that two arbitrary circles (and generally any two curves on the sphere) intersect at the same angle as their stereographic projection in the plane (and vice versa).

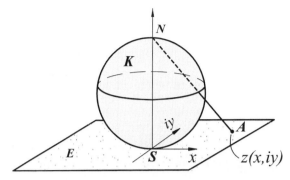

Fig. 2.12 Representation of the stereographic projection (complex sphere of Riemann)

2.11.6 Elementary Function

Complex functions are defined analogously to the introduction of real functions and can be given as follows:

> When $\mathbb{C}$ is an arbitrary set of complex numbers, $\mathbb{C}$ can be designated as the domain of the complex variables z. If one assigns to each complex variable z, within the domain $\mathbb{C}$, a complex quantity $F(z)$, then $F(z)$ is designated as the function of complex variables. The function $F(z)$ represents again a complex quantity:
> $$F(z) = \Phi + i\Psi \qquad (2.119)$$
> Here it is to be considered in general that the quantities Φ and Ψ again depend on x and iy, i.e. on the coordinates of the complex variable z.

When the definition of a complex function is compared with the often easier understandable real functions, the differentiation of a complex function has a significant difference compared to real functions. The existence of a derivative $f'(x)$ of a real function $f(x)$ does not say anything about the existence of possible higher-order derivatives, whereas from the existence of first-order derivative $f'(x)$ of a complex function $F(z)$, if automatically follows the existence of all higher derivatives, i.e.

> When a function $F(z)$, in a field $G \in \mathbb{C}$ is *holomorphic* (i.e. distinguishable in a complex manner) and exists and if the function posseses $F'(z)$, then there exist all higher-order derivatives $F''(z)$, $F'''(z)$,... also. Instead of the term holomorphic, the term analytical is often used.

The representation of $F(z)$ is often also treated as conformal mapping. The reason for this is based on the fact that, under certain restrictive conditions, the function $F(z)$ assigned to each point P in the plane z can map into another complex plane as a point Q in an imaginary plane W. In order to achieve this unequivocal assignment, a branch of an equivocal function is

2.11 Complex Numbers

often introduced as the main branch and only the latter is used for computing. The most important complex functions are as follows, see also refs. [2.2] and [2.7].

Polynomials of nth Order

$$F(z) = a_0 + a_1 z + a_2 z^2 + \cdots + a_n z^n \tag{2.120}$$

where $a_0, a_1, \ldots, a_n$ are complex constants and n a positive total number.

The transformation $F(z) = az + b$ is designated as a linear transformation in general.

Rational Algebraic Function

$$F(z) = \frac{P(z)}{Q(z)} \tag{2.121}$$

where $P(z)$ and $Q(z)$ are polynomials of arbitrary order. The special case

$$F(z) = \frac{az + b}{cz + d} \tag{2.122}$$

where $ad - bc \neq 0$ is often designated as a fractional linear function.

Exponential Function

$$F(z) = e^z = \exp(z) \tag{2.123}$$

where $e = 2.71828\ldots$ represents the basis of the (real) natural logarithm. Complex exponential functions have properties that are similar to those for real exponential functions. For example:

$$e^{z_1} \cdot e^{z_2} = e^{(z_1 + z_2)} \tag{2.124}$$

$$e^{z_1} / e^{z_2} = e^{(z_1 - z_2)} \tag{2.125}$$

Trigonometric Functions

The trigonometric functions for complex numbers are defined as follows:

$$\sin z = \frac{e^{iz} - e^{-iz}}{2i} \qquad \cos z = \frac{e^{iz} + e^{-iz}}{2} \tag{2.126}$$

$$\sec z = \frac{1}{\cos z} = \frac{2}{e^{iz} + e^{-iz}} \qquad \csc z = \frac{1}{\sin z} = \frac{2i}{e^{iz} - e^{-iz}} \tag{2.127}$$

$$\tan z = \frac{\sin z}{\cos z} = \frac{e^{iz} - e^{-iz}}{i(e^{iz} + e^{-iz})} \qquad \cot z = \frac{\cos z}{\sin z} = \frac{i(e^{iz} + e^{-iz})}{e^{iz} - e^{-iz}} \tag{2.128}$$

Many of the properties of the above functions are similar to those of real trigonometric functions. Thus it can be shown that

$$\sin^2 z + \cos^2 z = 1; \ 1 + \tan^2 z = \sec^2 z; \ 1 + \cot^2 z = \csc^2 z \quad (2.129)$$

$$\sin(-z) = -\sin z; \quad \cos(-z) = \cos z; \quad \tan(-z) = -\tan z \quad (2.130)$$

$$\sin(z_1 \pm z_2) = \sin z_1 \cos z_2 \pm \cos z_1 \sin z_2 \quad (2.131)$$

$$\cos(z_1 \pm z_2) = \cos z_1 \cos z_2 \mp \sin z_1 \sin z_2 \quad (2.132)$$

$$\tan(z_1 \pm z_2) = \frac{\tan z_1 \pm \tan z_2}{1 \mp \tan z_1 \tan z_2} \quad (2.133)$$

Hyperbolic Functions

The hyperbolic functions in the complex case are defined as follows:

$$\sinh z = \frac{e^z - e^{-z}}{2} \quad \cosh z = \frac{e^z + e^{-z}}{2} \quad (2.134)$$

$$\operatorname{sech} z = \frac{1}{\cosh z} = \frac{2}{e^z + e^{-z}} \quad \operatorname{csch} z = \frac{1}{\sinh z} = \frac{2}{e^z - e^{-z}} \quad (2.135)$$

$$\tanh z = \frac{\sinh z}{\cosh z} = \frac{e^z - e^{-z}}{e^z + e^{-z}} \quad \coth z = \frac{\cosh z}{\sinh z} = \frac{e^z + e^{-z}}{e^z - e^{-z}} \quad (2.136)$$

For these functions, the following relations apply:

$$\cosh^2 z - \sinh^2 z = 1; \ 1 - \tanh^2 z = \operatorname{sech}^2 z; \ \coth^2 z - 1 = \operatorname{csch}^2 z \quad (2.137)$$

$$\sinh(-z) = -\sinh z; \quad \cosh(-z) = \cosh z; \quad \tanh(-z) = -\tanh z \quad (2.138)$$

$$\sinh(z_1 \pm z_2) = \sinh z_1 \cosh z_2 \pm \cosh z_1 \sinh z_2 \quad (2.139)$$

$$\cosh(z_1 \pm z_2) = \cosh z_1 \cosh z_2 \pm \sinh z_1 \sinh z_2 \quad (2.140)$$

$$\tanh(z_1 \pm z_2) = \frac{\tanh z_1 \pm \tanh z_2}{1 \pm \tanh z_1 \tanh z_2} \quad (2.141)$$

From the above relations for trigonometric functions and hyperbolic functions, the following connections can be indicated:

$$\sin iz = i \sinh z \quad \cos iz = \cosh z \quad \tan iz = i \tanh z \quad (2.142)$$

$$\sinh iz = i \sin z \quad \cosh iz = \cos z \quad \tanh iz = i \tan z \quad (2.143)$$

2.11 Complex Numbers

Logarithmic Functions

As in the real case, the natural logarithm is the inverse function of the exponential function, i.e. it holds for complex cases that

$$F(z) = \ln z = \ln r + i(\varphi + 2k\pi) \qquad k = 0, \pm 1, \pm 2, \ldots \qquad (2.144)$$

where $z = re^{i\varphi}$ holds. It appears that the natural logarithm represents a non-equivocal function. By limitation to the so-called principal value of the function, an equivocalness can be produced. Here a certain arbitrariness is given. It can be eliminated by a specially desired branch, on which the equivocalness is guaranteed, which is also indicated, for example by $(\ln z)_0$.

The logarithmic functions can be defined for any real basis, i.e. also for values that differ from e. This means that the following can be stated:

$$F(z) = \log_a z \quad \Leftrightarrow \quad z = a^F \qquad (2.145)$$

where

$$a > 0 \quad \text{as well as} \quad a \neq 0 \quad \text{and} \quad a \neq 1 \qquad (2.146)$$

Inverse Trigonometric Functions

Inverse trigonometric functions for complex numbers can be stated as follows. These functions also are defined as non-equivocal, but show a periodicity:

$$\sin^{-1} z = i \ln\left(iz + \sqrt{1-z^2}\right) \quad \csc^{-1} z = \frac{1}{i} \ln\left(\frac{i + \sqrt{z^2 - 1}}{z}\right) \qquad (2.147)$$

$$\cos^{-1} z = i \ln\left(z + \sqrt{z^2 - 1}\right) \quad \sec^{-1} z = \frac{1}{i} \ln\left(\frac{1 + \sqrt{1-z^2}}{z}\right) \qquad (2.148)$$

$$\tan^{-1} z = \frac{1}{2i} \ln\left(\frac{1+iz}{1-iz}\right) \quad \cot^{-1} z = -\frac{1}{2i} \ln\left(\frac{iz+1}{iz-1}\right) \qquad (2.149)$$

Inverse Hyperbolic Functions

Analogous to the considerations of the trigonometric functions, the inverse functions of the hyperbolic functions can be formulated. These are as follows:

$$\sinh^{-1} z = \ln\left(z + \sqrt{z^2 + 1}\right) \quad \operatorname{csch}^{-1} z = \ln\left(\frac{i + \sqrt{z^2 + 1}}{z}\right) \qquad (2.150)$$

$$\cosh^{-1} z = \ln\left(z + \sqrt{z^2 - 1}\right) \quad \operatorname{sech}^{-1} z = \ln\left(\frac{1 + \sqrt{1 - z^2}}{z}\right) \qquad (2.151)$$

$$\tanh^{-1} z = \frac{1}{2} \ln\left(\frac{1+z}{1-z}\right) \quad \coth^{-1} z = \frac{1}{2} \ln\left(\frac{z+1}{z-1}\right) \qquad (2.152)$$

Differentiation of Complex Functions (Cauchy–Riemann Equations)

If the function $F(z)$ in a field $G \in \mathbb{C}$ is defined and the limiting value

$$F'(z) = \lim_{\Delta z \to 0} \frac{F(z+\Delta z) - F(z)}{\Delta z} \qquad (2.153)$$

is independent of the approximation $\Delta z \to 0$, then the function $F(z)$ in the field G is designated analytically.

A necessary condition so that the function $F(z) = \Phi + i\Psi$ represents a function analytically in $G \in \mathbb{C}$ is set by the Cauchy–Riemann differential equations:

$$\frac{\partial \Phi}{\partial x} = \frac{\partial \Psi}{\partial y} \qquad \frac{\partial \Phi}{\partial y} = -\frac{\partial \Psi}{\partial x} \qquad (2.154)$$

When the partial derivations of the Cauchy–Riemann equations in G are steady, then the Cauchy–Riemann equations are a sufficient condition to say that $F(z)$ is analytical in the field G.

From the Cauchy–Riemenn relations, it can be derived by differentiation that the real and imaginary parts of the function $F(z)$, i.e. the quantities $\Phi(x,y)$ and $\Psi(x,y)$, fulfill the Laplace equation, i.e.

$$\frac{\partial^2 \Phi}{\partial x^2} + \frac{\partial^2 \Phi}{\partial y^2} = 0 \qquad (2.155)$$

$$\frac{\partial^2 \Psi}{\partial x^2} - \frac{\partial^2 \Psi}{\partial y^2} = 0 \qquad (2.156)$$

Differentiation of Complex Functions

If $F(z)$, $G(z)$ and $H(z)$ are analytical functions of the complex variable z, then the differentiation laws of the functions result as indicated below. It is easy to see that they are analogus to the function of real variables.

$$\frac{d}{dz}[F(z) + G(z)] = \frac{d}{dz} F(z) + \frac{d}{dz} G(z) = F'(z) + G'(z) \qquad (2.157)$$

$$\frac{d}{dz}[F(z) - G(z)] = \frac{d}{dz} F(z) - \frac{d}{dz} G(z) = F'(z) - G'(z) \qquad (2.158)$$

$$\frac{d}{dz}[cF(z)] = c \frac{d}{dz} F(z) = cF'(z), \text{with } c \text{ as arbitrary constant} \qquad (2.159)$$

$$\frac{d}{dz}[F(z)G(z)] = F(z) \frac{d}{dz} G(z) + G(z) \frac{d}{dz} F(z) = F(z)G'(z) + G(z)F'(z) \qquad (2.160)$$

$$\frac{\mathrm{d}}{\mathrm{d}z}\left\{\frac{F(z)}{G(z)}\right\} = \frac{G(z)\dfrac{\mathrm{d}}{\mathrm{d}z}F(z) - F(z)\dfrac{\mathrm{d}}{\mathrm{d}z}G(z)}{[G(z)]^2}$$
$$= \frac{G(z)F'(z) - F(z)G'(z)}{[G(z)]^2} G(z) \neq 0 \quad (2.161)$$

When $W = F(\zeta)$ and $\zeta = G(z)$

$$\frac{\mathrm{d}W}{\mathrm{d}z} = \frac{\mathrm{d}W}{\mathrm{d}\zeta}\frac{\mathrm{d}\zeta}{\mathrm{d}z} = F'(\zeta)\frac{\mathrm{d}\zeta}{\mathrm{d}z} = F'[G(z)]G'(z) \quad (2.162)$$

The following table represents the important derivations of complex functions:

$$\begin{array}{lll}
\dfrac{\mathrm{d}}{\mathrm{d}z}(c) = 0 & \dfrac{\mathrm{d}}{\mathrm{d}z}z^n = nz^{n-1} & \dfrac{\mathrm{d}}{\mathrm{d}z}e^z = e^z \\[6pt]
\dfrac{\mathrm{d}}{\mathrm{d}z}a^z = a^z \ln a & \dfrac{\mathrm{d}}{\mathrm{d}z}\sin z = \cos z & \dfrac{\mathrm{d}}{\mathrm{d}z}\cos z = -\sin z \\[6pt]
\dfrac{\mathrm{d}}{\mathrm{d}z}\tan z = \sec^2 z & \dfrac{\mathrm{d}}{\mathrm{d}z}\cot z = -\csc^2 z & \dfrac{\mathrm{d}}{\mathrm{d}z}\sec z = \sec z \tan z \\[6pt]
\dfrac{\mathrm{d}}{\mathrm{d}z}\csc z = -\csc z \cot z & \dfrac{\mathrm{d}}{\mathrm{d}z}\log_e z = \dfrac{\mathrm{d}}{\mathrm{d}z}\ln z = \dfrac{1}{z} & \dfrac{\mathrm{d}}{\mathrm{d}z}\log_a z = \dfrac{1}{z \ln a} \\[6pt]
\dfrac{\mathrm{d}}{\mathrm{d}z}\sin^{-1} z = \dfrac{1}{\sqrt{1-z^2}} & \dfrac{\mathrm{d}}{\mathrm{d}z}\cos^{-1} z = \dfrac{-1}{\sqrt{1-z^2}} & \dfrac{\mathrm{d}}{\mathrm{d}z}\tan^{-1} z = \dfrac{1}{1+z^2} \\[6pt]
\dfrac{\mathrm{d}}{\mathrm{d}z}\cot^{-1} z = \dfrac{-1}{1+z^2} & \dfrac{\mathrm{d}}{\mathrm{d}z}\sec^{-1} z = \dfrac{1}{z\sqrt{z^2-1}} & \dfrac{\mathrm{d}}{\mathrm{d}z}\csc^{-1} z = \dfrac{-1}{z\sqrt{z^2-1}} \\[6pt]
\dfrac{\mathrm{d}}{\mathrm{d}z}\sinh z = \cosh z & \dfrac{\mathrm{d}}{\mathrm{d}z}\cosh z = \sinh z & \dfrac{\mathrm{d}}{\mathrm{d}z}\tanh z = \mathrm{sech}^2 z \\[6pt]
\dfrac{\mathrm{d}}{\mathrm{d}z}\coth z = \mathrm{csch}^2 z & \dfrac{\mathrm{d}}{\mathrm{d}z}\mathrm{sech}\, z = -\mathrm{sech}\, z \tanh z & \dfrac{\mathrm{d}}{\mathrm{d}z}\mathrm{csch}\, z = -\mathrm{csch}\, z \coth z \\[6pt]
\dfrac{\mathrm{d}}{\mathrm{d}z}\sinh^{-1} z = \dfrac{1}{\sqrt{1+z^2}} & \dfrac{\mathrm{d}}{\mathrm{d}z}\cosh^{-1} z = \dfrac{1}{\sqrt{z^2-1}} & \dfrac{\mathrm{d}}{\mathrm{d}z}\tanh^{-1} z = \dfrac{1}{1-z^2} \\[6pt]
\dfrac{\mathrm{d}}{\mathrm{d}z}\coth^{-1} z = \dfrac{1}{1-z^2} & \dfrac{\mathrm{d}}{\mathrm{d}z}\mathrm{sech}^{-1} z = \dfrac{-1}{z\sqrt{1-z^2}} & \dfrac{\mathrm{d}}{\mathrm{d}z}\mathrm{csch}^{-1} z = \dfrac{-1}{z\sqrt{z^2+1}}
\end{array}$$
$$(2.163)$$

The above derivations of important complex functions can be used for dealing with potential flows.

References

2.1. Richter EW, Joos G (1978) Höhere Mathematik für den Praktiker. Barth, Leipzig
2.2. Gradshteyn IS, Ryzhik IM (1980) Table of Integrals, Series, and Products. Academic press, San Diego
2.3. Großmann S (1988) Mathematischer Einführungskurs für die Physik. Teubner, Stuttgart

2.4. Aris R (1989) Vectors, Tensors and the Basic Equations of Fluid Mechanics, Dover, New York
2.5. Meyberg K, Vachenauer T (1989) Höhere Mathematik 1 und 2. Springer-Verlag, Berlin
2.6. Dallmann H, Elster KH (1991) Einführung in die höhere Mathematik 1–3. Gust Fischer Verlag, Jena
2.7. Rade L, Westergren B (2000) Springers Mathematische Formeln. 3. Auflage Springer-Verlag, Berlin

Chapter 3
Physical Basics

3.1 Solids and Fluids

All substances of our natural and technical environment can be subdivided into solid, liquid and gaseous media, on the basis of their state of aggregation. This subdivision is accepted in many fields of engineering in order to reveal important differences concerning the properties of the substances. This subdivision could also be applied to fluid mechanics, but, it would not be particularly advantageous. It is rather recommended to employ fluid mechanics aspects to achieve a subdivision of media, i.e. a subdivision appropriate for the treatment of fluid flow processes. To this end, the term fluid is introduced for designating all those substances that cannot be classified clearly as solids. Hence, from the point of view of fluid mechanics, all media can be subdivided into solids and fluids, the difference between the two groups being that solids possess *elasticity* as an important property, whereas fluids have *viscosity* as a characteristic property. Shear forces imposed on a solid from outside lead to inner elastic shear stresses which prevent irreversible changes of the positions of molecules of the solid. When, in contrast, external shear forces are imposed on fluids, they react with the build-up of velocity gradients, where the build-up of the gradient results via a molecule-dependent momentum transport, i.e. momentum transport through fluid viscosity. Thus elasticity (solids) and viscosity (liquids) are the properties of matter that are employed in fluid mechanics for subdividing media. However, there are a few exceptions to this subdivision, such as in the case of some of the materials in rheology exhibiting mixed properties. They are therefore referred to as visco-elastic media. Some of them behave such that for small deformations they behave like solids and for large deformations they behave like liquids.

At this point, attention is drawn to another important fact regarding the characterization of fluid properties. A fluid tries to evade the smallest external shear stresses by starting to flow. Hence it can be inferred from this that a fluid at rest is characterized by a state which is free of external shear stresses. Each area in a fluid at rest is therefore exposed to normal stresses only.

When shear stresses occur in a medium at rest, this medium is assigned to solids. The viscous (or the molecular) transport of momentum observed in a fluid, should not be mistaken to be similar to the elastic forces in solids. The viscous forces cannot even be analogously addressed as elastic force. This is the case for all liquids and gases as the two important subgroups of fluids which take part in the fluid motions considered in the book. Hence the present book is dedicated to the treatment of fluid flows of liquids and gases. On the basis of these explanations of fluid flows, the fluids in motion can simply be seen as media free from stresses and are therefore distinguished from solids. The "shear stresses" that are often introduced when treating fluid flows of common liquids and gases represent molecule-dependent momentum-transport terms in reality.

Neighboring layers of a flowing fluid, having a velocity gradient between them, do not interact with each another through "shear stresses" but through an exchange of momentum due to the molecular motion between the layers. This can be explained by simplified derivations aiming for a clear physical understanding of the molecular processes, as stated in the following section. The derivations presented below are carried out for an ideal gas, since they can be understood particularly well for this case of fluid motion. The results from these derivations therefore cannot be transferred in all aspects to fluids with more complex properties.

For further subdivision of fluids, it is recommended to make use of their response to normal stresses (or pressure) acting on fluid elements. When a fluid element reacts to pressure changes by adjusting its volume and consequently its density, the fluid is called compressible. When no volume or density changes occur with pressure or temperature, the fluid is regarded as incompressible although, strictly, incompressible fluids do not exist. However, such a subdivision is reasonable and moreover useful and this will be shown in the following derivations of the basic fluid mechanics equations. Indeed, this subdivision mainly distinguishes liquids from gases.

In general, as said above, fluids can be subdivided into liquids and gases. Liquids and some plastic materials show very small expansion coefficients (typical values for isobaric expansion are $\beta_P = 10 \times 10^{-6}\,\text{K}^{-1}$), whereas gases have much larger expansion coefficients (typical values are $\beta_P = 1{,}000 \times 10^{-6}\,\text{K}^{-1}$). A comparison of the two subgroups of fluids shows that liquids fulfill the condition of incompressibility with a precision that is adequate for the treatment of most flow problems. Based on the assumption of incompressibility, the basic equations of fluid mechanics can be simplified, as the following derivations show; in particular, the number of equations needed for the general description of fluid flow processes is reduced from 6 to 4. This simplification of the basic equations for incompressible fluid flows allows a considerable reduction in the complexity of the requested theoretical treatments for simple and complex geometries, e.g. in the case of problems without heat transfer the energy equation does not have to be solved.

The simplified basic equations of fluid mechanics, derived for incompressible media, can occasionally also be applied to flows of compressible fluids, such for cases where the density variations, occurring in the entire flow field, are small compared with the fluid density. This point is treated separately in Chap. 12, where conditions are derived under which density changes in gases can be neglected for the treatment of flow processes. Flows in gases can be treated like incompressible flows under the conditions indicated there.

For further characterization of a fluid, reference is made to the well-known fact that solids conserve their form, whereas a fluid volume has no form of its own, but takes the form of the container in which it is kept. Liquids differ from gases in terms of the available volume taken by the fluids, filling only part of the container, whereas the remaining part is either not filled or contains a gas and there exists a free surface between liquid and the gas. Such a surface does not exist when the container is filled only with a gas. As already said, a gas takes up the entire container volume.

Finally, it can be concluded that there are a number of media that can only be categorized, in a limited way, according to the above classification. They include media that consist of two-phase mixtures. These have properties that cannot be classified so easily. This holds also for a number of other media that can, as per the above classification, be assigned neither to solids nor to fluids and they start to flow only above a certain value of their internal "shear stress". Media of this kind will be excluded from this book, so that the above-indicated classifications of media into solids and fluids remain valid. Further restrictions on the fluid properties, that are applied in dealing with flow problems in this book, are clearly indicated in the respective sections. In this way, it should be possible to avoid mistakes that often arise in derivations of fluid mechanics equations valid only for simplified fluid properties.

3.2 Molecular Properties and Quantities of Continuum Mechanics

As all matter consists of molecules or aggregations of molecules, all macroscopic properties of matter can be described by molecular properties. Hence it is possible to derive all properties of fluids that are of importance for considerations in fluid mechanics, from properties of molecules, i.e. macroscopic properties of fluids can be described by molecular properties. However, a molecular description of the state of matter requires much effort owing to the necessary extensive formalism and moreover the treatment of macroscopic properties would remain unclear. A molecular-theoretical treatment of fluid properties would hardly be appropriate to supply application-oriented fluid mechanics information in an easily comprehensible form. For this reason, it is more advantageous to introduce quantities of continuum mechanics for describing fluid properties. The connection between continuum mechanics quantities, introduced in fluid mechanics, and the molecular properties should be considered, however, as the most important links between the two different ways of description and presentation of fluid properties. Every student should have a firm knowledge regarding this kind of considerations.

Some properties of the thermodynamic state of a fluid, such as density ρ, pressure P and temperature T, are essential for the description of fluid mechanics processes and these can easily be expressed in terms of molecular quantities. From the following derivations one can infer that the effects of molecules or molecular properties on fluid elements or control volumes are taken into consideration by introducing the properties the density ρ, pressure P, temperature T, viscosity μ, etc., in an "integral form". As will be seen, this integral consideration is sufficient for fluid mechanics. Therefore, continuum mechanics considerations do not neglect the molecular structure of the fluids, but take molecular properties into account in an integral form, i.e. averaged over a high number of molecules.

The mass per unit volume is called the specific density ρ of a material. For a fluid element this quantity depends on its position in space, i.e. on $x_i = (x_1, x_2, x_3)$, and also on time t, so that generally

$$\rho(x_i, t) = \lim_{\Delta V \to \delta V_\Re} \frac{\Delta M}{\Delta V} = \frac{\delta m_\Re}{\delta V_\Re} \qquad (3.1)$$

holds (Figs. 3.1 and 3.2). If n is considered to be the mean number of the molecules existing per unit volume and m the mass of a single molecule, the following connection between ρ, n and m holds:

$$\rho(x_i, t) = m n(x_i, t). \qquad (3.2)$$

The density of the matter is thus identical with the number of molecules available per unit volume, multiplied by the mass of a single molecule. Therefore, changes of density in space and in time correspond to spatial and temporal changes of the mean number of molecules available per unit volume, i.e. where ρ is large, n is large.

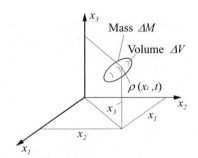

Fig. 3.1 Definition of the fluid density at a point in space, $\rho(x_i, t)$

3.2 Molecular Properties and Quantities of Continuum Mechanics

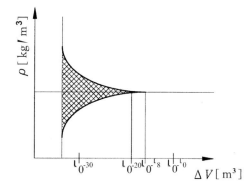

Fig. 3.2 Fluctuations with increasing ΔV while determining the density of fluids

Stochastic considerations of the thermal molecular motions in a fluid volume permit, having a very large number of molecules under normal conditions, a mean number of molecules to be specified at time t. Volumes of the order of magnitude of 10^{-18}–$10^{-20}\,\mathrm{m}^3$ are considered to be sufficiently large for arriving at a clear definition of density for gases. The treatments of flow processes in fluid mechanics are usually carried out for much larger volumes, therefore the specification of a "mean number of molecules" is appropriate in order to provide the needed mass in the considered volume. The local density $\rho(x_i, t)$ therefore describes a property of matter that is essential for fluid mechanics with a precision that is sufficient to treat fluid flows. The control volumes in fluid mechanics considerations are always selected such that the determination of a local density value is possible in spite of the molecular nature of matter. Its said above, a volume size of 10^{-18}–$10^{-20}\,\mathrm{m}^3$ fulfills the requirements of the considerations that are to be carried out from the fluid mechanics point of view. Hence, as said above, ρ can be defined in spite of the molecular structure of the fluids considered.

Similar considerations can also be made for the pressure that occurs in a fluid at rest and which is defined as the force acting per unit area (Figs. 3.3 and 3.4), i.e.:

$$P(x_i, t) = - \lim_{\Delta F_j \to \delta F_j} \frac{\Delta K_j}{\Delta F_j} \tag{3.3}$$

From the point of view of molecular theory, the pressure effect is defined as the momentum change per unit time felt per unit area, i.e. the force which the molecules experience and exert on a wall when colliding in an elastic way with the wall in the considered area. The following relation holds (see Sect. 3.3.2):

$$P = \frac{1}{3}mn\overline{u^2} = \frac{1}{3}\rho\overline{u^2}, \tag{3.4}$$

where m is the molecular mass, n the number of the molecules per unit volume and mean u the thermal velocity of the molecules.

Analogous to the above volume dimensions, it can be stated that most fluid mechanics considerations do not require area resolutions that fall below

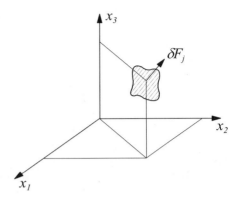

Fig. 3.3 Definition of the pressure in a fluid $P(x_i, t)$ by momentum exchange

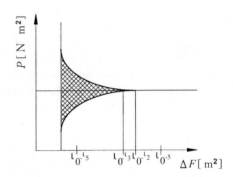

Fig. 3.4 Fluctuation while determining the pressure in a fluid

10^{-12} to 10^{-14} m² and therefore the mean numbers of molecules impinging on such an area are sufficient to have the force effect of the molecules per unit area. This, however, corresponds to a local definition of the pressure is permissible, $P(x_i, t)$ for a fluid.

Similarly to the above continuum mechanics quantities $\rho(x_i, t)$ and $P(x_i, t)$, there are other local field variables such as temperature, internal energy, and enthalpy of a fluid, etc., for which the above considerations can be repeated. Analogously to the above treatments for the density and pressure, it becomes apparent, that molecular properties define continuum properties. This again shows that it is possible for fluid mechanics considerations to neglect the complex molecular nature of fluids. It is sufficient to introduce continuum mechanics quantities into fluid mechanics considerations that correspond to mean values of corresponding molecular properties. Fluid mechanics considerations can therefore be carried out on the basis of continuum mechanics properties of fluids.

However, there are some important domains in fluid mechanics where continuum considerations are not appropriate, e.g. the investigation of flows in highly diluted gas systems. No clear continuum mechanics quantities can be defined there for the density and pressure with which fluid mechanics processes can be resolved. The required spatial resolution of the fluid mechanics considerations does not provide, due to the dilution, sufficient numbers of molecules for the necessary establishment of the mean values of the considered continuum properties. Hence there are insufficient molecules available in the considered δV for the introduction of the continuum mechanics quantities. When treating such fluid flows, priority has to be given to the molecular theory rather than continuum mechanics considerations. In the present introduction to fluid mechanics, the domain of flows of highly diluted gases is not dealt with, so that all required considerations can take place in the terminology of continuum mechanics. For continuum mechanics considerations, molecular effects, e.g. within the conservation laws for mass, momentum and energy, are presented in integral form, i.e. the molecular structure of the considered fluids is not neglected but taken into consideration in the form of integral quantities.

3.3 Transport Processes in Newtonian Fluids

3.3.1 General Considerations

When treating fluid motions including the transport of heat and momentum as well as mass transport, molecular transport processes occur that cannot be neglected and that, hence, have to be taken into account in the general transport equations. A physically correct treatment is necessary that orients itself on the general representations of these transport processes and this is indicated below. For explanation we refer to Figs. 3.5a–c. These figures show planes that lie parallel to the $x_1 - x_3$ plane of a Cartesian coordinate system. In each of these planes the temperature $T =$ constant (a), the concentration $c =$ constant (b) or the velocity $(U_j) =$ constant (c). There distributions in space are such that, when taking into account an increase in the quantities in the x_2 direction $= x_i$ direction, a positive gradient in each of these quantities exists. It is these gradients that result in the molecular transports of heat, mass and momentum.

In Fig. 3.5, the heat transport occurring, as a consequence of the molecular motion, is given by the Fourier law of heat conduction and the mass transport occurring analogously given by the Fick's law of diffusion. The Fourier law of heat conduction reads:

$$\dot{q}_i = -\lambda \frac{\partial T}{\partial x_i}, \tag{3.5}$$

where $\lambda =$ coefficient of heat conduction, and Fick's law of diffusion reads:

$$\dot{m}_i = -D \frac{\partial c}{\partial x_i}, \tag{3.6}$$

where $D =$ mass diffusion coefficient.

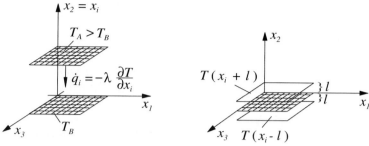

a) Explanations of heat transport

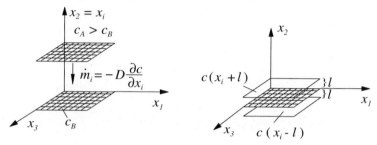

b) Explanations of transport of chemical species

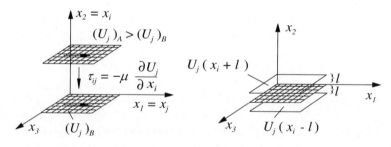

c) Explanations of momentum transport

Fig. 3.5 Analogy of the transport processes dependent on molecules for (**a**) heat transport, (**b**) mass transport, and (**c**) momentum transport

In an analogous way, the molecule-dependent momentum transport also has to be described by the Newtonian law, which in the presence of only one velocity component U_j can be stated as follows (see Bird et al. [3.1]):

$$\tau_{ij} = -\mu \frac{\partial U_j}{\partial x_i}, \tag{3.7}$$

where μ = dynamic viscosity.

In $\dot{q}_i$, $\dot{m}_i$, and τ_{ij} in the above three equations, the direction i indicates the "molecular transport direction", and j indicates the components of the

3.3 Transport Processes in Newtonian Fluids

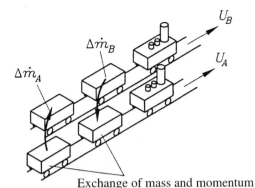

Fig. 3.6 Exchange of mass and momentum. Illustrative explanation of τ_{ij} as momentum transport

velocity vector for which momentum transport considerations are carried out. It will be shown in Chap. 5 that the complete equation for τ_{ij}, in the presence of a Newtonian medium, can be represented as follows:

$$\tau_{ij} = -\mu \left(\frac{\partial U_j}{\partial x_i} + \frac{\partial U_i}{\partial x_j} \right) + \frac{2}{3} \mu \delta_{ij} \left(\frac{\partial U_k}{\partial x_k} \right), \qquad (3.8^1)$$

where τ_{ij} represents the momentum transport per unit area and unit time and therefore represents a "stress", i.e. force per unit area. It is therefore often designated "shear stress" and the sign before the viscosity coefficient μ is chosen positive. This has to be taken into account when comparing treatments of momentum in this book with corresponding treatments in other books. The existing differences in the viewpoints are considered in following two annotations.

Annotation 1: The illustrative example in Fig. 3.6 shows how the viscosity-dependent momentum transport, introduced in continuum mechanics, is caused by motion of molecules. Two passenger trains may run next to one another at different speeds. In each of the trains, persons are assumed to travel carrying sacks along with them. These sacks are being thrown by the passengers in one train to the passengers in the other train, so that a momentum transfer takes place; it should be noted that the masses m_A and m_B of the trains do not change. Because the persons in the faster train catch the sacks that are being thrown to them from the slower train, the faster train is slowed down. In an analogous way, the slower train is accelerated. Momentum transfer (of the momentum in the direction of travel) takes place by a momentum transport perpendicular to the direction of travel. This idea transfers to the molecule-dependent momentum transport in fluids, is in accordance with the molecule-dependent transport processes of heat and mass that were stated above.

---

[1] τ_{ij} as molecule-dependent momentum transport, as introduced here, is to be differentiated, in principle, from the shear stress that is introduced in some books; see Sect. 3.3.3.

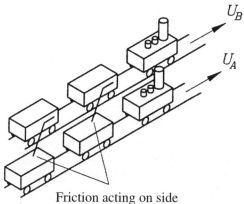

Fig. 3.7 Interaction of friction. Illustrative representation of the τ_{ij} term as a friction term

Friction acting on side walls of wagons of trains

Annotation 2: In continuum mechanics, the viscosity-dependent interaction between fluid layers is generally postulated as "friction forces" between layers. This would, in the above-described interaction between trains (Fig. 3.7), running alongside each another, correspond to passengers in each of the trains exerting a friction force of the other train with bars, by scratching along the other train's wall. This idea does not correspond to the concept of molecular-dependent transport processes between fluid layers of different speeds.

If one carries out physically correct considerations regarding the molecular-dependent momentum transport τ_{ij}, derivations have to be carried out as presented in Sect. 3.3.3. In addition, considerations are presented below concerning the existence of pressure and the occurrence of heat exchange and mass diffusion in gases in order to show the connection between molecular and continuum-mechanics quantities and transport processes.

3.3.2 Pressure in Gases

From the molecular theory point of view, the gaseous state of aggregation of a fluid is characterized by a random motion of the atoms and/or molecules. The properties that materials assume in this state of aggregation are described fairly well by the laws of an ideal gas. All the laws for ideal gases result from derivations that are based on mechanical laws for moving spheres. They interact by ideal elastic collisions and in the same way the moving molecules also interact with walls, e.g. with container walls. Between these collisions, the molecules move freely and in straight lines. In other words, no forces act between the molecules, except when their collisions take place. Likewise, container walls neither attract nor repel the molecules and the interactions of the walls with the moving molecules are limited to the moment of the collision. The most important properties of an ideal gas can be stated as follows:

3.3 Transport Processes in Newtonian Fluids

(a) The volume of the atoms and/or molecules is extremely small compared with the distances between them, so that the molecules can be regarded as material points.
(b) The molecules exert, except at the moment of their collisions, neither attractive nor repulsive forces on each other.
(c) For the collisions between two molecules or a molecule and a wall, the laws of perfect elastic collisions hold (collisions of two molecules take place exclusively).

When one takes into account the characteristic properties of an ideal gas listed in points (a)–(c), the derivations indicated below can be formulated to obtain the pressure. This fluid property represents a characteristic continuum mechanics quantity of the gas, but it can be derived by taking well-known molecular-theoretical considerations into account. The derivations in the theory only consider the known basic laws of mechanics and the molecule properties indicated above in (a)–(c).

In order to derive the "pressure effect" of the molecules on an area, the derivations are carried out by considering a control volume consisting of a cube with an edge length a as shown in Fig. 3.8. Regarding this control volume, the area standing perpendicular to the axis x_1 is hatched. All considerations are made for this area. For other areas of the control volume, the derivations have to be carried out in an analogous way, so that the considerations for the hatched area in Fig. 3.8 can be considered as generally valid, e.g. see Ref. [3.2].

In the control volume shown in Fig. 3.8, N molecules are present. With the introduction of n molecules per m^3 (molecular density), this number N is given by:

$$N = na^3. \tag{3.9}$$

From n molecules per unit volume, n_α molecules with a velocity component $(u_1)_\alpha$ may move in the direction of the axis x_1 and interact with the hatched area in Fig. 3.8. In a time Δt all molecules will hit the wall area that are at

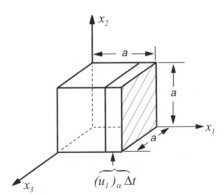

Fig. 3.8 Control volume for derivations of pressure from the molecular interaction with walls

a distance of $(u_1)_\alpha \Delta t$ from it

$$z_\alpha = n_\alpha a^2 (u_1)_\alpha \Delta t. \tag{3.10}$$

Each of the z_α molecules exerts a momentum on the wall that is formulated by the law of ideal elastic collision:

$$\Delta(i_1)_\alpha = -m\Delta(u_1)_\alpha = 2m(u_1)_\alpha \tag{3.11}$$

For the total momentum transferred by z_α molecules to the wall, we can write:

$$\Delta(J_1)_\alpha = z_\alpha \Delta(i_1)_\alpha = n_\alpha a^2 (u_1)_i \Delta t [2m(u_1)_\alpha] \tag{3.12}$$

$$\Delta(J_1)_\alpha = 2ma^2 \Delta t n_\alpha (u_1^2)_\alpha. \tag{3.13}$$

The wall experiences a force $(K_1)_\alpha$:

$$(K_1)_\alpha = \frac{\Delta(J_1)_i}{\Delta t} = 2ma^2 n_\alpha (u_1^2)_\alpha \tag{3.14}$$

and the following pressure $(P_1)_\alpha$ results:

$$(P_1)_\alpha = \frac{(K_1)_\alpha}{a^2} = 2m n_\alpha (u_1^2)_\alpha. \tag{3.15}$$

The total pressure which is exerted on the hatched area in Fig. 3.8 summarizes the pressure contributions $(P_1)_\alpha$ of all differing velocities $(u_1)_\alpha$. If one wants to calculate the total pressure, one has to sum up over all these contributions. Then one obtains from the above equation:

$$P_1 = \sum_{\alpha=1}^{n_x} (P_1)_\alpha = 2m \sum_{\alpha=1}^{n_x} n_\alpha (u_1^2)_\alpha. \tag{3.16}$$

The summation occurring in the above relation can be substituted by the following definition of the mean value of the velocity squared, $\overline{u_1^2}$:

$$\sum_{\alpha=1}^{n_x} n_\alpha (u_1^2)_\alpha = n_x \overline{(u_1^2)}, \tag{3.17}$$

If n_x is the total number of molecules per unit volume moving in the positive direction x_1, i.e.

$$n_x = \frac{1}{6} n, \tag{3.18}$$

where n_x represents the mean number of molecules present in a unit volume; $\overline{u_1^2}$ represents the square of the "effective value" of the molecular velocity, which, according to the above derivations, can be defined as follows:

$$\overline{(u_1^2)} = \frac{1}{n_\alpha} \sum_{\alpha=1}^{n_x} n_\alpha (u_1^2)_\alpha = \frac{6}{n} \sum_{\alpha=1}^{n_x} n_\alpha (u_1^2)_\alpha. \tag{3.19}$$

3.3 Transport Processes in Newtonian Fluids

The thermodynamic pressure P in a free fluid flow is defined generally as the mean value of the sum of the pressures in all three directions:

$$P = \frac{1}{3}(P_1 + P_2 + P_3) = \frac{2m}{3}\left[\sum_{\alpha=1}^{N_\alpha} n_\alpha \left(u_1^2\right)_\alpha + \sum_{\beta=1}^{N_\beta} n_\beta \left(u_2^2\right)_\beta + \sum_{\gamma=1}^{N_\gamma} n_\gamma \left(u_3^2\right)_\gamma\right]$$

$$= \frac{2m}{3}\left[n_x\overline{u_1^2} + n_y\overline{u_2^2} + n_z\overline{u_3^2}\right]$$

$$= \frac{2m}{3}\frac{1}{6}n\left[3\overline{u^2}\right]$$

$$= \frac{1}{3}mn\overline{u^2}$$

$$= \frac{1}{3}\rho\overline{u^2}$$

The pressure with which the hatched area is associated is therefore:

$$P = \frac{1}{3}mn\overline{(u^2)}. \qquad (3.20)$$

As m is the mass of a single molecule and n the mean number of molecules per unit volume, the expression mn corresponds to the density ρ in the terminology of continuum mechanics:

$$P = \frac{1}{3}mn\overline{(u^2)} = \frac{1}{3}\rho\overline{(u^2)} \qquad (3.21)$$

$$P = \frac{1}{3}\rho\overline{(u^2)}. \qquad (3.22)$$

This relationship contains the continuum property ρ and also the mean molecular velocity squared. The squared mean velocity can be eliminated by another quantity of continuum mechanics, namely the temperature T of an ideal gas (see Höfling [3.3]).

The mean kinetic energy of a molecule can be written according to the equipartition law of statistical physics:

$$e_k = \frac{1}{2}m\overline{(u^2)} = \frac{3}{2}kT, \qquad (3.23)$$

where $k = 1.380658 \times 10^{-23}\,\text{J K}^{-1}$ represents the Boltzmann constant. From (3.22) and (3.23) the following expression results:

$$P = \frac{1}{3}\rho(3\frac{k}{m}T) = \rho\frac{k}{m}T. \qquad (3.24)$$

Further,

$$k = \frac{\Re}{L} = \frac{\text{universal gas constant}}{\text{Loschmidts's number}}. \qquad (3.25)$$

Then,

$$P = \frac{\Re T}{Lm}\rho, \qquad (3.26)$$

where $\mathcal{M} = Lm$ is the mass per kmol of an ideal gas, so that $v = \mathcal{M}/\rho$ can be written:

$$Pv = \Re T, \tag{3.27}$$

where v represents the gas volume per kmol and $\Re$ is the general ideal gas constant (see to Bosnjakovic [3.4]).

Strictly, the above derivations can only be stated for a monatomic gas, with the assumption of ideal gas properties. However, the above law can be transferred to polyatomic gases with "ideal gas properties" if the additional degrees of freedom present in polyatomic gases and the corresponding constituents of the internal energy of a gas are taken into account. Generally, the energy content of a gas can be stated as follows:

$$e_{gas} = \frac{\alpha}{2} kT, \tag{3.28}$$

where α indicates the degrees of freedom of the molecular motion:

$\alpha = 3$ with monatomic gases,
$\alpha = 5$ with a diatomic gases,
$\alpha = 6$ with triatomic and polyatomic gases.

The above derivations have shown that the properties of an ideal gas, that are known from continuum mechanics, can be derived from molecular theory considerations. This means that the laws of continuum mechanics, at least for the pressure derived here, with the introduction of density and temperature, are consistent with the corresponding considerations of the mechanical theory of molecular motion.

3.3.3 Molecular-Dependent Momentum Transport

In Sect. 3.3.1, transport processes that are caused by the thermal motion of the molecules were considered in an introductory way. Attention was drawn to the analogy between momentum heat and mass transport and it was pointed out that the τ_{ij} terms used in fluid mechanics are not considered to be caused by friction, i.e. physically they represent no friction-caused "shear stress" but represent molecular-caused momentum transport terms occurring per unit area and time; the index i represents the considered molecular transport direction and j the direction of the considered momentum. In order to give an introduction into physically correct considerations of the molecular-dependent momentum transport terms, the following derivations for an ideal gas are made, where only an x_1 momentum transport in the direction x_2 is considered, i.e. the term τ_{21}.

For the derivations below, a velocity distribution has to be used that corresponds to the equilibrium distribution (Maxwell distribution), in which

3.3 Transport Processes in Newtonian Fluids

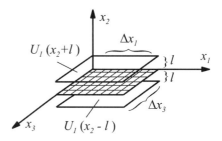

Fig. 3.9 Molecular motion and shear stress: consideration of transport in the x_2 direction of U_1 momentum

$\dot{m}_i = 0$.[2] The following simple model for a mean velocity distribution is used. One-sixth of all molecules at a time move with a velocity $(-\bar{u}, 0, 0)$, $(\bar{u}, 0, 0)$, $(0, -\bar{u}, 0)$, $(0, \bar{u}, 0)$, $(0, 0, -\bar{u})$, $(0, 0, \bar{u})$, with the directions of motion being perpendicular to the axis of the coordinates. When one assumes a molecular concentration per unit volume of n, i.e. n molecules per unit volume, one-third of them on an average move with a velocity $\bar{u}$ in the direction x_2 and of these again half, i.e. each $n/6$ molecules per unit volume, move in a negative and positive x_2 direction. On average, $\frac{1}{6} n \bar{u}$ molecules per unit time and unit area move through the area of the plane $x_2 = $ constant, which is indicated in Fig. 3.9.

The molecules which traverse the plane $x_2 = $ constant in the positive x_2 direction have, on average, collided the last time at a distance l with molecules below the plane, where l represents the mean free path of the molecular motion. The molecules coming from below thus possess, on average, the mean velocity which the flowing medium has in the plane $(x_2 - l)$. Consequently, the molecular transport in the positive direction x_2 posseses an x_1 momentum, which can be stated as follows:

$$\Delta J_1 = \frac{1}{6} n \bar{u} [m U_1(x_2 - l)] \Delta t \, \Delta x_1 \, \Delta x_3. \tag{3.29}$$

This is connected with an "effective force" per unit time and unit area, i.e. a force acting on the hatched area of Fig. 3.9, arising as a consequence of an x_1 momentum that is transported by molecular motion in the positive x_2 direction:

$$\tau_{21}^+ = \frac{\Delta J_1}{\Delta t} \frac{1}{\Delta x_1 \Delta x_3} = \frac{1}{6} m n \bar{u} U_1(x_2 - l). \tag{3.30}$$

In an analogous way, the molecular motion through the plane $x_2 = $ constant in the negative x_2 direction can be stated to carryout a x_1 momentum transport per unit time and unit area. For the latter the resultant stress can be calculated as:

$$\tau_{21}^- = -\frac{1}{6} n m \bar{u} U_1(x_2 + l). \tag{3.31}$$

Hence the entire momentum exchange per unit area and unit time which the hatched plane $x_2 = $ constant experiences can be expressed as:

---

[2] The derivations are only valid if no self diffusion of mass is present in a flow.

$$\tau_{21} = \tau_{21}^+ + \tau_{21}^- = \frac{1}{6}nm\bar{u}[U_1(x_2 - l) - U_1(x_2 + l)]. \tag{3.32}$$

The velocities $U_1(x_2 - l)$ and $U_1(x_2 + l)$ can be expressed by means of Taylor series expansions:

$$U_1(x_2 - l) = U_1(x_2) - \left(\frac{\partial U_1}{\partial x_2}\right) l + \cdots \tag{3.33}$$

$$U_1(x_2 + l) = U_1(x_2) + \left(\frac{\partial U_1}{\partial x_2}\right) l + \cdots \tag{3.34}$$

Thus one obtains:

$$\tau_{21} = \frac{1}{6}mn\bar{u}\left(-2\frac{\partial U_1}{\partial x_2}l\right), \tag{3.35}$$

i.e.

$$\tau_{21} = -\frac{1}{3}mn\bar{u}l\left(\frac{\partial U_1}{\partial x_2}\right) = -\mu\left(\frac{\partial U_1}{\partial x_2}\right). \tag{3.36}$$

Hence the derivation based on the molecualar theory of ideal gases resulted in a molecular transport of momentum. The resultant force per unit area τ_{21} turns out to be proportional to the local velocity gradient. The proportionality factor is the dynamic viscosity of the considered fluid.

For an ideal gas, this reads:

$$\mu = \frac{1}{3}mn\bar{u}l. \tag{3.37}$$

If one takes into account the following relationship:

$$\bar{u} = \sqrt{\frac{8kT}{\pi m}}; \quad l = \frac{1}{\sqrt{2}d^2\pi n}, \tag{3.38}$$

where d is the molecular diameter, one obtains:

$$\mu = \frac{2}{3\pi^{3/2}}\frac{\sqrt{mkT}}{d^2}. \tag{3.39}$$

This relationship tells us that, for an ideal gas, $\mu \sim \sqrt{T}$, i.e. the viscosity increases with increasing temperature. Furthermore, the viscosity of an ideal gas increases with increasing molecular mass, $\mu \sim \sqrt{m}$, whereas the viscosity decreases with increasing molecular size, $\mu \sim (1/d^2)$.

The above considerations were carried out to serve as an introduction to the derivations of continuum mechanics properties of fluids, using average molecular properties. Only one transport direction was taken into account and only the x_1 momentum term was included in the considerations. The complete term τ_{ij} is derived in Chap. 5 for Newtonian media, hence completing the considerations of the momentum transport in ideal gases. It is shown that the τ_{ij} term comprises essentially three terms, which can all be described physically with the help of considerations of the momentum transport by ideal gases. A general approach for momentum-transport processes in fluids is possible, i.e. of fluids different to an ideal gas, but this approach is not undertaken within the framework of this book.

3.3.4 Molecular Transport of Heat and Mass in Gases

When the temperature in a system is not constant spatially, this system is thermally not homogeneous and heat will be transferred from areas of higher temperature to areas of lower temperature. For the one-dimensional problem shown in Fig. 3.10, a heat flux $\dot{q}_{x_2} = \dot{Q}/(\Delta t \Delta x_1 \Delta x_3)$, i.e. per unit area and unit time, will take place, which is proportional to the temperature gradient existing at position $x_2 = 0$; this is known from experiments and is usually referred to as Fourier's law of heat conduction

$$\dot{q}_{x_2} = \dot{q}_2 = -\lambda \frac{\partial T}{\partial x_2}. \tag{3.40}$$

The proportionality constant λ is designated as the thermal conductivity of the fluid or the thermal conductivity coefficient of the fluid.

In this section, considerations will be summarized that are suitable for understanding the physical causes of heat conductivity from the point of view of the molecular theory of ideal gases. Hence the derivations are again given for the model of an ideal gas. Considerations of this kind are most suitable because the molecular motion, as a basis for heat transport considerations, can be presented in a simple way.

If one considers a plane A at $x_2 =$ constant in an ideal gas, where a temperature gradient exists that can be stated as the derivative of $T(x_2)$, the heat conduction through plane A can be explained in such a way that the molecules in both directions traverse plane A and carry the "thermal energy" with them. When $(\partial T/\partial x_2) > 0$, the molecules which move through the plane from the top to the bottom have a higher mean energy than the molecules which traverse plane A in the opposite direction. The heat flow

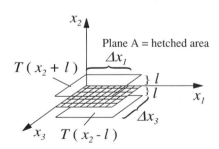

Fig. 3.10 Heat transport through a plane, caused by molecules

through plane A can thus be explained as the difference between the energy transports which stem from the upward and downward molecular motions. The following equations can be given to express this.

The energy flow in the positive x_2 direction can be given as:

$$\dot{q}_2^+ = \frac{1}{6} n \bar{u} e(x_2 - l), \qquad (3.41)$$

where n is the number of molecules per unit volume, $\bar{u}$ is the mean molecular velocity and $\frac{1}{6} n \bar{u}$ is the number of molecules that traverse the considered unit area per unit time in the positive x_2 direction. These molecules had, on average, the last contact with other molecules in a plane that has the distance of the mean free path l of the molecules. Concerning the "energy content" of the molecular flow through the area $A = 1$, it can be said that the molecules had an energy at the position $(x_2 - l)$, bring the same as that of all the molecules of the ideal gas, i.e. the energy $ne(x_2 - l)$.

In an analogous way, for the energy flow through the plane $A = 1$ constant in the negative x_2 direction [3.2]:

$$\dot{q}_2^- = -\frac{1}{6} n \bar{u} e(x_2 + l). \qquad (3.42)$$

The net heat flow results from the difference of the molecular-dependent energy transports in the positive and negative x_2 directions:

$$\dot{q}_2 = \dot{q}_2^+ + \dot{q}_2^- = \frac{1}{6} n \bar{u} [e(x_2 - l) - e(x_2 + l)]. \qquad (3.43)$$

By means of Taylor series expansion, one obtains:

$$e(x_2 - l) = e(x_2) - \left(\frac{\partial e}{\partial x_2}\right) l + \left(\frac{1}{2}\frac{\partial^2 e}{\partial x_2^2}\right) l^2 - \cdots \qquad (3.44)$$

$$e(x_2 + l) = e(x_2) + \left(\frac{\partial e}{\partial x_2}\right) l + \frac{1}{2}\left(\frac{\partial^2 e}{\partial x_2^2}\right) l^2 + \cdots \qquad (3.45)$$

For the difference $e(x_2 - l) - e(x_2 + l)$, one obtains in a first approximation, taking only first-order derivatives into account:

$$e(x_2 - l) - e(x_2 + l) = -2l \left(\frac{\partial e}{\partial x_2}\right) + \cdots \qquad (3.46)$$

and thus for the heat flow:

$$\dot{q}_2 = -\frac{1}{3} n \bar{u} l \left(\frac{\partial e}{\partial x_2}\right) = -\frac{1}{3} n \bar{u} l \left(\frac{\partial e}{\partial T}\right)\left(\frac{\partial T}{\partial x_2}\right). \qquad (3.47)$$

From the derivations given in Sect. 3.3.2 for the "heat energy" of molecules of an ideal gas:

$$e = e_k = \frac{3}{2} kT \qquad \frac{\partial e}{\partial T} = \frac{3}{2} k = c_v. \qquad (3.48)$$

3.3 Transport Processes in Newtonian Fluids

The Boltzmann constant k is understood to be a measure of the heat capacity of a molecule. When one considers again

$$\bar{u} = \sqrt{\frac{8kT}{\pi m}}; \qquad l = \frac{1}{\sqrt{2}d^2 \pi n} \qquad (3.49)$$

one obtains

$$\lambda = \frac{1}{d^2}\sqrt{\left(\frac{k}{\pi}\right)^3 \frac{T}{m}}. \qquad (3.50)$$

Analogous to the viscosity, the heat conductivity increases with increasing temperature, $\lambda \sim \sqrt{T}$, and decreases with increasing molecular size, $\lambda \sim (1/d^2)$. However, it also decreases with increasing molecular mass, $\lambda \sim (1/\sqrt{m})$.

Similarly to the considerations regarding the heat conductivity, where spatially different temperatures lead to temperature equilibrium by diffusion processes, spatially varying concentrations of certain molecule types cause concentration equalization processes that are to be understood analogously and in order to follow up such processes, and study them experimentally, gaseous radioactive particles could be used as tracers.

In equilibrium, these marked particles are distributed evenly over the available volume. However, if the concentration of the marked parts is position dependent where the entire number of the particles per unit volume is constant, this represents a non-equilibrium state which will try to equalize the concentration in the course of time by diffusion. This smoothing is possible by the temperature-dependent motion of the molecules. For the mathematical description of diffusion processes, the equation

$$\dot{m}_2 = -D\frac{\partial c}{\partial x_2} \qquad (3.51)$$

can be employed, which is known as Fick's law, where $\dot{m}_2$ is the mass flow per unit time and unit area that lies perpendicular to the direction x_2, i.e. the mass flow rate through a plane $x_2 =$ constant, D is the diffusion constant and c is the concentration of the marked substance. The minus sign expresses that the particles move from a position of higher concentration to a position of lower concentration.

Analogously to the molecular theory of heat conduction, the considerations below lead to the derivation of Fick's diffusion equation and to a relationship which states the diffusion coefficients in terms of molecular properties. It holds again that the mass flow of the molecules through a plane $x_2 =$ constant can be expressed as the difference in the mass flow of the molecules in the positive and negative x_2 directions (Fig. 3.11). In the positive x_2 direction, the area $\Delta x_1 \Delta x_3$ of the considered plane is passed by all the molecules, whose distance from the plane is not larger than $\bar{u}\Delta t$, i.e. $\frac{1}{6}\Delta x_1 \Delta x_3 \bar{u} \Delta t c(x_2)$.

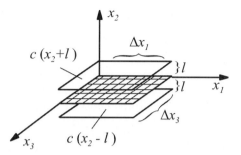

Fig. 3.11 Transport of marked molecules through a plane perpendicular to the x_2 direction

If one considers the particle flow per unit time and area [3.2], the following relationship results:

$$\dot{m}_2^+ = \frac{1}{6}\bar{u}c(x_2 - l). \tag{3.52}$$

The molecule concentration that exists at a distance l from the considered plane is represented by $c(x_2 - l)$.

Accordingly, for the mass flow rate in the negative x_2 direction, one can write:

$$\dot{m}_2^- = -\frac{1}{6}\bar{u}c(x_2 + l). \tag{3.53}$$

Through Taylor expansions of both c terms in (3.52) and (3.53), one obtains:

$$c(x_2 - l) = c(x_2) - \left(\frac{\partial c}{\partial x_2}\right)l + \cdots \tag{3.54}$$

$$c(x_2 + l) = c(x_2) + \left(\frac{\partial c}{\partial x_2}\right)l + \cdots \tag{3.55}$$

Hence this yields for the desired quantity $\dot{m}_2$:

$$\dot{m}_2 = \dot{m}_2^+ + \dot{m}_2^- \tag{3.56}$$

$$\dot{m} = -\frac{1}{3}\bar{u}l\frac{\partial c}{\partial x_2}. \tag{3.57}$$

A comparison with the diffusion equation (3.51) shows that the diffusion coefficient D is determined by molecular transport considerations as:

$$D = \frac{1}{3}\bar{u}l. \tag{3.58}$$

If, on the other hand, one sets:

$$\bar{u} = \sqrt{\frac{8kT}{\pi m}}; \quad l = \frac{1}{\sqrt{2}d^2\pi n} \tag{3.59}$$

then one obtains

$$D = \frac{2}{3}\frac{1}{nd^2}\sqrt{\frac{kT}{\pi^3 m}}. \tag{3.60}$$

From this equation, we can infer that the diffusion increases constantly with increasing temperature, $D \sim \sqrt{T}$, and decreases with increasing molecular mass, $D \sim 1/\sqrt{m}$. There exists a decrease with the molecular size, $D \sim (1/d^2)$, and also with increasing density of the gas $\rho = nm$, $D \sim (1/\rho)$.

3.4 Viscosity of Fluids

The molecular momentum transport in Newtonian fluid flow is given by

$$\tau_{ij} = -\mu\left(\frac{\partial U_j}{\partial x_i} + \frac{\partial U_i}{\partial x_j}\right) - \mu'\delta_{ij}\frac{\partial U_k}{\partial x_k}. \tag{3.61}$$

The material property μ is called the dynamic viscosity of a Newtonian medium, or the shear viscosity of the fluid. The coefficient μ' is defined as the expansion viscosity coefficient and can be formulated for a Newtonian medium as follows:

$$\mu' = -\frac{2}{3}\mu \tag{3.62}$$

with the same physical units for μ and μ', i.e. $[\mu] = [\mu']$. The dynamic shear viscosity is a thermodynamic property of a fluid and thus depends on temperature and pressure. For a Newtonian medium, μ is independent of $S_{ij} = \frac{1}{2}\left(\frac{\partial U_j}{\partial x_i} + \frac{\partial U_i}{\partial x_j}\right)$, i.e. τ_{ij} is linearly dependent on the velocity gradients occurring in a flow, or is connected with local fluid-element deformations. When the $\tau_{ij} = f(S_{ij})$, relationship (3.61) is non-linear, one speaks of fluids with non-Newtonian fluid viscosities. Figure 3.12 sketches some of these possible non-Newtonian fluid properties with pseudo-plastic behavior, i.e. with increasing shear rate the fluid tends to have lower viscosity. Dilatant fluids, on the other hand, show an increase in viscosity with an increase in the rate of deformation and one therefore defines them as "shear thickening fluids". In the diagram, a Bingham fluid also is shown that is characterized by a base value of τ_{ij} before the fluid moves. This book concentrates on the treatment of Newtonian fluids rather than on fluids with properties of non-Newtonian media. The non-Newtonian fluid properties indicated in Fig. 3.12 are presented to point out the existence of more complex fluid properties occuring in nature and in industry.

The dynamic viscosity of a Newtonian fluid depends indirectly on the molecular interactions and can therefore be regarded as a thermodynamic property that varies with temperature and pressure. A complete theory of this viscosity as a transport property in gases and liquids is still under development and the present state can be consulted in the book by Hirschfelder

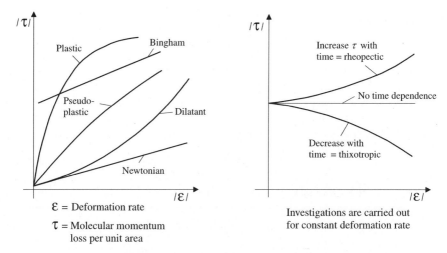

Fig. 3.12 Properties of Newtonian and non-Newtonian fluids

et al. [3.5]. For an entire class of fluids, the function $\mu[T, P]$ can be represented in a way that was presented by Kienan [3.6] and which makes use of a normalized expression such that all values are normalized with the viscosity at the critical point of the fluid. In this way the following expression is obtained:

$$\frac{\mu}{\mu_c} = f\left[\left(\frac{T}{T_c}\right), \left(\frac{P}{P_c}\right)\right] \tag{3.63}$$

which is illustrated in Fig. 3.13, showing that the viscosity of gases increases with pressure. The viscosity of liquids decreases with temperature. For gases, there is a very weak dependence of viscosity on pressure and this is generally neglected in gas-dynamic considerations. In short, Fig. 3.13 indicate that:

- The viscosity of liquids decreases rapidly with temperature.
- The viscosity of gases increases with temperature at moderate pressure values.
- The viscosity of all liquids increases more or less with pressure.
- The pressure dependence of viscosity of gases is negligible.

The above normalized property $\frac{\mu}{\mu_c}$ permits the conclusion that for most fluids the critical pressure is higher than 10 atm and hence conditions for low density are fulfilled very well under atmospheric pressure.

The theory of the physical properties of gases under pressure conditions $P < P_c$ is well established and has been advanced further on the basis of theories by Maxwell (1831–1879), as indicated by Hirschfelder et al. [3.5] and Present [3.7]. All of these theories are based on considerations yielding information given in Sect. 3.3. In this section, it is explained that the measured dynamic viscosity of a gas results from the statistical average of molecule-dependent momentum transport of the motion of fluids. In the case of gases,

3.4 Viscosity of Fluids

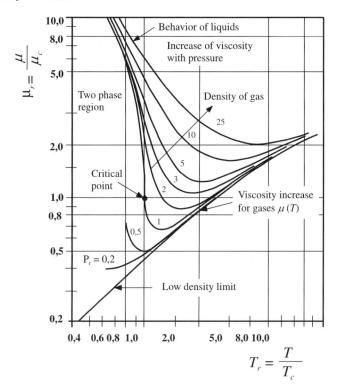

Fig. 3.13 Normalized viscosity as a function of pressure and temperature. Normalization is achieved with the viscosity at the critical point of the fluid

the dynamic viscosity can be given as follows:

$$\mu = \frac{2}{3}\rho l c, \qquad (3.64)$$

where ρ is the density of the fluid, l the mean free path length of the molecular motion, and c the speed of sound in a gas. For gases under normal pressure conditions $\rho l \approx$ constant. However, more precise considerations show that ρl increases slightly with temperature because of the so-called collision integral Ω_s. According to Chapman and Cowling [3.8]:

$$\mu = \frac{210^{-3}\sqrt{MT}}{\sigma^2 \Omega_s}, \qquad (3.65)$$

where M is the molecular weight of the gas, T the absolute temperature, and σ the collision cross-section of the molecules. For $\Omega_s = 1$ the molecules interact only in the form of collisions. When more complex molecular interactions exist, Ω_s has to be calculated according to following equation:

$$\Omega_s \approx 1.147 \left(\frac{T}{T_c}\right)^{-0.145} + \left(\frac{T}{T_c} + 0.5\right)^{-2}. \qquad (3.66)$$

Table 3.1 Stockmayer collision integral values for determining the viscosity of gases

$T^* = T/T_c$	Ω_s	Ω_s (3.66)
0.3	2.840	2.928
1.0	1.593	1.591
3.0	1.039	1.060
10.0	0.8244	0.8305
30.0	0.7010	0.7015
100.0	0.5887	0.5884
400.0	0.4811	0.4811

Table 3.2 Values for the calculation of the dynamic viscosity of gases according to the Sutherland equation (3.68)

Gas	T_0 (K)	μ_0 (mPa s^{-1})	n	Error (%)	Temperature range (K)	S (K)	Temperature range for ±2% error (K)
Air	887.65	0.01716	0.666	±4	745.65–4548.15	521.90	648.15–4548.15
Ar	887.65	0.02125	0.72	±3	723.15–3648.15	598.15	548.15–3648.15
CO_2	887.65	0.01370	0.79	±5	744.4–4098.15	773.15	700.65–4098.15
CO	887.65	0.01657	0.71	±2	790.65–3648.15	579.40	565.65–3648.15
N_2	887.65	0.01663	0.67	±3	773.15–3648.15	513.15	498.15–3648.15
O_2	887.65	0.01919	0.69	±2	790.65–4773.15	585.65	691.90–4773.15
H_2	887.65	0.008411	0.68	±2	453.15–2748.15	490.65	778.15–2748.15
Water vapor	1210.65	0.01703	1.04	±3	903.15–3648.15	2210.65	1085.65–3648.15

The values determined from the Stockmayer potential, e.g. see Bird et al. [3.1], are compared in Table 3.1 with the values from the above approximation relationship and they show good agreement.

For routine calculations, the following equation can be used:

$$\frac{\mu}{\mu_0} \approx \left(\frac{T}{T_0}\right)^n, \qquad (3.67)$$

where μ_0 and T_0 are corresponding reference values that were obtained from measurements or calculations on experimental findings.

In general, the value of n is found to be around 0.7. More precise values of n are given in Table 3.2 for different gases.

More extensive considerations were made by Sutherland [3.9], and these were based on an intermolecular potential for forces possessing an attractive part. The resulting Sutherland equation is

$$\frac{\mu}{\mu_0} \approx \left(\frac{T}{T_0}\right)^{\frac{3}{2}} \frac{T_0 + S}{T + S}, \qquad (3.68)$$

where S represents an effective temperature, the so-called "Sutherland constant". It is also given in Table 3.2 for different gases.

3.5 Balance Considerations and Conservation Laws

Before we conduct detailed considerations on fluid mechanics processes, some remarks need to be made on the acquisition of information in fluid mechanics, especially on the knowledge gained in analytical fluid mechanics, which is treated extensively in this book. Starting from conservation laws, analytical fluid mechanics employs deductive methods to solve various unsolved problems, i.e. to gain knowledge on existing flow problems. For analytical solutions, one makes use of derived relations that are based on balance considerations, as known from other fields of natural and engineering science and also from everyday observations. In many domains of daily life one acquires, starting from intuitive knowledge on the existence of conservation laws, useful information from balance considerations which one conducts over defined fields, domains, periods, etc. The way in which the changes of considered quantities of concern take place in detail is often not of interest. Rather, the "initial and end states of the considered quantities" need to be known. The changes within the considered domain are due to "inflows and outflows of the considered quantity". Relationships can be established between the changes within the considered fields, domains and periods knowing the "inflows and outflows". Considerations of the financial circumstances of a person or an institution are, for example, conducted by establishing balances of income and expenditure to obtain information on the development of the financial situation. Many more examples of this kind could be cited in order to make clear the importance of balances for obtaining information in daily life.

In fluid mechanics, we find balances of quantities such as mass, momentum, and energy in almost all fields of natural and engineering sciences. With these balances, basic equations are set up with the aid of existing conservation laws whose solution lead, in the presence of initial and boundary conditions, to the desired information on quantities. In order to obtain definite information, the balance considerations have to be based not only on valid conservation laws (mass conservation, energy conservation, momentum conservation, etc.) but also on definite specified domains. The field or the domain for which balances are formulated has to be defined precisely to guarantee the unambiguity of the derived basic laws. A relationship that was derived by considerations over a certain domain are, in general, not applicable when domain modifications have taken place and these are not included in the relationship.

Fluid mechanics is a subject based on the basic laws of mechanics and thermodynamics and, moreover, uses state equations in the derivations in order to establish relationships between thermodynamic state changes of a fluid. These state properties vary in the course of time or in space. However, the changes of state take place in accordance with the corresponding state equations while still observing the conservation laws. For the derivation of the basic equations of fluid mechanics, the basic laws of physics are:

- Mass conservation law (continuity equation)
- Momentum conservation law (equation of momentum)
- Energy conservation law (energy equation)
- Conservation of chemical species
- State equations

These basic laws can now be applied to several "balance domains". The size of the balance space is not important in general and it can include infinitesimally small balance domains (differential considerations) or finite volumes (integral considerations). Furthermore, the balance domains can lie at a fixed location in different coordinate systems and can undergo motions themselves (Euler and Lagrangian approaches). In general, once selected, domains for balance considerations are usually maintained for all quantities to be considered, i.e. mass, momentum, energy, etc.; however, this is not necessary. Changes in domains are admissible as long as they are known and thus can be included in the balance considerations.

Generally, to derive the basic fluid flow equations in fluid mechanics, only integral considerations are made, but balance considerations are made over different domains of interest. In the case of differential considerations, one finds balance considerations for moving fluid elements (Lagrangian approach) or space fixed elements (Eulerian approach). The two ways of fluid flow description have to be clearly distinguished and balances should always be set up separately for the Lagrangian and Eulerian balance spaces. Mixed balances leads to errors, in general. However, transformations of final equations are possible. It is usual, for example, in fluid mechanics to transform the balance relationships derived for a moving fluid element to space-fixed coordinate systems and thus to obtain balance relations for constant volumes. The connections between considerations for moving fluid elements and space-fixed control volume in this book are presented and the equations required for the transformation are derived. Particular attention is given to the physical understanding of the principal connections, so that advantages and disadvantages of the different approaches become clear. The advantages of the "Eulerian form" of the basic equations are brought out with respect to the imposed boundary conditions for obtaining solutions to flow problems. On the other hand, the Lagrangian considerations allow the transfer of often known physical knowledge of mechanics of moving bodies to fluid mechanics considerations.

When stating the basic equations in Lagrange variables, the following equations are valid for a fluid element:[3]

- Mass conservation: $\dfrac{\mathrm{d}(\delta m)_\Re}{\mathrm{d}t} = 0$

---

[3] The derivation of these equations is presented in Chap. 5. They are stated here beforehand to explain the basic physical knowledge that is taken from Physics.

3.5 Balance Considerations and Conservation Laws

- Newton's 2nd law: $\dfrac{d}{dt}[(\delta m)_\Re (U_j)_\Re] = \sum (\delta M_j)_\Re + (\delta \dot{J}_j)_\Re$
  $\qquad\qquad\qquad\qquad + \sum (\delta O_j)_\Re$

  (3.69)

- Energy conservation: $\dfrac{d}{dt}(e)_\Re = \dfrac{d}{dt}(\dot{q})_\Re - P_\Re \dfrac{1}{\delta V_\Re} \dfrac{d(\delta V)_\Re}{dt} + \phi_{diss}$

- State equation: $e_\Re = f(P_\Re, T_\Re)$ and $P_\Re = f(\rho_\Re, T_\Re)$

The above representations make it clear that, generally, in fluid mechanics, considerations agree with principles that are usually applied in thermodynamics, e.g. the energy equation (1st law of thermodynamics) and state equations for liquids and gases.

As is shown in Chap. 5, the above equations can also be expressed in field variables, such that the following set of differential equations for density $\rho(x_i, t)$, pressure $P(x_i, t)$, temperature $T(x_i, t)$, internal energy $e(x_i, t)$, and three velocity components $U_j(x_i, t)(j = 1, 2, 3)$ are obtained:

- Mass conservation: $\dfrac{\partial \rho}{\partial t} + \dfrac{\partial (\rho U_i)}{\partial x_i} = 0$

- Newton's 2nd law: $\rho \left[\dfrac{\partial U_j}{\partial t} + U_i \dfrac{\partial U_j}{\partial x_i}\right] = -\dfrac{\partial P}{\partial x_j} - \dfrac{\partial \tau_{ij}}{\partial x_i} + \rho g_i$

  with $\tau_{ij} = -\mu \left[\dfrac{\partial U_j}{\partial x_i} + \dfrac{\partial U_i}{\partial x_j}\right] + \dfrac{2}{3}\mu \delta_{ij} \dfrac{\partial U_k}{\partial x_k}$

- Energy conservation: $\rho \left[\dfrac{\partial e}{\partial t} + U_i \dfrac{\partial e}{\partial x_i}\right] = -\dfrac{\partial \dot{q}_i}{\partial x_i} - P \dfrac{\partial U_j}{\partial x_j} - \tau_{ij} \dfrac{\partial U_j}{\partial x_i}$

  with $\dot{q}_i = -\lambda \dfrac{\partial T}{\partial x_i}$

- State equations: $e = f(P, T)$ and $P = f(\rho, T)$

(3.70)

Thus five differential equations are available, if one inserts τ_{ij} and $\dot{q}_i$ through the above equations, for altogether seven unknowns and two thermodynamic equations. With this, the closed system of differential equations given above can be solved for specified initial and boundary conditions. Therefore, the given flow problem is defined by its initial and boundary conditions, needed for the solution of the above set of equations. The basic physical laws are identical for all flow problems. They comprise, as said above, the conservation and state equations that are usually treated in thermodynamics. Therefore, a repetition of thermodynamic fundamentals is essential as summarized in Sect. 3.6.

3.6 Thermodynamic Considerations

The thermodynamic state equations of fluids are often used as a supplement for the solution of flow problems. However, in the present text only "simple fluids", i.e. homogeneous liquids and gases for which the thermodynamic state can be expressed by a relation between pressure, temperature, and density, are considered. Hence the following statements are possible for both substantial and field quantities, i.e.:

$$P_\Re = f(T_\Re, \rho_\Re) \text{ or } P = f(T, \rho) \qquad (3.71)$$

The thermodynamic state equations for simple fluids are known to be:

$$\frac{P_\Re}{\rho_\Re} = RT_\Re \qquad \text{(thermodynamically ideal gases)}$$

$$\rho_\Re = \text{constant} \qquad \text{(thermodynamically ideal liquids)}.$$

If one defines $\alpha_\Re = P_\Re, T_\Re, \rho_\Re, e_\Re$ and $\alpha = P, T, \rho, e, \ldots$, the following relation holds when the fluid element $\Re$ is located at time t at position x_i (see Sect. 2.6):

$$\alpha_\Re(t) = \alpha(x_i, t) \rightsquigarrow \frac{d\alpha_\Re}{dt} = \frac{\partial \alpha}{\partial t} + U_i \frac{\partial \alpha}{\partial x_i} = \frac{D\alpha}{Dt}. \qquad (3.72)$$

The second part of (3.72) indicates how temporal changes of substantial, thermodynamic quantities can be computed from the substantial derivative of corresponding field quantities.

In addition to the above thermodynamic state properties $P_\Re, T_\Re, \rho_\Re, e_\Re, \ldots$, other state properties can be defined whose introduction is of advantage in certain thermodynamic considerations. Some of them are as follows:

- Specific volume: $v_\Re = \dfrac{1}{\rho_\Re}$
- Enthalpy: $h_\Re = e_\Re + P_\Re v_\Re$
- Free energy: $f_\Re = e_\Re - T_\Re s_\Re$ (Helmholtz potential)
- Free enthalpy: $g_\Re = h_\Re - T_\Re s_\Re$ (Gibb potential)

Accordingly, it is possible to apply certain mathematical operators in order to define "new thermodynamic quantities". However, their introduction makes sense, i.e. considerations become simpler or easier when carried out with newly introduced quantities. Advantages result from the introduction of newly defined quantities into thermodynamic considerations.

In one of the above definitions of thermodynamic potentials, the entropy $s_\Re$ was used, whose definition is given by a differential relationship:

$$T_\Re ds_\Re = de_\Re + P_\Re dv_\Re \qquad \text{(Gibbs relationship)}. \qquad (3.73)$$

3.6 Thermodynamic Considerations

Integrating, one obtains:

$$s_\Re = s_{(\Re)_0} + \int_{(e_\Re)_0}^{e_\Re} \frac{1}{T_\Re} de_\Re + \int_{(v_\Re)_0}^{v_\Re} \frac{P_\Re}{T_\Re} dv_\Re. \tag{3.74}$$

Equations (3.73) and (3.74) can be understood as identical definitions for the entropy $s_\Re$ of a fluid element. When employing (3.72) one obtains with:

$$\frac{ds_\Re}{dt} = \frac{Ds}{Dt} \qquad \frac{de_\Re}{dt} = \frac{De}{Dt} \quad \text{and} \quad \frac{dv_\Re}{dt} = v_\Re \frac{\partial U_i}{\partial x_i} \tag{3.75}$$

the following relationships:

$$T_\Re \frac{ds_\Re}{dt} = \frac{de_\Re}{dt} + P_\Re \frac{dv_\Re}{dt} \quad \rightsquigarrow \quad T\frac{Ds}{Dt} = \frac{De}{Dt} + P \frac{1}{\rho} \frac{\partial U_i}{\partial x_i} \tag{3.76}$$

or

$$T\left(\frac{\partial s}{\partial t} + U_i \frac{\partial s}{\partial x_i}\right) = \left(\frac{\partial e}{\partial t} + U_i \frac{\partial e}{\partial x_i}\right) + \frac{P}{\rho}\left(\frac{\partial U_i}{\partial x_i}\right). \tag{3.77}$$

When one applies the mass conservation equation (3.70), it can be rearranged further to yield:

$$\left(\frac{\partial U_i}{\partial x_i}\right) = -\frac{1}{\rho} \frac{D\rho}{Dt}. \tag{3.78}$$

Insertion in (3.76) yields

$$T\frac{Ds}{Dt} = \frac{De}{Dt} - \frac{P}{\rho^2} \frac{D\rho}{Dt}. \tag{3.79}$$

From this relation, further equations can be derived that are important in fluid mechanics, e.g. for $s_\Re = $ constant:

$$\left(\frac{De}{Dt}\right)_s = \frac{P}{\rho^2}\frac{D\rho}{Dt} \quad \rightsquigarrow \quad \left(\frac{de_\Re}{dt}\right)_{s_\Re} \frac{P_\Re}{\rho_\Re^2} = \left(\frac{de_\Re}{d\rho_\Re}\right)_{s_\Re}. \tag{3.80}$$

For $\rho_\Re = $ constant or $v_\Re = $ constant:

$$T\left(\frac{Ds}{Dt}\right)_\rho = \left(\frac{De}{Dt}\right)_\rho \quad \rightsquigarrow \quad T_\Re = \left(\frac{de_\Re}{ds_\Re}\right)_{\rho_\Re} \tag{3.81}$$

Further, for $e_\Re = $ constant:

$$T\left(\frac{Ds}{Dt}\right)_e = -\frac{P}{\rho^2}\left(\frac{D\rho}{Dt}\right)_e \quad \rightsquigarrow \quad P_\Re = T_\Re \left(\frac{\partial s_\Re}{\partial v_\Re}\right)_{e_\Re} = -T_\Re \rho_\Re^2 \left(\frac{\partial s_\Re}{\partial \rho_\Re}\right)_{e_\Re} \tag{3.82}$$

Further significant relationships, known from the field of thermodynamics, are needed in the following forms:

- Specific heat capacity of a fluid at constant volume:
$$c_v = \left(\frac{\partial e_\Re}{\partial T_\Re}\right)_{v_\Re} = T_\Re \left(\frac{\partial s_\Re}{\partial T_\Re}\right)_{v_\Re} \tag{3.83}$$

- Specific heat capacity of a fluid at constant pressure:
$$c_p = \left(\frac{\partial h_\Re}{\partial T_\Re}\right)_{P_\Re} = T_\Re \left(\frac{\partial s_\Re}{\partial T_\Re}\right)_{P_\Re}, \tag{3.84}$$

where $h_\Re = e_\Re + P_\Re v_\Re$.

- Isothermal compressibility coefficient:
$$\alpha = -\frac{1}{v_\Re}\left(\frac{\partial v_\Re}{\partial P_\Re}\right)_{T_\Re} = \frac{1}{\rho_\Re}\left(\frac{\partial \rho_\Re}{\partial P_\Re}\right)_{T_\Re} \tag{3.85}$$

- Thermal expansion coefficient:
$$\beta = \frac{1}{v_\Re}\left(\frac{\partial v_\Re}{\partial T_\Re}\right)_{P_\Re} = -\frac{1}{\rho_\Re}\left(\frac{\partial \rho_\Re}{\partial T_\Re}\right)_{P_\Re} \tag{3.86}$$

When one takes into account the following relationship:
$$d\rho_\Re = \left(\frac{\partial \rho_\Re}{\partial T_\Re}\right)_{P_\Re} dT_\Re + \left(\frac{\partial \rho_\Re}{\partial P_\Re}\right)_{T_\Re} dP_\Re \tag{3.87}$$

including (3.85) and (3.86), the following relation can be formulated for all fluids:
$$\frac{1}{\rho_\Re} d\rho_\Re = \alpha dP_\Re - \beta dT_\Re \tag{3.88}$$

or, rearranged in terms of field variables:
$$\frac{1}{\rho}\frac{D\rho}{Dt} = \alpha \frac{DP}{Dt} - \beta \frac{DT}{Dt}. \tag{3.89}$$

This relation allows the statement that all fluids of constant density, i.e. fluids having the property $\rho_\Re =$ constant or $(D\rho/Dt) = 0$, can be designated as incompressible. They react neither to pressure variations ($\alpha = 0$) nor to temperature variations ($\beta = 0$) with changes in volume or density.

For any fluid, the difference in the heat capacities is given by
$$(c_p - c_v) = \frac{T_\Re \beta^2}{\rho_\Re \alpha} = -\frac{T_\Re}{\rho_\Re} \cdot \beta \left(\frac{\partial P_\Re}{\partial T_\Re}\right)_{P_\Re} = -T_\Re \left(\frac{\partial P_\Re}{\partial T_\Re}\right)_{\rho_\Re} \left(\frac{\partial v_\Re}{\partial T_\Re}\right)_{P_\Re}. \tag{3.90}$$

The above general relationships (3.85) and (3.86) can now be employed to derive the special α and β equations below that hold for the two thermodynamic ideal fluids that receive special attention in this book, namely the

3.6 Thermodynamic Considerations

ideal gas and the ideal liquid. For an ideal gas:

$$\frac{P_\Re}{\rho_\Re} = RT_\Re \quad \text{and consequently} \quad \frac{P}{\rho} = RT \qquad (3.91)$$

and in addition $\left(\frac{\partial e_\Re}{\partial v_\Re}\right)_{T_\Re} = \left(\frac{\partial e_\Re}{\partial P_\Re}\right)_{T_\Re} = 0$ and $c_v = $ constant, i.e. the internal energy of an ideal gas is a pure function of the temperature. The isothermal compressibility coefficient α and the thermal expansion coefficient β are given by:

$$\begin{aligned} \alpha &= \frac{1}{\rho_\Re}\left(\frac{\partial \rho_\Re}{\partial P_\Re}\right)_{T_\Re} = \frac{1}{\rho_\Re}\frac{1}{RT_\Re} = \frac{1}{P_\Re} \\ \beta &= -\frac{1}{\rho_\Re}\left(\frac{\partial \rho_\Re}{\partial T_\Re}\right)_{P_\Re} = -\frac{1}{\rho_\Re}\left(-\frac{P_\Re}{RT_\Re^2}\right) = \frac{1}{T_\Re} \end{aligned} \qquad (3.92)$$

and therefore the difference in the specific heat capacities for an ideal gas is:

$$c_p - c_v = \frac{T_\Re}{\rho_\Re}\frac{1}{T_\Re^2}P_\Re = \frac{P_\Re}{\rho_\Re T_\Re} = R. \qquad (3.93)$$

It can further be formulated for the change in density:

$$\frac{\mathrm{d}\rho_\Re}{\rho_\Re} = \frac{\mathrm{d}P_\Re}{P_\Re} - \frac{\mathrm{d}T_\Re}{T_\Re}. \qquad (3.94)$$

If we introduce the thermodynamically ideal liquid that distinguishes itself by $\alpha = 0$ and $\beta = 0$, i.e. by:

$$\frac{\mathrm{d}\rho_\Re}{\rho_\Re} = 0 \qquad \text{(fluid of constant density)} \qquad (3.95)$$

we obtain the difference in the heat capacities and it can be computed that:

$$c_p - c_v = -\frac{T_\Re}{\rho_\Re}\beta\left(\frac{\partial P_\Re}{\partial T_\Re}\right)_{\rho_\Re} \quad \text{with } \beta = 0 \quad \leadsto \quad c_p = c_v. \qquad (3.96)$$

When one employs the Gibbs relationship, (3.73), $\mathrm{d}\rho_\Re = \frac{1}{dv_\Re} = 0$, one obtains:

$$\left(\frac{\partial s_\Re}{\partial e_\Re}\right)_{\rho_\Re} = \frac{1}{T_\Re}. \qquad (3.97)$$

Because $s_\Re \neq f(p_\Re)$, the pressure in an ideal liquid does not need to be taken into account as a thermodynamic quantity. It exists as a mechanical quantity, but for an ideal fluid it is not part of a thermodynamic state equation.

A further physical property of a fluid, which is of significance when dealing with some of the flow problems presented in this book, is the velocity

propagation of small pressure perturbations, the so-called sound velocity:
$$c^2 = \left(\frac{\partial P_\Re}{\partial \rho_\Re}\right)_{s_\Re}. \tag{3.98}$$

This quantity is defined as isentropic pressure change with density change, the entropy being maintained constant. Hence the propagation of small acoustic perturbations takes place isentropically.

If one takes into account the following relationship for the cited sequence of partial derivations:
$$1 = \left(\frac{\partial T_\Re}{\partial e_\Re}\right)_{\rho_\Re} \left(\frac{\partial e_\Re}{\partial s_\Re}\right)_{\rho_\Re} \left(\frac{\partial s_\Re}{\partial T_\Re}\right)_{\rho_\Re} \tag{3.99}$$

and if one considers:
$$T_\Re = \left(\frac{\partial e_\Re}{\partial s_\Re}\right)_{\rho_\Re} \quad c_v = \left(\frac{\partial e_\Re}{\partial T_\Re}\right)_{\rho_\Re} \tag{3.100}$$

then
$$c_v = T_\Re \left(\frac{\partial s_\Re}{\partial T_\Re}\right)_{\rho_\Re}. \tag{3.101}$$

Also taking into account the Maxwell relations:
$$\left(\frac{\partial T_\Re}{\partial \rho_\Re}\right)_{s_\Re} = \frac{1}{\rho_\Re^2} \left(\frac{\partial P_\Re}{\partial s_\Re}\right)_{\rho_\Re} \tag{3.102}$$

$$\left(\frac{\partial \rho_\Re}{\partial s_\Re}\right)_{T_\Re} = -\rho_\Re^2 \left(\frac{\partial T_\Re}{\partial P_\Re}\right)_{\rho_\Re} \tag{3.103}$$

it can be formulated that:
$$\left(\frac{\partial s_\Re}{\partial T_\Re}\right)_{\rho_\Re} \left(\frac{\partial T_\Re}{\partial \rho_\Re}\right)_{s_\Re} \left(\frac{\partial \rho_\Re}{\partial s_\Re}\right)_{T_\Re} = -1 \tag{3.104}$$

and one can also express the quantity c_v as:
$$c_v = -T_\Re \left(\frac{\partial \rho_\Re}{\partial T_\Re}\right)_{s_\Re} \left(\frac{\partial s_\Re}{\partial \rho_\Re}\right)_{T_\Re}. \tag{3.105}$$

Similarly, it can be derived that:
$$c_p = -T_\Re \left(\frac{\partial P_\Re}{\partial T_\Re}\right)_{s_\Re} \left(\frac{\partial s_\Re}{\partial P_\Re}\right)_{T_\Re}. \tag{3.106}$$

For the relationship of the heat capacities, it can be formulated that:
$$\kappa = \frac{c_p}{c_v} = \frac{\left(\frac{\partial P_\Re}{\partial T_\Re}\right)_{s_\Re} \left(\frac{\partial s_\Re}{\partial P_\Re}\right)_{T_\Re}}{\left(\frac{\partial \rho_\Re}{\partial T_\Re}\right)_{s_\Re} \left(\frac{\partial s_\Re}{\partial \rho_\Re}\right)_{T_\Re}} = \left(\frac{\partial P_\Re}{\partial \rho_\Re}\right)_{s_\Re} \left(\frac{\partial \rho_\Re}{\partial P_\Re}\right)_{T_\Re}. \tag{3.107}$$

With the definition equation for the speed of sound and the isothermal compressibility coefficient, one obtains

$$c^2 = \kappa \left(\frac{\partial P_\Re}{\partial \rho_\Re}\right)_{T_\Re} = \frac{\kappa}{\rho_\Re \alpha}. \qquad (3.108)$$

For an ideal gas with $\alpha = \frac{1}{P_\Re}$ and considering the ideal gas equation yields

$$c = \sqrt{\kappa R_\Re T_\Re}. \qquad (3.109)$$

For an ideal liquid with $\alpha \to 0$

$$c \to \infty, \qquad (3.110)$$

i.e. for a fluid with constant density an infinitely large sound velocity results.

References

3.1. Bird, R.B., Stewart, W.E., Lightfoot, E.N., Transport Phenomena, Wiley, New York, 1960.
3.2. Reif, F., Physikalische Statistik und Physik der Wärme, Walter de Gruyter, Berlin, 1976.
3.3. Höfling, O., Physik: Lehrbuch für Unterricht und Selbststudium, Dümmler, Bonn, 1994.
3.4. Bosnjakovic, F., Technische Thermodynamik, Theodor Steinkopf, Dresden, 1965.
3.5. Hirschfelder, J.O., Curtiss, C.F., Bird, R.B., Molecular Theory of Gases and Liquids, Wiley, New York, 1963.
3.6. Kienan
3.7. Present, R.D., Kinetic Theory of Gases, McGraw-Hill, New York, 1958.
3.8. Chapman, S., Cowling, T.G., The Mathematical Theory of Non-uniform Gases, Cambridge University Press, Cambridge, 1960.

Chapter 4
Basics of Fluid Kinematics

4.1 General Considerations

The previous chapters dealt with important basic knowledge and information of mathematics and physics as applied in the field of fluid mechanics. This knowledge is needed to describe fluid flows or derive and utilize the basic equations of fluid mechanics in order to solve flow problems. In this respect, it is important to know that fluid mechanics is primarily interested in the velocity field $U_j(x_i, t)$, for given initial and boundary conditions, and in the accompanying pressure field $P(x_i, t)$, i.e. fluid mechanics seeks to present and describe flow processes in field variables. This presentation and description result in the "Eulerian form" of considerations of fluid flows. This form is best suited for the solution of flow problems and is thus mostly applied in experimental, analytical and numerical fluid mechanics. Thanks to the introduction of field quantities also for the thermodynamic properties of a fluid, e.g. the pressure $P(x_i, t)$ and the temperature $T(x_i, t)$, the density $\rho(x_i, t)$ and the internal energy $e(x_i, t)$, and for the molecular transport quantities, e.g. the dynamic viscosity $\mu(x_i, t)$, the heat conductivity $\lambda(x_i, t)$ and the diffusion coefficients $D(x_i, t)$, a complete presentation of fluid mechanics is possible. With the inclusion of diffusive transport quantities, i.e. the molecular heat transport $\dot{q}_i(x_i, t)$, the molecular mass transport $\dot{m}_i(x_i, t)$ and the molecular momentum transport $\tau_{ij}(x_i, t)$, it is possible to formulate the conservation laws for mass, momentum and energy for general application. The basic equations of fluid mechanics can thus be formulated locally, as is shown in Chap. 5, and hold for all flow problems in the same form. The differences in the solutions of these equations result from the different initial and boundary conditions that define the actual flow problems. These enter into the solutions by the integration of the locally formulated basic fluid flow equations.

Experience shows that the derivation of the basic equations of fluid mechanics can be achieved in the easiest way by considering fluid elements, i.e. by employing the "Lagrangian considerations" for the derivation of the flow equations. The Lagrange approach starts from the assumption that a fluid can be split up, at a fixed time $t = 0$, into "marked elements" with mass $\delta m_\Re$, pressure $P_\Re$, temperature $T_\Re$, density $\rho_\Re$, internal energy $e_\Re$, etc. An element with the index $\Re$ possesses also a velocity $(U_j)_\Re$, which is defined as the Lagrangian velocity of the marked element. This velocity is always linked to the fluid element marked with $\Re$ and to all other quantities labelled with $\Re$. In fluid mechanics, these quantities are also designated as substantial properties and are best employed to derive the basic laws of fluid mechanics in an easily comprehensible way. As the following considerations will show, the basic knowledge of mechanics, gained in physics, can be transferred into fluid mechanics in the simplest way when it is introduced by way of Lagrangian considerations, yielding the basic equations for fluid flows, deriving them for a fluid element.

4.2 Substantial Derivatives

If one defines $\alpha_\Re$, a substantial quantity such as $P_\Re$, $T_\Re$, $\rho_\Re$ and $e_\Re$, the derivations of equations in fluid mechanics often require the total differential $d\alpha_\Re$ to be employed:

$$d\alpha_\Re = \frac{\partial \alpha}{\partial t}dt + \frac{\partial \alpha}{\partial x_1}(dx_1)_\Re + \frac{\partial \alpha}{\partial x_2}(dx_2)_\Re + \frac{\partial \alpha}{\partial x_3}(dx_3)_\Re \qquad (4.1)$$

where $\alpha(x_i, t)$ is the field variable corresponding to $\alpha_\Re$.

The fluid element motion in space (Fig. 4.1) can be described as follows:

$$\begin{aligned}(dx_1)_\Re &= (U_1)_\Re dt = U_1 dt \\ (dx_2)_\Re &= (U_2)_\Re dt = U_2 dt \\ (dx_3)_\Re &= (U_3)_\Re dt = U_3 dt\end{aligned} \qquad (4.2)$$

The replacement of the substantial velocities $(U_i)_\Re$ by the field quantities U_i in (4.2) is permissible, as at the time t we assume $(x_i)_\Re = x_i$ and thus $(U_i)_\Re(t) = U_i(x_i, t)$. Therefore, the following relationship holds:

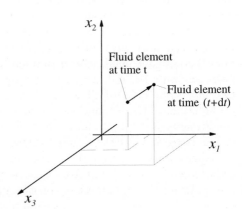

Fig. 4.1 Motion of a fluid element in space

4.3 Motion of Fluid Elements

$$d\alpha_\Re = \frac{\partial \alpha}{\partial t}dt + \frac{\partial \alpha}{\partial x_1}U_1 dt + \frac{\partial \alpha}{\partial x_2}U_2 dt + \frac{\partial \alpha}{\partial x_3}U_3 dt \qquad (4.3)$$

Hence the total time derivative of $\alpha_\Re(t)$ can be expressed as follows, if $x_\Re = x_i$ at time t:

$$\frac{d\alpha_\Re}{dt} = \frac{\partial \alpha}{\partial t} + U_i \frac{\partial \alpha}{\partial x_i} =: \frac{D\alpha}{Dt} \qquad (4.4)$$

where $(D\alpha/Dt)$ is the substantial derivative of the field quantity $\alpha(x_i, t)$ with respect to time and the operator:

$$\frac{D}{Dt} := \frac{\partial}{\partial t} + U_i \frac{\partial}{\partial x_i} \qquad (4.5)$$

indicates how the substantial derivative of a field quantity is to be calculated. The operator D/Dt may only be applied to field quantities. If one applies D/Dt to the velocity field $U_j(x_i, t)$, the substantial acceleration results, i.e. the local acceleration which a fluid element experiences in a flow field at a point x_i at time t where $U_j(x_i, t)$ exists:

$$\frac{DU_j}{Dt} = \frac{\partial U_j}{\partial t} + U_i \frac{\partial U_j}{\partial x_i} \qquad (4.6)$$

For further details about this subject, see refs. 4.1 to 4.3. The substantial derivative of the velocity plays an important role in deriving the momentum equation of fluid mechanics in Euler variables. In the acceleration term in Euler quantities, as shown by the subscript i in (4.6), four partial derivatives occur per momentum direction $j = 1, 2, 3$, namely one time derivative and three derivatives with respect to the space coordinates x_1, x_2 and x_3. Hence the spatial derivatives $(\partial U_j/\partial x_i)$, multiplied by U_i occur as three terms in each of the three momentum equations representing the substantial acceleration. These nonlinear terms lead to mathematical difficulties when flow problems are to be solved. They prevent the application of the superposition principle of solutions and result in solution bifurcations, i.e. in multiple solutions for equal initial and boundary conditions, and also in correlated velocity fluctuations, e.g. in turbulent flows. The treatment of these nonlinear terms is given good attention in several parts of this book. It is important that the significance of the non-linear terms is understood in detail as part of the acceleration term of fluid elements. It is important to realize that not only the temporal changes of the velocity field lead to accelerations of fluid elements, but also the motion of a fluid element in a non-uniform velocity field.

4.3 Motion of Fluid Elements

Flow kinematics is a vast field and a comprehensive treatment is beyond the scope of this book, which is meant to give only an introduction into various sub-domains of fluid mechanics, including fluid kinematics. To such an introduction belongs the treatment of path lines of fluid particles, i.e. the computation of space curves along which marked fluid elements move in a fluid. Further, the computation of streak lines will be treated, i.e. the "marked path". This is the line a tracer mark, in a fluid when it is added at a fixed point in the flow. The computation of both path lines and of streak lines is of importance for the entire field of experimental fluid mechanics, where it is often tried to gain an insight into a particular flow by observations, or also by quantitative measurements, of the temporal changes of positions of "flow markers". The basics for the evaluation of such measurements are stated in the following chapter.

4.3.1 Path Lines of Fluid Elements

If one subdivides, at time $t = 0$, the entire domain of a flow field, that is of interest for investigating, into defined fluid elements and if one defines the space coordinates of the mass centers of gravity of each element in a coordinate system at time $t = 0$, one achieves a marked fluid domain such that the position vector $\Re$ can be defined as follows:

$$\{x_i\}_{\Re,0} = \{x_i(t=0)\}_\Re \tag{4.7}$$

Hence $\{x_i\}_\Re$ at $t = 0$ is assigned to each marked fluid particle. Each of the moving fluid particles, defined by the subdivision of the fluid in space, and moving for $-\infty < t < +\infty$ is defined as a fluid element, that keeps its identity $0 \leq t < \infty$ i.e forever.

When kinematic considerations for each marked fluid element are carried out, only the motions of the individual fluid elements are of interest. These considerations result for each fluid element in a separate consideration and result for each marked element in a characteristic path line. The computation of these path lines will be explained in the following. In all kinematic considerations it can be assumed that the flow field determining the fluid element motions is known.

As the velocity of a fluid element is dependent only on time, it follows from $\mathrm{d}\{x_i\}_\Re/\mathrm{d}t = \{U_i\}_\Re$, that the path line of a fluid element $\Re$ can be calculated as follows:

$$\{x_i(t)\}_\Re = \{x_i\}_{\Re,0} + \int_0^t \{U_i(t')\}_\Re \mathrm{d}t' \tag{4.8}$$

The position vector $\{x_i(t)\}_\Re$, defined in this way for each instant in time t, contains as a parameter the position vector of the particle defined at time $t = 0$, i.e. $\Re$, i.e. $\{x_1\}_{\Re,0}$. The identity $\{U_i\}_\Re = \{U_i\}$ can be introduced into the considerations, i.e. at a certain moment in time t, the space change of a marked fluid element can be expressed as:

$$\frac{\mathrm{d}\{x_i\}_\Re}{\mathrm{d}t} = \{U_i\}_\Re = \{U_i\} \tag{4.9}$$

4.3 Motion of Fluid Elements

The equals sign between the substantial velocity $\{U_i\}_\Re$ and $\{U_i\}$, existing for a moment in time t, indicates that the identity $\{x_i\}_\Re = \{x_i\}$ which exists at time t justifies equating the substantial velocity $\{U_i\}_\Re$ with the field velocity $\{U_i\}$. For the time derivative of the components $\{x_i\}_\Re$ of the particle motion one can write:

$$\frac{d\{x_i\}_\Re}{dt} = \{U_i\} \quad \text{or} \quad \frac{d(x_i)_\Re}{dt} = U_i \qquad (4.10)$$

These differential equations have to be solved, for $i = 1, 2, 3$, in order to determine the path lines of fluid elements. The differential quotient in (4.10) states that, as a solution of the above differential equations, the path line of that fluid element $\Re$ is obtained whose position was defined at the moment in time $t = 0$ with $\{x_i\}_{\Re,0}$.

The general way of proceeding when defining path lines will be explained and made clear by the example given below. The components of the flow velocity field will be given as follows:

$$U_1 = x_1(1+t), \quad U_2 = -x_2 \quad \text{and} \quad U_3 = -x_3 t \qquad (4.11)$$

This flow case was also treated by Currie in ref. [4.4]. If one inserts these definitions of the components of the velocity field in the above differential equations for the path lines of a fluid element, one obtains

$$\begin{aligned} \frac{d(x_1)_\Re}{dt} &= x_1(1+t) \\ \frac{d(x_2)_\Re}{dt} &= -x_2 \\ \frac{d(x_3)_\Re}{dt} &= -x_3 t \end{aligned} \qquad (4.12)$$

This set of differential equations can be solved and results in the following solutions holding for the path lines of all fluid elements:

$$\begin{aligned} (x_1(t))_\Re &= C_1 \exp\left[t + \frac{t^2}{2}\right] \\ (x_2(t))_\Re &= C_2 \exp[-t] \\ (x_3(t))_\Re &= C_3 \exp\left[-\frac{t^2}{2}\right] \end{aligned} \qquad (4.13)$$

If one now considers a fluid element of interest which had the position coordinates $(1,1,1)$ at time $t = 0$, then from the initial conditions for each of the equations in (4.3), the introduced integration constants C_α result. They can be deduced to be

$$C_1 = C_2 = C_3 = 1 \qquad (4.14)$$

i.e. for the case considered here all integration constants are equal. Of course, other integration constants would have resulted from the choice of other marked fluid elements.

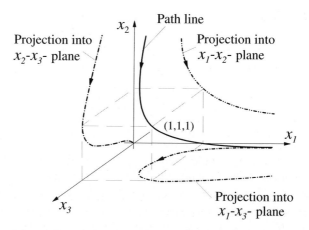

Fig. 4.2 Spatial path line of the considered fluid element and projections into planes of coordinate system

For the path lines of the selected fluid element, the introduction of C_α into (4.13) yields

$$x_1(t)_\Re = \exp\left[t + \frac{t^2}{2}\right]$$
$$x_2(t)_\Re = \exp\left[-t\right] \quad (4.15)$$
$$x_3(t)_\Re = \exp\left[-\frac{t^2}{2}\right]$$

This path line is presented in space in Fig. 4.2.

As said above, if one selects a particle whose position at time $t = 0$ showed different position coordinates, the integration constants C_1, C_2 and C_3 change accordingly and a different path line results. Thus the path line is an "individual" property of a fluid element. It is determined by the flow field and the position of the fluid element at time $t = 0$.

The general solution for the position coordinates $(x_i)_{\Re,0}$, which a fluid element takes at time $t = 0$, is obtained when one inserts these coordinates in the general solutions for the path line coordinates for the determination of the integration constants C_α ($\alpha = 1, 2, 3$). This results in

$$C_\alpha = (x_i)_{\Re,0} \quad (4.16)$$

and thus in the general solutions for the path line coordinates:

$$(x_1(t))_\Re = (x_1)_{\Re,0} \exp\left[t + \frac{t^2}{2}\right]$$
$$(x_2(t))_\Re = (x_2)_{\Re,0} \exp[-t] \quad (4.17)$$
$$(x_3(t))_\Re = (x_3)_{\Re,0} \exp\left[-\frac{t^2}{2}\right]$$

4.3 Motion of Fluid Elements

These coordinates yield the space curves, with the time t as parameter, which represent the path lines of fluid elements. Each fluid element has its own path line.

For further explanation, the following two-dimensional velocity field is also considered[1]:
$$U_1 = x_1, \quad U_2 = x_2(1+2t) \quad \text{and} \quad U_3 = 0. \tag{4.18}$$

With these data, the following law of differential equations for the coordinates of the path lines of fluid elements can be formulated:
$$\frac{d(x_1)_\Re}{dt} = x_1, \quad \frac{d(x_2)_\Re}{dt} = x_2(1+2t), \quad \frac{d(x_3)_\Re}{dt} = 0 \tag{4.19}$$

The solution of the third differential equation results in a constant that states in which plane the two-dimensional flow considerations need to be carried out. For the path line coordinates $x_1(t)$ and $x_2(t)$ it is computed from (4.10)
$$(x_1)_\Re = C_1 \exp[t], \quad (x_2)_\Re = C_2 \exp[t+t^2], \quad (x_3)_\Re = C_3 \tag{4.20}$$

When one computes the path line of the fluid element which at time $t=0$ took the coordinates (1,1,0), the result is
$$(x_1)_\Re = \exp[t], \quad (x_2)_\Re = \exp[t(t+1)], \quad (x_3)_\Re = 0 \tag{4.21}$$

When one resolves the equation obtained for $(x_1)_\Re$ with respect to time, the following can be deduced:
$$t = \ln(x_1)_\Re \tag{4.22}$$

Inserting this in the solution for $(x_2)_\Re$ for two-dimensional path lines in the plane x_1–x_2–, the following functional relation between $(x_1)_\Re$ and $(x_2)_\Re$ results:
$$(x_2)_\Re = (x_1)_\Re^{(1+\ln(x_1)_\Re)} \tag{4.23}$$

This path line is presented in Fig. 4.3.

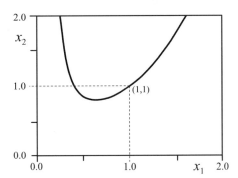

Fig. 4.3 Path line of the fluid flow in the plane x_1–x_2

---

[1] Attention is drawn to the fact that this flow field is not source-free, thus it violates the requirements of the continuity equation. This is, however, insignificant for the purely kinematic considerations mentioned here.

4.3.2 Streak Lines of Locally Injected Tracers

As mentioned before, it is usual in experimental fluid mechanics to gain qualitative insight into a flow process by the injection, at a fixed location, of a continuous fluid tracer. This leads to a marked "fluid thread" which is carried with the flow and thus marks the course of the flow. This is called a streak line. When the exact course of the flow is of interest, quantitative evaluations of the location coordinates of streak lines of locally installed tracer materials are required. These evaluations can, based on the derivations stated below, be carried out with methods of flow kinematics.

A fluid particle marked with a tracer, e.g. an air particle or any other gas particle marked with smoke, or a water or liquid particle marked with color, which at time t is located at the position $\{x_i\} = \{x_i(t)\}_\Re$ must have passed the injection point for the tracer at a moment in time $(t - \tau)$, in order to be present as a marked particle at the point $\{x_i\}$, i.e. the following relationship holds[2]

$$\{x_i(t)\}_\Re = \{x_i(t - \tau)\}_S \quad (4.24)$$

Hence the streak line covered by marked fluid elements up to time t can be computed as the path line of the fluid elements that fulfills the condition (4.24). A path line needs to be computed with the initial condition that for $t = \tau$ the fluid element held the position of the location coordinates of the injection point of the tracers. The streak line is thus composed of the sum of the path lines of individual particles. For each individual marked particle of a streak line a parameter τ is introduced, which for $0 \leq \tau \leq t$ covers all parts of a sweeping path. It is therefore important to vary the parameter τ in the solution equations in order obtain the entire streak line.

The above short explanations will be made clearer by way of an example, which is handled on the basis of the three-dimensional velocity field also used above:

$$U_1 = x_1(1+t), \quad U_2 = -x_2 \quad \text{and} \quad U_3 = -x_3 t \quad (4.25)$$

This velocity field yields the set of differential equations for the motion of a fluid element in space, i.e. the following differential equations:

$$\frac{d(x_1)_S}{dt} = x_1(1+t), \quad \frac{d(x_2)_S}{dt} = -x_2, \quad \frac{d(x_3)_S}{dt} = -x_3 t. \quad (4.26)$$

As a solution one obtains the components $(x_1)_S, (x_2)_S$ and $(x_3)_S$ according to (4.13):

$$(x_1)_S = C_1 \exp\left[t\left(1 + \frac{t}{2}\right)\right], \quad (x_2)_S = C_2 \exp[-t], \quad (x_3)_S = C_3 \exp\left[-\frac{t^2}{2}\right] \quad (4.27)$$

---

[2] The subscript s signifies that the location coordinate of the sweeping path is meant.

4.3 Motion of Fluid Elements

If one inserts the initial conditions, that $(x_1)_S = (x_1)_{t=\tau} = 1$, $(x_2)_S = (x_2)_{t=\tau} = 1$, $(x_3)_S = (x_3)_{t=\tau} = 1$ the injection condition for $t = \tau$, i.e. that the position $(1, 1, 1)$ serves as an injection point of the tracer, one obtains:

$$C_1 = \exp\left[-\tau\left(1 + \frac{\tau}{2}\right)\right], \quad C_2 = \exp[\tau] \quad \text{and} \quad C_3 = \exp\left[\frac{\tau^2}{2}\right] \tag{4.28}$$

Inserting in this result the solutions for $(x_1)_S, (x_2)_S$ and $(x_3)_S$ the equation for the space coordinates for the streak line results for all times[3]:

$$\begin{aligned}
(x_1)_S &= \exp\left[t\left(1 + \frac{t}{2}\right) - \tau\left(1 + \frac{\tau}{2}\right)\right], \\
(x_2)_S &= \exp[-(t - \tau)], \\
(x_3)_S &= \exp\left[-\frac{1}{2}(t^2 - \tau^2)\right].
\end{aligned} \tag{4.29}$$

When one wants to make the spatial course of a streak line visible at a moment in time t, one has to insert the value of t in the above equation. In this way one obtains the equation of a space curve, with τ as a parameter. Here τ is determined by the period of time $[\tau_1, \tau_2]$ of the tracer injection in $(1, 1, 1)$ with $-\infty < \tau_1 < \tau_2 < t$. For $\tau_1 \to -\infty, \tau_2 = t$ and $t = 0$ one can write:

$$\begin{aligned}
(x_1)_S &= \exp\left[-\tau\left(1 + \frac{\tau}{2}\right)\right] \\
(x_2)_S &= \exp[\tau] \qquad\qquad\qquad -\infty < \tau < 0 \\
(x_3)_S &= \exp\left[\frac{\tau^2}{2}\right]
\end{aligned} \tag{4.30}$$

The course of this space curve is shown in Fig. 4.4. It indicates the streak line existing at the moment in time $t = 0$ (made visible from $\tau = -\infty$ to $\tau = 0$). In Fig. 4.4 the projections of the streak line into the main level of the Cartesian coordinate system are also introduced.

When one compares the equation for the streak line fixed by the space point $(1, 1, 1)$ with the equations for the path line of a fluid element, stated for the same flow field, one realizes that path lines and streak lines are not identical for non-stationary flows. Only in the case of a stationary flow are path lines and streak lines identical, as can be shown easily by the following considerations.

Considering the stationary velocity field:

$$U_1 = 2x_1, \quad U_2 = -x_2, \quad U_3 = -x_3 \tag{4.31}$$

---

[3] As a space curve is involved here, the statement in x_1, x_2, x_3-coordinates is appropriate. The definition x_S indicates that the location coordinates of a streak line are meant.

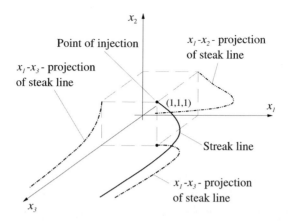

Fig. 4.4 Streak line for the time $t = 0$, with fluid tracer injections between $\tau = -\infty$ and $\tau = 0$ at the position $(1, 1, 1)$

one obtains for the path line of a fluid element the following differential equations:

$$\frac{d(x_1)_\Re}{dt} = 2x_1, \quad \frac{d(x_2)_\Re}{dt} = -x_2, \quad \frac{d(x_3)_\Re}{dt} = -x_3 \quad (4.32)$$

For $t = 0$ it will be assumed that $(x_1)_\Re = (x_2)_\Re = (x_3)_\Re = 1$ so that in the solution

$$(x_1)_\Re = C_1 \exp[2t], \quad (x_2)_\Re = C_2 \exp[-t], \quad (x_3)_\Re = C_3 \exp[-t] \quad (4.33)$$

holds and thus the path line for a marked tracer $\Re$ can be expressed as follows:

$$(x_1)_\Re = \exp[2t], \quad (x_2)_\Re = \exp[-t], \quad (x_3)_\Re = \exp[-t] \quad (4.34)$$

with $-\infty < t < \infty$. For the computation of the streak lines the solution in (4.27) can be employed again and C_1, C_2, and C_3 can be computed such that it is claimed that at time $t = \tau$ the following relationship holds:

$$(x_1(t = \tau))_S = 1, \quad (x_2(t = \tau))_S = 1, \quad (x_3(t = \tau))_S = 1 \quad (4.35)$$

Therefore, it can be deduced that

$$C_1 = \exp[-2\tau], \quad C_2 = \exp[\tau], \quad C_3 = \exp[\tau] \quad (4.36)$$

or as the coordinate for the streak line for each time t:

$$(x_1)_S = \exp[2(t - \tau)], \quad (x_2)_S = \exp[-(t - \tau)], \quad (x_3)_S = \exp[-(t - \tau)] \quad (4.37)$$

where the range of the values of τ is defined by the period of time of the tracer injection. In the case that the tracer substance is injected at all

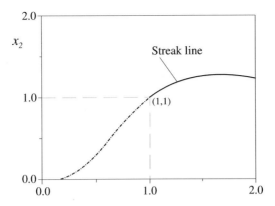

Fig. 4.5 Streak line in the plane x_1-x_2 (the full drawn line corresponds to $-\infty \leq \tau \leq 0$, broken line to $0 \leq \tau \leq \infty$)

times, i.e. $-\infty < \tau < \infty$, (4.34) and (4.37) yield the same curve. When the tracer injection is limited to certain time periods, one obtains as a visible streak line equation (4.37) only the corresponding part of the path line equation (4.34), see Fig. 4.5.

When one repeats the above derivations for the two-dimensional velocity field also stated in Sect. 4.3.1:

$$U_1 = x_1, \quad U_2 = x_2(1+2t), \quad U_3 = 0 \tag{4.38}$$

which leads to the differential equations (4.19):

$$\frac{d(x_1)_{\Re}}{dt} = x_{\Re}, \quad \frac{d(x_2)_{\Re}}{dt} = x_{\Re}(1+2t), \quad \frac{d(x_3)_{\Re}}{dt} = 0 \tag{4.39}$$

with the solution

$$(x_1)_{\Re} = C_1 \exp[t], \quad (x_2)_{\Re} = C_2 \exp[t(t+1)], \quad (x_3)_{\Re} = C_3 \tag{4.40}$$

On the other hand, when one demands that for the particle located at time t at the point x_1, x_2, x_3 passes the injection point $(1,1,0)$ of a tracer at the moment in time τ, the integration constants C_1, C_2 and C_3 can be obtained from the following conditional equations:

$$C_1 \exp[\tau] = 1, \quad C_2 \exp[\tau(\tau+1)] = 1, \quad C_3 = 0 \tag{4.41}$$

These "constants" can now be inserted in (4.40) again, where $C_3 = 0$ signifies that the streak line lies in the plane x_1-x_2, i.e. in the plane $x_3 = 0$, and is described there by the following equations for the position coordinates $(x_1)_S$ and $(x_2)_S$:

$$(x_1)_S = \exp[t-\tau], \quad (x_2)_S = \exp[t(t+1) - \tau(\tau+1)] \tag{4.42}$$

For the moment expressed by time $t = 0$, the course of the streak line is:

$$(x_1)_s = \exp[-\tau] \quad \text{and} \quad (x_2)_s = \exp[-\tau(\tau+1)] \quad \text{in the domain} \quad -\infty < \tau < \infty \tag{4.43}$$

From the equation for $(x_1)_s$, it follows that:

$$\tau = -\ln(x_1)_s \tag{4.44}$$

Inserted in the equation for $(x_2)_s$ at time $t = 0$, the existing course of the streak line made visible by the injection of tracer material in $-\infty < \tau < \infty$ results for the plane x_1–x_2:

$$\begin{aligned}(x_2)_s &= (x_1)_s^{(1-\ln(x_1)_s)} \\ (x_2)_s &= [(x_1)_s]^{(1-\ln(x_1)_s)}\end{aligned} \quad \text{in the domain} \quad 0 < (x_1)_s < \infty \tag{4.45}$$

4.4 Kinematic Quantities of Flow Fields

4.4.1 Stream Lines of a Velocity Field

The considerations in Sect. 4.3 for fluid elements, i.e. for the computation of path lines and streak lines have to be separated strictly from considerations for the determination of the stream lines of a flow field. Although for stationary flows stream lines, path lines and streak lines are identical, this does not justify neglecting the fundamental differences, or even worse, assuming that such differences do not exist. Only by a clear separation of the considerations does it become generally comprehensible why under the steady-state conditions of a flow field the above-stated identities of stream lines, path lines and streak lines exist.

If one considers the non-stationary flow field $U_j(x_i, t)$, stream lines can be defined for this field at any moment in time t, so that space curves run parallel to the velocity vectors. The latter are defined as stream lines at each point of the velocity field. In the case that one considers initially a two-dimensional flow field, for which all velocity vectors lie parallel to the plane x_1–x_2, then the above definition of the stream line leads to the following defining equation:

$$\frac{d(x_2)_\psi}{d(x_1)_\psi} = \frac{U_2}{U_1} \quad \text{for time } t = \text{constant} \tag{4.46}$$

This relationship means that the gradient of the stream line is equal to the tangent of the angle formed by the velocity vector with the axis x_1 at the instant in time t; see Fig. 4.6. The index ψ in (4.46) indicates that the stated coordinates x_1–x_2 describe the stream line $\psi = \text{const}$.

4.4 Kinematic Quantities of Flow Fields

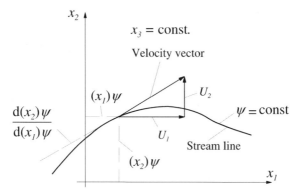

Fig. 4.6 Sketch for clarifying the defining equation for the stream line of a flow field

When one considers the defining equation (4.46) for the stream line of a two-dimensional flow field, it becomes understandable that, in general, for each moment in time t, the ratio of U_2 to U_1 is a function of x_1 and x_2. The resulting differential equation has to be solved in order to derive the fluid equation for the stream line. This will be explained on the basis of an example that starts from the following two-dimensional velocity field:

$$U_1 = x_1(1 + 2t), \quad U_2 = x_2 \quad U_3 = 0 \tag{4.47}$$

Introduced into the defining equation (4.46) for the stream line, one obtains

$$\frac{d(x_2)_\psi}{d(x_1)_\psi} = \frac{U_2}{U_1} = \frac{x_2}{x_1(1 + 2t)} \tag{4.48}$$

or, rewritten in the following form:

$$\frac{d(x_2)_\psi}{x_2} = \frac{d(x_1)_\psi}{x_1} \frac{1}{1 + 2t} \tag{4.49}$$

By integration, the following relationship for $(x_1)_\psi$ and $(x_2)_\psi$ results:

$$\ln(x_2)_\psi = C + \frac{1}{1 + 2t} \ln(x_1)_\psi \tag{4.50}$$

Considering the stream line, passing at time $t = 0$ the point $(1, 1, 0)$, $C = 0$ results and thus one can describe the stream line:

$$(x_2)_\psi = [(x_1)_\psi]^{\left(\frac{1}{1+2t}\right)} \tag{4.51}$$

In the case that a three-dimensional velocity field is considered, the above derivations, which were carried out for the projection of the stream lines into the planes x_1–x_2, can be performed in an analogous way. For the projections into the planes x_1-x_3 and x_2-x_3, relations analogous to the above used defining equation result:

$$\frac{\mathrm{d}(x_3)_\psi}{\mathrm{d}(x_1)_\psi} = \frac{U_3}{U_1} \qquad (4.52)$$

$$\frac{\mathrm{d}(x_3)_\psi}{\mathrm{d}(x_2)_\psi} = \frac{U_3}{U_2} \qquad (4.53)$$

Hence the defining equations of the stream lines of a velocity field can be stated as follows:

$$\frac{\mathrm{d}(x_1)_\psi}{U_1} = \frac{\mathrm{d}(x_2)_\psi}{U_2}, \quad \frac{\mathrm{d}(x_1)_\psi}{U_1} = \frac{\mathrm{d}(x_3)_\psi}{U_3}, \quad \frac{\mathrm{d}(x_2)_\psi}{U_2} = \frac{\mathrm{d}(x_3)_\psi}{U_3} \qquad (4.54)$$

or rewritten as:

$$\frac{\mathrm{d}(x_1)_\psi}{U_1} = \frac{\mathrm{d}(x_2)_\psi}{U_2} = \frac{\mathrm{d}(x_3)_\psi}{U_3} \qquad (4.55)$$

These differential equations for the stream line of a velocity field hold at each moment in time t. Their solution leads to a relationship $(x_3)_\psi = \psi(x_1, x_2)$, which describe a curve in space, the three-dimensional stream line.

Probably the simplest way to solve the set of differential equations (4.55) is to seek a parameter solution $(x_1)_\psi = x_1(s)$, where s is a parameter that varies along a streamline. The value of s at a certain reference point of the flow line is equal to zero. From there on it adopts increasing values along the flow line and in the flow direction. For all values $-\infty < s < \infty$ a presentation of the entire stream line is obtained.

By introducing s, one obtains:

$$\frac{\mathrm{d}(x_j)_\psi}{\mathrm{d}s} = U_j(x_i, t) \quad \text{for} \quad t = \text{constant} \quad \text{and} \quad j = s \qquad (4.56)$$

a relationship which represents, for each coordinate $(x_j)_\psi$, a differential equation and for $j = 1, 2, 3$ describing the stream lines in space for $t = $ constant. If the flow line passing through the space point $[x_0]_j$ at time t is sought, $s = 0$ results from integrating the three differential equations, when $x_j(t) = x_{j,0}$, then $s = 0$ is set. From this results the entire stream-line field as:

$$(x_j)_\psi = \psi_j(x_{0,j}, t, s) \qquad (4.57)$$

In order to demonstrate the above approached to obtaining the three-dimensional stream-line fields, the following velocity field is considered again:

$$U_1 = x_1(1+t), \quad U_2 = -x_2, \quad U_3 = -x_3 t \qquad (4.58)$$

A set of three differential equations for the stream lines of this velocity field results:

4.4 Kinematic Quantities of Flow Fields

$$\frac{\mathrm{d}(x_1)_\psi}{\mathrm{d}s} = x_1(1+t), \quad \frac{\mathrm{d}(x_2)_\psi}{\mathrm{d}s} = -x_2, \quad \frac{\mathrm{d}(x_3)_\psi}{\mathrm{d}s} = -x_3 t \quad (4.59)$$

Integration of these equations yields

$$\begin{aligned}(x_1)_\psi &= C_1 \exp[(1+t)s] \\ (x_2)_\psi &= C_2 \exp[-s] \\ (x_3)_\psi &= C_3 \exp[-txs]\end{aligned} \quad (4.60)$$

Searching for the stream line that passes through the point $(1,1,1)$, one can choose this point as reference point and set $s = 0$ for $(x_i)_\psi = 1$. From this one obtains the integration constants:

$$C_1 = C_2 = C_3 = 1 \quad (4.61)$$

Hence the following results for $(x_i)_\psi$ are obtained:

$$\begin{aligned}(x_1)_\psi &= \exp[(1+t)s] \\ (x_2)_\psi &= \exp[-s] \\ (x_3)_\psi &= \exp[-ts]\end{aligned} \quad (4.62)$$

For time $t = 0$, a stream line path results:

$$(x_1)_\psi = \exp[s], \quad (x_2)_\psi = \exp[-s], \quad (x_3)_\psi = 1 \quad (4.63)$$

Hence one obtains for $t = 0$ a stream line passing in the plane $x_3 = 1$, that is described by

$$(x_2)_\psi = \frac{1}{(x_1)_\psi} \quad (4.64)$$

The entire stream line field is obtained if one introduces for $s = 0$ arbitrary position coordinates $(x_{0,i})$ so that for any the position coordinates the following solution holds:

$$\begin{aligned}(x_1)_\psi &= (x_1)_{\psi,0} \exp[(1+t)s] \\ (x_2)_\psi &= (x_2)_{\psi,0} \exp[-s] \\ (x_3)_\psi &= (x_3)_{\psi,0} \exp[-txs]\end{aligned} \quad (4.65)$$

This set of equations indicates at each time t, the stream lines passing through the point $\{x_j\}_{\psi,0}$, see Fig. 4.7. There the parameter s is zero and for all other points of the considered stream line a value different to zero exists. This clear assignment is thus occurring which guarantees that stream lines never intersect, as otherwise two velocities would exist simultaneously at this point of intersection. This is precluded because of the existence of a well defined velocity fields (except for stagnation points and singularities).

In the preceding section, it was emphasized that, in general, stream lines, path lines and streak lines are not identical and the computed examples have confirmed this. For stationary flows all three lines are identical and are characteristic for each considered flow, i.e.

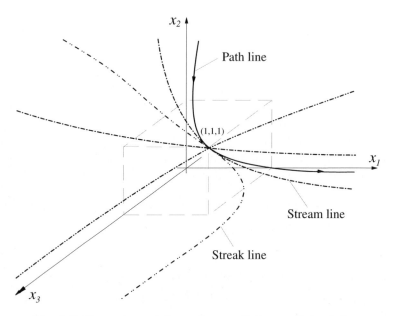

Fig. 4.7 Comparison of stream lines, path lines and streak lines

- For *stationary* flow fields marked fluid elements move along stream lines, i.e. stream lines are equal to their corresponding path lines
- For *stationary* flow fields stream lines can be made visible by locally injected tracer particles, i.e. stream lines are equal to streak lines

As already said, for *non-stationary* flows the corresponding stream lines, path lines and streak lines are different space curves, see Fig. 4.8.

4.4.2 Stream Function and Stream Lines of Two-Dimensional Flow Fields

For two-dimensional incompressible velocity fields $\{U_j\} = \{U_1, U_2, 0\}$, the stream function can be introduced as a field quantity. It is defined as follows, i.e the velocity components are thus defined as gradients of the stream function:

$$U_1 = \frac{\partial \psi}{\partial x_2} \quad \text{and} \quad U_2 = -\frac{\partial \psi}{\partial x_1} \tag{4.66}$$

Hence the stream function can be computed from the velocity field by the following line integral:

$$\psi - \psi_0 = \int_{x_{2,0}}^{x_2} U_1 \mathrm{d}x_2 \quad \text{for} \quad x_1 = \text{constant} \tag{4.67}$$

4.4 Kinematic Quantities of Flow Fields

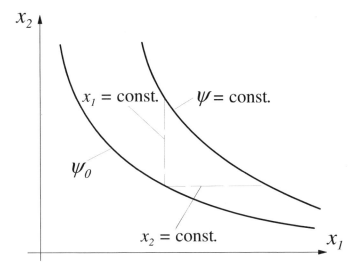

Fig. 4.8 Explanation of the defining equations of stream function

or also

$$\psi - \psi_0 = - \int_{x_{1,0}}^{x_1} U_2 \mathrm{d}x_1 \qquad \text{for } x_2 = \text{constant} \tag{4.68}$$

The above equations show that the difference that exists between the values of two stream functions is a measure of the volume flow rate that flows between two lines, each line corresponding to a constant stream function value. The depth of the considered area is taken as 1 perpendicular to the plane x_1–x_2. Accordingly the integrations along the line in (4.67) and (4.68) drawn in Fig. 4.8 and stated in the equations for $x_1 = $ constant and for $x_2 = $ constant, yield identical volume-flow values, when the upper integration limits x_1 and x_2 are chosen such that the differences $\psi - \psi_0$ are equally large in both integrals. For a fluid with $\rho = $ constant, i.e. for a thermodynamically ideal fluid, the identity of the integrals represents a mass-conservation relationship. When one carries out the integrations stated in (4.67) and (4.68) along a line $\psi = $ constant i.e. for $\mathrm{d}\psi = 0$, one obtains

$$U_1 \mathrm{d}(x_2)_\psi = U_2 \mathrm{d}(x_1)_\psi \tag{4.69}$$

or rewritten

$$\frac{\mathrm{d}(x_1)_\psi}{U_1} = \frac{\mathrm{d}(x_2)_\psi}{U_2} \tag{4.70}$$

From (4.70) it can be said by comparison with the definition equation for the stream line stated in (4.46) that the stream function defined for two-dimensional flow fields yields stream lines ψ for $\psi = $ constant.

Whereas for the kinematic considerations, e.g. in Chap. 2, arbitrary mathematical relationships for the velocity field could be used, the introduction of the stream function requires a limitation of the considerations to those velocity fields which have to fulfill physical conditions. Considerations in Chap. 5 show that physically existing flow fields have to fulfill the mass-conservation law, which can be formulated for ideal fluids (ρ = constant) as follows:

$$\frac{\partial U_i}{\partial x_i} = \frac{\partial U_1}{\partial x_1} + \frac{\partial U_2}{\partial x_2} + \frac{\partial U_3}{\partial x_3} = 0 \tag{4.71}$$

The stream function fulfills the mass-conservation law for two-dimensional flows automatically and can easily be checked by inserting the definition of ψ = in equation (4.66) and applying the Schwarz rule of differentiation.

When a prescribed velocity field does not fulfill the mass-conservation law, the integrations to be carried out according to (4.67) and (4.68) result in solutions that contradict one another. This can be shown for the two-dimensional flow field shown in Sect. 4.3.1 which does not fulfill the mass-conservation law (i.e. the continuity equation):

$$U_1 = x_1, \quad U_2 = x_2(1+2t), \quad U_3 = 0 \tag{4.72}$$

When one carries out the integration stated in (4.67), one obtains from the defining equations for the stream function, i.e. from (4.66) it results that:

$$\frac{\partial \psi}{\partial x_2} = x_1 \quad \text{and} \quad \frac{\partial \psi}{\partial x_1} = -x_2(1+2t) \tag{4.73}$$

By integration of these equations, one obtains for the stream function:

$$\psi = x_1 x_2 + F(x_1, t), \qquad \psi = -x_1 x_2 (1+2t) + G(x_2, t) \tag{4.74}$$

or, expressed otherwise,

$$x_1 x_2 + F(x_1, t) \neq -x_1 x_2 (1+2t) + G(x_2, t) \tag{4.75a}$$

or

$$2(t+1) x_1 x_2 \neq G(x_2, t) - F(x_1, t) \tag{4.75b}$$

A comparison of the results of both the integrations in equation (4.73), yields equation (4.74), which shows the contradiction resulting for the stream function ψ. The time dependence for the $x_1 x_2$ part in the right-hand side of equation (4.75a) is missing on the left-hand side. This results from the fact that the velocity field given in (4.72), although mathematically clearly defined, cannot exist physically; the velocity field in (4.72) does not fulfill the requirements determined by the mass-conservation law for ρ = constant. When one considers, on the other hand, the velocity field:

$$U_1 = \exp[x_1(1+t)], \quad U_2 = -x_2(1+t)\exp[x_1(1+t)], \quad U_3 = 0 \tag{4.76}$$

4.4 Kinematic Quantities of Flow Fields

for which equation (4.71) is fulfilled, since the following holds:

$$\frac{\partial U_1}{\partial x_1} = (1+t)\exp[x_1(1+t)] \quad \text{and} \quad \frac{\partial U_2}{\partial x_2} = -(1+t)\exp[x_1(1+t)] \quad (4.77)$$

one obtains from the defining equations for the stream function equation (4.66):

$$\frac{\partial \psi}{\partial x_2} = \exp[x_1(1+t)] \quad \text{and} \quad \frac{\partial \psi}{\partial x_1} = x_2(1+t)\exp[x_1(1+t)] \quad (4.78)$$

the following solution for the stream function ψ:

$$\psi = x_2 \exp[x_1(1+t)] + C \quad (4.79)$$

If one knows the value of the stream function for $x_{1,0}$ and $x_{2,0}$, then the following holds:

$$\psi_0 = x_{2,0} \exp[x_{1,0}(1+t)] + C \quad (4.80)$$

or, rewritten with ψ_0,

$$\psi - \psi_0 = x_2 \exp[x_1(1+t)] - x_{2,0} \exp[x_{1,0}(1+t)] \quad (4.81)$$

The result obtained can also be computed for $x_1 = x_{1,0} =$ constant from (4.66):

$$\psi - \psi_0 = (x_2 - x_{2,0}) \exp[x_{1,0}(1+t)] \quad \text{for} \quad x_1 = x_{1,0} \quad (4.82)$$

This expresses the distribution $\psi(x_2)$ at the location $x_{1,0}$. When one wants to determine the stream line in the x_1-x_2 plane, one has to consider ψ as a parameter in equation (4.81) and derive the relationship $(x_1-x_2)_\psi$, for ψ as a parameter from equation (4.81). Here $\psi_0, x_{1,0}$ and $x_{2,0}$ are constants that can be chosen freely.

4.4.3 Divergence of a Flow Field

In this section, mathematical operators will be explained that are known from vector analysis and that can be applied to flow fields. Their derivations will be repeated but also considered in some detail with respect to their physical meanings.

The divergence of the velocity field $\boldsymbol{U}$ can be expressed as:

$$\frac{\partial U_i}{\partial x_i} = \frac{\partial U_1}{\partial x_1} + \frac{\partial U_2}{\partial x_2} + \frac{\partial U_3}{\partial x_3} \quad (4.83)$$

The above-defined divergence of a velocity field is a scalar field which, in the presence of a steady velocity field, is defined at each point in space and can be computed from the velocity field if the latter can be assumed to be known.

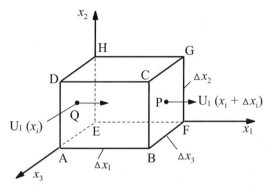

Fig. 4.9 Fluid element for explaining the physical meaning of $\frac{\partial U_i}{\partial x_i}$ = divergence of velocity field

When one wants to perceive the physical meaning of the operator $\frac{\partial}{\partial x_i}$ applied to the velocity field U_i, the consideration of a single fluid volume, as indicated in Fig. 4.9, is recommended.

The edge lengths Δx_i of the considered fluid element were assumed to be very small, so that a velocity vector can be assigned to each surface such that this vector indicates with which velocity the considered surface of a fluid element moves. Accordingly, the surface AEHD moves in the direction x_1 with the velocity component of the velocity field present at point Q, i.e. with $U_1(x_i) = U_1(x_1, x_2, x_3)$. In comparison, the surface BFGC moves with the velocity component $U_1(x_i + \Delta x_1, x_2, x_3)$, present at point P. This velocity component can be expressed by a Taylor series expansion as follows:

$$U_1(x_1 + \Delta x_1, x_2, x_3) = U_1(x_1) + \left(\frac{\partial U_1}{\partial x_1}\right) \Delta x_1 + \frac{1}{2}\left(\frac{\partial^2 U_1}{\partial x_1^2}\right) \Delta x_1^2 + \ldots \quad (4.84)$$

The difference velocity between the surfaces AEHD and BFGC can thus be computed as

$$\Delta U_1(x_1, \Delta x_1) = \left(\frac{\partial U_1}{\partial x_1}\right) \Delta x_1 + \frac{1}{2}\left(\frac{\partial^2 U_1}{\partial x_1^2}\right) \Delta x_1^2 + \ldots \quad (4.85)$$

As a consequence of this velocity difference, a volume increase or a volume decrease results, depending on the sign of the derivative of a considered velocity field. This value can be stated as follows, to a first approximation, by multiplication with the surface $(\Delta x_2 \Delta x_3)$, neglecting all the terms of second and higher order in Δx_1:

$$\frac{d}{dt}(\delta V_1)_\Re = \Delta U_1(x_i)(\Delta x_2 \Delta x_3) = \frac{\partial U_1}{\partial x_1} \Delta x_1 (\Delta x_2 \Delta x_3) \quad (4.86)$$

On the basis of simultaneously existing gradients of the velocity field in the directions x_2 and x_3, additional volume changes occur per unit time, which again can be stated to a first approximation as follows:

4.4 Kinematic Quantities of Flow Fields

$$\frac{d}{dt}(\delta V_2)_{\Re} = \frac{\partial U_2}{\partial x_2}\Delta x_2(\Delta x_1 \Delta x_3) \quad \text{and} \quad \frac{d}{dt}(\delta V_3)_{\Re} = \frac{\partial U_3}{\partial x_3}\Delta x_3(\Delta x_1 \Delta x_2) \tag{4.87}$$

so that the entire volume change that can be expected in a flow field for a fluid element per unit time can be expressed as follows:

$$\frac{d}{dt}(\delta V)_{\Re} = \sum_{\alpha=1}^{3}\frac{d}{dt}(\delta V_\alpha)_{\Re} = \frac{\partial U_i}{\partial x_i}(\delta V_{\Re}) \tag{4.88}$$

or can be rewritten as

$$\frac{\partial U_i}{\partial x_i} = \frac{1}{\delta V_{\Re}}\frac{d}{dt}(\delta V_{\Re}) \tag{4.89}$$

This relationship emphasizes the physical significance of the divergence of a velocity field.[4] The divergence of a velocity field states how large the volume change of a fluid element is that occurs per unit time and unit volume at a certain position in a flow field. At such locations of the flow field where the divergence of a velocity vector is equal to zero, there is no temporal volume change locally for a fluid element moving in the velocity field. When the divergence in sub-domains of the velocity field is computed to be negative, a fluid element experiences volume decreases in these domains.

When one considers the physical significance of the divergence of a velocity field for a stationary volume element of a fluid, inflows and outflows occur through the surfaces of the considered volume because of the existing velocity field. The volume flowing in per unit time can be stated as

$$\dot{V}_{\text{inflow}} = U_i \Delta x_j \Delta x_k \qquad i \neq j,k \tag{4.90}$$

For the volume flowing out, one can compute

$$\dot{V}_{\text{outflow}} = \left[U_i + \frac{\partial U_i}{\partial x_i}\Delta x_i\right]\Delta x_j \Delta x_k \qquad i \neq j,k \tag{4.91}$$

The difference of inflows and outflows, considering $\Delta V = \Delta x_1, \Delta x_2, \Delta x_3$ can be computed as

$$\Delta \dot{V} = \dot{V}_{\text{inflow}} - \dot{V}_{\text{outflow}} = -\frac{\partial U_i}{\partial x_i}\Delta V \tag{4.92}$$

This relation makes it clear that the presence of a positive divergence of the velocity field at a point in space is equal to a "volume source", as more "fluid volume" is flowing out of the considered control volume than is flowing in. When, however, the divergence of a velocity field is negative, a sink is present, as then the inflow in the "volume" has to be larger than the outflow.

---

[4] Equation (4.88) makes it clear that the summation for $i = 1 - 3$ is stated in a sufficiently comprehensible way by the subscript in $\partial U_i/\partial x_i$.

4.5 Translation, Deformation and Rotation of Fluid Elements

Analogous to considerations in solid-state mechanics, the deformations of fluid elements that occur due to existing velocity gradients are of interest in some fluid mechanics considerations. When one includes the translatory motion and the rotation of a fluid element in the fluid deformations, the entire local state of motion and deformation can be stated by four "geometrically easily separable" states of motion. The pure translatory motion sketched in Fig. 4.10 leads to a change in position of the fluid element marked $\Re$ to an extent that the following holds:

$$d(x_j)_\Re = (U_j)_\Re dt = U_j dt \qquad (4.93)$$

This relationship expresses that the locally existing velocity field is responsible for the translatory motion of a fluid element, i.e. fluid elements move at each moment in time with the locally existing velocity vector.

When one superimposes a fluid-element rotation upon the pure translatory motions sketched in Fig. 4.10, the image shown in Fig. 4.11 results.

In order to state or compute the rotation of a fluid element, one has to describe both the angles $\Delta\Theta_1$ and $d\Delta\Theta_2$:

$$\Delta\Theta_1 = \left| \tan \frac{\frac{\partial U_2}{\partial x_1}(\Delta x_1)_\Re \Delta t}{(\Delta x_1)_\Re} \right| = \frac{\partial U_2}{\partial x_1} \Delta t \qquad (4.94)$$

$$\Delta\Theta_2 = \left| \tan \frac{\frac{\partial U_1}{\partial x_2}(\Delta x_2)_\Re \Delta t}{(\Delta x_2)_\Re} \right| = -\frac{\partial U_1}{\partial x_2} \Delta t \qquad (4.95)$$

As the rotational speed of the fluid element, the positive change of angle of the diagonal of the element occurring per unit time is defined as:

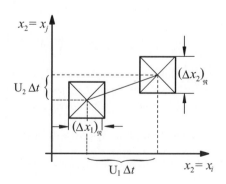

Fig. 4.10 Pure translatory motion; considerations of the projection into the x_1–x_2 planes

4.5 Translation, Deformation and Rotation of Fluid Elements

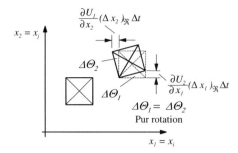

Fig. 4.11 Translation and rotation of a fluid element in a flow field, due to velocity gradients

$$\dot{\Theta}_{21}^R = \left(\frac{d\Theta^R}{dt}\right) = \frac{1}{2}\left(\frac{d\Theta_1}{dt} + \frac{d\Theta_2}{dt}\right) = \frac{1}{2}\left(\frac{\partial U_2}{\partial x_1} - \frac{\partial U_1}{\partial x_2}\right) \quad (4.96)$$

Generally, the components of the rotational velocity vector $\{\omega_k\}$ for the locally occurring rotation of a fluid element per unit time can be stated as follows:

$$\text{rot}(\boldsymbol{U}) = \epsilon_{ijk}\frac{\partial U_i}{\partial x_j} = 2\omega_k = 2\dot{\Theta}_{ij}^R = \frac{\partial U_j}{\partial x_i} - \frac{\partial U_i}{\partial x_j} \quad (4.97)$$

The quantity ω_k represents the double rotational speed around the axis of the fluid element, with the rotation occurring in the positive direction. The second diagonal of the considered fluid element in Fig. 4.11 rotates with the same angular speed. $\{\omega_k\}$ is an important kinematic quantity of the velocity field. It is defined mathematically as vorticity and is computed as half the rotational speed of the velocity field. Thus $\omega_k(x_i, t)$ is a field quantity of its own, for which we can easily show that $\omega_k(x_i, t) = 0$ when a flow field is free of rotation. When $\omega_k(x_i, t) \neq 0$, flows subjected to rotations are present, whose rotational properties can best be studied when expressing the conservation laws for mass, impulse and energy in terms of ω_k.

When one considers next the angular deformation of a fluid element shown in Fig. 4.12, one can see that for the angle deformation the following holds:

$$\dot{\Theta}_{21}^D = \frac{1}{2}\left(\frac{d\Theta_1}{dt} - \frac{d\Theta_2}{dt}\right) \quad \text{deformation angular speed}$$

Thus for the angular deformation in the plane x_1–x_2 the following expression holds:

$$\dot{\Theta}_{21}^D = \frac{1}{2}\left(\frac{\partial U_2}{\partial x_1} + \frac{\partial U_1}{\partial x_2}\right) \quad (4.98)$$

or, considering the angular deformation only,

$$\dot{\Theta}_{ji}^D = \frac{1}{2}\left(\frac{\partial U_j}{\partial x_i} + \frac{\partial U_i}{\partial x_j}\right) \quad i \neq j \quad (4.99)$$

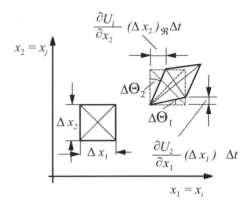

Fig. 4.12 Translation and angle deformation of a fluid element in a flow field due to velocity gradients

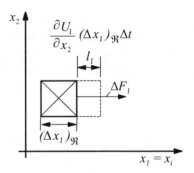

Fig. 4.13 Elongation of a volume element due to velocity gradients in the flow field

Analogous to considerations in solid-state mechanics, the symmetry of the deformation tensor holds:

$$\dot{\Theta}_{ij}^D = \dot{\Theta}_{ji}^D \tag{4.100}$$

Finally, one has to consider the dilatation of a fluid element which experiences strain rates due to the velocity gradient existing in a flow field, as is shown in Fig. 4.13. The linear deformation in length that occurs due to an existing velocity gradient in the direction x_1 can be stated as follows:

$$\frac{dl_1}{dt} = \lim_{\Delta t \to 0} \left(\frac{\partial U_1}{\partial x_1}\right) \frac{(\Delta x_1)_\Re \Delta t}{\Delta t} = \left(\frac{\partial U_1}{\partial x_1}\right)(\Delta x_1)_\Re \tag{4.101}$$

From this, the linear deformation that occurs per unit length and unit time is computed:

$$\frac{1}{(\Delta x_1)} \frac{d(l_1)}{dt} = \frac{\partial U_1}{\partial x_1} \tag{4.102}$$

4.5 Translation, Deformation and Rotation of Fluid Elements

On multiplying the linear deformation by the area perpendicular to the x_1 axis, the corresponding volume change results:

$$\frac{d(\delta V_1)_\Re}{dt} = \left(\frac{\partial U_1}{\partial x_1}\right)(\delta V_1)_\Re \qquad (4.103)$$

The same considerations hold for the x_2 and x_3 axes also. Summed over all three axis directions, one obtains for the entire volume change per unit time

$$\frac{1}{(\delta V)_\Re}\frac{d(\delta V)_\Re}{dt} = \frac{\partial U_i}{\partial x_i} \qquad (4.104)$$

i.e. the divergence of the velocity field indicates how the volume of a fluid element changes with time at a point in space. This was already shown in Sect. 4.3.

It is customary in the literature to combine elongations of fluid elements and their angular deformations with a deformation tensor in such a way that

$$\varepsilon_{ij} = \frac{1}{2}\left(\frac{\partial U_j}{\partial x_i} + \frac{\partial U_i}{\partial x_j}\right) \qquad (4.105)$$

so that for the deformation tensor

$$\{\varepsilon_{ij}\} = \begin{Bmatrix} \varepsilon_{11} & \varepsilon_{12} & \varepsilon_{13} \\ \varepsilon_{21} & \varepsilon_{22} & \varepsilon_{23} \\ \varepsilon_{31} & \varepsilon_{32} & \varepsilon_{33} \end{Bmatrix} \quad \text{and} \quad \varepsilon_{ij} = \varepsilon_{ji} \qquad (4.106)$$

From the above considerations, the following relationship results:

$$\frac{\partial U_j}{\partial x_i} = \frac{1}{2}\left(\frac{\partial U_j}{\partial x_i} + \frac{\partial U_i}{\partial x_j}\right) + \frac{1}{2}\left(\frac{\partial U_j}{\partial x_i} - \frac{\partial U_i}{\partial x_j}\right) \qquad (4.107)$$

i.e. the following kinematic relationship exists:

$$\frac{\partial U_j}{\partial x_i} = \varepsilon_{ij} + \frac{d\Theta_{ij}^R}{dt} = \varepsilon_{ij} + \frac{1}{2}\epsilon_{ijk}\frac{\partial U_j}{\partial x_i} \qquad (4.108)$$

Hence the gradients that existing in velocity fields are linked to deformations and rotations of fluid elements, the gradients yields corresponding rates of deformation and rotational angular velocities. This has to be taken into consideration when employing the analogy between solid-state mechanics and fluid mechanics. It is necessary to transfer considerations of deformations of elastic bodies, carried out in solid-state mechanics, to rates of deformation of fluid elements occurring in fluid mechanics. The latter occur due to gradients existing in velocity fields and the former due to existing internal surface forces.

4.6 Relative Motions

Considerations of the velocities at two points separated by (δx_i) result in the following relationship if one applies Taylor series expansion:

$$U_j(x_i + \delta x_i, t) = U_j(x_i, t) + \frac{\partial U_j}{\partial x_i} \delta x_i + \cdots \qquad (4.109)$$

or the relationship can be rewritten as follows:

$$U_j(x_i + \delta x_i, t) = \underbrace{U_j(x_i, t)}_{\text{Translation}} + \overbrace{\frac{1}{2}\left(\frac{\partial U_j}{\partial x_i} - \frac{\partial U_i}{\partial x_j}\right) \delta x_i}^{\text{Rotation}} + \underbrace{\frac{1}{2}\left(\frac{\partial U_j}{\partial x_i} + \frac{\partial U_i}{\partial x_j}\right) \delta x_i}_{\text{Deformation}} \qquad (4.110)$$

For further details, see Aris [4.5]. Considering the various terms in this relationship makes it clear that the velocity at the point adjacent to x_i, i.e. at the point $x_i + \delta x_i$, is composed of the translation by velocity at point $\mathcal{P}(x_i)$, a rotational velocity around this point and a deformation action in this point (Fig. 4.14). The components for $j = 1, 2, 3$ read

$$j = 1: \ U_1(x_i + \delta x_i, t) = \left[U_1 + \frac{\partial U_1}{\partial x_1}\delta x_1 + \frac{1}{2}\left(\frac{\partial U_1}{\partial x_2} - \frac{\partial U_2}{\partial x_1}\right)\delta x_2 \right.$$
$$\left. + \frac{1}{2}\left(\frac{\partial U_1}{\partial x_3} - \frac{\partial U_3}{\partial x_1}\right)\delta x_3 + \frac{1}{2}\left(\frac{\partial U_1}{\partial x_2} + \frac{\partial U_2}{\partial x_1}\right)\delta x_2 + \frac{1}{2}\left(\frac{\partial U_1}{\partial x_3} + \frac{\partial U_3}{\partial x_1}\right)\delta x_3\right]$$

$$(4.111)$$

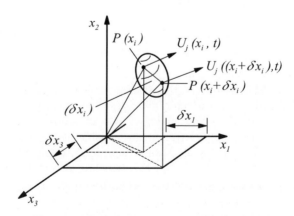

Fig. 4.14 Relative motions in a fluid element

4.6 Relative Motions

$$j = 2: \quad U_2(x_i + \delta x_i, t) = \left[U_2 + \frac{\partial U_2}{\partial x_2}\delta x_2 + \frac{1}{2}\left(\frac{\partial U_2}{\partial x_3} - \frac{\partial U_3}{\partial x_2}\right)\delta x_3 \right.$$
$$\left. + \frac{1}{2}\left(\frac{\partial U_2}{\partial x_1} - \frac{\partial U_1}{\partial x_2}\right)\delta x_1 + \frac{1}{2}\left(\frac{\partial U_2}{\partial x_3} + \frac{\partial U_3}{\partial x_2}\right)\delta x_3 + \frac{1}{2}\left(\frac{\partial U_2}{\partial x_1} + \frac{\partial U_1}{\partial x_2}\right)\delta x_1 \right] \quad (4.112)$$

$$j = 3: \quad U_3(x_i + \delta x_i, t) = \left[U_3 + \frac{\partial U_3}{\partial x_3}\delta x_3 + \frac{1}{2}\left(\frac{\partial U_3}{\partial x_1} - \frac{\partial U_1}{\partial x_3}\right)\delta x_1 \right.$$
$$\left. + \frac{1}{2}\left(\frac{\partial U_3}{\partial x_2} - \frac{\partial U_2}{\partial x_3}\right)\delta x_2 + \frac{1}{2}\left(\frac{\partial U_3}{\partial x_1} + \frac{\partial U_1}{\partial x_3}\right)\delta x_1 + \frac{1}{2}\left(\frac{\partial U_3}{\partial x_1} + \frac{\partial U_2}{\partial x_3}\right)\delta x_2 \right] \quad (4.113)$$

With the above equations, most general motions of fluid elements can now be described, i.e. the motion at any point of a fluid element can be stated as the sum of the translation of a reference point, a rotational motion around this point and an additional deformation. The motion is due to translation and rotation and a superimposed deformation.

The different components of the equations (4.111) to (4.113) can be obtained by regrouping as follows:

$$j = 1: \quad U_1(x_i) + \left[\frac{\partial U_1}{\partial x_1}\delta x_1 + \frac{1}{2}\left(\frac{\partial U_1}{\partial x_2} + \frac{\partial U_2}{\partial x_1}\right)\delta x_2 + \frac{1}{2}\left(\frac{\partial U_1}{\partial x_3} + \frac{\partial U_3}{\partial x_1}\right)\delta x_3\right]$$
$$+ \left\{ \frac{1}{2}\left(\frac{\partial U_1}{\partial x_2} - \frac{\partial U_2}{\partial x_1}\right)\delta x_2 + \frac{1}{2}\left(\frac{\partial U_1}{\partial x_3} - \frac{\partial U_3}{\partial x_1}\right)\delta x_3 \right\} \quad (4.114)$$

$$j = 2: \quad U_2(x_2) + \left[\frac{\partial U_2}{\partial x_2}\delta x_2 + \frac{1}{2}\left(\frac{\partial U_2}{\partial x_1} + \frac{\partial U_1}{\partial x_2}\right)\delta x_1 + \frac{1}{2}\left(\frac{\partial U_2}{\partial x_3} + \frac{\partial U_3}{\partial x_2}\right)\delta x_3\right]$$
$$+ \left\{ \frac{1}{2}\left(\frac{\partial U_2}{\partial x_1} - \frac{\partial U_1}{\partial x_2}\right)\delta x_1 + \frac{1}{2}\left(\frac{\partial U_2}{\partial x_3} - \frac{\partial U_3}{\partial x_2}\right)\delta x_3 \right\} \quad (4.115)$$

$$j = 3: \quad U_3(x_3) + \left[\frac{\partial U_3}{\partial x_3}\delta x_3 + \frac{1}{2}\left(\frac{\partial U_3}{\partial x_2} + \frac{\partial U_2}{\partial x_3}\right)\delta x_2 + \frac{1}{2}\left(\frac{\partial U_3}{\partial x_1} + \frac{\partial U_1}{\partial x_3}\right)\delta x_1\right]$$
$$+ \left\{ \frac{1}{2}\left(\frac{\partial U_3}{\partial x_2} - \frac{\partial U_2}{\partial x_3}\right)\delta x_2 + \frac{1}{2}\left(\frac{\partial U_3}{\partial x_1} - \frac{\partial U_1}{\partial x_3}\right)\delta x_1 \right\} \quad (4.116)$$

The expressions in front of the square brackets represent the translational velocity, which is given by the following velocity vector:

$$U_j(x_i, t) = \{U_1, U_2, U_3\}^T \quad (4.117)$$

In the square brackets the product of the deformation tensor:

$$D_{ij}(x_i,t) = \begin{Bmatrix} \dfrac{\partial U_1}{\partial x_1} & \dfrac{1}{2}\left(\dfrac{\partial U_1}{\partial x_2}+\dfrac{\partial U_2}{\partial x_1}\right) & \dfrac{1}{2}\left(\dfrac{\partial U_1}{\partial x_3}+\dfrac{\partial U_3}{\partial x_1}\right) \\ \dfrac{1}{2}\left(\dfrac{\partial U_2}{\partial x_1}+\dfrac{\partial U_1}{\partial x_2}\right) & \dfrac{\partial U_2}{\partial x_2} & \dfrac{1}{2}\left(\dfrac{\partial U_2}{\partial x_3}+\dfrac{\partial U_3}{\partial x_2}\right) \\ \dfrac{1}{2}\left(\dfrac{\partial U_3}{\partial x_1}+\dfrac{\partial U_1}{\partial x_3}\right) & \dfrac{1}{2}\left(\dfrac{\partial U_3}{\partial x_2}+\dfrac{\partial U_2}{\partial x_3}\right) & \dfrac{\partial U_3}{\partial x_3} \end{Bmatrix} \quad (4.118)$$

and of the "distance vector" from point x_i to $(x_i + \delta x_i)$:

$$\{\delta x_i\} = \{\delta x_1, \delta x_2, \delta x_3\} \quad (4.119)$$

is shown, and in the curly brackets the vector product of

$$\{\delta x_i\} = \{\delta x_1, \delta x_2, \delta x_3\} \quad (4.120)$$

and

$$2\omega_k = \left\{ \left(\dfrac{\partial U_3}{\partial x_2} - \dfrac{\partial U_2}{\partial x_3}\right); \ \left(\dfrac{\partial U_1}{\partial x_3} - \dfrac{\partial U_3}{\partial x_1}\right); \ \left(\dfrac{\partial U_2}{\partial x_1} - \dfrac{\partial U_1}{\partial x_2}\right) \right\} \quad (4.121)$$

is shown. Hence the entire motion can be written as

$$U_j(x + \mathrm{d}x_i, t) = U_j(x_i, t) + D_{ij}(x_i, t)\delta x_i + \varepsilon_{ijk}\omega_k(x_i, t)\delta x_i \quad (4.122)$$

This relationship again expresses the fact that the total motion of a point $P'(x_i + \mathrm{d}x_i)$ can be understood as the translational motion of the point $P(x_i)$,

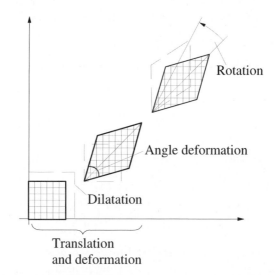

Fig. 4.15 Illustration of the translational motion, the deformation and the rotation of a fluid element

4.6 Relative Motions

superimposed by deformation motions and rotational motions around $P(x_i)$ (Fig. 4.14).

The different parts, i.e the translation, deformation and rotation, can be taken from the sequence of a fluid element which is shown in Fig. 4.15. For the two-dimensional case with $U_1 = u$ and $U_2 = v$ and with $x_1 = x$ and $x_2 = y$, the considerations for the different subjects are shown once more in Fig. 4.16. In Fig. 4.17 a fluid element is shown under translation and pure rotation once again. Fig. 4.18, on the other hand, shows a fluid element carrying out a translation motion in the presence of a pure deformation, i.e. without the presence of a rotation. The last manner of motion requires

$$\frac{\partial U_2}{\partial x_1} = \frac{\partial U_1}{\partial x_2} \qquad (4.123)$$

so that $\omega_3 = 0$.

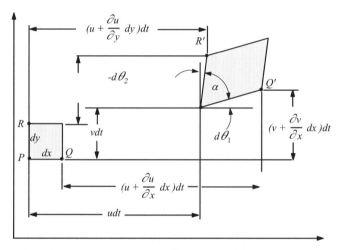

Fig. 4.16 Translation, deformation and rotation of a fluid element due to the velocity components u and v

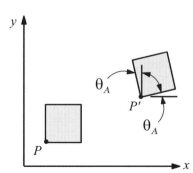

Fig. 4.17 Translation motion of a fluid element with rotation

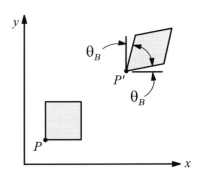

Fig. 4.18 Translation motion of a fluid element with deformation

References

4.1. Brodkey, R.S., The Phenomena of Fluid Motions, Dover, New York, 1967.
4.2. Spurk, J.H., Fluid Science/Technology: An Introduction to the Theory of Fluids, Springer, Berlin Heidelberg New York, 1966.
4.3. Hutter, K., Fluid Dynamics and Thermodynamics: An Introduction. Springer, Berlin Heidelberg New York, 1995.
4.4. Currie, I.G., Fundamental Mechanics of Fluids, McGraw-Hill, New York, 1974.
4.5. Aris, R., Vectors, Tensors and the Basic Equations of Fluid Mechanics, McGraw-Hill, New York, 1974.

Chapter 5
Basic Equations of Fluid Mechanics

5.1 General Considerations

Fluid mechanics considerations are applied in many fields, especially in engineering. Below a list is provided which clearly indicates the far-reaching applications of fluid-mechanics knowledge and their importance in various fields of engineering. Whereas it was usual in the past to carry out special fluid mechanics considerations for each of the areas listed below, today one strives increasingly at the development and introduction of generalized approaches that are applicable without restrictions to all of these fields. This makes it necessary to derive the basic equations of fluid mechanics so generally that they fulfill the requirements for the broadest applicability in areas of science and engineering, i.e. in those areas indicated in the list below. The objective of the derivations in this section is to formulate the conservation laws for mass, momentum, energy, chemical species, etc., in such a way that they can be applied to all the flow problems that occur in the following areas:

- Heat exchanger, cooling and drying technology
- Reaction technology and reactor layout
- Aerodynamics of vehicles and aeroplanes
- Semiconductor-crystal production, thin-film technology, vapor-phase deposition processes
- Layout and optimization of pumps, valves and nozzles
- Use of flow equipment parts such as pipes and junctions
- Development of measuring instruments and production of sensors
- Ventilation, heating and air-conditioning techniques, layout and tests, laboratory vents
- Problem solutions for roof ventilation and flows around buildings
- Production of electronic components, micro-systems analysis engineering
- Layout of stirrer systems, propellers and turbines
- Sub-domains of biomedicine and medical engineering
- Layout of baking ovens, melting furnaces and other combustion units

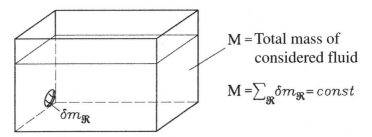

Fig. 5.1 Division of a fluid into $\Re$ fluid elements for mass conservation considerations

- Development of engines, catalyzers and exhaust systems
- Combustion and explosion processes, energy generation, environmental engineering
- Sprays, atomizing and coating technologies

Concerning the formulation of the basic equations of fluid mechanics, it is easy to formulate the conservation equations for mass, momentum, energy and chemical species for a fluid element, see Fig. 5.1, i.e. to derive the "Lagrange form" of the equations. In this way, the derivations can be represented in an easily comprehensible way and it is possible to build up the derivations upon the basic knowledge of physics. Derivations of the basic equations in the "Lagrange form" are usually followed by transformation considerations whose aim is to derive local formulations of the conservation equations and to introduce field quantities into the mathematical representations, i.e. the "Euler form" of the conservation equations is sought for solutions of fluid flow problems. This requires one to express temporal changes of substantial quantities as temporal changes of field quantities, which makes it necessary, partly, to repeat in this section the considerations in Chap. 2 but to explain them in a somewhat different and even deeper way.

The considerations to be carried in the sections below start out with the assumption that, at a certain point in time $t = 0$, the mass of a fluid is subdivided into fluid elements of the mass $\delta m_\Re$, i.e.

$$M = \sum_\Re \delta m_\Re.$$

Each fluid element $\delta m_\Re$ is chosen to be large enough to make the assumption $\delta m_\Re$ = constant possible, with sufficient precision, in spite of the molecular structure of the fluid. The assumption is also made to allow one to assign arbitrary thermodynamic and fluid mechanics properties $\alpha_\Re(x_\Re(t), t) = \alpha_\Re(t)$ to a fluid element to yield $\alpha_\Re$ = constant, with satisfactory precision for fluid mechanics considerations.

The term $\alpha_\Re(x_\Re, t)$, with $x_\Re = x_\Re(t)$, expresses the fact that the thermodynamic or fluid mechanics property, which is assigned to the considered fluid element, represents a substantial quantity that is only a function of time. This property of the element changes with time at a fixed position in space, but changes also due to the motion of the fluid element. For the description

5.2 Mass Conservation (Continuity Equation)

of these changes, it is important that one follows the mass $\delta m_\Re$, i.e. one takes $x_\Re(t)$ and also introduces it into consideration as known. It is assumed that the motion of sub-parts of $\delta m_\Re$ is the same for all parts of the considered fluid element. The fluid element is also assumed to consist at all times of the same fluid molecules, i.e. it is assumed that the considered fluid element does not split up during the considerations of its motion. This basically means that the fluid belonging to a considered fluid element, at time $t = 0$, remains also in the fluid element at all later moments in time. This signifies that it is not possible for two different fluid elements to take the same point in space at an arbitrary time: $x_\Re(t) \neq x_\mathcal{L}(t)$ for $\Re \neq \mathcal{L}$.

When a fluid element $\Re$ is at the position x_i at time t, i.e. $x_i = (x_\Re(t))_i$ at time t, then the substantial thermodynamic property, or any fluid mechanic property, $\alpha_\Re(t)$ is equal to the field quantity α at the point x_i at time t:

$$\alpha_\Re(t) = \alpha(x_i, t) \quad \text{when} \quad (x_\Re(t))_i = x_i \quad \text{at time } t. \tag{5.1}$$

For the temporal change of a quantity $\alpha_\Re(t)$ (see also Chaps. 2 and 3), one can write:

$$\frac{d\alpha_\Re}{dt} = \frac{\partial \alpha}{\partial t} + \frac{\partial \alpha}{\partial x_i}\left(\frac{dx_i}{dt}\right)_\Re. \tag{5.2}$$

With $(dx_i/dt)_\Re = (U_i)_\Re = U_i$:

$$\frac{d\alpha_\Re}{dt} = \frac{D\alpha}{Dt} = \frac{\partial \alpha}{\partial t} + U_i \frac{\partial \alpha}{\partial x_i}. \tag{5.3}$$

The operator $D t = \partial/\partial t + U_i \partial/\partial x_i$ applied to the field quantity $\alpha(x_i, t)$ is often defined as the substantial derivatve and will be applied in the subsequent derivations. The significance of individual terms are:

$\partial/\partial t = (\partial/\partial t)_{x_i}$ = change with time at a fixed location,
partial differentiation with respect to time

d/dt = total change with time (for a fluid element),
total differentiation with respect to time

e.g. for a fluid when $\alpha_\Re = \rho_\Re$ = constant, i.e. the density is constant, then:

$$\frac{d\rho_\Re}{dt} = \frac{D\rho}{Dt} = 0 \quad \text{or} \quad \frac{\partial \rho}{\partial t} = -U_i \frac{\partial \rho}{\partial x_i}. \tag{5.4}$$

When at a certain point in space $\partial/\partial t(\alpha)_{x_i} = 0$ indicates stationary conditions, the field property $\alpha(x_i, t)$ is stationary and thus has no time dependence. On the other hand, when $d(\alpha_\Re)/dt = D\alpha/Dt = 0$, then $\alpha_\Re(t) = \alpha(x_i, t) =$ constant, i.e. the field variable is independent of space and time.

5.2 Mass Conservation (Continuity Equation)

For fluid mechanics considerations, a "closed fluid system" can always be found, i.e. a system whose total mass $M =$ constant. This is easily seen for a fluid mass, which is stored in a container. For all other fluid flow

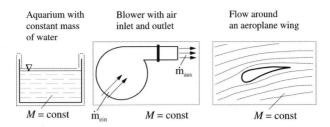

Fig. 5.2 Different fluid flow cases within control volumes for which $M = $ constant can be set

considerations, as shown in Fig. 5.2, control volumes can always be defined within which the system's total mass can be stated as constant. If necessary these control volumes can comprise the whole earth to reach $M = $ constant.

When one subdivides the fluid mass M within the considered system into $\Re$ fluid elements with sub-masses $\delta m_\Re$, then for the temporal change of the total mass one obtains:

$$0 = \frac{dM}{dt} = \frac{d}{dt}\sum_\Re (\delta m_\Re) = \sum_\Re \frac{d}{dt}(\delta m_\Re). \tag{5.5}$$

This equation expresses that the total mass conservation in the control volume of the fluid system is preserved when each individual fluid element conserves its mass $\delta m_\Re$. With this the balance equation for the mass conservation can be stated as follows, in Lagrange notation:

$$\frac{d(\delta m_\Re)}{dt} = 0. \tag{5.6a}$$

The basic molecular structure of matter and thermal motion connected with it indicates that to fulfill the above relationship absolutely, it is necessary that $\delta m_\Re \to 0$ is not taken into consideration. The derivations in this book therefore require that all the chosen $\delta m_\Re$ are considered as finite but nevertheless as very small. In Fig. 5.3 a fluid element with position coordinates $(x_i)_\Re$ is shown.[1]

The determination of the required size of $\delta m_\Re$ needs considerations that are given in Chap. 3, where it is shown what dimensions a volume of an ideal gas has to have in order to define "sufficiently clearly", e.g. the density of the gas within the volume. The considerations carried out there would have to be repeated here in order to ensure $\delta m_\Re \approx$ constant, in spite of the molecular structure of the fluid. With the choice of $\delta m_\Re =$ constant, the conditions are set to carry out continuum mechanics considerations for the motion of fluids, although the fluids show a molecular structure.

---

[1] See also the considerations in Sect. 3.2.

5.2 Mass Conservation (Continuity Equation)

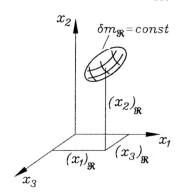

Fig. 5.3 $\delta m_\Re$ = constant, condition for the mass of a fluid element treated at $(x_i)_\Re = x_i$

It is often claimed that molecular motions are not considered when continuum assumptions are taken into fluid mechanics applications. Strictly, this means that the properties of the molecules, especially their transport properties, can only be introduced into continuum fluid mechanics considerations in an integral form.

Because of the above explanations, the mass conservation can be stated in Lagrange form as follows:

$$\frac{dM}{dt} = 0 \quad \text{and} \quad \frac{d\delta m_\Re}{dt} = 0. \tag{5.6b}$$

The above considerations confirm that it is very simple to formulate the mass-conservation law in Lagrange variables. Working practically with the law of mass conservation, however, requires its representation in field quantities, i.e. the Lagrange form of the mass-conservation law has to be brought into its corresponding Euler form.

Transformed into Euler variables (i.e. into field quantities), one obtains from (5.6a) for the mass conservation:

$$0 = \frac{d}{dt}(\delta m_\Re) = \frac{d}{dt}(\rho_\Re \delta V_\Re) = \underbrace{\rho_\Re \frac{d(\delta V_\Re)}{dt}}_{I} + \underbrace{\delta V_\Re \frac{d(\rho_\Re)}{dt}}_{II}. \tag{5.7}$$

For Term I in (5.7), using $\rho_\Re = \rho$ and $x_\Re(t)_i = x_i$ at time t, yields according to (4.89):

$$\rho_\Re \frac{d(\delta V_\Re)}{dt} = \rho \delta V_\Re \frac{\partial U_i}{\partial x_i}. \tag{5.8}$$

For Term II one obtains:

$$\delta V_\Re \frac{d(\rho_\Re)}{dt} = \delta V_\Re \left(\frac{\partial \rho}{\partial t} + U_i \frac{\partial \rho}{\partial x_i} \right). \tag{5.9}$$

With the above derivations, the substantial derivative of the corresponding field quantity $d(\rho_\Re)/dt$ could be applied for ρ, i.e. $D\rho/Dt$, in order to achieve the transformation of the substantial quantity $\rho_\Re(t)$ to the field quantity $\rho(x_i, t)$. The same procedure is not possible with $d(\delta V_\Re)/dt$ since there are no volume fields, i.e. a point has no volume. From Sect. 4.4.3, (4.89), it is known that the temporal change of a fluid element is equal to the divergence of the velocity field, however:

$$\rho_\Re \frac{d(\delta V_\Re)}{dt} = \delta V_\Re \frac{\partial U_i}{\partial x_i}. \tag{5.10}$$

When one inserts (5.8) and (5.9) into (5.7) one obtains:

$$\delta V_\Re \left[\frac{\partial \rho}{\partial t} + \rho \frac{\partial U_i}{\partial x_i} + U_i \frac{\partial \rho}{\partial x_i} \right] = 0. \tag{5.11}$$

As $\delta V_\Re \neq 0$, it follows for the continuity equation in field variables and in the most general form

$$\frac{\partial \rho}{\partial t} + \rho \frac{\partial U_i}{\partial x_i} + U_i \frac{\partial \rho}{\partial x_i} = \frac{\partial \rho}{\partial t} + \frac{\partial (\rho U_i)}{\partial x_i} = 0. \tag{5.12}$$

The equation can also be written as follows:

$$\frac{\partial \rho}{\partial t} + U_i \frac{\partial \rho}{\partial x_i} + \rho \frac{\partial U_i}{\partial x_i} = \frac{D\rho}{Dt} + \rho \frac{\partial U_i}{\partial x_i} = 0. \tag{5.13}$$

The right hand side of the continuity equation is not very useful for the solution of flow problems. However, it is very well suited for presentation of the basic equations of fluid mechanics in different ways, in order to bring out special physical facts. As an example, the special form of the continuity equation for $\rho_\Re = $ constant, i.e. $D\rho/Dt = 0$, from (5.13) results:

$$\frac{\partial U_i}{\partial x_i} = 0. \tag{5.14}$$

i.e. the divergence of the velocity field is zero for fields of constant fluid density. Since the divergence of the velocity field is zero, the change in volume is also zero from (5.10), this can also be obtained from equations (3.90) and (3.93):

$$\frac{1}{\rho} \frac{D\rho}{Dt} = \underbrace{\frac{1}{\rho} \left(\frac{\partial \rho}{\partial T} \right)_P}_{-\beta} \frac{DT}{Dt} + \underbrace{\frac{1}{\rho} \left(\frac{\partial \rho}{\partial P} \right)_T}_{+\alpha} \frac{DP}{Dt} = -\frac{\partial U_i}{\partial x_i}. \tag{5.15}$$

This relationship expresses for an ideal liquid, ideal in the thermodynamic sense, $\rho = $ constant, i.e. if the fluid density is constant, that the fluid has to be thermodynamically incompressible.

$$\frac{1}{\rho} \frac{D\rho}{Dt} = -\frac{\partial U_i}{\partial x_i} = 0 = \alpha \frac{DP}{Dt} - \beta \frac{DT}{Dt} \quad \text{with} \quad \alpha = 0 \quad \text{and} \quad \beta = 0, \tag{5.16}$$

5.3 Newton's Second Law (Momentum Equation)

where α and β are:

$$\alpha = -\frac{1}{v}\left(\frac{\partial v}{\partial P}\right)_T = \frac{1}{\rho}\left(\frac{\partial \rho}{\partial P}\right)_T = \text{isothermal compressibility coefficient,}$$

$$\beta = \frac{1}{v}\left(\frac{\partial v}{\partial T}\right)_P = -\frac{1}{\rho}\left(\frac{\partial \rho}{\partial T}\right)_P = \text{thermal expansion coefficient.}$$

Thus the continuity equation holds in one of the following two forms:

$$\frac{\partial \rho}{\partial t} + \frac{\partial(\rho U_i)}{\partial x_i} = 0 \quad \text{(compressible flows),} \tag{5.17}$$

$$\frac{\partial U_i}{\partial x_i} = 0 \quad \text{(incompressible flows).} \tag{5.18}$$

For further details of these derivations, see refs. [5.1] to [5.5].

5.3 Newton's Second Law (Momentum Equation)

The derivations of the momentum equations of fluid mechanics are usually given for the three coordinate directions $j = 1, 2, 3$. They express Newton's second law and are easiest formulated in their Lagrange forms. For a fluid element, it is stated that the time derivative of the momentum in the j direction is equal to the sum of the external forces acting in this direction on the fluid element, plus the molecular-dependent input of momentum per unit time. The forces can be stated as mass forces $(\delta M_j)_\Re$ caused by gravitation forces and electromagnetic forces[2], as well as surface forces caused by pressure, $(\delta O_j)_\Re$. After the addition of a temporal change of momentum introduced by the molecular movement input, the equation of motion can be formulated as follows:

$$\frac{\mathrm{d}(\delta J_j)_\Re}{\mathrm{d}t} = \underbrace{\sum(\delta M_j)_\Re}_{\text{mass forces}} + \underbrace{\sum(\delta O_j)_\Re}_{\text{surface forces}} + \underbrace{\left(\frac{\mathrm{d}}{\mathrm{d}t}(\delta J_M)_j\right)_\Re}_{\substack{\text{molecular-dependent} \\ \text{momentum input}}}. \tag{5.19}$$

Here, as shown in Fig. 5.4, $(\delta J_j)_\Re = \delta m_\Re (U_j)_\Re$.

Fluid elements act like rigid bodies. They do not change their state of motion, i.e. their momentum, if no mass or surfaces forces act on the fluid elements and no molecular-dependent momentum input is present. However, when forces are present, or when molecular momentum input occurs, a considered fluid element changes its momentum in accordance with (5.19). This equation represents the Lagrange form of the equations of momentum ($j = 1, 2, 3$) of fluid mechanics.

---

[2] The latter are not taken into consideration in the following.

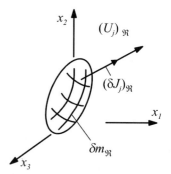

Fig. 5.4 The derivation of momentum equations are based on force considerations for a fluid element

In order to derive the Euler form of the equation of momentum, it is necessary to express each of the terms contained in (5.19) in field quantities. The left-hand side of (5.19) can be written as:

$$\frac{d(\delta J_j)_\Re}{dt} = \frac{d}{dt}[\delta m_\Re (U_j)_\Re] = \delta m_\Re \frac{d((U_j)_\Re)}{dt} + (U_j)_\Re \frac{d((\delta m)_\Re)}{dt}. \tag{5.20}$$

Because of the mass conservation for a fluid element expressed in (5.6a), the last term in (5.20) is equal to zero and hence one obtains:

$$\frac{d(\delta J_j)_\Re}{dt} = \delta m_\Re \frac{d((U_j)_\Re)}{dt} = \delta m_\Re \left(\frac{\partial U_j}{\partial t} + U_i \frac{\partial U_j}{\partial x_i} \right). \tag{5.21}$$

This relationship can be written as follows: $\delta m_\Re = \rho_\Re \delta V_\Re = \rho \delta V_\Re$ applying $\rho_\Re = \rho$ when $(x_\Re(t))_i = x_i$ at time t:

$$\frac{d(\delta J_j)_\Re}{dt} = \rho \delta V_\Re \left(\frac{\partial U_j}{\partial t} + U_i \frac{\partial U_j}{\partial x_i} \right). \tag{5.22}$$

In accordance with the above derivations, it is possible to state the left-hand side of the equation of momentum (5.19) in field quantities as shown in (5.22). For the right-hand side the considerations below can be carried out. The mass forces:

$$\sum_\Re (\delta M_j)_\Re = \text{mass forces acting on a fluid element}$$

acting on a fluid element can be expressed by means of the acceleration $\{g_j\} = \{g_1, g_2, g_3\}$ acting per unit mass (Fig. 5.5). The mass force acting on a fluid element in the j direction can therefore be stated as follows:

$$(\delta M_j)_\Re = (\delta m)_\Re g_j = \rho \delta V_\Re g_j. \tag{5.23}$$

Even when only gravitational acceleration is present, depending on the orientation of the coordinate system, several components of g_j may exist and

5.3 Newton's Second Law (Momentum Equation)

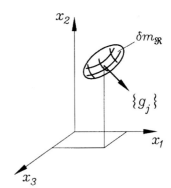

Fig. 5.5 Mass forces acting on a fluid element in the directions $j = 1, 2, 3$

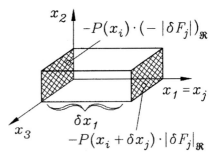

Fig. 5.6 Considerations concerning surface force on a fluid element in the directions $j = 1, 2, 3$

have to be taken into account:

$$\sum_{\Re}(\delta O_j)_{\Re} = \text{surface forces on a fluid element}.$$

Fluids, as they are treated in this book, i.e. liquids (e.g. water) and gases (e.g. air), are characterized by the way they apply surface forces on a fluid element (Fig. 5.6). The only surface forces that can exist are those imposed by the molecular pressure. The pressure force acting on a fluid element is calculated as the difference in the forces acting on the areas that stand vertically on the considered axes $j = 1, 2, 3$. For the pressure force, we can write:

$$\mathrm{d}\boldsymbol{K}_P = -P\,\mathrm{d}\boldsymbol{F}.$$

The surface force resulting for the motion of the fluid element in the j direction is the sum of the forces acting on the j planes of the element:

$$(\delta O_j)_{\Re} = -P(x_j)(-|\delta F_j|)_{\Re} - P(x_j + \delta x_j)(|\delta F_j|)_{\Re}. \tag{5.24a}$$

If one applies a Taylor series expansion for $P(x_i + \delta x_j)$, one obtains:

$$(\delta O_j)_{\Re} = +P(x_j)(|\delta F_j|)_{\Re} - [P(x_j) + \frac{\partial P}{\partial x_j}\delta x_j + \cdots](|\delta F_j|)_{\Re}. \tag{5.24b}$$

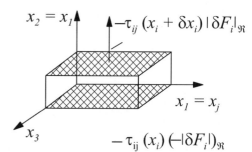

Fig. 5.7 Considerations on the molecular-dependent momentum input in the $j = 1, 2, 3$ directions

From this equation, one obtains for the surface force on a fluid element, neglecting all second and higher order terms of differentiations (Fig. 5.6),

$$(\delta O_j)_\Re = -\frac{\partial P}{\partial x_j} \delta V_\Re. \tag{5.25}$$

$\left(\dfrac{\mathrm{d}}{\mathrm{d}t}(\delta J_M)_j\right)_\Re =$ molecular-dependent momentum input per unit time into a fluid element.

When one defines the momentum j transported by molecules in the direction i per unit time and unit area as τ_{ij}, the input influencing the momentum j of a fluid element is calculated as an input at the position x_i and as an output at the position $(x_i + \Delta x_i)$, e.g. see Fig. 5.7:

$$\left(\frac{\mathrm{d}(\delta J_M)_j}{\mathrm{d}t}\right)_j = -\tau_{ij}(x_i)(-|\delta F_j|)_\Re - \tau_{ij}(x_i + \Delta x_i)(|\delta F_i|)_\Re. \tag{5.26a}$$

By a Taylor series expansion one obtains for the term $\tau_{ij}(x_i + \delta x_i)$

$$\left(\frac{\mathrm{d}(\delta J_M)_j}{\mathrm{d}t}\right)_\Re = +\tau_{ij}(x_i)(|\delta F_i|)_\Re - [\tau_{ij}(x_i) + \frac{\partial \tau_{ij}}{\partial x_i}\delta x_i \cdots](|\delta F_i|)_\Re. \tag{5.26b}$$

This results in:

$$\left(\frac{\mathrm{d}(\delta J_M)_j}{\mathrm{d}t}\right)_\Re = -\frac{\partial \tau_{ij}}{\partial x_i} \delta V_\Re. \tag{5.27}$$

When one inserts all these derived relationships (5.22), (5.23), (5.25) and (5.27) into (5.19) and after division by $\delta V_\Re$, the equation of momentum of fluid mechanics in the j direction results, i.e. for $j = 1, 2, 3$, three equations can be given:

$$\rho\left(\frac{\partial U_j}{\partial t} + U_i \frac{\partial U_j}{\partial x_i}\right) = \rho g_j - \frac{\partial P}{\partial x_j} - \frac{\partial \tau_{ij}}{\partial x_i}. \tag{5.28}$$

As in this equation the volume of the fluid element $\delta V_\Re$, appearing in all terms, was eliminated, the equations of momentum are given by (5.28) *per* unit volume. From the general equation (5.28), the momentum equations in the three coordinate directions result:

$$\rho\left[\frac{\partial U_1}{\partial t} + U_1\frac{\partial U_1}{\partial x_1} + U_2\frac{\partial U_1}{\partial x_2} + U_3\frac{\partial U_1}{\partial x_3}\right] = -\frac{\partial P}{\partial x_1} - \frac{\partial \tau_{11}}{\partial x_1} - \frac{\partial \tau_{21}}{\partial x_2} - \frac{\partial \tau_{31}}{\partial x_3} + \rho g_1$$

$$\rho\left[\frac{\partial U_2}{\partial t} + U_1\frac{\partial U_2}{\partial x_1} + U_2\frac{\partial U_2}{\partial x_2} + U_3\frac{\partial U_2}{\partial x_3}\right] = -\frac{\partial P}{\partial x_2} - \frac{\partial \tau_{12}}{\partial x_1} - \frac{\partial \tau_{22}}{\partial x_2} - \frac{\partial \tau_{32}}{\partial x_3} + \rho g_2$$

$$\rho\left[\frac{\partial U_3}{\partial t} + U_1\frac{\partial U_3}{\partial x_1} + U_2\frac{\partial U_3}{\partial x_2} + U_3\frac{\partial U_3}{\partial x_3}\right] = -\frac{\partial P}{\partial x_3} - \frac{\partial \tau_{13}}{\partial x_1} - \frac{\partial \tau_{23}}{\partial x_2} - \frac{\partial \tau_{33}}{\partial x_3} + \rho g_3$$

(5.29)

For fluids in general, $\tau_{ij} \neq 0$, but for *ideal* fluids, i.e. ideal in terms of fluid mechanics, the molecular momentum transport turns out to be $\tau_{ij} = 0$. Hence the following forms of the momentum equations can be stated:

$$\rho\left(\frac{\partial U_j}{\partial t} + U_i\frac{\partial U_j}{\partial x_i}\right) = -\frac{\partial P}{\partial x_j} - \frac{\partial \tau_{ij}}{\partial x_i} + \rho g_j \quad \text{(viscous fluids),} \quad (5.30)$$

$$\rho\left(\frac{\partial U_j}{\partial t} + U_i\frac{\partial U_j}{\partial x_i}\right) = -\frac{\partial P}{\partial x_j} + \rho g_j \quad \text{(ideal fluids).} \quad (5.31)$$

For further details of these derivations see refs. [5.1] to [5.5].

5.4 The Navier–Stokes Equations

In equation (5.30), the molecular-dependent momentum input τ_{ij} was introduced as an input as per unit surface area and unit time. It is an unknown term, i.e. it was introduced formally into the derivations without any details being considered as to how it can be formulated for various fluids. When one takes into consideration the symmetry of the term τ_{ij}, i.e. $|\tau_{ij}| = |\tau_{ji}|$, one finds that there are the following unknowns in the above equations:

$U_1, U_2, U_3, P, \tau_{11}, \tau_{12}, \tau_{13}, \tau_{22}, \tau_{23}, \tau_{33} = 10$ unknowns.

For these unknowns there are only four partial differential equations available to provide solutions to fluid flow problems, the continuity equation and three equations of momentum, i.e. an incomplete system of equations exists that does not permit the solution of flow problems. It is therefore necessary to state additional equations, i.e. to express the unknown terms τ_{ij} in a physically well-founded manner, as functions of $\partial U_j/\partial x_i$. This is done below for ideal gases, as their properties are usually well known to engineering students, from considerations in physics. From the derivations given below, relationships for $\tau_{ij} = f(\partial U_j/\partial x_i)$ result, that are valid also for non-ideal gases, and in fact for a whole class of fluids whose molecular momentum transport properties can be classified as "Newtonian". Thus the derived

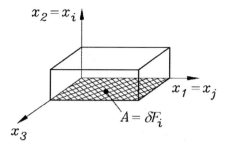

Fig. 5.8 Momentum input due to flow through the plane δF_i

relationships for τ_{ij} are valid far beyond ideal gases and represent in this book the basic equations to describe the molecular-dependent momentum transport in Newtonian fluids.

If one considers a fluid element, as shown in Fig. 5.8, with side walls parallel to the planes of a Cartesian coordinate system, one can see that the j momentum transported in the direction i by a velocity field U_i can be stated as follows:

$$\dot{I}_{ij} = \rho \hat{U}_i \hat{U}_j \delta F_i. \tag{5.32}$$

Assuming that the instantaneous velocity components are composed of the velocity components of the fluid flow U_i and the molecular velocity component u_i, one can write:

$$\begin{aligned}\rho \hat{U}_i \hat{U}_j &= \rho(U_i + u_i)(U_j + u_j) \\ &= \rho(U_i U_j + u_i U_j + u_j U_i + u_i u_j)\end{aligned} \tag{5.33}$$

By time averaging, one obtains for the time-averaged total momentum change of the fluid element:

$$\overline{\rho \hat{U}_i \hat{U}_j} = \rho[\underbrace{\overline{U_i U_j}}_{\text{I}} + \underbrace{\overline{u_i U_j}}_{\text{II}} + \underbrace{\overline{u_j U_i}}_{\text{III}} + \underbrace{\overline{u_i u_j}}_{\text{IV}}]. \tag{5.34}$$

The total momentum input consists of four terms that can be interpreted physically as follows:

Term I: j Momentum input in the i direction due to the velocity field U_i of the fluid.

Term II: j Momentum input in the i direction due to the molecular motion in the i direction, i.e. due to u_i.

Term III: j Momentum input in the i direction due to the molecular motion u_j in the j direction.

Term IV: For $i \neq j$ the $\overline{u_i u_j} = 0$, as the molecular motion in the three coordinate directions are not correlated. For $i = j$ the molecular caused pressure treated in Chap. 3 results.

5.4 The Navier–Stokes Equations

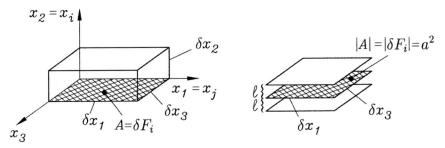

Fig. 5.9 Momentum input in the x_i direction caused by the molecular motion with mean velocity u_i

The molecular motion is characterized by the presence of the molecular free path lengths with finite dimensions, i.e. $l \neq 0$, and for this reason the time averages $\overline{u_i U_j}$ and $\overline{u_j U_i}$ are unequal to zero. In order to calculate the different contributions to the τ_{ij} terms making up the total molecular momentum transport in ideal gases, the considerations below are recommended. For the number of molecules moving in the direction x_i and passing the plane A in Fig. 5.9 in the time Δt, when $\delta x_1 = \delta x_2 = \delta x_3 = a$ we can be write:

$$z_i = \frac{1}{6} n a^2 u_i \Delta t, \tag{5.35}$$

where n is equal to the number of molecules per unit volume, a^2 is the magnitude of the area δF_i of the considered volume, oriented in the i direction, and u_i is the mean velocity of the molecules in the i direction. Connected with z_i, a mass transport through δF_i can be stated as follows:

$$m z_i = \frac{1}{6} \underbrace{(mn)}_{\rho} a^2 u_i \Delta t, \tag{5.36}$$

where m represents the mass of a molecule and thus $mn = \rho$ can be set.

If one considers now two parallel auxiliary planes in Fig. 5.9 located at a distances $\pm l$ above and below a main plane at δF_i and if one introduces for the derivations the mean flow field in the auxiliary planes to have the velocity components $U_j(x_i + l)$ and $U_j(x_i - l)$, the considerations below can be performed. In the positive and negative i directions, the j-directional momentum input and output can be stated as follows:

$$i_{ij}^+ = +z_i m U_j(x_i - l) \text{ momentum input over the area } |\delta F_i| = A = a^2$$

$$i_{ij}^- = -z_i m U_j(x_i + l) \text{ momentum output over the area } |\delta F_i| = A = a^2 \tag{5.37}$$

Therefore, for the net input of momentum the sum of the molecular-dependent input and output results:

$$\Delta i_{ij} = z_i m [U_j(x_i - l) - U_j(x_i + l)], \tag{5.38}$$

or, with z_i inserted from (5.35):

$$\Delta i_{ij} = \frac{1}{6} \underbrace{(mn)}_{\rho} a^2 u_i \Delta t [U_j(x_i - l) - U_j(x_i + l)]. \quad (5.39)$$

The net momentum input per unit area and unit time can be obtained by Taylor series expansion of the velocity terms around x_i. This can be expressed as given below by neglecting the higher order terms:

$$\tau_{ij}^{II} = \frac{1}{a^2} \frac{\Delta i_{ij}}{\Delta t} = \frac{1}{6} \rho u_i \left[U_j(x_i) - \frac{\partial U_j}{\partial x_i} l - U_j(x_i) - \frac{\partial U_j}{\partial x_i} l \right], \quad (5.40)$$

so that for Term II in (5.34) can be expressed as follows:

$$\tau_{ij}^{II} = -\underbrace{\frac{1}{3} \rho u_i l}_{\mu} \frac{\partial U_j}{\partial x_i} = -\mu \frac{\partial U_j}{\partial x_i}. \quad (5.41)$$

Analogous to this, considerations can be carried out on τ_{ij}^{III}, where for z_j it can be written

$$z_j = \frac{1}{6} n a^2 u_j \Delta t. \quad (5.42)$$

In accordance with Term III in (5.34), a j-momentum input results, see Fig. 5.10, which can be expressed as follows:

$$\begin{aligned} i_{ij}^+ &= z_j m U_i(x_j - l) \\ i_{ij}^- &= -z_j m U_i(x_j + l) \end{aligned}, \quad (5.43)$$

or

$$\Delta i_{ij} = z_j m [U_i(x_j - l) - U_i(x_j + l)]. \quad (5.44)$$

Analogous to the derivations in (5.38) to (5.41):

$$\tau_{ij}^{III} = -\underbrace{\frac{1}{3}(\rho u_j l)}_{\mu} \frac{\partial U_i}{\partial x_j} = -\mu \frac{\partial U_i}{\partial x_j}. \quad (5.45)$$

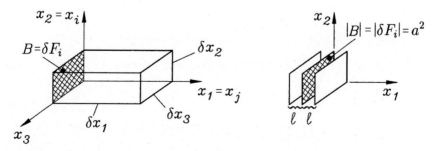

Fig. 5.10 Momentum input in the x_j direction with molecular velocity u_j and fluid velocity U_i

5.4 The Navier–Stokes Equations

For reasons of symmetry $\tau_{ij} = \tau_{ji}$, so that $u_i = u_j$ has to hold, i.e. the mean velocity field of the molecules is isotropic (no preferred velocity direction), so that the total j-momentum transport can be written as:

$$\tau_{ij} = \tau_{ij}^{\text{II}} + \tau_{ij}^{\text{III}} = -\mu\left[\frac{\partial U_j}{\partial x_i} + \frac{\partial U_i}{\partial x_j}\right]. \tag{5.46}$$

This is the total momentum input τ_{ij} for ρ = constant, i.e. when $d/dt(\delta V_\Re) = 0$, the thermodynamic state equation for a thermodynamically ideal liquid is assumed to be valid. For $\rho \neq$ constant, an additional term needs to be added to τ_{ij} which is caused by the volume increase of a fluid element. For the volume increase of a fluid element at point x_i and time t (see Chap. 4), one can write:

$$\frac{d(\delta V_\Re)}{dt} = (\delta V_\Re)\frac{\partial U_i}{\partial x_i}. \tag{5.47}$$

For the corresponding surface increase the following relationship holds:

$$\frac{d(\delta F_\Re)}{dt} = \frac{2}{3}(\delta F_\Re)\frac{\partial U_i}{\partial x_i}. \tag{5.48}$$

With this surface increase in time an increased momentum input results:

$$\tau_{ij} = +\mu\frac{2}{3}\delta_{ij}\frac{\partial U_k}{\partial x_k}. \tag{5.49}$$

This term has to be added to obtain the general τ_{ij}-relationship for the total momentum input per unit time and unit area for ideal gases. It can be stated as follows:

$$\tau_{ij} = -\mu\left(\frac{\partial U_j}{\partial x_i} + \frac{\partial U_i}{\partial x_j}\right) + \frac{2}{3}\delta_{ij}\mu\frac{\partial U_k}{\partial x_k}. \tag{5.50}$$

If one considers this equation for τ_{ij}, the basic equations of fluid mechanics can be written as follows:

Continuity equation:

$$\frac{\partial \rho}{\partial t} + \frac{\partial(\rho U_i)}{\partial x_i} = 0. \tag{5.51}$$

Momentum equations ($j = 1, 2, 3$):

$$\rho\left[\frac{\partial U_j}{\partial t} + U_i\frac{\partial U_j}{\partial x_i}\right] = -\frac{\partial P}{\partial x_j} - \frac{\partial \tau_{ij}}{\partial x_i} + \rho g_j. \tag{5.52}$$

For Newtonian fluids:

$$\tau_{ij} = -\mu\left[\frac{\partial U_j}{\partial x_i} + \frac{\partial U_i}{\partial x_j}\right] + \frac{2}{3}\delta_{ij}\mu\frac{\partial U_k}{\partial x_k}. \tag{5.53}$$

With τ_{ij} expressed by equation (5.53) there exist equations for the six unknown terms τ_{ij} in the momentum equations. The four differential equations, one continuity equation and three momentum equations, contain five remaining unknowns P, ρ, U_j, so that an incomplete system of partial differential equations still exists. With the aid of the thermal energy equation and the thermodynamic state equation, valid for the considered fluid, it is possible to obtain a complete system of partial differential equations that permits general solutions for flow problems, when initial and boundary conditions are present.

For $\rho = $ constant and $\mu = $ constant, using $\frac{\partial^2 U_i}{\partial x_i \partial x_j} = \frac{\partial^2 U_i}{\partial x_j \partial x_i} = \frac{\partial}{\partial x_j}\left(\frac{\partial U_i}{\partial x_i}\right) = 0$, the following set of equations can be stated:

Continuity equation:
$$\frac{\partial U_i}{\partial x_i} = 0. \tag{5.54}$$

Navier–Stokes equations ($j = 1, 2, 3$) (momentum equations):
$$\rho\left[\frac{\partial U_j}{\partial t} + U_i \frac{\partial U_j}{\partial x_i}\right] = -\frac{\partial P}{\partial x_j} + \mu \frac{\partial^2 U_j}{\partial x_i^2} + \rho g_j.$$

This system of equations comprises four equations for the four unknowns P, U_1, U_2, U_3. In principle, it can be solved for all flow problems to be investigated if suitable initial and boundary conditions are given. For thermodynamically ideal liquids, i.e. $\rho = $ constant, a complete system of partial differential equations exists through the continuity equation and the momentum equations, which can be used for solutions of flow problems.

5.5 Mechanical Energy Equation

In many fields in which fluid mechanics considerations are carried out, the mechanical energy equation is employed, which can, however, be derived from the momentum equation. For this purpose, one multiplies equation (5.52) by U_j:

$$\rho\left[U_j\frac{\partial U_j}{\partial t} + U_i U_j \frac{\partial U_j}{\partial x_i}\right] = -U_j\frac{\partial P}{\partial x_j} - U_j\frac{\partial \tau_{ij}}{\partial x_i} + U_j \rho g_j. \tag{5.55}$$

This equation can be rearranged to yield

$$\rho\left[\frac{\partial}{\partial t}\left(\frac{1}{2}U_j^2\right) + U_i \frac{\partial}{\partial x_i}\left(\frac{1}{2}U_j^2\right)\right] = -\frac{\partial(PU_j)}{\partial x_j} + P\frac{\partial U_j}{\partial x_j} - \frac{\partial(\tau_{ij}U_j)}{\partial x_i}$$
$$+ \tau_{ij}\frac{\partial U_j}{\partial x_i} + \rho g_j U_j. \tag{5.56}$$

This relationship expresses how the kinetic energy of a fluid element changes at a location due to energy production and dissipation terms that occur on the

5.5 Mechanical Energy Equation

right-hand side of (5.56). In order to discuss the significance of the different terms, the following modification of the last term is carried out, introducing a potential G from which the gravitational acceleration g_j is derived:

$$g_j = -\frac{\partial G}{\partial x_j} \quad \leadsto \quad \rho g_j U_j = -\rho \frac{\partial G}{\partial x_j} U_j. \tag{5.57}$$

Thus, employing $\frac{\partial G}{\partial t} = 0$, one can write:

$$\rho g_j U_j = -\rho \left[\frac{\partial G}{\partial t} + U_j \frac{\partial G}{\partial x_j} \right] = -\rho \frac{DG}{Dt}. \tag{5.58}$$

The combined equation (5.56) and (5.58) yield for the temporal change of the kinetic and potential energy of a fluid element

$$\rho \frac{D}{Dt}\left(\frac{1}{2}U_j^2 + G\right) = -\underbrace{\frac{\partial(PU_j)}{\partial x_j}}_{\text{I}} + \underbrace{P\frac{\partial U_j}{\partial x_j}}_{\text{II}} - \underbrace{\frac{\partial(\tau_{ij}U_j)}{\partial x_i}}_{\text{III}} + \underbrace{\tau_{ij}\frac{\partial U_j}{\partial x_i}}_{\text{IV}}, \tag{5.59}$$

where the terms I–IV have the following physical significance:

Term I: This term describes the difference between input and output of pressure energy. This refers to the considerations of ideal gases, in the framework of which it was shown that $P = \frac{1}{3}\rho \overline{u^2}$, i.e. the pressure expresses an energy per unit volume. Therefore, the following can be said:

$$PU_j(x_i) = \text{input of pressure energy per unit area}$$

$$-(PU_j(x_i + \Delta x_i)) = \text{output of pressure energy per unit area.}$$

Taylor series expansion and forming the difference yields for the energy per unit volume:

$$PU_j(x_i) - \left[PU_j(x_i) + \frac{(\partial PU_j)}{\partial x_j} + \cdots\right] \simeq -\frac{(\partial PU_j)}{\partial x_j}. \tag{5.60}$$

Term II: This term requires the following considerations:

with
$$\frac{\partial U_j}{\partial x_j} = \frac{1}{\delta V_\Re} \frac{d(\delta V_\Re)}{dt}, \tag{5.61}$$

the term
$$P \frac{\partial U_j}{\partial x_j} = \frac{P}{\delta V_\Re} \frac{d(\delta V_\Re)}{dt} \tag{5.62}$$

proves to be the work done during expansion, expressed per unit volume.

Term III: When taking into consideration that τ_{ij} represents the molecular-dependent momentum transport per unit area and unit time into a fluid element:

$-\dfrac{\partial(\tau_{ij}U_j)}{\partial x_j}$ represents the difference between the molecular input and output of the kinetic energy of the fluid.

Term IV: The term $\tau_{ij}\dfrac{\partial U_j}{\partial x_i}$ describes the dissipation of mechanical energy into heat.

The above derivations show that the mechanical energy equation can be deduced from the j momentum equation by multiplication by U_j. It is therefore not an independent equation and hence should not be employed along with the momentum equations for the solution of fluid mechanics problems.

A special form of the mechanical energy equation is the Bernoulli equation, which can be derived from the general form of the mechanical energy equation:

$$\rho \frac{\mathrm{D}}{\mathrm{D}t}\left(\frac{1}{2}U_j^2 + G\right) = -\frac{\partial P}{\partial x_j}U_j - \frac{\partial \tau_{ij}}{\partial x_i}U_j. \tag{5.63}$$

For $\tau_{ij} = 0$ and $\dfrac{\partial P}{\partial t} = 0$, and also $\rho = $ constant,

$$\rho \frac{\mathrm{D}}{\mathrm{D}t}\left[\frac{1}{2}U_j^2 + G\right] = -\rho\left[\frac{\partial\left(\frac{P}{\rho}\right)}{\partial t} + U_j\frac{\partial\left(\frac{P}{\rho}\right)}{\partial x_j}\right] = \frac{\mathrm{D}\frac{P}{\rho}}{\mathrm{D}t}, \tag{5.64}$$

$$\rho \frac{\mathrm{D}}{\mathrm{D}t}\left[\frac{1}{2}U_j^2 + \frac{P}{\rho} + G\right] = 0 \quad \leadsto \quad \frac{1}{2}U_j^2 + \frac{P}{\rho} + G = \text{constant}. \tag{5.65}$$

This form of the mechanical energy equation can be employed in many engineering applications to solve flow problems in an engineering manner.

5.6 Thermal Energy Equation

The derivations in Sect. 5.5 showed that the mechanical energy equation is derivable from the momentum equation, so that both equations have to be considered as not being independent of each other. From the derivations the following form of the energy equation was obtained:

$$\rho\left[\frac{\mathrm{D}}{\mathrm{D}t}\left(\frac{1}{2}U_j^2\right)\right] = -\frac{\partial(PU_j)}{\partial x_j} + P\frac{\partial U_j}{\partial x_j} - \frac{\partial(\tau_{ij}U_j)}{\partial x_i} + \tau_{ij}\frac{\partial U_j}{\partial x_i} + \rho g_j U_j. \tag{5.66}$$

5.6 Thermal Energy Equation

When one sets up the energy equation with the total energy balance, the considerations stated below result, which start from the entire internal, the kinetic and the potential energies of a fluid element and consider its evolution as a function of time:

$$\frac{d}{dt}\left(\delta m_\Re \underbrace{\left[\frac{1}{2}U_j^2 + e + G\right]}_{\cdots}\right) = \delta m_\Re \frac{d}{dt}[\ldots] + [\ldots]\frac{d\delta m_\Re}{dt}$$

For the temporal change of the total energy of a fluid element one obtains with $\delta m_\Re = $ constant, i.e. $\frac{d}{dt}(\delta m_\Re) = 0$:

$$\frac{d}{dt}\left(\delta m_\Re \left[\frac{1}{2}U_j^2 + e + G\right]\right) = \delta m_\Re \frac{D}{Dt}\left(\frac{1}{2}U_j^2 + e + G\right)$$

This is the total energy change with time of a fluid element which has to be considered concerning the derivation of the total energy equation.

The change in the total energy of the fluid element can emanate from the heat conduction, which yields the following inputs minus the output of heat:

$$-\frac{\partial \dot{q}_i}{\partial x_i}\delta V_\Re = \text{energy input into } \delta V_\Re \text{ per unit time by heat conduction}$$

An energy input can also originate from the convective transport of pressure energy:

$$-\frac{\partial}{\partial x_j}(PU_j)\delta V_\Re = \text{input of pressure energy into } \delta V_\Re \text{ through convection}$$

Also, the input of kinetic energy due to molecular transport into the fluid element has to be considered:

$$-\frac{\partial}{\partial x_i}(\tau_{ij}U_j)\delta V_\Re = \text{molecular-dependent input of kinetic energy}$$

The following total energy balance thus results:

$$\rho\delta V_\Re\frac{D}{Dt}\left[\frac{1}{2}U_j^2 + e + G\right] = -\frac{\partial \dot{q}_i}{\partial x_i}\delta V_\Re - \frac{\partial(PU_j)}{\partial x_j}\delta V_\Re - \frac{\partial(\tau_{ij}U_j)}{\partial x_i}\delta V_\Re. \quad (5.67)$$

As $\delta V_\Re \neq 0$, it follows that

$$\rho\frac{D}{Dt}\left[e + \left(\frac{1}{2}U_j^2 + G\right)\right] = -\frac{\partial \dot{q}_i}{\partial x_i} - \frac{\partial(PU_j)}{\partial x_j} - \frac{\partial(\tau_{ij}U_j)}{\partial x_i}. \quad (5.68)$$

When one deducts from this the derived mechanical parts of the energy, i.e. by subtracting from equation (5.68) the equation for the mechanical energy, given here once again:

$$\rho \frac{D}{Dt}\left[\frac{1}{2}U_j^2 + G\right] = -\frac{\partial(PU_j)}{\partial x_j} + P\frac{\partial U_j}{\partial x_j} - \frac{\partial(\tau_{ij}U_j)}{\partial x_i} + \tau_{ij}\frac{\partial U_j}{\partial x_i}, \tag{5.69}$$

one obtains the thermal energy equation:

$$\underbrace{\rho \frac{De}{Dt}}_{I} = \underbrace{-\frac{\partial \dot{q}_i}{\partial x_i}}_{II} \underbrace{- P\frac{\partial U_j}{\partial x_j}}_{III} \underbrace{- \tau_{ij}\frac{\partial U_j}{\partial x_i}}_{IV}. \tag{5.70}$$

Term I: Temporal change of the internal energy of a fluid per unit volume.
Term II: Heat supply per unit time and unit area.
Term III: Expansion work done per unit volume and unit time.
Term IV: Irreversible transfer of mechanical energy into heat, per unit volume and unit time.

Considering the energy equation of technical thermodynamics:

$$dq_\Re = de_\Re + P_\Re dv_\Re - dl_{\text{diss}}, \tag{5.71}$$

and the sign convention usually applied in technical thermodynamics, that the energy to be dissipated by a fluid element has to be regarded as negative, one obtains

$$\frac{de_\Re}{dt} = \frac{De}{Dt}; \quad \frac{dq_\Re}{dt} = -\frac{1}{\rho}\frac{\partial \dot{q}_i}{\partial x_i}; \quad P_\Re \frac{dv_\Re}{dt} = \frac{1}{\rho}P\frac{\partial U_j}{\partial x_j}$$

$$\text{and} \quad \frac{dl_{\text{diss}}}{dt} = \frac{1}{\rho}\tau_{ij}\frac{\partial U_j}{\partial x_i} \tag{5.72}$$

The above derivations thus lead to the form of energy equation used in thermodynamics but through the above derivation the energy per unit time results.

Different forms of the thermal energy equation can be derived from equation (5.70), I is advantageous for most fluid mechanics computations, to substitute the internal energy (e) by pressure and temperature relationships, the following relations being employed in most text books.

Generally, it can be written for thermodynamically simple fluids that

$$de_\Re = \left(\frac{\partial e}{\partial v}\right)_T dv + \left(\frac{\partial e}{\partial T}\right)_v dT = \left(\frac{\partial e}{\partial v}\right)_T dv + c_v\, dT. \tag{5.73}$$

Considering the Maxwell relationships of thermodynamics, one can be write

$$\left(\frac{\partial e}{\partial v}\right)_T = -P + T\left(\frac{\partial P}{\partial T}\right)_v, \tag{5.74}$$

5.6 Thermal Energy Equation

so that the following form of the energy equation can be given:

$$\rho \frac{De}{Dt} = \left[-P + T \left(\frac{\partial P}{\partial T} \right)_\rho \right] \frac{\partial U_i}{\partial x_i} + \rho c_v \frac{DT}{Dt}. \tag{5.75}$$

The thermal energy equation can thus also be written as:

$$\rho c_v \frac{DT}{Dt} = -\frac{\partial \dot{q}_i}{\partial x_i} - T \left(\frac{\partial P}{\partial T} \right)_\rho \frac{\partial U_i}{\partial x_i} - \tau_{ij} \frac{\partial U_j}{\partial x_i}. \tag{5.76}$$

For an ideal gas, as $\left(\frac{\partial P}{\partial T} \right)_\rho = \frac{P}{T}$ and $\dot{q}_i = -\lambda \frac{\partial T}{\partial x_i}$,

$$\rho c_v \frac{DT}{Dt} = \lambda \frac{\partial^2 T}{\partial x_i^2} - P \frac{\partial U_i}{\partial x_i} - \tau_{ij} \frac{\partial U_j}{\partial x_i}. \tag{5.77}$$

For a thermodynamically ideal liquid, as $\frac{\partial U_i}{\partial x_i} = 0$ and $c_v = c_p$, therefore:

$$\rho c_p \frac{DT}{Dt} = \lambda \frac{\partial^2 T}{\partial x_i^2} - \tau_{ij} \frac{\partial U_j}{\partial x_i}. \tag{5.78}$$

It was shown above that the equation for the change of the total energy can be derived by addition of the equations for the mechanical and thermal energies:

Equation for mechanical energy:

$$\rho \frac{D}{Dt} \left(\frac{1}{2} U_j^2 + G \right) = -\frac{\partial}{\partial x_j} (PU_j) + P \frac{\partial U_j}{\partial x_j} - \frac{\partial}{\partial x_i} (\tau_{ij} U_j) + \tau_{ij} \frac{\partial U_j}{\partial x_i}. \tag{5.79}$$

Equation for thermal energy:

$$\rho \frac{De}{Dt} = -\frac{\partial \dot{q}_i}{\partial x_i} - P \frac{\partial U_i}{\partial x_i} - \tau_{ij} \frac{\partial U_i}{\partial x_i}. \tag{5.80}$$

Equation for the total energy:

$$\rho \frac{D}{Dt} \left(\frac{1}{2} U_j^2 + G + e \right) = -\frac{\partial \dot{q}_i}{\partial x_i} - \frac{\partial}{\partial x_j} (PU_j) - \frac{\partial}{\partial x_i} (\tau_{ij} U_j)$$
$$= \lambda \frac{\partial^2 T}{\partial x_i^2} - \frac{\partial}{\partial x_j} (PU_j) - \frac{\partial}{\partial x_i} (\tau_{ij} U_j) \tag{5.81}$$

From this final relationship, the Bernoulli equation can be derived, which is often used for fluid mechanics considerations in engineering:

Ideal Liquid (ρ = constant): no heat conduction and viscous dissipation:

$$\rho \left[\frac{D}{Dt} \left(\frac{1}{2} U_j^2 + G \right) \right] = -U_j \frac{\partial P}{\partial x_j} = -U_i \frac{\partial P}{\partial x_i}. \tag{5.82}$$

For a steady flow:

$$\rho\left[\overbrace{\frac{\partial}{\partial t}\left(\frac{1}{2}U_j^2+G\right)}^{=0}+U_i\frac{\partial}{\partial x_i}\left(\frac{1}{2}U_j^2+G\right)\right]=-U_i\frac{\partial P}{\partial x_i},\quad(5.83)$$

$$\frac{\partial}{\partial x_i}\left(\frac{1}{2}U_j^2+G+\frac{P}{\rho}\right)=0$$

or after integration,

$$\frac{1}{2}U_j^2+G+\frac{P}{\rho}=\text{constant}.\quad(5.84)$$

Ideal Gas: $P/\rho = RT$, no heat conduction and neglecting viscous dissipation and not considering the potential energy:

$$\rho\frac{\mathrm{D}}{\mathrm{D}T}\left(\frac{1}{2}U_j^2+e\right)=-\frac{\partial}{\partial x_j}(PU_j)=-\frac{\partial}{\partial x_i}(PU_i).\quad(5.85)$$

For steady-state flows:

$$\rho\frac{\partial}{\partial x_i}\left(\frac{1}{2}U_j^2+e\right)=-\frac{\partial}{\partial x_i}(PU_i)=-P\frac{\partial U_i}{\partial x_i}-U_i\frac{\partial P}{\partial x_i}.\quad(5.86)$$

From the continuity equation, it follows for steady-state flows that

$$\rho\frac{\partial U_i}{\partial x_i}=-U_i\frac{\partial \rho}{\partial x_i}.\quad(5.87)$$

If one inserts into the considerations $e = c_v T$, the following equations result:

$$\rho\frac{\partial}{\partial x_i}\left(\frac{1}{2}U_j^2+e\right)=\rho\frac{\partial}{\partial x_i}\left(\frac{1}{2}U_j^2\right)+\rho c_v\frac{\partial T}{\partial x_i}=\frac{P}{\rho}\frac{\partial \rho}{\partial x_i}-\frac{\partial P}{\partial x_i},$$

$$\frac{\partial}{\partial x_i}\left(\frac{1}{2}U_j^2\right)=\frac{P}{\rho^2}\frac{\partial \rho}{\partial x_i}-\frac{1}{\rho}\frac{\partial P}{\partial x_i}-c_v\frac{\partial T}{\partial x_i}\quad(5.88)$$

Introducing

$$\frac{\partial T}{\partial x_i}=-\frac{P}{R\rho^2}\frac{\partial \rho}{\partial x_i}+\frac{1}{R\rho}\frac{\partial P}{\partial x_i},\quad(5.89)$$

$$\frac{\partial}{\partial x_i}\left(\frac{1}{2}U_j^2\right)=\frac{P}{\rho^2}\frac{\partial \rho}{\partial x_i}\left(1+\frac{c_v}{R}\right)-\frac{1}{\rho}\frac{\partial P}{\partial x_i}\left(1+\frac{c_v}{R}\right)$$

$$=-\frac{\kappa}{\kappa-1}\frac{\partial}{\partial x_i}\left(\frac{P}{\rho}\right)\quad(5.90)$$

yields the Bernoulli equation in its "compressible form":

$$\frac{\partial}{\partial x_i}\left[\frac{1}{2}U_j^2+\frac{\kappa}{\kappa-1}\left(\frac{P}{\rho}\right)\right]=0 \Rightarrow \frac{1}{2}U_j^2+\frac{\kappa}{\kappa-1}\left(\frac{P}{\rho}\right)=\text{constant}.\quad(5.91)$$

5.7 Basic Equations in Different Coordinate Systems

5.7.1 Continuity Equation

The derivations carried out for the continuity equation in *Cartesian coordinates* resulted in

$$\frac{\partial \rho}{\partial t} + \frac{\partial (\rho U_i)}{\partial x_i} = 0 \tag{5.92}$$

or, without the application of the summation convention,

$$\frac{\partial \rho}{\partial t} + \frac{\partial (\rho U_1)}{\partial x_1} + \frac{\partial (\rho U_2)}{\partial x_2} + \frac{\partial (\rho U_3)}{\partial x_3} = 0. \tag{5.93}$$

For $\rho =$ constant on obtains:

$$\frac{\partial U_1}{\partial x_1} + \frac{\partial U_2}{\partial x_2} + \frac{\partial U_3}{\partial x_3} = 0$$

In *cylindrical coordinates* (r, φ, z) with (U_r, U_φ, U_z), the following equation can be derived:

$$\frac{\partial \rho}{\partial t} + \frac{\partial (\rho U_r)}{\partial r} + \frac{1}{r}\frac{\partial (\rho U_\varphi)}{\partial \varphi} + \frac{\partial (\rho U_z)}{\partial z} + \frac{\rho U_r}{r} = 0, \tag{5.94}$$

and for $\rho =$ constant the equation reduces to:

$$\frac{\partial U_r}{\partial r} + \frac{1}{r}\frac{\partial U_\varphi}{\partial \varphi} + \frac{\partial U_z}{\partial z} + \frac{U_r}{r} = 0. \tag{5.95}$$

In *spherical coordinates* (r, θ, ϕ), the continuity equation can be stated as shown below for (U_r, U_θ, U_ϕ):

$$\frac{\partial \rho}{\partial t} + \frac{1}{r^2}\frac{\partial}{\partial r}(\rho r^2 U_r) + \frac{1}{r \sin \theta}\frac{\partial}{\partial \theta}(\rho U_\theta \sin \theta) + \frac{1}{r \sin \theta}\frac{\partial}{\partial \phi}(\rho U_\phi) = 0. \tag{5.96}$$

The coordinates mentioned below were employed in the derivations of the above relationships (Figs. 5.11 and 5.12).

The use of cylindrical coordinates in the derivations of the basic equations leads to the metric coefficients introduced in Sect. 2.10 for the transformation of the equations:

$$h_r = 1; \quad h_\varphi = r; \quad h_z = 1$$

for the general continuity equation one can also write:

$$\frac{\partial (\rho)}{\partial t} + \frac{1}{r}\frac{\partial}{\partial r}(r \rho U_r) + \frac{1}{r}\frac{\partial}{\partial \varphi}(\rho U_\varphi) + \frac{\partial (\rho U_z)}{\partial z} = 0, \tag{5.97}$$

or for the continuity equation with $\rho =$ constant:

$$\frac{\partial U_r}{\partial r} + \frac{1}{r}\frac{\partial U_\varphi}{\partial \varphi} + \frac{\partial U_z}{\partial z} + \frac{U_r}{r} = 0. \tag{5.98}$$

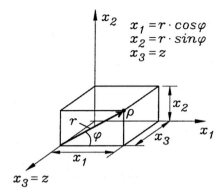

Fig. 5.11 Coordinate systems and transformation equations for cylindrical coordinates

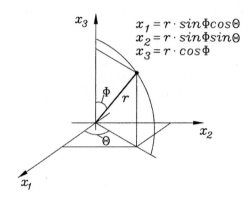

Fig. 5.12 Coordinate systems and transformation equations for spherical coordinates

Analogously to the above derivations of the continuity equation in cylindrical coordinates, one obtains for spherical coordinates the following metric coefficients for the transformations:

$$h_r = 1; \quad h_\upsilon = r; \quad h_\varphi = r\sin\theta, \tag{5.99}$$

and thus for the continuity equation in spherical coordinates reads:

$$\frac{\partial(\rho)}{\partial t} \frac{1}{r^2}\frac{\partial}{\partial r}(\rho r^2 U_r) + \frac{1}{r\sin\theta}\frac{\partial}{\partial\theta}(\rho U_\theta \sin\theta) + \frac{1}{r\sin\theta}\frac{\partial}{\partial\phi}(\rho U_\phi) = 0. \tag{5.100}$$

The continuity equation in spherical coordinates with $\rho = $ constant results in:

$$\frac{1}{r^2}\frac{\partial}{\partial r}(r^2 U_r) + \frac{1}{r\sin\theta}\frac{\partial}{\partial\theta}(U_\theta \sin\theta) + \frac{1}{r\sin\theta}\frac{\partial}{\partial\phi}(U_\phi) = 0. \tag{5.101}$$

5.7.2 Navier–Stokes Equations

Analogously to the transformation of the continuity equation into cylindrical and spherical coordinates, the different terms of the Navier–Stokes equations

5.7 Basic Equations in Different Coordinate Systems

can also be transformed. These equations can be stated, for Newtonian fluids, and in Cartesian coordinates as follows:

$$\rho \frac{DU_j}{Dt} = \rho \left[\frac{\partial U_j}{\partial t} + U_i \frac{\partial U_j}{\partial x_i} \right]$$
$$= -\frac{\partial P}{\partial x_j} + \frac{\partial}{\partial x_i} \left[\mu \left(\frac{\partial U_j}{\partial x_i} + \frac{\partial U_i}{\partial x_j} \right) - \frac{2}{3} \mu \delta_{ij} \frac{\partial U_k}{\partial x_k} \right] + \rho g_j. \quad (5.102)$$

Written out for $j = 1, 2, 3$:

$$\rho \frac{DU_1}{Dt} = -\frac{\partial P}{\partial x_1} + \frac{\partial}{\partial x_1} \left[2\mu \frac{\partial U_1}{\partial x_1} - \frac{2}{3}\mu(\frac{\partial U_k}{\partial x_k}) \right] + \frac{\partial}{\partial x_2} \left[\mu \left(\frac{\partial U_1}{\partial x_2} + \frac{\partial U_2}{\partial x_1} \right) \right]$$
$$+ \frac{\partial}{\partial x_3} \left[\mu \left(\frac{\partial U_3}{\partial x_3} + \frac{\partial U_1}{\partial x_1} \right) \right] + \rho g_1, \quad (5.103)$$

$$\rho \frac{DU_2}{Dt} = -\frac{\partial P}{\partial x_2} + \frac{\partial}{\partial x_1} \left[\mu \left(\frac{\partial U_2}{\partial x_1} + \frac{\partial U_1}{\partial x_2} \right) \right] + \frac{\partial}{\partial x_2} \left[2\mu \frac{\partial U_2}{\partial x_2} - \frac{2}{3}\mu(\frac{\partial U_k}{\partial x_k}) \right]$$
$$+ \frac{\partial}{\partial x_3} \left[\mu \left(\frac{\partial U_3}{\partial x_3} + \frac{\partial U_2}{\partial x_1} \right) \right] + \rho g_2, \quad (5.104)$$

$$\rho \frac{DU_3}{Dt} = -\frac{\partial P}{\partial x_3} + \frac{\partial}{\partial x_1} \left[\mu \left(\frac{\partial U_3}{\partial x_1} + \frac{\partial U_1}{\partial x_3} \right) \right] + \frac{\partial}{\partial x_2} \left[\mu \left(\frac{\partial U_3}{\partial x_2} + \frac{\partial U_2}{\partial x_3} \right) \right]$$
$$+ \frac{\partial}{\partial x_3} \left[2\mu \frac{\partial U_3}{\partial x_3} - \frac{2}{3}\mu(\frac{\partial U_k}{\partial x_k}) \right] + \rho g_3. \quad (5.105)$$

- *Momentum Equations in Cartesian Coordinates*
  - Momentum equations with τ_{ij} terms:

x_1 Component: $\rho \left(\frac{\partial U_1}{\partial t} + U_1 \frac{\partial U_1}{\partial x_1} + U_2 \frac{\partial U_1}{\partial x_2} + U_3 \frac{\partial U_1}{\partial x_3} \right)$
$$= -\frac{\partial P}{\partial x_1} - \left(\frac{\partial \tau_{11}}{\partial x_1} + \frac{\partial \tau_{21}}{\partial x_2} + \frac{\partial \tau_{31}}{\partial x_3} \right) + \rho g_1, \quad (5.106)$$

x_2 Component: $\rho \left(\frac{\partial U_2}{\partial t} + U_1 \frac{\partial U_2}{\partial x_1} + U_2 \frac{\partial U_2}{\partial x_2} + U_3 \frac{\partial U_2}{\partial x_3} \right)$
$$= -\frac{\partial P}{\partial x_2} - \left(\frac{\partial \tau_{12}}{\partial x_1} + \frac{\partial \tau_{22}}{\partial x_2} + \frac{\partial \tau_{32}}{\partial x_3} \right) + \rho g_2, \quad (5.107)$$

x_3 Component: $\rho \left(\frac{\partial U_3}{\partial t} + U_1 \frac{\partial U_3}{\partial x_1} + U_2 \frac{\partial U_3}{\partial x_2} + U_3 \frac{\partial U_3}{\partial x_3} \right)$
$$= -\frac{\partial P}{\partial x_3} - \left(\frac{\partial \tau_{13}}{\partial x_1} + \frac{\partial \tau_{23}}{\partial x_2} + \frac{\partial \tau_{33}}{\partial x_3} \right) + \rho g_3. \quad (5.108)$$

– Navier–Stokes equations for ρ and μ both being constant:

$$x_1 \text{ Component: } \rho\left(\frac{\partial U_1}{\partial t} + U_1\frac{\partial U_1}{\partial x_1} + U_2\frac{\partial U_1}{\partial x_2} + U_3\frac{\partial U_1}{\partial x_3}\right)$$
$$= -\frac{\partial P}{\partial x_1} + \mu\left(\frac{\partial^2 U_1}{\partial x_1^2} + \frac{\partial^2 U_1}{\partial x_2^2} + \frac{\partial^2 U_1}{\partial x_3^2}\right) + \rho g_1, \qquad (5.109)$$

$$x_2 \text{ Component: } \rho\left(\frac{\partial U_2}{\partial t} + U_2\frac{\partial U_2}{\partial x_1} + U_2\frac{\partial U_2}{\partial x_2} + U_3\frac{\partial U_2}{\partial x_3}\right)$$
$$= -\frac{\partial P}{\partial x_2} + \mu\left(\frac{\partial^2 U_2}{\partial x_1^2} + \frac{\partial^2 U_2}{\partial x_2^2} + \frac{\partial^2 U_2}{\partial x_3^2}\right) + \rho g_2, \qquad (5.110)$$

$$x_3 \text{ Component: } \rho\left(\frac{\partial U_3}{\partial t} + U_1\frac{\partial U_3}{\partial x_1} + U_2\frac{\partial U_3}{\partial x_2} + U_3\frac{\partial U_3}{\partial x_3}\right)$$
$$= -\frac{\partial P}{\partial x_3} + \mu\left(\frac{\partial^2 U_3}{\partial x_1^2} + \frac{\partial^2 U_3}{\partial x_2^2} + \frac{\partial^2 U_3}{\partial x_3^2}\right) + \rho g_3. \qquad (5.111)$$

- **Momentum Equations in Cylindrical Coordinates**

  – Momentum equations with τ_{ij} terms:

$$r \text{ Component: } \rho\left(\frac{\partial U_r}{\partial t} + U_r\frac{\partial U_r}{\partial r} + \frac{U_\varphi}{r}\frac{\partial U_r}{\partial \varphi} - \frac{U_\varphi^2}{r} + U_z\frac{\partial U_r}{\partial z}\right)$$
$$= -\frac{\partial P}{\partial r} - \left(\frac{1}{r}\frac{\partial}{\partial r}(r\tau_{rr}) + \frac{1}{r}\frac{\partial \tau_{r\varphi}}{\partial \varphi} - \frac{\tau_{\varphi\varphi}}{r} + \frac{\partial \tau_{rz}}{\partial z}\right) + \rho g_r, \qquad (5.112)$$

$$\varphi \text{ Component: } \rho\left(\frac{\partial U_\varphi}{\partial t} + U_r\frac{\partial U_\varphi}{\partial r} + \frac{U_\varphi}{r}\frac{\partial U_\varphi}{\partial \varphi} + \frac{U_r U_\varphi}{r} + U_z\frac{\partial U_\varphi}{\partial z}\right)$$
$$= -\frac{1}{r}\frac{\partial P}{\partial \varphi} - \left(\frac{1}{r^2}\frac{\partial}{\partial r}(r^2\tau_{r\varphi}) + \frac{1}{r}\frac{\partial \tau_{\varphi\varphi}}{\partial \varphi} + \frac{\partial \tau_{\varphi z}}{\partial z}\right) + \rho g_\varphi, \qquad (5.113)$$

$$z \text{ Component: } \rho\left(\frac{\partial U_z}{\partial t} + U_r\frac{\partial U_z}{\partial r} + \frac{U_\varphi}{r}\frac{\partial U_z}{\partial \varphi} + U_z\frac{\partial U_z}{\partial z}\right)$$
$$= -\frac{\partial P}{\partial z} - \left(\frac{1}{r}\frac{\partial}{\partial r}(r\tau_{rz}) + \frac{1}{r}\frac{\partial \tau_{\varphi z}}{\partial \varphi} + \frac{\partial \tau_{zz}}{\partial z}\right) + \rho g_z. \qquad (5.114)$$

  – Navier–Stokes equations for ρ and μ both being constant:

$$r \text{ Component: } \rho\left(\frac{\partial U_r}{\partial t} + U_r\frac{\partial U_r}{\partial r} + \frac{U_\varphi}{r}\frac{\partial U_r}{\partial \varphi} - \frac{U_\varphi^2}{r} + U_z\frac{\partial U_r}{\partial z}\right)$$
$$= -\frac{\partial P}{\partial r} + \mu\left[\frac{\partial}{\partial r}\left(\frac{1}{r}\frac{\partial}{\partial r}(rU_r)\right) + \frac{1}{r^2}\frac{\partial^2 U_r}{\partial \varphi^2} - \frac{2}{r^2}\frac{\partial U_\varphi}{\partial \varphi} + \frac{\partial^2 U_r}{\partial z^2}\right] + \rho g_r, \qquad (5.115)$$

5.7 Basic Equations in Different Coordinate Systems

φ Component: $\rho \left(\dfrac{\partial U_\varphi}{\partial t} + U_r \dfrac{\partial U_\varphi}{\partial r} + \dfrac{U_\varphi}{r} \dfrac{\partial U_\varphi}{\partial \varphi} + \dfrac{U_r U_\varphi}{r} + U_z \dfrac{\partial U_\varphi}{\partial z} \right)$

$= -\dfrac{1}{r}\dfrac{\partial p}{\partial \varphi} + \mu \left[\dfrac{\partial}{\partial r}\left(\dfrac{1}{r}\dfrac{\partial}{\partial r}(rU_\varphi) \right) + \dfrac{1}{r^2}\dfrac{\partial^2 U_\varphi}{\partial \varphi^2} + \dfrac{2}{r^2}\dfrac{\partial U_r}{\partial \varphi} + \dfrac{\partial^2 U_\varphi}{\partial z^2} \right] + \rho g_\varphi,$ (5.116)

z Component: $\rho \left(\dfrac{\partial U_z}{\partial t} + U_r \dfrac{\partial U_z}{\partial r} + \dfrac{U_\varphi}{r}\dfrac{\partial U_z}{\partial \varphi} + U_z \dfrac{\partial U_z}{\partial z} \right)$

$= -\dfrac{\partial P}{\partial z} + \mu \left[\dfrac{1}{r}\dfrac{\partial}{\partial r}\left(r\dfrac{\partial U_z}{\partial r} \right) + \dfrac{1}{r^2}\dfrac{\partial^2 U_z}{\partial \varphi^2} + \dfrac{\partial^2 U_z}{\partial z^2} \right] + \rho g_z.$ (5.117)

- *Momentum Equations in Spherical Coordinates*
  - Momentum equations with τ_{ij} terms

r Component: $\rho \left(\dfrac{\partial U_r}{\partial t} + U_r \dfrac{\partial U_r}{\partial r} + \dfrac{U_\theta}{r}\dfrac{\partial U_r}{\partial \theta} + \dfrac{U_\phi}{r\sin\theta}\dfrac{\partial U_r}{\partial \phi} - \dfrac{U_\theta^2 + U_\phi^2}{r} \right)$

$= -\dfrac{\partial P}{\partial r} - \left(\dfrac{1}{r^2}\dfrac{\partial}{\partial r}(r^2 \tau_{rr}) + \dfrac{1}{r\sin\theta}\dfrac{\partial}{\partial \theta}(\tau_{r\theta}\sin\theta) \right.$

$\left. + \dfrac{1}{r\sin\theta}\dfrac{\partial \tau_{r\phi}}{\partial \phi} - \dfrac{\tau_{\theta\theta} + \tau_{\phi\phi}}{r} \right) + \rho g_r,$ (5.118)

θ Component: $\rho \left(\dfrac{\partial U_\theta}{\partial t} + U_r \dfrac{\partial U_\theta}{\partial r} + \dfrac{U_\theta}{r}\dfrac{\partial U_\theta}{\partial \theta} + \dfrac{U_\phi}{r\sin\theta}\dfrac{\partial U_\theta}{\partial \phi} + \dfrac{U_r U_\theta}{r} \right.$

$\left. - \dfrac{U_\phi^2 \cot\theta}{r} \right) = -\dfrac{1}{r}\dfrac{\partial p}{\partial \theta} - \left(\dfrac{1}{r^2}\dfrac{\partial}{\partial r}(r^2 \tau_{r\theta}) + \dfrac{1}{r\sin\theta}\dfrac{\partial}{\partial \theta} \right.$

$\left. (\tau_{\theta\theta}\sin\theta) + \dfrac{1}{r\sin\theta}\dfrac{\partial \tau_{\theta\phi}}{\partial \phi} + \dfrac{\tau_{r\theta}}{r} - \dfrac{\cot\theta}{r}\tau_{\phi\phi} \right) + \rho g_\theta,$ (5.119)

ϕ Component: $\rho \left(\dfrac{\partial U_\phi}{\partial t} + U_r \dfrac{\partial U_\phi}{\partial r} + \dfrac{U_\theta}{r}\dfrac{\partial U_\phi}{\partial \theta} + \dfrac{U_\phi}{r\sin\theta}\dfrac{\partial U_\phi}{\partial \phi} + \dfrac{U_\phi U_r}{r} \right.$

$\left. + \dfrac{U_\theta U_\phi}{r}\cot\theta \right) = -\dfrac{1}{r\sin\theta}\dfrac{\partial p}{\partial \phi} - \left(\dfrac{1}{r^2}\dfrac{\partial}{\partial r}(r^2 \tau_{r\phi}) \right.$

$\left. + \dfrac{1}{r}\dfrac{\partial \tau_{\theta\phi}}{\partial \theta} + \dfrac{1}{r\sin\theta}\dfrac{\partial \tau_{\phi\phi}}{\partial \phi} + \dfrac{\tau_{r\phi}}{r} + \dfrac{2\cot\theta}{r}\tau_{\theta\phi} \right) + \rho g_\phi.$ (5.120)

- *Navier–Stokes Equations for ρ and μ both being constant*

r Component: $\rho \left(\dfrac{\partial U_r}{\partial t} + U_r \dfrac{\partial U_r}{\partial r} + \dfrac{U_\theta}{r} \dfrac{\partial U_r}{\partial \theta} + \dfrac{U_\phi}{r \sin \theta} \dfrac{\partial U_r}{\partial \phi} - \dfrac{U_\theta^2 + U_\phi^2}{r} \right)$

$= -\dfrac{\partial P}{\partial r} + \mu \left(\nabla^2 U_r - \dfrac{2}{r^2} U_r - \dfrac{2}{r^2} \dfrac{\partial U_\theta}{\partial \theta} - \dfrac{2}{r^2} U_\theta \cot \theta \right.$

$\left. - \dfrac{2}{r^2 \sin \theta} \dfrac{\partial U_\phi}{\partial \phi} \right) + \rho g_r,$ (5.121)

θ Component: $\rho \left(\dfrac{\partial U_\theta}{\partial t} + U_r \dfrac{\partial U_\theta}{\partial r} + \dfrac{U_\theta}{r} \dfrac{\partial U_\theta}{\partial \theta} + \dfrac{U_\phi}{r \sin \theta} \dfrac{\partial U_\theta}{\partial \phi} + \dfrac{U_r U_\theta}{r} \right.$

$\left. - \dfrac{U_\phi^2 \cot \theta}{r} \right) = -\dfrac{1}{r} \dfrac{\partial P}{\partial \theta} + \mu \left(\nabla^2 U_\theta + \dfrac{2}{r^2} \dfrac{\partial U_r}{\partial \theta} \right.$

$\left. - \dfrac{U_\theta}{r^2 \sin^2 \theta} - \dfrac{2 \cos \theta}{r^2 \sin^2 \theta} \dfrac{\partial U_\phi}{\partial \phi} \right) + \rho g_\theta,$ (5.122)

ϕ Component: $\rho \left(\dfrac{\partial U_\phi}{\partial t} + U_r \dfrac{\partial U_\phi}{\partial r} + \dfrac{U_\theta}{r} \dfrac{\partial U_\phi}{\partial \theta} + \dfrac{U_\phi}{r \sin \theta} \dfrac{\partial U_\phi}{\partial \phi} + \dfrac{U_\phi U_r}{r} \right.$

$\left. + \dfrac{U_\theta U_\phi}{r} \cot \theta \right) = -\dfrac{1}{r \sin \theta} \dfrac{\partial P}{\partial \phi} + \mu \left(\nabla^2 U_\phi - \dfrac{U_\phi}{r^2 \sin^2 \theta} \right.$

$\left. + \dfrac{2}{r^2 \sin^2 \theta} \dfrac{\partial U_r}{\partial \phi} + \dfrac{2 \cos \theta}{r^2 \sin^2 \theta} \dfrac{\partial U_\theta}{\partial \phi} \right) + \rho g_\phi.$ (5.123)

In these equations the operator ∇^2 corresponds to:

$$\nabla^2 = \dfrac{1}{r^2} \dfrac{\partial}{\partial r} \left(r^2 \dfrac{\partial}{\partial r} \right) + \dfrac{1}{r^2 \sin \theta} \dfrac{\partial}{\partial \theta} \left(\sin \theta \dfrac{\partial}{\partial \theta} \right) + \dfrac{1}{r^2 \sin^2 \theta} \left(\dfrac{\partial^2}{\partial \phi^2} \right). \quad (5.124)$$

- *Components of the molecular momentum transport tensor in Cartesian coordinates*

$\tau_{11} = -\mu \left[2 \dfrac{\partial U_1}{\partial x_1} - \dfrac{2}{3} \left(\dfrac{\partial U_k}{\partial x_k} \right) \right]; \quad \tau_{22} = -\mu \left[2 \dfrac{\partial U_2}{\partial x_2} - \dfrac{2}{3} \left(\dfrac{\partial U_k}{\partial x_k} \right) \right],$

$\tau_{33} = -\mu \left[2 \dfrac{\partial U_3}{\partial x_3} - \dfrac{2}{3} \left(\dfrac{\partial U_k}{\partial x_k} \right) \right],$ (5.125)

$\tau_{12} = \tau_{21} = -\mu \left[\dfrac{\partial U_1}{\partial x_2} + \dfrac{\partial U_2}{\partial x_1} \right], \quad \tau_{23} = \tau_{32} = -\mu \left[\dfrac{\partial U_2}{\partial x_3} + \dfrac{\partial U_3}{\partial x_2} \right],$

$\tau_{31} = \tau_{13} = -\mu \left[\dfrac{\partial U_3}{\partial x_1} + \dfrac{\partial U_1}{\partial x_3} \right].$ (5.126)

5.7 Basic Equations in Different Coordinate Systems

- Components of the molecular m on in time transport tensor in cylindrical coordinates

$$\tau_{rr} = -\mu \left[2\frac{\partial U_r}{\partial r} - \frac{2}{3}(\nabla \cdot \boldsymbol{U}) \right],$$

$$\tau_{\varphi\varphi} = -\mu \left[2\left(\frac{1}{r}\frac{\partial U_\varphi}{\partial \varphi} + \frac{U_r}{r}\right) - \frac{2}{3}(\nabla \cdot \boldsymbol{U}) \right], \tau_{zz} = -\mu \left[2\frac{\partial U_z}{\partial x_z} - \frac{2}{3}(\nabla \cdot \boldsymbol{U}) \right], \tag{5.127}$$

$$\tau_{r\varphi} = \tau_{\varphi r} = -\mu \left[r\frac{\partial}{\partial r}\left(\frac{U_\varphi}{r}\right) + \frac{1}{r}\frac{\partial U_r}{\partial \varphi} \right],$$

$$\tau_{\varphi z} = \tau_{z\varphi} = -\mu \left[\frac{\partial U_\varphi}{\partial z} + \frac{1}{r}\frac{\partial U_z}{\partial \varphi} \right], \quad \tau_{zr} = \tau_{rz} = -\mu \left[\frac{\partial U_z}{\partial r} + \frac{\partial U_r}{\partial z} \right], \tag{5.128}$$

with

$$(\nabla \cdot \boldsymbol{U}) = \frac{1}{r}\frac{\partial}{\partial r}(rU_r) + \frac{1}{r}\frac{\partial U_\varphi}{\partial \varphi} + \frac{\partial U_z}{\partial z}. \tag{5.129}$$

- Components of the molecular momentum transport tensor in spherical coordinates

$$\tau_{rr} = -\mu \left[2\frac{\partial U_r}{\partial r} - \frac{2}{3}(\nabla \cdot \boldsymbol{U}) \right], \tag{5.130}$$

$$\tau_{\theta\theta} = -\mu \left[2\left(\frac{1}{r}\frac{\partial U_\theta}{\partial \theta} + \frac{U_r}{r}\right) - \frac{2}{3}(\nabla \cdot \boldsymbol{U}) \right], \tag{5.131}$$

$$\tau_{\phi\phi} = -\mu \left[2\left(\frac{1}{r\sin\theta}\frac{\partial U_\phi}{\partial \phi} + \frac{U_r}{r} + \frac{U_\theta \cot\theta}{r}\right) - \frac{2}{3}(\nabla \cdot \boldsymbol{U}) \right], \tag{5.132}$$

$$\tau_{r\theta} = \tau_{\theta r} = -\mu \left[r\frac{\partial}{\partial r}\left(\frac{U_\theta}{r}\right) + \frac{1}{r}\frac{\partial U_r}{\partial \theta} \right], \tag{5.133}$$

$$\tau_{\theta\phi} = \tau_{\phi\theta} = -\mu \left[\frac{\sin\theta}{r}\frac{\partial}{\partial \theta}\left(\frac{U_\phi}{\sin\theta}\right) + \frac{1}{r\sin\theta}\frac{\partial U_\theta}{\partial \phi} \right], \tag{5.134}$$

$$\tau_{\phi r} = \tau_{r\phi} = -\mu \left[\frac{1}{r\sin\theta}\frac{\partial U_r}{\partial \phi} + r\frac{\partial}{\partial r}\left(\frac{U_\phi}{r}\right) \right], \tag{5.135}$$

$$(\nabla \cdot \boldsymbol{U}) = \frac{1}{r^2}\frac{\partial}{\partial r}(r^2 U_r) + \frac{1}{r\sin\theta}\frac{\partial}{\partial \theta}(U_\theta \sin\theta) + \frac{1}{r\sin\theta}\frac{\partial U_\phi}{\partial \phi}. \tag{5.136}$$

- Dissipation function $\tau_{ij}\frac{\partial U_j}{\partial x_i} = \mu\Phi_\mu$
  Cartesian coordinates:

$$\Phi_\mu = 2\left[\left(\frac{\partial U_1}{\partial x_1}\right)^2 + \left(\frac{\partial U_2}{\partial x_2}\right)^2 + \left(\frac{\partial U_3}{\partial x_3}\right)^2\right] + \left[\frac{\partial U_2}{\partial x_1} + \frac{\partial U_1}{\partial x_2}\right]^2$$

$$+ \left[\frac{\partial U_3}{\partial x_2} + \frac{\partial U_2}{\partial x_3}\right]^2 + \left[\frac{\partial U_1}{\partial x_3} + \frac{\partial U_3}{\partial x_1}\right]^2 - \frac{2}{3}\left[\frac{\partial U_1}{\partial x_1} + \frac{\partial U_2}{\partial x_2} + \frac{\partial U_3}{\partial x_3}\right]^2. \tag{5.137}$$

Cylindrical coordinates:

$$\Phi_\mu = 2\left[\left(\frac{\partial U_r}{\partial r}\right)^2 + \left(\frac{1}{r}\frac{\partial U_\varphi}{\partial \varphi} + \frac{U_r}{r}\right)^2 + \left(\frac{\partial U_z}{\partial x_z}\right)^2\right]$$
$$+ \left[r\frac{\partial}{\partial r}\left(\frac{U_\varphi}{r}\right) + \frac{1}{r}\frac{\partial U_z}{\partial \varphi}\right]^2 + \left[\frac{1}{r}\frac{\partial U_z}{\partial \varphi} + \frac{\partial U_\varphi}{\partial x_z}\right]^2 + \left[\frac{\partial U_r}{\partial x_z} + \frac{\partial U_z}{\partial r}\right]^2$$
$$- \frac{2}{3}\left[\frac{1}{r}\frac{\partial}{\partial r}(rU_r) + \frac{1}{r}\frac{\partial U_\varphi}{\partial \varphi} + \frac{\partial U_z}{\partial x_z}\right]^2. \tag{5.138}$$

Spherical coordinates:

$$\Phi_\mu = 2\left[\left(\frac{\partial U_r}{\partial r}\right)^2 + \left(\frac{1}{r}\frac{\partial U_\theta}{\partial \theta} + \frac{U_r}{r}\right)^2 \left(\frac{1}{r\sin\theta}\frac{\partial U_\phi}{\partial \phi} + \frac{U_r}{r} + \frac{U_\theta \cot\theta}{r}\right)^2\right]$$
$$+ \left[r\frac{\partial}{\partial r}\left(\frac{U_\theta}{r}\right) + \frac{1}{r}\frac{\partial U_r}{\partial \theta}\right]^2 + \left[\frac{\sin\theta}{r}\frac{\partial}{\partial \theta}\left(\frac{U_\phi}{\sin\theta}\right) + \frac{1}{r\sin\theta}\frac{\partial U_\theta}{\partial \phi}\right]^2$$
$$+ \left[\frac{1}{r\sin\theta}\frac{\partial U_r}{\partial \phi} + r\frac{\partial}{\partial r}\left(\frac{U_\phi}{r}\right)\right]^2$$
$$- \frac{2}{3}\left[\frac{1}{r^2}\frac{\partial}{\partial r}(r^2 U_r) + \frac{1}{r\sin\theta}\frac{\partial}{\partial \theta}(U_\theta \sin\theta) + \frac{1}{r\sin\theta}\frac{\partial U_\phi}{\partial \phi}\right]^2. \tag{5.139}$$

The above equations can be solved in connection with the initial and boundary conditions describing the actual flow problems. Very often the boundary conditions define the coordinate system chosen for solving a particular flow problem.

5.8 Special Forms of the Basic Equations

Due to the multitude of fluid mechanical considerations, special forms of the equations have crystallized out of those treated in the preceding sections. Some of these equations will be derived and also explained in this section. These are the vorticity equation, the Bernoulli equation and the Crocco equation, some of which have already been treated. The following derivations will also treat the Kelvin theorem as a basis for explanations of its physical significance. The objective of the considerations is to bring out clearly the prerequisites under which the derived special forms of the basic equations are valid. Only when a sound basis for the simplified treatments of flow problems exists can the special forms of the basic equations lead to valuable results. In case the insight into a particular flow is not present, the general form of the basic equations of fluid mechanics should be employed to slove the considered flow problems.

5.8 Special Forms of the Basic Equations

5.8.1 Transport Equation for Vorticity

The vorticity ω_i is a local property of the flow field which can be employed advantageously in considerations of rotating fluid motions. It can be computed from the velocity field as follows:

$$2\omega_k = \nabla \times \boldsymbol{U} = -\epsilon_{ijk}\frac{\partial U_j}{\partial x_i} = \left(\frac{\partial U_j}{\partial x_i} - \frac{\partial U_i}{\partial x_j}\right). \tag{5.140}$$

For a fluid with the properties ρ = constant and μ = constant, the Navier–Stokes equation can be written in the following way:

$$\rho\left[\frac{\partial U_j}{\partial t} + U_i\frac{\partial U_j}{\partial x_i}\right] = -\frac{\partial P}{\partial x_j} + \mu\frac{\partial^2 U_j}{\partial x_i^2} + \rho g_j, \tag{5.141}$$

or in vector form:

$$\left[\frac{\partial \boldsymbol{U}}{\partial t} + (\boldsymbol{U}\cdot\nabla)\boldsymbol{U}\right] = -\frac{1}{\rho}\nabla P + \nu\nabla^2\boldsymbol{U} + \boldsymbol{g}. \tag{5.142}$$

This vector form of the Navier–Stokes equation can also be written as:

$$\frac{\partial \boldsymbol{U}}{\partial t} + \nabla\left(\frac{1}{2}\boldsymbol{U}\cdot\boldsymbol{U}\right) - \boldsymbol{U}\times(\nabla\times\boldsymbol{U}) = -\frac{1}{\rho}\nabla P + \nu\nabla^2\boldsymbol{U} + \boldsymbol{g}. \tag{5.143}$$

When one applies the operator $\nabla\times(\ldots)$ to each of the terms appearing in the above equation, one obtains:

$$\frac{\partial \boldsymbol{\omega}}{\partial t} - \nabla\times(\boldsymbol{U}\times\boldsymbol{\omega}) = \nu\nabla^2\boldsymbol{\omega}. \tag{5.144}$$

Making use of the following relationship, valid for vectors:

$$\nabla\times(\boldsymbol{U}\times\boldsymbol{\omega}) = \boldsymbol{U}(\nabla\cdot\boldsymbol{\omega}) - \boldsymbol{\omega}(\nabla\cdot\boldsymbol{U}) - (\boldsymbol{U}\cdot\nabla)\boldsymbol{\omega} + (\boldsymbol{\omega}\cdot\nabla)\boldsymbol{U},$$

where $\nabla\cdot\boldsymbol{\omega} = 0$ as the divergence of the rotation of each vector is equal to zero, and where at the same time ρ = constant and also $\nabla\cdot\boldsymbol{U} = 0$ holds due to the continuity equation. When one introduces all this into the above equations, the transport equation for vorticity results:

$$\frac{\partial \boldsymbol{\omega}}{\partial t} + (\boldsymbol{U}\nabla)\boldsymbol{\omega} = (\boldsymbol{\omega}\nabla)\boldsymbol{U} + \nu\nabla^2\boldsymbol{\omega}, \tag{5.145}$$

or written in tensor notation:

$$\frac{D\omega_j}{Dt} = \frac{\partial \omega_j}{\partial t} + U_i\frac{\partial \omega_j}{\partial x_i} = \omega_j\frac{\partial U_i}{\partial x_j} + \nu\frac{\partial^2 \omega_j}{\partial x_i^2}. \tag{5.146}$$

Equation (5.146) does not contain the pressure term, hence it is apparent that the vorticity field can be determined without knowledge of the pressure distribution. To be able to compute the pressure, one forms the divergence of the Navier–Stokes equation and obtains for $g_j = 0$:

$$\frac{\partial^2}{\partial x_i^2}\left(\frac{P}{\rho}\right) = \omega_j^2 + U_j \frac{\partial^2 U_j}{\partial x_i^2} - \frac{1}{2}\frac{\partial^2 U_j^2}{\partial x_i^2}, \qquad (5.147)$$

thus yielding a Poisson equation for the computation of the pressure. For two-dimensional flows, for which the vorticity vector stands vertical on the plane of the flow, $(\boldsymbol{\omega} \cdot \nabla)\boldsymbol{U} = 0$. The transport equation for the vorticity therefore reads:

$$\frac{\partial \omega_j}{\partial t} + U_i \frac{\partial \omega_j}{\partial x_i} = \nu \frac{\partial^2 \omega_j}{\partial x_i^2}. \qquad (5.148)$$

5.8.2 The Bernoulli Equation

The general momentum equations can be transferred into the Euler equations by assuming a fluid mechanically ideal fluid to exist. This is the equation with which to start to derive the Bernoulli equation:

$$\rho \frac{DU_j}{Dt} = \rho\left[\frac{\partial U_j}{\partial t} + U_i \frac{\partial U_j}{\partial x_i}\right] = -\frac{\partial P}{\partial x_j} + \rho g_j. \qquad (5.149)$$

Multiplying this equation by U_j, one obtains the mechanical energy equation valid for dissipation-free fluid flows:

$$\rho \frac{D}{Dt}\left(\frac{1}{2}U_j^2\right) = \rho\left[\frac{\partial}{\partial t}\left(\frac{1}{2}U_j^2\right) + U_i \frac{\partial}{\partial x_i}\left(\frac{1}{2}U_j^2\right)\right] = -U_j \frac{\partial P}{\partial x_j} + \rho g_j U_j. \qquad (5.150)$$

When one introduces the potential field G for the presentation of g_j as follows:

$$g_j = -\frac{\partial G}{\partial x_j}, \qquad (5.151)$$

the last term of (5.151) reads:

$$\rho g_j U_j = -\rho U_j \frac{\partial G}{\partial x_j} = -\rho \frac{DG}{Dt} + \rho \frac{\partial G}{\partial t}, \qquad (5.152)$$

and one obtains for $\frac{\partial G}{\partial t} = 0$:

$$\rho \frac{D}{Dt}\left[\left(\frac{1}{2}U_j^2\right) + G\right] = -U_j \frac{\partial P}{\partial x_j}. \qquad (5.153)$$

5.8 Special Forms of the Basic Equations

Considering that

$$\rho \frac{D}{Dt}\left(\frac{P}{\rho}\right) = \rho \frac{\partial}{\partial t}\left(\frac{P}{\rho}\right) + \rho U_j \frac{\partial}{\partial x_j}\left(\frac{P}{\rho}\right), \qquad (5.154)$$

and that moreover the following conversions of terms are possible:

$$\rho \frac{\partial}{\partial t}\left(\frac{P}{\rho}\right) = \frac{\partial P}{\partial t} - \frac{P}{\rho}\frac{\partial \rho}{\partial t}, \qquad (5.155)$$

$$\rho U_j \frac{\partial}{\partial x_j}\left(\frac{P}{\rho}\right) = U_j \frac{\partial P}{\partial x_j} - \frac{P}{\rho} U_j \frac{\partial \rho}{\partial x_j}, \qquad (5.156)$$

then (5.154) can be written as:

$$\rho \frac{D}{Dt}\left(\frac{P}{\rho}\right) = \frac{\partial P}{\partial t} + U_j \frac{\partial P}{\partial x_j} - \frac{P}{\rho}\frac{D\rho}{Dt}. \qquad (5.157)$$

From (5.154) and (5.157) one obtains

$$\rho \frac{D}{Dt}\left[\left(\frac{1}{2}U_j^2\right) + G\right] = -U_j \frac{\partial P}{\partial x_j} = -\rho \frac{D}{Dt}\left(\frac{P}{\rho}\right) + \frac{\partial P}{\partial t} - \frac{P}{\rho}\frac{D\rho}{Dt}, \qquad (5.158)$$

or, after conversion of some terms,

$$\rho \frac{D}{Dt}\left[\left(\frac{1}{2}U_j^2\right) + \frac{P}{\rho} + G\right] = \frac{\partial P}{\partial t} - \frac{P}{\rho}\frac{D\rho}{Dt}. \qquad (5.159)$$

or

$$\rho \frac{D}{Dt}\left[\left(\frac{1}{2}U_j^2\right) + \frac{P}{\rho} + G\right] = \frac{\partial P}{\partial t} + \frac{\partial U_j}{\partial x_j} P. \qquad (5.160)$$

For stationary pressure fields $\frac{\partial P}{\partial t} = 0$, and, for $\rho =$ constant, the Bernoulli equation can be stated as follows:

$$\frac{1}{2}U_j^2 + \frac{P}{\rho} + G = \frac{1}{2}U_j^2 + \frac{P}{\rho} - x_j g_j = \text{constant}. \qquad (5.161)$$

The above derivations make it clear under which conditions the well-known Bernoulli (5.161) holds.

From the above derivations, the general form of the mechanical energy equation, by introducing dissipation into the considerations, can be written in the following form:

$$\rho \frac{D}{Dt}\left[\frac{1}{2}U_j^2 + \frac{P}{\rho} + G\right] = \frac{\partial P}{\partial t} + P\frac{\partial U_j}{\partial x_j} + \frac{\partial}{\partial x_i}(\tau_{ij}U_j) - \tau_{ij}\frac{\partial U_j}{\partial x_i}. \qquad (5.162)$$

The left-hand side of this form of the mechanical energy equation contains all terms of the Bernoulli equation.

5.8.3 Crocco Equation

The Crocco equation is a special form of the momentum equation which shows in an impressive manner how purely fluid mechanics considerations can be supplemented by thermodynamic considerations, yielding new insights into fluid flows. The Crocco equation connects the vorticity of a flow field to the entropy of the considered fluid. It can be shown from this equation that isotropic flows are free of rotation and vice versa, at least under certain conditions. So, when one recognizes a flow field to be isentropic, the simplified rotation-free flow field considerations can be applied.

For the derivation of the Crocco equation, one starts from the Navier–Stokes equation, as stated in (5.143) supplemented by $\nu = 0$, i.e. one introduces an ideal fluid into the considerations, by neglecting viscous forces:

$$\frac{\partial \boldsymbol{U}}{\partial t} + \nabla\left(\frac{1}{2}\boldsymbol{U} \cdot \boldsymbol{U}\right) - \boldsymbol{U} \times (\nabla \times \boldsymbol{U}) = -\frac{1}{\rho}\nabla P. \qquad (5.163)$$

In Sect. 3.6, it was shown that:

$$T_\Re \, \mathrm{d}s_\Re = \mathrm{d}e_\Re + P_\Re \, \mathrm{d}v_\Re = \mathrm{d}e_\Re + P_\Re \, \mathrm{d}\left(\frac{1}{\rho_\Re}\right). \qquad (5.164)$$

With $e_\Re = h_\Re - P_\Re/\rho_\Re$, the following relation holds:

$$\mathrm{d}h_\Re - \mathrm{d}\left(\frac{P_\Re}{\rho_\Re}\right) = -P_\Re \, \mathrm{d}\left(\frac{1}{\rho_\Re}\right) + T_\Re \, \mathrm{d}s_\Re. \qquad (5.165)$$

Because $\mathrm{d}\left(\dfrac{P_\Re}{\rho_\Re}\right) = P_\Re \, \mathrm{d}\left(\dfrac{1}{\rho_\Re}\right) + \dfrac{1}{\rho_\Re}\,\mathrm{d}P_\Re$, it holds that

$$-\frac{1}{\rho_\Re} \, \mathrm{d}P_\Re = T_\Re \, \mathrm{d}s_\Re - \mathrm{d}h_\Re. \qquad (5.166)$$

This relation can also be written in field variables as:

$$-\frac{1}{\rho}\nabla P = T\nabla s - \nabla h. \qquad (5.167)$$

Equation (5.167) is inserted in (5.163) to yield:

$$\frac{\partial \boldsymbol{U}}{\partial t} + \nabla\left(\frac{1}{2}\boldsymbol{U} \cdot \boldsymbol{U}\right) - \boldsymbol{U} \times (\nabla \times \boldsymbol{U}) = T\nabla s - \nabla h. \qquad (5.168)$$

5.8 Special Forms of the Basic Equations

For stationary adiabatic processes, the thermal energy equation can be written in the following form:

$$\rho \frac{Dh}{Dt} = \frac{DP}{Dt}. \tag{5.169}$$

From the momentum equation, it follows further that:

$$\rho \frac{D}{Dt}\left(\frac{1}{2}UU\right) = -U\nabla P. \tag{5.170}$$

Hence

$$\rho \frac{D}{Dt}\left(h + \frac{1}{2}UU\right) = \frac{DP}{Dt} - U\nabla P, \tag{5.171}$$

$$\rho \frac{D}{Dt}\left(h + \frac{1}{2}UU\right) = \frac{DP}{Dt}. \tag{5.172}$$

Equation (5.172), inserted in (5.163) under stationary flow conditions, yields the following relationship

$$U \times \omega + T\nabla s = \nabla\left(h + \frac{1}{2}UU\right). \tag{5.173}$$

If a flow is considered along a flow line, then $\nabla(h + 1/2\,U \cdot U)$ is a vector perpendicular to the considered flow line. $U \times \omega$ is also a vector and also lies perpendicular to the flow line. Hence $T\nabla s$ lies vertical to the fluid motion along a flow line, and therefore it can be stated that:

$$U_n \omega_n + T\frac{ds}{dn} = \frac{d}{dn}\left(h + \frac{1}{2}UU\right), \tag{5.174}$$

when $\left(h + \frac{1}{2}UU\right)$ is constant along a flow field, then $\frac{d}{dn}\left(h + \frac{1}{2}UU\right) = 0$ and thus

$$U_n \omega_n + T\frac{ds}{dn} = 0. \tag{5.175}$$

If $\omega_n = 0$ then $ds/dn = 0$, hence rotation-free flows are isentropic and vice versa. If the flow is assumed to be stationary and in the absence of viscosity, the inertial forces turn out to be zero.

5.8.4 Further Forms of the Energy Equation

The close connection between fluid mechanics and thermodynamics becomes clear from different forms of the energy equation, summarized in the following table, as introduced by Bird, Steward and Lightfoot [5.1] with the notation adapted in this book.

Symbol	Explanation	Dimensions
c_p	Heat capacity at constant pressure, per unit mass	$L^2/(Tt^2)$
c_v	Heat capacity at constant volume, per unit mass	$L^2/(Tt^2)$
e_{total}	Total energy of the fluid, per unit mass	L^2/t^2
e	Internal energy, per unit mass	L^2/t^2
$\boldsymbol{g}, g_i$	External mass acceleration	L/t^2
G	Potential energy, potential of G	ML^2/t^2
h	Enthalpy	L^2/t^2
P	Pressure field	$M/(Lt^2)$
$\dot{\boldsymbol{q}}, \dot{q}_i$	Heat flow per unit area	M/t^3
T	Absolute temperature	T
$\boldsymbol{U}, U_i$	Velocity field	L/t
V	Volume	L^3
x_i	Cartesian coordinates	L
β	Thermal expansion coefficient	$1/T$
ρ	Fluid density field	M/L^3
$\boldsymbol{\tau}, \tau_{ij}$	Molecular momentum transport	$M/(Lt^2)$

Mass conservation (continuity equation)

Equations in vector and tensor notation	Special forms
$\frac{D\rho}{Dt} = -\rho(\nabla \cdot \boldsymbol{U})$	For $\frac{D\rho}{Dt} = 0;\ (\nabla \cdot \boldsymbol{U}) = 0$
$\frac{D\rho}{Dt} = -\rho \frac{\partial U_i}{\partial x_i}$	or $\frac{\partial U_i}{\partial x_i} = 0$

Equation of motion (momentum equation)

Special form	Equations in vector and tensor notation	Special forms
Imposed convection	$\rho \frac{D\boldsymbol{U}}{Dt} = -\nabla P - [\nabla \cdot \boldsymbol{\tau}] + \rho \boldsymbol{g}$ $\rho \frac{DU_j}{Dt} = -\frac{\partial P}{\partial x_j} - \frac{\partial \tau_{ij}}{\partial x_i} + \rho g_j$	For $\nabla \cdot \boldsymbol{\tau} = 0$ one obtains the Euler equations
Free convection	$\rho \frac{D\boldsymbol{U}}{Dt} = -[\nabla \cdot \boldsymbol{\tau}] - \rho \beta \boldsymbol{g} \Delta T$ $\rho \frac{DU_j}{Dt} = -\frac{\partial \tau_{ij}}{\partial x_i} - \rho \beta g_j \Delta T$	This equation comprises approximations by Boussinesque assumptions

5.8 Special Forms of the Basic Equations

Energy equations

Special form	Equations in vector and tensor notation	Special forms
Written for $e_{\text{total}} = e + \frac{1}{2}\boldsymbol{U}^2 + G$	$\rho\frac{De_{\text{total}}}{Dt} = -(\nabla \cdot \boldsymbol{q}) - (\nabla \cdot \rho\boldsymbol{U}) - (\nabla \cdot [\boldsymbol{\tau} \cdot \boldsymbol{U}])$ $\rho\frac{De_{\text{total}}}{Dt} = -\frac{\partial \dot{q}_i}{\partial x_i} - \frac{\partial (PU_i)}{\partial x_i} - \frac{\partial (\tau_{ij}U_j)}{\partial x_i}$	Exact only for G time independent
$e + \frac{1}{2}\boldsymbol{U}^2$	$\rho\frac{D(e+\frac{1}{2}\boldsymbol{U}^2)}{Dt} = -(\nabla \cdot \boldsymbol{q}) - (\nabla \cdot \rho\boldsymbol{U})$ $\qquad -(\nabla \cdot [\boldsymbol{\tau} \cdot \boldsymbol{U}]) + \rho(\boldsymbol{U} \cdot \boldsymbol{g})$ $\rho\frac{D(e+\frac{1}{2}U_i^2)}{Dt} = -\frac{\partial \dot{q}_i}{\partial x_i} - \frac{\partial (PU_i)}{\partial x_i} - \frac{\partial (\tau_{ij}U_j)}{\partial x_i} + \rho U_i g_i$	
$\frac{1}{2}\boldsymbol{U}^2$	$\rho\frac{D\frac{1}{2}\boldsymbol{U}^2}{Dt} = -(\boldsymbol{U} \cdot \nabla P) - (\boldsymbol{U} \cdot [\nabla \cdot \boldsymbol{\tau}])$ $\qquad + \rho(\boldsymbol{U} \cdot \boldsymbol{g})$ $\rho\frac{D\frac{1}{2}U_i^2}{Dt} = -U_i\frac{\partial P}{\partial x_i} - U_i\frac{\partial \tau_{ij}}{\partial x_j} + \rho U_i g_i$	
e	$\rho\frac{De}{Dt} = -(\nabla \cdot \boldsymbol{q}) - P(\nabla \cdot \boldsymbol{U}) - (\boldsymbol{\tau} : \nabla \boldsymbol{U})$ $\rho\frac{De}{Dt} = -\frac{\partial \dot{q}_i}{\partial x_i} - P\frac{\partial U_i}{\partial x_i} - \tau_{ij}\frac{\partial U_i}{\partial x_j}$	The term containing P is zero for $\frac{D\rho}{Dt} = 0$
h	$\rho\frac{Dh}{Dt} = -(\nabla \cdot \boldsymbol{q}) - (\boldsymbol{\tau} : \nabla \boldsymbol{U}) + \frac{DP}{Dt}$ $\rho\frac{Dh}{Dt} = -\frac{\partial \dot{q}_i}{\partial x_i} - \tau_{ij}\frac{\partial U_i}{\partial x_j} + \frac{DP}{Dt}$	
Written for c_v and T	$\rho c_v\frac{DT}{Dt} = -(\nabla \cdot \boldsymbol{q}) - T(\frac{\partial P}{\partial T})_\rho(\nabla \cdot \boldsymbol{U})$ $\qquad -(\boldsymbol{\tau} : \nabla \boldsymbol{U})$ $\rho c_v\frac{DT}{Dt} = -\frac{\partial \dot{q}_i}{\partial x_i} - T(\frac{\partial P}{\partial T})_\rho(\frac{\partial U_i}{\partial x_i}) - \tau_{ij}\frac{\partial U_i}{\partial x_j}$	For an ideal gas $(\frac{\partial P}{\partial T})_\rho = \frac{P}{T}$
Written for c_p and T	$\rho c_p\frac{DT}{Dt} = -(\nabla \cdot \boldsymbol{q}) + (\frac{\partial \ln V}{\partial \ln T})_\rho\frac{DP}{Dt}$ $\qquad -(\boldsymbol{\tau} : \nabla \boldsymbol{U})$ $\rho c_p\frac{DT}{Dt} = -\frac{\partial \dot{q}_i}{\partial x_i} + (\frac{\partial \ln V}{\partial \ln T})_\rho\frac{DP}{Dt} - \tau_{ij}\frac{\partial U_i}{\partial x_j}$	For an ideal gas $(\frac{\partial \ln V}{\partial \ln T})_\rho = 1$

5.9 Transport Equation for Chemical Species

In many domains of engineering science, investigations of fluids with chemical reactions are required, which make it necessary to extend the considerations carried out so far. It is necessary to state the basic equations of fluid mechanics for the different chemical components:

- Local change of the mass per unit time of the chemical component A $\qquad \dfrac{\partial \rho_A}{\partial t} \rho V_\Re$

- Change of the mass of component A by inflow and outflow of A $\qquad -\dfrac{\partial}{\partial x_i} \rho_A (U_A)_i \delta V_\Re$

- Production of the chemical component A by chemical reactions in $V_\Re$ $\qquad r_A \delta V_\Re$

This yields a mass balance:

$$\frac{\partial \rho_A}{\partial t} \delta V_\Re = -\frac{\partial}{\partial x_i} [\rho_A (U_A)_i] \delta V_\Re + r_A \delta V_\Re, \tag{5.176}$$

and the equation for the mass conservation for the chemical component A of a fluid is

$$\frac{\partial \rho_A}{\partial t} + \frac{\partial}{\partial x_i} [\rho_A (U_A)_i] = r_A. \tag{5.177}$$

For a chemical component B, as a consequence of identical considerations,

$$\frac{\partial \rho_B}{\partial t} + \frac{\partial}{\partial x_i} [\rho_B (U_B)_i] = r_B. \tag{5.178}$$

The addition of these equations yields

$$\frac{\partial \rho}{\partial t} + \frac{\partial (\rho U_i)}{\partial x_i} = 0, \tag{5.179}$$

i.e. the total mass conservation equation for a mixture of different components is equal to the continuity equation for a fluid which consists of one chemical component only. By considering Fick's law of diffusion, it can be stated that

$$\frac{\partial \rho_A}{\partial t} + \frac{\partial}{\partial x_i} (\rho_A U_i) = \frac{\partial}{\partial x_i} \left[\rho D_{AB} \frac{\partial (C_A/C)}{\partial x_i} \right] + r_A. \tag{5.180}$$

For ρ = constant and D_{AB} = constant, one obtains

$$\frac{\partial \rho_A}{\partial t} + \overbrace{\rho_A \frac{\partial U_i}{\partial x_i}}^{=0} + U_i \frac{\partial \rho_A}{\partial x_i} = D_{AB} \frac{\partial^2 \rho_A}{\partial x_i} + r_A, \tag{5.181}$$

or, expressed in terms of concentration, C_A,

$$\frac{DC_A}{Dt} = \left[\frac{\partial C_A}{\partial t} + U_i \frac{\partial C_A}{\partial x_i}\right] = D_{AB}\frac{\partial^2 C_A}{\partial x_i^2} + R_A \tag{5.182}$$

with $r_A = AR_A$ [5.1].

References

5.1. Bird, R.B., Stewart, W.E. and Lightfoot, E.N., Transport Phenomena, John Wiley and Sons, New York, 1960.
5.2. Spurk, J.H., Strömungslehre – Einführung in die Theorie der Strömungen, Springer-Verlag, Berlin, 4. Aufl., 1996.
5.3. Brodkey, R.S., The Phenomena of Fluid Motions, Dover, New York, 1967.
5.4. Sherman, F.S., Viscous Flow, McGraw-Hill, Singapore, 1990.
5.5. Schlichting, H., Boundary Layer Theory, 6th edition, McGraw-Hill, New York, 1968.

Chapter 6
Hydrostatics and Aerostatics

6.1 Hydrostatics

Hydrostatics deals with the laws to which fluids are subjected that do not show motions in the coordinate system in which the considerations are carried out, i.e. fluids which are at rest in the coordinate system employed for the considerations. As the relationships derived in the preceding chapter represent general laws of fluid motions, they are also applicable to the cases of fluids at rest, i.e. non-flowing fluids. Thus, from the continuity equation,

$$\frac{\partial \rho}{\partial t} + \frac{\partial}{\partial x_i}(\rho U_i) = 0 \tag{6.1}$$

it can be shown that for $\rho = $ constant and $U_i \neq f(x_i)$ the continuity equation is given by

$$\underbrace{\frac{\partial \rho}{\partial t} + U_i \frac{\partial \rho}{\partial x_i}}_{D\rho/Dt=0} + \rho \underbrace{\frac{\partial U_i}{\partial x_i}}_{=0} = 0 \tag{6.2}$$

This means that for $U_i = 0$ the following simple partial differential equation holds:

$$\frac{\partial \rho}{\partial t} = 0 \tag{6.3}$$

whose general solution can be stated as follows:

$$\rho = F(x_i) \tag{6.4}$$

The density ρ in a fluid at rest is thus only a function of the spatial coordinates x_i. When time variations of the density of the fluid occur, these lead inevitably to motions within the fluid because of the relationship between the flow and density fields attributable to the continuity equation, i.e. because of (6.4).

The general equations of momentum can be expressed as

$$\rho\left[\frac{\partial U_j}{\partial t} + U_i\frac{\partial U_j}{\partial x_i}\right] = -\frac{\partial P}{\partial x_j} - \frac{\partial \tau_{ij}}{\partial x_i} + \rho g_j \tag{6.5}$$

and its special form is deduced for a fluid at rest ($U_j = 0$ and the molecular-dependent momentum transport τ_{ij}) as the following system of partial differential equations, which represents the set of basic equations of hydrostatics and aerostatics:

$$\frac{\partial P}{\partial x_j} = \rho g_j \qquad (j = 1, 2, 3) \tag{6.6}$$

or, written out for all three directions $j = 1, 2, 3$,

$$\frac{\partial P}{\partial x_1} = \rho g_1, \qquad \frac{\partial P}{\partial x_2} = \rho g_2, \qquad \frac{\partial P}{\partial x_3} = \rho g_3 \tag{6.7}$$

In this section, the pressure distribution in a fluid, mainly defined by the field of gravity, will be considered more closely. Restrictions are made concerning the possible fluid properties; the fluid is assumed to be incompressible for hydrostatics, i.e. $\rho = $ constant. This condition is in general fairly well fulfilled by liquids, so that the following derivations can be considered as valid for liquids, see also refs. [6.1] to [6.8].

For the derivation of the pressure distribution in a liquid at rest, a rectangular Cartesian coordinate system is introduced, whose position is chosen such that the mass acceleration $\{g_i\}$ given by the field of gravity shows only one component in the negative x_2 direction, i.e. the following vector holds:

$$\{g_i\} = \begin{Bmatrix} 0 \\ -g \\ 0 \end{Bmatrix}. \tag{6.8}$$

Then the differential equations (6.7), given above generally for the pressure, can be written as follows:

$$\frac{\partial P}{\partial x_1} = 0, \qquad \frac{\partial P}{\partial x_2} = -\rho g, \qquad \frac{\partial P}{\partial x_3} = 0 \tag{6.9}$$

From $\partial P/\partial x_1 = 0$, it follows that $P = f(x_2, x_3)$ and from $\partial P/\partial x_3 = 0$ it follows that $P = f(x_1, x_2)$. Thus a comparison yields $P = f(x_2)$ and this shows that the pressure of a fluid within a plane is constant, $P(x_2) = $ const, when it is perpendicular to the direction of the field of gravity. The free surface of a fluid stored in a container is a plane of constant pressure and all planes parallel to it are also planes of constant pressure. The pressure increases in the direction that was defined by g_j, i.e. in the direction of the gravitational acceleration.

6.1 Hydrostatics

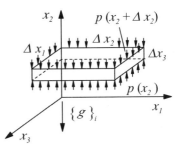

Fig. 6.1 Coordinate system for the derivation of the pressure distribution in fluids of constant density

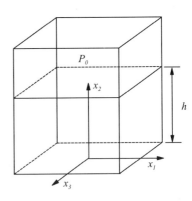

Fig. 6.2 Fluid at rest in a container with a free interface at height h

For the physical understanding of hydrostatics, it is also important to recognize that equation (6.9) expresses that the increase in pressure in the negative x_2 direction is caused by the weight of the fluid element plotted in Fig. 6.1, i.e. the following force balance holds:

$$-\rho g \overbrace{\Delta x_1 \Delta x_2 \Delta x_3}^{\Delta V} + p \overbrace{\Delta x_1 \Delta x_3}^{\Delta A} - (p + \frac{\partial p}{\partial x_2}\Delta x_2) \overbrace{\Delta x_1 \Delta x_3}^{\Delta A} = 0 \qquad (6.10)$$

Employing the above physical insights and the resultant equations, the following statements can be made for a liquid of constant density located in a container (Fig. 6.2). In the case that the field of gravity acts in the negative x_2 direction, i.e.

$$g_1 = 0, \qquad g_2 = -g, \qquad g_3 = 0 \qquad (6.11)$$

the differential equations stated in (6.9) with the solution $P = f(x_2)$ hold for this case. Thus the partial differential $\partial P/\partial x_2$ in (6.9) can be written as a total differential and one obtains for constant density fluids ($\rho =$ constant)

$$\frac{dP}{dx_2} = -\rho g \longrightarrow P = -\rho g x_2 + C \qquad (6.12)$$

which can be rewritten as:

$$\frac{P}{\rho} + g x_2 = C$$

This relationship expresses that the sum of the "pressure energy" P/ρ and the potential energy ($gx_2 = -g_j x_j$) is constant at each point of a fluid at rest. As all points of different fluid elements possess the same "total energy", the driving energy gradient for motion is absent. Hence also from the energy point of view the condition for hydrostatic fluid behavior exists.

When a fluid with a height h has a free surface on which an equally distributed pressure P_0 acts at all points, it represents, because of the relation $P = f(x_2)$, a plane $x_2 = h$ constant, i.e. a horizontal plane.

For the pressure distribution, one obtains with the boundary condition $P = P_0$ for $x_2 = h \leadsto C = P_0 + \rho g h$

$$P = P_0 + \rho g (h - x_2) \qquad 0 \leq x_2 \leq h \qquad (6.13)$$

This relationship expresses the known hydrostatic law, according to which the pressure in a liquid at rest increases in a linear way with the depth below the free surface.

When one rewrites equation (6.13), one obtains:

$$\frac{P_0}{\rho} + gh = \frac{P}{\rho} + gx_2 = \text{constant} \qquad (6.14)$$

The laws of hydrostatics are often applicable also to fluids in moving containers when one treats these in "accelerated reference systems". The externally imposed accelerative forces are then to be introduced as inertia forces. Figure 6.3 shows as an example, a "container lorry" filled with a fluid which is at rest at the time $t < t_0$; it then increases its speed linearly for all times $t \geq t_0$, i.e. the fluid experiences a constant acceleration.

In a state of rest or in non-accelerating motion, the fluid surface in the container forms a horizontal level. When the container experiences a constant acceleration b, the fluid surface will adopt a new equilibrium position, provided that one disregards the initially occurring "swishing motions". When one now wants to compute the new position of the fluid surface, the introduction of a coordinate system x_i is recommended which is closely connected with the container. For this coordinate system, the hydrostatic equations read as follows:

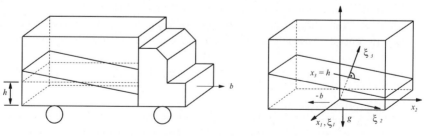

Fig. 6.3 Position of the fluid level under constant acceleration

6.1 Hydrostatics

$$\frac{\partial P}{\partial x_1} = 0; \qquad \frac{\partial P}{\partial x_2} = -\rho b; \qquad \frac{\partial P}{\partial x_3} = -\rho g \qquad (6.15)$$

From this, the general solution results:

$$\frac{\partial P}{\partial x_1} = 0 \qquad\qquad p = f_1(x_2, x_3) \qquad (6.16a)$$

$$\frac{\partial P}{\partial x_2} = -\rho b \qquad\qquad P = -\rho b x_2 + f_2(x_1, x_3) \qquad (6.16b)$$

$$\frac{\partial P}{\partial x_3} = -\rho g \qquad\qquad P = -\rho g x_3 + f_3(x_1, x_2) \qquad (6.16c)$$

By comparing the solutions, one obtains that f_1, f_2, f_3 can only be the sum of the terms obtained by partial integration plus a constant:

$$P = C - \rho(bx_2 + gx_3) \qquad (6.17)$$

Along the free surface of the fluid the pressure $P = P_0$ exists and thus the equation of the plane in which the free surface lies reads

$$x_3 = -\frac{b}{g}x_2 + \frac{1}{g\rho}(C - P_0) \qquad \text{for} \qquad -\infty < x_1 < +\infty \qquad (6.18)$$

The integration constant C is determined by the condition that the fluid volume before and after the onset of the acceleration is the same. Therefore, the same volume holds for the moving fluid and the fluid at rest:

$$C = g\rho h + P_0 \qquad (6.19)$$

Hence the equation for the plane of the free surface reads

$$x_3 = h - \frac{b}{g}x_2 \qquad \text{for} \qquad -\infty < x_1 < +\infty \qquad (6.20)$$

As the solution of the problem has to be independent of the chosen coordinate system, a coordinate system ξ_i can be introduced which is rotated against the system x_i in such a way that the following equations for the coordinate transformations hold:

$$\xi_1 = x_1 \text{ (axis of rotation)}$$

$$\xi_2 = \frac{1}{\sqrt{b^2 + g^2}}(gx_2 + bx_3) \qquad (6.21)$$

and

$$\xi_3 = \frac{1}{\sqrt{b^2 + g^2}}(-bx_2 + gx_3)$$

This is equivalent to the introduction of a resulting acceleration of the quantity $\sqrt{b^2 + g^2}$ in the direction ξ_3. Hence the basic hydrostatic equations read

$$\frac{\partial P}{\partial \xi_1} = 0; \qquad \frac{\partial P}{\partial \xi_2} = 0; \qquad \frac{\partial P}{\partial \xi_3} = -\sqrt{b^2 + g^2}\,\rho \qquad (6.22)$$

Thus $P = F(\xi_3)$ holds and $P = C - \rho\sqrt{b^2 + g^2}\,\xi_3$.

The integration constant C results from the boundary condition

$$P = P_0 + g\rho h \quad \text{for} \quad \xi_3 = 0 \qquad (6.23)$$

$$P = P_0 + \rho g \left(h - \sqrt{1 + \left(\frac{b}{g}\right)^2}\,\xi_3 \right) \qquad (6.24)$$

All further statements concerning the problem of the accelerated fluid container can also be made in the coordinate system ξ_i. Along the free surface $P = P_0$ and

$$\xi_3 = \frac{h}{\sqrt{1 + \left(\dfrac{b}{g}\right)^2}} = \text{constant} \qquad (6.25)$$

Hence this is the equation of the plane in which the free surface of the fluid in the moving container lies.

By the above treatment, it becomes clear that it is possible to employ the basic hydrostatic laws also in accelerated reference systems, provided the inertia forces that are attributable to the external motions are taken into consideration. The accelerations that occur (inertial and gravitational) are to be added in a vectorial manner to yield a total acceleration. Through this one obtains the direction and magnitude of the total acceleration. The free surface occurs perpendicular to the vector of the total acceleration.

The example given in Fig. 6.4 can also be categorized into the group of examples that can be treated by means of the basic laws of hydrostatics. This figure shows a water container which is sliding down an inclined plane with an angle of inclination α with respect to the horizontal plane. The container at rest possesses a water surface which is horizontal, as only the gravitational

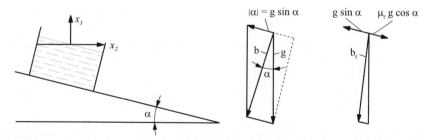

Fig. 6.4 Water container sliding down an inclined plane. Motion with and without friction

6.1 Hydrostatics

acceleration appears as inertia force per kg of fluid. When the fluid container is released and when the acceleration directed downwards is $|b| = g \sin \alpha$, the body starts to move and thus experiences an acceleration which is parallel to the inclined plane. The resulting acceleration component acting on the fluid is composed of the component directed upwards with $|b| = g \sin \alpha$ and the component directed downwards with $\mu_r g \cos \alpha$. Here μ_r is the friction coefficient which characterizes the interaction between the container bottom and the surface of the inclined plane.

When one treats first the accelerated motion occurring downwards on the inclined plane without friction, one obtains in the coordinate system, indicated in Fig. 6.4, the following set of basic hydrostatic equations:

$$\frac{\partial P}{\partial x_1} = 0 \tag{6.26a}$$

$$\frac{\partial P}{\partial x_2} = -\rho g \sin \alpha \cos \alpha \tag{6.26b}$$

$$\frac{\partial P}{\partial x_3} = -\rho g (1 - \sin^2 \alpha) \tag{6.26c}$$

The pressure distribution in the container sliding downwards in Fig. 6.4 and thus also the solution for the position of the fluid surface can be obtained by the solution of (6.26).

From $\dfrac{\partial P}{\partial x_1} = 0$, it follows on the one hand that $P = f(x_2, x_3)$ and thus the following solution holds:

$$\frac{\partial P}{\partial x_2} = -\frac{1}{2} \rho g \sin(2\alpha) \longrightarrow P = f_1(x_3) - \frac{1}{2} \rho g \sin(2\alpha) x_2 \tag{6.27a}$$

and also the solution

$$\frac{\partial P}{\partial x_3} = -\rho g \cos^2 \alpha \longrightarrow P = f_2(x_2) - \rho g (\cos^2 \alpha) x_3 \tag{6.27b}$$

By comparing the solutions, one obtains:

$$P = C - \frac{1}{2} \rho g (\sin(2\alpha) x_2 + 2(\cos^2 \alpha) x_3) \tag{6.28}$$

Along the free surface, $P = P_0$ holds and thus one obtains as the relationship for the location of the free surface:

$$x_3 = -(\tan \alpha) x_2 + \frac{1}{\rho g \cos^2 \alpha} (C - P_0) \quad \text{for} \quad -\infty < x_1 < +\infty \tag{6.29}$$

As the origin of the coordinates also lies on the free surface, $C = P_0$ follows and thus for the plane in which the free surface lies one obtains

$$x_3 = -(\tan \alpha) x_2 \quad \text{for} \quad -\infty < x_1 < +\infty \tag{6.30}$$

This equation shows that for a friction-free sliding along the inclined plane, the free surface lies parallel to the plane along which the container slides. This can also be derived from considerations of the left-hand acceleration diagram in Fig. 6.4, in which it can be seen that the resulting acceleration b is located vertically with respect to the inclined plane.

When one adds for the downward motion the occurring frictional force, one obtains the following set of basic hydrostatic equations:

$$\frac{\partial P}{\partial x_1} = 0 \tag{6.31a}$$

$$\frac{\partial P}{\partial x_2} = -\rho g (\sin \alpha - \mu_r \cos \alpha) \cos \alpha \tag{6.31b}$$

$$\frac{\partial P}{\partial x_3} = -\rho g [1 - (\sin \alpha - \mu_r \cos \alpha) \sin \alpha] \tag{6.31c}$$

Thus the solution corresponding to (6.31) reads:

$$P = C - \rho g [(\sin \alpha - \mu_r \cos \alpha) \cos \alpha] x_2 - \rho g [1 - (\sin \alpha - \mu_r \cos \alpha) \sin \alpha] x_3 \tag{6.32}$$

If one puts on the one hand $P = P_0$, for the free surface, one obtains the equation for the plane in which the free surface lies. When one takes further into consideration that the origin of the coordinates lies again on the free surface, i.e. $C = P_0$, one obtains as the final equation for the plane of the free fluid surface

$$x_3 = -\left[\frac{(\sin \alpha - \mu_r \cos \alpha) \cos \alpha}{1 - (\sin \alpha - \mu_r \cos \alpha) \sin \alpha}\right] x_2 \tag{6.33}$$

For this general case of motion with friction along the inclined plane of the fluid container, shown in Fig. 6.4, a free liquid surface appears which is less inclined with respect to the horizontal plane without surface friction. Attention has to be paid, however, to the fact that the derivations only hold when $\mu_r \leq \tan \alpha$. For $\mu_r \geq \tan \alpha$ one obtains the limiting case of a container at rest, i.e. the frictional force is higher than the forward accelerating force.

As a last example to show the employment of hydrostatic laws in accelerated reference systems, the problem presented in Fig. 6.5 is considered. It shows a rotating cylinder closed at the top and bottom, which is partly filled with a liquid. When the cylinder is at rest, the free surface of this liquid assumes a horizontal position, as the different liquid particles only experience the gravitational force as mass force. When the cylinder is put into rotation, one observes a deformation of the liquid surface which progresses until finally it becomes parabolic, as shown in Fig. 6.5.

When now on this rotating motion an additional accelerated vertical motion is superimposed, one detects that the hyperboloid can assume different shapes, depending on the magnitude of the vertical acceleration and on the direction (upwards or downwards). In the following it will be shown that the

6.1 Hydrostatics

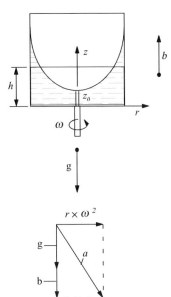

Fig. 6.5 Treatment of the "fluid flows" in a rotating vertically moved and partly filled cylinder

issue of the shape of the hyperboloid can be answered on the basis of the basic equations of hydrostatics. For this purpose, a coordinate system is chosen which is firmly coupled to the walls of the rotating and vertically accelerated cylinder and which thus experiences both rotating motion and accelerated vertical motion.

The above examples have shown that the basic hydrostatic equations are applicable, provided that no fluid motion occurs in the chosen coordinate system and that the external acceleration forces are taken into consideration as inertia forces. In the acceleration diagram in Fig. 6.5, it is shown that, for the following derivations, the horizontally occurring centrifugal acceleration $\omega^2 r$, as well as the "vertical acceleration" b, have been taken into account.

If one considers the fluid flow problem sketched in Fig. 6.5, in a coordinate system (r, φ, z), rotating with the cylinder, one finds that all fluid particles are at rest after having reached the steady final state of motion. With reference to the chosen coordinate system, the prerequisites for the employment of the basic hydrostatic equation are fulfilled, which in cylindrical coordinates adopt the following form:

$$\frac{\partial P}{\partial r} = \rho g_r; \qquad \frac{1}{r}\frac{\partial P}{\partial \varphi} = \rho g_\varphi; \qquad \frac{\partial P}{\partial z} = \rho g_z \qquad (6.34)$$

For $g_r = r\omega^2$, $g_\varphi = 0$ and $g_z = -(g + b)$ one obtains, for the problem to be treated, the following set of basic equations and their general solution:

$$\frac{\partial P}{\partial r} = \rho r \omega^2 \qquad \longrightarrow \qquad P = \frac{1}{2}\rho \omega^2 r^2 + f_1(\varphi, z) \qquad (6.35a)$$

$$\frac{1}{r}\frac{\partial P}{\partial \varphi} = 0 \qquad \longrightarrow \qquad P = f_2(r,z) \qquad (6.35b)$$

$$\frac{\partial P}{\partial z} = -\rho(g+b) \qquad \longrightarrow \qquad P = -\rho(g+b)z + f_3(r,\varphi) \qquad (6.35c)$$

Comparison of the solutions (6.35a–c) results in:

$$P = C + \frac{\rho}{2}\omega^2 r^2 - \rho(g+b)z \qquad \text{for} \qquad 0 \le \varphi \le 2\pi \qquad (6.36)$$

When one introduces on the axis $r = 0$, for the position of the parabolic apex $z = z_0$, $P = P_0$ holds at the location $r = 0$ and $z = z_0$. This yields for the integration constant:

$$C = P_0 + \rho(g+b)z_0$$

Therefore, the equation for the pressure distribution in the liquid body in Fig. 6.5 reads:

$$P = P_0 + \frac{\rho}{2}\omega^2 r^2 - \rho(g+b)(z - z_0) \qquad \text{for} \qquad 0 < \varphi < 2\pi \qquad (6.37)$$

Along the free surface of the liquid, the following holds for the pressure $P = P_0$, so that the free surface employing (6.37) can be represented as follows:

$$z = z_0 + \frac{\omega^2}{2(g+b)}r^2 \qquad \text{for} \qquad 0 \le \varphi \le 2\pi \qquad (6.38)$$

The coordinate z_0 of the introduced apex position can be determined from the condition that the liquid volume before the rotations starts, i.e. $\pi R^2 h$, has to be equal to the liquid volume which exists in rotation between the free surface of the liquid and the cylinder walls. Thus the following holds:

$$\pi R^2 h = 2\pi \int_0^R rz\,\mathrm{d}r = 2\pi \int_0^R r\left[z_0 + \frac{\omega^2}{2(g+b)}r^2\right]\mathrm{d}r \qquad (6.39)$$

and carrying out the integration yields:

$$\frac{1}{2}R^2 h = \left[\frac{1}{2}z_0 r^2 + \frac{\omega^2}{8(g+b)}r^4\right]_0^R = \frac{1}{2}R^2\left[z_0 + \frac{\omega^2}{4(g+b)}R^2\right] \qquad (6.40)$$

$$z_0 = h - \frac{\omega^2}{4(g+b)}R^2 \qquad (6.41)$$

Inserting (6.41) in (6.38) yields:

$$z = h - \frac{\omega^2}{4(g+b)}(R^2 - 2r^2) \qquad (6.42)$$

On the basis of this, the different forms of the free liquid surface can now be looked at. Some typical cases are shown in Fig. 6.6. These will be discussed

6.2 Connected Containers and Pressure-Measuring Instruments

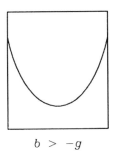
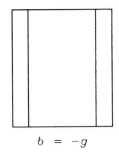
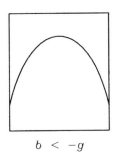

$b > -g$ $b = -g$ $b < -g$

Fig. 6.6 Examples of possible forms of the fluid surface in a rotating vertically accelerated cylinder

in the following on the basis of the above derivations and the derived final relationship. It is hoped that it will be clear to the reader how physical information can be obtained by analytical derivations employing the basic equations of fluid mechanics, e.g. in the present example the form of the free surface of the liquids in the containers can be calculated.

The positions of the liquid surface indicated in Fig. 6.6 can be stated by the indicated relationship of the relative magnitudes of b and g:

$b > -g$: When the vertical acceleration of the container takes place upwards and the resultant acceleration b points downwards, with $0 > b > -g$, the "opening" of the parabola is positive according to (6.42). The liquid touches the bottom and side areas of the container.

$b = -g$: When the vertical acceleration of the container takes place upwards with $b = -g$, the entire fluid rests at the side wall of the container.

$b < -g$: When the vertical acceleration of the container takes place downwards with $b < -g$, the "opening" of the parabola is negative according to (6.42). The fluid touches the ceiling and side areas of the container.

All this can be deduced from (6.42) and all intermediate forms, not shown in Fig. 6.6 can also be completed.

6.2 Connected Containers and Pressure-Measuring Instruments

6.2.1 Communicating Containers

In many fields of engineering one has to deal with fluid systems that are connected to one another by pipelines. Special systems are those in which

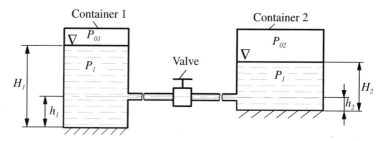

Fig. 6.7 Sketch for explanation of the pressure conditions with communicating containers

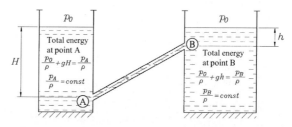

Fig. 6.8 Communicating containers with inclined communication tube

the fluid is at rest, i.e. in which the fluid does not flow. Figure 6.7 represents schematically such a system, which consists of two containers with "fluids at rest" that are connected with one another by a pipeline with a valve.

When the valve is opened, both systems can interact with one another in such a way that a flow takes place from the container with higher pressure, at the entrance of the pipe connecting the containers, with lower pressure. When the flow through the pipe has stopped, the same fluid pressure exists on both sides of the valve, i.e.

$$p_{01} + \rho_1 g (H_1 - h_1) = p_{02} + \rho_2 g (H_2 - h_2) \tag{6.43}$$

When there is the same fluid in both containers with $\rho_1 = \rho_2 = \rho$ the following relationship holds:

$$p_{02} - p_{01} = \rho g [(H_1 - H_2) - (h_2 - h_1)] \tag{6.44}$$

For the containers shown in Fig. 6.8, the pressure P_0 acts on both of the top surfaces.

Hence, introducing that the pressure over both of the free surfaces is equal, one obtains

$$p_{02} = p_{01} = p_0 \tag{6.45}$$

and thus

$$(H_1 - h_1) = (H_2 - h_2) \tag{6.46}$$

i.e. in open communicating containers filled with the same fluid, the fluid levels take the same height with respect to a horizontal plane.

6.2 Connected Containers and Pressure-Measuring Instruments

This is the basic principle according to which simple liquid level indicators which are installed outside liquid containers operate. They consist of a vertical tube connected with a container in which the liquid filled in the container can rise. The fluid level indicated in the connecting tube shows the fluid level in the container.

As a last example, open containers are considered that are connected to one another by means of an inclined tube that is directed upwards. For these containers, one finds that the fluid surfaces in both containers adopt the same level. When this final state is reached (equilibrium state), no equalizing flow takes place between the containers, although the deeper lying end of the pipe shows a higher hydrostatic pressure at the connecting point. The reasons for the fact that an equalizing flow does not occur, in spite of a higher hydrostatic pressure at the deeper lying end of the pipe were stated in Sect. 6.1. The energy considerations carried out there show that the total energies of the fluid particles are the same at both ends of the pipe and thus the basic prerequisite for the start of fluid flows is missing.

The behavior of communicating containers that are filled with fluids at rest can often be understood easily by making it clear to oneself that the pressure influence of a fluid on walls is identical at each point with the pressure influence on fluid elements which one installs instead of walls. For example, the pressure distributions in the fluid container shown in Fig. 6.9 are identical with those of the same container when components are installed to obtain two partial containers connected with one another, in the case that the fluid surfaces are kept at the same level.

Owing to the installed walls, the pressure conditions do not change in the right container as compared with the left container. The container areas installed on the left replace the pressure influence of the fluid particles acting on the walls.

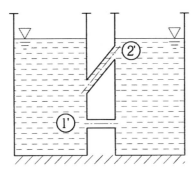

Fig. 6.9 Sketch for the consideration of the pressure influence on liquids at rest

6.2.2 Pressure-Measuring Instruments

The insights into the pressure distribution in containers gained in Sect. 6.2.1 were obtained through pressure relationships that were described for communicating systems. From the derived relationships for the pressure distribution in the containers, information could be obtain for the location of the free liquid surfaces. In return, it is possible, in the case that the established fluid levels are known, to employ the general pressure relationships, indicated in Sect. 6.1, in order to obtain information on the pressures occurring in containers.

This is illustrated in Fig. 6.10, which shows a sketch for explaining the basic principle according to which pressure measurements are carried out by communicating systems.

To measure the pressure at point A in the container to which a "U-tube manometer" is connected, the latter is filled with a measuring liquid (dotted part of the U tube) and partly also with the liquid which enters into the U tube from the container. The two liquids are assumed not to mix. For the location of the separating plane between the two fluids, the following pressure equilibrium holds:

$$p_A + \rho_A g \Delta h = p_0 + \rho_F g h \qquad (6.47)$$

For the pressure to be measured at point A, it follows that:

$$p_A = p_0 + \rho_F g h - \rho_A g \Delta h \qquad (6.48)$$

Equation (6.48) makes it clear that it is possible to determine the pressure at point A in the container by measurements of h and Δh when the fluid densities ρ_F and ρ_A are known. In Fig. 6.10, it was assumed that the pressure in the container is high compared with the ambient pressure p_0. When there is a negative pressure in the container with respect to the outside pressure, the conditions presented in Fig. 6.11 will exist for the fluid level in the U-tube manometer.

Thus, for the pressure equilibrium at the parting surface of the two fluids, the following relationship holds:

$$p_A - \rho_A g \Delta h = p_0 - \rho_F g h \qquad (6.49)$$

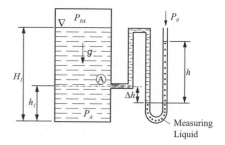

Fig. 6.10 Diagram for explaining the basic principle of pressure measurements by communicating systems

6.2 Connected Containers and Pressure-Measuring Instruments

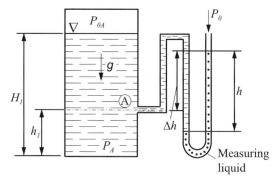

Fig. 6.11 Fluid columns in the U-tube manometer for "negative pressure" giving an upward rise of the measuring liquid

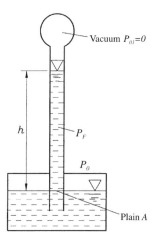

Fig. 6.12 Basic principle of barometric pressure measurements

For the pressure at point A one then obtains the following expression:

$$p_A = p_0 - \rho_F g h + \rho_A g \Delta h \tag{6.50}$$

On the basis of communicating containers, measuring devices can also be designed and employed to measure the atmospheric pressure, i.e. to carry out barometric measurements. To explain this, reference is made to the pressure measuring equipment shown in Fig. 6.12.

A system can in principle be produced as follows:

(a) A glass tube with a length of more than 1 m is filled to the top with mercury; at the lowest end of the tube a spherical extension of the tube section has been made.
(b) The glass tube, filled with mercury, is turned upside down and its lower end is displaced into a container also filled with mercury. This needs to be achieved without mercury running out of the tube, as shown in Fig. 6.12.

(c) The level of the mercury column in the glass tube over the surface of the mercury in the external container is a measure of the barometric pressure:
$$p_0 = \rho_F g h \qquad (6.51)$$
A barometer, as shown in Fig. 6.12, can be employed to verify experimentally the pressure distributions in the atmosphere stated in Sect. 6.4.1.

6.3 Free Fluid Surfaces

6.3.1 Surface Tension

In Chap. 1, it was emphasized that a special characteristic of fluids is that, in contrast to solids, they have no form of their own, but always adopt the form of the container in which they are stored. While doing this, a free surface forms and it was shown in Sect. 6.1 that this surface adopts a position which is perpendicular to the vector of the gravitational acceleration.

In this way, the fluid properties under gravitational influence were formulated which are known from phenomena of every day life. It was always assumed that the fluid at disposal possesses a total volume having the same order of magnitude as the larger container in which it is stored. The fluid properties hold only when these conditions are met. This is known from observations of small quantities of liquids which form drops when put on surfaces, as sketched in Fig. 6.13. It is seen that different shapes of drops can form, depending on which surface and which fluid for forming drops are used. More detailed considerations show, moreover, that the gas surrounding the fluid and the solid surface all have an influence on the shape of a drop. The latter is often neglected and one differentiates considerations of fluid–solid combinations with reference to their "wetting possibility", depending on whether the angle of contact established between the edge of the fluid surface and solid surface is smaller or larger than $\pi/2$.

The surface is classified as non-wetting by the fluid when
$$\gamma_{gr} > \pi/2 \qquad (6.52)$$

Fig. 6.13 (a) Shape of a drop in the case of non-wetting fluid surfaces; (b) shape of a drop in the case of wetting fluid surfaces

6.3 Free Fluid Surfaces

It holds furthermore that for

$$\gamma_{gr} < \pi/2 \qquad (6.53)$$

the surface is classified as wetting for the fluid.

Surfaces covered by a layer of fat are examples of surfaces that cannot be wetted by water. Cleaned glass surfaces are to be classified as wetting for many fluids.

The above phenomena can be explained by the fact that different "actions of forces" are experienced by fluid elements. Equivalent physical considerations can be made also by referring to the surface energy that can be attributed to free fluid surfaces. The equivalence of forces and energy considerations in mechanical systems is explained in Sect. 5.5. When a fluid element is located in a layer that is far away from a free fluid surface, it is surrounded from all sides by homogeneous fluid molecules and one can assume that the cohesion forces occurring between the molecules cancel each other. This is, however, no longer the case when one considers fluid elements in the proximity of free surfaces. As the forces exerted by gas molecules on the water particles are negligible in comparison with the interacting forces of the liquid, a particle lying at the free surface experiences an action of forces in the direction of the fluid. "Lateral forces" also act on the fluid element, which thus finds itself, in the considered free interphase surface, in a state of tension that attributes special characteristics to the free surface. It is therefore possible, for example, to deposit carefully applied flat metal components on free surfaces without the metal piece penetrating the liquid. The floating of razor blades on water surfaces is an experiment that is often presented in basic courses of physics to demonstrate this. In nature, animals like "pond skaters" make use of this particular property of the water surface in order to cross pools and ponds skillfully and quickly.

When a drop of liquid comes into contact with a firm support, attracting forces also occur in addition to the internal interacting forces. When these attracting forces are stronger than the internal forces that are typical for the fluid, we have the case of a wetting surface and water drops form as shown in Fig. 6.13b. If, however, the forces attributed to a thin layer of the free surface are stronger, we have the case of a non-wetting surface and the shapes of the drops correspond to those in Fig. 6.13a.

More detailed considerations of the processes in the proximity of the free surface of a liquid show that we have to deal with a complicated layer (with finite extension vertical to the fluid surface) from a liquid area to a gas area. It suffices, for many considerations to be made in fluid mechanics however, to introduce the surface as a layer with a thickness $\delta \to 0$. To the same are attributed the properties that comprise the complex layer between fluid and gas. The property that is of particular importance for the considerations to be carried out here is the surface tension. This surface tension can be demonstrated by immersing a strap, as shown in Fig. 6.14, in a fluid. When pulling the strap through the free surface upwards, one observes that this

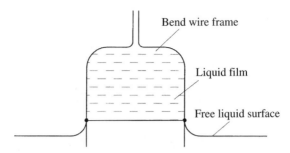

Fig. 6.14 Wire frame experiment to prove the action of forces as a consequence of surface tension

requires an action of forces which is proportional to the distance between the strap arms. The proportionality constant describing this fact is defined as the surface constant.

The surface tension thus represents a force of the free surface per unit linear length. It can also be introduced as the energy that is required to build up the tension in the liquid film in Fig. 6.14. The two definitions are identical. In both ways of looking at the energy equation, the length of the liquid film in the direction in which the wire frame is pulled can be understood as a process to introduce energy to produce free surfaces. One can also look at the force that is needed to pull the wire frame up to produce the free surface. This makes it clear that both possibilities of introduction of the surface tension, one as the action of forces per unit length and the other as the energy per unit area, are identical.

In concluding these introductory considerations, the effect of the surface tension on the areas above and below a free surface will be looked at more closely. From observations of free surfaces in the middle of large containers one can infer that the surface tension there has no influence on the fluid and the gas area lying above it, as the free surface forms vertically to the field of gravity of the earth, as stated in Fig. 6.1. From this it follows that considerations of liquids with free surfaces can be carried out far away from solid boundaries (container walls), without consideration of the wall effects.

When one considers a curved surface element, as shown in Fig. 6.15, one understands that, as a consequence of the surface tensions that occur, actions of forces are directed to the side of the surface on which the center points of the curvatures are located. The forces acting on sides AD and BC of the surface element are computed for each element ds_1 and the forces resulting from them, in the direction of the center points of the circles of curvature, are:

$$dK_1 = 2\sigma \, ds_1 \sin\beta = 2\sigma \, ds_1 \frac{ds_2}{2R_1}$$
$$dK_1 = \frac{\sigma}{R_2} ds_1 \, ds_2 = \frac{\sigma}{R_1} dO \qquad (6.54)$$

6.3 Free Fluid Surfaces

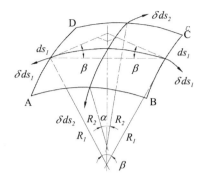

Fig. 6.15 Schematic representation of a curved surface

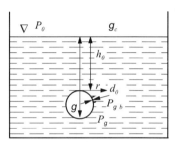

Fig. 6.16 Diagram for the consideration of pressure in bubbles

Similarly, the action of forces dK_2 is computed as

$$dK_2 = \frac{\sigma}{R_2} ds_2\, ds_1 = \frac{\sigma}{R_2} dO \qquad (6.55)$$

This shows that as a consequence of the surface tension, pressure effects occur that are directed towards the center points of the circles of curvature. This pressure effect is computed as force per unit area. As the considerations show, a differential pressure results, which is caused by the surface tension:

$$\Delta p_\sigma = \frac{dK}{dO} = \sigma \left(\frac{1}{R_1} + \frac{1}{R_2} \right), \quad \text{with } dK = dK_1 + dK_2 \qquad (6.56)$$

If a spherical surface is considered, the following relationship holds:

$$R_1 = R_2 = R \quad \longrightarrow \quad \Delta p_\sigma = \frac{2\sigma}{R} \qquad (6.57)$$

This relation means that the gas pressure in a spherical bubble is higher than the liquid pressure imposed from outside:

$$p_F + \frac{2\sigma}{R} = P_g \qquad (6.58)$$

For very small bubbles this pressure difference can be very large.

When one considers the equilibrium state of a surface element of a bubble (see Fig. 6.16), the following relationship can be written for the pressure in the upper apex:

$$p_0 + \rho_F g h_0 + \sigma \left(\frac{2}{R_0}\right) = p_{g,0} \tag{6.59}$$

For a surface element of any height, the following pressure equilibrium holds:

$$p_0 + \rho_F g (h_0 + y) + \sigma \left(\frac{1}{R_1} + \frac{1}{R_2}\right) = p_{g,0} + \rho_g g y \tag{6.60}$$

When one now forms the difference of these pressure relationships, one obtains:

$$\left(\frac{1}{R_1} + \frac{1}{R_2}\right) - \frac{2}{R_0} + \frac{1}{\sigma}(\rho_F - \rho_g) g y = 0 \tag{6.61}$$

Hence the characteristic quantity for the normalization of equation (6.61) is to be introduced:

$$a = \sqrt{\frac{2\sigma}{g(\rho_F - \rho_g)}} \tag{6.62}$$

which is known as the Laplace constant or capillary constant. It has the dimensions of length and indicates the order of magnitude when a perceptible influence of the surface tension on the surface shape of a medium exists.

- When the Laplace constant of a free surface of a liquid is comparable to the dimensions of the liquid body, an influence of the surface tension on the free surface shape is to be expected.
- In the proximity of liquid rims (container walls), an influence of the surface tension on the shape of the fluid surface is to be expected with linear dimensions that are of the order of magnitude of the Laplace constant.

6.3.2 Water Columns in Tubes and Between Plates

From the final statements in Sect. 6.3.1, consequences result for considerations of heights of liquid columns in pipes and channels of small dimensions. Considerations, have been carried out already in Sect. 6.2, but influences of the interactions between liquid, solid and gaseous phases remained unconsidered there, i.e. the influence of the surface tension was not taken into account. Considering the insights into the properties of connected containers filled with liquid gained in Sect. 6.3.1, one sees that the considerations stated for "communicating systems" in Sect. 6.2 hold only when the dimensions of the systems are larger than the Laplace constant of the fluid interface. Moreover, the considerations only hold far away from fluid rims. In the immediate proximity of the rim there exists an influence of the surface tension which was neglected in the considerations in Sect. 6.2.

6.3 Free Fluid Surfaces

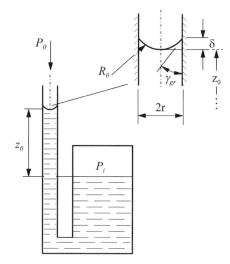

Fig. 6.17 Diagram for consideration of heights of liquid column in tubes and between plates

The processes taking place in fluid containers of small dimensions can be treated easily if one carries out a division of the properties of container walls into "wetting" ones and "non-wetting" ones. When making the considerations first for wetting walls, experiments show that for such surfaces, in small tubes and between plates with small distances forming small gaps, the fluid in the tube or between the plates assumes a height which is above the height of the surface of a connected larger container, as indicated in Fig. 6.17.

From pressure equilibrium considerations it follows that

$$\text{Pressure between plates: } p_0 - \frac{\sigma}{R_0} = p_F = p_i - \rho_F g z_0 \quad (6.63\text{a})$$

$$\text{Pressure in tubes: } p_0 - \frac{2\sigma}{R_0} = p_F = p_i - \rho_F g z_0 \quad (6.63\text{b})$$

or, written in another form,

$$\text{Height of water surface between plates: } z_0 = \frac{1}{\rho_F g}(p_i - p_0) + \frac{\sigma}{\rho_F g R_0}, \quad (6.64\text{a})$$

$$\text{Height of water surface in tubes: } z_0 = \frac{1}{\rho_F g}(p_i - p_0) + \frac{2\sigma}{\rho_F g R_0} \quad (6.64\text{b})$$

In these relationships, the radius of curvature R_0 has to be considered as an unknown before its determination, for which two possibilities exist. To simplify the corresponding derivations one can assume, with a precision that is sufficient in practice, that the surface in the rising pipe adopts the form of a partial sphere for the tube and that of a partial cylinder for the gap of the plate. The angle of contact between the fluid surface and the tube wall or plate wall has to be known from information regarding the wetting properties of the wall materials. When one defines this angle as γ_{gr}, one obtains the following relation:

$$r = R_0 \cos \gamma_{gr} \tag{6.65}$$

For the final relationships of the height z_0 of the water surface for plates and tubes, the following expressions hold:

$$\text{Plates:} \quad z_0 = \frac{1}{\rho_F g} (p_i - p_0) + \frac{\sigma}{\rho_F g r} \cos \gamma_{gr} \tag{6.66a}$$

$$\text{Tube:} \quad z_0 = \frac{1}{\rho_F g} (p_i - p_0) + \frac{2\sigma}{\rho_F g r} \cos \gamma_{gr} \tag{6.66b}$$

These final relationships now show that even in the case of pressure equality, i.e. $p_i = p_0$, the height z_0 of the free liquid surface assumes finite values if $\gamma_{gr} < \frac{\pi}{2}$ exists. This has to be considered when employing communicating systems for measurements of the height of the water level in containers and when measuring pressures.

The second possibility for computing the pressure difference below and above interphases is given by the fact that it is experimentally possible, although with greater inaccuracy, to determine the quantity δ in Fig. 6.17 by means of the following considerations:

$$r^2 + (R_0 - \delta)^2 = R_0^2 \qquad R_0 = \frac{r^2 + \delta^2}{2\delta} \tag{6.67}$$

The height z_0 of the liquid surface is calculated from this as follows:

$$z_0 = \frac{1}{\rho g} (p_i - p_0) + \frac{4\sigma\delta}{\rho_F g (r^2 + \delta^2)} \tag{6.68}$$

This proves that for $\sigma = 0$ no height difference exists and a liquid level increase due to surface effects is not to be expected in tubes or between plates. Under such conditions, for considerations of wetting of a surface, the relations derived in Sect. 6.2 hold also for small tube diameters and small gaps between plates.

In the case of non-wetting surfaces, it is observed that the fluid in the interior of a rising tube or the gap between plates does not reach the height which the fluid outside the tube (or the gap of the plate) assumes, as indicated in Fig. 6.18. Analogous to the preceding considerations for wetting fluids, it can be stated that the following relationship holds:

$$z_0 = -\frac{2\sigma}{R_0 \rho g} \tag{6.69}$$

where R_0 can be introduced again as indicated above.

6.3 Free Fluid Surfaces

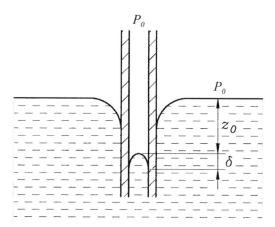

Fig. 6.18 Considerations of the height of the meniscus in tubes and between plates for non-wetting surfaces

The equation thus obtained indicates that the final relationships derived for the wetting surfaces can often be applied also to non-wetting media, if one considers the sign of γ_{gr} and δ, Thus δ needs, for example, to be introduced positively in the above relationships for wetting fluids, whereas for non-wetting surfaces δ has to be inserted negatively.

6.3.3 Bubble Formation on Nozzles

The injection of gases into fluids for chemical reactions or for an exchange of gases represents a process which is employed in many fields of process engineering. Thus bubble formation on nozzles as an introduction process for the injected gases is of interest for these applications. Moreover, the simulation of boiling processes, where steam bubbles are replaced by gas bubbles, represents another field where precise knowledge of bubble formation is required.

When gas bubbles form on nozzles during the gassing of liquids, the pressure in the interior of bubbles changes. For the theoretically conceivable static bubble formation, this is attributed to different curvatures of the bubble interface during the formation of bubbles and thus to changes of the pressure difference caused by capillary effects . Superimposed upon these are changes in pressure which have their origin in the upward movement of the bubble vertex taking place during the formation. With the dynamic formation of bubbles, additional changing pressure effects are to be expected which are essentially caused by accelerative and frictional forces.

By the term "static bubble formation", one understands the formation of bubbles under pressure conditions, which allow one to neglect the pressure effects on an element of the interface due to accelerative and frictional forces. Although in practice this kind of bubble formation exists only to a very limited extent, static bubble formation has a certain importance. As it is analytically treatable, some important basic knowledge can be gained from it which contributes to the general understanding of bubble formation, also under dynamic conditions. Furthermore, knowledge is required on static bubble formation in order to investigate the influences of the accelerative and frictional forces in the case of the dynamic formation of bubbles.

The essential basic equations of static bubble formation can be derived from the equilibrium conditions for the pressure forces at a free surface element of the bubble.

For the pressure equilibrium at an element, the following relationship of the interface boundary surface holds, in accordance with Fig. 6.19. This means that the gas pressure in the bubble p_G has to be equal to the sum of the hydrostatic pressure p_h and the capillary pressure p_σ:

$$p_G = p_\sigma + p_h = \left(\frac{1}{R_1} + \frac{1}{R_2}\right)\sigma + p_0 + \rho_F g(h_0 + y) \qquad (6.70)$$

Here the gas pressure is

$$p_G = p_{G,0} + \rho_G g y \qquad (6.71)$$

When one considers the definition for the normalized radii of curvature, employing the Laplace constant, one obtains $\bar{R}_j = R_j/a$, $\bar{r} = r/a$, $\bar{y} = y/a$. Hence the following differential equation can be derived:

$$\frac{\bar{y}''}{(1+\bar{y}'^2)^{3/2}} + \frac{\bar{y}'}{\bar{r}(1+\bar{y}'^2)^{1/2}} = 2\left(\frac{1}{\bar{R}_0} - \bar{y}\right) \qquad (6.72)$$

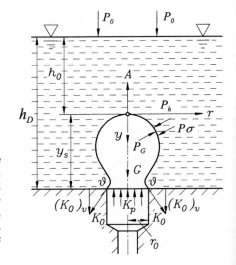

Fig. 6.19 Equilibrium of forces at the bubble interface (A, buoyancy force; G, gravity; h_D, distance of the nozzle from the fluid surface; h_0, distance of the bubble vertex from the fluid surface; K_0, surface forces; K_P, pressure forces; P_h, hydrostatic pressure; P_0, atmospheric pressure on the fluid surface)

6.3 Free Fluid Surfaces

By the substitution of

$$\bar{z} = \frac{\bar{y}'}{\sqrt{1+\bar{y}'^2}} = \sin\vartheta \tag{6.73}$$

the differential equation of second order can be replaced by a system of two differential equations of first order:

$$\frac{d}{d\bar{r}}(\bar{r}\bar{z}) = 2\bar{r}\left(\frac{1}{R_0} - \bar{y}\right) \tag{6.74}$$

$$\frac{d\bar{y}}{d\bar{r}} = \frac{\bar{z}}{\sqrt{1-\bar{z}^2}} = \tan\vartheta \tag{6.75}$$

which are used for integration. The desired bubble volume $\bar{V}$ is obtained in dimensionless form by the following partial integration:

$$\bar{V} = \pi \int_0^{\bar{y}} \bar{r}^2 \, d\bar{y} = \pi \bar{r}^2 \bar{y} - 2\pi \int_0^{\bar{y}} \bar{r}\bar{y} \, d\bar{r} \tag{6.76}$$

and, with the use of (6.74), one obtains:

$$\bar{V} = \pi\bar{r}\left[\bar{z} + \bar{r}\left(\bar{y} - \frac{1}{R_0}\right)\right] \tag{6.77}$$

If one introduces again dimensional quantities, equation (6.77) can be written as follows:

$$\frac{V}{a^3} = \pi\left(\frac{r}{a}\right)\left[\bar{z} + \frac{r}{a}\left(\frac{y}{a} - \frac{a}{R_0}\right)\right] \tag{6.78}$$

$$V = a^2\pi r\left[\bar{z} + \frac{r}{a^2}\left(y - \frac{a^2}{R_0}\right)\right] \tag{6.79}$$

With Laplace constant a and equation (6.73), the bubble volume V can be written as c'

$$V = \frac{2\sigma}{g(\rho_F - \rho_G)}\pi r \left\{\sin\vartheta + \frac{r}{2\sigma}g(\rho_F - \rho_G)\left[y - \frac{2\sigma}{g(\rho_F - \rho_G)R_0}\right]\right\} \tag{6.80}$$

Equation (6.80) represents an integral property of the differential equation system (6.74) and (6.75) which was obtained from considerations of the equilibrium of forces on a bubble surface element.

For the forces acting on a bubble as a whole (Fig. 6.19), the equilibrium condition can be written in the form:

$$Vg\rho_F - Vg\rho_G + \pi r^2\left[\frac{2\sigma}{R_0} - g(\rho_F - \rho_G)y\right] = 2\pi r\sigma\sin\vartheta \tag{6.81}$$

where the first two terms represent the buoyancy force and the weight of the bubble and the third term on the left-hand side is the pressure force on the bubble cross-section πr^2 at the height y. The surface forces are indicated on the right-hand side. Equation (6.81) should be employed in cases where the bubble volume is to be computed from the conditions of the equilibrium of forces.

For the computation of the pressure changes during bubble formation, the pressure in the bubble vertex can be expressed using equations (6.70) and (6.71) and can be expressed as:

$$p_{G,0} = \frac{2\sigma}{R_0} + p_0 + \varrho_F g (h_D - y_s) \tag{6.82}$$

The pressure at the nozzle mouth varies according to (6.71) and (6.82) as the following relationship shows:

$$p_{G,D} = \frac{2\sigma}{R_0} + p_0 + \varrho_F g h_D - g(\varrho_F - \varrho_g) y_s \tag{6.83}$$

Equation (6.83) can be written in dimensionless form:

$$\Delta \bar{p}_D = \frac{1}{\sqrt{2g\sigma(\varrho_F - \varrho_G)}} [p_{G,D} - p_0 - \varrho_F g h_D] = \frac{1}{R_0} - \bar{y}_s \tag{6.84}$$

Although the differential equation system (6.74) and (6.75) permits the computation of all bubble forms of static bubble formation and by means of equations (6.77) and (6.84), the corresponding bubble volumes and pressure differences can be obtained as important quantities of the bubbles, the problem with regard to the single steps of bubble formation is not determined. The solution of the equations only allows the computation of a one-parameter set of bubble shapes, where the vertex radius R_0 is introduced into the derivations as a parameter. It does not permit one to predict in which order the different values of the parameters are determined during bubble formation. This has to be introduced into the considerations as additional information in order to obtain a set of bubble forms that are generated in the course of the bubble formation.

Theoretically, it is now possible to choose any finite set of quantities of $R_{0,i}$ values and to compute for these the corresponding bubble forms. Of practical importance, however, is only one $R_{0,i}$ variation, which is given by most of the experimental conditions. For these conditions, the set of $R_{0,i}$ values can be selected as follows:

(a) All bubbles form above a nozzle with radius $\bar{r}_D$.
(b) $\bar{R}_{0,i} = \infty$ is the starting shape of static bubble formation; the horizontal position of the interface boundary surface above the nozzle is chosen to start computations.
(c) All further vertex radii are selected according to the condition $\bar{V}_D [\bar{R}_{0,i+1}] \geq \bar{V}_D [R_{0,i}]$.

6.3 Free Fluid Surfaces

This means that the theoretical investigations are restricted to the bubble formation which comes about through slow and continuous gas feeding through nozzles having a radius $\bar{r}_D$. Gas flows through the nozzles, and thus a decrease in the bubble volume with the selected vertex radius, as (6.77) or (6.79) would permit, are excluded from the consideration by the imposed relationship (c). The consequent application of this relationship leads to the formation of a maximum bubble volume. The same has to be considered to be the volume of the bubble at the start of the separation process, i.e. the lift-off occurs for

(d) $\bar{V}_A = (\bar{V}_D)_{max}$.

In the computations, the differential equation system (6.74) and (6.75) was solved numerically for different vertex radii, considering the indicated conditions, and thus a sequence of bubble forms was ascertained. The results of these computations are summarized in Figs. 6.20–6.25, which can be consulted in order to understand the static bubble formation on nozzles in liquids.

Figure 6.20 shows bubble forms that represent different stages of bubble formation with slow gas feeding through nozzles. The results are reproduced for $\bar{r}_D = 0.4$ and this corresponds to a nozzle radius of roughly $r_D \approx 1.6\,\text{mm}$ in the case of air bubble formation in water.

Figure 6.21 shows the change in the bubble volume during the formation of gas bubbles on nozzles of different radii $\bar{r}_D$, where the vertex radius $\bar{R}_0$ was chosen for designating the respective formation stage. From this diagram, it can be inferred that a large part of the bubble forms at an almost constant vertex radius and this is an important property for larger nozzle radii. For

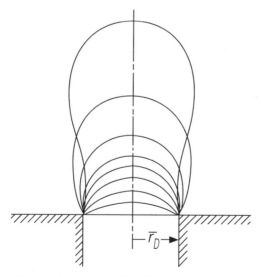

Fig. 6.20 Bubble forms of the static bubble formation for $\bar{r}_D = 0.4$ computed through integration of the equation system (6.74) and (6.75)

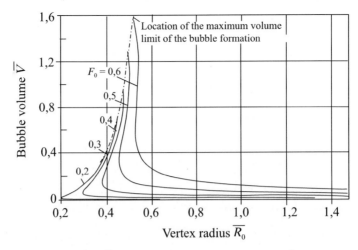

Fig. 6.21 Bubble volume $\bar{V}$ as a function of the vertex radius $\bar{R}_0$ for different nozzle radii $\bar{r}_D$

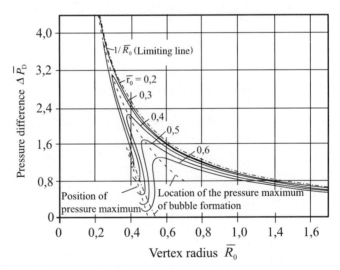

Fig. 6.22 Pressure difference $\Delta \bar{P}_D$ as a function of the vertex radius $\bar{R}_0$ for different nozzle radii $\bar{r}_D$

smaller nozzle radii, larger changes in the vertex radius are to be expected during the formation of the gas bubbles. The vertex radius $\bar{R}_0$ first decreases and then increases again before the bubble separates from the nozzle.

In Fig. 6.22, the pressure difference $\Delta \bar{P}_D$ as a function of the vertex radius $\bar{R}_0$ is represented for different nozzle radii $\bar{r}_D$. From this representation, it can be gathered that, for static bubble formation on nozzles, initially a continuous pressure increase at the nozzle mouth is necessary. After having

6.3 Free Fluid Surfaces

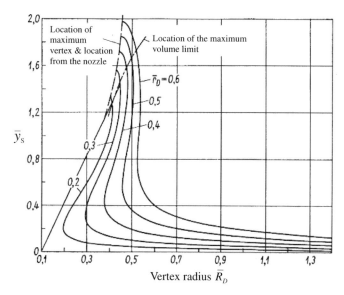

Fig. 6.23 Distance of the bubble vertex $\bar{y}_s$ from the nozzle top as a function of the vertex radius $\bar{R}_0$ for different nozzle radii $\bar{r}_D$

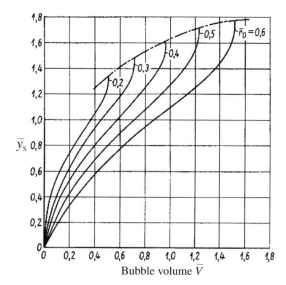

Fig. 6.24 Distance of the bubble vertex $\bar{y}_s$ from the nozzle top as a function of the nozzle volume $\bar{V}$ for different nozzle radii $\bar{r}_D$

reached a maximum, distinct for all nozzle radii, the pressure decreases again. This continuously increasing and then decreasing pressure change, which is required for static bubble formation, makes it difficult to investigate experimentally the static formation of gas bubbles on nozzles in liquids.

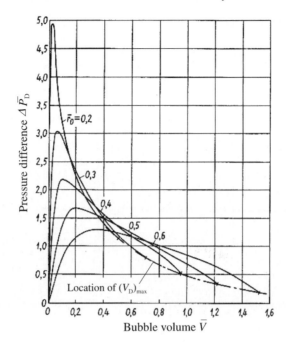

Fig. 6.25 Pressure difference $\Delta \bar{P}_D$ as a function of the nozzle volume for different nozzle radii $\bar{r}_D$

Figures 6.23 and 6.24 show the change in the vertex distance from the nozzle during bubble formation for different nozzle radii. In Fig. 6.23, the vertex radius was chosen for designating the respective stage of bubble formation, whereas in Fig. 6.24 the bubble volume was employed.

Figure 6.25 represents the dimensionless pressure difference $\Delta \bar{P}_D$ as a function of the bubble volume for different nozzle radii $\bar{r}_D$. The final points of the different curves are given by the existence of a maximum bubble volume. As can be seen from Fig. 6.22, the bubble of maximum volume above a nozzle is not identical with that having a minimum pressure. However, the latter is excluded from the possible bubble forms. This exclusion is due to the definition of static bubble formation given above. It is stated that a continuous gas flow takes place through the nozzles towards the bubble.

For a general understanding of static bubble formation, two further facts can be referred to:

(a) For the static bubble formation, a radius $\bar{r}_{D,gr} = 0.648$ exists which splits up the static bubble formation into two different domains. For larger $\bar{r}_D$ bubbles form that differ in their shape from those shown in Fig. 6.20. Theoretical investigations for nozzle radii $\bar{r}_D \geq \bar{r}_{D,gr}$ were not carried out here. They turned out to be unimportant for the introduction of

6.4 Aerostatics

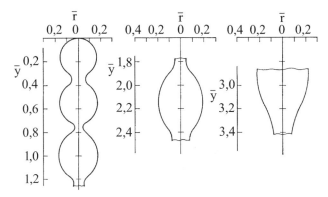

Fig. 6.26 Bubble chains $\bar{R}_0 = 1.60$ ascertained from the differential equation system (6.74) and (6.75)

the results intended here for a better understanding of the principles of bubble formation.

(b) The differential equation system (6.74) and (6.75) allows the computation of bubbles, as shown in Fig. 6.26. These bubble columns were not investigated further, as they are not in accordance with the above-stated air flow through the nozzle for the examined static bubble formation.

Whereas static nozzle formation can theoretically be understood essentially with simple mathematical means, there are considerable difficulties with similar investigations of dynamic bubble formation. This is attributable mainly to the fact that no coordinate system could be found in which dynamic bubble formation could be described as a stationary process. Moreover, for dynamic bubble formation, the pressure on an element of the interface boundary surface is dependent on the fluid motions during the bubble formation and thus is computable only by solving the non-stationary Navier–Stokes equation. This, however, is solvable only with difficulty and great computing effort using numerical methods. Solutions of this fluid are not treated in this book.

6.4 Aerostatics

6.4.1 Pressure in the Atmosphere

Aerostatics differs from hydrostatics in the treatment of the properties of the fluid. The partial differential equations for fluids at rest, derived from the general Navier–Stokes equations, are applied along with the equation of state for an ideal gas:

$$\frac{P}{\rho} = RT \qquad (6.85)$$

rather than for an ideal liquid, $\rho =$ constant.

Hence the valid partial differential equations in aerostatics read:

$$\frac{\partial P}{\partial x_j} = \rho \hat{g}_j = \frac{P}{RT}\hat{g}_j \tag{6.86}$$

or written out for $j = 1, 2, 3$:

$$\frac{\partial P}{\partial x_1} = \frac{P}{RT}\hat{g}_1, \quad \frac{\partial P}{\partial x_2} = \frac{P}{RT}\hat{g}_2, \quad \frac{\partial P}{\partial x_3} = \frac{P}{RT}\hat{g}_3 \tag{6.87}$$

This set of partial differential equations can now be employed for the computation of the pressure distribution in such gases whose thermodynamic state equation is given by (6.85). For most gases, and definitely for air under atmospheric conditions, considerations can be carried out with the help of this ideal gas equation. This will be made clear in the sections below with the help of some examples.

When one considers the atmosphere of the Earth as consisting of a compressible fluid at rest, whose thermodynamic state can be described by the ideal gas equation (6.85), with a precision sufficient for the derivations at issue, an approximate relationship between the height above the surface of the Earth and the pressure of the atmosphere at a considered height H can be derived, which in general is defined as the barometric height-pressure relationship. In particular, the relationship known as Babinets approximation equation will be derived here.

When one uses the coordinate system indicated in Fig. 6.27, in which the plane x_1, x_2 forms a horizontal area at the level of the sea surface, the partial differential equations (6.87) can be written as follows:

$$\frac{\partial P}{\partial x_1} = 0 \rightsquigarrow P = f(x_2, x_3)$$
$$\frac{\partial P}{\partial x_2} = 0 \rightsquigarrow P = f(x_1, x_3) \tag{6.88}$$

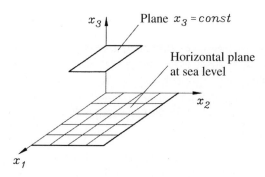

Fig. 6.27 Coordinate system for the derivation of Babinet's approximation equation for the atmospheric pressure above the Earth's surface

6.4 Aerostatics

i.e. $P = f(x_3)$ and thus the relation $\frac{\partial P}{\partial x_3} \longrightarrow \frac{dP}{dx_3}$ holds:

$$\frac{dP}{dx_3} = -\frac{P}{RT}\hat{g} \qquad (6.89)$$

The differential equation which is to be solved can be written as follows:

$$\frac{dP}{dx_3} = -\frac{P}{RT}\hat{g} \qquad \frac{dP}{P} = -\frac{1}{RT}\hat{g}dx_3 \qquad (6.90)$$

with the general solution:

$$\ln\frac{P_2}{P_1} = -\frac{1}{R}\int_{(x_3)_1}^{(x_3)_2}\left(\frac{\hat{g}}{T}\right)dx_3 \qquad (6.91)$$

The integral in the above equation can be solved only when it is known how the gravitational acceleration g changes with height and when further the temperature variation as a function of x_3 can be stated. For the gravitational acceleration $\hat{g}$, it is known that it changes with the height x_3, exactly speaking, with the square of the distance from the center of the Earth. This follows directly from Newton's law of gravitation, when the influence of the rotation of the Earth is neglected. When one designates the radius of the Earth R and when g is the gravitational acceleration at sea level, the following relation holds:

$$\hat{g} = g\frac{1}{\left(1+\frac{x_3}{R}\right)^2} \qquad (6.92)$$

Taking into consideration the linear decrease in temperature with height which often exists in the atmosphere, i.e. if one introduces:

$$T = T_0(1 - \alpha x_3) \qquad (6.93)$$

one obtains the following final relation:

$$\ln\frac{P_H}{P_0} = -\frac{g}{RT_0}\int_0^H\frac{dx_3}{(1-\alpha x_3)\left(1+\frac{x_3}{R}\right)^2} \qquad (6.94)$$

When one now imposes restrictions concerning the height above which the above integration is to take place and when one chooses the height such that the following relationship holds:

$$\alpha\,x_3 \ll 1 \qquad \frac{x_3}{R} \ll 1 \qquad (6.95)$$

one can obtain by series development of the terms in brackets:

$$\ln\frac{P_H}{P_0} = -\frac{g}{RT_0}\int_0^H(1+\alpha x_3 + \ldots)\left(1 - \frac{2x_3}{R} + \ldots\right)dx_3 \qquad (6.96)$$

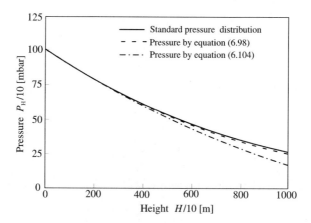

Fig. 6.28 Standard pressure distribution in the atmosphere and distributions computed on the basis of approximate equation (6.98)

or the following approximate equation by neglecting the terms of higher order:

$$\ln \frac{P_H}{P_0} = -\frac{g}{RT_0} \int_0^H \left[1 + \left(\alpha - \frac{2}{R}\right) x_3\right] dx_3 \qquad (6.97)$$

By solving this equation, the following final relationship results for the pressure distribution in the lower atmosphere:

$$P_H = P_0 \exp\left\{-\frac{gH}{RT_0}\left[1 + \frac{1}{2}\left(\alpha - \frac{2}{R}\right)H\right]\right\} \qquad (6.98)$$

As Fig. 6.28 shows, this pressure relationship describes, with good precision, the standard pressure distribution existing in the atmosphere.

The approximate equations stated by Babinet for the pressure distribution in the atmosphere can be derived from the general differential equation

$$\frac{dP}{P} = -\frac{\hat{g}}{RT} dx_3 \qquad (6.99)$$

by introducing the following hypotheses:

$$P = \frac{P_H + P_0}{2} = \text{constant} \qquad \hat{g} = g = \text{constant}$$
$$T = \frac{T_0 + T_H}{2} = \text{constant} \qquad (6.100)$$

When one introduces these simplifications, the following solution for the differential equation (6.99) results:

6.4 Aerostatics

$$\frac{2}{P_H + P_0}(P_H - P_0) = -\int_0^H \frac{g}{RT}\,dx_3 \tag{6.101}$$

or resolved:

$$\frac{P_H - P_0}{P_H + P_0} = -\frac{gH}{R(T_0 + T_H)} \tag{6.102}$$

$$H = -\frac{R}{g}T_0\left(1 + \frac{T_H}{T_0}\right)\left(\frac{P_H - P_0}{P_H + P_0}\right) \tag{6.103}$$

When one rearranges this equation to obtain P_H relative to P_0:

$$P_H = P_0 \frac{\frac{R}{g}T_0\left(1 + \frac{T_H}{T_0}\right) - H}{\frac{R}{g}T_0\left(1 + \frac{T_H}{T_0}\right) + H} \tag{6.104}$$

Figure 6.28 shows the pressure distribution by the two relationships (6.98) and (6.104) in comparison with the standard pressure distribution in the atmosphere.

6.4.2 Rotating Containers

As a further example of the employment of the basic equation of aerostatics, the problem indicated in Fig. 6.29 will be considered. By a single pressure measurement in the upper center point of the cylinder top surface and by employment of the partial differential equation of aerostatics, the entire pressure

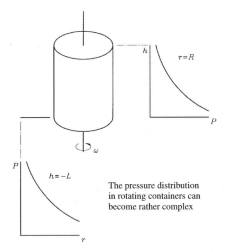

The pressure distribution in rotating containers can become rather complex

Fig. 6.29 Rotating cylinder with a compressible medium

load on the circumferential surface of a rotating cylinder, filled with a compressible fluid, can be defined. In contrast to the example treated in Sect. 6.1, here the cylinder will experience a pure rotational motion only, so that the partial differential equations of aerostatics can be written for $T =$ constant as follows:

$$\frac{\partial P}{\partial r} = \frac{P}{RT} r\omega^2 \longrightarrow \ln P = \frac{\omega^2}{2RT} r^2 + F(\varphi, z) \tag{6.105}$$

$$\frac{1}{r}\frac{\partial P}{\partial \varphi} = 0 \longrightarrow P = F(r, z) \tag{6.106}$$

$$\frac{\partial P}{\partial z} = -\frac{P}{RT} g \longrightarrow \ln P = -\frac{g}{RT} z + F(r, \varphi) \tag{6.107}$$

By comparing the solutions one obtains for the logarithm of the pressure:

$$\ln P = C + \frac{\omega^2}{2RT} r^2 - \frac{g}{RT} z \tag{6.108}$$

The pressure P_0, measured at the coordinate origin, permits the definition of the integration constant C as:

$$C = \ln P_0 \tag{6.109}$$

Introducing C into the (6.108), one obtains the pressure distribution. The pressure P is given by:

$$P = P_0 \exp\left\{ \frac{\omega^2}{2RT} r^2 - \frac{g}{RT} z \right\} \tag{6.110}$$

Along the floor space $z = -L$ the pressure distribution is computed as:

$$P(r, z = -L) = P_0 \exp\left\{ \frac{\omega^2}{2RT} r^2 + \frac{gL}{RT} \right\} \tag{6.111}$$

Along the vertical circumferential surface $r = R$:

$$P(r = R, z) = P_0 \exp\left\{ \frac{\omega^2}{2RT} R^2 - \frac{g}{RT} z \right\} \tag{6.112}$$

Pressure measurements show a very good agreement of $P(R, z)$ of equation (6.112) with the experiments.

6.4.3 Aerostatic Buoyancy

When employing aerostatic laws, mistakes are often made by using relations that are valid strictly only in hydrostatics. An example is the buoyancy of bodies, which is computed according to Fig. 6.30 as the difference in the

6.4 Aerostatics

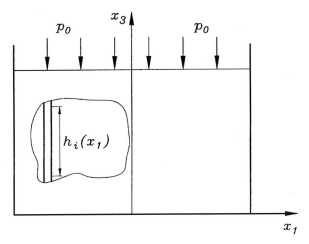

Fig. 6.30 Illustration of the hydrostatic buoyancy of immersed bodies

pressure forces on the lower and upper sides of a immersed body. The basic equations for these computations are stated in Fig. 6.30. They lead to the generally known relation which expresses that the buoyancy force experienced by a body immersed in a fluid is equal to the weight of the fluid displaced by the body. The derivations on the basis of Fig. 6.30 make it clear that the simple relationship holds only for fluids with $\rho = $ constant.

For the buoyancy force on a body element:

$$\Delta A_i = \Delta P_i \Delta F_i$$
$$\Delta P_i = \rho_F g h_i(x_1)$$

Thus for the buoyancy force:

$$A = \sum_{i1}^{N} \rho_F g h_i(x_1) \Delta F_i = \rho_F g V$$

When one wants to compute the buoyancy forces in gases, the laws of aerostatics have to be employed, i.e. the derivations shown in Fig. 6.30 are to be modified as follows.

When one applies the basic equations (6.87) to an "isothermal atmosphere", in which the axis x_3 of a rectangular Cartesian coordinate system points in the upward direction, one obtains

$$\frac{dP}{dx_3} = -\frac{P}{RT}g \qquad (6.113)$$

When one sets $g = $ constant and $T = T_0 = $ constant, one obtains

$$\frac{dP}{P} = -\frac{g}{RT_0} dx_3 \qquad (6.114)$$

or when integrated:

$$\ln \frac{P}{P_0} = -\frac{g}{RT_0} x_3 \tag{6.115}$$

In this relation, the pressure P_0 prevails at the level $x_3 = 0$. For the pressure path in an isothermal atmosphere, one can therefore write:

$$P = P_0 \exp\left\{-\frac{g}{RT_0} x_3\right\} \tag{6.116}$$

Employing now again the relation for the buoyancy occurring on a body element:

$$\Delta A_i = \Delta P_i \Delta F_i \tag{6.117}$$

and rewriting using (6.116), we obtain:

$$\Delta P_i = P_0 \exp\left\{-\frac{g x_{3i}}{RT_0}\right\} - P_0 \exp\left\{-\frac{g}{RT_0}(x_{3i} + h_i)\right\} \tag{6.118}$$

i.e.

$$\Delta P_i = P_0 \exp\left\{-\frac{g x_{3i}}{RT_0}\right\}\left[1 - \exp\left\{-\frac{g}{RT_0} h_i\right\}\right] \tag{6.119}$$

and it quickly becomes evident that, for the total buoyancy in the present case, a simple relation stated for $\rho = $ constant for the total buoyancy force does not result. The Taylor series expansion for P_i can be written as:

$$\Delta P_i = \frac{\partial P}{\partial x_3} h_i + \frac{1}{2}\frac{\partial^2 P}{\partial x_3^2} h_i^2 + \frac{1}{6}\frac{\partial^3 P}{\partial x_3^3} h_i^3 + \ldots \tag{6.120}$$

$$= -\frac{P_0 g}{RT_0}\exp\left\{-\frac{g x_3}{RT_0}\right\} h_i + \frac{P_0 g^2}{2R^2 T_0^2}\exp\left\{-\frac{g x_3}{RT_0}\right\} h_i^2 - \tag{6.121}$$

$$\Delta P_i \approx \rho(x_3) g h_i \left[1 - \frac{g h_i}{2RT_0} + \frac{1}{6}\left(\frac{g h_i}{RT_0}\right)^2 - + \ldots\right] \tag{6.122}$$

Thus one obtains for the buoyancy force:

$$A \simeq \rho(x_3) g \left[V - \sum_{i=1}^{N} \frac{g h_i^2 \Delta F_i}{2RT_0} + - \ldots\right] \tag{6.123}$$

The first term in (6.123) corresponds to the buoyancy force of hydrostatics. The second terms takes the compressibility of the considered gas into account.

6.4.4 Conditions for Aerostatics: Stability of Layers

A fluid can find itself in mechanical equilibrium (i.e. show no macroscopic motion) without being in thermal equilibrium. An equation similar to (6.7), for the condition of mechanical equilibrium, can also be given when the temperature in the fluid is not constant. Here the question arises, however, of whether an established mechanical equilibrium is stable. It appears from simple considerations that the equilibrium shows the stability only under certain conditions. When this condition is not met, the static state of the fluid is unstable and in the fluid flows of random motions occur, which tend to mix the fluid such that the condition of constant temperature is achieved. This motion is defined as free convection. The stability condition for mechanical equilibrium is, in other words, the condition for the absence of natural convection for a stratified fluid. It can be derived as follows.

We consider a fluid at a level z and with a specific volume $v(P, s)$, where P and s are the equilibrium pressure and the equilibrium entropy of the fluid at this level. We assume that the considered fluid element is displaced upwards by a small stretch ξ adiabatically. Its specific volume thus becomes $v(P', s)$, where P' is the pressure at the level $z + \xi$. For the stability of the chosen equilibrium state it is necessary (although in general not sufficient) that the force occurring here drives the element back to the starting position. The considered volume element therefore has to be heavier than the fluid in which it was displaced into the new position. The specific volume of the fluid is $v(P', s')$, where s' is the equilibrium entropy of the fluid at the level $z + \xi$. Thus we have as a stability condition

$$v(P', s') - v(P', s) > 0$$

This difference is expanded in powers of $s' - s = \frac{ds}{dz} \xi$ and we obtain

$$\left(\frac{\partial v}{\partial s}\right)_P \frac{\partial s}{\partial z} > 0 \tag{6.124}$$

In accordance with thermodynamic relations, because $T\,ds = dh - v\,dP$, the following holds, namely:

$$\left(\frac{\partial v}{\partial s}\right)_P = \frac{T}{c_p}\left(\frac{\partial v}{\partial T}\right)_P$$

c_p being the specific heat at constant pressure. The specific heat c_p is always positive like the temperature T and therefore we can transform (6.124) into

$$\left(\frac{\partial v}{\partial T}\right)_P \frac{ds}{dz} > 0 \tag{6.125}$$

Most materials expand with warming, i.e. $\left(\frac{\partial v}{\partial T}\right)_p > 0$. The condition for the absence of the convection is then reduced to the inequality:

$$\frac{ds}{dz} > 0 \tag{6.126}$$

i.e. the entropy has to increase with the level.

From this, one can easily find a condition for the temperature gradient dT/dz. From the derivation of ds/dz we can write

$$\frac{ds}{dz} = \left(\frac{\partial s}{\partial T}\right)_P \frac{dT}{dz} + \left(\frac{\partial s}{\partial p}\right)_T \frac{dp}{dz} = \frac{c_p}{T}\frac{dT}{dz} - \left(\frac{\partial V}{\partial T}\right)_P \frac{dp}{dz} > 0$$

Finally, we insert $dP/dz = -\rho g$ according to (6.7) and obtain:

$$\frac{dT}{dz} > -\frac{gT\rho}{c_p}\left(\frac{\partial V}{\partial T}\right)_P \tag{6.127}$$

Hence convection will occur when the temperature in the direction from top to bottom decreases and the temperature gradient is larger in value than $\frac{gT}{c_p}\rho\left(\frac{\partial v}{\partial T}\right)_p$.

When one investigates the equilibrium of a gas column, one can assume the gas stratification for (6.127) to be fulfilled. The limiting value $\frac{T}{V}\left(\frac{\partial V}{\partial T}\right)_p = 1$ exists. Hence the stability condition for the equilibrium simply reads

$$\frac{dT}{dz} > -\frac{g}{c_p} \tag{6.128}$$

When this stability requirement in the atmosphere is not met, the established temperature stratification is unstable and it will give way to a convective temperature-driven flow as soon as the smallest disturbances occur. The motion will eliminate the unstable stratification.

References

6.1. Schade, H., Kunz, E., Flow Mechanics, Walter de Gruyter, Berlin, 1989.
6.2. Potter, M.C., Foss, J.F., Fluid Mechanics, Wiley, New York, 1975.
6.3. Hutter, K., Fluid- and Thermodynamics, Springer, Berlin Heidelberg Newyork, 1995.
6.4. Pnueli, D., Gutfinger, C., Fluid Mechanics, Cambridge University Press, Cambridge, 1992.
6.5. Spurk, J.H., Stroemungslehre, Springer, Berlin, Heidelberg Newyork, 1996.
6.6. Höfling, O., Physik: Lehrbuch für Unterrichtund Selbststudium, Dümmler-Verlag, Bonn, 1994.
6.7. Yuan, S.W., Foundations of Fluid Mechanics, Prentice-Hall, Taiwan, 1971.
6.8. Bergmann, L., Schaefer, C.-L., Lehrbuch der Experimentalphysik I: Mechanik-, Akustik-, Waermelehre, Walter de Gruyter, Berlin, New York, 1961.

Chapter 7
Similarity Theory

7.1 Introduction

The knowledge gained from similarity theory is applied in many fields of natural and engineering science, among others in fluid mechanics. In this field, similarity considerations are often used for providing insight into the flow phenomenon and for *generalization of results*. The importance of similarity theory rests on the recognition that it is possible to gain important new insights into flows from the similarity of conditions and processes without having to seek direct solutions for posed problems. This will be known to most readers of this book from geometry, where geometrically similar figures and bodies, e.g. similar triangles, are introduced and employed extend considerations. In this way, the height of a tower or the width of a river can, for example, be found by means of similarity considerations, without directly determining the height or the width. From similarity considerations in the field of geometry, it is known that a sufficient condition for the presence of geometrically similar triangles, quadrangles, etc., is the equality of corresponding angles or equality of the ratios of the corresponding sides, i.e. it holds for similar triangles:

$$\frac{L_1}{l_1} = \frac{L_2}{l_2} = \frac{L_3}{l_3} = \text{constant}. \tag{7.1}$$

For a channel with a step:

$$\frac{L_1}{l_1} = \frac{L_2}{l_2} = \frac{D}{d} = \frac{H}{h} = \text{constant}. \tag{7.2}$$

ensures geometrical similarity (see Fig. 7.1). The geometric similarity of fluid flow boundaries is always an important assumption for extended similarity considerations in fluid mechanics, i.e. the geometric similarity, is an inherent part of the subject of this chapter.

The similarity relationships indicated above for the field of geometry can be transferred to other domains, e.g. to mathematics and physics, and can

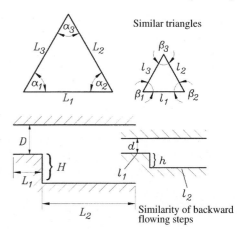

Fig. 7.1 Explanation of geometric similarity

be introduced in engineering science with the objective of achieving *indirect* solutions for problems. A similar methodology can be applied to obtain the *kinematic similarity* of flows, which is considered as given when, e.g. two fluid flows exhibit similar spatial motions. This requires, in general, similarity of the forces acting on individual fluid elements, i.e. the existence of *dynamic* similarity is a precondition for kinematic similarity of fluid motions. More detailed considerations show, in addition, that a further prerequisite for kinematic similarity of fluid motions is geometric similarity and the presence of similar boundary conditions.

When one extends the similarity considerations also to heat transport problems, it makes sense to introduce the term *thermal* similarity. Here again, similar temperature fields can be expected only when similar heat flows exist in the considered temperature fields, i.e. when the *caloric* similarity of the considered heat-transport problems is given. However, it is important here to separate the thermal and caloric similarity from one another.

In similarity considerations, strictly, only quantities with the same physical units can be included. The "dimensionless proportionality factors" of the different terms of a physical relationship computed from it by dividing all terms by one term in the equation are designated similarity numbers or dimensionless characteristic numbers of the physical problem. Physical processes of all kinds can thus be categorized as similar only when the corresponding dimensionless characteristic numbers, defining the physical problem, are equal. This requires, in addition, that geometric similarity exists and the boundary conditions for the considered problems are similar. The concept of similarity can therefore only be applied to physical processes of the same kind, i.e. to fluid flows *or* heat transport processes separately. When certain relationships apply both to flow processes and to heat transfer process, one talks of an *analogy* between the two processes.

7.1 Introduction

The above general considerations on theoretical similarity deliberations in fluid mechanics make it clear that two fundamentally different process exist concerning the solution of flow problems:

- Solutions of concrete flow problems using dimensional physical quantities permit flow systems with all parameters to be considered providing a solution of a specific flow problem. Hence a solution with all dimensions of the problem is provided.
- Generalization of solutions of flow problems, with the help of dimensionless "characteristic numbers", in order to present the solution obtained for a specific flow problem as a generally valid solution for similar flows.

In Chap. 2, physical quantities were introduced that consisted of a sign, a numerical value and a unit. It was made clear that according to the chosen system of basic units, the magnitude of a considered physical quantity can vary. When one chooses the SI system as a basis of units (m, kg, s) for some mechanical quantities, other numerical values result for the value of a considered quantity then when the basic units (cm, g, s) are introduced, as is sometimes still the case. The international system of basic units, comprising the following units, is gaining more and more acceptance, however:

- Length L (m = metre)
- Time t (s = second)
- Amount of substance N (mol = mole)
- Light intensity S (cd = candela)
- Mass M (kg = kilogram)
- Temperature θ (K = kelvin)
- Current intensity I (A = ampere)

Thanks to the introduction of an international system of units, the communication of physical processes and, above all, the comparison of analytical, numerical and experimental results have been simplified. Nevertheless, in principle, there is a multitude of different basic systems of units that can be employed for the presentation of the same physical quantity. With each system of units the considered physical quantity has another "absolute value", i.e. the physical quantity changes its numerical value according to the chosen system of units. This indicates the significance of the Table 7.1 of dimensions and units of the most important physical quantities in fluid mechanics. At the same time the Table 7.1 points out the principle differences between the *dimensions* and *units* of physical quantities.

In the presentation of fluid mechanics, in this book, transport processes by molecular motions (diffusive transport) or by flows (convective transport) are of particular interest. In order to guarantee the similarities between the transport processes, different dimensionless characteristic numbers are employed, which are subdivided below into four groups:

Table 7.1 Dimensions and units of important physical quantities of fluid mechanics

Quantity, designation	Dimensions		Units
	F, L, t, ϑ	M, L, t, ϑ	
Length	L	L	metre, m
Force	F	MLt^{-2}	newton, N
Mass	$FL^{-1}t^2$	M	kilogram, kg
Time	t	t	second, s
Temperature	Θ	Θ	kelvin, K
Velocity	Lt^{-1}	Lt^{-1}	$\mathrm{m\,s^{-1}}$
Mass flow $\dot{m}$	$FL^{-1}t$	Mt^{-1}	$\mathrm{kg\,s^{-1}}$
Volume flow	$L^3 t^{-1}$	$L^3 t^{-1}$	$\mathrm{m^3\,s^{-1}}$
Pressure, stress	FL^{-2}	$ML^{-1}t^{-2}$	pascal, $\mathrm{Pa = N\,m^{-2}}$
Work, energy	FL	$ML^2 t^{-2}$	joule, $\mathrm{J = W\,s = N\,m}$
Power	FLt^{-1}	$ML^2 t^{-3}$	watt, $\mathrm{W = N\,m\,s^{-1}}$
Density ρ	$FL^{-4}t^2$	ML^{-3}	$\mathrm{kg\,m^{-3}}$
Dynamic viscosity μ	$FL^{-2}t$	$ML^{-1}t^{-1}$	$\mathrm{Pa\,s = N\,s\,m^{-2}}$
Kinematic viscosity ν	$L^2 t^{-1}$	$L^2 t^{-1}$	$\mathrm{m^2\,s^{-1}}$
Thermal expansion coefficient β	ϑ^{-1}	ϑ^{-1}	$\mathrm{L\,K^{-1}}$
Compressibility coefficient α	$F^{-1}L^2$	$M^{-1}Lt^2$	$\mathrm{m\,s^{-2}\,kg^{-1}}$
Specific heat capacity c_p, c_v	$L^2 t^{-2} \vartheta^{-1}$	$L^2 t^{-2} \vartheta^{-1}$	$\mathrm{J\,kg^{-1}\,K^{-1}}$
Thermal conductivity λ	$Ft^{-1}\vartheta^{-1}$	$MLt^{-3}\vartheta^{-1}$	$\mathrm{W\,m^{-1}\,K^{-1}}$
Surface tension σ	FL^{-1}	Mt^{-2}	$\mathrm{N\,m^{-1}}$
Thermal diffusivity $a = \lambda/\rho c_p$	$L^2 t^{-1}$	$L^2 t^{-1}$	$\mathrm{m^2\,s^{-1}}$
Heat transfer coefficient α	$FL^{-1}t^{-1}\vartheta^{-1}$	$Mt^{-3}\vartheta^{-1}$	$\mathrm{W\,m^{-2}\,K^{-1}}$
Gas constant R	$L^2 t^{-2} \vartheta^{-1}$	$L^2 t^{-2} \vartheta^{-1}$	$\mathrm{J\,kg^{-1}\,K^{-1}}$
Entropy s	$L^2 t^{-2} \vartheta^{-1}$	$L^2 t^{-2} \vartheta^{-1}$	$\mathrm{J\,kg^{-1}\,K^{-1}}$

1. *Similarity of molecular transport processes*

    $Pr = \nu/a = (\mu c_p/\lambda) =$ Prandtl number
    $Sc = \nu/D = \mu/(\rho D) =$ Schmidt number

2. *Similarity of flow processes*

    $Re = LU/\nu =$ Reynolds number
    $Fr = U^2 g/L =$ Froude number
    $Ma = U/c =$ Mach number
    $Eu = \Delta P/\rho U^2 =$ Euler number
    $St = Lf/U =$ Strouhal number, f is the frequency
    $Gr = L^3 g \beta \rho^2 \Delta T/\mu^2 =$ Grashof number

3. *Similarity of heat transfer processes*

    $Pe = RePr = UL/a =$ Peclet number
    $E_c = U^2/c_p \Delta T =$ Eckert number

4. Similarity of integral quantities of heat and mass transfer

$Nu = \alpha L / \lambda$ = Nusselt number
$Sh = \beta L / D$ = Sherwood number

where α is introduced as heat transfer coefficient and β as mass transfer coefficient.

It is recommended to bear these groups in mind when in the following sections similarity theory and its application in fluid mechanics are treated.

7.2 Dimensionless Form of the Differential Equations

7.2.1 General Remarks

In Chap. 5 it was shown that fluid mechanics is a field whose physical basics not only exist in complete form, but can also be formulated as a complete set of partial differential equations. Closing the momentum and energy equations, molecular transport properties of fluids were included, i.e. the molecular structure of the fluids was introduced with regard to its integral effect on momentum, heat and mass transport. This resulted in a set of transport equations which are summarized below for Newtonian fluids:

- Continuity equation:

$$\frac{\partial \rho}{\partial t} + \frac{\partial (\rho U_i)}{\partial x_i} = 0 \tag{7.3}$$

- Momentum equations ($j = 1, 2, 3$):

$$\rho \left(\frac{\partial U_j}{\partial t} + U_i \frac{\partial U_j}{\partial x_i} \right) = -\frac{\partial P}{\partial x_j} + \frac{\partial}{\partial x_i} \left[\mu \left(\frac{\partial U_j}{\partial x_i} + \frac{\partial U_i}{\partial x_j} \right) - \frac{2}{3} \delta_{ij} \mu \frac{\partial U_k}{\partial x_k} \right] + \rho g_j \tag{7.4}$$

or for ρ = constant and μ = constant:

$$\rho \left(\frac{\partial U_j}{\partial t} + U_i \frac{\partial U_j}{\partial x_i} \right) = -\frac{\partial P}{\partial x_j} + \mu \frac{\partial^2 U_j}{\partial x_i^2} + \rho g_j \tag{7.5}$$

- Energy equation for ρ = constant, μ = constant and λ = constant:

$$\rho c_p \left(\frac{\partial T}{\partial t} + U_i \cdot \frac{\partial T}{\partial x_i} \right) = \lambda \frac{\partial^2 T}{\partial x_i^2} + \mu \Phi_\mu \tag{7.6}$$

with

$$\Phi_\mu = 2\left[\left(\frac{\partial U_1}{\partial x_1}\right)^2 + \left(\frac{\partial U_2}{\partial x_2}\right)^2 + \left(\frac{\partial U_3}{\partial x_3}\right)^2\right] + \left[\left(\frac{\partial U_1}{\partial x_2} + \frac{\partial U_2}{\partial x_1}\right)^2 \right.$$
$$\left. + \left(\frac{\partial U_2}{\partial x_3} + \frac{\partial U_3}{\partial x_2}\right)^2 + \left(\frac{\partial U_3}{\partial x_1} + \frac{\partial U_1}{\partial x_3}\right)^2\right] \quad (7.7)$$

- State equations:

$$\frac{P}{\rho} = RT \quad \text{(ideal gas)} \quad \text{and} \quad \rho = \text{constant} \quad \text{(ideal liquid)} \quad (7.8)$$

Through the above set of partial differential equations there exist excellent conditions for all fields of fluid mechanics for applying similarity considerations. Similarity considerations can be introduced into the differential equations in many ways, depending on the problem and the solution sought or on the solution path that one wants to take. This is explained in Sects. 7.2.2–7.2.4, by way of examples, see also refs. [7.1] to [7.4].

Looking at the above set of partial differential equations, one finds that the solution requires the specification of initial and boundary conditions. In similarity analysis, these boundary conditions must meet strict similarity requirements. Geometric similarity of boundaries is an important prerequisite for similar solutions of the above differential equations. As further considerations show, the continuity equation can be employed to formulate the conditions as to how the characteristic temporal changes of flow fields are to be coupled with the characteristic dimensions of flow geometries and characteristic fluid velocities, in order to deduce the conditions for the similarity of solutions of the basic fluid-mechanical equations. Through the momentum equations, for all three velocity components, which are formulated as equations of the forces acting on a fluid element (per unit volume), strict statements can be made on the requirements that have to exist for the dynamic similarity of flows. Moreover, dynamic similarity is a prerequisite for the existence of kinematic similarity of flow fields that is often necessary for transferring the insights on structural observations in one flow to another flow.

Finally, it should be pointed out that with the above locally formulated energy equation, all information is available to ensure that for heat transfer problems the conditions for caloric similarity are given. This then is again a prerequisite for the existence of thermal similarity. Employing also the state equations, the conditions required to transfer the temperature field of a gas flow to the temperature field of a liquid flow can then be derived. All these possibilities, to extend a solution from a special flow or temperature field, through similarity considerations, into generally valid knowledge, make similarity theory an important section of fluid mechanics. The following considerations show how one uses the insights, gained by similarity considerations, of the differential equations, in different applications in fluid mechanics.

7.2.2 Dimensionless Form of the Differential Equations

The deliberations in Sect. 7.1 show that insights into the existence of similarity could be obtained by forming fixed relationships from momentum changes per unit time and corresponding force effects. From this result, one can obtain dimensionless characteristic numbers that are employed as a basis for the desired generalizations of certain insights into fluid flows. Such knowledge can also be gained when transferring the partial differential equations summarized in Sect. 7.2.1 from their dimensional into a dimensionless form. For this purpose, one introduces "characteristic quantities" that are designated below with the subscript c. All quantities marked with asterisks (*) are dimensionless.

$$U_j = U_c U_j^*; \quad t = t_c t^*; \quad \rho = \rho_c \rho^*; \quad P = \Delta P_c P^*; \quad \tau_{ij} = \tau_c \tau^*;$$

$$g_j = g_c g^*; \quad \mu = \mu_c \mu^*; \quad \text{etc.}$$

When one inserts the dimensionless quantities into the continuity equation (7.3), one obtains:

$$\frac{\partial \rho}{\partial t} + \frac{\partial (\rho U_i)}{\partial x_i} = \frac{\rho_c}{t_c} \frac{\partial \rho^*}{\partial t^*} + \frac{\rho_c U_c}{L_c} \frac{\partial (\rho^* U_i^*)}{\partial x_i^*} = 0 \qquad (7.9)$$

or resulting:

$$\frac{L_c}{t_c U_c} \frac{\partial \rho^*}{\partial t^*} + \frac{\partial (\rho^* U_i^*)}{\partial x_i^*} = 0. \qquad (7.10)$$

$$\underbrace{St}_{} = \text{Strouhal number}$$

The above derivations make it clear that similar solutions for flow problems can follow from the continuity equation only when the Strouhal numbers for two flow problems A and B, to which the continuity equation is applied, are equal, i.e. when the following holds:

$$\left(\frac{L_c}{t_c U_c}\right)_A = \left(\frac{L_c}{t_c U_c}\right)_B. \qquad (7.11)$$

Normalizing also the momentum equations in a similar way, one obtains:

$$\rho_c \rho^* \left(\frac{U_c}{t_c} \frac{\partial U_j^*}{\partial t^*} + \frac{U_c^2}{L_c} U_i^* \cdot \frac{\partial U_j^*}{\partial x_i^*} \right) = -\frac{\Delta P_c}{L_c} \frac{\partial P^*}{\partial x_j^*} - \frac{\tau_c}{L_c} \frac{\partial \tau_{ij}^*}{\partial x_i^*} + \rho_c g_c \rho^* g_j^* \qquad (7.12)$$

or once again rewritten in dimensionless form:

$$\rho^* \left(\underbrace{\frac{L_c}{t_c U_c}}_{St} \frac{\partial U_j^*}{\partial t^*} + U_i^* \frac{\partial U_j^*}{\partial x_i^*} \right) = - \underbrace{\frac{\Delta P_c}{\rho_c U_c^2}}_{Eu} \frac{\partial P^*}{\partial x_j^*} - \underbrace{\frac{\tau_c}{\rho_c U_c^2}}_{1/Re} \frac{\partial \tau_{ij}^*}{\partial x_i^*} + \underbrace{\frac{g_c L_c}{U_c^2}}_{1/Fr} \rho^* g_j^*. $$

$$(7.13)$$

From (7.13), one can see that three new dimensionless characteristic numbers were created by normalization of the momentum equation: from the continuity equation and the momentum equation, four dimensionless characteristic numbers can thus be derived for flow problems:

$$
\begin{aligned}
St &= L_c/t_c U_c &&= \text{Strouhal number} &&= \frac{\text{local acceleration forces}}{\text{spatial acceleration forces}} \\
Eu &= \Delta P_c/\rho_c U_c^2 &&= \text{Euler number} &&= \frac{\text{pressure forces}}{\text{acceleration forces}} \\
Re &= \rho_c U_c^2/\tau_c &&= \text{Reynolds number} &&= \frac{\text{acceleration forces}}{\text{molecular momentum transport}} \\
Fr &= U_c^2/(g_c L_c) &&= \text{Froude number} &&= \frac{\text{acceleration forces}}{\text{mass forces}}.
\end{aligned}
\tag{7.14}
$$

When one wants to obtain a general solution of (7.13), it is necessary that the above-stated dimensionless characteristic numbers of the considered flow problems are equal. Naturally, this is only a necessary, but not sufficient, requirement for the existence of a uniform solution for the considered flow problems. Similarity of the boundary conditions has to exist also and this presupposes, in general, geometric similarity for the introduction of the boundary conditions.

When one includes also into the consideration the momentum transport relationship valid for a Newtonian medium:

$$
\tau_{ij} = -\mu \left(\frac{\partial U_j}{\partial x_i} + \frac{\partial U_i}{\partial x_j} \right) + \frac{2}{3} \mu \delta_{ij} \frac{\partial U_k}{\partial x_k} \tag{7.15}
$$

one obtains, by introducing dimensionless quantities,

$$
\frac{\tau_c L_c}{\mu_c U_c} \tau_{ij}^* = -\mu^* \left(\frac{\partial U_j^*}{\partial x_i^*} + \frac{\partial U_i^*}{\partial x_j^*} \right) + \frac{2}{3} \mu^* \delta_{ij} \frac{\partial U_k^*}{\partial x_k^*}. \tag{7.16}
$$

Setting now $\tau_c = \frac{\mu_c U_c}{L_c}$, one obtains $Re = \frac{U_c L_c}{\nu_c}$ with $\nu_c = \frac{\mu_c}{\rho_c}$, i.e. for Newtonian media the following holds:

$$
Re = \frac{\rho_c U_c^2}{\tau_c} = \frac{U_c L_c}{\nu_c} = \text{Reynolds number, or } L_c = \frac{\mu_c U_c}{\tau_c}. \tag{7.17}
$$

The dimensionless characteristic numbers stated above in (7.14) indicate the conditions under which the dynamic similarity between flows is given. It means that the dimensionless characteristic numbers, introduced as force relationships, adopt the same values. In the case of geometric similarity and similar boundary conditions, i.e. for similar flow problems, $Re_A = Re_B$, $St_A = St_B$, $Eu_A = Eu_B$ and $Fr_A = Fr_B$ must hold. In this way, for example, horizontal, stationary pipe flows of gases and fluids can be compared in all their flow properties when the same Reynolds numbers exist. The pressure losses occurring in the pipe flows can generally be plotted as

7.2 Dimensionless Form of the Differential Equations

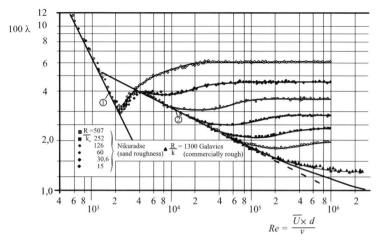

Fig. 7.2 Pressure loss factors for smooth and rough pipes for Newtonian fluids. The dimensionless presentation makes the generalization of measured values possible

$$c_f = \frac{\Delta P_c}{\frac{\rho_c}{2} U_c^2} = 2Eu = f(Re). \tag{7.18}$$

Thus c_f values that were measured for an airstream can be employed for the computation of pressure losses of flows of different Newtonian media, e.g. water, oils, etc. When one considers also the influence of the "sand roughness" k_s on the pressure losses, see Fig. 7.2, pressure losses in smooth and rough pipes results, with λ being equal to $4c_f$.

Extending the above similarity considerations also to the energy equation, one obtains

$$St\frac{\partial T^*}{\partial t^*} + U_i^* \frac{\partial T^*}{\partial x_i^*} = \underbrace{\frac{1}{RePr}}_{Pe} \frac{\partial}{\partial x_i^*}\left[\lambda^*\left(\frac{\partial T^*}{\partial x_i^*}\right)\right]$$

$$+ Ec\left(St\frac{\partial P^*}{\partial t^*} + U_i^*\frac{\partial P^*}{\partial x_i} + \frac{1}{Re}\Phi^*\right). \tag{7.19}$$

The normalization of the energy equation adds the following new dimensionless characteristic numbers to the similarity considerations valid for momentum and heat transfers; $\Delta T_c = T_c - T_\infty$ is introduced for the normalization of the temperatures:

$$Pr = \frac{\nu_c}{a_c} = \frac{\mu_c(c_p)_c}{\lambda_c} = \text{Prandtl number}$$

$$Pe = RePr = \frac{U_c L_c}{\nu_c}\frac{\nu_c}{a_c} = \frac{U_c L_c}{a_c} = \text{Peclet number} \tag{7.20}$$

$$Ec = \frac{U_c^2}{(c_p)_c(T_c - T_\infty)} = \text{Eckert number.}$$

The above dimensionless characteristic numbers can be grouped, as indicated in Sect. 7.1:

1. Similarity of molecular transport processes: $Pr, (Sc), \ldots$
2. Similarity of flow processes: $St, Re, Eu, Fr, (Gr)$
3. Similarity of energy-transport processes: $Pe, Ec, \ldots$

Finally, in the framework of the similarity of momentum and heat transport processes, one should consider the dimensionless characteristic numbers that result from experimental and theoretical investigations of *heat transfer* processes. Extensions can easily be made to mass transfer processes. These considerations result in the introduction of the Nusselt number (Nu) for the heat transfer and in the introduction of the Sherwood number (Sh) for the mass transfer. The considerations are limited at this point to the Nusselt number. The heat transfer depends on the following parameters:

$|U_\infty|; |\rho|; |\mu|; |L|; |D|$ with $\dot{Q}(U_\infty, L, D, \text{Fluid})$ results from measurements of the heat transfer:
$\dot{Q}$ = heat supply to the cylinder shown in Fig. 7.3.

Sought is a method for the reduction of the number of measurements, given by the set of parameters. Without similarity considerations, one has to carry out many measurements $\dot{Q}(U_\infty, L, D, \text{Fluid})$.

The introduction of the Nusselt number is possible in many different ways, e.g. through similarity considerations in fluid mechanics with heat transfer. The introduction favored here starts from the following considerations of the heat transfer from a cylinder (see Fig. 7.3).

In experimental investigations, the transfer of heat is measured per unit area that can be expressed by a heat transfer coefficient α and a temperature difference between the surface and surrounding fluid:

$$\frac{\dot{Q}}{L} = \alpha(\pi D)(T_D - T_\infty), \tag{7.21}$$

where $\dot{Q}$ represents the transferred heat per unit time, L the length of the cylinder, T_D the temperature of the cylinder and T_∞ the temperature of the oncoming fluid far away from the cylinder.

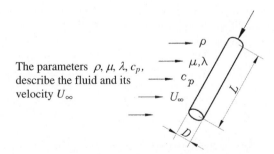

Fig. 7.3 Heat transfer considerations for a two-dimensional cylinder in a cross flow

7.2 Dimensionless Form of the Differential Equations

α is defined from experiments and without using similarity insights it would have to be determined for all diameters D, for all fluids of interest and as a function of the incoming flow velocity U_∞. A generalization of the information obtained through one series of measurements, i.e. for a single fluid, can be achieved by recognizing that the heat transfer rate per unit length $\dot{Q}/L$ supplied to the cylinder in Fig. 7.3 can be set equal to the removal of heat conduction at the cylinder surface:

$$\frac{\dot{Q}}{L} = -\int_F \lambda_D \left(\frac{\partial T}{\partial r}\right)_{r=R} \frac{\mathrm{d}F}{L}. \tag{7.22}$$

Thus the following relationship holds:

$$\alpha(\pi D)(T_D - T_\infty) = -\int_F \lambda_D \left(\frac{\partial T}{\partial r}\right)_{r=R} \frac{\mathrm{d}F}{L}. \tag{7.23}$$

When one introduces for the derivation of the dimensionless form of this relationship the following quantities:

$$T = (T_D - T_\infty)T^*; \quad r = Rr^*; \quad \frac{\mathrm{d}F}{L} = (\pi R)\,\mathrm{d}\varphi^*; \quad \varphi = 2\pi\varphi^*$$

one obtains

$$\alpha(\pi D)(T_D - T_\infty) = \frac{1}{R}\lambda_D(T_D - T_\infty)(\pi R)\left[-\int_0^1 \lambda^*\left(\frac{\partial T^*}{\partial r^*}\right)_{r^*=1} \mathrm{d}\varphi^*\right] \tag{7.24}$$

or, rewritten:

$$Nu = \frac{\alpha D}{\lambda_D} = -\int_0^1 \lambda^*\left(\frac{\partial T^*}{\partial r^*}\right)_{r^*=1} \mathrm{d}\varphi^*. \tag{7.25}$$

The heat transfer obtained from a single series of measurements and for a single fluid can be generalized by an $Nu(Re)$ representation of the data. Hilpert carried out experiments in an airstream and could thus cover an Re domain of almost 10^6, see ref. [7.5].

Equation (7.25) shows that heat transfer measurements can be generalized when the results are given in terms of the Nusselt number and not by the measured heat transfer coefficient α. For the heat transfer problem considered in Fig. 7.3, a correlation of the measuring results in the form $Nu = f(Re)$ is shown in Fig. 7.4. From a single series of measurements for an airstream, a relationship can be derived that is valid for heat transfers from cylinders, independent of the fluid and the cylinder diameter.

Analogous to the introduction of the Nusselt number, considerations on mass transfer can be carried out that lead to the introduction of the Sherwood number as a fourth group of dimensionless characteristic numbers:

4. Similarity of integral heat and mass transfer: Nu, Sh

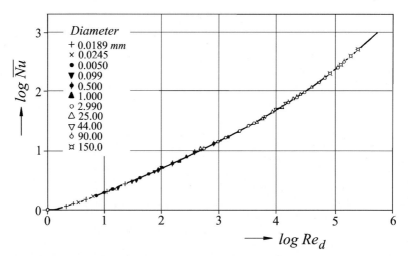

Fig. 7.4 Measured Nusselt numbers for the heat transfer into flows around a cylinder as a function of the Reynolds number

This classification of dimensionless characteristic numbers represents the basis for deriving generally valid laws in fluid mechanics from experiments, analytical and numerical computations, carried out for special fluids, individual flow velocities and a limited set of geometric parameters. The dimensionless presentation of results obtained, e.g. for one fluid and one flow geometry by variation of the flow velocity, can thus be transferred to other fluids and other similar flow geometries of different dimensions.

7.2.3 Considerations in the Presence of Geometric and Kinematic Similarities

When solving fluid mechanical problems, the question often arises of the conditions under which the results of flow investigations, obtained in a test section, can be transferred to a second flow with another fluid and for the flow in a geometrically similar test section. When considering in this context the geometrically similar test section for step flows sketched in Fig. 7.5 and postulating stationary flow conditions, geometric similarity demands that the ratio of corresponding geometric dimensions of the test sections yields a constant value

$$\frac{d_A}{D_B} = \frac{D_A}{D_B} = \frac{h_A}{h_B} = \frac{\xi_1}{x_1} = \text{constant}. \tag{7.26}$$

When one considers as characteristic linear measures of the flow, the step heights $l_A = h_A$ and $l_B = h_B$ and the corresponding characteristic flow velocities $(U_c)_A = (U_1)_A$ and $(U_c)_B = (U_1)_B$, one obtains for the dimensionless

7.2 Dimensionless Form of the Differential Equations

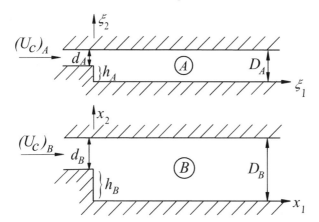

Fig. 7.5 Diagram for similarity considerations at a step flow

momentum equation, when assuming stationary flow conditions, the following form:

$$\rho^* U_i^* \frac{\partial U_j^*}{\partial x_i^*} = -\frac{\Delta P}{\rho_c U_c^2} \frac{\partial P^*}{\partial x_j^*} + \frac{\nu_c}{U_c L_c} \frac{\partial}{\partial x_i} \mu^* \frac{\partial U_j^*}{\partial x_i^*} + \frac{L_c g_c}{U_c^2} \rho^* g_j^*, \quad (7.27)$$

i.e. for the dynamic similarity it is necessary that the following dimensionless characteristic numbers are the same for both flows:

$$Eu = \frac{\Delta P}{\rho_c U_c^2}; \quad Re = \frac{U_c L_c}{\nu}; \quad \text{and } Fr = \frac{U_c^2}{L_c g_c}. \quad (7.28)$$

Thus for similar flows one has to expect

$$\frac{\Delta P_A}{\rho_A (U_1)_A^2} = \frac{\Delta P_B}{\rho_B (U_1)_B^2}; \quad \frac{(U_1)_A h_A}{\nu_A} = \frac{(U_1)_B h_B}{\nu_B}; \quad \frac{(U_1)_A^2}{h_A g} = \frac{(U_1)_B^2}{h_B g}. \quad (7.29)$$

Large Froude numbers, i.e. high $(U_1)_A^2$ or $(U_1)_B^2$ values and small values of h_A or h_B, result in a single dependency of all flow quantities on the Reynolds number. For $(Re)_A = (Re)_B$, the differential pressures obtained in a test section can be transferred from one test section to the other:

$$\text{for:} \quad \frac{\xi_1}{h_A} = \frac{x_1}{h_B} \quad \text{we have} \quad \frac{\Delta P_A}{\rho_A U_A^2} = \frac{\Delta P_B}{\rho_B U_B^2}. \quad (7.30)$$

For the velocity profile measured in the two test sections, the following holds:

$$\frac{\left[U_1\left(\frac{\xi_2}{h_A}\right)\right]_A}{(U_1)_A} = \frac{\left[U_1\left(\frac{x_2}{h_B}\right)\right]_B}{(U_1)_B} \quad \text{for} \quad \frac{\xi_1}{h_A} = \frac{x_1}{h_B}. \quad (7.31)$$

The velocity profiles obtained in the two test sections are therefore similar when they are measured in positions corresponding to similarity considerations and when the measurements are taken with equal Reynolds numbers

Fig. 7.6 Streamlines for the flow field of sudden channel expansion as a function of the Reynolds number

in test sections that are similar in the strictly geometric sense. This holds, of course, not only for the results from experimental investigations but also for results from numerical flow computations. The streamlines of the flow fields computed for different Reynolds numbers and represented in Fig. 7.6 for a sudden channel expansion are identical for all Newtonian fluids and for large and small dimensions as long as always only the corresponding Reynolds number is present.

If the flow represented in Fig. 7.6 by its streamlines for $Re = 610$ is started from rest, a temporal path of the streamlines results as depicted in Fig. 7.7. Here the time until the flow reaches its stationary final state is a multiple of the characteristic time of the flow, i.e. $t_{\text{stat}} \sim t_c \sim \frac{L_c}{U_c}$. Thus, when one wants to reach the stationary state of the flow quickly, i.e to reach a certain final Reynolds number, one has to choose small dimensions and high velocities, both being chosen to yield $Re = 610$.

To what extent a numerically computed flow is stable depends on the influence of disturbances which one can impose on the flow without changing its stationary final state. The stability of the flows for the sudden channel expansion for $Re = 70$ and 610 is shown in Fig. 7.8. Disturbances imposed at the inlet of the test section yield temporal changes of the rate of inflow and these lead to strong temporal changes of the spatial dimensions of the separation regions behind the step of the sudden channel expansion.

7.2 Dimensionless Form of the Differential Equations

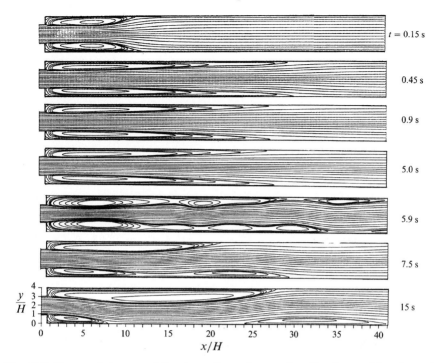

Fig. 7.7 Temporal changes of the streamlines of a transient flow in a sudden channel expansion for $Re = 610$

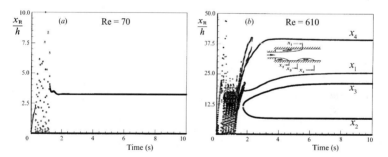

Fig. 7.8 Time variation of the length of separation regions in channel flow with sudden expansion because of imposed disturbances. The flow is stable with respect to the imposed disturbances as it reverts to its original condition after the disturbance has faded away

7.2.4 Importance of Viscous Velocity, Time and Length Scales

In fluid mechanics, there is a multitude of problems, outside the scope of civil engineering, where gravitational forces are unimportant. This is equivalent to the statement that large Froude numbers exist and that therefore the normalized momentum equation (7.13) can be written as follows:

$$\rho^* \left[\overbrace{\frac{L_c}{Z_c \cdot U_c}}^{St} \frac{\partial U_j^*}{\partial t^*} + U_i^* \frac{\partial U_j^*}{\partial x_i^*} \right] = - \overbrace{\frac{\Delta P_c}{P_c U_c^2}}^{Eu} \frac{\partial P^*}{\partial x_j^*} - \overbrace{\frac{\tau_c}{\rho_c U_c^2}}^{1/Re} \frac{\partial \tau_{ij}^*}{\partial x*_i}. \quad (7.32)$$

Completing this equation with the dimensionless molecular-dependent momentum transport from (7.16):

$$\overbrace{\frac{\tau_c L_c}{\mu_c U_c}}^{c} \tau_{ij}^* = -\mu^* \left[\frac{\partial U_j^*}{\partial x_i^*} + \frac{\partial U_i^*}{\partial x_j^*} \right] + \frac{2}{3} \mu^* \delta_{ij} \frac{\partial U_k^*}{\partial x_k^*} \quad (7.33)$$

one obtains a dependence of the solutions of flow problems on the following dimensionless characteristic numbers:

$$St = \frac{L_c}{t_c \cdot U_c} = \text{Strouhal number} \quad Eu = \frac{\Delta P_c}{\rho_c U_c^2} = \text{Euler number}$$
$$Re = \frac{\rho_c U_c^2}{\tau_c} = \text{Reynolds number} \quad C = \frac{\tau_c L_c}{\mu_c U_c} = \text{Shear number}. \quad (7.34)$$

Considering these dimensionless characteristic numbers, it becomes understandable that the basic equations of fluid mechanics deliver uniform solutions also in the case that one chooses the characteristic quantities of flow problems such that all characteristic numbers produce the value 1, i.e.

$$Re = \frac{\rho_c U_c^2}{\tau_c} = 1 \quad \rightsquigarrow \quad U_c = \sqrt{\frac{\tau_c}{\rho_c}} \quad (7.35)$$

$$Eu = \frac{\Delta P_c}{\rho_c U_c^2} = 1 \quad \rightsquigarrow \quad \Delta P_c = \tau_c. \quad (7.36)$$

With the quantities

$$St = \frac{L_c}{t_c U_c} = 1 \quad \text{and} \quad C = \frac{\tau_c L_c}{\mu_c U_c} = 1 \quad (7.37)$$

one obtains the following characteristic time and length scales:

$$t_c = \frac{\mu_c}{\tau_c} = \frac{\nu_c}{U_c^2} \quad L_c = t_c U_c = \frac{\nu_c}{U_c}. \quad (7.38)$$

These characteristic quantities suggest that fluid flows in *different* flow geometries can be grouped in a uniform representation. Thus the characteristic flow properties can be described by the following set of differential equations that are free from dimensionless characteristic numbers:

7.2 Dimensionless Form of the Differential Equations

continuity equation: $\dfrac{\partial \rho^*}{\partial t^*} + \dfrac{\partial (\rho^* U_i^*)}{\partial x_i^*} = 0$ (7.39)

momentum equations $(j = 1, 2, 3)$: $\rho^* \left(\dfrac{\partial U_j^*}{\partial t^*} + U_i^* \dfrac{\partial U_j^*}{\partial x_i^*} \right) = -\dfrac{\partial P^*}{\partial x_j^*} - \dfrac{\partial \tau_{ij}^*}{\partial x_i^*}$

(7.40)

molecular momentum transport: $\tau_{ij}^* = -\mu^* \left(\dfrac{\partial U_j^*}{\partial x_i^*} + \dfrac{\partial U_i^*}{\partial x_j^*} \right) + \dfrac{2}{3}\mu^* \delta_{ij} \dfrac{\partial U_k^*}{\partial x_k^*}.$

(7.41)

The dependence of flow results on the dimensionless characteristic numbers occurs in the solutions of (7.39)–(7.41) via the imposed boundary conditions. Thus for all flows in Fig. 7.9, a uniform representation of measuring results near the wall is achieved.

All these flows are characterised by a momentum loss to walls. This is generally nominated as τ_w, the momentum loss.

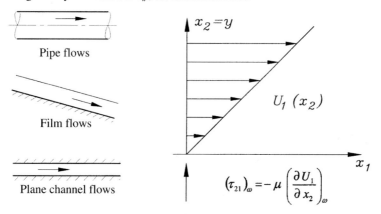

There is a general representation of data possible using normalised velocities u^+ and distances y^+ as defined below.

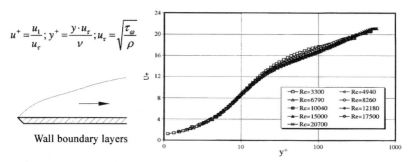

Fig. 7.9 Standardized flow profiles near the wall as a function of the standardized distance from the wall for channel, tube, film and flat plate boundary layer flows

The above representations have shown that for fluid flow measurements for Newtonian media, the occurrence of velocity gradients are controlled by the presence of characteristic viscous velocity, time and length scales. These can be stated for $\tau_c = \tau_w$, $\rho_c = \rho$ and $\mu_c = \mu$ as follows:

$$u_\tau = \sqrt{\frac{\tau_w}{\rho}}; \quad t_\tau = \frac{\nu}{u_\tau^2}; \quad \text{and} \quad L_\tau = \frac{\nu}{u_\tau}. \tag{7.42}$$

When employing these characteristic quantities, the general representation of velocity profiles in turbulent wall boundary layer flows results, indicated in Fig. 7.9. In this figure the following quantities are plotted:

$$u^+ = \frac{U_1(y)}{u_\tau} \quad y^+ = \frac{y u_\tau}{\nu},$$

where y is the distance from the wall.

The importance of the quantities in (7.42) becomes obvious when one tries to solve some typical flow problems, e.g. the one-dimensional diffusion problem which is sketched in Fig. 7.10, described by the following differential equation:

$$\rho \frac{\partial U_1}{\partial t} = -\mu \frac{\partial^2 U_1}{\partial x_2^2}. \tag{7.43}$$

After normalization, the equation can be written as

$$\left(\frac{\rho_c L_c^2}{\mu_c}\right) \frac{\partial U_1^*}{\partial t^*} = +\mu^* \frac{\partial^2 U_1^*}{\partial x_2^{*2}}. \tag{7.44}$$

From this one can deduce for the time which is required by the molecules to transport the momentum, entering at the position $x_1 = 0$, and transporting it over the distance D, the following relationship:

$$t_{\text{Diff}} = \frac{D^2}{\nu}. \tag{7.45}$$

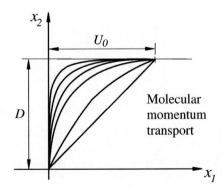

Fig. 7.10 Considerations of molecular-dependent momentum transport in fluid flows

7.2 Dimensionless Form of the Differential Equations

Forming the relationship $t_{\text{Diff}}/t_\tau = (D^2/\nu)/(u_\tau^2/\nu)$, one obtains the diffusion time expressed as a multiple of viscous time units:

$$\frac{t_{\text{Diff}}}{t_\tau} = \frac{D^2}{\nu}\frac{u_\tau^2}{\nu} = \frac{D^2}{\nu^2}\frac{U_0^2}{U_0^2}U_\tau^2 = \underbrace{\left(\frac{DU_0}{\nu}\right)^2}_{Re^2}\underbrace{\frac{u_\tau^2}{U_0^2}}_{c_f}, \tag{7.46}$$

i.e. for the momentum diffusion problem sketched in Fig. 7.10 the following holds:

$$\frac{t_{\text{Diff}}}{t_\tau} = Re^2 c_f = Re^2(2Eu). \tag{7.47}$$

This relationship makes it clear that the dimensionless characteristic numbers can be understood also as relationships for characteristic times, e.g.,

$$Re = \frac{U_0 D}{\nu} = \frac{1}{(D/U_0)}\frac{D^2}{\nu} = \frac{t_{\text{Diff}}}{t_{\text{Conv}}} = \frac{\text{Diffusion time}}{\text{Convection time}}. \tag{7.48}$$

The molecular-dependent heat transfer occurs in an analogous way to the momentum transport, as shown in Fig. 7.11. From this results the temporal formation of the temperature profile between the planes $x_2 = 0$ and $x_2 = D$:

$$\left(\frac{c_p \rho D^2}{\lambda}\right)\frac{\partial T^*}{\partial t^*} = \lambda^* \frac{\partial^2 T^*}{\partial x_2^{*2}} \tag{7.49}$$

and the diffusion time thus ensures that for the temperature propagation problem, the following holds:

$$(t_{\text{Diff}})_T = \frac{D^2}{\left(\frac{\lambda}{\rho c_p}\right)} = \frac{D^2}{a}, \tag{7.50}$$

where a is the thermal diffusion constant.

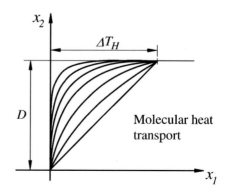

Fig. 7.11 Considerations of the molecular-dependent heat transfer in fluid flows

The above representations show that the momentum and heat diffusion processes in fluids occur in an analogous way. The ratio of the resulting diffusion times is

$$\frac{(t_{\text{Diff}})_T}{(t_{\text{Diff}})_U} = \frac{\left(\dfrac{D^2}{a}\right)}{\left(\dfrac{D^2}{\nu}\right)} = \frac{\nu}{a} = \frac{\mu c_p}{\lambda} = Pr. \tag{7.51}$$

For Prandtl numbers larger than 1, the resulting linear temperature distribution between plates at $x_1 = 0$ and $x_1 = H$ in Fig. 7.11 is formed more slowly than the analogous linear velocity distribution in Fig. 7.10. In contrast, for Prandtl numbers smaller than 1, the development of the linear temperature profile is quicker than that of the velocity profile.

7.3 Dimensional Analysis and π-Theorem

The formal tool of the similarity theory, illustrated in many ways in the preceding sections, is *dimensional analysis, if the differential equations describing the flow problem are not known*. Its special importance lies in the fact that it can also be employed when the physical relationships between quantities are not known at all. Dimensional analysis proves to be a generally valid method to recognize the information structure in the relationships between physical quantities in a precise and clear way. It starts from the fact that in quantitative natural science the descriptive quantities, as illustrated before, have dimensions and can be divided correspondingly into *basic quantities* and *derived quantities*. In the framework of fluid mechanics, one could regard length, time and mass as (dimensional) basic quantities and, e.g. area, volume, velocity, acceleration, pressure (or shear stresses), energy, density and (dynamic and kinematic) viscosity, in relation to them, as derived quantities. This classification has an important consequence that the units in which the basic quantities are measured can be chosen independently, and those of the dependent quantities are determined by this choice. Thus, with the units metre (m), second (s) and kilogram (kg) for the basic quantities, the derived quantities are:

Area:	m^2
Volume:	m^3
Velocity:	$m\,s^{-1}$
Acceleration:	$m\,s^{-2}$
Pressure (or shear stress):	$kg\,m^{-1}\,s^{-2}$
Energy:	$kg\,m^2\,s^{-2}$
Density:	$kg\,m^{-3}$
Dynamic viscosity:	$kg\,m^{-1}\,s^{-1}$
Kinematic viscosity:	$m^{-2}\,s^{-1}$

7.3 Dimensional Analysis and π-Theorem

In particular the rule then follows that *a modification* of the units of the basic quantities entails also *a modification* of the units of the derived quantities. This principle determines formally the dimensions of the derived quantities from the dimensions length (L), time (t) and mass (M) of the basic quantities. The dimensions of the above-derived quantities are:

Area:	L^2
Volume:	L^3
Velocity:	Lt^{-1}
Acceleration:	Lt^{-2}
Pressure (or shear stress):	$ML^{-1}t^{-2}$
Energy:	ML^2t^{-2}
Density:	ML^{-3}
Dynamic viscosity:	$ML^{-1}t^{-1}$
Kinematic viscosity:	L^2t^{-1}

Each physical quantity is characterized quantitatively by its unit and the numerical value related to this unit. When one modifies the unit by a factor λ, the numerical value changes by the inverse factor λ^{-1}.

The interdependent relationships between physical quantities, shown many times in examples, relate to their numerical values. *As a generally valid statement, based on the dimensional analysis*, the interdependent relationships between physical quantities are dimensionally homogeneous, i.e. they are valid independent of the choice of the units. This rule can also be expressed in the following way: *The relationships are invariant towards all changes of units, i.e. changes of scales of the basic quantities, although the quantities appearing individually in them possess units, i.e. scales.*

The overall consequence of this statement becomes clear by a mathematical observation: the set of all modifications of scale of the basic quantities meets the conditions, not described here in detail, of the (mathematical) group concept. The latter is often associated with the concept of symmetry: the elements of the group are operations on a certain object which do not change this object. Just as the reflection of a circle along one of its diameters leaves the circle unchanged (invariant) and thus is a symmetry operation of the circle; all scale transformations of the physical basic quantities can be understood as symmetry operations of these relationships, as they do not change the interdependent relationships. The formal objective of the dimensional analysis is to work out these circumstances and their consequences. The consequence can be stated as follows:

The scale-invariant relationships between scale-possessing physical quantities can be represented in the form of relationships between scale-invariant quantities.

The direct objective of dimensional analysis is to develop the methodology for determining from a given relation the number and form of the scale-invariant quantities, the so-called **characteristic numbers,** to which this relation can be attributed. *This objective is summarized in the π-theorem.*

A deeper reason for the occupation with dimensional analysis, which in its nucleus is represented in the π-theorem, lies in the benefit to be gained from it. **The practical benefit of dimensional analysis for fluid mechanics lies in the possibility of scale transfer.** Dimensional analysis has its origin in the variety of (passive) considerations on the same physical situation, arising from the choice of the scales concerning the units and the numerical values of the basic quantities. However, owing to dimensional homogeneity only, the numerical values of the physical quantities enter into the physical interdependent relationships, and their scaling can be separated from the units of these quantities. *Dimensional analysis therefore can be considered also as scaling of the numerical values of fixed units, i.e. as an instrument of (active) scale transfer.*

Dimensional analysis assures, however, that the physical relationships are always attributed to relationships that comprise dimensionless quantities (so-called characteristic numbers). When a physical relation is described by a differential equation, the method illustrated in Sect. 7.2 can be applied and by normalization of the equation a set of characteristic numbers can be determined. When this form of equation of the physical relation does not exist, one has to make use of the π-theorem in order to determine a set of characteristic numbers describing the physical problem. This theorem makes a statement on the relevant number of characteristic numbers: it is equal to the number of variables minus the maximum number of variables with which no dimensionless characteristic number can be formed. The theorem also gives a recipe for the construction of the characteristic numbers. For both aspects the so-called dimensional matrix is employed, which can be formed from the quantities of the problem. Expressed by this matrix:

The number of the dimensionless characteristic numbers of a physical problem, for which a complete set of dimensional quantities is available, is equal to the total number of the dimensional quantities minus the rank of the dimensional matrix.

The setting up of the dimensional matrix is shown below for some typical examples, starting from the following (mechanical) basic quantities:

$$M = \text{mass (kg)}; \quad L = \text{length (m)}; \quad t = \text{time (s)}$$

The chosen units are stated in parentheses. Each mechanical quantity can now be traced in its dimension to the above basic quantities mass, length and time, so the following holds:

$$[Q] = M^{\alpha_1}, L^{\alpha_2}, t^{\alpha_3}. \tag{7.52}$$

When a mechanical problem depends on the quantities $Q_1, Q_2, Q_3, \ldots, Q_h, \ldots, Q_{n-1}, Q_n$, it holds that

$$[Q_k] = M^{\alpha_{1k}} L^{\alpha_{2k}} t^{\alpha_{3k}}. \tag{7.53}$$

7.3 Dimensional Analysis and π-Theorem

Example 1: Fluid Flowing out of a Container

For this physical problem, the following dimensional matrix can be given:

	Q_1	Q_2	Q_3	Q_k	Q_{n-2}	Q_{n-1}	Q_n
M	α_{11}	α_{12}	α_{13}	α_{1k}	$\alpha_{1(n-2)}$	$\alpha_{1(n-1)}$	α_{1n}
L	α_{21}	α_{22}	α_{23}	α_{2k}	$\alpha_{2(n-2)}$	$\alpha_{2(n-1)}$	α_{2n}
T	α_{31}	α_{32}	α_{33}	α_{3k}	$\alpha_{3(n-2)}$	$\alpha_{3(n-1)}$	α_{3n}

Matrix with rank $r = 3$, and n influencing parameters, yield the quantity of π-numbers: $\pi = (n - r)$

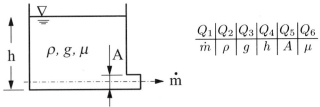

Q_1	Q_2	Q_3	Q_4	Q_5	Q_6
$\dot{m}$	ρ	g	h	A	μ

$n = 6$ influencing parameters, i.e. $\dot{m} = f(\rho, g, h, A, \mu)$.

With the above determination of the quantities relevant for a fluid flowing out of a container, the following dimensional matrix can be set up:

	$\dot{m}$	A	g	ρ	h	μ
M	1	0	0	1	0	1
L	0	2	1	-3	1	-1
T	-1	0	-2	0	0	-1

From the below-stated considerations with rank $r = 3$, three dimensionless characteristic numbers result, π_1, π_2 and π_3.

When one chooses as a first determinant variable $\dot{m}$, one obtains

$$[\dot{m}][h]^\alpha [\mu]^\beta = (MT^{-1})(L^\alpha)(M^\beta L^{-\beta} T^{-\beta}) = M^0 L^0 T^0, \quad (7.54)$$

i.e. $(1 + \beta) = 0; (-1 - \beta) = 0; \alpha - \beta = 0$

or $\beta = -1$ and $\alpha = \beta \quad \leadsto \quad$ and thus $\quad \pi_1 = \dfrac{\dot{m}}{h\mu}. \quad (7.55)$

where π_1 is the Reynolds number.

Choosing as a second determinant variable A, one obtains

$$[A][h]^\alpha = (L^2)(h)^\alpha = M^0 L^0 T^0, \quad (7.56)$$

i.e. $(2 + \alpha) = 0 \quad \leadsto \quad \alpha = -2 \quad \leadsto \quad$ and thus $\quad \pi_2 = \dfrac{A}{h^2}, \quad (7.57)$

where π_2 is the geometric similarity number.

When one chooses as a third determinant variable g, it can be stated that

$$[g][h]^\alpha [\mu]^\beta [\rho]^\gamma = 1 , \quad (7.58)$$

i.e. $(\beta + \gamma) = 0;\ 1 + \alpha - \beta - 3\gamma = 0;\ -2 - \gamma = 0$.

Thus one obtains $\alpha = 3;\ \beta = -2;\ \gamma = 2$, and therefore $\pi_3 = \dfrac{gh^3}{\nu^2}$ (7.59)

$$\pi_1 = \frac{\dot m}{h\mu} = [\frac{kg}{s}]\left[\frac{m\cdot s}{kg}\right]\left[\frac{1}{m}\right] = Re$$
$$(\pi_1 \text{ is the Reynolds number of the problem}) \quad (7.60)$$

$$\pi_2 = \frac{A}{h^2} = [m^2]\left[\frac{1}{m^2}\right] \quad (7.61)$$

$$\pi_3 = \frac{gh^3}{\nu^2} \qquad \pi_1 = f(\pi_2, \pi_3). \quad (7.62)$$

Hence a general representation of measured results can be achieved for $\dot m = f(\rho,\ g,\ h,\ A,\ \mu)$, such that one applies π_1 and chooses π_2, π_3 as parameters.

Example 2: Flow Through Rough Pipes

This is an example of high relevance to engineering:

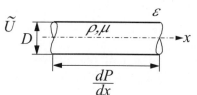

number of the variables

Q_1	Q_2	Q_3	Q_4	Q_5	Q_6
$\tilde U$	D	ρ	μ	ϵ	dP/dx

$\epsilon =$ roughness of the pipe

The parameters relevant for the representation of the problem of flow through rough pipes were determined above. Thus the dimensional matrix can be determined as follows:

	$\tilde U$	ϵ	dP/dx	D	ρ	μ
M	0	0	1	0	1	1
L	1	1	−2	1	−3	−1
T	−1	0	−2	0	0	−1

$r = 3$ and $n = 6$

When choosing $\tilde U$ as a first determinant variable, one obtains

$$[\tilde U][D]^\alpha [\rho]^\beta [\mu]^\gamma = (LT^{-1})(L^\alpha)(M^\beta L^{-3\beta})(M^\gamma L^{-\gamma}T^{-\gamma}) = M^0 L^0 T^0 \quad (7.63)$$

i.e. $(\beta + \gamma) = 0;\ (1 + \alpha - 3\beta - \gamma) = 0;\ (-1 - \gamma) = 0$.

7.3 Dimensional Analysis and π-Theorem

Thus one obtains: $\gamma = -1$; $\beta = 1$; $\alpha = 1$ and therefore $\pi_1 = \dfrac{UD\rho}{\mu}$. (7.64)

When choosing $\mathrm{d}P/\mathrm{d}x$ as a second determinant constant, it can be stated that

$$\left[\frac{\mathrm{d}P}{\mathrm{d}x}\right][D]^\alpha [\rho]^\beta [\mu]^\gamma = (MT^{-2}L^{-1})(L^\alpha)(M^\beta L^{-3\beta})(M^\gamma L^{-\gamma}T^{-\gamma}),\quad (7.65)$$

i.e. $1 + \beta + \gamma = 0$; $-2 + \alpha - 3\beta - \gamma = 0$; $-2 - \gamma = 0$.

Thus it can be computed that $\alpha = 3$; $\beta = 1$; $\gamma = -2$; $\rightsquigarrow$

$$\pi_2 = \frac{(\mathrm{d}P/\mathrm{d}x)D^3 \rho}{\mu^2}.\quad (7.66)$$

When one chooses ϵ as a third determinant variable, it yields

$$[D]^\alpha [\rho]^\beta [\mu]^\gamma [\epsilon] = (L^\alpha)(M^\beta L^{-3\beta})(M^\gamma L^{-\gamma} T^{-\gamma})(L) = 0,\quad (7.67)$$

i.e. it holds that $(\alpha - 3\beta - \gamma + 1) = 0$; $\beta + \gamma = 0$; $(-\gamma) = 0$.

Thus it results that $\gamma = 0$; $\beta = 0$; $\alpha = -1$; $\rightsquigarrow$ $\pi_3 = \dfrac{\epsilon}{D}$ (7.68)

$$\pi_1 = \frac{U \cdot D\rho}{\mu} = \left[\frac{m}{s}\right][m]\left[\frac{kg}{m^3}\right]\left[\frac{ms}{kg}\right] = Re$$
(π_1 is the Reynolds number of the problem) (7.69)

$$\pi_2 = \frac{D^3\left(\frac{\mathrm{d}P}{\mathrm{d}x}\right)\rho}{\mu^2} = \text{pressure-drop number} \quad (7.70)$$

$$\pi_3 = \frac{\epsilon}{D} = [m]\left[\frac{1}{m}\right] = \text{"relative roughness"} \quad \pi_2 = f(\pi_1; \pi_3). \quad (7.71)$$

However, instead of the characteristic number π_2, the product

$$\frac{\pi_2}{\pi_3^2} = \frac{D^3 \rho \frac{\mathrm{d}P}{\mathrm{d}x}}{\mu^2} \cdot \frac{\mu^2}{D^2 \tilde{U}^2 \rho^2} = \frac{D \frac{\mathrm{d}P}{\mathrm{d}x}}{\rho U^2}$$

can be used, as it was presented in Fig. 7.2.

Example 3: Pumping Capacity for an Incompressible Fluid

Q_1	Q_2	Q_3	Q_4	Q_5	Q_6	Q_7	Q_8
P_E	η	gH	ρ	ω	D	$\dot{V}$	μ

As the electrical pumping capacity is of special interest, it was also included in the list of the relevant parameters. Thus the dimensional matrix can be set up as follows:

	P_E	η	g_H	ρ	ω	D	$\dot{V}$	μ
M	1	0	0	1	0	0	0	1
L	2	0	2	−3	0	1	3	−1
T	−3	0	−2	0	−1	0	−1	−1

$r = 3$ and $n = 8$

Consequent use of the methods indicated in Examples 1 and 2 leads to the following characteristic numbers:

$$\pi_1 = \frac{P_E}{\rho \omega^3 D^5}; \quad \pi_2 = \eta; \quad \pi_3 = \frac{g_H}{\omega^2 D^2}; \quad \pi_4 = \frac{\dot{V}}{\omega D^3}; \quad \pi_5 = \frac{\rho \omega D^2}{\mu}. \quad (7.72)$$

The quantities to be ascertained as important for a pump, such as the electrical power capacity P_E, the efficiency factor η and the pump head g_H, can be represented via the following characteristic numbers:

$$\frac{P_E}{\rho \omega^3 D^5} = f_1\left(\frac{\dot{V}}{\omega D^3}; \frac{\rho \omega D^2}{\mu}\right) = \pi_1 \quad (7.73)$$

$$\eta = f_2\left(\frac{\dot{V}}{\omega D^3}; \frac{\rho \omega D^2}{\mu}\right) = \pi_2 \quad (7.74)$$

$$\frac{g_H}{\omega^2 D^2} = \left(\frac{\dot{V}}{\omega D^3}; \frac{\rho \omega D^2}{\mu}\right) = \pi_3. \quad (7.75)$$

These characteristic numbers are dependent on the remaining characteristic numbers π_4 and π_5. *From experiments*, it is known, however, that the dependence of the standardized pump properties does not depend on $\pi_5 = (\rho \omega D^2)/\mu$, or that the dependence is very small. Hence it can definitively be stated that

$$\frac{P_E}{\rho \omega^3 D^5} = f_1\left(\frac{\dot{V}}{\omega D^3}\right); \quad \eta = f_2\left(\frac{\dot{V}}{\omega D^3}\right); \quad \frac{g_H}{\omega^2 D^2} = f_3\left(\frac{\dot{V}}{\omega D^3}\right). \quad (7.76)$$

The experimental results of pump investigations are generally also plotted as indicated above, in diagrams with $\dot{V}/(\omega D^3)$ as abscissa.

References

7.1. Hutter, K., Fluid- und Thermodynamik: Eine Einführung, Springer, Berlin Heidelberg, 1995
7.2. Spurk, J.H., Dimensionsanalyse in der Strömungslehre, Springer, Berlin Heidelberg, 1992
7.3. Zierep, J., Ähnlichkeitsgesetze und Modellregeln der Strömungslehre, Braun, Karlsruhe, 3. überarb. Auflage, 1991
7.4. Zlokarnik, M., Dimensional Analysis and Scale-Up in Chemical Engineering, Springer, Berlin Heidelberg, 1991
7.5. Hilpert, R., Wärmeübergang von geheizten Drähten und Rohren in Luftstorm. Forsch. G. Ing, **4**(5), pp 215–224, 1933

Chapter 8
Integral Forms of the Basic Equations

In Chap. 5, the basic equations of fluid mechanics were derived in a form valid for all flow problems of Newtonian fluids. In order to obtain this generally valid form of the equations, they were formulated as differential equations for field quantities. They represent local formulations of mass, momentum and energy conservations. Applying these equations to special flow problems, it is advantageous and often essential to derive and employ the *integral forms* of these equations. These are derived in this chapter from the basic differential equations stated in Chap. 5 for a point in space, i.e. they are locally formulated and are valid per unit volume. The integral form of these equations is derived by integration over a pre-defined control volume. In the preceding chapters these derivations take place separately for the continuity equation, the momentum equation in direction j and the mechanical and caloric forms of the energy equation. In the following sections exemplary applications are described to make it clear how the applications of the derived integral forms of the fluid-mechanical basic equations take place. It will thus be shown how it is possible to solve fluid-mechanical problems in a somewhat engineering manner. As in this book only an introduction to the solution of problems is given, simplified assumptions are made in the course of the solutions. Attention is drawn to these simplifications in order to ensure that the reader is aware of the limits of the validity of the derived results. On the basis of the exemplary applications, independent solutions of more extensive and complicated problems should be possible by readers of this book. Depending on the problem, the integral form of the momentum equation or the mechanical energy equation can be employed, since the latter results from the former, as shown in Chap. 5.

8.1 Integral Form of the Continuity Equation

In Sect. 5.2, the continuity equation, i.e. the mass-conservation equation in the local formulation expressed in field variables, see equation (5.17), was stated as follows:

$$\frac{\partial \rho}{\partial t} + \frac{\partial (\rho U_i)}{\partial x_i} = 0. \tag{8.1}$$

Applying to (8.1) the integral operator $\int_{V_K}() \, dV$, i.e. integrating this equation over a given control volume $V = V_K$, one obtains

$$\int_{V_K} \left(\frac{\partial \rho}{\partial t}\right) dV + \int_{V_K} \left(\frac{\partial (\rho U_i)}{\partial x_i}\right) dV = 0, \tag{8.2}$$

where V_K is an arbitrary control volume which is to be selected for the solution of flow problems in such a way that a simple solution path for each solved problem can be found.

Considering that the integration applied to $(\partial \rho / \partial t)$ and the partial differentiation carried out for ρ can be done in any sequence, one obtains

$$\frac{\partial}{\partial t}\left(\int_{V_K} \rho \, dV\right) + \int_{V_K} \left(\frac{\partial (\rho U_i)}{\partial x_i}\right) dV = 0. \tag{8.3}$$

Applying now, to the second term of the above equation, Gauss's integral theorem (see Sect. 2.9), the following equation results:

$$\frac{\partial}{\partial t}\left(\int_{V_K} \rho \, dV\right) + \int_{O_K} \rho U_i \, dA_i = 0. \tag{8.4}$$

Here the second integral is to be carried out over the entire outer surface of the control volume, where the direction of dF_i is considered positive from the inside of the volume to its outside.

In (8.2), the following consideration was applied:

$$\int_{V_K} \frac{\partial (\rho U_i)}{\partial x_i} \, dV \iff \int_{O_K} (\rho U_i) \, dA_i$$

- In this relationship the surface vector dF_i represents a directed quantity, i.e. it contains the normal vector $\boldsymbol{n}$ of the surface element with its absolute value $|dA_i|$.
- Because of the double index we have a scalar product of the velocity vector U_i with the surface vector dA_i.

In (8.4), the resulting integrals have the following meaning:

$$M = \int_{V_K} \rho \, dV = \text{total mass in the control volume},$$

8.1 Integral Form of the Continuity Equation

$$\dot{m}_{\text{out}} - \dot{m}_{\text{in}} = \int_{O_K} \rho U_i \, \mathrm{d}A_i = \frac{\text{difference of the mass outflows and inflows}}{\text{over the surface of the control volume}}.$$

(8.5)

Thus the integral form of the continuity equation yields:

$$\frac{\partial M}{\partial t} = \dot{m}_{\text{in}} - \dot{m}_{\text{out}}.$$

(8.6)

For the volume flow through a surface with the velocity component U_i normal to this surface, one obtains, because of the above sign convention for the surface vector $\mathrm{d}A_i$ (outer normal to the surface), for the inflow in the control volume or the outflow from it:

$$\dot{V}_{\text{in}} = -\int_F |U_i| |\mathrm{d}A_i| \qquad \dot{V}_{\text{out}} = +\int_F |U_i| |\mathrm{d}A_i|.$$

(8.7)

The surface-averaged flow velocity can therefore be computed for any point in time at

$$\tilde{U} = \frac{\dot{V}_{\text{in}}}{|A|} = \frac{1}{|A|} \int_F U_i \, \mathrm{d}F_i.$$

(8.8)

When one considers, however, that for the area-averaged density $\tilde{\rho}$

$$\tilde{\rho} = \frac{1}{|A|} \int_F \rho |\mathrm{d}A_i|,$$

(8.9)

the mass flow through a surface can also be written:

$$|\dot{m}| = \tilde{\rho} \tilde{U} F \quad \text{mit} \quad F = |F|.$$

(8.10)

For moderate velocities, where ρ is hardly changing, for internal flows, a surface decrease is connected with an increase in velocity.

When carrying out considerations in fluid mechanics, the above derivations, taking into account the physical conditions of mass conservation, lead to the following mathematical representations of mass conservation:

Differential form:
$$\frac{\partial \rho}{\partial t} + \frac{\partial (\rho U_i)}{\partial x_i} = 0.$$

(8.11)

Integral form:
$$\frac{\partial M}{\partial t} = \dot{m}_{\text{in}} - \dot{m}_{\text{out}}.$$

(8.12)

Internal flows:
$$\frac{\partial M}{\partial t} = 0 \quad \leadsto \quad \dot{m} = \tilde{\rho} \tilde{U} A = \text{constant}$$

(8.13)

From (8.13) one can derive by differentiating

$$\frac{d}{dx}(\dot{m}) = \frac{d}{dx}(\tilde{\rho}\tilde{U}F) = 0 \quad \rightsquigarrow \quad \frac{d\tilde{\rho}}{\tilde{\rho}} + \frac{d\tilde{U}}{\tilde{U}} + \frac{dF}{F} = 0 \qquad (8.14)$$

All of the above equations represent different forms of mass conservation that can be employed for the solution of flow problems; which form should be chosen depends on the method that is applied to solve a particular flow problem.

8.2 Integral Form of the Momentum Equation

In Sect. 5.3, the momentum equation for local considerations of flow problems was formulated and derived for each component j of the momentum as follows, i.e. for $j = 1, 2, 3$:

$$\rho\left(\frac{\partial U_j}{\partial t} + U_i \frac{\partial U_j}{\partial x_i}\right) = -\frac{\partial P}{\partial x_j} - \frac{\partial \tau_{ij}}{\partial x_i} + \rho g_j. \qquad (8.15)$$

When one adds to this equation the continuity equation in the form (8.1), multiplied by U_j, i.e. adding the following terms:

$$U_j \frac{\partial \rho}{\partial t} + U_j \frac{\partial (\rho U_i)}{\partial x_i} = 0, \qquad (8.16)$$

one obtains

$$\frac{\partial (\rho U_j)}{\partial t} + \frac{\partial (\rho U_i U_j)}{\partial x_i} = -\frac{\partial P}{\partial x_j} - \frac{\partial \tau_{ij}}{\partial x_i} + \rho g_j. \qquad (8.17)$$

When integrating (8.17) over a given control volume, i.e. when applying the operator $\int_{V_K} (\,) \, dV$ to all terms of the equation, one obtains the integral form of the component j of the momentum equation of fluid mechanics:

$$\int_{V_K} \frac{\partial (\rho U_j)}{\partial t} dV + \int_{V_K} \frac{\partial (\rho U_i U_j)}{\partial x_i} dV = -\int_{V_K} \frac{\partial P}{\partial x_j} dV - \int_{V_K} \frac{\partial \tau_{ij}}{\partial x_i} dV + \int_{V_K} \rho g_j dV + \sum F_j. \qquad (8.18)$$

The term $\sum K_j$ is added as "integration constant", i.e. all those forces in the direction j have to be included in the equation that act as external forces on the boundaries of the chosen control volume. Considering that the integration and differentiation represent in their sequence exchangeable mathematical operators, and also employing Gauss's integration theorem, the following form of the integral momentum equation can be derived:

$$\underbrace{\frac{\partial}{\partial t} \int_{V_K} \rho U_j \, dV}_{\text{I}} + \underbrace{\int_{O_K} \rho U_i U_j \, dF_i}_{\text{II}} = \underbrace{-\int_{O_K} P \, dF_j}_{\text{III}} - \underbrace{\int_{O_K} \tau_{ij} \, dF_i}_{\text{IV}} + \underbrace{\int_{V_K} \rho g_j \, dV}_{\text{V}} + \underbrace{\sum F_j}_{\text{VI}}.$$

$$(8.19)$$

8.3 Integral Form of the Mechanical Energy Equation

This equation comprises six terms whose physical meanings are stated below:

I: Temporal change of the j momentum in the interior of a control volume.
II: Sum of inflows and outflows of flow momentum per unit time in the direction j summed up over the entire surface surrounding the considered control volume. The momentum over $i = 1, 2, 3$ represents this.
III: Resulting pressure force in the direction j, obtained by integration over the entire j components of the surface elements surrounding the considered control volume.
IV: Sum of the j momentum inflows and outflows occurring per unit time by molecular momentum transport over the entire surface of the control volume. The double index i expresses the summation over $1, 2, 3$.
V: The j component of the mass force acting on the control volume.
VI: Sum of all external (not fluid mechanically induced) forces acting in the j direction on the boundaries of the control volume.

This integral form of the momentum equation can be employed for a large number of flow problems in fluid mechanics, in order to determine the cause of forces on fluid motions on walls, flow aggregates, etc. Their application is explained in the examples that are dealt with in Sects. 8.5.1–8.5.9. On the basis of selected representative samples, it will be made clear how the above-derived integral form of the momentum equation can be used to solve flow problems. Here it is important to recognize the universal validity of the integral form of the momentum equation to ensure its general use in solving flow problems, beyond the examples considered in the following sections.

8.3 Integral Form of the Mechanical Energy Equation

In Sect. 5.5, it was shown that the momentum equation j:

$$\text{in differential form}: \rho\left[\frac{\partial U_j}{\partial t} + U_i\frac{\partial U_j}{\partial x_i}\right] = -\frac{\partial P}{\partial x_j} - \frac{\partial \tau_{ij}}{\partial x_i} + \rho g_j \quad (8.20)$$

can be transferred to the mechanical energy equation by multiplication by U_j, [see (5.56)], to yield the following relationship:

$$\rho\left[\frac{\partial\left(\frac{1}{2}U_j^2\right)}{\partial t} + U_i\frac{\partial\left(\frac{1}{2}U_j^2\right)}{\partial x_i}\right] = -\frac{\partial(PU_j)}{\partial x_j} + P\frac{\partial U_j}{\partial x_j} - \frac{\partial(\tau_{ij}U_j)}{\partial x_i} + \tau_{ij}\frac{\partial U_j}{\partial x_i} + \rho g_j U_j. \quad (8.21)$$

Multiplying the continuity equation by $\left(\frac{1}{2}U_j^2\right)$, the following results:

$$\left(\frac{1}{2}U_j^2\right)\frac{\partial \rho}{\partial t} + \left(\frac{1}{2}U_j^2\right)\frac{\partial(\rho U_i)}{\partial x_i} = 0, \quad (8.22)$$

which can be added to (8.21) so that one obtains

$$\frac{\partial}{\partial t}\left(\frac{1}{2}\rho U_j^2\right)+\frac{\partial}{\partial x_i}\left(\rho U_i \frac{1}{2} U_j^2\right) = -\frac{\partial (P U_j)}{\partial x_j}+P\frac{\partial U_j}{\partial x_j}-\frac{\partial (\tau_{ij} U_j)}{\partial x_i}+\tau_{ij}\frac{\partial U_j}{\partial x_i}+\rho g_j U_j. \tag{8.23}$$

When one integrates this equation over a given control volume, one obtains, by employing Gauss's integral theorem and taking into account the mathematically possible inversion of the integration and differentiation sequence:

$$\underbrace{\frac{\partial}{\partial t}\int_{V_K}\frac{1}{2}\rho U_j^2\,\mathrm{d}V}_{\text{I}} + \underbrace{\int_{O_K}\rho U_i \frac{1}{2}U_j^2\,\mathrm{d}A_i}_{\text{II}} = -\underbrace{\int_{O_K} P U_j\,\mathrm{d}A_j}_{\text{III}} + \underbrace{\int_{V_K} P\frac{\partial U_j}{\partial x_j}\,\mathrm{d}V}_{\text{IV}} - \underbrace{\int_{O_K} \tau_{ij} U_j\,\mathrm{d}A_i}_{\text{V}}$$

$$+ \underbrace{\int_{V_K}\tau_{ij}\frac{\partial U_j}{\partial x_i}\,\mathrm{d}V}_{\text{VI}} + \underbrace{\int_{V_K}\rho g_j U_j\,\mathrm{d}V}_{\text{VII}} + \underbrace{\sum \dot{E}}_{\text{VIII}}. \tag{8.24}$$

This equation comprises eight terms having the following physical meanings:

I: Temporal change of the entire kinetic energy of the flowing fluid within the limits determining the control volume.

II: Outflow minus inflow of the kinetic energy of the flowing fluid per unit time over the entire surface of the considered control volume.

III: Inflow minus outflow of "pressure energy" per unit time over the entire surface of the considered control volume.

IV: Work done by expansion of the flowing fluid per unit time integrated over the entire control volume.

V: Molecule-dependent input minus output of kinetic energy of the considered flowing fluid per unit time integrated over the entire surface of the control volume.

VI: The kinetic energy per unit time dissipated within the entire control volume. This energy is transferred into heat.

VII: Potential energy per unit time of the total mass then integrated over the entire control volume.

VIII: Energy input per unit time over the surface of the control volume or the power supplied to the fluid by flow machines.

In Sect. 5.5, attention was drawn to the fact that the differential form of the j momentum equation and the differential form of the mechanical energy equation do not represent independent equations. The latter emanated from the first by multiplication by U_j, followed by various mathematical derivations and rearrangements of the different terms. This statement holds only in a restricted way for the integral form of the basic equations. By addition of the term $\sum F_j$ in (8.19) and the term $\sum \dot{E}$ in (8.24), it is possible that independent forms of the momentum equation and the mechanical energy equation come about. This is known from the treatment of impacts

8.3 Integral Form of the Mechanical Energy Equation

of spheres treated in mechanics, for which the known momentum and energy equations from (8.15) or (8.19) and (8.20) or (8.24) can be derived as follows:

- The left side of (8.15) yields for $\rho = $ const. for an integration over the entire sphere volume

$$\int_{V_K} \left[\rho \frac{\partial U_j}{\partial t} + \rho U_i \frac{\partial U_j}{\partial x_i} \right] dV = \int_{V_K} \frac{D}{Dt}(\rho U_j)\, dV = \frac{D}{Dt}\int_{V_K} \rho U_j\, dV, \qquad (8.25)$$

and thus

$$\frac{D}{Dt}\int_{V_K} \rho U_j\, dV = \frac{d}{dt}(m_K U_j). \qquad (8.26)$$

- With (8.25) and (8.26) for spheres 1 and 2, (8.15) can be written:

$$\frac{d}{dt}(m_K U_j)_1 = (K_j)_1 \quad \text{and} \quad \frac{d}{dt}(m_K U_j)_2 = (K_j)_2, \qquad (8.27)$$

or rewritten, because $(K_j)_1 = -(K_j)_2$:

$$\frac{d}{dt}\left[(m_K U_j)_1 + (m_K U_j)_2\right] = 0 \rightsquigarrow (m_K U_j)_1 + (m_K U_j)_2 = \text{constant} \qquad (8.28)$$

- The left-hand side of (8.20) yields for $\rho = $ constant

$$\int_{V_K} \left[\rho \frac{\partial}{\partial t}\left(\frac{1}{2}U_j^2\right) + \rho U_i \frac{\partial}{\partial x_i}\left(\frac{1}{2}U_j\right)^2 \right] dV = \int_{V_K} \frac{D}{Dt}\left(\rho \frac{1}{2}U_j^2\right)$$

$$= \frac{D}{Dt}\int_{V_K} \rho \frac{1}{2}U_j^2\, dV \qquad (8.29)$$

and thus

$$\frac{d}{dt}\left(m_K \frac{1}{2}U_j^2\right) = \frac{D}{Dt}\int_{V_K} \rho \frac{1}{2}U_j^2\, dV. \qquad (8.30)$$

- With (8.29) and (8.30), (8.20) yields for spheres

$$\frac{d}{dt}\left(m_K \frac{1}{2}U_j^2\right)_1 = \left(\dot{E}\right)_1 \quad \text{and} \quad \frac{d}{dt}\left(m_K \cdot \frac{1}{2}U_j^2\right)_2 = \left(\dot{E}\right)_2, \qquad (8.31)$$

or rewritten because $\left(\dot{E}\right)_1 + \left(\dot{E}\right)_2 = 0$:

$$\frac{d}{dt}\left[\left(m_K \frac{1}{2}U_j^2\right)_1 + \left(m_K \frac{1}{2}U_j^2\right)_2\right] = \left(\dot{E}\right)_1 + \left(\dot{E}\right)_2 = 0, \qquad (8.32)$$

$$\left(m_K \frac{1}{2}U_j\right)_1 + \left(m_K \frac{1}{2}U_j\right)_2 = \text{constant} \qquad (8.33)$$

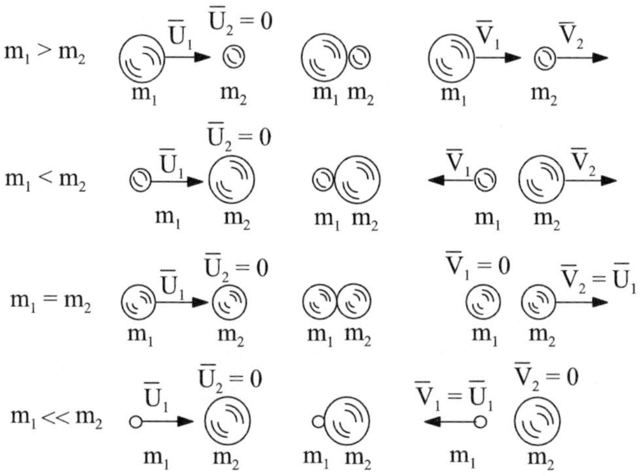

Fig. 8.1 Possible motions of spheres following an elastic collision

The insights gained on the elastic collision by employing (8.28) and (8.33) are sketched in Fig. 8.1. The representations are stated for different mass ratios of the spheres. From the integral forms of the basic equations of flow mechanics result the collision laws for spheres which are known from applications of mechanics in physics. This makes clear the general applicability of the integral form of the mechanical energy equation stated in (8.24).

8.4 Integral Form of the Thermal Energy Equation

In Sect. 5.6, the thermal energy equation was derived and stated for an ideal gas in (5.77) as follows:

$$\rho c_v \left[\frac{\mathrm{D}T}{\mathrm{D}t}\right] = \lambda \frac{\partial^2 T}{\partial x_i^2} - P\frac{\partial U_i}{\partial x_i} - \tau_{ij}\frac{\partial U_j}{\partial x_i}, \tag{8.34}$$

For an ideal liquid with $\rho = \mathrm{const.}$ it was stated with (5.78):

$$\rho c_v \left[\frac{\mathrm{D}T}{\mathrm{D}t}\right] = \lambda \frac{\partial^2 T}{\partial x_i^2} - \tau_{ij}\frac{\partial U_j}{\partial x_i}. \tag{8.35}$$

When one chooses (8.34) for further considerations, this equation can also be written:

$$\rho c_v \left[\frac{\partial T}{\partial t} + U_i \frac{\partial T}{\partial x_i}\right] = -\frac{\partial \dot{q}_i}{\partial x_i} - P\frac{\partial U_i}{\partial x_i} - \tau_{ij}\frac{\partial U_j}{\partial x_i}. \tag{8.36}$$

Adding to (8.36) the continuity equation multiplied by $c_v T$:

$$c_v T \frac{\partial \rho}{\partial t} + c_v T \frac{\partial(\rho U_i)}{\partial x_i} = 0,$$

8.4 Integral Form of the Thermal Energy Equation

one obtains the initial equation for the derivation of the integral form of the thermal energy equation:

$$\frac{\partial(\rho c_v T)}{\partial t} + \frac{\partial(\rho c_v T U_i)}{\partial x_i} = -\frac{\partial \dot{q}_i}{\partial x_i} - P\frac{\partial U_i}{\partial x_i} - \tau_{ij}\frac{\partial U_j}{\partial x_i}. \tag{8.37}$$

With $c_v T = e$ (inner energy), one obtains

$$\frac{\partial(\rho e)}{\partial t} + \frac{\partial(\rho e U_i)}{\partial x_i} = -\frac{\partial \dot{q}_i}{\partial x_i} - P\frac{\partial U_i}{\partial x_i} - \tau_{ij}\frac{\partial U_j}{\partial x_i}. \tag{8.38}$$

The integration of (8.38) over a control volume yields

$$\int_{V_K}\frac{\partial(\rho e)}{\partial t}\,\mathrm{d}V + \int_{V_K}\frac{\partial(\rho e U_i)}{\partial x_i}\,\mathrm{d}V = -\int_{V_K}\frac{\partial \dot{q}_1}{\partial x_i}\,\mathrm{d}V - \int_{V_K} P\frac{\partial U_i}{\partial x_i}\,\mathrm{d}V - \int_{V_K}\tau_{ij}\frac{\partial U_j}{\partial x_i}\,\mathrm{d}V$$
$$+ \sum(\dot{Q} + \dot{E}). \tag{8.39}$$

Rewriting (8.39) with consideration of Gauss's integral theorem and the reversibility of the sequence of integration and differentiation, one obtains

$$\underbrace{\frac{\partial}{\partial t}\left(\int_{V_K}\rho e\,\mathrm{d}V\right)}_{\text{I}} + \underbrace{\int_{O_K}\rho e U_i\,\mathrm{d}A_i}_{\text{II}} = -\underbrace{\int_{O_K}\dot{q}_i\,\mathrm{d}A_i}_{\text{III}} - \underbrace{\int_{V_K} P\cdot\frac{\partial U_i}{\partial x_i}\,\mathrm{d}V}_{\text{IV}} - \underbrace{\int_{V_K}\tau_{ij}\frac{\rho U_j}{\partial x_i}\,\mathrm{d}V}_{\text{V}}$$
$$+ \underbrace{\sum(\dot{Q} + \dot{E})}_{\text{VI}}. \tag{8.40}$$

The terms of the resulting equation can be interpreted as follows:

I: Temporal change of the inner energy of the fluid within the control volume V_K.
II: Convective outflow and inflow of inner energy per unit time over the surface O_K of the control volume.
III: Molecular heat flow per unit time, i.e. the sum of the outflow and inflow, over the surface O_K of the control volume.
IV: The work carried out during expansion by the total volume per unit time.
V: The mechanical energy dissipated per unit time in the entire control volume.
VI: External heat and energy supply per unit time which is added to the entire control volume.

The above equation holds likewise for an ideal liquid, but for this term IV is equal to zero, as no work can be done during expansion because of $\rho = $ constant.

8.5 Applications of the Integral Form of the Basic Equations

The importance of the integral forms of the basic equations of fluid mechanics becomes clear from applications that are listed below. Many books on the basics of fluid mechanics treat flow problems of this kind, so that the considerations carried out in the following sections can be brief. Typical examples are treated that make it clear that the derived integral forms of the basic equations represent the basis for a variety of engineering problem solutions. However, attention has to be paid to the fact that solutions often can be derived only by employing simplifications to the general form of the equations. Reference is made to these simplifications for each of the treated flow problems and their implications for the obtained solutions in the framework of the derivations.

In order to introduce the reader to the methodically of the correct handling of the integral form of the equations, each of the problems treated below is solved by starting from the employed basic equations. Starting with the general form of the integral form of the equations, terms are deleted which are equal to zero for the treated flow problem. In addition, by introducing simplifications, terms in the equations are also removed which are small and therefore have very little influence on the treated flow problem, so that easily comprehensible solutions are obtained. Below only examples for the applications of the integral forms of the basic equations are given. More examples are found in refs. [8.1] to [8.5].

8.5.1 Outflow from Containers

In Fig. 8.2, a simple container is sketched, having a diameter D, which is partly filled with a fluid and is assumed to be closed at the top. Between the fluid surface and the container lid there is a gas having a constant pressure P_H. The fluid height is H and at the bottom of the container there is an opening with diameter d. Sought is the outflow velocity from the container, i.e. the velocity U_d.

From Fig. 8.2, it can be seen that the water surface is moving downwards with a velocity U_D, because of the fluid flowing out, which exits with U_d from

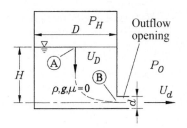

Fig. 8.2 Diagram for the treatment of outflows from containers

8.5 Applications of the Integral Form of the Basic Equations

the container opening. Through the integral form of the continuity equation one obtains

$$\tilde{\rho}\tilde{U}F = \text{constant} \quad \rightsquigarrow \quad \rho U_D \frac{\pi}{4} D^2 = \rho U_d \frac{\pi}{4} d^2 \quad \rightsquigarrow \quad U_D = \frac{d^2}{D^2} U_d \quad (8.41)$$

By employing the Bernoulli equation between the points (A) and (B), one obtains

$$\frac{1}{2} U_D^2 + \frac{P_H}{\rho} + gH = \frac{1}{2} U_d^2 + \frac{P_0}{\rho}. \quad (8.42)$$

Hence

$$\frac{1}{2} U_d^2 = \frac{1}{2} U_D^2 + gH + \frac{1}{\rho}(P_H - P_0), \quad (8.43)$$

or, after insertion of (8.41)

$$\frac{1}{2} U_d^2 = \frac{1}{2} \left(\frac{d^4}{D^4} \right) U_d^2 + gH + \frac{1}{\rho}(P_H - P_0), \quad (8.44)$$

the following relationship for U_d results:

$$U_d = \sqrt{\frac{2gH + \frac{2}{\rho}(P_H - P_0)}{1 - \frac{d^4}{D^4}}}. \quad (8.45)$$

For $P_H = P_0$ and $d \ll D$, the well known equation $U_d = \sqrt{2gH}$ results.

8.5.2 Exit Velocity of a Nozzle

In fluid mechanics, it is necessary to calibrate indirectly operating measuring processes (e.g. stagnation-pressure tubes, hot-wire anemometers) in flow fields in which the flow velocity is known. By letting a fluid flow through a nozzle, it can be achieved that at the nozzle exit the flow velocity required for calibration can be adjusted via the pressure in the input pipe of the nozzle (Fig. 8.3).

With the statements made in Sect. 8.1, the integral form of the continuity equation holds in the following form:

$$\dot{m} = \tilde{\rho}\tilde{U}F = \text{constant} \quad \rightsquigarrow \quad \rho U_A \frac{\pi}{4} D^2 = \rho U_B \frac{\pi}{4} d^2, \quad (8.46)$$

i.e. one can write for U_A

$$U_A = \frac{d^2}{D^2} U_B. \quad (8.47)$$

For the planes (A) and (B) it can be written as a result of the Bernoulli equation:

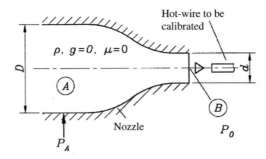

Fig. 8.3 Diagram of a nozzle-calibrating test rig for velocity-measuring sensors

$$\frac{1}{2}U_A^2 + \frac{P_A}{\rho} = \frac{1}{2}U_B^2 + \frac{P_B}{\rho} = \frac{1}{2}\frac{d^4}{D^4} \cdot U_B^2 + \frac{P_A}{\rho}. \tag{8.48}$$

From this, the following results:

$$U_B = \sqrt{\frac{2(P_A - P_B)}{\rho\left(1 - \frac{d^4}{D^4}\right)}}. \tag{8.49}$$

When one chooses $D \gg d$, one obtains for U_B, to a good approximation:

$$U_B = \sqrt{\frac{2}{\rho}(P_A - P_B)} = \sqrt{\frac{2}{\rho}(P_A - P_0)}. \tag{8.50}$$

By adjusting different P_A values, the entire velocity regime required for the calibration of measuring tubes can be set. Hence, measuring $P_A = P_0$ and knowing ρ yields U_B against which velocity sensors can be calibrated.

8.5.3 Momentum on a Plane Vertical Plate

When flowing fluid jets are decelerated, forces appear which are used in many fields of technology. For the flow problem shown in Fig. 8.4, the question arises as to what force needs to be applied in the x_1 direction to prevent the deflection of the plate due to the momentum impact of the plane fluid jet. The plane jet has a thickness H in the x_2 direction and a width b in the x_3 direction. The jet velocity far away from the plate is known and is U_A. The density ρ is known and $g_1 = 0$, as the x_1 direction axis is horizontal.

Employing the integral form of the continuity equation gives

$$\rho U_A H b = \rho U_C H_C b + \rho U_B H_B b. \tag{8.51}$$

From the Bernoulli equation, one obtains

$$\frac{1}{2}U_A^2 + \frac{P_A}{\rho} = \frac{1}{2}U_B^2 + \frac{P_B}{\rho} \quad \text{and} \quad P_A = P_B = P_0, \tag{8.52}$$

8.5 Applications of the Integral Form of the Basic Equations

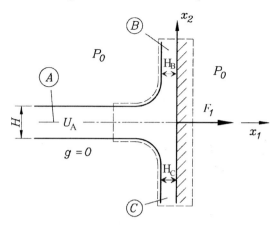

Fig. 8.4 Diagram for consideration of the momentum impact on a vertical plate

and thus $U_A = U_B$. Analogous considerations yield $U_A = U_C$. Owing to the symmetry of the problem, one obtains

$$H = 2H_C = 2H_B; \quad \text{hence} \quad H_C = H_B.$$

For the solution of the problem to yield K, the integral form of the momentum equation can be employed, as it is stated in (8.19):

$$\frac{\partial}{\partial t} \int_{V_K} \rho U_j \, dV + \int_{O_K} \rho U_i U_j \, dA_i = -\int_{O_K} P \, dA_j - \int_{O_K} \tau_{ij} \, dA_i + \int_{V_K} \rho g_i \, dV + \sum F_j.$$

(8.53)

For the considered flow problem, the following simplifications of the above universally valid equation hold:

$\frac{\partial}{\partial t} \int_{V_K} \rho U_j \, dV = 0$, stationary flow problem

$\int_{O_K} P \, dA_j = 0$, as $P = P_0$ on all surfaces of the chosen control volume

$\int_{O_K} \tau_{ij} \, dA_i = 0$, absence of viscosity in the fluid

$\int_{\dot{V}} \rho g_j \, dV = 0$, gravitation term, here $g_j = g_1 = 0$.

(8.54)

Therefore, the following holds for the simplified form of (8.53):

$$\int_{O_K} \rho U_i U_j \, dA_i = F_j,$$

(8.55)

and thus one obtains by integration for $j = 1$:

$$F_1 = -\rho U_A^2 HB. \tag{8.56}$$

The result of the above derivations shows that F_1 must act in the negative x_1 direction, in order to prevent deflection of the plane plate by the incoming plane fluid jet. This gives an example to make clear the kind of force terms $\sum F_j$ that occur in (8.19). All forces need to be included for a particular flow problem that act on the considered control volume.

8.5.4 Momentum on an Inclined Plane Plate

For a fluid jet hitting an inclined plane plate, the jet behavior is shown in Fig. 8.5. Because of the inclination of the plate, which encloses the angle α with the axis of the incoming plane fluid jet, the jet splits into two parts of unequal thickness. The thicker jet goes upwards and has a height $h_B = \varepsilon H_A$ and the thinner jet goes downwards and has a height $h_C = (1-\varepsilon)H_A$. This results from the continuity equation. Applying this equation, $U_A = U_B = U_C$ results when $g = 0$ is introduced into the Bernoulli equation.

When employing the continuity equation in integral form, i.e. when considering that for the solution of the problem the mass conservation law can be used, it follows that

$$\rho U_A H_A b = \rho U_B h_B b + \rho U_C h_C b, \tag{8.57}$$

or, because $U_A = U_B = U_C$ from the Bernoulli equation:

$$U_A H_A = U_B h_B + U_C h_C \quad \leadsto \quad H_A = h_B + h_C. \tag{8.58}$$

For the two split jets forming on the plate, it can therefore be stated that

$$h_B = \varepsilon H_A \quad \text{and} \quad h_C = (1-\varepsilon)\,H_A. \tag{8.59}$$

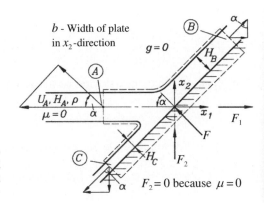

Fig. 8.5 Diagram for explaining the independence of fluid flow considerations on the chosen coordinate system

8.5 Applications of the Integral Form of the Basic Equations

When one employs the integral momentum equation:

$$\frac{\partial}{\partial t}\int_{V_K} \rho U_j \, dV + \int_{O_K} \rho U_i U_j \, dA_i = -\int_{O_K} P \, dA_j - \int_{O_K} \tau_{ij} \, dA_i + \int_{V_K} \rho g_j \, dV + \sum F_j. \tag{8.60}$$

For the problems sketched in Fig. 8.5, the following simplifications can be introduced:

$$\frac{\partial}{\partial t}\int_{V_K} \rho U_j \, dV = 0, \text{ stationary problem}$$

$$\int_{O_K} P \, dA_j = 0, \text{ as } P = P_O \text{ on all surfaces of the chosen control volume}$$

$$\int_{O_K} \tau_{ij} \, dA_i = 0, \text{ viscosity-free fluid}$$

$$\int_{V_K} \rho g_j \, dV = 0, \text{ insignificant term or } g_j = 0.$$

$$\tag{8.61}$$

For the simplified form of the above integral momentum equation it thus holds that

$$\int_{O_K} \rho U_i U_j \, dA_i = F_j. \tag{8.62}$$

When one chooses a coordinate system oriented along the plate, then for K_p, integration over the planes (A), (B) and (C), yields three contributions:

$$F_P = -\rho U_A^2 H_A b \sin\alpha + \rho U_A^2 \varepsilon H_A b - \rho U_A^2 (1-\varepsilon) H_A b. \tag{8.63}$$

As in the present problem $\mu = 0$ was set, the following results for the force $F_P = 0$, due to the moving fluid along the plate, can be deduced from (8.63):

$$-1 - \sin\alpha + 2\varepsilon = 0, \tag{8.64}$$

or one obtains for ε

$$\varepsilon = \frac{1}{2}(1 + \sin\alpha). \tag{8.65}$$

For the force acting vertically on the plate, it can be computed that K_S is

$$F_S = \rho U_A^2 H_A b \cos\alpha. \tag{8.66}$$

It is evident that the above considerations have to be independent from the chosen coordinate system. When one chooses the coordinate system indicated by x_1 and x_2 in Fig. 8.6, one obtains for K_1 the following contributions, derived by integration over the planes (A), (B) and (C):

$$F_1 = -\rho U_A^2 H_A b + \rho U_A^2 \varepsilon H_A b \sin\alpha - \rho U_A^2 (1-\varepsilon) H_A b \sin\alpha. \tag{8.67}$$

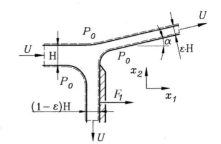

Fig. 8.6 Jet deflection at a knife edge and its cause

For the force K_2 similar derivations yield:

$$F_2 = \rho U_A^2 \varepsilon H_A b \cos\alpha - \rho U_A^2 (1-\varepsilon) H_A b \cos\alpha. \tag{8.68}$$

Since, because $\mu = 0$, the total force on the plate resulting from F_1 and F_2 has to act on the plate K vertically, the following holds:

$$\tan\alpha = \frac{F \sin\alpha}{-F_1} = \frac{F_2}{F \cos\alpha} = \frac{2\varepsilon \cos\alpha - \cos\alpha}{1 - 2\varepsilon \sin\alpha + \sin\alpha}. \tag{8.69}$$

From this it can be derived by introduction of $2\varepsilon \cos\alpha - \cos\alpha = \sin\alpha - 2\varepsilon \sin\alpha + \sin\alpha$ or for ε:

$$\varepsilon = \frac{1}{2}(1 + \sin\alpha), \tag{8.70}$$

which is the same result as (8.65).

8.5.5 Jet Deflection by an Edge

When a fluid jet (height H, width b) hits, with part of its cross-sectional area, a plate standing vertically to the jet, the arriving fluid is partitioned into two partial jets. One of the two partial jets runs, vertically to the original jet direction, downwards along the plate, and the other partial jet is deflected upwards by an angle α with respect to the original jet direction. Neglecting viscosity forces and gravitational forces and assuming a constant ambient pressure, it results from the Bernoulli equation that the two partial jets have the same velocity, which is equal to the velocity of the fluid in the original jet. Because of the continuity equation, the two partial jets have jet heights εH and $(1-\varepsilon)H$. In the integral form of the momentum equation

$$\frac{\partial}{\partial t}\int_{V_K} \rho U_j \, dV + \int_{O_K} \rho U_i U_j \, dA_i = -\int_{O_K} P \, dA_j - \int_{O_K} \tau_{ij} \, dA_i + \int_{V_K} \rho g_j \, dV + \sum F_j, \tag{8.71}$$

8.5 Applications of the Integral Form of the Basic Equations

the following simplifications can be introduced:

$$\frac{\partial}{\partial t} \int_{V_K} \rho U_j \, dV = 0, \text{stationary fluid flow problem}$$

$$\int_{O_K} P \, dF_j = 0, \text{constant pressure along the surface of the control volume}$$

$$\int_{O_K} \tau_{ij} \, dF_i = 0, \text{viscosity forces are neglected}$$

$$\int_{O_K} \rho g_j \, dV = 0, \text{gravitation is neglected.}$$

(8.72)

Hence the following simplified momentum equation results:

$$\int_{O_K} \rho U_i U_j dA_i = \sum F_j. \tag{8.73}$$

The force exerted on the fluid can be determined by the equation for the x_1 components of the total force

$$-\rho UU(bH) + \rho(U\cos\alpha)U(\varepsilon bH) = F_1, \tag{8.74}$$

$$F_1 = -\rho U^2 bH(1 - \varepsilon \cos\alpha). \tag{8.75}$$

The negative value of the force K_1 results from the fact that the force exerted on the plate is computed.

From the equation for K_2, one obtains:

$$\rho U(1-\epsilon)Hb(-U) + \rho U \varepsilon bH(U \sin\alpha) = 0, \tag{8.76}$$

and, hence, for $K_2 = 0$, the connection between the deflection angle α and the ratio ε can be computed as

$$-\rho U^2 bH \left[(1-\varepsilon) - \varepsilon \sin\alpha\right] = 0, \tag{8.77}$$

$$\varepsilon = \frac{1}{1 + \sin\alpha}. \tag{8.78}$$

Hence the ratio of splitting the jet in Fig. 8.6 can be determined from the deflection of the jet from the horizontal position, i.e. by measuring the angle α.

8.5.6 Mixing Process in a Pipe of Constant Cross-Section

In a pipe, two fluids flow at constant velocities U_A, U_B. These fluids mix with one another as they move downstream in a channel Fig. 8.7. The pressure at point 1 and the partial areas in which the velocities U_A, U_B hold, will be

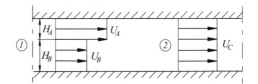

Fig. 8.7 Diagram for explaining the mixing process

given. Sought is the pressure P_2 at point 2 where a constant velocity $\bar{U}_C$ over the pipe cross-section has been reached.

From the integral form of the continuity equation, one obtains

$$b(-\rho H_A U_A - \rho H_B U_B + \rho(H_A + H_B)\bar{U}_C) = 0, \quad (8.79)$$

or, rearranged for $\bar{U}_C$,

$$\bar{U}_C = \frac{U_A H_A + U_B H_B}{H_A + H_B}. \quad (8.80)$$

The momentum equation:

$$\frac{\partial}{\partial t}\int_{V_K} \rho U_j \, dV + \int_{O_K} \rho U_i U_j \, dA_i = -\int_{O_K} P \, dF_j - \int_{O_K} \tau_{ij} \, dA_i + \int_{V_K} \rho g_j \, dV + \sum F_j \quad (8.81)$$

can be simplified as follows:

$$\frac{\partial}{\partial t}\int_{V_K} \rho U_j \, dV = 0, \text{ stationary flow problem}$$

$$\int_{O_K} \tau_{ij} \, dA_i = 0, \ \mu = 0, \text{ i.e. the assumption of absence of viscosity is made}$$

$$\int_{V_K} \rho g_j \, dV = 0, \text{ no component of gravitation exists in horizontal direction}$$

$$\sum F_j = 0, \text{ no external forces act on the control volume.} \quad (8.82)$$

From these simplifying assumptions it follows that

$$\int_{O_K} \rho U_i U_j \, dA_i = -\int_{O_K} P \, dA_j. \quad (8.83)$$

For the present problem this results in:

$$-\rho U_A^2 H_A - \rho U_B^2 H_B + \rho \bar{U}^2(H_A + H_B) = (P_1 - P_2)(H_A + H_B). \quad (8.84)$$

When one now inserts the above expression for $\bar{U}_c$ and solves the equation for P_2, one obtains:

$$P_2 = P_1 + \rho(U_A - U_B)^2 \frac{H_A H_B}{(H_A + H_B)^2}. \quad (8.85)$$

The pressure therefore increases as a consequence of mixing the two flows from position (1) to position (2).

8.5.7 Force on a Turbine Blade in a Viscosity-Free Fluid

In flow machines, wheels with blades are used to exploit the momentum of fluid flows for propulsion purposes, i.e. to drive the rotating wheels. A jet from a rectangular nozzle (plane jet having a width H and depth b) hits a stationary blade which deflects the jet symmetrically to two sides around the angle $180° - \beta$ (Fig. 8.8). The information on the inflowing and outflowing fluid flows is given. The pressure along the surface of the marked control volume is equal all over and can be assumed to be the ambient pressure, so that the integral form of the momentum equation can be given as

$$\frac{\partial}{\partial t} \int_{V_K} \rho U_j \, dV + \int_{O_K} \rho U_i U_j \, dA_i = - \int_{O_K} P \, dA_j - \int_{O_K} \tau_{ij} \, dA_i$$
$$+ \int_{V_K} \rho g_j \, dV + \sum F_j. \qquad (8.86)$$

This equation can be simplified in the following way, if gravitation is negligible:

$$\frac{\partial}{\partial t} \int_{V_K} \rho U_j \, dV = 0, \quad \text{stationary flow problem}$$

$$\int_{O_K} P \, dA_j = 0, \quad \text{as } P = P_0 \text{ along all surfaces of the control volume}$$

$$\int_{O_K} \tau_{ij} \, dA_i = 0, \quad \text{as } \mu = 0 \text{ was set in the assumptions}$$

$$\int_{V_K} \rho g_j \, dV = 0, \quad \text{as gravitation is negligible.} \qquad (8.87)$$

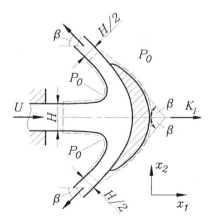

Fig. 8.8 Diagram for explaining the force on a turbine blade moving in the x_1 direction

Hence the following final equation results:

$$\int_{O_K} \rho U_i U_j \, dA_i = \sum F_j. \tag{8.88}$$

As the problem is symmetrical, only the horizontal component $j = 1$ has to be considered, i.e. in the $j = 2$ direction no resultant force appears, so that the conservation of momentum can be written as follows:

$$-\rho H b U^2 + 2\frac{1}{2}\rho H b U(-U \cdot \cos \beta) = F_1. \tag{8.89}$$

The resultant force on the blade thus results in:

$$F_S = -F_1 = \rho H b U^2 (1 + \cos \beta). \tag{8.90}$$

This equation makes it clear that by deflecting the jets in direction of the incoming flow, an increase in the force F_1 acting on the turbine blade can be obtained.

8.5.8 Force on a Periodical Blade Grid

Assumptions: $g = 0$, $\mu = 0$ for the flow problem given below.

Two-dimensional grids of blades are used in flow machines in order to exploit forces caused by flows to drive rotating wheels.

Figure 8.9 shows such a set of blades arranged periodically in the $x_1 - x_2$ plane. These blades are approached by a flowing fluid with an approach

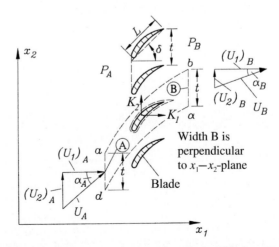

Fig. 8.9 Diagram explaining the effect of blade grids

8.5 Applications of the Integral Form of the Basic Equations

velocity $\boldsymbol{U}_A = \{(U_1)_A, (U_2)_A, 0\}$ comprising two components, i.e. the U_3 component of the oncoming flow field is assumpt to be zero. Because of the deflection of the flow by the blade, the departing velocity differs from the approach velocity, so that the following holds:

$$\boldsymbol{U}_B = \{(U_1)_B, (U_2)_B, 0\} \neq \boldsymbol{U}_A. \tag{8.91}$$

The assumed features of $\boldsymbol{U}_A$ and $\boldsymbol{U}_B$, for which it always holds that $(U_3)_A = (U_3)_B = 0$, make it clear that the flow field remains "two-dimensional" and also has only two components. Thus for the inflow and outflow the following equation can be derived from the continuity equation, where b is the width of the blade in the x_3 direction:

$$\rho(U_1)_A tb = \rho(U_1)_B tb. \tag{8.92}$$

For the inflow and outflow the same velocity components in the x_1 direction thus result. This makes it clear that the purpose of the blades is to change the velocity component $(U_2)_A$ of the approaching flow.

For the computation of the force on a row of blades, a control volume (a, b, c, d). is chosen, as shown in Fig. 8.9. In the flow direction, two flow lines are chosen as boundaries of the control volume which are positioned along $(a - b)$ and $(d - c)$, with the distance of the blades being t. Hence the inflow and outflow to the chosen control volume take place only over the areas $(d - a)b$ and $(c - d)b$.

When one neglects gravitational effects, the Bernoulli equation yields the following relationship between the velocity and pressure fields:

$$\underbrace{\frac{P_A}{\rho} + \frac{1}{2}\left[(U_1)_A^2 + (U_2)_A^2\right]}_{\text{entire kinetic energy in (A)}} = \underbrace{\frac{P_B}{\rho} + \frac{1}{2}\left[(U_1)_B^2 + (U_2)_B^2\right]}_{\text{entire kinetic energy in (B)}}. \tag{8.93}$$

For the computation of the force on the blade, the integral form of the momentum equation is used:

$$\frac{\partial}{\partial t}\left(\int_{V_K} \rho U_j \, dV\right) + \int_{O_K} \rho U_i U_j \, dA_i = -\int_{O_K} P \, dA_j - \int_{O_K} \tau_{ij} \, dA_i + \int_{V_K} \rho g_j \, dV + \sum F_j. \tag{8.94}$$

With the following assumptions, one obtains from (8.94)

$$\frac{\partial}{\partial t}\int_{V_K} \rho U_j \, dV_p = 0, \quad \text{as there is a stationary flow}$$

$$\int_{O_K} \tau_{ij} \, dA_i = 0, \quad \text{as the flow is that of a viscosity-free fluid} \tag{8.95}$$

$$\int_{V_K} \rho g_j \, dV = 0, \quad \text{as gravitational forces are negligible,}$$

and the momentum equation in simplified form:

$$\int_{O_K} \rho U_i U_j \, dA_i = - \int_{O_K} P \, dA_j + \sum F_j \quad \text{oder} \quad F_j = \int_{O_K} \rho U_i U_j \, dA_i + \int_{O_K} P \, dA_j. \tag{8.96}$$

Hence for the forces in the directions $j = 1$ and $j = 2$:

$$F_1 = \dot{m} \underbrace{[(U_1)_B - (U_1)_A]}_{=0} + (P_B - P_A) \, bt = (P_B - P_A) \, bt. \tag{8.97}$$

$$F_2 = \dot{m} \left[(U_2)_B - (U_2)_A\right]. \tag{8.98}$$

From the Bernoulli (8.93), one obtains

$$(P_B - P_A) = \frac{\rho}{2} \left[(U_2)_A^2 - (U_2)_B^2\right]. \tag{8.99}$$

Hence for F_1 the following results:

$$F_1 = \frac{\rho}{2} Bt \left[(U_2)_A^2 - (U_2)_B^2\right], \tag{8.100}$$

or, expressed in terms of the in-flow angle and the out-flow angle:

$$F_1 = \frac{\rho}{2} Bt \left[U_A^2 \sin^2 \alpha_A - U_B^2 \sin^2 \alpha_B\right], \tag{8.101}$$

and

$$F_2 = -\rho \cdot Bt \left[U_A \sin \alpha_A - U_B \sin \alpha_B\right], \tag{8.102}$$

where F_1 and F_2 are the forces acting on the control volume. The forces acting on the blades are

$$(F_1)_s = -F_1 = -\frac{\rho}{2} bt \left[U_A^2 \sin^2 \alpha_A - U_B^2 \sin^2 \alpha_B\right]. \tag{8.103}$$

$$(F_2)_s = -F_2 = \rho bt \left[U_A \sin \alpha_A - U_B \sin \alpha_B\right]. \tag{8.104}$$

The blade thus experiences a force $(F_1)_s$ in the negative x_1 direction and a force $(F_2)_s$ in the positive x_2 direction. It is the force in the x_2 direction that drives the set of blades in Fig. 8.9.

8.5.9 Euler's Turbine Equation

The considerations in Sect. 8.5.8 related to a set of blades arranged in a plane, which is located in a $x_1 - x_2$ plane. Arranging the blades radially, a rotating wheel results as shown in Fig. 8.10, one obtains the basic arrangement of a radial turbine. The term "radial" designates the main flow direction in which the flow through the turbine blades takes place, namely radially from

8.5 Applications of the Integral Form of the Basic Equations

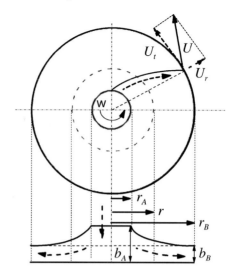

Fig. 8.10 Diagram showing schematically the flow through a radial turbine

the inside of the impeller to the outside. When employing the continuity equation, one obtains

$$\rho(U_r)_A (2\pi r_A) b = \rho(U_r)_B (2\pi r_B) b, \qquad (8.105)$$

or, stated differently, the demand for mass conservation results in the following relationship:

$$(U_r)_B = \frac{r_A}{r_B}(U_r)_A. \qquad (8.106)$$

From the Bernoulli equation, the following results:

$$\frac{P_A}{\rho} + \frac{1}{2}\left[(U_r)_A^2 + (U_t)_A^2\right] = \frac{P_B}{\rho} + \frac{1}{2}\left[(U_r)_B^2 + (U_t)_B^2\right]. \qquad (8.107)$$

From the integral form of the momentum equation, one can determine for the force in the radial direction, if $b_A = b_B = b$ is introduced:

$$F_r = \dot{m}\left[(U_r)_B - (U_r)_A\right] + 2\pi(r_B P_B - r_A P_A)b. \qquad (8.108)$$

Furthermore, for the force in the tangential direction the following holds:

$$F_t = \dot{m}\left[(U_t)_B - (U_t)_A\right]. \qquad (8.109)$$

For the moment imposed on the control volume by the running wheel, one can derive:

$$M_t = \dot{m}\left[r_B(U_t)_B - r_A(U_t)_A\right]. \qquad (8.110)$$

The mechanical power output, resulting from the turbine, amounts to:

$$P_{\text{turb}} = -M_t \omega = -\dot{m}\omega\left[r_B(U_t)_B - r_A(U_t)_A\right]. \qquad (8.111)$$

Equations (8.110) and (8.111) are referred to in the literature as Euler's turbine equation. Here, for $r_A = r_i$, the "inflow radius" of the blade rim is assumed, and for $r_B = r_0 =$, the "outflow radius" is set.

The resulting equation holds not only for turbines, but also generally for flow machines, such as compressors, air blowers (ventilators), pumps, etc. Here pumps and turbines differ in the considerations carried out only with regard to the sign of the energy exchange between running wheel and flowing fluid. In a turbine, energy is extracted from the fluid flow, so that one can collect a usable moment at the shaft of this power engine, or the corresponding energy can be extracted. In a pump, on the other hand, energy is supplied to the fluid flow via the "running wheel", i.e. a torque for driving the machine is exerted at the shaft. Thus the energy necessary for pumping is supplied in pumps to the fluid.

Finally, it is mentioned that Euler's turbine equation can also be applied to flow machines through which flows pass axially, as Figs. 8.11 and 8.12 show them.

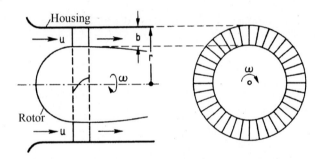

Fig. 8.11 Diagram of the course of the flow through an axial turbine

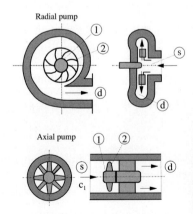

Fig. 8.12 Schematic representation of radial and axial pumps

8.5.10 Power of Flow Machines

In Fig. 8.13, a typical application of a flow machine is shown schematically. In the chosen case, the application of a pump is shown as an example. It sucks on the intake side a certain quantity of water $\dot{m}$ in order to transport it upwards at its discharge side. Here, differences exist in the pipe diameter between the intake side and the discharge side of the pump.

The suction of the pumped fluid takes place from a container (A) and the transport is carried out into a container (B), as shown schematically in Fig. 8.13. All quantities located at the suction side of the considered problem are designated by the subscript A, the quantities on the discharge side of the pump by B.

From the integral form of the continuity equation, the following results for the considered pumping fluid problem:

$$\rho_A \frac{\pi}{4} D_A^2 \tilde{U}_A = \rho_B \frac{\pi}{4} D_B^2 \tilde{U}_B. \tag{8.112}$$

For $\rho_A = \rho_B = \rho = $ constant, the following results:

$$D_A^2 \tilde{U}_A = D_B^2 \tilde{U}_B = \frac{\dot{m}}{\rho} = \dot{V}. \tag{8.113}$$

For the computation of the required pumping power, it is recommended to employ the integral form of the mechanical energy equation, as stated in (8.24):

$$\frac{\partial}{\partial t} \int_{V_K} \frac{1}{2} \rho U_j^2 \, dV + \int_{O_K} \rho U_i \left(\frac{1}{2} U_j^2\right) dA_i = -\int_{O_K} P U_j \, dA_j + \int_{V_K} P \frac{\partial U_j}{\partial x_j} \, dV$$
$$- \int_{O_K} \tau_{ij} U_j \, dA_i + \int_{V_K} \tau_{ij} \frac{\partial U_j}{\partial x_i} \, dV + \int_{V_K} \rho g_j \dot{V} \, dx_j + \sum \dot{E}. \tag{8.114}$$

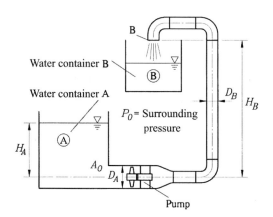

Fig. 8.13 Diagram for the computation of the pumping capacity

Reducing this equation by simplifying assumptions that apply to the present flow problem:

$$\frac{\partial}{\partial t} \int_{V_K} \frac{1}{2} \rho U_j^2 \, dV = 0 \text{ stationary pumping conditions} \qquad (8.115)$$

$$\int_{V_K} P \frac{\partial U_j}{\partial x_j} \, dV = 0 \text{ no work done during expansion, as } \rho = \text{ constant and thus}$$

$$\frac{\partial U_j}{\partial x_j} = 0, \qquad (8.116)$$

and neglecting the τ_{ij} terms in the above integral energy equation, one obtains:

$$\int_{O_K} \rho U_i \left(\frac{1}{2} U_j^2\right) dA_i = -\int_{O_K} P U_j \, dA_j + \int_{V_K} \rho g_j U_j \, dV + \sum \dot{E}, \quad (8.117)$$

or

$$-\dot{m} \left[\frac{1}{2} \tilde{U}_A^2 + \frac{1}{2} \tilde{U}_B^2\right] = \dot{V}_A P_A - \dot{V}_B P_B + \rho g \dot{V} H_C + p_m, \qquad (8.118)$$

where P_m represents the power transferred by the pump into the fluid. For the required mechanical power of the pump, the following holds:

$$P_m = \dot{V} \left\{ \left(\frac{\rho}{2} \tilde{U}_B^2 + P_B + \rho g H_C\right) - \left(\frac{\rho}{2} \tilde{U}_A^2 + P_A\right) \right\}. \qquad (8.119)$$

When one now considers P_B and P_A in more detail, the following can be said:

$P_B = P_0 \quad$ and hence $\quad P_A = P_0 + \rho g H_A - \frac{\rho}{2} U_A^2$ and $\tilde{U}_A \approx 0$,
and
$U_A = \dfrac{\dot{V}}{F_A} \quad$ and hence $\quad U_B = \dfrac{\dot{V}}{F_B}$

Therefore, one obtains the following relationship:

$$P_m = \dot{V} \left\{ \frac{\rho}{2} \left(\frac{\dot{V}^2}{F_B^2} - \frac{\dot{V}^2}{F_A^2}\right) - \rho g H_A + \frac{\rho}{2} \left(\frac{\dot{V}}{F_A}\right)^2 + \rho g H_C \right\} \qquad (8.120)$$

or, summarized:

$$P_m = \frac{\rho}{2} \left(\frac{\dot{V}^3}{F_B^2}\right) + \rho g (H_C - H_A) \dot{V}. \qquad (8.121)$$

For the electrical power p_e of the pump, the following results, with η being the efficiency factor:

$$p_e = \frac{1}{\eta} p_m = \frac{1}{\eta} \left[\frac{\rho \dot{V}^3}{2F_B^2} + \rho g \dot{V} (H_C - H_A) \right]. \tag{8.122}$$

For the pumping power it results that the electrical power stated in (8.122) is required to supply the kinetic energy of the fluid leaving the pipe per unit time, plus the power required per unit time for overcoming the hydrostatic pressure level.

References

8.1. Becker, E., Technische Strömungslehre, Teubner, Stuttgart, 6. überarb. Auflage, 1986.
8.2. Hutter, K., Fluid- und Thermodynamik – Eine Einführung, Springer, Berlin, Heidelberg, New York, 1995.
8.3. Potter, M.C., Foss, J.F., Fluid Mechanics, Wiley, New York, 1975.
8.4. Spurk, J.H., Strömungslehre – Einführung in die Theorie der Strömungen, Springer, Berlin, Heidelberg, 3. überarb. Auflage, 1993.
8.5. Zierep, J., Bühler, K., Strömungsmechanik, Springer, Berlin, Heidelberg, New York, 1991.

Chapter 9
Stream Tube Theory

9.1 General Considerations

The preceding considerations, that covered the derivation of the integral form of the basic equations of fluid mechanics, can also be used advantageously to derive simplified equations applicable to so-called flow filaments or also called stream tubes. The latter can be applied to solve some flow problems. For this purpose, one starts the considerations from flow lines that are introduced as lines of a flow which, at a certain point in time, possess the direction of the flow at each point of the flow field. One can imagine a so-called flow filament to be built up from a bundle of such flow lines and one can make a subdivision of the entire flow field into a multitude of flow filaments. Furthermore, it is possible to bundle flow filaments to obtain stream tube, as indicated in Fig. 9.1. For the suggested approach, one has to consider the properties of flows applied to flow lines, filaments and stream tubes because this concept can only be employed advantageously when the flow quantities assigned to each area of the flow filament can be considered to be constant over the cross-section of the flow filament. This makes it necessary occasionally to choose the cross-sectional area of a flow filament sufficiently small that, for the considered problem, the assumption of uniform state and flow quantities over the cross-sectional area of the flow filament can be fulfilled sufficiently precisely.

For the stationary flow tube theory, it results that the fluid elements constituting a flow filament constitute this flow filament permanently. Fluid particles that are located outside a flow filament at a certain point in time can never become components of the considered filament. Each fluid particle of a stationary flow area belongs to a certain flow filament, so that it is possible to describe the properties of the flow area by the properties of the considered flow filaments (Fig. 9.2).

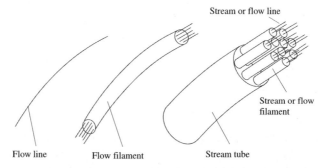

Fig. 9.1 Flow or stream line, flow or stream filament and stream tube

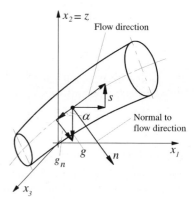

Fig. 9.2 Stream with introduced n–s coordinate system

To simplify the considerations on flow filaments, the following assumptions for flow filaments are introduced:

- A flow filament is always completely filled with the fluid for which the flow considerations are carried out.
- The cross-sectional area changes along a flow filament are small
- A flow filament is assumed to be only slightly curved in the flow direction

Although the assumptions introduced above for flow filaments have considerable limitations, the derivations given in the following sections show that the introduction of flow filaments, with the above properties, into fluid mechanical considerations, leads to equations through which physically very illustrative solutions of flow problems can be derived.

The considerations carried out on the basis of flow filaments show that in some cases the properties of entire flow fields can be described by the properties of flow filaments. When the flow quantities change only slightly over the entire cross-sections of internal flows, the basic equations derived for flow filaments of small dimensions can also be employed to acquire the most important properties of internal flows by a one-dimensional flow theory. For this purpose, the internal flow is treated as a single stream tube. The justification

9.2 Derivations of the Basic Equations

9.2.1 Continuity Equation

The derivation of the continuity equation for a flow filament builds up on the differential form of mass conservation as derived in Sect. 5.2, and which, after integration over a control volume, and having employed Gauss's integral theorem, can be stated as:

$$\frac{\partial \rho}{\partial t} + \frac{\partial (\rho U_i)}{\partial x_i} = 0 \quad \leadsto \quad \int_{V_c} \left(\frac{\partial \rho}{\partial t}\right) dV + \int_{O_c} \rho U_i dA_i = 0 \quad (9.1)$$

where V_c is identical with the considered control volume and O_c is its outer surface. Exchanging in the first term of this equation integration and differentiation, one obtains:

$$\frac{\partial}{\partial t} \int_{V_c} \rho dV + \int_{O_c} \rho U_i dA_i = 0 \quad \leadsto \quad \frac{\partial M_c}{\partial t} = -\int_{O_c} \rho U_i dA_i \quad (9.2)$$

Applying this form of the mass conservation equation to a flow filament and considering that the same mass flux passes through all the cross-sectional areas of the flow filament, $\frac{\partial M_c}{\partial t} = 0$ (stationary flow conditions) the following results, i.e. the mass inflows and outflows for a flow filament are the same:

$$\int_{A_A} \rho U_i df_i = \int_{A_B} \rho U_i df_i \quad \leadsto \quad A_A U_{s,A} \rho_A = A_B U_{s,B} \rho_B \quad (9.3)$$

where the plane of the area A stands perpendicular to the flow direction s. Therefore, one can conclude that the mass flow $\dot{m} = \rho A U_s$ through the cross-sectional area along a flow filament is constant.

In the derivations carried out above, it was already said that because of small cross-sectional area changes in the flow filaments A, ρ and U_s can be set to be constant over A. When one wants to apply the considerations also to stream tubes, as shown in Fig. 9.3, a more refined approach is necessary. It must be taken into account that the assumption of constant density and velocity U_s in the presence of large cross-sectional areas is only permitted conditionally. The introduction of cross-sectionally averaged quantities is necessary, as will be shown below.

When carrying out the following averaging, with the use of the mean value theorem of integration:

$$\widetilde{\rho U_s} = -\frac{1}{A_s} \int_{A_s} \rho U_s df_s$$

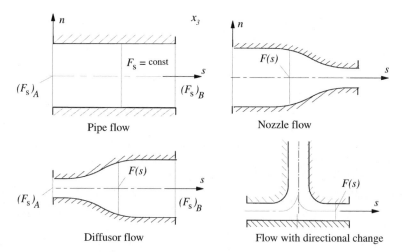

Fig. 9.3 Flows which can be computed one-dimensionally by the approximation method introduced by stream tubes

it can be stated for stationary flow conditions that

$$\overbrace{\frac{\partial M_c}{\partial t}}^{=0} = -\int_{V_c} \rho U_i \mathrm{d}A_i \quad \rightsquigarrow \quad \left(\widetilde{\rho U_s}\right)_A A_A = \left(\widetilde{\rho U_s}\right)_B A_B \quad (9.4)$$

This relationship for the momentum flows through the cross-sectional areas A and B is often simplified further in stream tube theory, by assuming $\widetilde{\rho U_s} = \tilde{\rho}\tilde{U}_s$, i.e. (9.4) is employed as follows:

$$\tilde{\rho}_A \tilde{U}_{s,A} A_A = \tilde{\rho}_B \tilde{U}_{s,B} A_B = \rho U_s A_s \quad (9.5)$$

where $\tilde{\rho}_A$ and $\tilde{\rho}_B$, and also $\tilde{U}_{s,A}$ and $\tilde{U}_{s,B}$ are defined as follows:

$$\tilde{\rho}_A = \frac{1}{A_s} \iint_{A_s} \rho \mathrm{d}f \quad \text{and} \quad \tilde{U}_s = \frac{1}{A_s} \iint_{f_s} U_s \mathrm{d}f \quad (9.6)$$

The above derivations make it clear that the employment of the simplified integral form of the continuity equation $\tilde{\rho}\tilde{U}_s A_s$ = constant is only justified for flows that have no strong variation in density or velocity over the flow cross-section of a considered stream tube. Taking this assumption into account, (9.5) is used in the following derivations. Since in the following considerations, strong variations of the quantities ρ and U_1 over the cross-section are excluded, it is also justified to introduce local quantities in (9.5) as a valid approximation.

In fluid mechanics, a number of questions arise relating to infinitesimal changes of a thermodynamic state or a flow quantity when infinitesimal

9.2 Derivations of the Basic Equations

changes of other parameters take place. For this reason, the continuity equation is often applied in a form derived below:

Differentiation of (9.5) gives:

$$\tilde{U}_s A_s d\tilde{\rho} + \tilde{\rho} A_s d\tilde{U}_s + \tilde{\rho}\tilde{U} dA_s = 0 \tag{9.7}$$

The division of (9.7) by (9.5) leads to a further form of the continuity equation which is employed in some cases in the following sections:

$$\frac{d\tilde{\rho}}{\tilde{\rho}} + \frac{d\tilde{U}_s}{\tilde{U}_s} + \frac{dA_s}{A_s} = 0 \tag{9.8}$$

The equation expresses how, e.g., the velocity of a fluid will change in a relative manner when common relative changes in density and cross-sectional area occur.

9.2.2 Momentum Equation

Solutions of flow problems on the basis of the stream tube theory require the inclusion of the momentum equations. However, these have to be transformed to the flow filament or stream tube coordinates (Fig. 9.4). Starting from the general momentum equation:

$$\rho \left(\frac{\partial U_j}{\partial t} + U_i \frac{\partial U_j}{\partial x_i} \right) = -\frac{\partial P}{\partial x_j} - \frac{\partial \tau_{ij}}{\partial x_i} + \rho g_j \tag{9.9}$$

and neglecting the molecular momentum transport terms acting on the flow filament, the following form of the momentum equations results:

$$\rho \left(\frac{\partial U_s}{\partial t} + U_s \frac{\partial U_s}{\partial s} \right) = -\frac{\partial P}{\partial s} + \rho g_s \tag{9.10}$$

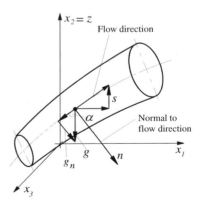

Fig. 9.4 Expression of the momentum equation for the flow filament

with $g_s = -g\cos\alpha$ and $\cos\alpha = \mathrm{d}z/\mathrm{d}s$, so that:

$$\rho\left(\frac{\partial U_s}{\partial t} + U_s\frac{\partial U_s}{\partial s}\right) = -\frac{\partial P}{\partial s} - \rho g\frac{\mathrm{d}z}{\mathrm{d}s} \qquad (9.11)$$

In an analogous way, it can be derived that the momentum equation in the n direction is

$$\rho\frac{U_s^2}{R} = -\frac{\partial P}{\partial n} - g\frac{\mathrm{d}z}{\mathrm{d}n} \qquad (9.12)$$

where the direction of z needs to be chosen in the negative direction of the gravitational field.

This equation expresses that for straight flow filaments, i.e. $R \to \infty$, the pressure variation vertical to the flow direction is given only by the gravitation. When the gravitational forces are negligible, the pressure over the cross-section of a non-curved flow filament is constant.

Starting from the general momentum equation in Eulerian form, i.e. neglecting the molecular momentum loss terms:

$$\frac{\partial(\rho U_j)}{\partial t} + \frac{\partial(\rho U_i U_j)}{\partial x_i} = -\frac{\partial P}{\partial x_j} + \rho g_j \qquad (9.13)$$

and integrating these over a control volume corresponding to the entire space of a flow filament, one obtains for stationary flow conditions:

$$\int_{A_c} \rho U_i U_j \mathrm{d}f = -\int_{A_c} P \mathrm{d}f_j + \int_{V_c} \rho g_j \mathrm{d}V \qquad (9.14)$$

For the special inflow and outflow conditions at the areas A_A and A_B of a flow filament, it can thus be stated that

$$-\rho_A U_{s,A}^2 A_A + \rho U_{s,B}^2 A_B = +P_A A_A - P_B A_B - \int_V \rho g \frac{\mathrm{d}z}{\mathrm{d}s}(A\mathrm{d}s) \qquad (9.15)$$

or rewritten for $g = 0$, i.e. neglecting gravity:

$$\rho U_s^2 A_s + P A_s = \text{constant} \qquad (9.16)$$

This form of momentum equation is employed for many problem solutions in fluid mechanics and will also be employed in the sections to follow in this book.

9.2.3 Bernoulli Equation

When carrying out fluid mechanics considerations, often the pressure and the velocity changes in the flow direction are of interest. Such changes can, when resulting only from mechanical energy changes, be determined from the

9.2 Derivations of the Basic Equations

mechanical energy equation. This equation can be stated in general form as follows (see Sect. 5.5):

$$\rho \frac{D}{Dt}\left(\frac{1}{2}U_j^2 + G\right) = -\frac{\partial(PU_j)}{\partial x_j} + P\frac{\partial U_j}{\partial x_j} - \frac{\partial(\tau_{ij}U_j)}{\partial x_i} + \tau_{ij}\frac{\partial U_j}{\partial x_i} \quad (9.17)$$

When carrying out considerations neglecting the molecular-dependent energy transport on molecular dissipations terms, i.e. setting $\tau_{ij} = 0$, then one obtains for $\rho = $ constant and thus $\frac{\partial U_j}{\partial x_j} = 0$:

$$\rho \frac{D}{Dt}\left(\frac{1}{2}U_j^2 + G\right) = -\frac{\partial(PU_j)}{\partial x_j} = -U_j \frac{\partial P}{\partial x_j} \quad (9.18)$$

In the presence of only stationary pressure fields the following holds:

$$\frac{DP}{Dt} = \underbrace{\frac{\partial P}{\partial t}}_{=0} + U_i\frac{\partial P}{\partial x_i} = U_i\frac{\partial P}{\partial x_i} \quad (9.19)$$

so that under these conditions one can write:

$$\rho \frac{D}{Dt}\left(\frac{1}{2}U_j^2 + G\right) = -\frac{DP}{Dt} \quad (9.20)$$

For $G = -x_j g_j$, (9.20) can be written as:

$$\rho \frac{D}{Dt}\left(\frac{1}{2}U_j^2 + \frac{P}{\rho} - x_j g_j\right) = 0 \quad (9.21)$$

which leads for $j = s$ to the statement of the Bernoulli equation for a flow filament:

$$\frac{1}{2}U_s^2 + \frac{P}{\rho} - g_s s = \text{constant} \quad (9.22)$$

Considering that $-g_s s = gh$, one obtains the final form of the Bernoulli equation:

$$\frac{1}{2}U_s^2 + \frac{P}{\rho} + gh = \text{constant} \quad (9.23)$$

This equation can be interpreted physically such that the mass flux $\dot{m}$, flowing into a flow filament per unit time, introduces the kinetic energy $\dot{m}\frac{1}{2}U_s^2$, the pressure energy $\dot{m}Pv = \dot{m}\frac{P}{\rho}$ and the potential energy $\dot{m}gh$ as total energy. The sum of these three parts along the flow filament cannot change, i.e. the total energy is constant along the flow filament. As at the same time $\dot{m} = $ constant holds, (9.23) results from all these considerations of mass and energy conservations.

9.2.4 The Total Energy Equation

The above considerations of the Bernoulli equation must also include the expansion work when carrying out energy considerations for compressible media. Considering the derivations in Chap. 5, the equation for the total energy has to be employed instead of the above treated mechanical energy equation. According to (5.68), the equation of the total energy can be stated as follows:

$$\rho \frac{D}{Dt}\left(e + \frac{1}{2}U_j^2 + G\right) = -\frac{\partial \dot{q}_j}{\partial x_i} - \frac{\partial (PU_j)}{\partial x_j} - \frac{(\tau_{ij}U_j)}{x_i} \quad (9.24)$$

Neglecting the contributions in the equation due to the molecular-dependent heat and momentum transport, i.e. $\dot{q}_i = 0$ and $\tau_{ij} = 0$, the following equation results for $g \approx 0$:

$$\frac{\partial}{\partial t}\left(\rho e + \frac{1}{2}\rho U_j^2\right) + \frac{\partial}{\partial x_i}\left[U_i\left(\rho e + \frac{1}{2}\rho U_j^2\right)\right] = \frac{\partial}{\partial x_j}(PU_j) \quad (9.25)$$

For stationary flow processes, i.e. neglecting the time derivative terms, the following equation holds:

$$\frac{\partial}{\partial x_i}\left[\rho U_i\left(e + \frac{1}{2}U_j^2 + \frac{P}{\rho}\right)\right] = 0 \quad (9.26)$$

Introducing the enthalpy $h = e + P/\rho$, one obtains for the energy equation for stationary flows of compressible media:

$$h + \frac{1}{2}U_j^2 = \text{constant} \quad (9.27)$$

When carrying out the above considerations for stream tubes, one obtains also the relationship stated in (9.27) for area-averaged quantities:

$$\tilde{h} + \frac{1}{2}\tilde{U}_j^2 = \text{constant} \quad (9.28)$$

The sum of the area-averaged enthalpy of a flowing fluid and the area-averaged kinetic energy per unit mass of the fluid is a constant for adiabatic flows that are free of viscosity and when gravity influences are negligible. With this, the following equations for the computation of flows in stream tubes result:

Flows of incompressible fluids:

- Mass conservation: $\quad \tilde{\rho}\tilde{A}_s\tilde{U}_s = \text{constant} \quad (9.29)$

- Momentum conservation: $\quad \tilde{\rho}\tilde{U}_s^2 A_s + PA_s = \text{constant} \quad (9.30)$

- Mechanical energy equation: $\quad \left\{\frac{1}{2}\tilde{U}_s^2 + \frac{\tilde{P}}{\tilde{\rho}} + g\tilde{h}\right\} = \text{constant} \quad (9.31)$

9.3 Incompressible Flows

Flows of compressible fluids:

- Mass conservation: $\tilde{\rho}\tilde{A}_s\tilde{U}_s = \text{constant}$ (9.32)

- Momentum conservation: $\tilde{\rho}\tilde{U}_s^2 A_s + PA_s = \text{constant}$ (9.33)

- Thermal energy equation: $\frac{1}{2}\tilde{U}_s^2 + \tilde{h} = \text{constant}$ (9.34)

9.3 Incompressible Flows

9.3.1 Hydro-Mechanical Nozzle Flows

The flow problem shown in Fig. 9.5 can be solved, neglecting the friction forces, with the aid of the flow tube theory, in order to obtain a first overview of the flow processes taking place in converging pipes.

When applying one-dimensional flow computations, for the flow in cross-section A, a velocity can be stated which is constant over the entire tube diameter:

$$U_D = \frac{\dot{m}}{\rho} \frac{1}{\frac{\pi}{4}D^2} \quad (9.35)$$

Since the continuity equation holds:

$$\frac{\pi}{4}D^2 U_D = \frac{\pi}{4}d^2 U_d \quad \leadsto \quad U_d = \frac{D^2}{d^2} U_D \quad (9.36)$$

From the Bernoulli equation:

$$\frac{P_D}{\rho} + \frac{1}{2}U_D^2 = \frac{P_d}{\rho} + \frac{1}{2}U_d^2 = \frac{P_0}{\rho} + \frac{1}{2}U_2^2 \quad (9.37)$$

From this P_D is computed as:

$$P_D = P_0 + \frac{\rho}{2}\left(U_d^2 - U_D^2\right) = P_0 + \frac{\rho}{2}\left[\left(\frac{D}{d}\right)^4 - 1\right]U_D^2 \quad (9.38)$$

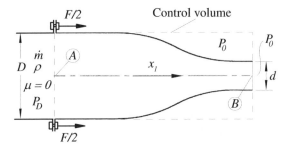

Fig. 9.5 Nozzle flow at the end of a pipe

The flange force F on the control volume in Fig. 9.5 is computed from the integral momentum equation as follows:

$$-\rho U_D^2 \frac{\pi}{4} D^2 + \rho U_d^2 \frac{\pi}{4} d^2 \underbrace{- P_D \frac{\pi}{4} D^2 + P_O \frac{\pi}{4} D^2}_{-\frac{\pi}{4} D^2 (P_D - P_O)} = F \qquad (9.39)$$

or, after rearrangement, in consideration of (9.38):

$$-\frac{\rho}{2} \frac{\pi}{4} D^2 U_D^2 \left(2 - 2 \frac{d^2}{D^2} \frac{U_d^2}{U_D^2} \right) - \frac{\pi}{4} D^2 \frac{\rho}{2} \left(\frac{D^2}{d^4} - 1 \right) U_D^2 = F \qquad (9.40a)$$

and after further rearrangement:

$$-\frac{\rho}{2} \frac{\pi}{4} D^2 U_D^2 \left(1 - \frac{D^2}{d^2} \right)^2 = F \qquad (9.40b)$$

On inserting the corresponding relationships for U_D from (9.35), one obtains for the flange force F

$$A = -\frac{\dot{m}^2}{\rho \frac{\pi}{2} D^2} \left(1 - \frac{D^2}{d^2} \right)^2 \qquad (9.41)$$

The force applied by the flange on the examined nozzle part proves to be positive, so that the supporting surface of the flange receives a negative force F. The screws in the flange can therefore be regarded to be force free as far as any contribution from the flow is concerned. The nozzle is pressed on to the flange.

9.3.2 Sudden Cross-Sectional Area Extension

In practical fluid mechanics, often pipes of different cross-sections are lined up and flows proceed through them. In this way, viewed in the flow direction, internal flows result that are exposed to sudden cross-section widenings, as shown in Fig. 9.6. In this manner separation areas are generated whose influence on the flow can be understood from the following considerations. When treating flows with sudden changes in their cross-sectional area, mass conservation between the planes A and B of the pipe flow exists, and thus yields from the continuity equation:

$$\frac{\dot{m}}{\rho} = U_d \frac{\pi}{4} d^2 = U_D \frac{\pi}{4} D^2 \qquad (9.42)$$

Hence the velocities U_d and U_D can be determined by the given mass flow $\dot{m}$, if the density of the fluid is known, and also the cross sectional areas.

In the case of flows passing through the considered region without losses, the following difference in pressure would result between the planes A and B, which can be computed from the Bernoulli equation:

9.3 Incompressible Flows

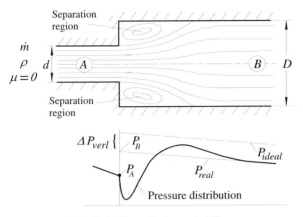

Fig. 9.6 Carnot's impact diffusor

$$\Delta P_{\text{ideal}} = (P_B - P_A)_{\text{ideal}} = \frac{\rho}{2}U_d^2 \left(1 - \frac{d^4}{D^4}\right) \tag{9.43}$$

Under real conditions, as indicated in Fig. 9.6 by the occurrence of the separation areas, the following momentum equation results:

$$F = \rho U_d^2 \frac{\pi}{4} - \rho U_D^2 \frac{\pi}{4}D^2 + P_A \frac{\pi}{4}d^2 - P_B \frac{\pi}{4}D^2 \tag{9.44}$$

When one neglects the contributions to the force F by momentum losses at the pipe walls, then the force F can be computed as the pressure force on the ring surface after the sudden expansion of the pipe, i.e. as

$$F = P_A \frac{\pi}{4}\left(D^2 - d^2\right) \tag{9.45}$$

Thus one obtains for the pressure difference

$$\Delta P_{\text{real}} = (P_A - P_B)_{\text{real}} = \rho U_d^2 \frac{d^2}{D^2}\left(1 - \frac{d^2}{D^2}\right), \tag{9.46}$$

so that a pressure loss (Carnot's momentum loss) can be determined as follows:

$$\Delta P_{\text{loss}} = \Delta P_{\text{ideal}} - \Delta P_{\text{real}} = \frac{\rho}{2}U_d^2\left(1 - \frac{d^4}{D^4}\right) = \frac{\rho}{2}\left(U_d^2 - U_D^2\right) \tag{9.47}$$

For $D \to \infty$ results as a maximum value for $\Delta P_{\text{loss}} = \frac{\rho}{2}U_d^2$, the discharge pressure loss. This means that there is no diffusor available to convert the "dynamic pressure" $\frac{1}{2}\rho U_d^2$ back into "static pressure".

9.4 Compressible Flows

9.4.1 Influences of Area Changes on Flows

The general treatment here provides information on what effect cross-sectional changes in flow channels have on fluid flows, i.e. to what extent and in which way area changes determine the distribution of velocity, pressure, density and temperature along the channel. The equations that were derived in Sect. 9.2 are employed, i.e. the continuity equation reads:

$$\tilde{\rho}\tilde{U}_1 A = \text{constant} \qquad (9.48)$$

Equation (9.48) can be written in differential form as:

$$\frac{dA}{A} + \frac{d\tilde{\rho}}{\tilde{\rho}} + \frac{d\tilde{U}_1}{\tilde{U}_1} = 0 \qquad (9.49)$$

According to the considerations in Sect. 9.2, the variation of the velocity in the flow direction can be described, as a first approximation, by Euler's equation, reduced for one-dimensional flows, i.e.

$$\tilde{\rho}\tilde{U}_1 \frac{d\tilde{U}_1}{dx_1} = -\frac{d\tilde{P}}{dx_1} = -\frac{d\tilde{P}}{d\tilde{\rho}} \frac{d\tilde{\rho}}{dx_1} \qquad (9.50)$$

On the basis of the energy equation, written for reversible adiabatic fluid flows, the following relationship holds:

$$\frac{\tilde{P}}{\tilde{\rho}^{\kappa}} = \text{constant} \qquad (9.51)$$

Under these adiabatic conditions, the differentiation of $\tilde{P}$ with respect to ρ yields the sound velocity c:

$$\tilde{c}^2 = \left(\frac{d\tilde{P}}{d\tilde{\rho}}\right)_{ad} \qquad (9.52)$$

Equation (9.52) introduced into (9.50) yields the following relationship:

$$\tilde{\rho}\tilde{U}_1 \frac{d\tilde{U}_1}{dx_1} = -\tilde{c}^2 \frac{d\tilde{\rho}}{dx_1} \qquad (9.53)$$

When one introduces the Mach number Ma of the flow as:

$$\widetilde{Ma} = \frac{\tilde{U}_1}{\tilde{c}} \qquad (9.54)$$

9.4 Compressible Flows

(9.53) can be written as:

$$\frac{d\tilde{\rho}}{\tilde{\rho}} = -\widetilde{Ma}^2 \frac{d\tilde{U}_1}{\tilde{U}_1} \tag{9.55}$$

Inserting the density variation in (9.55) in (9.49), one obtains:

$$\frac{dA}{A} - \widetilde{Ma}^2 \frac{d\tilde{U}_1}{\tilde{U}_1} + \frac{d\tilde{U}_1}{\tilde{U}_1} = 0 \tag{9.56}$$

or

$$\frac{d\tilde{U}_1}{\tilde{U}_1} = \frac{-1}{(1-\widetilde{Ma}^2)} \frac{dA}{A} \tag{9.57}$$

When one takes into consideration that subsonic flows are given by $\widetilde{Ma} < 1$ and supersonic flows by $\widetilde{Ma} > 1$, the above relationship expresses:

- In the presence of a *subsonic flow* ($\widetilde{Ma} < 1$), a *decrease* in the cross-sectional area of a flow channel in the flow direction is linked to an increase in the flow velocity. An *increase* in the channel cross-sectional area in the flow direction results in a *decrease* in the flow velocity (see Fig. 9.7).
- In the presence of a *supersonic flow* ($\widetilde{Ma} > 1$), a *decrease* in the cross-sectional area of a flow channel in the flow direction is linked to a *decrease* in the flow velocity. An *increase* in the flow cross-section in the flow direction results in an *increase* in the flow velocity (see Fig. 9.8).

In addition to the changes in the flow velocity, caused by changes in the cross-sectional areas, the changes in pressure, density and temperature of the flowing fluid are also of interest. From (9.51), it can be seen that the relative change in density always has the opposite sign to the change in velocity, i.e. the density increases in the flow direction when the velocity decreases and vice versa. In the region of subsonic flow, the locally present relative

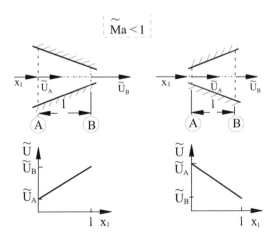

Fig. 9.7 Influence of a change in the flow cross-section on a subsonic flow

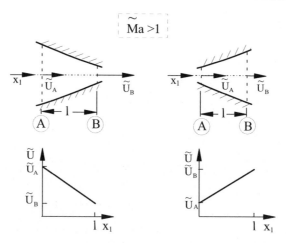

Fig. 9.8 Influence of a change in the flow cross-section on a supersonic flow

change in density is smaller than the local relative change in velocity. In the region of supersonic flow, the locally present relative change in density is larger than the relative change in velocity. The changes in the density for the corresponding changes in cross-sectional area changes of the flow channel are given by the relationship:

$$\frac{d\tilde{\rho}}{\tilde{\rho}} = \frac{\widetilde{Ma}^2}{(1 - \widetilde{Ma}^2)} \frac{dA}{A} \tag{9.58}$$

With regard to the pressure variation, the following considerations can be carried out. From the adiabatic pressure-density relationship (9.51), the following results:

$$d\tilde{P} = \frac{\tilde{P}}{\tilde{\rho}^\kappa} \kappa \tilde{\rho}^{(\kappa-1)} d\tilde{\rho} = \kappa \frac{\tilde{P}}{\tilde{\rho}} d\tilde{\rho} \tag{9.59}$$

Therefore, for the local relative change in pressure one can derive:

$$\frac{d\tilde{P}}{\tilde{P}} = \kappa \widetilde{Ma}^2 \frac{d\tilde{U}_1}{\tilde{U}_1} \tag{9.60}$$

or with regard to the local relative change in the cross-sectional area of the flow, the following relative change in pressure results:

$$\frac{d\tilde{P}}{\tilde{P}} = \frac{\kappa \widetilde{Ma}^2}{(1 - \widetilde{Ma}^2)} \frac{dA}{A} \tag{9.61}$$

Finally, it is necessary to consider the variations in temperature. For this purpose, the state equation for ideal gases is differentiated:

$$-\tilde{P}\frac{d\tilde{\rho}}{\tilde{\rho}^2} + \frac{d\tilde{P}}{\tilde{\rho}} = Rd\tilde{T}\frac{dA}{A} \tag{9.62}$$

9.4 Compressible Flows

or rewritten in the following form:

$$-\frac{\mathrm{d}\tilde{\rho}}{\tilde{\rho}} + \frac{\mathrm{d}\tilde{P}}{\tilde{P}} = \frac{\mathrm{d}\tilde{T}}{\tilde{T}} \tag{9.63}$$

Hence, knowing $\frac{\mathrm{d}\tilde{\rho}}{\tilde{\rho}}$ and $\frac{\mathrm{d}\tilde{P}}{\tilde{P}}$, the following relationship for the temperature changes results:

$$\frac{\mathrm{d}\tilde{T}}{\tilde{T}} = -(\kappa - 1)\widetilde{Ma}^2 \frac{\mathrm{d}\tilde{U}_1}{\tilde{U}_1} \tag{9.64}$$

The locally occurring relative change in temperature has the opposite sign to the local relative change in velocity. The relative changes in temperature are weaker than the corresponding relative changes in density. With regard to the relative area change of the flow cross-section, it results that

$$\frac{\mathrm{d}\tilde{T}}{\tilde{T}} = \frac{(\kappa - 1)\widetilde{Ma}^2}{(1 - \widetilde{Ma}^2)} \frac{\mathrm{d}A}{A} \tag{9.65}$$

The considerations stated for the flow velocity variations in supersonic and subsonic flows, which are sketched in Figs. 9.7 and 9.8, can also be carried out for the variations in pressure, density and temperature with the aid of the above equations.

Another important result of the above derivations can be stated through a rearrangement of the relationships derived above, such that the following equation holds:

$$\frac{\mathrm{d}A}{\mathrm{d}\tilde{U}_1} = \frac{A}{\tilde{U}_1}(1 - \widetilde{Ma}^2) \tag{9.66}$$

This relationship expresses that the condition for achieving the sound velocity is given by $\mathrm{d}A = 0$, i.e. $\widetilde{Ma} = 1$. Since for the second derivative of A

$$\frac{\mathrm{d}^2 A}{\mathrm{d}\tilde{U}_1^2} = \frac{A}{\tilde{U}_1^2}\widetilde{Ma}^2(\widetilde{Ma}^2 - 2) \tag{9.67}$$

for $\widetilde{Ma} = 1$ the condition for some flow to exist is given by a minimum of the flow cross-section. Further considerations of changes if errors sectional area are given in refs. [9.1] to [9.5].

9.4.2 Pressure-Driven Flows Through Converging Nozzles

In many technical plants, flows of gases occur which are to be classified into a group of flows that take place between reservoirs with differing pressure levels. Gases, for example, are often stored under high pressure in large storage reservoirs, in order to be discharged through conduits for the intended purpose when need arises. This discharge can be idealized as a "equalization flow" between two reservoirs or two chambers of which one represents the storage reservoir under pressure, while the environment represents the second reservoir. In the following considerations it is assumed that both reservoirs are very large, so that constant reservoir conditions exist during the entire "equalization flow" under investigation. These are assumed to be known and are given by the pressure P_H, the temperature T_H, etc., in the high-pressure reservoir, and also through the pressure P_N and temperature T_N for the low-pressure reservoir. The compensating flow takes place via a continually converging nozzle as indicated in Fig. 9.9, whose largest cross-section thus represents the discharge opening of the high pressure reservoir, whereas the smallest nozzle cross-section represents the inlet opening into the low-pressure reservoir.

When one wants to investigate the fluid flows taking place in the above equalization flow in more detail, the final equations for flows through channels, pipes, etc., derived in Sect. 9.2 can be used:

$$\tilde{\rho}\tilde{U}_1 A = \text{constant} \tag{9.68}$$

$$\tilde{h} + \frac{1}{2}\tilde{U}_1^2 = \text{constant}; \quad \frac{\tilde{P}}{\tilde{\rho}^\kappa} = \text{constant} \tag{9.69}$$

$$\frac{\tilde{P}}{\tilde{\rho}} = R\tilde{T} \tag{9.70}$$

With (9.68)–(9.70), a sufficient number of equations exists to determine the changes in the area-averaged velocity and the area-averaged thermodynamic state quantities of the flowing gas. Hence the velocity, pressure, temperature and density along the x_1 axis, shown in Fig. 9.9, can be found by solving this set of equations.

When one considers that, based on the assumption of a large high-pressure reservoir there is the constant pressure P_H and the velocity $(U_1)_H = 0$, then for the velocity U_1 at each point x_1 of the nozzle the following relationship can be stated to be valid:

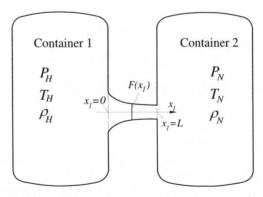

Fig. 9.9 Flow between two reservoirs through a converging nozzle

9.4 Compressible Flows

$$\tilde{h} + \frac{1}{2}\tilde{U}_1^2 = h_H \tag{9.71}$$

Taking into account that the enthalpy for an ideal gas can be stated as $c_P T$ and moreover that the ideal gas equation (9.70) holds, (9.71) can be rewritten as follows:

$$c_P \frac{\tilde{P}}{R\tilde{\rho}} + \frac{1}{2}\tilde{U}_1^2 = \frac{\kappa}{\kappa-1} \frac{\tilde{P}}{\tilde{\rho}} + \frac{1}{2}\tilde{U}_1^2 = \frac{\kappa}{\kappa-1} \frac{P_H}{\rho_H} \tag{9.72}$$

The velocity U_1 is thus linked to the change in the pressure along the axis of the nozzle as follows:

$$\tilde{U}_1 = \sqrt{\frac{2\kappa}{\kappa-1}\left(\frac{P_H}{\rho_H} - \frac{\tilde{P}}{\tilde{\rho}}\right)} \tag{9.73}$$

The above equation indicates that for $\tilde{P} = 0$, i.e. for the outflow into a vacuum, a maximum possible flow velocity develops which is given by the state of the reservoir only:

$$U_{\max} = \sqrt{\frac{2\kappa}{\kappa-1}\frac{P_H}{\rho_H}} = \sqrt{2c_P T_H} \tag{9.74}$$

Standardizing the flow velocity U_1, existing at a point x_1, with $U_{\max}$, one obtains:

$$\frac{\tilde{U}_1}{U_{\max}} = \sqrt{1 - \frac{\tilde{P}\cdot\rho_H}{P_H \tilde{\rho}}} \tag{9.75}$$

or rewritten by taking the ideal gas equation into account:

$$\frac{\tilde{U}_1}{U_{\max}} = \sqrt{1 - \frac{\tilde{T}}{T_H}} \tag{9.76}$$

Linking the adiabatic equation (9.86) to the state (9.70) leads to the following relationships:

$$\frac{\tilde{T}}{T_H} = \left(\frac{\tilde{\rho}}{\rho_H}\right)^{\kappa-1} \quad \text{and} \quad \frac{\tilde{T}}{T_H} = \left(\frac{\tilde{P}}{P_H}\right)^{\frac{\kappa-1}{\kappa}} \tag{9.77}$$

Thus the following equations hold:

$$\frac{\tilde{U}_1}{U_{\max}} = \sqrt{\left[1 - \left(\frac{\tilde{\rho}}{\rho_H}\right)^{\kappa-1}\right]} \tag{9.78}$$

and

$$\frac{\tilde{U}_1}{U_{\max}} = \sqrt{\left[1 - \left(\frac{\tilde{P}}{P_H}\right)^{\frac{\kappa-1}{\kappa}}\right]} \qquad (9.79)$$

On choosing the normalized velocity $(\tilde{U}_1/U_{\max})$ as a parameter for the representation of the flow in the nozzle, the distributions of pressure, density and temperature along the nozzle can be stated as follows:

$$\frac{\tilde{P}}{P_H} = \left[1 - \left(\frac{\tilde{U}_1}{U_{\max}}\right)^2\right]^{\frac{\kappa}{\kappa-1}} \qquad (9.80)$$

$$\frac{\tilde{\rho}}{\rho_H} = \left[1 - \left(\frac{\tilde{U}_1}{U_{\max}}\right)^2\right]^{\frac{1}{\kappa-1}} \qquad (9.81)$$

$$\frac{\tilde{T}}{T_H} = \left[1 - \left(\frac{\tilde{U}_1}{U_{\max}}\right)^2\right] \qquad (9.82)$$

These relationships are shown in Fig. 9.10 as functions of $(\tilde{U}_1/U_{\max})$. Also, along the $(\tilde{U}_1/U_{\max})$ axis, the corresponding Mach number of the flow is plotted, which under consideration of the relationship $c = \sqrt{(\mathrm{d}P/\mathrm{d}\rho)_{\mathrm{ad}}} = \sqrt{\kappa R T}$ can be shown to be identical with

$$\frac{\tilde{U}_1^2}{U_{\max}^2} = \frac{\tilde{U}_1^2}{2c_P T_H} \frac{\kappa R \tilde{T}}{\kappa R \tilde{T}} = \widetilde{Ma}_1^2 \frac{\kappa-1}{2}\left(\frac{\tilde{T}}{T_H}\right) \qquad (9.83)$$

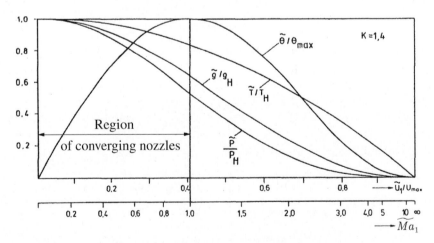

Fig. 9.10 Distributions of the pressure, density and temperature as a function of the local normalized velocity or as a function of the local Mach number

9.4 Compressible Flows

When one considers the relationship derived above for (T/T_H), in (9.77), one obtains for the Mach number the following dependence on $(\tilde{U}_1/U_{\max})$:

$$\widetilde{Ma} = \sqrt{\frac{2}{\kappa-1}\left\{\frac{\left(\dfrac{\tilde{U}_1}{U_{\max}}\right)^2}{\left[1-\left(\dfrac{\tilde{U}_1}{U_{\max}}\right)^2\right]}\right\}} \qquad (9.84)$$

Hence a Mach number of the flow can be assigned to each value of an area-averaged velocity normalized with the maximum velocity.

All quantities which are stated in (9.80)–(9.82) can also be written as functions of the Mach number $\widetilde{Ma}$. This in turn can be considered as an area-averaged flow quantity describing the distributions of the flow along the x_1 axis.

For the derivation showing the dependence of the pressure, density and temperature on the Mach number of the flow, shown graphically in Fig. 9.10, (9.71) is written as follows:

$$c_P \tilde{T} + \frac{1}{2}\tilde{U}_1^2 = c_P T_H \qquad (9.85)$$

By division with $c_P \tilde{T}$ one obtains

$$\frac{T_H}{\tilde{T}} = 1 + \frac{\tilde{U}_1^2}{2c_P \tilde{T}} \frac{\kappa R}{\kappa R} = 1 + \frac{\kappa-1}{2}\widetilde{Ma}_1^2 \qquad (9.86)$$

or for the reciprocal

$$\frac{\tilde{T}}{T_H} = \frac{2}{2+(\kappa-1)\widetilde{Ma}_1^2} \qquad (9.87)$$

This equation makes it clear that there is a relationship between the area-averaged temperature at a location on the x_1 axis and the Mach number existing at the same point of the flow. Hence it becomes clear that for each point x_1 the temperature can be computed when the high-pressure reservoir temperature is given and the Mach number of the flow is known.

Taking into account the adiabatic equation, the relationship between the pressure $\tilde{P}$ and reservoir pressure P_H is given by

$$\frac{\tilde{P}}{P_H} = \left(\frac{\tilde{T}}{T_H}\right)^{\frac{\kappa}{\kappa-1}} = \left[\frac{2}{2+(\kappa-1)\widetilde{Ma}_1^2}\right]^{\frac{\kappa}{\kappa-1}} \qquad (9.88)$$

and the corresponding relationship for the density is

$$\frac{\tilde{\rho}}{\rho_H} = \left(\frac{\tilde{T}}{T_H}\right)^{\frac{1}{\kappa-1}} = \left[\frac{2}{2+(\kappa-1)\widetilde{Ma}_1^2}\right]^{\frac{1}{\kappa-1}} \qquad (9.89)$$

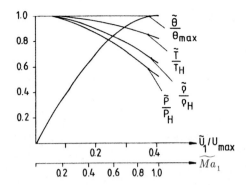

Fig. 9.11 Distribution of the pressure, density, temperature and mass flow density for converging nozzles

Figure 9.11 also contains the distribution of the flux density $\theta = \dot{m}/A = \tilde{\rho}\tilde{U}_1$, i.e. the mass flowing per unit area and unit time through the cross-section of the flow. The equation for this quantity can be written as follows, using the relationships for $\tilde{U}_1$ and $\tilde{\rho}$:

$$\tilde{\rho}_1 \tilde{U}_1 = \rho_H \left[1 - \left(\frac{\tilde{U}_1}{U_{\max}}\right)^2\right]^{\frac{1}{\kappa-1}} \tilde{U}_1^2 \qquad (9.90)$$

or for the normalized mass flow density:

$$\frac{\tilde{\rho}_1 \tilde{U}_1}{\rho_H U_{\max}} = \frac{\tilde{U}_1}{U_{\max}} \left[1 - \left(\frac{\tilde{U}_1}{U_{\max}}\right)^2\right]^{\frac{1}{\kappa-1}} \qquad (9.91)$$

The above relationship for the mass flow density makes it clear that for $U_1 = 0$, the mass flow $\theta = 0$ is achieved. The mass flow density, however, assumes the value zero also for $U_1 = U_{\max}$. The reason for this is that at the maximum possible velocity the density of the fluid, also determining the mass flow density, has dropped to $\tilde{\rho} = 0$. Between these two minimum values the mass flow density has to traverse a maximum which can be computed by differentiation of the above functions and by setting the derivation to zero. The value obtained by solving the resulting equation has to be inserted for $\tilde{U}_1/U_{\max}$ in the above equation for the mass flow density in order to achieve the maximum value. We have:

$$\theta_{\max} = \rho_H U_{\max} \sqrt{\frac{\kappa-1}{\kappa+1}} \left(\frac{2}{\kappa+1}\right)^{\frac{1}{\kappa-1}} \qquad (9.92)$$

where for the velocity value:

$$\frac{\tilde{U}_1}{U_{\max}} = \sqrt{\frac{\kappa-1}{\kappa+1}} \qquad \text{for } \theta = \theta_{\max} \qquad (9.93)$$

9.4 Compressible Flows

With this, the mass flow density normalized with the maximum value can be written as:

$$\frac{\theta}{\theta_{max}} = \sqrt{\frac{\kappa+1}{\kappa-1}} \frac{\tilde{U}_1}{U_{max}} \left[\frac{\kappa+1}{2}\left(1 - \frac{\tilde{U}_1^2}{U_{max}}\right)\right]^{\frac{1}{\kappa-1}} \quad (9.94)$$

The distribution of this quantity with $\tilde{U}_1/U_{max}$ is also represented in Fig. 9.10. The significance of the maximum of the mass flow density for the distribution of pressure-driven flows is dealt with more in detail later. Its appearance prevents a steady increase of the mass flow with increase in the pressure difference between the pressure reservoirs when the compensating flow takes place via steadily converging nozzles.

A representation of the flows through converging nozzles, often regarded as simpler, is achieved by relating the quantities designating the flow to the corresponding quantities of the "critical state", which is designated by $\widetilde{Ma} = 1$. To this state corresponds not only a certain Mach number, i.e. $\widetilde{Ma}_1 = 1$, but also certain values of the thermodynamic state quantities: These can be determined from (9.87)–(9.89) by setting $\widetilde{Ma}_1 = 1$. From this the following values for thermodynamic state quantities of the fluid result at the critical state of the flow, i.e. for $\widetilde{Ma}_1 = 1$:

$$\frac{\tilde{P}^*}{P_H} = \left(\frac{2}{\kappa+1}\right)^{\frac{\kappa}{\kappa-1}} \quad (9.95)$$

$$\frac{\tilde{\rho}^*}{\rho_H} = \left(\frac{2}{\kappa+1}\right)^{\frac{1}{\kappa-1}} \quad (9.96)$$

$$\frac{\tilde{T}^*}{T_H} = \frac{2}{\kappa+1} \quad (9.97)$$

With these equations, the pressure, density and temperature of a flowing medium can be determined in that cross-section of a converging nozzle in which the velocity of the fluid takes on the local sound velocity. According to the considerations at the end of Sect. 9.4.1, a minimum of the cross-section has to exist at this point. As here the Mach number assumes the value $\widetilde{Ma}_1 = 1$, (9.83) can be written as follows:

$$\frac{\tilde{U}_1^{*2}}{U_{max}^2} = \frac{\kappa-1}{2}\left(\frac{\tilde{T}}{T_H}\right) = \frac{\kappa-1}{\kappa+1} \quad (9.98)$$

On comparing the values for $\tilde{U}_1/U_{max}$ in 9.98 and 9.93, one finds that they are identical, i.e. the maximum mass flow density can only occur in the narrowest cross-section of a nozzle, where the sound velocity then also applies.

In accordance with the above derivations of the basic equations for pressure-driven flows between large reservoirs, the flow which occurs in a steadily converging nozzle, as sketched in Fig. 9.7, will be discussed. The considerations will be carried out in such a way that the mass flow is computed which results when a certain pressure relationship (P_N/P_H) between the reservoirs applies. Here two pressure ranges are of interest:

$$\frac{P_N}{P_H} > \frac{P^*}{P_H}\qquad\text{The ratio of the normalized reservoir pressures is larger than the critical pressure ratio}$$

$$\frac{P_N}{P_H} < \frac{P^*}{P_H}\qquad\text{The ratio of the reservoir pressures is smaller than the critical pressure ratio}$$

When the pressure ratio is larger than the critical value, a steady decrease in the ratio of the reservoir pressures leads to a steady increase in the mass flow density, as indicated in Fig. 9.11. The latter represents part of the total diagram stated in Fig. 9.10, namely up to $Ma = 1$. For the variation of the state quantities, namely for the pressure and the density the diagram is given. On the assumption that in the narrowest cross-section of the steadily converging nozzle the pressure of the low-pressure reservoir sets in, the pressure ratio P_N/P_H can be determined from the known values P_N and P_H. Via the same approach the mass flow density in this cross-section can be determined in the following manner and thus also the total mass flowing through the nozzle:

$$\dot{m}_H = A_H \tilde{\theta}_H = A_H (\tilde{\rho}\tilde{U}_1)_H \tag{9.99}$$

For reasons of continuity, this total mass flow is constant in all cross-section planes of the nozzle, so that

$$\dot{m}_H = \dot{m} \quad\text{i.e.}\quad A_N \tilde{\theta}_N = A_{x_1}\tilde{\theta}_{x_1} \tag{9.100}$$

Starting from the assumption that the specified distribution of the cross-sectional area of the nozzle along the x_1 axis is known, then the mass flow density distribution along the x_1 axis can be determined using (9.100). Via the same approach, one can then compute, as indicated in Fig. 9.12, the pressure distribution along the nozzle, or the resulting distributions of the density and the temperature, but also of the Mach number and the flow velocity.

The approach to determining the pressure distribution along the nozzle, indicated in Fig. 9.13, can be applied analogously also to define the density distribution and the temperature distribution. To determine the distribution of the Mach number and the velocity, the approach indicated in Fig. 9.12 holds.

It follows from the above considerations that the velocity $(U_1)_N$ in the entrance cross-section of the nozzle indicated in Fig. 9.7 is finite and that there the mass flow density

$$\tilde{\theta}_H = \frac{A_N}{A_H}\left(\tilde{\rho}\tilde{U}_1\right)_N \tag{9.101}$$

9.4 Compressible Flows

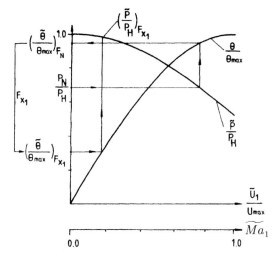

Fig. 9.12 Determining the pressure distribution along the nozzle axis for $(P_N/P_H) > (P_N^*/P_H)$

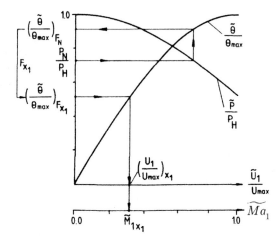

Fig. 9.13 Determining the Mach number and the velocity distribution along a converging nozzle for $(P_N/P_H) < (P_N^*/P_H)$

is present. Also in this cross-section a pressure, a density and a temperature exist which do not correspond to the values in the high-pressure reservoir. It is necessary always to take this into consideration when computing pressure-driven flows through nozzles. The quantities designating the flows that exist at the nozzle entrance are to be determined via the above diagrams from the mass flow density computed for the entrance cross-section in accordance with (9.101).

When carrying out the above computations for determining the flow quantities and the thermodynamic quantities, with a decrease in the pressure ratio P_N/P_H an increase in the mass flow density in each cross-section of the nozzle is obtained, as long as the pressure ratio is larger than the critical value. When the critical value itself is reached:

$$\frac{P_N}{P_H} = \left(\frac{2}{\kappa+1}\right)^{\frac{\kappa}{\kappa-1}} = \frac{P^*}{P_H} \qquad (9.102)$$

This value cannot be exceeded in the case of a further decrease in the pressure ratio P_N/P_H i.e. for all pressure ratios smaller than the critical value:

$$\frac{P_N}{P_H} < \frac{P^*}{P_H} = \left(\frac{2}{\kappa+1}\right)^{\frac{\kappa}{\kappa-1}} \qquad (9.103)$$

in the steadily converging nozzle, a flow comes about which is identical for all pressure relationships. At the exit cross-section of the nozzle, i.e. in the entrance cross-section to the low-pressure reservoir, the pressure P_N no longer applies. In this cross-section, the maximum mass flow density rather is reached:

$$\tilde{\theta}_{\max} = \rho_H \sqrt{\frac{2\kappa}{\kappa-1}\frac{P_H}{\rho_H}} = \sqrt{\frac{2\kappa}{\kappa-1}P_H\rho_H} \qquad (9.104)$$

or

$$\tilde{\theta}_{\max} = \rho_H \sqrt{2c_P T_H} \qquad (9.105)$$

The total mass flow thus is computed as:

$$\dot{m} = \dot{m}_{\max} = A_N \tilde{\theta}_{\max} \qquad (9.106)$$

Starting again from the assumption that the nozzle form is known, then the mass flow distribution existing along the axis x_1 can be computed via the continuity equation. When this distribution is known, the corresponding distributions of the pressure, density, temperature, Mach number and flow velocity can be determined as stated above. Of importance is that for all pressure ratios P_N/P_H that are equal to or smaller than the critical ratio, the same flow occurs in the nozzle. In the exit cross-section of the nozzle for

$$\frac{P_N}{P_H} < \frac{\tilde{P}^*}{P_H} = \left(\frac{2}{\kappa+1}\right)^{\frac{\kappa}{\kappa-1}} \qquad (9.107)$$

an area-averaged pressure exists which is larger than the pressure P_N existing in the low-pressure reservoir. The pressure compensation takes place via fluid flows that form in the open jet flow, stretching from the nozzle tip to the interior of the low-pressure reservoir (Fig. 9.14).

Finally, attention is drawn to important facts that arise when considering pressure-driven flow. The above representations started from the state often

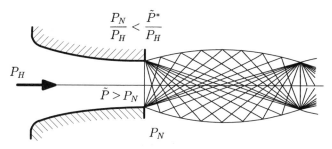

Fig. 9.14 Pressure compensation at the nozzle exit via density impacts

existing in practice that pressure-driven flows are controlled via pressure differences between reservoirs. This means that it was assumed that P_H, ρ_H or T_H are known and constant and that they have an influence on how the flow forms. In the low-pressure reservoir it was only assumed that P_N is given and can be forced upon the flow in the narrowest cross-section of the nozzle (for P_N/P_H larger than the critical value P^*/P_H). The density of the flowing gas that occurs for these conditions in the exit cross-section of the nozzle or the temperature that arises are not identical with the corresponding values for the fluid in the low-pressure reservoir.

Equalization of these values and the corresponding values for the low-pressure reservoir takes place in the open jet flow following the nozzle flow. For pressure conditions

$$\frac{P_N}{P_H} < \frac{P^*}{P_H} = \left(\frac{2}{\kappa+1}\right)^{\frac{\kappa}{\kappa-1}} \tag{9.108}$$

the equalization takes place between the pressure in the nozzle exit cross-section and the pressure in the low-pressure reservoir, and likewise in the open jet flow following the nozzle flow. Further details of one-dimensional compressible flows are provided in refs. [9.1] to [9.2].

References

9.1. E. Becker: "Technische Thermodynamik", Teubner Studienbuecher Mechanik, B.G. Teubner, Stuttgart (1985)
9.2. K. Hutter: "Fluid- und Thermodynamik - Eine Einfuehrung", Springer, Berlin Heidelberg New York (1995)
9.3. K. Oswatitsch: "Gasdynamik," Springer, Berlin Heidelberg New York (1952)
9.4. J.H. Spurk: "Stroemungslehre", 4. Auflage, Springer, Berlin Heidelberg New York (1996)
9.5. S.W. Yuan: "Foundations of Fluid Mechanics", Civil Engineering and Mechanics Series, Mei Ya Publications, Taipei, Taiwan (1971)

Chapter 10
Potential Flows

10.1 Potential and Stream Functions

In order to make the integration of the partial differential equation of fluid mechanics possible by simple mathematical means, the introduction of irrotationality of the flow field is necessary. The introduction of irrotationality is necessary to yield a replacement for the momentum equations and it is this fact that permits simpler mathematical methods to be applied. In Sect. 5.8.1, a transport equation equivalent to the momentum equation was derived for the vorticity which for viscosity-free flows is reduced to the simple form $D\omega/Dt = 0$. From this equation, two things follow. On the one hand, it becomes evident that irrotational fluids obey automatically a simplified form of the momentum equation. On the other hand, Kelvin's theorem results immediately, according to which all flows of viscosity-free fluids are irrotational, when at any point in time the irrationality of the flow field was detected. This can be understood graphically by considering that all surface forces acting on a non-viscous fluid element act normal to the surface and as a resultant go through the center of mass of the fluid element. At the same time, the inertia forces also act on the center of mass, so that no resultant momentum comes about which can lead to a rotation. Hence the conclusion is possible that rotating fluid elements cannot receive an additional rotation due to pressure and inertia forces acting on ideal fluids. This is indicated in Fig. 10.1.

In addition to the above requirement for irrotationality, a further restriction will now be introduced regarding the properties of the flows that are dealt with in this chapter, namely the exclusive consideration of two-dimensional flows. This restriction imposed on the allowable properties of flows is not a condition resulting from irrotationality; one can, on the contrary, well imagine three-dimensional flows of viscosity-free fluids that are irrotational. For two-dimensional, irrotational flows there exists, however, a very elegant solution method which is based on the employment of complex analytical functions and which is used exclusively in the following.

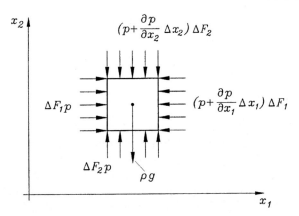

Fig. 10.1 Graphical representation of the physical cause of irrotationality of ideal flows (Kelvin's theorem)

When considering two-dimensional flow fields with flow property dependences on x_1 and x_2, the only remaining component of the rotational vector is:

$$\omega_3 = \frac{1}{2}\left(\frac{\partial U_2}{\partial x_1} - \frac{\partial U_1}{\partial x_2}\right). \tag{10.1}$$

When one assumes the considered two-dimensional flow fields to be irrotational, it holds that $\omega_3 = 0$ or:

$$\frac{\partial U_1}{\partial x_2} = \frac{\partial U_2}{\partial x_1}. \tag{10.2}$$

This condition has to be fulfilled in addition to the continuity equation when irrotational two-dimensional flow problems are to be solved.

Disregarding singularities, for irrotational flow fields the above relationship has to be fulfilled in all points of the flow field. This is tantamount to the statement that, for two-dimensional irrotational flows, a velocity potential $\Phi(x_1, x_2)$ driving the flow exists, to such an extent that the following relationships hold:

$$U_1 = \frac{\partial \Phi}{\partial x_1} \quad \text{and} \quad U_2 = \frac{\partial \Phi}{\partial x_2}. \tag{10.3}$$

Equation (10.3) inserted in (10.2) leads to the following relations:

$$\frac{\partial U_1}{\partial x_2} = \frac{\partial^2 \Phi}{\partial x_1 \partial x_2} \quad \text{and} \quad \frac{\partial U_2}{\partial x_1} = \frac{\partial^2 \Phi}{\partial x_1 \partial x_2}, \tag{10.4}$$

which for irrotational flow fields, i.e. for $\omega_3 = 0$ [see (10.1)], confirm the reasonable introduction of a potential driving the velocity field. When one

10.1 Potential and Stream Functions

inserts (10.3) into the two-dimensional continuity equation (5.18) for $\rho = $ constant then one obtains the Laplace equation for the velocity potential:

$$\frac{\partial^2 \Phi}{\partial x_1^2} + \frac{\partial^2 \Phi}{\partial x_2^2} = 0. \tag{10.5}$$

For determining two-dimensional potential fields, it is sufficient to solve (10.3) and (10.5), i.e. for determining the velocity field it is not necessary to solve the Navier–Stokes equation, formulated in velocity terms. These equations, or the equation derived in Sect. 5.8.1, have to be employed, however, for determining the pressure field.

The solution of the partial differential (10.5) for the velocity potential requires at the boundary of the flow the boundary condition

$$\frac{\partial \Phi}{\partial n} = 0, \tag{10.6}$$

where n is the normal unit vector at each point of the flow boundary.

When the velocity potential Φ, or potential field Φ, has been obtained as a solution of (10.5), the velocity components U_1 and U_2 can be determined for each point of the flow field by partial differentiations, according to (10.3). After that, the determination of the pressure via Euler's equations, i.e. via the momentum equations for viscosity-free fluids, can be computed. Determining the pressure can also be done, however, via the integrated form of Euler's equations, which leads to the "non-stationary Bernoulli equation".

The above treatments make it clear that the introduction of the irrotationality of the flow field has led to considerable simplifications of the solution ansatz for the basic equations for flow problems. The equations that have to be solved for the flow field are linear and they can be solved decoupled from the pressure field. The linearity of the equations to be solved is an essential property as it permits the superposition of individual solutions of the equations in order to obtain also solutions of complex flow fields. This solution principle will be used extensively in the following sections.

In the derivations of the above equations for two-dimensional potential flows, the potential function was introduced in such a way that the irrotationality of the flow field was fulfilled by definition. The introduction of the potential function Φ into the continuity equation then led to the two-dimensional Laplace equation; only such functions Φ which fulfil this equation can be regarded as solutions of the basic equations of irrotational flows.

Via a procedure similar to the above introduction of the potential function Φ, it is possible to introduce a second important function for two-dimensional flows of incompressible fluids, the so-called stream function Ψ. The latter is defined in such a way that through the stream function the two-dimensional continuity equation is automatically fulfilled, i.e.

$$U_1 = \frac{\partial \Psi}{\partial x_2} \quad \text{and} \quad U_2 = -\frac{\partial \Psi}{\partial x_1}. \tag{10.7}$$

This relationship, inserted into the continuity equation, shows directly that the stream function Ψ (introduced according to (10.7)) fulfills this equation; by definition this is the case for rotational and irrotational flow fields.

When one wants to define analytically or numerically the stream function of an irrotational flow the function Ψ has to be a solution of the Laplace equation:

$$\frac{\partial^2 \Psi}{\partial x_1^2} + \frac{\partial^2 \Psi}{\partial x_2^2} = 0. \tag{10.8}$$

This equation can be derived by inserting equations (10.7) into the condition for the irrotationality of the flow:

$$\frac{\partial U_2}{\partial x_1} - \frac{\partial U_1}{\partial x_2} = 0.$$

The stream function for two-dimensional potential flows fulfils the two-dimensional Laplace equation, similar to the potential function Φ.

The stream function has a number of properties that prove useful for the treatment of two-dimensional flow problems. Lines of constant stream-function values, for example, are path lines of the flow field when stationary conditions exist for the flow. This can be derived by stating the total differential of Ψ:

$$d\Psi = \frac{\partial \Psi}{\partial x_1} dx_1 + \frac{\partial \Psi}{\partial x_2} dx_2. \tag{10.9}$$

For $\Psi = \text{constant}$ $d\Psi = 0$ and therefore

$$\left(\frac{dx_2}{dx_1}\right)_{\Psi=\text{constant}} = -\frac{\frac{\partial \psi}{\partial x_1}}{\frac{\partial \Psi}{\partial x_2}} = \frac{U_2}{U_1}. \tag{10.10}$$

This is the relationship for the gradient of the tangent of the stream line, but also for the gradient of the path line of a fluid element. Accordingly, the total family of stream lines of a velocity field is described by all of Ψ values from 0 to infinity.

A further essential property of the stream function becomes clear from the fact that the difference of the stream-function values of two flow lines indicates the volume flow rate that flows between the flow lines. This can be derived with the aid of Fig. 10.2, which shows two flow lines that are connected to one another by a control line AB.

When computing the volume flow that passes the control area AB in the flow direction passing perpendicular to the $x_1 - x_2$ plane of depth 1, one obtains:

$$\dot{Q} = \int_A^B U_1 dx_2 - \int_A^B U_2 dx_1 = \int_A^B (U_1 dx_2 - U_2 dx_1). \tag{10.11}$$

10.1 Potential and Stream Functions

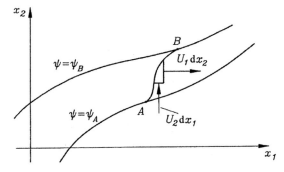

Fig. 10.2 Schematic representation of the flow between flow lines

It holds, however, that $d\Psi = U_1 dx_2 - U_2 dx_1$, so that the following can be written:

$$\dot{Q} = \int_A^B (U_1 dx_2 - U_2 dx_1) = \int_A^B d\Psi = \Psi_B - \Psi_A. \tag{10.12}$$

It should be mentioned that from the statement that Ψ = constant are stream lines of the considered flow field, it follows immediately that solid walls have to run tangentially to lines Ψ = constant. From the orthogonality of equipotential lines and stream lines, which is shown in the following section, it results at once that equipotential lines always have to stand vertically on solid walls. When one considers the stream lines of a flow field, which can be given for two-dimensional potential flows in connection with the potential lines of the same flow field, i.e. lines with Φ = constant, one finds that Ψ = constant and Φ = constant lines lie orthogonally to one another. This can be shown by stating the total differential $d\Phi$:

$$d\Phi = \frac{\partial \Phi}{\partial x_1} dx_1 + \frac{\partial \Phi}{\partial x_2} dx_2, \tag{10.13}$$

or writing the same with consideration of (10.3) as follows:

$$d\Phi = U_1 dx_1 + U_2 dx_2. \tag{10.14}$$

The lines Φ = constant are thus given by:

$$\left(\frac{dx_2}{dx_1}\right)_{\Phi=\text{constant}} = -\frac{U_1}{U_2}. \tag{10.15}$$

A comparison of relationships (10.10) and (10.15) yields:

$$\left(\frac{dx_2}{dx_1}\right)_\Phi = -\frac{1}{\left(\frac{dx_2}{dx_1}\right)_\Psi}. \tag{10.16}$$

As the gradient of the equipotential lines is equal to the negative reciprocal of the gradient of the flow lines, these lines form an orthogonal net. The velocity along a stream can be computed as

$$U_s = \left(\frac{\partial \Phi}{\partial s}\right)_{\Psi=\text{constant}}. \tag{10.17}$$

This relationship is often used in investigations of flow fields for which values of flow lines and equipotential lines have been computed or have to be obtained from measurements.

From the above derivations, it is apparent that a stream function Ψ can be computed when the potential function Φ is known and that also inversely the potential function Φ can be determined when the stream function Ψ is available. The procedure for determining one function from the other is to be considered in accordance with the following single steps for determining the stream function:

- The known potential function $\Phi(x_1, x_2)$ is examined with regard to whether it represents a solution of (10.5).
- By partial differentiation of the function $\Phi(x_1, x_2)$ with respect to x_1 and x_2 the velocity components U_1 and U_2 are determined, in accordance with relations (10.3).
- From this the gradient of the equipotential line can be determined [see (10.15)]:

$$\left(\frac{\mathrm{d}x_2}{\mathrm{d}x_1}\right)_\Phi = -\frac{U_1}{U_2}.$$

- From (10.16) it follows for the gradient of the stream lines that

$$\left(\frac{\mathrm{d}x_2}{\mathrm{d}x_1}\right)_\Psi = \frac{U_2}{U_1}.$$

- By integration of this relationship, the course of the stream lines is determined. These are lines of constant Ψ values.

10.2 Potential and Complex Functions

The considerations in Sect. 10.1 have shown that the velocities U_1 and U_2 can be stated as partial derivatives of the stream function *and* the potential function for irrotational two-dimensional flows of incompressible and viscosity-free fluids:

$$U_1 = \frac{\partial \Phi}{\partial x_1} = \frac{\partial \Psi}{\partial x_2}, \tag{10.18}$$

and

$$U_2 = \frac{\partial \Phi}{\partial x_2} = -\frac{\partial \Psi}{\partial x_1}. \tag{10.19}$$

10.2 Potential and Complex Functions

On the basis of their definition, the stream and potential functions satisfy the Cauchy–Rieman differential equations:

$$\frac{\partial \Phi}{\partial x_1} = \frac{\partial \Psi}{\partial x_2}, \tag{10.20}$$

$$\frac{\partial \Phi}{\partial x_2} = -\frac{\partial \Psi}{\partial x_1}. \tag{10.21}$$

These relationships provide the basis to deduce that a complex analytical function $F(z)$ (see Sect. 2.11.6) can be introduced in which $\Phi(x, y)$ represents the real part and $\Psi(x, y)$ the imaginary part of the function $F(z)$. The latter being refered to as the complex potential of the velocity field. This function is usually written as:

$$F(z) = \Phi(x, y) + i\Psi(x, y), \tag{10.22}$$

where $x = x_1$ and $y = x_2$ and $z = x + iy$ indicates a point in the considered complex number plane. Conversely, it can be said that for any analytical function it holds that its real part represents automatically the potential of a velocity field whose stream lines are described by the corresponding imaginary part of the complex function $F(z)$. As a consequence, it results that each real part of an analytical function, and also the corresponding imaginary part of $F(z)$, separately fulfil the two-dimensional Laplace equation. Analytical functions, as they are dealt with in functional theory, can thus be employed for describing potential flows. When setting their real part $\Re(x, y)$ equal to the potential function $\Phi(x, y)$ and the imaginary part $\mathrm{Im}(x, y)$ equal to the stream function $\Psi(x, y)$, it is possible to state these as the equipotential and the stream lines. By proceeding in this way, solutions to flow problems are obtained without partial differential equations having to be solved. The inverse way of proceeding, which is sought in this chapter for the solution of flow problems, namely interpreting a known solution of the potential equation as a flow, is regarded as acceptable because of the evident advantages of proceeding in this way for introducing students to the subject of potential flows.

From a complex potential $F(z)$, a complex velocity can be derived by differentiation. As $F(z)$ represents an analytical function, and therefore is continuous and can be continiously differentiated. The differentiation has to be independent of the direction in which it is carried out, as is shown in the following. Since the smoothness of $F(z)$ holds, we can derive:

$$\begin{aligned}\frac{\mathrm{d}F}{\mathrm{d}z} &= \lim_{\Delta z \to 0} \frac{\Delta F}{\Delta z} = \lim_{\Delta z \to 0} \frac{\Delta F}{(z + \Delta z) - z} \\ &= \lim_{\Delta z \to 0} \frac{\Delta F}{(x + \Delta x) + i(y + \Delta y) - (x + iy)},\end{aligned}$$

and as one is free to choose the way in which Δz goes towards zero (the differentiation has to be independent of the approach selected), the following special ways can be taken into consideration:

$$\Delta y = 0: \quad \frac{\mathrm{d}F}{\mathrm{d}z} = \lim_{\Delta x \to 0} \frac{\Delta F}{(x + \Delta x) + iy - (x + iy)} = \lim_{\Delta x \to 0} \frac{\Delta F}{\Delta x} = \frac{\partial F}{\partial x},$$

$$\Delta x = 0: \quad \frac{\mathrm{d}F}{\mathrm{d}z} = \lim_{\Delta y \to 0} \frac{\Delta F}{x + i(y + \Delta y) - (x + iy)} = \lim_{\Delta y \to 0} \frac{\Delta F}{i \Delta y}$$

$$= \frac{\partial F}{i \partial y} = -i \frac{\partial F}{\partial y}.$$

The result of differentiation of the complex potential $F(z)$ is thus for $x = x_1$:

$$w(z) = \frac{\mathrm{d}F(z)}{\mathrm{d}z} = \frac{\partial \Phi}{\partial x_1} + i \frac{\partial \Psi}{\partial x_1}, \tag{10.23}$$

or, expressed in velocity components:

$$w(z) = U_1 - iU_2. \tag{10.24}$$

Based on the above considerations, the following also holds:

$$w(z) = \frac{\mathrm{d}F(z)}{\mathrm{d}z} = \frac{\partial \Phi}{i \partial x_2} + i \frac{\partial \Psi}{i \partial x_2}, \tag{10.25}$$

or after transformation, considering that $i^2 = -1$, one can write:

$$w(z) = \frac{\partial \Psi}{\partial x_2} - i \frac{\partial \Phi}{\partial x_2} = U_1 - iU_2. \tag{10.26}$$

The above relationships are used in the following to investigate different potential flows. For these investigations, occasionally use is made of the fact that the complex number z can also be stated in cylindrical coordinates (r, φ):

$$z = re^{(i\varphi)} = r\cos\varphi + ir\sin\varphi. \tag{10.27}$$

Between the velocity components in Cartesian coordinates and in cylindrical coordinates, the known relationships:

$$U_1 = U_r \cos\varphi - U_\varphi \sin\varphi, \tag{10.28}$$

$$U_2 = U_r \sin\varphi + U_\varphi \cos\varphi \tag{10.29}$$

hold. Thus, for the complex velocity the following expressions result:

$$w(z) = \frac{\mathrm{d}F(z)}{\mathrm{d}z} = U_1 - iU_2 = (U_r \cos\varphi - U_\varphi \sin\varphi) - i(U_r \sin\varphi + U_\varphi \cos\varphi)$$
$$= U_r(\cos\varphi - i\sin\varphi) - iU_\varphi(\cos\varphi - i\sin\varphi), \tag{10.30}$$

$$w(z) = (U_r - iU_\varphi)e^{(-i\varphi)}. \tag{10.31}$$

10.3 Uniform Flow

Probably the simplest analytical function $F(z)$, disregarding a constant, is a function which is directly proportional to z and whose proportionality constant is a real number:

$$F(z) = U_0 z = U_0(x + iy). \qquad (10.32)$$

This analytical function describes a flow with the following potential and stream functions:

$$\Phi(x,y) = U_0 x \quad \text{and} \quad \Psi(x,y) = U_0 y. \qquad (10.33)$$

Via the relationship for the complex velocity, one obtains:

$$w(z) = \frac{dF(z)}{dz} = U_0 = U_1 - iU_2, \qquad (10.34)$$

or for $U_1 = U_0$ and $U_2 = 0$, the complex potential $F(z)$ describes a uniform flow parallel to the x_1-axis or the x-axis. This flow is sketched in Fig. 10.3a. For the velocity field it can be deduced that in every point of the flow field, the velocity components are $U_1 = U_0$ and $U_2 = 0$.

This figure shows the stream lines $\Psi = $ constant, where the arrows indicate the direction of the velocity. The potential lines $\Phi = $ constant are not indicated in Fig. 10.3a. In Fig. 10.3b, stream lines of another flow are

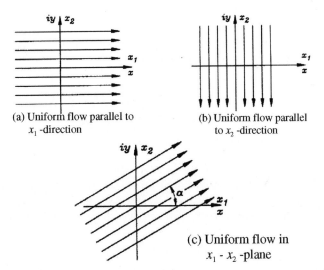

Fig. 10.3 Uniform flow in the (a) x_1 and (b) the x_2 direction and (c) in the direction of the angle α relative to the x_1 direction

shown, representing the lines parallel to the x_2-axis. When the proportionality constant is imaginary, i.e. it holds that:

$$F(z) = iV_0 z = V_0(-y + ix), \qquad (10.35)$$

then one obtains for the potential and stream functions:

$$\Phi(x,y) = -V_0 y \quad \text{and} \quad \Psi(x,y) = V_0 x. \qquad (10.36)$$

For the complex velocity it is computed that:

$$w(z) = iV_0 = U_1 - iU_2, \qquad (10.37)$$

or $U_1 = 0$ and $U_2 = -V_0$, i.e. in this case the complex potential describes a flow parallel to the x_2-axis or the y-axis which takes place in the direction of the negative axis (see Fig. 10.3b).

When there is a flow in the direction indicated in Fig. 10.3c, the complex potential is:

$$\begin{aligned} F(z) &= (U_0 - iV_0)z \\ &= (U_0 - iV_0)(x + iy). \end{aligned} \qquad (10.38)$$

From this result, the following relationships for $\Phi(x,y)$ and $\Psi(x,y)$ can be obtained:

$$\Phi(x,y) = U_0 x + V_0 y \quad \text{and} \quad \Psi(x,y) = U_0 y - V_0 x.$$

Via the complex velocity, one obtains:

$$w(z) = U_0 - iV_0 = U_1 - iU_2. \qquad (10.39)$$

$U_1 = U_0$ and $U_2 = V_0$, The components give a velocity field which is sketched in Fig. 10.3c.

10.4 Corner and Sector Flows

Potential flows around corners and or in sectors of defined angles are described by a complex potential $F(z)$ which is proportional to z^n, where for $n \leq 1$ flows around corners are described, and for $n \geq 1$, flows in sectors of angles $\frac{\pi}{n}$ are obtained. This will be derived and explained through the following considerations. The derivations below are based on the following complex potential:

$$F(z) = Cz^n. \qquad (10.40)$$

When one replaces z by $z = re^{(i\varphi)}$ and divides the complex potential into real and imaginary parts, one obtains:

$$F(z) = C\left[r^n \cos(n\varphi) + ir^n \sin(n\varphi)\right]. \qquad (10.41)$$

10.4 Corner and Sector Flows

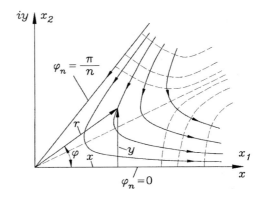

Fig. 10.4 General representation of corner flows and sector flows

From this relationship and taking (10.22) into account, the potential and stream function can be stated as follows:

$$\Phi(r,\varphi) = Cr^n \cos(n\varphi) \quad \text{and} \quad \Psi(r,\varphi) = Cr^n \sin(n\varphi). \tag{10.42}$$

The resulting relationship for the stream function in (10.42) makes it clear that $\Psi(r,\varphi)$ assumes the values $\Psi = 0$ for $\varphi = 0$ and for $\varphi = \pi/n$. This means that the lines $\varphi = 0$ and $\varphi = \pi/n$ represent the flow line $\Psi = 0$ and are regarded here as walls of the flow field. Between them the stream lines for $\Psi = r^n \sin(n\varphi) = $ constant are stated. These result for $\Psi = $ constant in stream lines as they are sketched in Fig. 10.4. The velocity components that are to be assigned to this flow field can be expressed in cylindrical coordinates as follows:

$$w(z) = \frac{dF(z)}{dz} = nCz^{(n-1)} = nCr^{(n-1)} e^{\{i(n-1)\varphi\}} \tag{10.43}$$

or, rewritten, one obtains:

$$w(z) = \left[nCr^{(n-1)}(\cos(n\varphi) + i\sin(n\varphi)) \right] e^{(-i\varphi)}, \tag{10.44}$$

so that one can state [see (10.31)]:

$$U_r = nCr^{(n-1)} \cos(n\varphi) \quad \text{and} \quad U_\varphi = -nCr^{(n-1)} \sin(n\varphi). \tag{10.45}$$

For $\frac{1}{2} < n < 1$, one obtains fluid flows around corners as sketched in Fig. 10.5. Flows around corners are of concern here. They are designated here, in short, as corner flows. For $\frac{2}{3} < n < 1$ flows around obtuse-angled corners are described by (10.40) and for $\frac{1}{2} < n \le \frac{2}{3}$ a representation of flows is achieved which comprises the flow around acute-angled corners.

For $1 < n < \infty$, flows in angle sectors result from the complex potential $F(z) = Cz^n$ as sketched for obtuse-angled angle sectors $(1 < n < 2)$ in Fig. 10.6a and for acute-angled ones $(2 \le n \le \infty)$ in Fig. 10.6b. As for $0 < \varphi < (\pi/2n)U_r$ is always positive, whereas U_φ assumes negative values

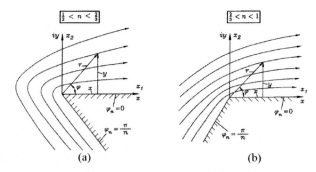

Fig. 10.5 Flow around (a) acute-angled and (b) obtuse-angled corners

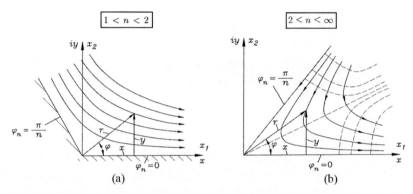

Fig. 10.6 Flow in the (a) obtuse-angled and (b) acute-angled angle sector

in this domain, and as for $(\pi/2n) < \varphi < (\pi/n) U_r$ becomes negative and U_φ remains negative, the courses of the stream and potential lines result as sketched in Fig. 10.6.

The planes $\varphi_n = 0$ and $\varphi_n = \pi/n$ represent a stream line. Along this stream line there are no velocity components in direction normal to the wall. The velocity changes along the boundary stream line, i.e. the wall boundary of the flow. The flow in an angle sector with acute angle differs from the flow in an obtuse-angle flow domain only by the exponent n in the complex velocity potential.

From the above derivations, it can be seen that the complex potential (10.40) includes for $n = 1$ also the uniform flow dealt with in Sect. 10.3. Another important special case is the flow around a thin plate, which can be treated as flow around a border with the angle 360°, i.e. this flow is described by the complex potential:

$$F(z) = Cz^{(\frac{1}{2})}. \tag{10.46}$$

10.4 Corner and Sector Flows

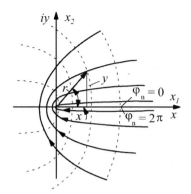

Fig. 10.7 Potential flow around the front of an infinitely thin plate

The proportionality constant is real and the angular area occupied by the flow is:
$$0 \leq \varphi \leq 2\pi.$$

In cylindrical coordinates the complex potential can be written as:
$$F(z) = Cr^{(\frac{1}{2})} e^{(i\frac{\varphi}{2})}. \tag{10.47}$$

The potential and stream functions can be stated as follows:
$$\Phi(r,\varphi) = Cr^{\frac{1}{2}} \cos\left(\frac{\varphi}{2}\right) \quad \text{and} \quad \Psi(r,\varphi) = Cr^{\frac{1}{2}} \sin\left(\frac{\varphi}{2}\right). \tag{10.48}$$

From the relationship for the stream function, it can be derived that the lines $\varphi = 0$ and $\varphi = 2\pi$ correspond to the stream line $\Psi = 0$. The stream lines for other φ values are described by the stream function in (10.48) and are sketched in Fig. 10.7. Also indicated are the equipotential lines, which are also computable according to (10.44). The complex flow velocity is obtained by differentiation of the complex potential $F(z)$ to yield:

$$w(z) = \frac{dF(z)}{dz} = \frac{C}{2z^{(\frac{1}{2})}} \tag{10.49}$$
$$= \frac{C}{2r^{(\frac{1}{2})}} e^{(-i\frac{\varphi}{2})}.$$

One can rewrite this relationship as:
$$w(z) = \frac{C}{2r^{(\frac{1}{2})}} \left[\cos\left(\frac{\varphi}{2}\right) + i\sin\left(\frac{\varphi}{2}\right)\right] e^{(-i\varphi)}. \tag{10.50}$$

The velocity components U_r and U_Φ can therefore be computed as:

$$U_r = \frac{C}{2r^{(\frac{1}{2})}} \cos\left(\frac{\varphi}{2}\right) \quad \text{and} \quad U_\varphi = -\frac{C}{2r^{(\frac{1}{2})}} \sin\left(\frac{\varphi}{2}\right). \tag{10.51}$$

These relationships make it clear that the velocity component U_φ for $0 < \varphi < 2\pi$ is negative, whereas U_r for $0 < \varphi < \pi$ is positive and for $\pi < \varphi < 2\pi$ negative. This leads to the stream lines of the flow sketched in Fig. 10.7.

As an important result of the above derivations, one can deduce that the velocity field possesses a singularity at the origin of the coordinate system. This is caused by the flow around the front corner of the flat plate. This corner is characterized by extreme values of the velocity field. The values of both velocity components approach ∞ for $r \to 0$.

10.5 Source or Sink Flows and Potential Vortex Flow

When one chooses a complex potential $F(z)$ which is proportional to the natural logarithm of z, one obtains the complex potential of a source or a sink flow selecting a real proportionality constant, and depending on whether one chooses a positive or negative sign, the source flow (+sign) and the sink ($-$sign) flow results:

$$F(z) = \pm C \ln z, \tag{10.52}$$

or, with $z = re^{(i\varphi)}$:

$$F(z) = \pm C \left[\ln r + i\varphi\right] = \Phi + i\Psi. \tag{10.53}$$

For the potential and stream functions of the source and sink flow, one thus obtains:

$$\begin{aligned}\Phi(r,\varphi) &= \pm C \ln r & \Psi(r,y) &= \pm C \varphi \\ \Phi(x,y) &= \pm C \ln \sqrt{x^2 + y^2} & \Psi(x,y) &= \pm C \arctan \frac{y}{x}\end{aligned}. \tag{10.54}$$

These equations show that the equipotential lines represent circles with $r =$ constant whereas the stream lines represent radial lines with $\Phi =$ constant. When computing the complex velocity:

$$w(z) = \frac{dF(z)}{dz} = \pm C \frac{1}{z} = \pm C \frac{(x - iy)}{x^2 + y^2}, \tag{10.55}$$

one obtains for the velocity components:

$$w(z) = \frac{\pm C}{x^2 + y^2}(x - iy) = U_1 - iU_2, \tag{10.56}$$

or, written for U_1 and U_2,

$$U_1 = \frac{\pm Cx}{x^2 + y^2} \quad \text{and} \quad U_2 = \frac{\pm Cy}{x^2 + y^2}. \tag{10.57}$$

10.5 Source or Sink Flows and Potential Vortex Flow

In (10.55), $w(z)$ can also be written in $r - \Phi$ coordinates:

$$w(z) = \frac{\pm C}{z} = \pm \frac{C}{r} e^{(-i\varphi)}. \tag{10.58}$$

A comparison of (10.58) with (10.31) shows that the following relationships hold:

$$U_r = \pm \frac{C}{r} \quad \text{and} \quad U_\varphi = 0. \tag{10.59}$$

The velocity component U_r decreases with $1/r$; however, this velocity has a singularity in origin at $r = 0$ in the selected coordinate system.

A flow thus comes about which is sketched in Fig. 10.8 for the source flow and which is purely radial.

The volume flow released per unit time and unit depth by the source, characterizing the strength of the source, is given by:

$$\dot{Q} = \int_0^{2\pi} U_r r \, d\varphi = C 2\pi, \tag{10.60}$$

so that the complex potential for the source or the sink flow can be written as follows:

$$F(z) = \pm \frac{\dot{Q}}{2\pi} \ln z \quad \begin{array}{l} (+) = \text{source flow} \\ (-) = \text{sink flow} \end{array}. \tag{10.61}$$

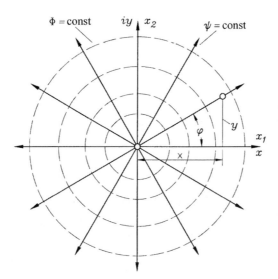

Fig. 10.8 Representation of the potential and stream lines for source flows

When the source or sink does not lie at the origin of the coordinate system but at the point z_0, one obtains:

$$F(z) = \pm \frac{\dot{Q}}{2\pi} \ln(z - z_0). \tag{10.62}$$

When considering a potential z proportional to the natural logarithm, in which the proportionality constant is imaginary, one obtains $F(z)$ of a potential vortex:

$$F(z) = iC \ln z = C(-\varphi + i \ln r). \tag{10.63}$$

For the potential and stream functions one can deduce from this that:

$$\Phi(r, \varphi) = -C\varphi \quad \text{and} \quad \Psi(r, \varphi) = C \ln r, \tag{10.64}$$

or:

$$\Phi(x, y) = -C \arctan \frac{y}{x} \quad \text{and} \quad \Psi(x, y) = C \ln \sqrt{x^2 + y^2}. \tag{10.65}$$

These relationships show that the equipotential radially and outward going lines represented by $\Phi = $ constant whereas the stream lines are circles with $r = $ constant (Fig. 10.9). For the complex velocity, one can derive:

$$w(z) = \frac{\mathrm{d}F(z)}{\mathrm{d}z} = iC \frac{1}{z} = i \frac{C}{r} e^{(-i\varphi)}. \tag{10.66}$$

By comparing (10.66) and (10.31), one can deduce that:

$$U_r = 0 \quad \text{and} \quad U_\varphi = -\frac{C}{r}. \tag{10.67}$$

This resulting flow field is that of a potential vortex with a characteristic decrease of the circumferential velocity with distance from the vortex center. When defining the strength of the potential vortex by the circulation Γ, one

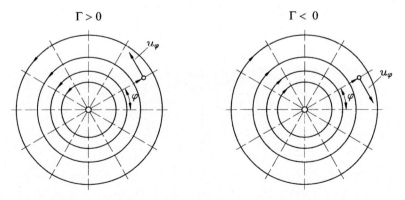

Fig. 10.9 Stream lines and equipotential lines of the potential vortex

can derive:

$$\Gamma = \oint U_s \, ds = \int_0^{2\pi} U_\varphi r \, d\varphi = -2\pi C. \tag{10.68}$$

With this the potential vortex rotating in the mathematically positive direction (Γ is positive), the complex potential can be stated as follows:

$$F(z) = -\frac{\Gamma}{2\pi} i \ln z. \tag{10.69}$$

When the sign is positive, a potential vortex rotating in the mathematically negative direction results with Γ being positive.

A strict distinction has to be made between the potential vortex and vortex motions whose flow fields possess rotations, e.g. vortices result where the entire flow field rotates analogously to the rotation of a solid body. The flow field of the potential vertex is irrotational. The entire circulation of a potential vertex is limited to the vortex-center line where the total circulation is located.

10.6 Dipole-Generated Flow

In this section, a potential flow will be discussed which is defined as dipole-generated flow and results as a limiting case of the superposition of a source flow with a sink flow. Considered is a source with a strength $\dot{Q}$ which is located on the x-axis at a distance $-a$ from the origin of a coordinate system and a sink of the same strength, which has been arranged on the x-axis at a distance $+a$ as shown in Fig. 10.10a.

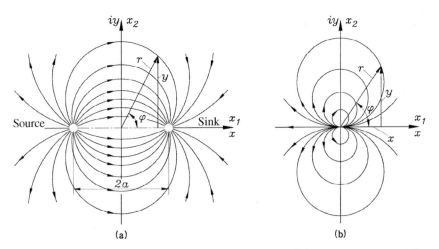

Fig. 10.10 Flow lines of (a) source and sink flows and (b) a dipole-generated flow

When the distances $\pm a$ are reduced, the source and the sink of the considered potential flow move closer together until, for the limiting case $a \to 0$, they coincide in the coordinate origin and thus result in the dipole-generated flow sketched in Fig. 10.10b.

It is the task of the following derivations to find the complex potential of the dipole-generated flow and to derive and discuss, based on the carried out derivations, the flow field of the dipole generated flow.

The complex potential of the combined source and sink flow sketched in Fig. 10.10 can be stated as the sum of the complex potential of both flows:

$$F(z) = \frac{\dot{Q}}{2\pi} \ln(z+a) - \frac{\dot{Q}}{2\pi} \ln(z-a), \tag{10.70}$$

or rewritten in the following form:

$$F(z) = \frac{\dot{Q}}{2\pi} \left[\ln\left(\frac{z+a}{z-a}\right) \right] = \frac{\dot{Q}}{2\pi} \ln\left[\frac{1+a/z}{1-a/z}\right]. \tag{10.71}$$

When carrying out a series expansion for the term $\left(\frac{1}{1-a/z}\right)$, one obtains:

$$F(z) = \frac{\dot{Q}}{2\pi} \ln\left[\left(1+\frac{a}{z}\right)\left(1+\frac{a}{z}+\frac{a^2}{z^2}+\frac{a^3}{z^3}+\cdots\right)\right], \tag{10.72}$$

or, after performing multiplication and truncation after the linear terms, one obtains:

$$F(z) = \frac{\dot{Q}}{2\pi} \ln\left(1+2\frac{a}{z}\right). \tag{10.73}$$

When one carries out another series expansion:

$$\ln\left(1+2\frac{a}{z}\right) = 2\frac{a}{z} - 2\frac{a^2}{z^2} + \frac{8a^3}{3z^3} \mp \cdots, \tag{10.74}$$

one obtains for small values (a/z):

$$F(z) = \frac{\dot{Q}}{2\pi} 2\frac{a}{z}. \tag{10.75}$$

With the strength of the dipole generated flow being characterized as:

$$D = \frac{\dot{Q}a}{\pi},$$

the following complex potential results for the dipole-generated flows:

$$F(z) = \frac{D}{z} = \frac{D}{(x+iy)}. \tag{10.76}$$

For the potential and stream functions, the following expressions can be derived:

$$\Phi(r,\varphi) = \frac{Dx}{x^2 + y^2} \quad \text{and} \quad \Psi(r,\varphi) = \frac{-Dy}{x^2 + y^2},$$

$$\Phi(r,\varphi) = \frac{D}{r}\cos\varphi \quad \text{and} \quad \Psi(r,\varphi) = \frac{-D}{r}\sin\varphi. \tag{10.77}$$

The flow lines and equipotential lines are indicated in Fig. 10.10b. For the complex velocity, one can derive:

$$w(z) = \frac{dF(z)}{dz} = -\frac{D}{z^2} = -\frac{D}{r^2}e^{(-i2\varphi)}, \tag{10.78}$$

or rewritten in the following form:

$$w(z) = -\frac{D}{r^2}(\cos\varphi - i\sin\varphi)e^{(-i\varphi)}. \tag{10.79}$$

From this result, the following expressions for the velocity components result:

$$U_r = -\frac{D}{r^2}\cos\varphi \quad \text{and} \quad U_\varphi = -\frac{D}{r^2}\sin\varphi. \tag{10.80}$$

The signs of these velocity components confirm the direction of the flow indicated in Fig. 10.10b.

10.7 Potential Flow Around a Cylinder

The significance of the dipole-generated flow discussed above lies in the fact that its complex potential can be superimposed with the complex potential of the uniform flow parallel to the x-axis; in this way, a complex potential arises which describes the flow around a cylinder. The simple superposition of the $F(z)$ functions of these two kinds of flows is permitted as the partial differential equations, derived from the basic equations of fluid mechanics, are linear for the potential and stream function. By addition of the complex potentials for the constant flow parallel to the x-axis and for the dipole-generated flow, one obtains the following relationship:

$$F(z) = U_0 z + \frac{D}{z} = U_0 r e^{(i\varphi)} + \frac{D}{r}e^{(-i\varphi)}, \tag{10.81}$$

which is equivalent to:

$$F(z) = U_0 r(\cos\varphi + i\sin\varphi) + \frac{D}{r}(\cos\varphi - i\sin\varphi). \tag{10.82}$$

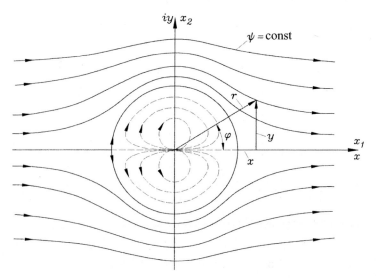

Fig. 10.11 Flow lines of the flow around a cylinder

For the potential and stream functions the following relationships can thus be found:

$$\Phi(r,\varphi) = \left(U_0 r + \frac{D}{r}\right)\cos\varphi \quad \text{and} \quad \Psi(r,\varphi) = \left(U_0 r - \frac{D}{r}\right)\sin\varphi. \quad (10.83)$$

When one now inserts the radius $r = R$ of a cylinder, the stream function along a cylinder wall results as

$$\Psi(r,\varphi) = \left(U_0 R - \frac{D}{R}\right)\sin\varphi. \quad (10.84)$$

When choosing the strength of the dipole-generated flow $D = U_0 R^2$, one obtains for the stream function $\Psi = 0$ along the cylinder wall ($r = R$ for all φ). The resulting stream lines of this flow are shown in Fig. 10.11. From this representation, it can be seen that the stream line representing the cylinder wall is a dividing line between an internal flow caused by the dipole-generated flow and an external flow coming from the flow parallel to the x-axis. We thus have an external flow that can be interpreted as the flow resulting from two-dimensional considerations of the flow of an incompressible viscosity-free fluid around a cylinder. When one takes into consideration the relationship $D = U_0 R^2$, derived for the strength of the dipole-generated flow, for the complex potential of the flow around a cylinder with $r \geq R$, the following final equation can be given:

$$F(z) = U_0\left(z + \frac{R^2}{z}\right). \quad (10.85)$$

10.7 Potential Flow Around a Cylinder

In addition, for the potential and stream functions the following relationships hold:

$$\Phi(r,\varphi) = U_0 \left(r + \frac{R^2}{r}\right) \cos\varphi \quad \text{and} \quad \Psi(r,\varphi) = U_0 \left(r - \frac{R^2}{r}\right) \sin\varphi. \tag{10.86}$$

For the complex velocity, one can derive:

$$w(z) = \frac{dF(z)}{dz} = U_0 \left(1 - \frac{R^2}{z^2}\right) = U_0 \left[1 - \frac{R^2}{r^2} e^{(-i2\varphi)}\right]. \tag{10.87}$$

Further conversions yield:

$$w(z) = U_0 \left[e^{(i\varphi)} - \frac{R^2}{r^2} e^{(-i\varphi)}\right] e^{(-i\varphi)} \tag{10.88}$$

$$= U_0 \left[(\cos\varphi + i\sin\varphi) - \frac{R^2}{r^2}(\cos\varphi - i\sin\varphi)\right] e^{(-i\varphi)}$$

and lead to the following velocity components:

$$U_r = U_0 \left(1 - \frac{R^2}{r^2}\right) \cos\varphi \quad \text{and} \quad U_\varphi = -U_0 \left(1 + \frac{R^2}{r^2}\right) \sin\varphi. \tag{10.89}$$

For the outer cylinder area ($r = R$) one obtains $U_r = 0$; along the actual cylinder there is only a flow along the cylinder wall. For the latter a velocity component results:

$$U_\varphi = -2U_0 \sin\varphi \quad \text{for} \quad r = R. \tag{10.90}$$

For $\varphi = \frac{\pi}{2}$, a velocity component therefore exists which is equal to twice the value of the velocity parallel to the x-axis.

The indicated potential flow around a cylinder results in a solution having outflow conditions which are equal to the inflow conditions, so that no force resulting from the flow acts on the cylinder. This can also be derived from the solution for the velocity field itself. As concerns the quantity of the U_φ component, there exists a symmetry to the x-axis, so that the pressure distribution is also symmetrical and therefore no resulting buoyancy force comes about. Because of a likewise existing symmetry of the pressure distribution to the y-axis, no resulting resistance force is produced either. As this result is contradictory to our experience (d'Alambert's paradox), this investigation shows clearly the significance of the viscosity terms in the basic equations of fluid mechanics. When these terms are not considered in fluid-technical considerations, for obtaining relevant information with regard to fluid physics, fluid forces on bodies can only be dealt with to a limited extent.

10.8 Flow Around a Cylinder with Circulation

In the previous section, the potential functions of the two potential flows were added to yield a new flow. Determing the strength of the dipole-generated flow, the flow around a circular cylinder resulted. In a similar way, one can add complex potentials to give the following complex potential:

$$F(z) = U_0 \left(z + \frac{R^2}{z} \right) + \frac{i\Gamma}{2\pi} \ln z + Ci. \tag{10.91}$$

This complex potential results from the summation of the complex potential of the flow around a cylinder and the complex potential of a vortex, where the centers of both flows lie at the origin of the coordinate system. The constant C was included in the above equation to be able to choose the quantity of the stream function again in such a way that $\Psi = 0$ when $r = R$, i.e. the outer cylinder is to represent the flow line $\Psi = 0$ in the finally derived relation. For determining now the constant C, we insert in the above equation $z = re^{(i\varphi)}$:

$$F(z) = U_0 \left[re^{(i\varphi)} + \frac{R^2}{r} e^{(-i\varphi)} + \frac{i\Gamma}{2\pi} \ln(re^{(i\varphi)}) \right] + Ci. \tag{10.92}$$

Making use of the relation $e^{i\varphi} = \cos\varphi + i\sin\varphi$ we obtain

$$F(z) = U_0 \left[\left(r + \frac{R^2}{r} \right) \cos\varphi + i \left(r - \frac{R^2}{r} \right) \sin\varphi \right] - \frac{\Gamma}{2\pi}\varphi + i\frac{\Gamma}{2\pi} \ln r + Ci \tag{10.93}$$

from which one can deduce the following relationship for the potential and stream functions:

$$\Phi(r,\varphi) = U_0 \left(r + \frac{R^2}{r} \right) \cos\varphi - \frac{\Gamma}{2\pi}\varphi, \tag{10.94}$$

and

$$\Psi(r,\varphi) = U_0 \left(r - \frac{R^2}{r} \right) \sin\varphi + \frac{\Gamma}{2\pi} \ln r + C. \tag{10.95}$$

In order to obtain $\Psi = 0$ for $r = R$ and for all values of φ, one has to choose the constant $C = -\left(\frac{\Gamma}{2\pi}\right) \ln R$. In this way, for the complex potential of the flow around a cylinder with circulation the following complex potential results:

$$F(z) = U_0 \left(z + \frac{R^2}{z} \right) + i\frac{\Gamma}{2\pi} \ln \frac{z}{R}. \tag{10.96}$$

This potential describes the plane flow parallel to the x-axis, made up of a dipole-generated flow and a potential vortex located at the origin of the coordinate system. For this flow, the potential and stream functions can be

10.8 Flow Around a Cylinder with Circulation

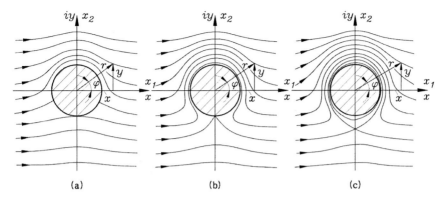

Fig. 10.12 Stream lines for the flow around a cylinder with rotation: (a) circulation $0 \leq \frac{\Gamma}{4\pi U_0 R} < 1$; (b) circulation $\frac{\Gamma}{4\pi U_0 R} = 1$; (c) circulation $\frac{\Gamma}{4\pi U_0 R} > 1$

stated as follows:

$$\Phi(r, \varphi) = U_0 \left(r + \frac{R^2}{r}\right) \cos \varphi - \frac{\Gamma}{2\pi}\varphi.$$

$$\Psi(r, \varphi) = U_0 \left(r - \frac{R^2}{r}\right) \sin \varphi + \frac{\Gamma}{2\pi} \ln \frac{r}{R}. \tag{10.97}$$

The corresponding flow and equipotential lines are shown in Fig. 10.12 for three typical domains of the normalized circulation. The velocity components of the flow field can be computed with the help of the complex velocity:

$$\begin{aligned} w(z) &= U_0 \left[1 - \frac{R^2}{r^2} e^{(-i2\varphi)}\right] + \frac{i\Gamma}{2\pi r} e^{(-i\varphi)} \\ &= \left[U_0 \left(e^{(i\varphi)} - \frac{R^2}{r^2} e^{(-i\varphi)}\right) + i\frac{\Gamma}{2\pi r}\right] e^{(-i\varphi)}. \end{aligned} \tag{10.98}$$

By comparing this relationship with (10.31), the following velocity components result:

$$U_r = U_0 \left(1 - \frac{R^2}{r^2}\right) \cos \varphi \quad \text{and} \quad U_\varphi = -U_0 \left(1 + \frac{R^2}{r^2}\right) \sin \varphi - \frac{\Gamma}{2\pi r}. \tag{10.99}$$

For $\Gamma = 0$ the equations given in Sect. 10.7, resulting for the potential flow around a cylinder without circulation, can be deduced from (10.99).

By setting $r = R$ in the above relationship, one obtains the velocity components U_r and U_φ along the circumferential area of the cylinder:

$$U_r = 0 \quad \text{and} \quad U_\varphi = -2U_0 \sin \varphi - \frac{\Gamma}{2\pi R}. \tag{10.100}$$

As expected, the stream line $\Psi = 0$ fulfils the boundary condition employed with all potential flows for solid boundaries. The U_φ component of the velocity has finite values along the cylinder surface. However, a stagnation point forms in which also $U_\varphi = 0$. These are the stagnation points of the flow with positions on the surface of the cylinder. These locations are obtained from (10.100) for $U_\varphi = 0$.

It should be noted that the position of the stagnation points on the cylinder surface are only given for $\Gamma \leq 4\pi U_0 R$. For $\Gamma = 0$ the stagnation points are located at $\varphi_s = 0$ and $\varphi_s = \pi$, i.e. on the x-axis. For finite Γ values in the range $0 < \Gamma/(4\pi U_0 R) < 1$, φ_s is computed as negative, so that the stagnation points come to lie in the third and fourth quadrants of the cylinder area, as shown in Fig. 10.12. For $\Gamma/(4\pi U_0 R) = 1$, the stagnation points are located in the lowest point of the cylinder surface area. For this location, $\varphi_s = -\frac{\pi}{2}$ and $\frac{3}{2\pi}$ is computed (see Fig. 10.12).

When the circulation of the flow is increased further, so that $\Gamma > 4\pi U_0 R$ holds, stagnation points of the flow cannot form any more along the cylinder surface area. The formation of a "free stagnation point" in the flow field comes about. The position of this point for $U_r = 0$ and $U_\varphi = 0$ can be computed from the above equations for the velocity components, i.e. from:

$$U_0 \left(1 - \frac{R^2}{r_s^2}\right) \cos \varphi_s = 0, \qquad (10.101)$$

and

$$U_0 \left(1 + \frac{R^2}{r_s^2}\right) \sin \varphi_s = -\frac{\Gamma}{2\pi r_s}. \qquad (10.102)$$

As $r_s \neq R$, i.e. the formation of the free stagnation point on the circumferential area is excluded, the first of the above two equations can only be fulfilled for $\varphi_s = \frac{\pi}{2}$ or $\frac{3}{2\pi}$. Hence the second conditional equation for the position coordinate of the "free stagnation point" is:

$$U_0 \left(1 + \frac{R^2}{r_s^2}\right) = \pm \frac{\Gamma}{2\pi r_s}. \qquad (10.103)$$

As $\Gamma > 0$ can be assumed in the above equation, and as the left-hand side of the equation can only adopt positive values, only the positive sign of the above equation yields values consistent with the flow field, i.e. the conditional equation for r_s reads:

$$U_0 \left(1 + \frac{R^2}{r_s^2}\right) = \frac{\Gamma}{2\pi r_s}, \qquad (10.104)$$

or rewritten in the following form to compute r_s:

$$r_s^2 - \frac{\Gamma}{2\pi U_0} r_s + R^2 = 0. \qquad (10.105)$$

As a solution of this equation, one obtains:

$$r_s = \frac{\Gamma}{4\pi U_0} \pm \sqrt{\left(\frac{\Gamma}{4\pi U_0}\right)^2 - R^2}. \qquad (10.106)$$

With this the position coordinates of the free stagnation point result as:

$$\varphi_s = \frac{3\pi}{2} \quad \text{and} \quad \frac{r_s}{R} = \frac{\Gamma}{4\pi U_0 R}\left[1 + \sqrt{1 - \left(\frac{4\pi U_0 R}{\Gamma}\right)^2}\right]. \quad (10.107)$$

The negative sign of the root in the solution for r_s was omitted in the statement of the position coordinates for the free stagnation point, as this would lead to a radius which is located inside the cylinder surface area. As only the flow around the cylinder is of concern, this second solution of the square equation for r_s is of no interest in the considerations presented here.

Moreover, it was also excluded from the solution for the position coordinates of the free stagnation point that the angle φ_s also has a solution for $\frac{\pi}{2}$. The reason for this is that for $\frac{\Gamma}{4\pi U_0 R} = 1$ the stagnation point appears as a solution only in the lower half of the cylinder surface area. Inclusion of the solution for $\varphi_s = \frac{\pi}{2}$ would mean that a small increase in the circulation, to an extent that the standardized circulation is given a value larger than 1, would lead to a jump of the stagnation point from the lower to the upper half of the flow. Considerations on the stability of the position of the stagnation points show, however, that only the lower stagnation point, i.e $\varphi_s = \frac{3\pi}{2}$, can exist as a stable solution. Because of the superposition of the flow around a cylinder with a potential vortex, a flow field has come about, which again is symmetrical concerning the y-axis. With this outcome of the above considerations it is in turn determined that, owing to the flow around the cylinder, the cylinder surface area obtains no resulting force acting in the flow direction, i.e. no resistance force occurs because of the flow. Owing to the imposed circulation, an asymmetric flow with respect to the x-axis, has come about, however, and this leads to a buoyancy force, i.e. to a resulting force on the cylinder, directed upwards. As the velocity component on the upper side of the cylinder is larger than that on the lower side, because of the Bernoulli equation a pressure difference results, with low pressure on the upper side. This causes a flow force directed upwards. The quantitative determination of this force requires integral relationships to be applied as derived in Sect. 10.10.

10.9 Summary of Important Potential Flows

In the preceding sections, a number of potential flows were discussed which are known as basic potential flows and whose treatment give an insight into the fluid flow processes that occur. In Table 10.1, further analytical functions are stated, in addition to the already extensively discussed examples, which can be used for the derivation of potential and stream functions and the corresponding velocity fields of potential flows. By equating the indicated

Table 10.1 Examples of complex functions, potentials ϕ and stream functions ψ for two-dimensional potential flows

Complexes potential $F(z)$	Potential $\Phi(x,y)$	Stream function $\Psi(x,y)$	Velocity u	v	c	Stream lines $\Psi = \text{const}$
$(u_\infty - iv_\infty)z$ Parallel flow	$u_\infty x + v_\infty y$	$u_\infty y - v_\infty x$	u_∞	v_∞	$c_\infty = \sqrt{u_\infty^2 + v_\infty^2}$	
Parallel flow $u_\infty z$ in x-direction	$u_\infty x$	$u_\infty y$	u_∞	0	u_∞	
$\frac{a}{2}z^2$ Stagnation point a real > 0	$\frac{a}{2}(x^2 - y^2)$	axy	ax	$-ay$	ar	
$\frac{\dot Q}{2\pi}\ln z$ Source $\dot Q > 0$, Sink $\dot Q < 0$	$\frac{\dot Q}{2\pi}\ln r = \frac{\dot Q}{2\pi}\ln\sqrt{x^2+y^2}$	$\frac{\dot Q}{2\pi}\varphi = \frac{\dot Q}{2\pi}\arctan\frac{y}{x}$	$\frac{\dot Q}{2\pi}\frac{x}{x^2+y^2}$	$\frac{\dot Q}{2\pi}\frac{y}{x^2+y^2}$	$\frac{\dot Q}{2\pi r}$	
Vortex $\frac{\Gamma}{2\pi}i\ln z$ $\Gamma > 0$ turning right $\Gamma < 0$ turning left	$-\frac{\Gamma}{2\pi}\arctan\frac{y}{x}$	$\frac{\Gamma}{2\pi}\ln\sqrt{x^2+y^2}$	$\frac{\Gamma}{2\pi}\frac{y}{x^2+y^2}$	$-\frac{\Gamma}{2\pi}\frac{x}{x^2+y^2}$	$\frac{\Gamma}{2\pi r}$	

10.9 Summary of Important Potential Flows

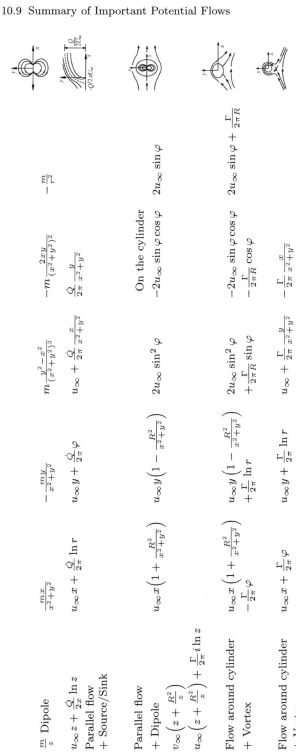

$\frac{m}{z}$ Dipole	$\frac{mx}{x^2+y^2}$	$-\frac{my}{x^2+y^2}$	$m\frac{y^2-x^2}{(x^2+y^2)^2}$	$-m\frac{2xy}{(x^2+y^2)^2}$	$-\frac{m}{r^2}$
$u_\infty z + \frac{\dot{Q}}{2\pi}\ln z$ Parallel flow + Source/Sink	$u_\infty x + \frac{\dot{Q}}{2\pi}\ln r$	$u_\infty y + \frac{\dot{Q}}{2\pi}\varphi$	$u_\infty + \frac{\dot{Q}}{2\pi}\frac{x}{x^2+y^2}$	$\frac{\dot{Q}}{2\pi}\frac{y}{x^2+y^2}$	
Parallel flow + Dipole $v_\infty\left(z + \frac{R^2}{z}\right)$ $u_\infty\left(z + \frac{R^2}{z}\right) + \frac{\Gamma}{2\pi}i\ln z$	$u_\infty x\left(1 + \frac{R^2}{x^2+y^2}\right)$	$u_\infty y\left(1 - \frac{R^2}{x^2+y^2}\right)$	$2u_\infty \sin^2\varphi$	$-2u_\infty \sin\varphi\cos\varphi$	On the cylinder $2u_\infty \sin\varphi$
Flow around cylinder + Vortex	$u_\infty x\left(1 + \frac{R^2}{x^2+y^2}\right)$ $-\frac{\Gamma}{2\pi}\varphi$	$u_\infty y\left(1 - \frac{R^2}{x^2+y^2}\right)$ $+\frac{\Gamma}{2\pi}\ln r$	$2u_\infty \sin^2\varphi$ $+\frac{\Gamma}{2\pi R}\sin\varphi$	$-2u_\infty \sin\varphi\cos\varphi$ $-\frac{\Gamma}{2\pi R}\cos\varphi$	$2u_\infty \sin\varphi + \frac{\Gamma}{2\pi R}$
Flow around cylinder + Vortex	$u_\infty x + \frac{\Gamma}{2\pi}\varphi$	$u_\infty y + \frac{\Gamma}{2\pi}\ln r$	$u_\infty + \frac{\Gamma}{2\pi}\frac{y}{x^2+y^2}$	$-\frac{\Gamma}{2\pi}\frac{x}{x^2+y^2}$	

potential or stream-function values to a constant, the equipotential or flow lines of the potential flow can be stated.

The procedure concerning the derivations of fluid-mechanically interesting quantities will be represented here once again briefly with the aid of the source-sink flow taken from the table.

Example:
$$F(z) = \frac{\dot{Q}}{2\pi} \cdot \ln z = \frac{\dot{Q}}{2\pi} (\ln r + i\varphi) \; ; \quad z = x + iy = re^{i\varphi}$$

Potential:
$$\Phi = \frac{\dot{Q}}{2\pi} \ln r = \frac{\dot{Q}}{2\pi} \ln \sqrt{x^2 + y^2}$$

Stream function:
$$\Psi = \frac{\dot{Q}}{2\pi} \varphi = \frac{\dot{Q}}{2\pi} \arctan \frac{y}{x}$$

Velocity:
$$u = \frac{\partial \Phi}{\partial x} = \frac{\dot{Q}}{2\pi} \frac{x}{x^2 + y^2} = \frac{\partial \Psi}{\partial y}$$

$$v = \frac{\partial \Phi}{\partial y} = \frac{\dot{Q}}{2\pi} \frac{y}{x^2 + y^2} = -\frac{\partial \Psi}{\partial x}$$

$$c = \sqrt{u^2 + v^2} = \frac{\dot{Q}}{2\pi} \sqrt{\frac{x^2 + y^2}{(x^2 + y^2)^2}} = \frac{\dot{Q}}{2\pi r}$$

Equipotential lines: $y = \sqrt{e^{\frac{2\pi}{c} \cdot K_\Phi} - x^2}$
$\Phi = K_\Phi$

Stream lines: $y = x \tan\left(\frac{2\pi}{\dot{Q}}\right) K_\Psi$
$\Psi = K_\Psi$

10.10 Flow Forces on Bodies

In Sects. 10.1 and 10.2 the possibility was already mentioned of computing from the pressure distribution along a body contour the forces acting on bodies that are caused by potential flows. When one has determined the velocity field of a potential flow according to the preceding sections, the velocity distribution is also known along the body contour. This contour represents a flow line of the flow field (as the reader will hopefully remember). In each point of the flow holds the Bernoulli equation in the following form:

$$P + \frac{\rho}{2}(U_s^2 + U_n^2) = \text{constant.} \tag{10.108}$$

10.10 Flow Forces on Bodies

For the stream line $\Psi = 0$ and thus the body contour, $U_n = 0$ holds, i.e.

$$P + \frac{\rho}{2}U_s^2 = \text{constant}. \tag{10.109}$$

The quantity U_s^2 can be computed from U_1 and U_2 or from U_r and U_φ as follows:

$$U_s^2 = U_1^2 + U_2^2 = U_r^2 + U_\varphi^2. \tag{10.110}$$

Along the contour of a flow body, the following integrations can be carried out:

$$F_1 = -\oint P\cos\varphi ds = -\oint Pdx_2 \quad \text{and} \quad F_2 = -\oint P\sin\varphi ds = -\oint Pdx_1, \tag{10.111}$$

in order to conserve the flow forces in the x_1 or the x_2 direction of a Cartesian coordinate system (here φ is the angle between body contour and x_2-axis).

On referring the directions of the forces to the inflow direction and choosing the latter such that it is identical with the x_1 direction, F_1 results in the resisting force on the body, while F_2 yields the buoyancy force.

In the present section, an attempt is made to derive the forces directly through appropriate equations which use the complex velocity. To carry out the necessary derivations, a control volume around the flow body is taken with the height 1 vertical to the plane of flow, as indicated in Fig. 10.13.

In this way, a control volume comes about which is determined by an internal and an external contour. The fluid forces attacking in the center of gravity of the submerged body and given in the directions of the x_1- and x_2-axis, respectively, are likewise indicated in Fig. 10.13. Also sketched is the moment which a body can experience by the flow forces that occur.

When now applying to the control volume, indicated in Fig. 10.13, the momentum equations in integral form, as they were treated in Chap. 8,

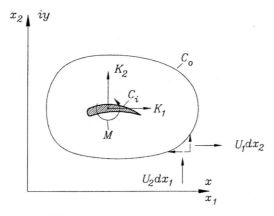

Fig. 10.13 Fluid element and surrounding control volume

the consideration can be expressed in words that the increase of x_1 or x_2 momentum of the flow can only be caused by the flow forces acting on the body in the x_1 or x_2 direction. In the x_1 direction the following force results:

$$-F_1 - \oint_{C_0} P \mathrm{d}x_2 = \oint_{C_0} \rho U_1 (U_1 \mathrm{d}x_2 - U_2 \mathrm{d}x_1). \tag{10.112}$$

This relationship considers that the internal contour of the control volume represents the surface of an emerged body, so that the fluid does not flow through it. The pressure forces acting on the internal contour C_i in the x_1 direction were combined into the resulting force F_1. The force acts in the positive direction on the body and thus in the negative direction on the fluid; this explains the negative sign in front of F_1.

A similar relationship can be written for the x_2 direction:

$$-F_2 + \oint_{C_0} P \mathrm{d}x_1 = \oint_{C_0} \rho U_2 (U_1 \mathrm{d}x_2 - U_2 \mathrm{d}x_1). \tag{10.113}$$

By integrating the two equations in terms of the forces and solving them, one obtains

$$F_1 = \oint_{C_0} \left[-(P + \rho U_1^2) \mathrm{d}x_2 + \rho U_1 U_2 \mathrm{d}x_1 \right], \tag{10.114}$$

and

$$F_2 = \oint_{C_0} \left[(P + \rho U_2^2) \mathrm{d}x_1 - \rho U_1 U_2 \mathrm{d}x_2 \right]. \tag{10.115}$$

Applying the Bernoulli equation:

$$P + \frac{\rho}{2}(U_1^2 + U_2^2) = \text{constant}, \tag{10.116}$$

and taking into consideration that the line integrals $\oint_{C_0} (\text{constant}) \mathrm{d}x_1$ and $\oint_{C_0} (\text{constant}) \mathrm{d}x_2$ are both equal to zero along a closed contour of the control volume, one obtains for the forces in the x_1 and x_2 directions the following terms:

$$F_1 = \rho \oint_{C_0} \left[U_1 U_2 \mathrm{d}x_1 - \frac{1}{2}(U_1^2 - U_2^2) \mathrm{d}x_2 \right], \tag{10.117}$$

$$F_2 = -\rho \oint_{C_0} \left[U_1 U_2 \mathrm{d}x_2 + \frac{1}{2}(U_1^2 - U_2^2) \mathrm{d}x_1 \right].$$

Considering the quantity:

$$i \frac{\rho}{2} \oint_{C_0} w^2(z) \mathrm{d}z = i \frac{\rho}{2} \oint_{C_0} (U_1 - iU_2)^2 (\mathrm{d}x + i \mathrm{d}y) \tag{10.118}$$

one obtains:
$$i\frac{\rho}{2}\oint_{C_0} w^2(z)dz = \rho\oint_{C_0}\left[\left(U_1U_2 dx_1 - \frac{1}{2}(U_1^2 - U_2^2)dx_2\right)\right.$$
$$\left. + i\left(U_1U_2 dx_2 + \frac{1}{2}(U_1^2 - U_2^2)dx_1\right)\right]. \quad (10.119)$$

This equation shows that the flow forces in the x_1 and x_2 directions that act on a body can be computed as follows:
$$i\frac{\rho}{2}\oint_{C_0} w^2(z)dz = F_1 - iF_2. \quad (10.120)$$

Through this relationship, the Blasius integral for flow forces, the flow forces on bodies submerged in potential flows can be computed easily.

Employing the above relationship to compute the resulting force components on the cylinder with circulation, one obtains, beginning with the complex potential:
$$F(z) = U_0\left(z + \frac{R^2}{z}\right) + i\frac{\Gamma}{2\pi}\ln\frac{z}{R} \quad (10.121)$$

for the complex velocity:
$$w(z) = \frac{dF(z)}{dz} = U_0\left(1 - \frac{R^2}{z^2}\right) + \frac{i\Gamma}{2\pi z}. \quad (10.122)$$

For $w^2(z)$, one can compute:
$$w^2(z) = U_0^2 - \frac{2U_0^2 R^2}{z^2} + \frac{U_0^2 R^4}{z^4} + \frac{iU_0\Gamma}{\pi z} - \frac{iU_0\Gamma R^2}{\pi z^3} - \frac{\Gamma^2}{4\pi^2 z^2}, \quad (10.123)$$

or, rewritten:
$$w^2(z) = U_0^2 + \frac{U_0^2 R^4}{z^4} - \frac{1}{z^2}\left(2U_0^2 R^2 + \frac{\Gamma^2}{4\pi^2}\right) - i\left[\frac{U_0\Gamma R^2}{\pi z^3} - \frac{U_0\Gamma}{\pi z}\right]. \quad (10.124)$$

Inserting this into the relationship for the components K_1 and K_2 of the flow force, given above, one obtains for the integration along the cylinder surface area
$$F_1 - iF_2 = i\frac{\rho}{2}\oint w^2(z)dz = i\frac{\rho}{2}\oint\left[U_0^2 + \frac{U_0^2 R^4}{z^4} - \frac{1}{z^2}\left(2U_0^2 R^2 + \frac{\Gamma}{4\pi^2}\right)\right.$$
$$\left. - i\left(\frac{U_0\Gamma R^2}{\pi z^3} - \frac{U_0\Gamma}{\pi z}\right)\right]dz \quad (10.125)$$

On introducing into this integral $z = re^{(i\varphi)}$ and considering that for the cylinder surface area $r = R$ holds, then integration can be carried out and leads to the result:
$$F_1 - iF_2 = -i\rho U_0\Gamma \quad (10.126)$$

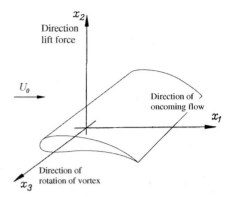

Fig. 10.14 Determination of the direction of the buoyancy forces

or $F_1 = 0$ and $F_2 = \rho U_0 \Gamma$. This is the Kutta–Joukowski equation for the lift force. This equation indicates that the flow force occurring through a potential flow around a cylinder is equal to zero, when there is no circulation.

When there is circulation present, no resisting force occurs but a buoyancy force, which is proportional to the fluid density, to the inflow velocity and the circulation:

$$K_2 = \rho U_0 \Gamma \tag{10.127}$$

As the sign of this force is positive, there is a buoyancy force acting on the cylinder. The rule holding for the direction of the bouyancy is stated in Fig. 10.14. The inflow direction, the direction of rotation of the vortex and the direction of the resulting buoyancy represent the directions of the axes of a rectangular coordinate. Hence, the force orientation is that of the "right hand rule".

The positive force in the case of the flow around a cylinder with circulation comes about as a result of the mathematically positive direction of rotation of the potential vortex at the origin of the coordinate system.

Flow forces acting on bodies can also lead to moments of rotation. There, computation can again be carried out in a conventional way, i.e. by integration of the moment contributions generated by pressure effects on areas. When again assuming the moment acting on the body to be positive, the following equation holds for the moment acting on the fluid:

$$M + \oint_{C_0} [Px_1 \mathrm{d}x_1 + Px_2 \mathrm{d}x_2 + \rho U_1 x_2 (U_1 \mathrm{d}x_2 - U_2 \mathrm{d}x_1) \tag{10.128}$$

$$- \rho U_2 x_1 (U_1 \mathrm{d}x_2 - U_2 \mathrm{d}x_1)] = 0$$

On solving in terms of M, one obtains:

$$M = -\oint_{C_0} \left[Px_1 \mathrm{d}x_1 + Px_2 \mathrm{d}x_2 + \rho (U_1^2 x_2 \mathrm{d}x_2 + U_2^2 x_1 \mathrm{d}x_1 \right.$$

$$\left. - U_1 U_2 x_2 \mathrm{d}x_1 - U_1 U_2 x_1 \mathrm{d}x_2) \right] \tag{10.129}$$

By eliminating the pressure with the help of the Bernoulli equation:

$$P + \frac{\rho}{2}(U_1^2 + U_2^2) = \text{constant}, \qquad (10.130)$$

and considering that the integrals are $\oint_{C_0}(\text{constant})x_1 dx_1 = \oint_{C_0}(\text{constant})x_2 dx_2 = 0$, one obtains:

$$M = \frac{\rho}{2}\oint_{C_0}\left[(U_1^2 - U_2^2)(x_1 dx_1 - x_2 dx_2) + 2U_1 U_2(x_1 dx_2 + x_2 dx_1)\right], \qquad (10.131)$$

and it can be shown that the following holds (second Blasius integral):

$$M = \frac{\rho}{2}\text{Re}\left(\oint_{C_0} z w^2(z) dz\right). \qquad (10.132)$$

An evaluation of the integral yields:

$$M = \text{Re}\left[\frac{\rho}{2}\oint_{c_0}(x+iy)(U_1-iU_2)^2(dx+idy)\right], \qquad (10.133)$$

and considering that $x_1 = x$ and $x_2 = y$, one obtains:

$$M = \text{Re}\left\{\frac{\rho}{2}\oint\left[(U_1^2 - U_2^2)(x_1 dx_1 - x_2 dx_2) + 2U_1 U_2 \cdot (x_1 dx_2 + x_2 dx_1)\right]\right.$$
$$\left. + i\left[(U_1^2 - U_2^2)(x_1 dx_2 + x_2 dx_1) - 2U_1 U_2(x_1 dx_1 - x_2 dx_2)\right]\right\} \qquad (10.134)$$

The real part of (10.134) corresponds to the term (10.131), which was to be proved.

On applying relation (10.131) to the flow around a cylinder with circulation, one obtains:

$$M = \text{Re}\left[\frac{\rho}{2}\oint_{C_0}\left(U_0^2 z - \frac{2U_0^2 R^2}{z} + \frac{U_0^2 R^4}{z^3} + \frac{iU_0\Gamma}{\pi} - \frac{iU_0\Gamma R^2}{\pi z^2} - \frac{\Gamma^2}{4\pi^2 z}\right) dz\right]. \qquad (10.135)$$

On inserting $z = re^{(i\varphi)}$ and $r = R$ in (10.135), one obtains as a solution $M = 0$. The flow around a cylinder does not furnish a hydrostatic moment on the cylinder, even when the flow has circulation.

References

10.1. Yuan, S.W., Foundations of Fluid Mechanics. Mei Ya Publications, Taipei, 1967.
10.2. Allen, T. Jr., Ditsworth, R.L., Fluid Mechanics, McGraw-Hill, New York, 1972.
10.3. Schade, H., Kunz, E., Strömungslehre mit einer Einführung in die Strömungsmestechnik von Jorg-Dieter Vagt. 2, Auflage, Walter de Greyter, Berlin, 1989.

10.4. Zierep, J., Grundzüge der Strömungslehre. 6. Auflage, Springer, Berlin, Heidelberg, New York, 1997.
10.5. Siekmann, H.E., Strömungslehre für den Maschinenbau. Springer, Berlin, Heidelberg, New York, 2001.
10.6. Spurk, J.H., Strömungslehre. 5. Auflage, Springer, Berlin, Heidelberg, New York, 2004.

Chapter 11
Wave Motions in Non-Viscous Fluids

11.1 General Considerations

In Chaps. 9 and 10, fluid flows were considered whose analytical treatment was possible by employing simplified forms of the generally valid basic equations of fluid mechanics. The solution methods required for this are known, i.e. they are at everybody's disposal, and it is known that they can be successfully employed to solve flow problems. Thus in Chap. 10, for example, the application of methods was shown which permit the solutions of the basic equations of fluid mechanics in order to obtain solutions to one- and two-dimensional flow problems. In particular, in Chap. 10 potential flows were dealt with whose given properties were chosen such that methods of functional theory can be employed to treat analytically two-dimensional and irrotational flow problems. Hence, the special properties of potential flows made it possible to take a fully developed domain of mathematics into fluid mechanics and to employ it for computing potential flows and their potential and streamlines. From these computed quantities, velocity fields of the treated potential flows could be derived. The employment of the mechanical energy equation, in its integral form, finally led to pressure distributions in the considered flow fields. The latter again led to the computations of forces and moments for pre-chosen control volumes. Lift and drag forces were considered that are of particular interest for the solution of engineering problems. Simplifications of the flow properties by introducing two-dimensionality and irrotationality have thus permitted a closed treatment of flow problems with known mathematical methods.

A similar solution procedure is adopted in this chapter, in which an introduction to the treatment of wave motions in fluids is attempted. As with all mechanical wave motions, they are usually treated as fluid motions in a medium at rest and around a mean location, i.e. the fluid particles involved in the wave motion experience no change of position when considered over long times. Thus, in the case of wave motions in fluids, only the energy in the

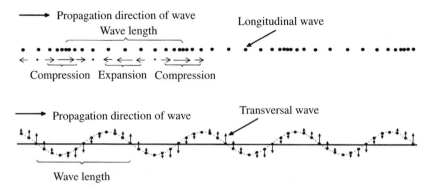

Fig. 11.1 Instantaneous image of progressing longitudinal and transversal waves

wave propagates and not the fluid itself. This holds independently of whether the wave motions in fluids are longitudinal or transversal waves.

Figure 11.1 shows the oscillation motion of fluid particles for both wave modes. From the diagrams one can infer that the considered *wave motions* are periodical, with regard to both space and time. *Oscillations*, on the other hand, are periodical with respect to either time or space.

It can be seen from Fig. 11.1 that mechanical longitudinal waves, which are characterized by compressions and dilatations i.e. by changes of the specific volume or density of a fluid, can exist in all media having "volume elasticity", i.e. react with elastic counter forces to the occurring volume changes. Such counter forces form in gases, and their volume changes are coupled to pressure changes, so that for an ideal gas at $T =$ constant, the following holds:

$$P \, \mathrm{d}v = -v \, \mathrm{d}P \tag{11.1}$$

and therefore, due to compressibility, longitudinal waves can occur in isothermal gases, which are not possible in thermodynamically ideal liquids because of $\rho = 1/v =$ constant.

Figure 11.1 also makes it clear that the formation of transversal waves is dependent on the presence of "shear forces", i.e. lateral forces must exist in order to permit the wave motion of "particles" perpendicular to the direction of propagation. Hence, these mechanical transversal waves only occur in solid matter which can build up elastic transversal forces. This makes it clear that in purely viscose fluids no transversal waves are possible. At first sight, this statement seems to be a contradiction to observations of water waves whose development and propagation can be observed easily when one throws an object into a water container. A transversal wave develops which, however, proves to be a wave motion restricting itself to a small height perpendicular to the water surface. In the interior of the fluid, the wave motion cannot be observed. Moreover, it can be seen that the observed wave on the surface does not form due to "shear forces", but that the presence of gravity or the occurrence of surface tensions is responsible for the wave motion.

11.1 General Considerations

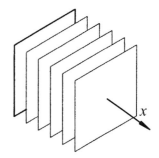

Fig. 11.2 Diagram of a two-dimensional plane wave in its direction of propagation

In fluids, many different wave motions are possible, whose initiation and existence are connected with an energy input into the fluid. For the generation of a wave and its maintenance, a certain energy input is necessary which then propagates in space as the energy of the wave. With this, two different types of energy modes must exist and are essential for a wave to occur. Between the two types of energy an exchange of energy can take place in a periodical sequence. This makes it clear that an essential characteristic of a wave motion in a fluid is that energy is transported without mass transport taking place. Depending on the form of the wave fronts, i.e. also the form of the source of the wave motion, one distinguishes different wave namely plane waves, spherical waves and cylindrical waves. For the velocity field of such waves the following can be stated:

Plane waves: $\qquad u'(x,t) = u_A \sin\left[2\pi\left(\dfrac{t}{T} \mp \dfrac{x}{\lambda}\right)\right]$

Plane waves (Fig. 11.2) are of particular importance for the considerations in this chapter. In the case of a plane wave, the mean energy density is constant, as a considered surface of a wave does not change in area along the propagation direction x. In the above equation, T is the time of the oscillation period of the wave motion and λ is the wavelength. The periodicity of the plane wave in the propagation direction x and the time t can be seen from the sinusoidal term.

Spherical wave: $\qquad u'(x,t) = \dfrac{u_A}{r} \sin\left[2\pi\left(\dfrac{t}{T} \mp \dfrac{r}{\lambda}\right)\right]$

As far as spherical waves in fluids are concerned (Fig. 11.3), the energy density decreases with the square of the distance from the point $r = 0$, as the surface of the sphere increases with the square of the distance. At point $r = 0$ the generator of the spherical wave is located; the entire origin of the energy of the wave is concentrated at this location. Hence the above equation for a spherical wave only holds for $r \neq 0$. A negative sign in front of the r/λ term indicates a diverging wave that moves away from the wave centre of origin and a positive sign indicates a converging wave, moving towards the center $r = 0$.

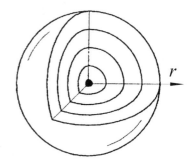

Fig. 11.3 Diagram of a spherical wave showing its radial propagation

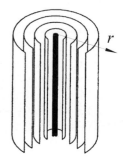

Fig. 11.4 Diagram of a cylindrical wave showing plane radical propagation

Cylindrical wave: $\quad u'(x,t) = \dfrac{u_A}{\sqrt{r}} \sin\left[2\pi\left(\dfrac{t}{T} \mp \dfrac{r}{\lambda}\right)\right]$

Cylindrical waves (Fig. 11.4) propagate radially from the line of the wave generation located in the center, i.e. at $r = 0$. Hence, for cylindrical waves the wave surface increases linearly with the distance r. Thus the energy density decreases linearly with the distance r. Therefore, the amplitude of the wave is inversely proportional to the square root of the distance r from the wave-generating line. Again, a negative sign in front of the r/λ term indicates a wave moving from the generating line in the positive r direction, whereas a positive sign describes a wave moving towards the wave origin.

Many general properties of wave motions, known from physics, can be transferred to wave propagations in fluids. Nevertheless, in a book meant as an introduction to fluid mechanics, special considerations are required; in particular, it is necessary to create a deeper understanding of the causes of the considered wave motions. Especially it is necessary to show how to deal with the wave motion on the basis of the Navier–Stokes equations. In the following sections, important wave motions in fluids are considered. The derivations of the properties of these waves will show the way in which to proceed in fluid mechanics to derive the properties from the basic equations of fluid mechanics. The aim of the derivations is therefore not to provide broad considerations about different wave motions in fluids, but an introduction to the mathematical treatment of longitudinal and transverse waves in fluids.

11.2 Longitudinal Waves: Sound Waves in Gases

In order to be able to deal theoretically with the properties of longitudinal, e.g. with sound waves, in ideal gases, the basic equations of fluid mechanics derived in Chap. 5 can be used. These can be stated for ideal gases as follows:

Continuity equation:
$$\frac{\partial \rho}{\partial t} + \frac{\partial (\rho U_i)}{\partial x_i} = 0 \tag{11.2}$$

Momentum equations: $(j = 1, 2, 3)$
$$\rho \left(\frac{\partial U_j}{\partial t} + U_i \frac{\partial U_j}{\partial x_i} \right) = -\frac{\partial P}{\partial x_j} - \frac{\partial \tau_{ij}}{\partial x_i} + \rho g_j \tag{11.3}$$

Thermal energy equation:
$$\rho c_v \left(\frac{\partial T}{\partial t} + U_i \frac{\partial T}{\partial x_i} \right) = -\frac{\partial q_i}{\partial x_i} - P \frac{\partial U_i}{\partial x_i} - \tau_{ij} \frac{\partial U_j}{\partial x_i} \tag{11.4}$$

State equation:
$$\frac{P}{\rho} = RT \quad \text{and} \quad e = c_v T \tag{11.5}$$

The above system of partial differential equations and thermodynamic state equations comprises seven unknowns, namely U_1, U_2, U_3, P, ρ, e and T, for the determination of which, seven equations are available. Thus we have a closed system of equations which, with sufficient initial and boundary conditions, can be solved, at least in principle. It definitely therefore allows one to treat fluid motions caused by longitudinal waves.

The above system of equations is considerably simplified when one neglects the diffusive heat and momentum transport terms, so that all terms of the momentum and energy equation provided by q_i and τ_{ij} can be dropped. Mass forces can also be neglected, i.e. $g_j = 0$. Maintaining the tensor approach, the equations, after introduction of the suggested simplifications, can be written as follows:

Continuity equation:
$$\frac{\partial \rho}{\partial t} + \frac{\partial (\rho U_i)}{\partial x_i} = 0 \tag{11.6}$$

Momentum equations: $(j = 1, 2, 3)$
$$\rho \frac{DU_j}{Dt} = \rho \left(\frac{\partial U_j}{\partial t} + U_i \frac{\partial U_j}{\partial x_i} \right) = -\frac{\partial P}{\partial x_j} \tag{11.7}$$

Energy equation:
$$\rho \frac{De}{Dt} = \rho c_v \frac{DT}{Dt} = \rho c_v \left(\frac{\partial T}{\partial t} + U_i \frac{\partial T}{\partial x_i} \right) = -P \frac{\partial U_i}{\partial x_i} \tag{11.8}$$

State equation:

$$\frac{P}{\rho} = RT \quad \text{and} \quad e = c_v T \tag{11.9}$$

Taking into consideration that the continuity equation can be written as

$$\frac{\partial \rho}{\partial t} + U_i \frac{\partial \rho}{\partial x_i} + \rho \frac{\partial U_i}{\partial x_i} = \frac{D\rho}{Dt} + \rho \frac{\partial U_i}{\partial x_i} = 0 \tag{11.10}$$

the following relationship holds:

$$\frac{\partial U_i}{\partial x_i} = -\frac{1}{\rho} \frac{D\rho}{Dt}. \tag{11.11}$$

Inserting (11.11) into the energy equation (11.8) and considering the state equation (11.9), the energy equation can be written in the following form:

$$\rho c_v \left[\frac{D}{Dt} \left(\frac{P}{\rho R} \right) \right] = P \left[\frac{1}{\rho} \frac{D\rho}{Dt} \right] \tag{11.12}$$

or

$$\frac{\rho c_v}{R} \left[\frac{1}{\rho} \frac{DP}{Dt} - \frac{P}{\rho^2} \frac{D\rho}{Dt} \right] = \frac{P}{\rho} \frac{D\rho}{Dt} \tag{11.13}$$

$$\frac{1}{P} \frac{DP}{Dt} = \left(\frac{R + c_v}{c_v} \right) \frac{1}{\rho} \frac{D\rho}{Dt}. \tag{11.14}$$

Considering $R = (c_P - c_v)$ and $\kappa = (c_P / c_v)$ permits the following relationship to be derived:

$$\frac{1}{P} \frac{DP}{Dt} = \kappa \frac{1}{\rho} \frac{D\rho}{Dt}. \tag{11.15}$$

Equation (11.15) allows the following general solution to be derived by integration:

$$\frac{D}{Dt}(\ln P) = \frac{D}{Dt}(\ln \rho^\kappa) \tag{11.16}$$

or

$$\frac{D}{Dt}\left[\ln \left(\frac{P}{\rho^\kappa} \right) \right] = 0 \rightsquigarrow \frac{P}{\rho^\kappa} = \text{constant.} \tag{11.17}$$

The above relationship shows that the energy equation can be reduced to the adiabatic equation of property changes known in thermodynamics. This implies that no molecular heat and momentum transport takes place. The relationship was derived from the energy equation, taking into account the continuity equation and the state equation for ideal gases. Thus, along a stream line of a flow the following relation holds for the indicated conditions of the considered fluid motion:

$$\frac{P}{\rho^\kappa} = \text{constant.} \tag{11.18}$$

11.2 Longitudinal Waves: Sound Waves in Gases

There exist a number of fluid mechanics processes in compressible media that can be dealt with by means of reduced equations, which result from the above equations by further simplifications. Assuming that there are flow processes that take place in such a way that the velocity field depends only on one spatial coordinate, then we can write $U_1 = U_1(x_1)$, $U_2 = U_2(x_1)$ and $U_3 = U_3(x_1)$. Moreover, the simplifications $\frac{\partial U_2}{\partial x_1} \ll \frac{\partial U_1}{\partial x_1}$ and $\frac{\partial U_3}{\partial x_1} \ll \frac{\partial U_1}{\partial x_1}$ are introduced so that the following equations can be stated:

Continuity equation:
$$\frac{\partial \rho}{\partial t} + \frac{\partial}{\partial x_1}(\rho U_1) = 0 \tag{11.19}$$

Momentum equation:
$$\rho \left[\frac{\partial U_1}{\partial t} + U_1 \frac{\partial U_1}{\partial x_1} \right] = -\frac{\partial P}{\partial x_1} \tag{11.20}$$

Energy equation:
$$\frac{P}{\rho^\kappa} = \text{constant} \tag{11.21}$$

By sound waves one understands the propagation of small disturbances in gases. The sound velocity is, therefore, the velocity with which small disturbances propagate in a fluid at rest. Whereas for an incompressible fluid an infinitely large propagation velocity results for small disturbances. For compressible fluids a finite propagation velocity results, defined by the considered kind of gas and its temperature. The quantitative determination of the propagation velocity of small disturbances in fluids at rest can be derived as stated below by employing equations (11.19)–(11.21), which hold for unsteady, one-dimensional non-viscous, compressible fluids.

Computing the pressure gradient $(\partial P/\partial x_1)$ in (11.20), one obtains:

$$\frac{\partial P}{\partial x_1} = \left(\frac{\mathrm{d}P}{\mathrm{d}\rho} \right)_{\text{ad}} \frac{\partial \rho}{\partial x_1}. \tag{11.22}$$

This relationship results since the pressure is a function of only one thermodynamic quantity, such as the density, as shown by (11.21). The differentiation $\mathrm{d}P/\mathrm{d}\rho$ has to be applied for the adiabatic conditions as required by (11.21). The continuity equation and the momentum equation can thus be written for one-dimensional fluid flows as:

$$\frac{\partial \rho}{\partial t} + U_1 \frac{\partial \rho}{\partial x_1} + \rho \frac{\partial U_1}{\partial x_1} = 0 \tag{11.23}$$

$$\frac{\partial U_1}{\partial t} + U_1 \frac{\partial U_1}{\partial x_1} + \frac{1}{\rho} \left(\frac{\mathrm{d}P}{\mathrm{d}\rho} \right)_{\text{ad}} \frac{\partial \rho}{\partial x_1} = 0. \tag{11.24}$$

These two equations are available for the two unknowns U_1 and ρ, whose analytical solution is sought on the hypothesis that through the wave motion small pressure and density fluctuations exist, i.e. the following holds:

$$U_1 = 0 + u'(x_1, t) \qquad P = P_0 + p'(x_1, t) \qquad (11.25)$$

$$\rho = \rho_0 + \rho'(x_1, t).$$

These relationships inserted in the above equations (11.23) and (11.24) lead to:

$$\frac{\partial}{\partial t}(\rho_0 + \rho') + u'\frac{\partial}{\partial x_1}(\rho_0 + \rho') + (\rho_0 + \rho')\frac{\partial u'}{\partial x_1} = 0 \qquad (11.26)$$

$$\frac{\partial u'}{\partial t} + u'\frac{\partial u'}{\partial x_1} + \left(\frac{dP}{d\rho}\right)_{ad}\frac{1}{(\rho_0 + \rho')}\frac{\partial}{\partial x_1}(\rho_0 + \rho') = 0. \qquad (11.27)$$

On considering that the variables ρ' and u' depend on the location and time, whereas the quantities P_0 and ρ_0 do not depend on either location or time, the following set of partial differential equations results. To derive these, the assumption was introduced that the products of fluctuating quantities are negligible with reference to the linear terms:

$$\frac{\partial \rho'}{\partial t} + \rho_0 \frac{\partial u'}{\partial x_1} = 0 \quad \text{and} \quad \frac{\partial u'}{\partial t} + \left(\frac{dP}{d\rho}\right)_{ad}\frac{1}{\rho_0}\frac{\partial \rho'}{\partial x_1} = 0. \qquad (11.28)$$

There are now two differential equations for ρ' and u' that can be employed to yield solutions for ρ' and u'. In order to obtain the solution for the propagation of sound waves, the first of the above two equations is differentiated with respect to t:

$$\frac{\partial^2 \rho'}{\partial t^2} + \rho_0 \frac{\partial^2 u'}{\partial x_1 \partial t} = 0. \qquad (11.29)$$

The second equation multiplied by ρ_0 and differentiated with respect to x_1 yields:

$$\rho_0 \frac{\partial^2 u'}{\partial x_1 \partial t} + \left(\frac{dP}{d\rho}\right)_{ad}\frac{\partial^2 \rho'}{\partial x_1^2} = 0. \qquad (11.30)$$

The subtraction of (11.30) from (11.29) results in a differential equation for ρ':

$$\frac{\partial^2 \rho'}{\partial t^2} - \left(\frac{dP}{d\rho}\right)_{ad}\left(\frac{\partial^2 \rho'}{\partial x_1^2}\right) = 0. \qquad (11.31)$$

Furthermore, the differentiation of (11.28) with respect to x_1 yields:

$$\frac{\partial^2 \rho'}{\partial t \partial x_1} + \rho_0 \frac{\partial^2 u'}{\partial x_1^2} = 0 \qquad (11.32)$$

11.2 Longitudinal Waves: Sound Waves in Gases

and multiplication of (11.28) by $\rho/(\frac{dP}{d\rho})_{ad}$ and differentiation with respect to t leads to:

$$\frac{\partial^2 \rho'}{\partial t \partial x_1} + \frac{\rho_0}{\left(\frac{dP}{d\rho}\right)_{ad}} \frac{\partial^2 u'}{\partial t^2} = 0. \tag{11.33}$$

The subtraction of (11.33) from (11.28) and multiplication by $\frac{1}{\rho} \left(\frac{dP}{d\rho}\right)_{ad}$ results in:

$$\frac{\partial^2 u'}{\partial t^2} - \left(\frac{dP}{d\rho}\right)_{ad} \frac{\partial^2 u'}{\partial x_1^2} = 0 \tag{11.34}$$

which is the differential equation for the velocity fluctuation u'.

By comparing the differential equations for ρ' and u', one sees that both have the same form, i.e. ρ' and u' will show the same dependence on location and time. The solutions for both quantities are stated by the one-dimensional wave equation, i.e. there exists a wave motion for ρ' and u' with a propagation velocity which generally reads:

$$c = \sqrt{\left(\frac{dP}{d\rho}\right)_{ad}} = \sqrt{\kappa \rho^{\kappa-1} \frac{P}{\rho^\kappa}} = \sqrt{\kappa \frac{P}{\rho}} = \sqrt{\kappa RT}. \tag{11.35}$$

The general solutions of the differential equations for $\rho'(x_1, t)$ and $u'(x_1, t)$ can be stated as follows:

$$\rho' = f_\rho(x_1 - ct) + g_\rho(x_1 + ct) \quad \text{and} \quad u' = f_u(x_1 - ct) + g_u(x_1 + ct), \tag{11.36}$$

where $f_{\rho,u}(x_1 - ct)$ represents the respective wave which propagates in the positive x_1 direction and $g_{\rho,u}(x_1 + ct)$ the wave moving in the negative x_1 direction.

Further considerations on the propagation of disturbances in compressible fluids at rest can now be made on the basis of the above results. To this effect, one computes from the general solution for u' (wave in the positive x_1 direction):

$$\frac{\partial u'}{\partial t} = -c \left(\frac{\partial f}{\partial \eta}\right) \quad \text{with} \quad \eta = x_1 - ct \tag{11.37}$$

and

$$\frac{\partial u'}{\partial t} = -c \frac{\partial u'}{\partial x_1}. \tag{11.38}$$

From the momentum equation it follows that

$$\frac{\partial u'}{\partial t} = -\frac{c^2}{\rho_0} \frac{\partial \rho'}{\partial x_1} = -c \frac{\partial u'}{\partial x_1} \tag{11.39}$$

or rewritten

$$\frac{\partial u'}{\partial x_1} = \frac{c}{\rho_0} \frac{\partial \rho'}{\partial x_1} \quad \Longrightarrow \quad \frac{u'}{c} = \frac{\rho'}{\rho_0}. \tag{11.40}$$

When we have a disturbance in the form of a compression wave, i.e. $\rho' > 0$, then also $u' > 0$, and this means that the fluid particles move in the direction of the disturbance when a compression disturbance occurs.

When, on the other hand, an expansion disturbance occurs, i.e. $\rho' < 0$, then also $u' < 0$, and in this case the fluid particles move opposite to the direction of the propagation of the disturbance.

The most important result obtained from the above derivations was that small disturbances in non-viscous and compressible fluids at rest propagate with the sound velocity of the fluid that can be computed as follows:

$$c = \sqrt{\left(\frac{\mathrm{d}P}{\mathrm{d}\rho}\right)_{\mathrm{ad}}} = \sqrt{\kappa R T}. \tag{11.41}$$

This relationship will find extensive employment in the derivations in Chap. 12.

11.3 Transversal Waves: Surface Waves

11.3.1 General Solution Approach

On free surfaces of fluids, transversal wave appearances and wave propagations can occur, i.e. propagation of transversal waves introduced by disturbances. These can be two- or three-dimensional; however, the analytical treatment of surface waves presented here concentrates on two-dimensional surface waves. By linearization of the basic equations written in potential form, one obtains the partial differential equations, normally solved for surface waves. Looking at these indicates that the treatment of the propagation of surface waves belongs to the field of the potential theory. The treatment of waves takes place separately, as a special problem, i.e. with surface waves, a special class of flow problems occurs whose treatment correspondingly requires a special methodology. The latter is shown below in an introductory way.

The relationships stated in the following can again be derived from the basic equations of fluid mechanics, which can be stated as follows for a fluid-mechanically ideal fluid, i.e. a non-viscous fluid:

$$\frac{\partial U}{\partial t} + U_i \frac{\partial U_j}{\partial x_i} = -\frac{1}{\rho}\frac{\partial P}{\partial x_j} + g_j. \tag{11.42}$$

By integrating this equation over a period of time τ, one obtains

$$\bar{U}_j + \int_0^\tau U_i \frac{\partial U_j}{\partial x_i}\,\mathrm{d}t = -\frac{1}{\rho}\frac{\partial}{\partial x_j}\int_0^\tau P\,\mathrm{d}t + \int_0^\tau g_j\,\mathrm{d}t. \tag{11.43}$$

11.3 Transversal Waves: Surface Waves

This equation can now be interpreted, with $\Pi = \int_0^\tau P dt$, as the pressure impulse during the time interval τ. For small time intervals τ and for $\rho =$ constant the following results:

$$\bar{U}_j = -\frac{\partial}{\partial x_j}\frac{P}{\rho} \quad \text{with} \quad \int_0^\tau U_i \frac{\partial U_j}{\partial x_i} dt \approx 0 \quad \text{and} \quad \int_0^\tau g_j dt \approx 0. \tag{11.44}$$

Hence the fluid motion generated as a result of pressure impulses on free surfaces is described by a velocity potential. By setting $\bar{U}_j = U_j$:

$$U_j = \bar{U}_j = -\frac{\partial \phi}{\partial x_j} \quad \text{with} \quad \phi = \frac{P}{\rho}. \tag{11.45}$$

The motion is therefore irrotational. Strictly, all this holds only at the free surface and the determination of ϕ in the entire flow area requires further considerations.

The continuity equation can be written in terms of ϕ:

$$\frac{\partial^2 \phi}{\partial x_i \partial x_i} = 0 = \frac{\partial^2 \phi}{\partial x_1^2} + \frac{\partial^2 \phi}{\partial x_2^2} + \frac{\partial^2 \phi}{\partial x_3^2}. \tag{11.46}$$

The momentum (11.42) can be written as:

$$\frac{DU_j}{Dt} = -\frac{1}{\rho}\frac{\partial P}{\partial x_j} + g_j \tag{11.47}$$

and can be rewritten, after multiplication of the equation by U_j, as follows:

$$\frac{D}{Dt}\left(\frac{1}{2}U_j^2\right) = -\frac{1}{\rho}\left[\frac{DP}{Dt} - \frac{\partial P}{\partial t}\right] + g_j U_j. \tag{11.48}$$

With $g_j = -\rho \frac{DG}{Dt}$ for $\frac{\partial G}{\partial t} = 0$ (see (5.57) and (5.58)), we obtain:

$$\frac{D}{Dt}\left(\frac{1}{2}U_j^2\right) = -\frac{1}{\rho}\frac{DP}{Dt} - \frac{1}{\rho}\frac{\partial P}{\partial t} - \frac{DG}{Dt} \tag{11.49}$$

or, rewritten:

$$\frac{\partial \phi}{\partial t} + \frac{P}{\rho} + \frac{1}{2}U_j^2 + G = F(t). \tag{11.50}$$

The function $F(t)$ introduced by the integration can be included in the potential ϕ, so that the final relationship is:

$$\frac{\partial \phi}{\partial t} + \frac{P}{\rho} + \frac{1}{2}U_j^2 + G = 0. \tag{11.51}$$

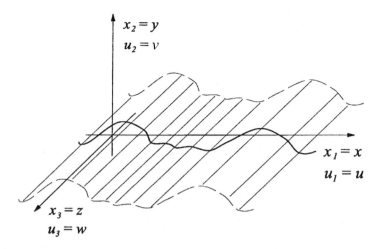

Fig. 11.5 Two-dimensional surface wave

Figure 11.5 represents a two-dimensional surface wave whose deflection, measured from the position of rest $x_2 = 0$, can be expressed as follows:

$$x_2 = y = \eta(x_1, t) = \eta(x, t).$$

The kinematic boundary condition of the flow problem to be solved can therefore be stated as follows:

$$y = \eta(x, t) = 0. \tag{11.52}$$

This means that a fluid particle which belonged to the fluid surface at an instant in time, will always belong to the free surface. From (11.52), the following results, with u_i as fluid velocity of the considered wave motion:

$$\frac{D}{Dt}(y - \eta) = 0 = \frac{\partial}{\partial t}(y - \eta) + u_i \frac{\partial}{\partial x_i}(y - \eta) = 0 \tag{11.53}$$

or after differentiation:

$$-\frac{\partial \eta}{\partial t} - u_1 \frac{\partial \eta}{\partial x_1} + u_2 - u_3 \frac{\partial \eta}{\partial x_3} = 0. \tag{11.54}$$

On now introducing the potential function ϕ, for which the following relationship holds:

$$u_1 = \frac{\partial \phi}{\partial x_1}, \quad u_2 = \frac{\partial \phi}{\partial x_2} \quad \text{and} \quad u_3 = \frac{\partial \phi}{\partial x_3} \tag{11.55}$$

one obtains for the free surface with $x_1 = x$, $x_2 = y$ and $x_3 = z$:

$$\frac{\partial \phi}{\partial y} = \frac{\partial \eta}{\partial t} + \frac{\partial \phi}{\partial x}\frac{\partial \eta}{\partial x} + \frac{\partial \phi}{\partial z}\frac{\partial \eta}{\partial z}. \tag{11.56}$$

11.3 Transversal Waves: Surface Waves

In the entire area of the flow, the potential function fulfils the continuity equation, which can be stated in its two-dimensional form as follows:

$$\frac{\partial^2 \phi}{\partial x^2} + \frac{\partial^2 \phi}{\partial y^2} = 0. \tag{11.57}$$

Under the assumption of absence of viscosity, the Bernoulli equation can be employed in the form indicated by (11.51). Hence we can write:

$$\frac{\partial \phi}{\partial t} + \frac{P}{\rho} + \frac{1}{2}U_j^2 + G = 0. \tag{11.58}$$

This equation is equivalent to the assumption that typically the pressure along a free surface is constant and corresponds to the atmospheric pressure over the surface. If one now includes in the considerations the solid bottom in a certain position $y = -h$, one obtains as a boundary condition at this distance

$$\frac{\partial \phi}{\partial y} = 0 \quad \text{for} \quad y = -h. \tag{11.59}$$

Hence one can write the following set of equations, which are to be fulfilled in order to treat the propagation of waves on free surfaces analytically:

$$\begin{aligned} \frac{\partial^2 \phi}{\partial x^2} + \frac{\partial^2 \phi}{\partial y^2} &= 0 \\ \frac{\partial \eta}{\partial t} + \frac{\partial \phi}{\partial x}\frac{\partial \eta}{\partial x} + \frac{\partial \phi}{\partial z}\frac{\partial \eta}{\partial z} &= \frac{\partial \phi}{\partial y} \quad \text{for } y = \eta \\ \frac{\partial \phi}{\partial t} + \frac{P}{\rho} + \frac{1}{2}U_j^2 + g\eta &= 0 \quad \text{for } y = \eta \\ \frac{\partial \phi}{\partial y} &= 0 \quad \text{for } y = -h. \end{aligned} \tag{11.60}$$

Here the last equations are to be understood as boundary conditions. Hence it becomes clear that the problem, when solving wave problems for fluids with free surfaces, is heavily determined by the imposed kinematic and dynamic boundary conditions. It proves to be a peculiarity of the treatment of wave motion in fluids with free surfaces that the main problem is the introduction of the boundary conditions and not the solution of the differential equations describing the fluid motion.

Considerable simplifications of the system of equations result from the assumption of surface waves of small amplitudes. Assuming that the amplitude of the wave is smaller than all other linear dimensions of the problem, i.e. smaller than the depth of the water h and the wavelength λ, it results that η is small and $\frac{\partial \eta}{\partial x}$ also can be assumed to be small. The latter is the gradient of the shape of the water surface. Moreover, it holds that $\frac{\partial \phi}{\partial x}$ can also be assumed to be small. Surface waves have no high frequencies and the assumption of small amplitudes is also valid for their propagation. Thus, for two-dimensional waves we can write:

$$\frac{\partial \eta}{\partial t} = \frac{\partial \phi}{\partial \eta} \quad \text{for } y = \eta. \tag{11.61}$$

This equation still contains the problem that the boundary condition, applied for surface waves, has to be imposed at the point $y = \eta$. However, when one expands $\frac{\partial \phi}{\partial \eta}$ in a Taylor series:

$$\frac{\partial \phi}{\partial y}(x, \eta, t) = \frac{\partial \phi}{\partial y}(x, 0, t) + \eta \frac{\partial^2 \phi}{\partial \eta^2}(x, 0, t) + \cdots \tag{11.62}$$

it can be seen that the second term on the right-hand side can be neglected because of the assumed small η values. In an analogous way, we can write:

$$\frac{\partial \phi}{\partial t}(x, \eta, t) + \frac{P(x, t)}{\rho} + g \cdot \eta(x, t) = F(t) \tag{11.63}$$

and for small velocities the following relation is valid:

$$\frac{\partial \phi}{\partial t}(x, 0, t) + \frac{P(x, t)}{\rho} + g \cdot \eta(x, t) = 0, \tag{11.64}$$

where the function $F(t)$ was included in the potential $\phi(x, y, z)$. Differentiation of (11.64) with respect to t yields:

$$\frac{\partial^2 \phi}{\partial t^2} + \frac{1}{\rho}\frac{\partial P}{\partial t} + g\frac{\partial \eta}{\partial t} = \frac{\partial^2 \phi}{\partial t^2}(x, 0, z) + \frac{1}{\rho}\frac{\partial P(x, z)}{\partial t} + g \cdot \frac{\partial \phi}{\partial y}(x, 0, z) = 0 \tag{11.65}$$

so that one obtains the following simplified set of equations for the treatment of surface waves of small amplitudes:

$$\frac{\partial^2 \phi}{\partial x^2} + \frac{\partial^2 \phi}{\partial y^2} = 0$$

$$\frac{\partial \phi}{\partial y}(x, 0, t) = \frac{\partial \eta}{\partial t}(x, t) \quad (\text{for } y = \eta)$$

$$\frac{\partial^2 \phi}{\partial t^2}(x, 0, t) + \frac{1}{\rho}\frac{\partial P(x, t)}{\partial t} + g\frac{\partial \phi}{\partial y}(x, 0, t) = 0 \quad (\text{for } y = \eta)$$

$$\frac{\partial \phi}{\partial y}(x, -h, t) = 0 \quad (\text{for } y = -h). \tag{11.66}$$

With the above equations, gravitational waves and capillary waves can be treated, which usually represent waves with small amplitudes.

11.4 Plane Standing Waves

When considering wave motions, where the fluid particles move only parallel to the $x_1 - x_2$ plane, i.e. where the pressure P and the velocity U_j are independent of x_3, so that the fluid motions in all areas parallel to the $x_1 - x_2$ plane take place in the same way, a plane wave motion with the following potential results:

$$\phi(x, y, t) = \phi(x, y) \cos(\varphi t + \epsilon). \tag{11.67}$$

For the case of a standing wave to be dealt with in this section, it can be stated that:

$$\phi(x, y) = \frac{P(y)}{\rho} \sin[k(x - \xi)]. \tag{11.68}$$

The potential ϕ fulfils the Laplace equation:

$$\frac{\partial^2 \phi}{\partial x^2} + \frac{\partial^2 \phi}{\partial y^2} = 0. \tag{11.69}$$

With $\rho \frac{\partial^2 \phi}{\partial x^2} = -P(y)k^2 \sin[k(x - \xi)]$ and $\rho \frac{\partial^2 \phi}{\partial y^2} = P \frac{d^2 P}{dy^2} \sin[k(x - \xi)]$, one obtains the following differential equation:

$$\frac{d^2 P}{dy^2} - k^2 P = 0 \tag{11.70}$$

the solution of which is:

$$P(y) = C_1 \exp(ky) + C_2 \exp(-ky). \tag{11.71}$$

From more precise considerations, the integration constant C_2, results as $C_2 = 0$, as otherwise for large depths $y = \to -\infty$ the $P(y)$ term would become very large, so that the solution:

$$\phi(x, y) = \frac{C_1}{\rho} \exp(ky) \sin[k(x - \xi)] \tag{11.72}$$

can be obtained, or:

$$\phi(x, y, t) = \frac{C_1}{\rho} \exp(ky) \sin[k(x - \xi)] \cos(\varphi t + \epsilon). \tag{11.73}$$

By starting from the assumption that the occurring fluid motion is slow, the equation:

$$\frac{\partial \phi}{\partial t} + \frac{P}{\rho} + \frac{1}{2} U_j^2 + g\eta = 0 \quad \text{for } y = \eta \tag{11.74}$$

can be written as follows:

$$\frac{\partial \phi}{\partial t} + \frac{P}{\rho} + g\eta = 0 \quad \text{for } y = \eta. \tag{11.75}$$

Differentiation with respect to time yields the differential equation indicated below, as the pressure along the free surface does not change:

$$\frac{\partial^2 \phi}{\partial t^2} + g \frac{\partial \eta}{\partial t} = \frac{\partial^2 \phi}{\partial t^2} + g u_2. \tag{11.76}$$

With $u_2 = \frac{\partial \phi}{\partial y}$, one can finally write:

$$\frac{\partial^2 \phi}{\partial t^2} = -g \frac{\partial \phi}{\partial y}. \tag{11.77}$$

Employing (11.77) to treat (11.73), one obtains:

$$\frac{\partial^2 \phi}{\partial t^2} = -\frac{C_1}{\rho} \exp(ky) \sin[k(x-\xi)] \varphi^2 \cos(\varphi + \epsilon) \tag{11.78}$$

and

$$\frac{\partial \phi}{\partial y} = +\frac{C_1}{\rho} k \exp(ky) \sin[k(x-\xi)] \cos(\varphi + \epsilon) \tag{11.79}$$

$$\varphi^2 = kg. \tag{11.80}$$

Hence, for the remaining considerations, the following fluid motion has to be examined, which, for the sake of simplicity, is considered for $\xi = 0$ and $\epsilon = 0$:

$$\phi(x, y, t) = \frac{C_1}{\rho} \exp(ky) \sin(kx) \cos(\varepsilon t). \tag{11.81}$$

For the free surface one can compute from (11.75):

$$\eta = -\frac{1}{g} \frac{\partial \phi}{\partial t} = -\frac{1}{g} \frac{\partial}{\partial t} \phi(x, 0, t) \tag{11.82}$$

or

$$\eta = \frac{C_1 \varphi}{g} \sin(kx) \sin(\varphi t). \tag{11.83}$$

With $A = \frac{C_1 \varphi}{\rho g} \sin(\varphi t)$ it holds that $\eta = A \sin(kx)$.

For $x = \frac{m\pi}{k}$, for $m = 0, \pm 1, \pm 2, \pm \cdots$, nodal points of a standing wave result. In the middle between these nodes are the "antinodal points" of the wave motion. The wavelength of the sinusoidal fluid motion can be computed as:

$$\lambda = \frac{2\pi}{k}. \tag{11.84}$$

The amplitude of the wave motion is $\frac{C_1 \varphi}{g} \sin(\varphi t) = A$, where for the frequency of the wave motion the following holds:

$$f = \frac{\varphi}{2\pi} = \frac{1}{T}. \tag{11.85}$$

Taking into consideration (11.80), (11.84) and (11.85), one obtains

$$T = \frac{1}{f} = \sqrt{\frac{2\pi k}{g}} \quad \text{or} \quad \lambda = \frac{g\tau^2}{2\pi} \tag{11.86}$$

or

$$\lambda = \frac{g}{2\pi f^2}. \tag{11.87}$$

The above relationships show that the wavelength of standing fluid waves decreases with increasing frequency of the motion.

11.5 Plane Progressing Waves

For the derivations given below, it is assumed that the fluid considered takes up the space as follows (see Fig. 11.6) for the $x-y$ coordinate arrangements:

$$-\infty \leq y \leq 0 \quad \text{and} \quad -\infty \leq x \leq +\infty$$

and the fluid is assumed, at the point $y = 0$, to occupy a free surface. For the considerations carried out it represents a finite surface. The equations required for the treatment of progressing waves can be stated as follows:

$$\frac{\partial^2 \phi}{\partial x^2} + \frac{\partial^2 \phi}{\partial y^2} = 0. \tag{11.88}$$

With $y = \eta(x_1, t)$ for the free surface it holds that

$$\frac{D}{Dt}(y - \eta) = 0 \quad \leadsto \quad u_2 = \left(\frac{\partial}{\partial t} + u_1 \frac{\partial}{\partial x_1}\right)\eta. \tag{11.89}$$

Neglecting the term of second order, the following equation results:

$$\frac{\partial \phi}{\partial y} = \frac{\partial \eta}{\partial t}. \tag{11.90}$$

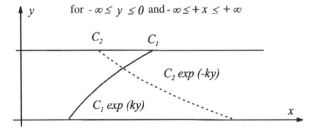

Fig. 11.6 Illustration of the decrease for $y \to -\infty$

For the pressure at the free surface it can be stated that:

$$P = -\sigma \left[\frac{1}{R_1} + \frac{1}{R_2}\right], \qquad (11.91)$$

where R_1 and R_2 represent the main radii of curvature of the free surface and σ is the surface tension. Linearized, this relationship can be written as:

$$P = -\sigma \frac{\partial^2 \eta}{\partial x^2}, \qquad (11.92)$$

where the pressure above the free surface is taken to be $P = 0$, otherwise P is to be replaced by $P = P_0$.

For plane progressing waves, the following potential can be stated:

$$\phi(x, y, t) = C \exp(ky) \cos\left[k\left(x - ct\right)\right] \qquad (11.93)$$

with $\phi = 0$ for $y = -\infty$. The formulation for $\phi(x, y, z)$ fulfils the continuity equation in the form of (11.69).

The Bernoulli equation can be stated as follows:

$$\frac{P}{\rho} = -\frac{\partial \phi}{\partial t} - gy \qquad (11.94)$$

or rewritten:

$$\frac{\partial \phi}{\partial t} = -\frac{P}{\rho} - g\eta = \frac{\sigma}{\rho}\frac{\partial^2 \eta}{\partial x^2} - g\eta. \qquad (11.95)$$

From this, the following relationship results:

$$\frac{\partial^2 \phi}{\partial t^2} = \frac{\sigma}{\rho}\frac{\partial^2}{\partial x^2}\left(\frac{\partial \eta}{\partial t}\right) - g\frac{\partial \eta}{\partial t} \qquad (11.96)$$

and with consideration of equations (11.90), (11.95) can be written as:

$$\frac{\partial^2 \phi}{\partial t^2} = \left[\frac{\sigma}{\rho}\frac{\partial^2}{\partial x^2} - g\right]\frac{\partial \phi}{\partial g}. \qquad (11.97)$$

For the left-hand side of (11.95), one can write using (11.93):

$$\frac{\partial^2 \phi}{\partial t^2} = -Ck^2c^2 \exp(ky) \cos\left[k(x - ct)\right] = -k^2c^2\phi \qquad (11.98)$$

so that the following holds:

$$k^2c^2\phi = \left[g - \frac{\sigma}{\rho}\frac{\partial^2}{\partial x^2}\right]\frac{\partial \phi}{\partial y}. \qquad (11.99)$$

11.5 Plane Progressing Waves

With $\frac{\partial \phi}{\partial y} = k\phi$ and $\frac{\partial^2 \phi}{\partial x^2} = -k^2\phi$, the following relationship results from (11.99) for the velocity of the progressing wave:

$$c^2 = \frac{g}{k} + \frac{k\sigma}{\rho}. \tag{11.100}$$

With $k = \frac{2\pi}{\lambda}$, it can be seen that for long waves the influence of gravity dominates:

$$c = \sqrt{\frac{g}{k}} \qquad \text{shear waves.} \tag{11.101}$$

For waves with small wavelengths, the capillary effects dominate:

$$c = \sqrt{\frac{k\sigma}{\rho}} \qquad \text{capillary waves.} \tag{11.102}$$

Concerning wavelengths, often the path lines of the fluid particles are also of interest, which occur close to the water surface or at certain depths below the water surface. In this respect, the following can be carried out, where x_0 and y_0 are introduced as the coordinates which, with the help of:

$$u_x = \frac{\partial \phi}{\partial x} = \frac{dx}{dt} = Ck \exp(ky) \sin[k(x - ct)] \tag{11.103}$$

$$u_y = \frac{\partial \phi}{\partial y} = \frac{dy}{dt} = Ck \exp(ky) \cos[k(x - ct)] \tag{11.104}$$

fulfil the following equations:

$$x = x_0 + C \cdot k \exp(ky) \cos[k(x - ct)] \left(\frac{+1}{Ck}\right) \tag{11.105}$$

$$y = y_0 + ck \exp(ky) \sin[k(x - ct)] \left(\frac{-1}{ck}\right) \tag{11.106}$$

or rewritten:

$$(x - x_0)^2 + (y - y_0)^2 = \left(\frac{C}{c}\right)^2 \exp(2ky_0). \tag{11.107}$$

The path lines of the fluid particles are derived as circles whose radii become smaller with increasing water depth. For the water surface, the radius of the circular path is equal to the amplitude of the surface wave, while at a certain depth it has already decreased to 1/535th of the wave amplitude at the water surface. This makes it clear that the considered wave motion of fluid particles remains limited to an area in the immediate proximity of the water surface.

Figure 11.7 shows the circular paths described by fluid particles. These will run in an anticlockwise direction, so that the following expressions for the x and y motion holds:

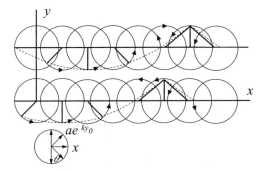

Fig. 11.7 Circular paths of fluid particle motion

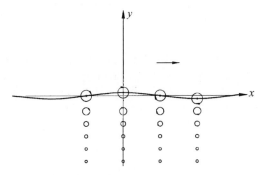

Fig. 11.8 Path lines in a plane gravity wave

$$(x - x_0) = a \exp(ky_0) \sin \Theta \tag{11.108}$$

$$(y - y_0) = -a \exp(ky_0) \cos \Theta. \tag{11.109}$$

Hence we can write for Θ:

$$\Theta = kx_0 + kct \cos(yt + \epsilon). \tag{11.110}$$

The changes of the motions of the fluid particles with water depth are sketched in Fig. 11.8. The strong decrease of the radius of the circular fluid motion with depth was not taken into consideration in Fig. 11.7.

In order to be able to investigate gravity waves with free surfaces in fluids with a finite depth h, a mean surface position at point $y = 0$ is assumed. At position $y = -h$ a wall is considered as being given, so that a mean fluid film thickness with height h occurs. To fulfill now the continuity equation:

$$\frac{\partial^2 \phi}{\partial x^2} + \frac{\partial^2 \phi}{\partial y^2} = 0 \tag{11.111}$$

by a wave with wavenumber k, the following potential formulation is carried out:
$$\phi = C \cosh k(y+h) \cos k(x-ct). \tag{11.112}$$

This formulation not only fulfils the continuity equation, but also permits the boundary condition at the bottom of the fluid layer to be fulfilled:
$$\frac{\partial \phi}{\partial y} = 0 \quad \text{for } y = -h. \tag{11.113}$$

The procedure for deriving the required relationship is now similar to that in Sect. 11.5. It then results in a condition for the free surface that can be stated as:
$$kc^2 \cosh kh = \left(g + \frac{k^2 T}{\rho}\right) \sinh kh \tag{11.114}$$

or resolved for the wave velocity, one obtains:
$$c^2 = \left(g + \frac{k^2 T}{\rho}\right) \frac{\tanh kh}{k}. \tag{11.115}$$

For waves with long wave lengths, i.e. for small values of the wavenumber k, one obtains for the wave velocity:
$$c^2 = gh. \tag{11.116}$$

The waves moving with this velocity are essentially gravitation waves, as the surface curvature is so small that the influences of the surface tension at the wave motion are not felt in the wave velocity.

For very short waves, i.e. for large values of the wavenumber k, one obtains on the other hand
$$c^2 = \frac{kT}{\rho}. \tag{11.117}$$

This is the propagation velocity of the capillary waves. This equation shows for the velocity of the capillary waves that they are waves of small amplitudes, so that their propagation velocity is not influenced by the height of the fluid layer.

In this section only an introduction to the treatment of wave motions in fluids is given. Further treatments of waves in fluids are given in refs. [11.1] to [11.7].

11.6 References to Further Wave Motions

The wave motions dealt with in Sects. 11.1–11.5 need considerations that require extensions with emphasis on other kind of wave motions, e.g. see Yih [11.6]. Nonetheless, good text books with general considerations on wave motions in fluids are lacking, i.e. the treatment of wave motions in books is always limited to the treatment of very special wave motions. In Yih [11.6], for example, the following wave motions in fluids are dealt with:

- Gerstner waves
- Solitary waves
- Rossby waves
- Stokes waves
- Cnoidal waves
- Axisymmetric waves

If one wants, however, to find the introductory literature on the mathematical treatments of wave motion observed in nature, it is necessary to have a clear understanding of the physical cause of the considered wave motion. Thus one observes, for example, that a long body which is moved perpendicular to its linear expansion near the free surface of a liquid forms waves mainly in its wake. In front of the body one observes, with respect to the amplitude, smaller surface waves, when the dimensions of the bodies in flow direction are smaller than $(\sigma/\rho g)^{1/2}$. Otherwise the gravity waves occurring behind the body dominate and the capillary waves that can be observed in front of the body are negligible. Hence, when one has recognized the nature of the observed wave motions, the appertaining analytic treatment can be found in the tables of contents, listed in references.

References

11.1. Batchelor, G.K.: An Introduction to Fluid Dynamics, Cambridge University Press, Cambridge, 1970
11.2. Bergmann, L. and Schaefer, Cl.: Lehrbuch der Experimentalphysik, Band I, 6, Walter de Gruyter, Berlin, 1961
11.3. Currie, I.G.: Fundamental Mechanics of Fluids, McGraw-Hill, New York, 1974
11.4. Lamb, H.: Hydrodynamics, Dover, New York, 1945
11.5. Spurk, J.H.: Strömungslehre, Springer, Berlin Heidelberg New York, 4. Aufl., 1996
11.6. Yih, C.S.: Fluid Mechanics: A Concise Introduction to the Theory, West River, Ann Arbor, MI, 1979
11.7. Yuan, S.W.: Foundations of Fluid Mechanics, Prentice-Hall, Englewood Cliffs, NJ, 1971

Chapter 12
Introduction to Gas Dynamics

12.1 Introductory Considerations

Gas dynamics is a branch of fluid mechanics which deals with the motion of gases at high velocities. Gravitational forces and their influence on the state of the gas can be neglected in the majority of gas dynamic flow cases. Considerations of the pressure differences to be expected by gravitational forces in a gas show that this is justified. According to the force balance shown in a previous chapter, the following relationship holds for the pressure:

$$\Delta P = -\rho g_j x_j = \rho g \Delta z, \tag{12.1}$$

which can be determined for an ideal gas ($P = \rho RT$) such as air as:

$$\frac{\Delta P}{P} = g\frac{\Delta z}{RT} \approx 9{,}81\frac{\Delta z}{287T}\left[\frac{\text{m}}{\text{s}^2}\frac{\text{ms}^2\text{K}}{\text{m}^2\text{K}}\right]. \tag{12.2}$$

On inserting $T \approx 293$ K it can be seen that the relative pressure changes due to gravitation assume values around 1% only when vertical displacements of about 100 m occur. As gas dynamic considerations are usually restricted to installations of flow equipment of much smaller dimensions, it is justified to simplify the fluid mechanical equations in gas dynamics by neglecting the gravitational forces. For many fluid mechanical considerations in gas dynamics, it is permissible to regard gaseous fluids also as *incompressible* when the fluid velocities occurring are small compared with the sound velocity of the fluid. This can be explained for a stagnation point flow by the following:

$$P_S = P_\infty + \rho\frac{U_\infty^2}{2} \text{ with } \rho = \text{constant}. \tag{12.3}$$

For a compressible flow, the stagnation point pressure may be obtained using the stream line relationship derived in Chap. 9 for adiabatic changes of thermodynamic state of a gas, i.e.

$$P_S = P_\infty \left(1 + \frac{\kappa-1}{2\kappa}\frac{\rho_\infty}{P_\infty}U_\infty^2\right)^{\frac{\kappa}{\kappa-1}}, \tag{12.4}$$

or rewritten as a series expansion:

$$P_S = P_\infty \left[1 + \frac{\rho_\infty}{2P_\infty}U_\infty^2 + \frac{1}{2\kappa}\left(\frac{\rho_\infty U_\infty^2}{2P_\infty}\right)^2 + \cdots\right]. \tag{12.5}$$

where $\kappa = c_p/c_v$, the ratio of the heat capacitances. If we compose the stagnation pressure for an incompressible flow in (12.3) with the result from compressible flows, we observe a difference of about 2% for velocities of around $70\,\mathrm{m\,s^{-1}}$ (assuming standard state conditions in the free stream). This corresponds to a free stream Mach number $Ma \approx 0.2$. Thus, it may be concluded that compressibility effects in gases have to be taken into account for velocities well above $Ma \approx 0.2$.

$$Ma = \frac{U}{c} = \frac{U}{\sqrt{\kappa RT}} \geq 0.2. \tag{12.6}$$

From this consideration, a second conclusion may be derived concerning the viscous effects if the flow velocity is fairly large, namely the Reynolds number also takes on large values. Hence viscous effects may be neglected and the Euler equations are usually used as a starting point for a mathematical treatment of gas flows. For a ideal gas, that flows under gas dynamic conditions, the equations are as follows:

Continuity equation:
$$\frac{\partial \rho}{\partial t} + \frac{\partial(\rho U_i)}{\partial x_i} = 0. \tag{12.7}$$

Momentum equation ($j = 1, 2, 3$):
$$\rho\left[\frac{\partial U_j}{\partial t} + U_i \frac{\partial U_j}{\partial x_i}\right] = -\frac{\partial P}{\partial x_j}. \tag{12.8}$$

Energy equation:
$$\rho c_v \left[\frac{\partial T}{\partial t} + U_i \frac{\partial T}{\partial x_i}\right] = -P \frac{\partial U_j}{\partial x_j}, \tag{12.9}$$

where the energy (12.9) is given for *adiabatic fluid flows*. Together with the thermodynamic equation of state for ideal gases, a closed system of differential equations exists which can be solved, in principle, for given boundary conditions. The possible solutions require special considerations; however, the appearance of high flow velocities is linked to specific phenomena which differentiate gas dynamics sharply from other areas of fluid mechanics. As the following considerations will show, the presence of high Mach numbers, $Ma = U/c$, leads to the emergence of "discontinuity surfaces" (compression

12.1 Introductory Considerations

shocks) in which the pressure (and other flow quantities) experience a sudden jump. This makes special procedures necessary when solving flow problems.

The employment of the differential form of the basic equations usually requires that the quantities describing a flow are steady in the flow area. There is also the fact that when treating fluid flows at high Mach numbers, processes occur that are linked to different time scales, namely the time scales of the diffusion Δt_{Diff}, the convection Δt_{Conv} and the sound propagation Δt_{Sound}:

$$\Delta t_{\text{Diff}} = \frac{L_c^2}{\nu_c}; \quad \Delta t_{\text{conv}} = \frac{L_c}{U_c}; \quad \Delta t_{\text{Sound}} = \frac{L_c}{c}. \quad (12.10)$$

For $\Delta t_{\text{Conv}} \ll \Delta t_{\text{Diff}}$, the following results:

$$Re = \frac{\Delta t_{\text{Diff}}}{\Delta t_{\text{Conv}}} = \frac{U_c L_c}{\nu} \gg 1, \quad (12.11)$$

i.e. during the time that a flow needs to cover a certain distance, the molecular transport, at high flow velocities, manages only to overcome a negligible distance, i.e. at high Reynolds numbers the formation of thin boundary layers takes place. However, in gas dynamics, considerations of boundary layers are neglected, especially in the introductory considerations presented here. From the point of view of characteristic times, the Mach number is represented by the following ratio:

$$Ma = \frac{\Delta t_{\text{Conv}}}{\Delta t_{\text{Sound}}} = \frac{U_c}{c}, \quad (12.12)$$

i.e. the Mach number shows how fast a fluid element is transported in comparison with the disturbances arising from the motion of this fluid element. The disturbances vary with the velocity c:

$$c = \sqrt{\kappa RT} \quad \kappa = c_p/c_v = \text{relation of the heat capacities}, \quad (12.13)$$

where R = specific gas constant and T = absolute temperature.

This relationship between the sound velocity and the thermodynamic state quantities pressure and density can be presented as follows: we consider the propagation of a small (i.e. isentropic) disturbance at the velocity c in a fluid at rest. This is a non-stationary process which, by changing the reference system (the observer moves together with the flow), can be modified into a stationary problem, as shown in Fig. 12.1. Now the momentum (12.8)

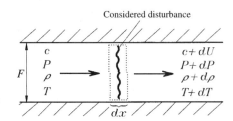

Fig. 12.1 Propagation of a disturbance in a compressible fluid

can be employed as a balance of forces at a control volume around the disturbance:
$$F[P - (P + \mathrm{d}P)] = \rho F c[(c + \mathrm{d}U) - c]. \tag{12.14}$$

Equation (12.14) gives the following relation:
$$-\mathrm{d}P = \rho c \,\mathrm{d}U. \tag{12.15}$$

For the mass conservation, it can be stated that
$$\rho F c = (\rho + \mathrm{d}\rho)(c + \mathrm{d}U) F, \tag{12.16}$$

so that
$$\mathrm{d}\rho = -\rho \frac{\mathrm{d}U}{c}, \tag{12.17}$$

and from (12.15) and (12.17):
$$\frac{\mathrm{d}P}{\mathrm{d}\rho} = c^2 = \left(\frac{\partial P}{\partial \rho}\right)_{\mathrm{ad}} \tag{12.18}$$

as no heat exchange is included in the present considerations.

The sound velocity is therefore a local quantity, i.e. it depends on the local pressure changes under adiabatic conditions. With the local value $c(x_i, t)$, the local Mach number can be computed at each point of a flow field $U_j(x_i, t)$, so that the corresponding Mach number field can also be assigned to the flow field, i.e. $Ma(x_i, t)$. This local Mach number expresses essentially how quickly at each point of the flow field disturbances propagate relative to the existing flow velocity.

From a historical point of view, it is interesting to note that Newton was the first scientist to compute the sound velocity for gases, although on the assumption of an isothermal process in which no temperature changes occur due to the sound propagation. He obtained
$$c_{\mathrm{Newton}} = \sqrt{\frac{P}{\rho}} = \sqrt{RT} < c. \tag{12.19}$$

Only a century later, Marquis de Laplace corrected the result of Newton's computations by recognizing that the temperature fluctuations produced by sound disturbances and also the temperature gradients connected with them are very small. Laplace recognized that it is not possible to transport the heat produced by the compression of a pressure disturbance to the environment. The $\sqrt{\kappa}$ correction of Newton's equation introduced by Laplace led to the correct propagation velocity of sound waves in ideal gases:
$$c = \sqrt{\kappa R T}. \tag{12.20}$$

Attention is drawn once again to the fact that via this equation a sound velocity field also $c(x_i, t)$ is assigned to each temperature field $T(x_i, t)$ of an ideal gas.

12.2 Mach Lines and Mach Cone

When considering a disturbance originating from a point source at the origin of a coordinate system, it will propagate radially at a velocity c, if the point source does not undergo any motion, i.e. the surfaces of disturbance of the same phase represent spherical surfaces when the propagation takes place in a field of constant temperature. Whereas on the other hand, there is a temperature field with variations of temperature, these variations are reflected as deformations of the spherical surfaces shown in Fig. 12.2. The propagation takes place more rapidly in the direction of high temperatures, as predicted by equation (12.20). Possible temperature distributions thus impair the symmetry of the propagation of sound waves.

When one now extends the considerations of the propagation of sound to moving disturbance sources of small dimensions, propagation phenomena result as shown in Fig. 12.3 for $U < c$, i.e. $Ma < 1$ and $U > c$, i.e. $Ma > 1$. By moving the sound source at a velocity lower than the propagation velocity of

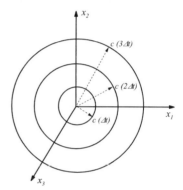

Fig. 12.2 Propagation of disturbances with a stationary source of disturbance

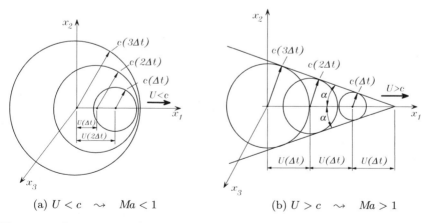

Fig. 12.3 Propagation of disturbances caused by a moving sound source for (a) $Ma < 1$ and (b) $Ma > 1$

the disturbances as shown in Fig. 12.3a, a propagation image results which does not show the symmetry seen in Fig. 12.2. Instead, a concentration of the emitted waves is observed in the direction of prorogation of the source. As a consequence, an observer standing upstream of the disturbance will recognize a frequency increase as compared to the disturbances originating from a source at rest. In the opposite direction, on the other hand, a frequency decrease takes place with respect to the emitted disturbance.

When one computes this frequency change for the frequency increase in the positive x_1 direction, one obtains according to Fig. 12.4 for $Ma < 1$: $\lambda' = \frac{c - U_i \ell_i}{f}$, or $U_i \ell_i = U$ where $\lambda' = \frac{(c-U)}{f}$ and ℓ_i is the unit vector in the

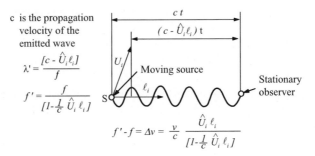

Fig. 12.4 Frequency change by moving the sound source (Doppler effect by moving source)

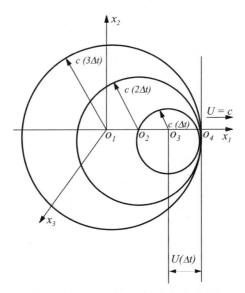

Fig. 12.5 Propagation of disturbances with a source moving at sound velocity

12.2 Mach Lines and Mach Cone

direction of propagation (f = frequency of the disturbance). Thus for f' we have

$$f' = \frac{f}{1 - U_i \ell_i / c} = \frac{f}{1 - (U/c)}. \tag{12.21}$$

$$f' = \frac{c}{\lambda'} = \frac{f}{1 - U/c} = \frac{f}{1 - Ma}. \tag{12.22}$$

In the negative x_1 direction, the following relationship holds:

$$f'' = \frac{f}{1 + U/c} = \frac{f}{1 + Ma}. \tag{12.23}$$

Thus the Mach number proves to be an important quantity for characterizing wave propagations in fluids.

In the case that the velocity of the sound source exceeds the propagation velocity of the sound, a characteristic propagation image develops which is shown in Fig. 12.3b. This illustrates that the propagation of the disturbances in relation to the moving sound source takes place within a cone, the so-called Mach cone. In front of the cone a disturbance-free area results, which is strictly separated from the area with disturbances within the Mach cone. From considerations, shown in Fig. 12.6, it results for the half-angle of the aperture α of the cone:

$$\sin \alpha = \frac{c \Delta t}{U \Delta t} = \frac{1}{Ma}. \tag{12.24}$$

The above equation, employing Fig. 12.6, is derived from the following quantities:

$c \Delta t$ = propagation distance of the disturbance in the time Δt,

$U \Delta t$ = propagation distance of the disturbance source in the time Δt.

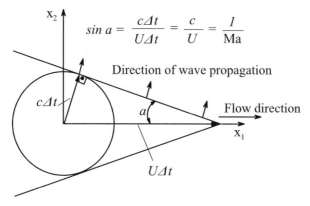

The angle α depends on the Mach numbers

Fig. 12.6 Formation of the Mach cone with typical angle

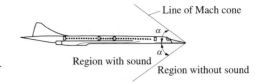

Fig. 12.7 Explanation for perception of aeroplanes

In two dimensions, the Mach cone consists of two planes representing the *Mach planes* or *Mach waves*. The considerations stated above for spatial motions can easily be employed for one-dimensional problems also. They show that propagations of disturbances occur in the form of plane waves. The propagation takes place vertically to the wave planes.

With the aid of the above considerations, observations can be explained that can be made in connection to the flight of supersonic aeroplanes (see Fig. 12.7). Aeroplanes of this kind show a region in which they cannot be heard, i.e. observers can perceive an aeroplane flying towards them at supersonic speed much earlier with the eye than they can hear it. Only when the observers are within the Mach cone do they succeed in seeing *and* hearing the aeroplane.

12.3 Non-Linear Wave Propagation, Formation of Shock Waves

The considerations in Sects. 12.1 and 12.2 concentrated on disturbances of small amplitudes which can be treated as disturbances through linearized equations, as was shown in Chap. 9. There it was explained that small disturbances of the fluid properties ρ', P', T' or of the flow velocity u' can be treated through linearizations of the basic equations of fluid mechanics. Based on assumptions for this fluid, a constant wave velocity resulted. The resultant propagation is such that a given wave form does not change. The implied assumptions no longer hold for wave motions of larger amplitudes, so that wave velocities may change locally and wave fronts may develop that deform with propagation. In order to understand such processes, it is better to consider the one-dimensional form of the continuity and momentum equations with $U = U_1, x = x_1$:

Continuity equation:

$$\frac{\partial \rho}{\partial t} + U \frac{\partial \rho}{\partial x} + \rho \frac{\partial U}{\partial x} = 0. \tag{12.25}$$

12.3 Non-Linear Wave Propagation

Momentum equation:

$$\frac{\partial U}{\partial t} + U \frac{\partial U}{\partial x} = -\frac{1}{\rho}\frac{\partial P}{\partial x}. \tag{12.26}$$

From (12.25) the following relation results for $\rho = \rho(U)$:

$$\frac{d\rho}{dU}\frac{\partial U}{\partial t} + U\frac{d\rho}{dU}\frac{\partial U}{\partial x} + \rho\frac{\partial U}{\partial x} = 0. \tag{12.27}$$

Analogously, (12.26) can be written:

$$\frac{\partial U}{\partial t} + U\frac{\partial U}{\partial x} + \frac{1}{\rho}\left(\frac{dP}{d\rho}\right)\left(\frac{d\rho}{dU}\right)\frac{\partial U}{\partial x} = 0. \tag{12.28}$$

On multiplying (12.28) by $d\rho/dU$ and subtracting it from (12.27), one obtains:

$$\rho\frac{\partial U}{\partial x} = \frac{1}{\rho}\left(\frac{dP}{d\rho}\right)\left(\frac{d\rho}{dU}\right)^2 \frac{\partial U}{\partial x}, \tag{12.29}$$

or rewriting:

$$\frac{dU}{d\rho} = \pm\frac{1}{\rho}\sqrt{\left(\frac{dP}{d\rho}\right)} = \pm\frac{1}{\rho}\sqrt{\left(\frac{\partial P}{\partial \rho}\right)_{ad}}. \tag{12.30}$$

This equation can now be integrated:

$$\int_0^U dU = \pm\int_{\rho_\infty}^{\rho}\sqrt{\left(\frac{dP}{d\rho}\right)}\frac{d\rho}{\rho}. \tag{12.31}$$

For isentropic flows, i.e. $P/\rho^\kappa = $ constant, (12.31) can be integrated:

$$U = \pm\int_{\rho_\infty}^{\rho}\sqrt{\kappa \text{ constant}}\,\rho^{\frac{\kappa-1}{2}}\frac{d\rho}{\rho} = \pm\frac{2}{\kappa-1}\left[\sqrt{\kappa\rho^{\kappa-1}\text{ constant}}\right]_{\rho_\infty}^{\rho}$$

$$U = \pm\frac{2}{(\kappa-1)}(a-c). \tag{12.32}$$

Thus for the propagation velocity of a wave of large amplitude:

$$a = c \pm \frac{(\kappa-1)}{2}U, \tag{12.33}$$

a propagation velocity a results, which depends on the local flow velocity. Here c is the computed sound velocity for the undisturbed fluid.

By inserting (12.30) into (12.28), one obtains the following relationship:

$$\frac{\partial U}{\partial t} + U\frac{\partial U}{\partial x} \pm \sqrt{\left(\frac{dP}{d\rho}\right)}\frac{\partial U}{\partial x} = 0, \tag{12.34}$$

or rewritten:
$$\frac{\partial U}{\partial t} + (U \pm a) \frac{\partial U}{\partial x} = 0. \tag{12.35}$$

From the continuity equation, one obtains:
$$\frac{\partial \rho}{\partial t} + (U \pm a) \frac{\partial \rho}{\partial x} = 0, \tag{12.36}$$

so that for ρ the following general solution of the differential (12.36) can be stated:
$$\rho = F_\rho \left(x_1 - (U_1 \pm a) \right) = F_\rho \left(x_1 - \left(c \pm \frac{\kappa + 1}{2} U_1 \right) t \right), \tag{12.37}$$

where $F_\rho()$ can be any function. Analogously for the velocity:
$$U = F_u \left(x - (U \pm a) \right) = F_u \left(x - \left(c \pm \frac{\kappa + 1}{2} U \right) t \right). \tag{12.38}$$

Equations (12.37) and (12.38) allow one to explain the propagation of a disturbance with a propagation velocity $c \pm \frac{\kappa+1}{2} U$. Because of this propagation velocity, which depends on the local flow velocity, wave deformations develop as they are indicated in Fig. 12.8. On considering the propagating

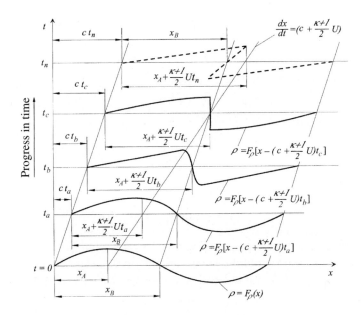

Fig. 12.8 Wave deformations and formation of compression shocks

12.4 Alternative Forms of the Bernoulli Equation

part with a + sign, then characteristic position changes in times t can be stated as

$$x'_A = c \cdot t_\omega; \quad x'_B = x_A + ct_a + \frac{\kappa+1}{2}Ut_a; \quad x'_C = x_B + ct_b. \quad (12.39)$$

The developing and progressive deformation of the wave is apparent. Thus the formation of compression shocks comes about.

The local ambiguity of the density stated in Fig. 12.8 for t_n cannot occur, of course. When the wave front has built up in such a way that all thermodynamic quantities of the fluid and also the velocity experience sudden changes, then the maximum deformation possible of the propagating flow is reached. A compression shock has built up.

12.4 Alternative Forms of the Bernoulli Equation

In Sect. 9.4.2, the stream tube theory was used to consider one-dimensional isentropic flows leading to the Bernoulli equation for compressible flows:

$$\frac{1}{2}U^2 + \frac{\kappa}{(\kappa-1)}\frac{P}{\rho} = \frac{\kappa}{(\kappa-1)}\frac{P_H}{\rho_H}. \quad (12.40)$$

The thermodynamically possible maximum velocity was determined for $(P/\rho) \to 0$:

$$(U_{max})^2 = \frac{2\kappa}{(\kappa-1)}\frac{P_H}{\rho_H} = \frac{2\kappa}{(\kappa-1)}RT_H. \quad (12.41)$$

Thus (12.40) may be expressed as

$$\frac{1}{2}U^2 = \frac{1}{2}U^2_{max} - \frac{\kappa}{(\kappa-1)}\frac{P}{\rho}. \quad (12.42)$$

As the Mach number represents a fundamental quantity in the treatment of gas-dynamic flow problems, we can write

$$1 = \left(\frac{U_{max}}{U}\right)^2 - \frac{2\kappa}{(\kappa-1)}\frac{RT}{U^2} = \left(\frac{U_{max}}{U}\right)^2 - \frac{2}{(\kappa-1)}\frac{1}{Ma^2}. \quad (12.43)$$

or rewritten:

$$\frac{1}{Ma^2} = \frac{\kappa-1}{2}\left[\left(\frac{U_{max}}{U_1}\right)^2 - 1\right]. \quad (12.44)$$

The basis for the above considerations was an expanding flow, as it is indicated in Fig. 12.9. For this flow the so-called critical state results, when the local velocity reaches the speed of sound, i.e. $U_1 = c = U_c$. Then, from (12.40):

$$\frac{1}{2}U_c^2 + \frac{U_c^2}{(\kappa-1)} = \frac{c_H^2}{(\kappa-1)} \quad U_c^2 = \frac{2\kappa}{(\kappa+1)}RT_H. \quad (12.45)$$

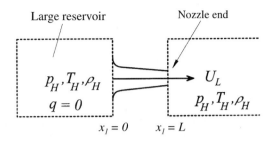

Fig. 12.9 Flow between two pressure tanks of different pressures

The critical pressure can be computed according to (12.40), considering (12.45) and assuming isentropy:

$$U_c^2 = \frac{2\kappa}{(\kappa-1)} RT_H \left[1 - \left(\frac{P_c}{P_H}\right)^{\frac{\kappa-1}{\kappa}}\right] = \frac{2\kappa}{(\kappa+1)} RT_H. \tag{12.46}$$

$$\frac{P_c}{P_H} = \frac{P^*}{P_H} = \left[\frac{2}{(\kappa+1)}\right]^{\frac{\kappa}{(\kappa-1)}}. \tag{12.47}$$

Employing the relationships for isentropic density and temperature changes, one obtains:

$$\left(\frac{\rho_c}{\rho_H}\right) = \frac{\rho^*}{\rho_H} = \left(\frac{P^*}{P_H}\right)^{1/\kappa} = \left[\frac{2}{(\kappa+1)}\right]^{\frac{1}{(\kappa-1)}}. \tag{12.48}$$

$$\left(\frac{T_c}{T_H}\right) = \frac{T^*}{T_H} = \left(\frac{P^*}{P_H}\right)^{\frac{\kappa-1}{\kappa}} = \frac{2}{(\kappa+1)}. \tag{12.49}$$

The above results may now be expressed in terms of the Mach number. From Bernoulli's equation for compressible fluids it follows that:

$$\frac{1}{2}U_1^2 + \frac{c^2}{(\kappa-1)} = \frac{c_H^2}{(\kappa-1)} \rightsquigarrow \frac{\kappa-1}{2} Ma^2 + 1 = \frac{T_H}{T}, \tag{12.50}$$

or rewritten for the temperature ratio:

$$\left(\frac{T}{T_H}\right) = \left[1 + \frac{(\kappa-1)}{2} Ma^2\right]^{-1}. \tag{12.51}$$

For the density and pressure variations, the following relations can be derived:

$$\left(\frac{\rho}{\rho_H}\right) = \left(\frac{T}{T_H}\right)^{\frac{1}{\kappa-1}} = \left[1 + \frac{(\kappa-1)}{2} Ma^2\right]^{\frac{-1}{(\kappa-1)}}, \tag{12.52}$$

$$\left(\frac{P}{P_H}\right) = \left(\frac{T}{T_H}\right)^{\frac{\kappa}{(\kappa-1)}} = \left[1 + \frac{(\kappa-1)}{2} Ma^2\right]^{\frac{-\kappa}{(\kappa-1)}}, \tag{12.53}$$

12.4 Alternative Forms of the Bernoulli Equation

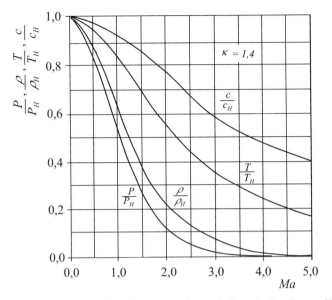

Fig. 12.10 Diagram representing the parameter variations in the Bernoulli equation

For the sound velocity relation c/c_H, the following results:

$$\frac{c}{c_H} = \left(\frac{T}{T_H}\right)^{\frac{1}{2}} = \left[1 + \frac{\kappa - 1}{2} Ma^2\right]^{-2}. \quad (12.54)$$

The above relationships can be plotted as shown in Fig. 12.10. Thus, as a result of the Bernoulli equation for isentropic flows, the figure shows the change of pressure, density, temperature and speed of sound, each normalized with its stagnation value. All data are represented as functions of Mach number changes. This figure corresponds to Fig. 9.10 in Chap. 9, where the temperature, density and pressure variations with (U_1/U_{max}) were employed as a parameter for the representation of different forms of the Bernoulli equation.

It is characteristic for compressible flows that the local dynamic pressure

$$\frac{1}{2}\rho U_1^2 = \frac{1}{2}\rho c^2 Ma^2 = \frac{1}{2}\rho \left(\frac{\kappa P}{\rho}\right) Ma^2 = \frac{1}{2}\kappa P Ma^2 \quad (12.55)$$

depends on the local pressure and the local Mach number.

For the normalized pressure difference, the following holds:

$$\frac{P_H - P}{\frac{1}{2}\rho U_1^2} = \frac{2}{\kappa Ma^2} \frac{P_H - P}{P} = \frac{2}{\kappa Ma^2}\left[\frac{P_H}{P} - 1\right], \quad (12.56)$$

and with (P_H/P) from (12.53) one obtains:

$$\frac{P_H - P}{\frac{1}{2}\rho U_1^2} = \frac{2}{\kappa Ma^2}\left[\left(1 + \frac{\kappa - 1}{2}Ma^2\right)^{\frac{\kappa}{(\kappa-1)}} - 1\right]. \qquad (12.57)$$

Through a series expansion for $Ma^2 < \frac{2}{(\kappa-1)}$, the following results:

$$C_p = \frac{P_H - P}{\frac{1}{2}\rho U_1^2} = 1 + \frac{1}{4}Ma^2 + \frac{2-\kappa}{24}Ma^4 + \frac{(2-\kappa)(3-2\kappa)}{192}Ma^6 + \cdots. \qquad (12.58)$$

For incompressible flows, the Mach number goes to zero so that only the first term of the series expansion remains. For compressible flows, with increasing Mach number a substantial deviation of C_p from the incompressible result is obtained. However, for Mach numbers below 0.3. this deviation is below 1%. Therefore, compressibility effects may be neglected up to this Mach number. This is the basis for treating low-velocity gas flows as incompressible.

12.5 Flow with Heat Transfer (Pipe Flow)

Each chapter in this book tries to give an introduction into a sub-domain of fluid mechanics and in particular each chapter aims at a deepening of the *physical* understanding of the fluid flows treated there. For this purpose, often simplifications were introduced into the considerations of an analytical problem. In the preceding chapter, for example, adiabatic, reversible (dissipation-free) and one-dimensional fluid flows were treated, i.e. isentropic flow processes of compressible media which depend on only one space coordinate. These considerations need some supplementary explanation in order to be able to understand special phenomena in the case of high-speed flows *with heat transfer*. For dealing with such flows, which can be considered as stationary and one-dimensional, i.e. experience changes only in the flow direction $x_1 = x$, the following basic equations are available, which are stated by $U_1 = U$:

- Mass conservation:

$$\rho F U = \dot{m} = \text{ constant}. \qquad (12.59)$$

- Momentum equation:

$$\rho U \frac{dU}{dx} = -\frac{dP}{dx}. \qquad (12.60)$$

- Energy equation:

$$(dq) = c_v dT + Pd\left(\frac{1}{\rho}\right) = c_p dT - \frac{1}{\rho}dP. \qquad (12.61)$$

12.5 Flow with Heat Transfer (Pipe Flow)

- State equation for ideal gases:
$$\frac{P}{\rho} = RT. \tag{12.62}$$

From the mass conservation (12.59), one obtains:
$$\frac{d\rho}{\rho} + \frac{dU}{U} + \frac{dF}{F} = 0. \tag{12.63}$$

or for pipe flows with $\frac{dF}{F} = 0$:
$$\frac{dU}{U} = -\frac{d\rho}{\rho}. \tag{12.64}$$

From the ideal gas (12.62), it can be derived that:
$$\frac{dP}{P} = \frac{d\rho}{\rho} + \frac{dT}{T}, \tag{12.65}$$

and from the momentum equation one obtains $-\frac{dP}{\rho} = U\,dU$ or:
$$-\frac{dP}{P} = \frac{\rho}{P} U\,dU = \frac{1}{RT} U^2 \frac{dU}{U}. \tag{12.66}$$

With $\kappa RT = c^2$ and from the momentum (12.66), one obtains:
$$-\frac{dP}{P} = \frac{\kappa}{c^2} U^2 \frac{dU}{U} = \kappa Ma^2 \frac{dU}{U}. \tag{12.67}$$

On finally including the energy equation into the considerations, it can be stated that the following relationship holds:
$$(dq) = c_p dT - \frac{dP}{\rho} = c_p dT + U\,dU, \tag{12.68}$$

or rewritten:
$$\frac{dU}{U} = \frac{(dq)}{U^2} - \frac{c_p dT}{U^2} = \frac{1}{Ma^2}\left(\frac{c_p}{\kappa RT}\right)\frac{(dq)}{c_p} - \frac{1}{Ma^2}\frac{c_p}{\kappa RT}dT, \tag{12.69}$$

i.e. for the relative velocity change in a pipe flow as a result of heat supply, the following can be written:
$$\frac{dU}{U} = \frac{1}{(\kappa-1)Ma^2}\left(\frac{(dq)}{h} - \frac{dT}{T}\right), \tag{12.70}$$

where $h = c_p T$ was set. From (12.65), it follows that:

$$\frac{dT}{T} = \frac{dP}{P} - \frac{d\rho}{\rho} = -\kappa Ma^2 \frac{dU}{U} + \frac{dU}{U}, \tag{12.71}$$

or rewritten:

$$\frac{dT}{T} = (1 - \kappa Ma^2)\frac{dU}{U}. \tag{12.72}$$

When this relationship is inserted in (12.70), the following results:

$$\frac{dU}{U} = \frac{1}{(\kappa - 1)Ma^2}\left(\frac{(dq)}{h} - (1 - \kappa Ma^2)\frac{dU}{U}\right). \tag{12.73}$$

Solving in terms of $\frac{dU}{U}$, one obtains:

$$\frac{dU}{U} = \frac{1}{(1 - Ma^2)}\frac{(dq)}{h}. \tag{12.74}$$

This relationship inserted in (12.64) yields for the relative density change:

$$\frac{d\rho}{\rho} = \frac{-1}{(1 - Ma^2)}\frac{(dq)}{h}. \tag{12.75}$$

or for the relative changes in pressure and temperature:

$$\frac{dP}{P} = \frac{-\kappa Ma^2}{(1 - Ma^2)}\frac{(dq)}{h} \quad \text{and} \quad \frac{dT}{T} = \frac{(1 - \kappa Ma^2)}{(1 - Ma^2)}\frac{(dq)}{h}. \tag{12.76}$$

For the local change of the Mach number, it can also be derived that:

$$\frac{d(Ma^2)}{Ma^2} = \frac{d(U^2/c^2)}{(U^2/c^2)} = \frac{T}{U^2}d\left(\frac{U^2}{T}\right) = 2\frac{dU}{U} - \frac{dT}{T}. \tag{12.77}$$

Thus for the change of the Mach number with heat supply, the following holds:

$$\frac{dMa^2}{Ma^2} = \frac{(1 + \kappa Ma^2)}{(1 - Ma^2)}\frac{(dq)}{h}. \tag{12.78}$$

As $(dq) = T \cdot ds$ and $h = c_p \cdot T$ it holds furthermore that:

$$\frac{dMa^2}{Ma^2} = \frac{(1 + \kappa Ma^2)}{(1 - Ma^2)}\frac{ds}{c_p}. \tag{12.79}$$

The above relations can now be employed for understanding how P, T, ρ, U and Ma change locally when one introduces heat to a pipe flow, i.e. $dq/h > 0$:

12.5.1 Subsonic Flow

$\dfrac{dU}{U} > 0$; the flow velocity increases with heat supply.

$\dfrac{d\rho}{\rho} < 0$ and $\dfrac{dP}{P} < 0$; density and pressure decrease with heat supply.

$\dfrac{dT}{T} > 0$; the temperature increases with heat supply for $Ma < \sqrt{\dfrac{1}{\kappa}}$.

$\dfrac{dT}{T} < 0$; the temperature decreases in spite of heat supply for $Ma > \sqrt{\dfrac{1}{\kappa}}$.

$\dfrac{dMa^2}{Ma^2} > 0$; the local Mach number increases with heat supply.

The above relationships indicate that, in spite of heat supply, there is a decrease in temperature for $\sqrt{1/\kappa} < Ma < 1$. This is not expected from simple energy considerations that do not take the above details into account.

12.5.2 Supersonic Flow

$\dfrac{dU}{U} < 0$; the flow velocity decreases with heat supply.

$\dfrac{d\rho}{\rho} > 0$ and $\dfrac{dP}{P} > 0$; density and pressure increase with heat supply.

$\dfrac{dT}{T} > 0$; the temperature increases with heat supply.

$\dfrac{dMa^2}{Ma^2} < 0$; the local Mach number decreases with heat transfer.

Thus, in a heated pipe, the change of the thermo-fluid dynamic state differs substantially, depending on the Mach number of the flow.

If one considers to deepen the physical insight into pipe flows with heat supply, the processes that occur in the T–s diagram for an ideal gas, one obtains:

$$(dq)_v = c_v dT_v = T ds_v \quad \leadsto \quad \left(\dfrac{\partial T}{\partial s}\right)_v = \dfrac{T}{c_v}, \tag{12.80}$$

$$(dq)_P = c_p dT_P = T ds_P \quad \leadsto \quad \left(\dfrac{\partial T}{\partial s}\right)_P = \dfrac{T}{c_p}, \tag{12.81}$$

From (12.76), one obtains for the temperature change in a pipe flow with heat supply:

$$\dfrac{dT}{T} = \dfrac{(1 - \kappa Ma^2)}{(1 - Ma^2)} \dfrac{dq}{h} = \dfrac{(1 - \kappa Ma^2)}{(1 - Ma^2)} \dfrac{T\, ds_R}{c_p T}. \tag{12.82}$$

From this it can be computed that:

$$\left(\frac{\partial T}{\partial s}\right)_{\text{pipe}} = \frac{T(1-\kappa Ma^2)}{c_p(1-Ma^2)} = \left(\frac{\partial T}{\partial s}\right)_R, \qquad (12.83)$$

On now introducing an effective heat capacity $c_{\text{pipe}} = c_R$, the following holds:

$$(\mathrm{d}q)_R = c_R\,\mathrm{d}T_R = T\mathrm{d}s_R \quad \leadsto \quad \left(\frac{\partial T}{\partial s}\right)_R = \frac{T}{c_R}, \qquad (12.84)$$

and c_R is computed as:

$$c_R = c_p \frac{(1-Ma^2)}{(1-\kappa Ma^2)} = T \frac{1}{\left(\frac{\partial T}{\partial s}\right)_R}. \qquad (12.85)$$

With $\kappa = \dfrac{c_p}{c_v}$, one can also write:

$$c_R = c_v \frac{(1-Ma^2)}{\left(\frac{1}{\kappa}-Ma^2\right)} = T\frac{1}{\left(\frac{\partial T}{\partial s}\right)_R}. \qquad (12.86)$$

Hence the following relationship holds:

$$\frac{\left(\frac{\partial T}{\partial s}\right)_P - \left(\frac{\partial T}{\partial s}\right)_R}{\left(\frac{\partial T}{\partial s}\right)_V - \left(\frac{\partial T}{\partial s}\right)_R} = \frac{\frac{T}{c_p} - \frac{T}{c_R}}{\frac{T}{c_v} - \frac{T}{c_R}} = \frac{c_R - c_p}{c_R - c_v}, \qquad (12.87)$$

and further rewritten:

$$\frac{\left[\left(\frac{c_R}{c_p}\right) - 1\right]}{\left[\frac{c_R}{c_p} - \frac{1}{\kappa}\right]} = \frac{1 - Ma^2 - 1 + \kappa Ma^2}{\kappa - \kappa Ma^2 - 1 + \kappa Ma^2} = \frac{(\kappa-1)Ma^2}{(\kappa-1)} = Ma^2. \qquad (12.88)$$

$$Ma^2 = \frac{\left(\frac{\partial T}{\partial s}\right)_p - \left(\frac{\partial T}{\partial s}\right)_R}{\left(\frac{\partial T}{\partial s}\right)_v - \left(\frac{\partial T}{\partial s}\right)_R} = \frac{A}{B}. \qquad (12.89)$$

In Fig. 12.11, the relationships expressed by (12.89) are shown graphically. Here $(\partial T/\partial s)_p$ signifies the gradient of the isobars in the T–s state diagram and $(\partial T/\partial s)_v$ the gradient of the isochors and $(\partial T/\partial s)_R$ the change of the thermodynamic state of a gas in a pipe flow with heat supply. It can be shown

12.5 Flow with Heat Transfer (Pipe Flow)

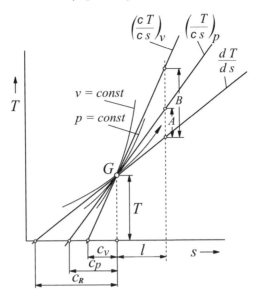

Fig. 12.11 Change of state in the T–s diagram for pipe flows with heat supply

that (12.89) holds not only for flows of ideal gases, generally treated in gas dynamics, but also for the flows of real gases.

In conclusion, it can be remarked that the relationships for $\frac{dU}{U}$ (12.74), $\frac{d\rho}{\rho}$ (12.75), $\frac{dP}{P}$ (12.76), $\frac{dT}{T}$ (12.76) and $\frac{dMa^2}{Ma^2}$ (12.78) for $Ma = 1$ lose their validity if $(dq) \neq 0$. In order to accelerate a subsonic flow to supersonic flow speeds through heat supply, the heat supply has to be stopped once $Ma = 1$ is reached. After that, it is necessary to cool the flow in order to obtain a further velocity increase.

Extended considerations show that the heat supply in the subsonic region leads to accelerating the flow and in the supersonic region to decelerating the flow. For pipe flows with a radius $R =$ constant, a subsonic flow cannot be turned into a supersonic flow with constant heat supply.

By considering the course of the effective heat capacity of the pipe flow, the behaviour shown in Fig. 12.12 results:

$$\frac{c_R}{c_v} = \frac{(Ma^2 - 1)}{(Ma^2 - 1/\kappa)}. \tag{12.90}$$

For $0 \leq Ma < \sqrt{1/\kappa}$ and $1 \leq Ma < \infty$ the heat capacity is positive and in the range $\sqrt{1/\kappa} < Ma < 1$ a negative heat capacity results. At $Ma = \sqrt{1/\kappa}$ the local flow velocity has the value of the isothermal sound velocity.

According to (12.88) for the effective heat capacity c_R/c_v in heated and cooled pipe flows, the thermodynamic state developes as shown in Fig. 12.13.

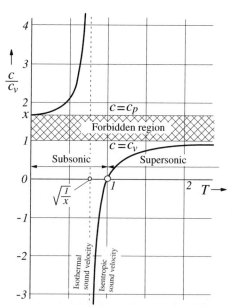

Fig. 12.12 Behavior of the effective heat capacity of a gas with heat supply in a pipe flow

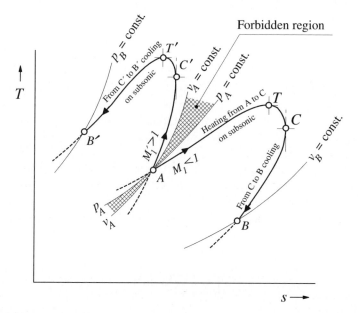

Fig. 12.13 Thermodynamic changes of state at subsonic and supersonic pipe flows according to Bošnjaković

Starting for subsonic flow from state A, one reaches state C by heating and subsequently state B by cooling, where a supersonic flow is achieved. If, on the other hand, state A is supersonic, heating would decelerate the flow towards C' and finally cooling would further decelerate to subsonic flow B'.

12.6 Rayleigh and Fanno Relations

The considerations in the preceding section concentrated on the investigation of infinitesimal changes of fluid-mechanical and thermodynamic state quantities in pipe flows, i.e. on the changes in the case of an infinitesimal heat supply to the fluid, with the assumption that no dissipative processes occur. For the pressure and Mach number changes that occur, it was derived that:

$$\frac{dP}{P} = \frac{-\kappa Ma^2}{(1 - Ma^2)} \frac{(dq)}{h} \quad \text{and} \quad \frac{dMa^2}{Ma^2} = \frac{(1 + \kappa Ma^2)}{(1 - Ma^2)} \frac{(dq)}{h}. \tag{12.91}$$

From these, the following relationship between the relative changes of pressure and Mach number for the pipe flow is obtained:

$$\frac{dP}{P} = -\frac{\kappa Ma^2}{(1 + \kappa Ma^2)} \frac{(dMa^2)}{Ma^2}. \tag{12.92}$$

This differential equation can be integrated between two states 1 and 2, yielding:

$$\int_1^2 \frac{dP}{P} = -\int_1^2 \frac{\kappa Ma^2}{(1 + \kappa Ma^2)} \frac{dMa^2}{Ma^2} \rightsquigarrow \ln\left(\frac{P_2}{P_1}\right) = \ln\left(\frac{1 + \kappa Ma_1^2}{1 + \kappa Ma_2^2}\right), \tag{12.93}$$

and thus:

$$\frac{P_2}{P_1} = \frac{1 + \kappa (Ma)_1^2}{1 + \kappa (Ma)_2^2} \quad \text{etc..} \tag{12.94}$$

From (12.53), a thermodynamically achievable maximum pressure P_H follows:

$$P_H = P \left(1 + \frac{\kappa - 1}{2} Ma^2\right)^{\frac{\kappa}{(\kappa-1)}}. \tag{12.95}$$

With this relationship, also known as the Rayleigh flow, the following case can be computed:

$$\frac{(P_H)_2}{(P_H)_1} = \frac{(1 + \kappa Ma_1^2)}{(1 + \kappa Ma_2^2)} \left[\frac{1 + \frac{(\kappa-1)}{2} Ma_2^2}{1 + \frac{(\kappa-1)}{2} Ma_1^2}\right]^{\frac{\kappa}{(\kappa-1)}}. \tag{12.96}$$

Analogously for the temperature relationship T_2/T_1, the following can be derived:

$$\frac{T_2}{T_1} = \frac{(Ma)_2^2}{(Ma)_1^2} \left[\frac{1 + \kappa (Ma)_1^2}{1 + \kappa (Ma)_2^2}\right]^2, \tag{12.97}$$

and for the corresponding relationship of the stagnation temperature ratio:

$$\frac{(T_H)_2}{(T_H)_1} = \frac{Ma_2^2}{Ma_1^2}\left(\frac{1+\kappa Ma_1^2}{1+\kappa Ma_2^2}\right)^2 \left[\frac{1+\frac{(\kappa-1)}{2}Ma_2^2}{1+\frac{(\kappa-1)}{2}Ma_A^2}\right]. \quad (12.98)$$

For the density and velocity relationship one can write:

$$\frac{\rho_2}{\rho_1} = \frac{U_1}{U_2} = \frac{P_2 T_1}{P_1 T_2} = \frac{Ma_1^2}{Ma_2^2}\left(\frac{1+\kappa Ma_2^2}{1+\kappa Ma_1^2}\right). \quad (12.99)$$

Finally, the following relationship can also be derived:

$$ds = c_p \left[\frac{1-Ma^2}{1+\kappa \cdot Ma^2}\right] \frac{d(Ma^2)}{Ma^2} \quad (12.100)$$

in order to compute the entropy change of the flowing gas in the pipe flow with heat supply, the following holds:

$$s_2 - s_1 = \underbrace{\frac{\kappa R}{(\kappa-1)}}_{c_p} \cdot \ln\left[\frac{(Ma)_2^2}{(Ma)_1^2}\left(\frac{1+\kappa(Ma)_1^2}{1+\kappa(Ma)_2^2}\right)^{\frac{(\kappa+1)}{\kappa}}\right]. \quad (12.101)$$

The above equations can now be employed to determine in a T–s diagram the thermodynamically possible states with the Mach numbers as parameters, e.g. for Rayleigh flow. We start here from a state 1, for which T_1 and s_1 are known, as well as U_1 and therefore also $(Ma)_1$. For each value $(Ma)_2$, T_2 an s_2 can be computed and thus the Rayleigh curve, as shown in Fig. 12.14 can be obtained. For the direct connection between s and T one obtains:

$$\frac{s_2 - s_1}{c_p} = \ln\left(\frac{T_2}{T_1}\right)^{\frac{\kappa+1}{2\kappa}}$$

From Fig. 12.14, it can be seen that for the subsonic part of the Rayleigh curve the temperature increases, together with an increase in the Mach number up to $Ma = \sqrt{1/\kappa}$. After that the temperature decreases until $Ma = 1$. On moving on the branch of the supersonic flow, the Mach number decreases with increasing entropy until $Ma = 1$ is achieved.

It is also usual in gas dynamics to employ values for $(Ma)_1 = 1$, which are usually designated with an asterisk $(*)$ as reference quantities for the standardized representation of P, P_H, T, T_H and ρ. For such a representation of the above results, the following holds:

$$\frac{P}{P^*} = \frac{1+\kappa}{1+\kappa Ma^2}; \quad \frac{T}{T^*} = \frac{(1+\kappa)^2 Ma^2}{(1+\kappa Ma^2)^2}, \quad (12.102)$$

12.6 Rayleigh and Fanno Relations

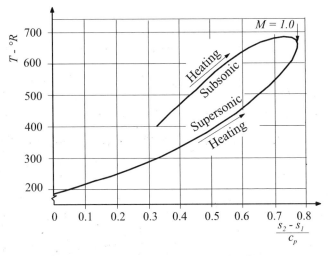

Fig. 12.14 Rayleigh curves on a T–s diagram

$$\frac{\rho}{\rho^*} = \frac{1}{(U_1/U_1^*)} = \frac{1 + \kappa Ma^2}{(1+\kappa)Ma^2}. \quad (12.103)$$

A generalization of the considerations above for flows in pipes with change in cross-section, which are furthermore exposed to externally imposed forces, leads to the relationships below for the fluid mechanical and thermodynamic changes of state caused in compressible flows.

Continuity equation:
$$\frac{d\rho}{\rho} + \frac{dU}{U} + \frac{dF}{F} = 0. \quad (12.104)$$

Momentum equation:
$$\rho U \, dU = -dP + d\Pi, \quad (12.105)$$

where $d\Pi$ is an externally applied pressure gradient which can be forced on the flow by a compressor. The energy equation can be stated for the extended considerations as follows:
$$c_p \, dT + U \, dU_1 = dq. \quad (12.106)$$

By division by P and after introduction of $c^2 = \kappa \frac{P}{\rho}$, the momentum (12.105) can be written as
$$\kappa Ma^2 \frac{dU}{U} + \frac{dP}{P} = \frac{d\Pi}{P}. \quad (12.107)$$

For the energy equation, the following rearrangements of terms are possible:
$$\frac{dT}{T} + \frac{U \, dU}{c_p T} = \frac{dq}{c_p T}, \quad (12.108)$$

or rewritten:
$$\frac{dT}{T} + (\kappa - 1)Ma^2 \frac{dU}{U} = \frac{dq}{c_p T}. \tag{12.109}$$

Finally, the state equation for ideal gases is employed for the considerations to be carried out:
$$\frac{P}{\rho} = RT \quad \rightsquigarrow \quad \frac{dP}{P} - \frac{d\rho}{\rho} - \frac{dT}{T} = 0. \tag{12.110}$$

The above set of equations can now be employed to express the quantities dU/U, $d\rho/\rho$, dT/T, etc. as a function of the local Mach number and the local relative area change (dF/F), the heat supplied (dq/h) and the applied external forces $(d\Pi/P)$:

$$\frac{dU}{U} = \frac{1}{(Ma^2 - 1)} \left(\frac{dF}{F} + \frac{d\Pi}{P} - \frac{dq}{h} \right), \tag{12.111}$$

$$\frac{dP}{P} = -\frac{\kappa Ma^2}{(Ma^2 - 1)} \frac{dF}{F} - \frac{1 + (\kappa - 1)Ma^2}{(Ma^2 - 1)} \frac{d\Pi}{P} + \frac{\kappa Ma^2}{(Ma^2 - 1)} \frac{dq}{h}, \tag{12.112}$$

$$\frac{dT}{T} = -\frac{(\kappa - 1)Ma^2}{(Ma^2 - 1)} \frac{dF}{F} - \frac{(\kappa - 1)Ma^2}{(Ma^2 - 1)} \frac{d\Pi}{P} + \frac{(\kappa Ma^2 - 1)}{(Ma^2 - 1)} \frac{dq}{h}. \tag{12.113}$$

From the above general equations, the preceding derivations, that are referred solely to the area changes (see Chap. 9), can now be derived for $d\Pi/P = 0$ and $dq/h = 0$. Furthermore, one obtains for $dF/F = 0$ and $d\Pi/P = 0$ the relationships derived at the beginning of this chapter for heated pipes. When one now sets $dF/F = 0$ and $dq/h = 0$, one obtains:

$$\frac{dU}{U} = \frac{1}{(Ma^2 - 1)} \frac{d\Pi}{P}; \quad \frac{dP}{P} = -\frac{1 + (\kappa - 1)Ma^2}{(Ma^2 - 1)} \frac{d\Pi}{P}, \tag{12.114}$$

and finally also:
$$\frac{dT}{T} = -\frac{(\kappa - 1)Ma^2}{(Ma^2 - 1)} \frac{d\Pi}{P}. \tag{12.115}$$

On considering now for a viscous flow the molecular momentum transport as an external *action of forces* $(d\Pi_R/P) < 0$, one realizes that the following temperature changes are connected with it:

$$\left(\frac{dT}{T} \right)_R > 0 \text{ for } Ma < 1, \tag{12.116}$$

or
$$\left(\frac{dT}{T} \right)_R < 0 \text{ for } Ma > 1. \tag{12.117}$$

Analogous to the considerations that were based on (12.91) and (12.92), all derivations that lead to the relationships for the Rayleigh flow, i.e. for the flow through pipes having constant cross-sections with heat supply, can now be repeated for pipe flow under the influence of friction without heat supply. From the derivations, similar relationships result as for the Rayleigh flow in (12.94)–(12.102). From, this the "Fanno curve" in the T–s diagram results, which indicates the possible states of the thermodynamic state that develop in adiabatic pipe flow with internal friction. The "Rayleigh curve" in the T–s diagram, on the other hand, represents the thermodynamic change of state which develops with heat supply in the case of friction-free flow of an ideal gas in a pipe. With this the Fanno curve indicates the influence of friction in a pipe flow with constant cross-section, whereas the Rayleigh curve shows the influence of the heat supply.

12.7 Normal Compression Shock (Rankine–Hugoniot Equation)

In Sect. 12.3, the formation of compression shocks was explained as a phenomenon of wave motions with a state-dependent wave velocity. The discontinuity surface formed in this way shows a thickness which can be considered to be of the order of magnitude of the free path length of the molecules of an ideal gas. It is thus possible to describe, within the assumptions chosen in this book, the compression shock shown in Fig. 12.15 in a medium at rest by the fluid mechanical and thermodynamic state quantities before and after the compression shock. As before, the analysis may be simplified by considering a stationary problem, i.e. assuming that the shock is at rest. The flow velocity and the variables describing the thermodynamic state upstream of the shock are denoted by $P_A, \rho_A, T_A, e_A, s_A$ and downstream by $P_B, \rho_B, T_B, e_B, s_B$. Thus the transformation is $U_A = U_s$ and $U_B = U_A - U_g$, respectively.

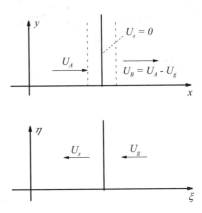

Fig. 12.15 Plotted normal compression shocks and selection of coordinate systems

The integral conservation laws formulated with these variables are expressed as:

$$\rho_A U_A = \rho_B (U_A - U_g) = \rho_B U_B, \tag{12.118}$$

$$\rho_A U_A^2 + P_A = \rho_B U_B^2 + P_B, \tag{12.119}$$

or rewritten:

$$\rho_A U_A (U_A - U_B) = P_B - P_A. \tag{12.120}$$

Furthermore, the energy equation holds:

$$\frac{1}{2} U_A^2 + \frac{P_A}{\rho_A} + c_v T_A = \frac{1}{2} U_B^2 + \frac{P_B}{\rho_B} + c_v T_B. \tag{12.121}$$

The left-hand side of (12.121) can be described by the quantities P, ρ and U of the mass conservation and momentum equations as one obtains:

$$\frac{1}{2} U_A^2 + \frac{P_A}{\rho_A} + c_v \frac{P_A}{R \rho_A} = \frac{1}{2} U_A^2 + \frac{\kappa}{(\kappa - 1)} \frac{P_A}{\rho_A} = \frac{\kappa}{(\kappa - 1)} \frac{P_H}{\rho_H}. \tag{12.122}$$

From the momentum equation and the continuity equation, it follows that:

$$U_A - U_B = \frac{P_B}{\rho_B U_B} - \frac{P_A}{\rho_A U_A}. \tag{12.123}$$

Multiplication by $U_A + U_B$ yields:

$$U_A^2 - U_B^2 = \left(\frac{P_B}{\rho_B U_B} - \frac{P_A}{\rho_A U_A} \right) (U_A + U_B), \tag{12.124}$$

$$U_A^2 - U_B^2 = \left(\frac{P_B}{\rho_A} + \frac{P_B}{\rho_B} - \frac{P_B}{\rho_A} - \frac{P_A}{\rho_A} \right), \tag{12.125}$$

or rewritten:

$$U_A^2 - U_B^2 = (P_B - P_A) \left[\frac{1}{\rho_A} + \frac{1}{\rho_B} \right]. \tag{12.126}$$

From the energy equation it follows:

$$U_A^2 - U_B^2 = \frac{2\kappa}{\kappa - 1} \left(\frac{P_B}{\rho_B} - \frac{P_A}{\rho_A} \right). \tag{12.127}$$

Equations (12.125) and (12.126) set equal yields:

$$(P_B - P_A) \left[\frac{1}{\rho_A} + \frac{1}{\rho_B} \right] = \frac{2\kappa}{\kappa - 1} \left(\frac{P_B}{\rho_B} - \frac{P_A}{\rho_A} \right). \tag{12.128}$$

From this, the following relationship is obtained:

$$\frac{P_B}{P_A} \left[\frac{\rho_B}{\rho_A} - \frac{(\kappa + 1)}{(\kappa - 1)} \right] = 1 - \frac{\rho_B}{\rho_A} \frac{(\kappa + 1)}{(\kappa - 1)}. \tag{12.129}$$

12.7 Normal Compression Shock (Rankine–Hugoniot Equation)

Hence one can write:

$$\frac{P_B}{P_A} = \frac{1 + \frac{(\kappa+1)}{2}\left(\frac{\rho_B}{\rho_A} - 1\right)}{1 - \frac{(\kappa-1)}{2}\left(\frac{\rho_B}{\rho_A} - 1\right)} \quad \text{and} \quad \frac{\rho_B}{\rho_A} = \frac{1 + \frac{(\kappa+1)}{2\kappa}\left(\frac{P_B}{P_A} - 1\right)}{1 + \frac{(\kappa-1)}{2\kappa}\left(\frac{P_B}{P_A} - 1\right)}. \tag{12.130}$$

The above relations $P_A/P_B = f(\rho_B/\rho_A)$ or $(\rho_B/\rho_A) = g(P_B/P_A)$ are known as Rankine–Hugoniot equations. They state the pressure and density changes through the normal compression shock. As the compression shock is linked to a dissipation of mechanical energy into heat, the compression shock is a non-isentropic process.

Derivations of somewhat different nature, using the energy equation yield:

$$U_A + \frac{2\kappa}{(\kappa-1)} \frac{P_A}{\rho_A U_A} = \frac{2\kappa}{(\kappa-1)} \frac{P_H}{\rho_H U_A} \tag{12.131}$$

and

$$U_B + \frac{2\kappa}{(\kappa-1)} \frac{P_B}{\rho_B U_B} = \frac{2\kappa}{(\kappa-1)} \frac{P_H}{\rho_H U_B}. \tag{12.132}$$

From this, the following results:

$$(U_A - U_B) + \frac{2\kappa}{(\kappa-1)} \underbrace{\left(\frac{P_A}{\rho_A U_A} - \frac{P_B}{\rho_B U_B}\right)}_{U_B - U_A} = \frac{2\kappa}{(\kappa-1)} \frac{P_H}{\rho_H} \left(\frac{1}{U_A} - \frac{1}{U_B}\right),$$

$$\tag{12.133}$$

$$(U_A - U_B) - \frac{2\kappa}{(\kappa-1)}(U_A - U_B) = \frac{2c_H^2}{(\kappa-1)} \frac{U_B - U_A}{U_A U_B}, \tag{12.134}$$

$$1 - \frac{2\kappa}{(\kappa-1)} = -\frac{2c_H^2}{(\kappa-1)U_A U_B}. \tag{12.135}$$

From this, the Prandtl compression relationship can be computed:

$$U_A U_B = \frac{2c_H^2}{(\kappa+1)}; \quad (Ma)_{A,H} (Ma)_{B,H} = \frac{2}{(\kappa+1)}. \tag{12.136}$$

From the energy equation:

$$\frac{1}{2}U_A^2 + c_p T_A = c_p T_H \quad \text{and} \quad \frac{1}{2}U_B^2 + c_p T_B = c_p T_H. \tag{12.137}$$

It then follows that:

$$\left[(\kappa-1) + \frac{2}{(Ma)_A^2}\right] = \frac{2c_p T_H}{U_A^2}(\kappa-1), \tag{12.138}$$

$$\left[(\kappa-1) + \frac{2}{(Ma)_B^2}\right] = \frac{2c_p T_H}{U_B^2}(\kappa-1). \tag{12.139}$$

Multiplying these two equations yields:

$$\left[(\kappa-1)+\frac{2}{(Ma)_A^2}\right]\left[(\kappa-1)+\frac{2}{(Ma)_B^2}\right] = \frac{4c_p^2 T_H^2}{U_A^2 U_B^2}(\kappa-1)^2. \quad (12.140)$$

With $U_A U_B = \dfrac{2\kappa R T_H}{(\kappa+1)}$, the following equation results:

$$\left[\left(\frac{\kappa-1}{2}\right)+\frac{1}{(Ma)_A^2}\right]\left[\left(\frac{\kappa-1}{2}\right)+\frac{1}{(Ma)_B^2}\right] = \left[\frac{(\kappa+1)}{2}\right]^2. \quad (12.141)$$

It is usual to express the state quantities, after the vertical compression shock, scaled with the corresponding quantity before the shock, as a function of the Mach number before the shock. These normalized quantities can be written as follows:

$$\frac{P_B}{P_A} = \frac{2\kappa}{(\kappa+1)}(Ma)_A^2 - \frac{(\kappa-1)}{(\kappa+1)} \quad\rightsquigarrow\quad \frac{P_B}{P_A} = 1 + \frac{2\kappa}{\kappa+1}\left(Ma_A^2-1\right), \quad (12.142)$$

$$\frac{\rho_B}{\rho_A} = \frac{(\kappa+1)(Ma)_A^2}{(\kappa-1)(Ma)_A^2+2}, \quad (12.143)$$

$$\frac{T_B}{T_A} = \frac{P_B \rho_A}{P_A \rho_B} = \frac{1}{(\kappa+1)^2 (Ma)_A^2}\left[2\kappa(Ma)_A^2-(\kappa-1)\right]\left[(\kappa-1)(Ma)_A^2+2\right]. \quad (12.144)$$

Furthermore, it can be stated for the Mach number after the compression shock:

$$(Ma)_B^2 = \frac{[(\kappa-1)(Ma)_A^2+2]}{[(2(Ma)_A^2-1)\kappa+1]}, \quad (12.145)$$

and for the pressure difference $\Delta p/p_A$ as a measure for the strength of the compression shock:

$$\frac{(P_B-P_A)}{P_A} = \frac{2\kappa}{(\kappa+1)}\left[(Ma)_A^2-1\right]. \quad (12.146)$$

For the change of entropy linked to the shock, one can compute:

$$s_B - s_A = c_v \ln\left(\left[\frac{2\kappa}{\kappa+1}(Ma)_A^2 - \frac{\kappa-1}{\kappa+1}\right]\left[\frac{(\kappa-1)(Ma)_A^2+2}{(\kappa+1)(Ma)_A^2}\right]^\kappa\right). \quad (12.147)$$

The changes are shown in Fig. 12.17, where as the abscissa the Mach number before the shock was chosen.

12.7 Normal Compression Shock (Rankine–Hugoniot Equation)

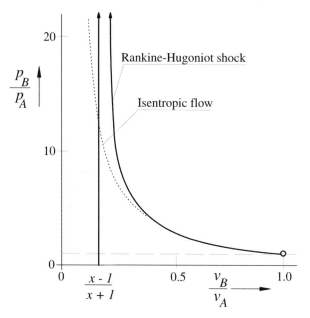

Fig. 12.16 Changes of state for vertical compression shock

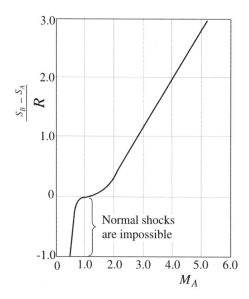

Fig. 12.17 Change of entropy as a result of vertical compression shocks in an ideal gas with $\kappa = 1.4$

Figure 12.16 shows further that the fluid is compressed when it is moving through the compression shock. The pressure, density and temperature increase on passing through the compression shock, i.e.

$$\frac{P_B}{P_A} \geq 1; \quad \frac{\rho_B}{\rho_A} \geq 1; \quad \text{and} \quad \frac{T_B}{T_A} \geq 1$$

When considering the difference $s_B - s_A$, using (12.46), it is evident (see Fig. 12.17) that $s_B - s_A$ can only be larger than zero for $M_A \geq 1$. This expresses that dilution shocks cannot occur in ideal gases, as they are not permitted by the second law of thermodynamics, which demands $s_B - s_A \geq 0$.

In this section only an introduction into gas dynamics was given providing treatments of compressible flows in a manner also applied in refs. [12.1] and [12.2] as well as in [12.4] to [12.6]. More advanced treatments are provided in [12.3].

References

12.1. Bošnjaković, F., Technische Thermodynamik I, Verlag von Theodor Steinkopff, Dresden, Leipzig, 1960.
12.2. Currie, I.G., Fundamental Mechanics of Fluids, McGraw-Hill, New York, 1974.
12.3. Oswatitsch, K., Gasdynamik, Springer, Berlin Heidelberg New York Vienna, 1952.
12.4. Yuan, S.W., Foundations of Fluid Mechanics, Prentice-Hall, Englewood Cliffs, NJ, 1971.
12.5. Becker, E., Technische Thermodynamik, Teubner Studienbuecher, Mechanik, Stuttgart, 1985.
12.6. Spurk, J.H., Stroemungslehre, Springer, Berlin, Heidelberg, New York, 4. Aufl., 1996.

Chapter 13
Stationary, One-Dimensional Fluid Flows of Incompressible, Viscous Fluids

13.1 General Considerations

In this chapter, flows of viscous fluids ($\mu \neq 0$) are considered which are stationary and two-dimensional. They are assumed to occur in fluids of constant density and, in addition, the fluid is assumed to be fully developed in the flow direction. The simplified equations determining this class of flow can be derived from the general equations of fluid mechanics and the resultant equations are basically one-dimensional. They are, moreover, for a number of boundary conditions, accessible to analytical solutions and thus well suited for students of natural and engineering sciences to provide to them an introduction into fluid mechanics of viscous fluids. The basic knowledge gained by studying these fluid flows can then be deepened in specialized lectures. In this way, the knowledge of how flows of viscous fluids behaves in one-dimensional flow cases can be extended and used for the solution of practical flow problems.

As shown below, the problems discussed in this chapter can be tackled by analytical solutions. Hence their properties with regard to the physics of fluid flows can be described with a few terms of the Navier–Stokes equations and solutions become possible due to the existence of simple boundary conditions. In addition, it is assumed that stationarity exists for all flow quantities and that fluids with a constant density are treated, i.e. fluids with $\rho =$ constant. This property holds not only for thermodynamically ideal liquids but also, as shown in Sect. 12.1, for thermodynamically ideal gases when they flow at moderate velocities.

Simple considerations show that gas flows with Mach numbers $Ma \leq 0.2$ can be treated as incompressible with a precision which is sufficient for practical application, i.e. gas flows at low Mach numbers can be treated as fluids of constant density. For such fluids, the basic equations in the form stated below hold, for Newtonian media when τ_{ij} is introduced as follows:

$$\tau_{ij} = -\mu \left[\frac{\partial U_j}{\partial x_i} + \frac{\partial U_i}{\partial x_j} \right] + \underbrace{\frac{2}{3} \delta_{ij} \mu \frac{\partial U_k}{\partial x_k}}_{=0 \text{ because } \rho = \text{constant}}. \tag{13.1}$$

- Continuity equation:

$$\frac{\partial U_1}{\partial x_1} + \frac{\partial U_2}{\partial x_2} + \frac{\partial U_3}{\partial x_3} = 0. \tag{13.2}$$

- Momentum equations:
  - x_1-component:

$$\rho \left[\frac{\partial U_1}{\partial t} + U_1 \frac{\partial U_1}{\partial x_1} + U_2 \frac{\partial U_1}{\partial x_2} + U_3 \frac{\partial U_1}{\partial x_3} \right] = -\frac{\partial P}{\partial x_1} + \mu \left[\frac{\partial^2 U_1}{\partial x_1^2} + \frac{\partial^2 U_1}{\partial x_2^2} + \frac{\partial^2 U_1}{\partial x_3^2} \right] + \rho g_1. \tag{13.3}$$

  - x_2-component:

$$\rho \left[\frac{\partial U_2}{\partial t} + U_1 \frac{\partial U_2}{\partial x_1} + U_2 \frac{\partial U_2}{\partial x_2} + U_3 \frac{\partial U_2}{\partial x_3} \right] = -\frac{\partial P}{\partial x_2} + \mu \left[\frac{\partial^2 U_2}{\partial x_1^2} + \frac{\partial^2 U_2}{\partial x_2^2} + \frac{\partial^2 U_2}{\partial x_3^2} \right] + \rho g_2. \tag{13.4}$$

  - x_3-component:

$$\rho \left[\frac{\partial U_3}{\partial t} + U_1 \frac{\partial U_3}{\partial x_1} + U_2 \frac{\partial U_3}{\partial x_2} + U_3 \frac{\partial U_3}{\partial x_3} \right] = -\frac{\partial P}{\partial x_3} + \mu \left[\frac{\partial^2 U_3}{\partial x_1^2} + \frac{\partial^2 U_3}{\partial x_2^2} + \frac{\partial^2 U_3}{\partial x_3^2} \right] + \rho g_3. \tag{13.5}$$

13.1.1 Plane Fluid Flows

As a further simplification for the subsequent considerations, the flow field is assumed to be two-dimensional, i.e. for all quantities of the velocity and pressure fields $[\partial(\cdots)/\partial x_3] = 0$ can be introduced. It is further assumed that in the x_3 direction there is no flow component or that it is always possible to introduce a coordinate system in such a way that only in the directions of the two coordinate axes x_1 and x_2 do velocity components occur. Thus one obtains the final equations for two-dimensional and two-directional flow problems, which are employed in the following analytical solutions:

$$\frac{\partial U_1}{\partial x_1} + \frac{\partial U_2}{\partial x_2} = 0, \tag{13.6}$$

$$\rho \left[\frac{\partial U_1}{\partial t} + U_1 \frac{\partial U_1}{\partial x_1} + U_2 \frac{\partial U_1}{\partial x_2} \right] = -\frac{\partial P}{\partial x_1} + \mu \left[\frac{\partial^2 U_1}{\partial x_1^2} + \frac{\partial^2 U_1}{\partial x_2^2} \right] + \rho g_1, \tag{13.7}$$

$$\rho \left[\frac{\partial U_2}{\partial t} + U_1 \frac{\partial U_2}{\partial x_1} + U_2 \frac{\partial U_2}{\partial x_2} \right] = -\frac{\partial P}{\partial x_2} + \mu \left[\frac{\partial^2 U_2}{\partial x_1^2} + \frac{\partial^2 U_2}{\partial x_2^2} \right] + \rho g_2. \tag{13.8}$$

The above equations are employed in subsequent sections for analytical computations of fluid flows. It is assumed here that the flow causing effects are known and that they fulfil the assumptions made to yield the above-stated simplified form of the basic equations, i.e. (13.6)–(13.8).

Further restrictions which are made concerning the subsequently treated flow problems should be mentioned with regard to the boundary conditions. It is assumed that these boundary conditions are known and that they fulfil the condition of stationarity, i.e. temporal changes do not occur. Because of another restriction in the following considerations, only solutions of the above equations are listed which are laminar. The perturbations acting on fluid flows in practice constitute, in general, boundary conditions that depend on time. Moreover, the disturbances have to be considered as unknown. Their effects on flows are therefore not treated in the subsequent considerations in this section.

13.1.2 Cylindrical Fluid Flows

For a large number of flow problems, boundary conditions exist which originate from axi-symmetric flow geometries and which can be introduced more easily into solutions of the basic equations of fluid mechanics, when these equations are written in cylindrical coordinates. To provide these equations, $\rho =$ constant and $\mu =$ constant are also assumed.

- Continuity equation:
$$\frac{\partial \rho}{\partial t} + \rho \left[\frac{1}{r}\frac{\partial}{\partial r}(rU_r) + \frac{1}{r}\frac{\partial}{\partial \varphi}(U_\varphi) + \frac{\partial}{\partial z}(U_z) \right] = 0. \quad (13.9)$$

- Momentum equations:
  - r-component:
$$\rho \left[\frac{\partial U_r}{\partial t} + U_r \frac{\partial U_r}{\partial r} + \frac{U_\varphi}{r}\frac{\partial U_r}{\partial \varphi} - \frac{U_\varphi^2}{r} + U_z \frac{\partial U_r}{\partial z} \right]$$
$$= -\frac{\partial P}{\partial r} + \mu \left[\frac{\partial}{\partial r}\left(\frac{1}{r}\frac{\partial (rU_r)}{\partial r}\right) + \frac{1}{r^2}\frac{\partial^2 U_r}{\partial \varphi^2} - \frac{2}{r^2}\frac{\partial U_\varphi}{\partial \varphi} + \frac{\partial^2 U_r}{\partial z^2} \right] + \rho g_r. \quad (13.10)$$

  - φ-component:
$$\rho \left[\frac{\partial U_\varphi}{\partial t} + U_r \frac{\partial U_\varphi}{\partial r} + \frac{U_\varphi}{r}\frac{\partial U_\varphi}{\partial \varphi} + \frac{U_r U_\varphi}{r} + U_z \frac{\partial U_\varphi}{\partial z} \right]$$
$$= -\frac{1}{r}\frac{\partial P}{\partial \varphi} + \mu \left[\frac{\partial}{\partial r}\left(\frac{1}{r}\frac{\partial (rU_\varphi)}{\partial r}\right) + \frac{1}{r^2}\frac{\partial^2 U_\varphi}{\partial \varphi^2} + \frac{2}{r^2}\frac{\partial U_r}{\partial \varphi} + \frac{\partial^2 U_\varphi}{\partial z^2} \right] + \rho g_\varphi. \quad (13.11)$$

– z-component:

$$\rho\left[\frac{\partial U_z}{\partial t} + U_r\frac{\partial U_z}{\partial r} + \frac{U_\varphi}{r}\frac{\partial U_z}{\partial \varphi} + U_z\frac{\partial U_z}{\partial z}\right]$$
$$= -\frac{\partial P}{\partial z} + \mu\left[\frac{1}{r}\frac{\partial}{\partial r}\left(r\frac{\partial U_z}{\partial r}\right) + \frac{1}{r^2}\frac{\partial^2 U_z}{\partial \varphi^2} + \frac{\partial^2 U_z}{\partial z^2}\right] + \rho g_z. \quad (13.12)$$

For stationary, incompressible (ρ = constant) fluid flows of Newtonian fluids, assuming axi-symmetry $\partial(\cdots)/\partial\varphi = 0$ and $U_\varphi = 0$, one can obtain the following final equations:

$$\frac{1}{r}\frac{\partial(rU_r)}{\partial r} + \frac{\partial U_z}{\partial z} = 0, \quad (13.13)$$

$$\rho\left[U_r\frac{\partial U_r}{\partial r} + U_z\frac{\partial U_r}{\partial z}\right] = -\frac{\partial P}{\partial r} + \mu\left[\frac{\partial}{\partial r}\left(\frac{1}{r}\frac{\partial(rU_r)}{\partial r}\right) + \frac{\partial^2 U_r}{\partial z^2}\right] + \rho g_r, \quad (13.14)$$

$$\rho\left[U_r\frac{\partial U_z}{\partial r} + U_z\frac{\partial U_z}{\partial z}\right] = -\frac{\partial P}{\partial z} + \mu\left[\frac{1}{r}\frac{\partial}{\partial r}\left(r\frac{\partial U_z}{\partial r}\right) + \frac{\partial^2 U_z}{\partial z^2}\right] + \rho g_z. \quad (13.15)$$

These equations can be employed for solutions of fluid flow problems for stationary axially symmetric fluid flows and for $U_\varphi = 0$.

13.2 Derivations of the Basic Equations for Fully Developed Fluid Flows

13.2.1 Plane Fluid Flows

The basic equations for stationary, two-dimensional and fully developed fluid flows can be derived from the equations for incompressible Newtonian media on the assumption that the resulting fluid flow in the x_1 direction fulfils the following relationships:

$$\frac{\partial U_1}{\partial x_1} = 0 \quad \text{and} \quad \frac{\partial U_2}{\partial x_2} = 0. \quad (13.16)$$

Thus the continuity equation is reduced to:

$$\underbrace{\frac{\partial U_1}{\partial x_1}}_{=0} + \frac{\partial U_2}{\partial x_2} = 0 \quad \rightsquigarrow \quad \frac{\partial U_2}{\partial x_2} = 0 \quad \text{and therefore} \quad U_2 = f(x_1). \quad (13.17)$$

Based on the assumption of a fully developed fluid flow, the relationships (13.16) hold and from (13.17) we can derive:

13.2 Derivations of the Basic Equations

$$U_2 = \text{constant} \quad U_2 = 0 \quad \begin{pmatrix} U_2 = 0 \text{ holds for fluid flows with} \\ \text{impermeable walls} \end{pmatrix}, \tag{13.18}$$

i.e. stationary, incompressible and internal flows are unidirectional. They flow only in the x_1 direction, i.e. only one U_1 component of the flow field exists.

This is a statement for the flow field that was obtained from the continuity equation for fluid flows which are fully developed in the flow direction x_1. The momentum equations are simplified for this class of fluid flows as follows:

x_1 direction:

$$0 = -\frac{\partial P}{\partial x_1} + \mu \frac{\partial^2 U_1}{\partial x_2^2} + \rho g_1. \tag{13.19}$$

x_2 direction:

$$0 = -\frac{\partial P}{\partial x_2} + \rho g_2. \tag{13.20}$$

From (13.20), one obtains a general solution for the pressure field:

$$P = \rho g_2 x_2 + \Pi(x_1). \tag{13.21}$$

The pressure field $P(x_1, x_2)$ comprises an externally imposed pressure, $\Pi(x_1)$, which can be applied along the x_1 axis. The implementation of $\Pi(x_1)$ usually takes place in practice with pumps and blowers. On introducing this general pressure relationship into the momentum equation x_1, taking $U_1(x_2)$ into consideration, one obtains:

$$0 = -\frac{d\Pi}{dx_1} + \mu \frac{d^2 U_1}{dx_2^2} + \rho g_1, \tag{13.22}$$

i.e. a differential equation for the unknown flow field $U_1(x_2)$. This is the basic equation which holds for incompressible, stationary and one-dimensional, i.e. fully developed, fluid flows, if the flow medium has Newtonian properties and the fluid can be regarded as incompressible and the viscosity as constant. Physically, the equation can be interpreted in such a way that the pressure gradient imposed externally in the x_1 direction counteracts the viscosity and mass forces of the flow field

$$\frac{d\Pi}{dx_1} = \mu \frac{d^2 U_1}{dx_2^2} + \rho g_1. \tag{13.23}$$

Here it is important that the pressure gradient, in accordance with (13.21), can assume any externally imposed value, which for the flow problems treated here must depend only on x_1. Considering however, (13.18), and admitting only constant mass forces, i.e. $g_1 = \text{constant}$, the right-hand side of (13.22) is a function only of x_2. Thus the pressure gradient in the x_1 direction assumes a constant value in the case of stationary, incompressible and one-dimensional fluid flows.

13.2.2 Cylindrical Fluid Flows

Analogous to the above derivations of plane fluid flows, the derivations of the basic equations for stationary, one-dimensional fluid flows can be made for axi-symmetric flow cases also. For the following derivation, it is assumed that in the z direction the fluid flow is fully developed, i.e. all derivatives of the velocity components are zero in the z direction, as stated below

$$\frac{\partial U_r}{\partial z} = 0, \qquad \frac{\partial U_z}{\partial z} = 0. \tag{13.24}$$

With these assumptions, one obtains from the continuity equation:

$$\frac{\partial}{\partial r}(rU_r) = 0 \quad \leadsto \quad \rho r U_r = \text{constant} \tag{13.25}$$

and because of the assumption of impermeable walls for the fluid (see (13.18)) one obtains:

$$U_r = 0 \quad \begin{pmatrix} \text{for the presence of impermeable walls} \\ \text{for the considered fluid flow} \end{pmatrix} \tag{13.26}$$

and thus the momentum equations hold:

$$0 = -\frac{\partial P}{\partial r} + \rho g_r. \tag{13.27}$$

$$0 = -\frac{\partial P}{\partial z} + \mu \left[\frac{1}{r}\frac{\partial}{\partial r}\left(r\frac{\partial U_z}{\partial r}\right) \right] + \rho g_z \tag{13.28}$$

by integration of:

$$P(r,z) = \rho g_z r + \Pi(z), \quad \text{i.e.} \quad \frac{\partial P}{\partial z} = \frac{d\Pi}{dz}. \tag{13.29}$$

Finally, the above derivations result in:

$$0 = -\frac{d\Pi}{dz} + \mu \left[\frac{1}{r}\frac{\partial}{\partial r}\left(r\frac{\partial U_z}{\partial r}\right) \right] + \rho g_z. \tag{13.30}$$

This last equation represents the conditional equation for the velocity field, which has to be employed for solutions of one-dimensional (fully developed) flow problems in axi-symmetric geometries.

13.3 Plane Couette Flow

In chemical process engineering, it is common practice, e.g. when coating sheet metals, foils, plates, etc., to employ coating systems of the kind shown in Fig. 13.1. This figure shows that the actual material to be coated is moved

13.3 Plane Couette Flow

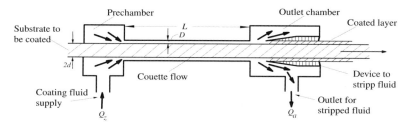

Fig. 13.1 Schematic representation of a coating system

through a pre-chamber filled with the coating fluid. From there the material enters a channel with plane parallel walls that end in a fluid collecting chamber, where the coating thickness is brought to the required final value by scrubbers installed on both sides. The wiped off coating material is collected in the discharge chamber and is fed back through a discharge pipe to the coating fluid supply system.

The flow forming in the slots between the pre-chamber and the discharge chamber, after a certain distance from the inlet, is called Couette flow. It is characterized by the fact that no pressure gradients are used for driving the flow, i.e. for Couette flow the following holds:

$$\frac{\mathrm{d}\Pi}{\mathrm{d}x_1} = 0 \quad \text{and therefore} \quad 0 = \mu \frac{\mathrm{d}^2 U_1}{\mathrm{d}x_2{}^2} + \rho g_1. \qquad (13.31)$$

In the case of a horizontal flow direction with respect to the vertical direction of the field of gravity, no mass forces are active which could drive the fluid flow, as the x_1 direction is vertical to the direction of the gravitational acceleration, i.e. for the Couette flow in Fig. 13.1 the following holds:

$$g_1 = 0. \qquad (13.32)$$

Thus the basic equation stated in Sect. 13.2 for plane flows is reduced to the differential equation describing the Couette flow:

$$-\underbrace{\frac{\mathrm{d}\Pi}{\mathrm{d}x_1}}_{=0} + \mu \frac{\mathrm{d}^2 U_1}{\mathrm{d}x_2^2} + \underbrace{\rho g_1}_{=0} = 0 \quad \Longrightarrow \quad \frac{\partial^2 U_1}{\partial x_2{}^2} = 0. \qquad (13.33)$$

From the resulting equation for U_1, i.e. from (13.33), it can be seen that the velocity profile $U_1(x_2)$ occurring in the slot is independent of the viscosity of the coating fluid. Thus also the required quantity of coating material is independent of the viscosity of the coating medium, a property which is often regarded to be desirable for well-designed coating systems. The system thus becomes equally applicable for all fluid properties and results in velocity profiles that are independent of the fluid properties (Fig. 13.2).

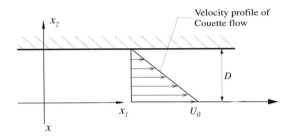

Fig. 13.2 Basic geometry of the upper slit in the coating system shown in Fig. 13.1 (width of the slit in the x_3 direction is B)

When considering that, for the assumptions made, the velocity U_1 can depend on the coordinate x_2 only, the final equation can be written as follows:

$$\frac{d^2 U_1}{dx_2^2} = 0 \quad \leadsto \quad U_1 = C_1 x_2 + C_2. \tag{13.34}$$

Because of the boundary conditions existing due to the operation of the coating system, the integration constants C_1 and C_2 result in values as shown below

$$\begin{aligned} x_2 = 0 &: U_1 = U_0 = C_1 0 + C_2; \quad C_2 = U_0, \\ x_2 = D &: U_1 = 0 = C_1 D + C_2; \quad C_1 = -U_0/D. \end{aligned} \tag{13.35}$$

Thus for the velocity profile one obtains

$$U_1 = \frac{U_0}{D}(D - x_2) \quad \text{for} \quad 0 \leq x_2 \leq D. \tag{13.36}$$

The required fluid volume of the coating material that has to be supplied per unit time results from integration over the entire slot having the width B in the x_3 direction, i.e. the integration has to be taken over both slot openings, on the top and at the bottom of the substrate. For the system in Fig. 13.1, $\dot{Q}_z$ needed for coating results from the following integration:

$$\dot{Q}_z = 2\dot{Q} = 2B \int_0^D U_1 \, dx_2 = 2\frac{U_0 B}{D}\left[Dx_2 - \frac{1}{2}x_2^2\right]_0^D \tag{13.37}$$

$$= 2\frac{U_0 B}{D}\left[\frac{1}{2}D^2\right] \tag{13.38}$$

or as the final relationship:

$$\dot{Q} = \frac{1}{2} B U_0 D. \tag{13.39}$$

The force exerted on one side of the material to be coated in the slot can be computed as follows:

$$F = BL\tau_w = -BL\mu \frac{dU_1}{dx_2} = BL\mu \frac{U_0}{D}. \tag{13.40}$$

Finally, attention is drawn to the fact that the Couette flow is characterized in such a way that in the entire flow field, the same molecular-dependent momentum transport takes place at every location x_2. For this reason the Couette flow is often sought as a fluid flow for basic investigations, in order to examine experimentally the influence of the shear stresses on fluid properties of non-Newtonian fluids.

13.4 Plane Fluid Flow Between Plates

In Sect. 13.2, the generally valid basic equation for an incompressible ($\rho =$ constant), stationary and one-dimensional (fully developed) flow of a Newtonian medium with constant viscosity ($\mu =$ constant) was derived. This equation:

$$\frac{d\Pi}{dx_1} = \mu \frac{d^2 U_1}{dx_2{}^2} + \rho g_1 \tag{13.41}$$

holds also for the fluid flow between two infinitely long plane plates arranged as shown in Fig. 13.3.

This figure shows two plates which are placed at a distances $x_2 = +D$ and $x_2 = -D$ with the planes in $x_3 =$ constant as surfaces located in a Cartesian coordinate system. The fluid flow takes place between these two plates and the flow velocity is equal to zero at the surfaces of the plates (non-slip condition).

If one selects x_1 perpendicular to the gravity field, then (13.41) reduces to the following form:

$$\frac{d\Pi}{dx_1} = \mu \frac{d^2 U_1}{dx_2{}^2}. \tag{13.42}$$

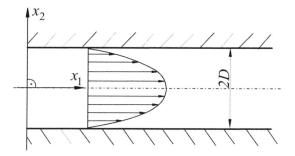

Fig. 13.3 Fully developed fluid flow between two plane and parallel plates

This relationship expresses the fact that the motion of the flow between the plates is caused by an imposed external pressure gradient. Pressure and viscosity forces for these kinds of flows are in equilibrium for a fluid element.

As the pressure distribution $d\Pi/dx_1$ can only be a function of x_1 (see Sect. 13.2) and the right-hand side of the above equation depends only on x_2, i.e. $U_1(x_2)$, $d\Pi/dx_1$ has to be constant for the flow between parallel plates. Thus, we have a simple linear differential equation of second order which has to be solved to obtain the velocity profile of the plane channel flow.

By a first integration one obtains:

$$\frac{dU_1}{dx_2} = \frac{1}{\mu}\left(\frac{d\Pi}{dx_1}\right)x_2 + C_1. \tag{13.43}$$

This differential equation has as a general solution obtained by a second integration:

$$U_1 = \frac{1}{2\mu}\left(\frac{d\Pi}{dx_1}\right)x_2^2 + C_1 x_2 + C_2. \tag{13.44}$$

Due to the following boundary conditions:

$$x_2 = +D \longrightarrow U_1 = 0 = \frac{1}{2\mu}\left(\frac{d\Pi}{dx_1}\right)D^2 + C_1 D + C_2, \tag{13.45}$$

$$x_2 = -D \longrightarrow U_1 = 0 = \frac{1}{2\mu}\left(\frac{d\Pi}{dx_1}\right)D^2 - C_1 D + C_2 \tag{13.46}$$

one obtains the values for the integration constants:

$$C_1 = 0 \quad \text{and} \quad C_2 = -\frac{1}{2\mu}\left(\frac{d\Pi}{dx_1}\right)D^2 \tag{13.47}$$

and thus the solution for the velocity distribution between the plates can be given as follows:

$$U_1 = -\frac{1}{2\mu}\left(\frac{d\Pi}{dx_1}\right)D^2\left[1 - \left(\frac{x_2}{D}\right)^2\right] \quad \text{for} \quad -D \leq x \leq D. \tag{13.48}$$

This relationship for the flow velocity U_1 shows that the velocity profile between the plates represents a parabola. The maximum velocity is at the center of the channel. At the surfaces of both the plates the flow velocity is zero and in the entire flow field U_1 is positive, because for the flow region $|x_2| \leq D$ holds and, hence, $[1 - (x^2/D)^2]$ is always positive. However, the pressure gradient in the x_1 direction decreases, i.e. the resultant pressure gradient is negative, so that the velocity U_1 in the x_1 direction, according to (13.43), is positive. Assuming that the plates in the x_3 direction have a width B, the volumetric flow rate per unit time can be computed for the flow in Fig. 13.3 as follows:

13.4 Plane Fluid Flow Between Plates

$$\dot{Q} = 2B \int_0^D U_1 \, dx_2 = \frac{2B}{2\mu} \left(\frac{d\Pi}{dx_1} \right) \left[\frac{1}{3} x_2^3 - D^2 x_2 \right]_0^D. \tag{13.49}$$

Thus the following results are valid for the flow rate $\dot{Q}$ and the mean velocity:

$$\dot{Q} = -\frac{B}{\mu} \left(\frac{d\Pi}{dx_1} \right) \frac{2}{3} D^3 \quad \longrightarrow \quad \tilde{U} = \frac{\dot{Q}}{2DB} = -\frac{1}{3\mu} \left(\frac{d\Pi}{dx_1} \right) D^2. \tag{13.50}$$

For the velocity $U_{\max}$ one can compute

$$U_{\max} = U(x_2 = 0) = -\frac{1}{2\mu} \left(\frac{d\Pi}{dx_1} \right) D^2 \quad \rightsquigarrow \quad \tilde{U} = \frac{2}{3} U_{\max}. \tag{13.51}$$

From $\dot{Q}$ one obtains for the pressure gradient

$$\frac{d\Pi}{dx_1} = \frac{\Delta P}{\Delta L} = \frac{3\mu \dot{Q}}{2BD^3}. \tag{13.52}$$

From this it can be seen that the pressure drop is linear and is directly proportional to the dynamic viscosity and to the volume flow rate and inversely proportional to the cube of half the channel height. The action of forces on the plate due to the molecular momentum transport results from the product of the shear stress at the wall τ_w and the area of the plates

$$\tau_w = -\mu \left(\frac{dU_1}{dx_2} \right)_{x_2 = x_w}, \tag{13.53}$$

$$\left(\frac{dU_1}{dx_2} \right)_w = \frac{1}{\mu} \left(\frac{d\Pi}{dx_1} \right) (x_2)_w ; \quad (x_2)_w = D; \quad \tau_w = -\left(\frac{d\Pi}{dx_1} \right) D. \tag{13.54}$$

The force acting on one of the plates having length L and width B is given by

$$F = \tau_w A = \left(\frac{d\Pi}{dx_1} \right) DLB. \tag{13.55}$$

As a further quantity, which is often used in fluid mechanics, the friction coefficient of the flow can be computed

$$c_f = \frac{\tau_w}{\frac{\rho}{2} \tilde{U}^2} = \frac{\tau_w 2D}{\frac{\mu}{2} \left(\frac{\tilde{U} 2D}{\mu/\rho} \right) \tilde{U}} = \frac{1}{Re} \frac{4D\tau_w}{\mu \tilde{U}}. \tag{13.56}$$

With

$$\tau_w = \left(\frac{d\Pi}{dx_1} \right) D \quad \text{and} \quad \tilde{U} = \frac{\dot{Q}}{2DB} = \frac{1}{3\mu} \left(\frac{d\Pi}{dx_1} \right) D^2, \tag{13.57}$$

one obtains

$$c_f = \frac{12}{Re} \quad \text{with} \quad Re = \frac{\tilde{U} 2D}{\nu}. \tag{13.58}$$

On plotting the friction coefficient as a function of the Reynolds number in a diagram with double-logarithmic scales, one obtains a straight line with a gradient of -1.

13.5 Plane Film Flow on an Inclined Plate

In this section, fluid flows which are generally called film flows will be considered. They find applications in many fields of chemical engineering. Such flows can be extremely complex, when the base plates of the flow show irregularities or waviness. To simplify the considerations to be carried out here, only smooth surfaces are considered. In addition, the considerations are only carried out for incompressible fluids with constant viscosity. Furthermore, the assumption of two-dimensionality of the fluid flow is introduced into the derivations and extended by the assumption of fully developed film flows finally yielding the one-dimensionality of the flow, so that the following basic equation holds:

$$0 = -\left(\frac{\mathrm{d}\Pi}{\mathrm{d}x_1}\right) + \mu \frac{\mathrm{d}^2 U_1}{\mathrm{d}x_2^2} + \rho g_1. \tag{13.59}$$

In the examples shown in Fig. 13.4, the film motion is caused by the mass forces occurring in the flow direction and not, as in the case of the plane channel flow, by an externally imposed pressure gradient, i.e. for the film flow the following holds for the pressure gradient:

$$\frac{\mathrm{d}\Pi}{\mathrm{d}x_1} = 0, \tag{13.60}$$

which finally results in the following simple basic equation for gravity-driven film flows:

$$\mu \frac{\mathrm{d}^2 U_1}{\mathrm{d}x_2^2} + \rho g_1 = 0. \tag{13.61}$$

In the case of film flows which are caused by mass forces on the fluid, the mass and the viscous forces at a fluid element are in equilibrium. The diagram in Fig. 13.4 shows a film flow which can be treated analytically as will be shown later.

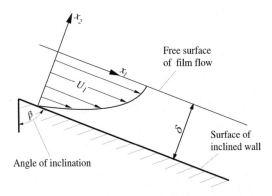

Fig. 13.4 A fluid film on a plane, inclined wall

13.5 Plane Film Flow on an Inclined Plate

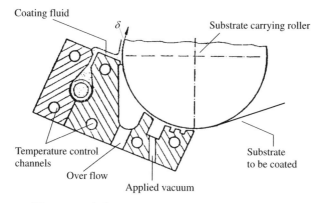

Fig. 13.5 A film-coating system for a single layer

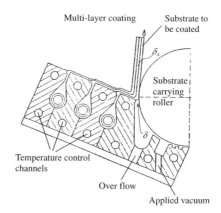

Fig. 13.6 A film-coating system for several layers

As an example of film flows that occur in the practice of chemical engineering, different coating procedures are mentioned here which are applied in industry in order to coat photographic papers and foils of all kinds. Current coating procedures are presented in Figs. 13.5 and 13.6, which show that a characteristic of the customary coating procedures is that the material used for coating is supplied in fluid films. The fluid-volume flow supplied in the films is, at a given geometry of the actual coating apparatus, controlled by the supplied volume flow only. In this way, the supplied volume flow $\dot{Q}$ controls the film thickness δ and through the latter it also controls the velocity distribution in the wet film.

For the design and construction of coating systems of the kind shown in Figs. 13.5 and 13.6, it is important to know the relationship $\delta(\dot{Q})$ The latter can be found by solving the above differential equation for the boundary conditions of the film flow. This needs then to be integrated to obtain the

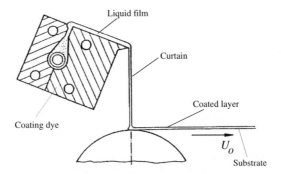

Fig. 13.7 Curtain coating procedure and equipment

volume flow rate ($\dot{Q}$). In addition, the solution of the differential equation also renders details of the velocity field that establishes itself in the fluid film.

In the simple coating system represented in Fig. 13.7, the coating material is supplied through a slot opening leading it on to a plane, inclined surface where, due to gravitation, a film flow forms. The film falling downwards impinges on to the substrate to be coated which, moved by rollers, carries away the fluid film.

For the actual coating dye, after the fluid has reached the inclined flat plate, a plane film flow develops which can be treated analytically. After a short entrance length, the conditions for a stationary, fully developed film flow exist. The component of the gravity acting in the x_1 direction is:

$$g_1 = g \cos \beta. \tag{13.62}$$

Thus, the differential equation describing the flow field reads:

$$\frac{\mathrm{d}^2 U_1}{\mathrm{d} x_2^2} = -\frac{\rho g \cos \beta}{\mu}. \tag{13.63}$$

By a first integration one obtains from the above differential equation:

$$\frac{\mathrm{d} U_1}{\mathrm{d} x_2} = -\frac{\rho g \cos \beta}{\mu} x_2 + C_1 \tag{13.64}$$

and by another integration the final relationship for the velocity distribution in the film results:

$$U_1 = -\frac{\rho g \cos \beta}{2\mu} x_2^2 + C_1 x_2 + C_2. \tag{13.65}$$

As boundary conditions are available (see Fig. 13.4):

$$x_2 = 0: \quad \frac{\mathrm{d} U_1}{\mathrm{d} x} = 0, \quad \text{i.e. } C_1 = 0 \text{ because of the free surface,} \tag{13.66}$$

13.5 Plane Film Flow on an Inclined Plate

$$x_2 = -\delta: \quad U_1 = 0, \quad \text{i.e.} \quad C_2 = \frac{\rho g \cos \beta}{2\mu}\delta^2. \tag{13.67}$$

Thus for the velocity distribution of the film, U_1 can be expressed as:

$$U_1 = \frac{\rho g \cos \beta \delta^2}{2\mu}\left[1 - \left(\frac{x_2}{\delta}\right)^2\right]. \tag{13.68}$$

This equation describes the parabolic velocity profile which is characteristic for film flows with the maximum velocity being at the free surface of the film, i.e. for the coordinate system chosen in Fig. 13.4 at the location $x_2 = 0$.

When the velocity profile in the fluid film is known (see (13.68)), the volume flow $\dot{Q}$ can be computed by the following integration, where B is the width of the film perpendicular to the $x_1 - x_2$ plane:

$$\dot{Q} = B\int_{-\delta}^{0} U_1 dx_2 = B\frac{\rho g \cos \beta \delta^2}{2\mu}\int_{-\delta}^{0}\left[1 - \left(\frac{x_2}{\delta}\right)^2\right]dx_2, \tag{13.69}$$

$$\dot{Q} = B\frac{\rho g \cos \beta \delta^2}{2\mu}\left[x_2 - \frac{1}{3\delta^2}x_2^3\right]_{-\delta}^{0}. \tag{13.70}$$

From this, $\dot{Q}$ can be computed as:

$$\dot{Q} = B\frac{\rho g \cos \beta \delta^3}{3\mu}. \tag{13.71}$$

The volume flow running in a fluid film is inversely proportional to the dynamic viscosity and directly proportional to the cubic power of the film thickness. The mean velocity results as:

$$\tilde{U} = \frac{\dot{Q}}{B\delta} = \frac{\rho g \cos \beta \delta^2}{3\mu}. \tag{13.72}$$

When the force acting on the film carrying surface in the x_1 direction is of interest, it can be computed for a surface having the dimensions L and B as follows:

$$F_1 = \tau_B LB = -\mu\left(\frac{dU_1}{dx_2}\right)_{x_2=-\delta} LB = -\delta LB\rho g \cos \beta. \tag{13.73}$$

This value corresponds to the component of the weight of the total film acting in the x_1 direction over the length. This final result expresses that the film as a whole adheres to the plate and thus the momentum transport to the wall compensates the weight of the film.

In connection with the motion of the plane fluid film, the energy dissipation in viscous fluids will be considered in more detail. In a fluid film, as shown in Fig. 13.3, a fluid volume ($LB\,dx_2$) having the mass ($\rho LB\,dx_2$) is flowing downwards, per unit time, over a distance $U_1 \cos \beta$ in the direction of the gravitational acceleration. In this way, the following potential energy per unit time is set free:

$$\mathrm{d}\dot{E}_{\mathrm{pot}} = \rho L B \, \mathrm{d}x_2 U_1 \cos\beta g. \tag{13.74}$$

Hence the potential energy $\dot{E}_{\mathrm{pot}}$ for the entire fluid film results as:

$$\dot{E}_{\mathrm{pot}} = \int_{-\delta}^{0} \rho L B \frac{\rho g \cos\beta \delta^2}{2\mu} \left[1 - \left(\frac{x_2}{\delta}\right)^2\right] \cos\beta g \, \mathrm{d}x_2, \tag{13.75}$$

$$\dot{E}_{\mathrm{pot}} = L B \frac{\rho^2 g^2 \cos^2\beta \delta^2}{2\mu} \left[x_2 - \frac{x_2^3}{3\delta^2}\right]_{-\delta}^{0}, \tag{13.76}$$

$$\dot{E}_{\mathrm{pot}} = L B \frac{\rho^2 g^2 \cos^2\beta \delta^3}{3\mu}. \tag{13.77}$$

This energy, set free per unit time by moving in the direction of gravity along the length L, dissipates due to the viscosity of the flow medium. The dissipated energy E_{diss} per unit time and unit volume for a fluid layer of width 1 can be given as follows:

$$\frac{\mathrm{d}E_{\mathrm{diss}}}{\mathrm{d}V} = \mu \left(\frac{\mathrm{d}U_1}{\mathrm{d}x_2}\right)^2 = \rho^2 g^2 \cos^2\beta x_2^2 \frac{1}{\mu}. \tag{13.78}$$

For the considered volume of the entire film, the dissipated energy per unit time is computed by integration

$$\dot{E}_{\mathrm{diss}} = L B \frac{\rho^2 g^2 \cos^2\beta}{\mu} \int_{-\delta}^{0} x_2^2 \, \mathrm{d}x_2, \tag{13.79}$$

$$\dot{E}_{\mathrm{diss}} = -L B \frac{\rho^2 g^2 \cos^2\beta}{3\mu} \delta^3. \tag{13.80}$$

Thus $\dot{E}_{\mathrm{pot}} + \dot{E}_{\mathrm{diss}} = 0$ holds, i.e. the total potential energy of the falling film is dissipated due to the viscosity of the flowing fluid, i.e. potential energy is converted into heat. Because of the generally very high heat capacity of fluids, this means, e.g. for water, only a very small increase of the fluid temperature.

13.6 Axi-Symmetric Film Flow

In addition to the description of plane film flows in Sect. 13.5, fluid films that develop on axi-symmetric surfaces are also of interest in chemical engineering. As an example, a film is shown in Fig. 13.8 which flows down on the outside of a cylindrical body. The volume flow needed for the stationary fluid film is conveyed upwards in the inner space of the cylindrical body, flows outwards

13.6 Axi-Symmetric Film Flow

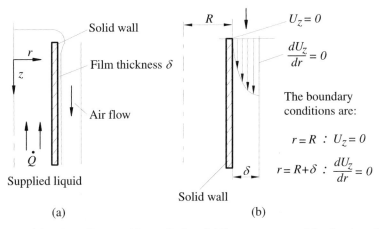

Fig. 13.8 (a) Falling film outside a cylinder. (b) Important quantities for the solution of the differential equation for the velocity of the fluid film

at the upper edge and forms there, after a short development length, an axi-symmetric, stationary fluid film which is fully developed in the flow directions. The fluid volume running down in the film per unit time corresponds to the volume flow transported upwards in the inner space of the cylinder.

Film producing systems, such as schematically indicated in Fig. 13.8 are often employed in chemical engineering. The fluid film running downward has a large surface area, when compared with its volume, which can be brought in contact with the surrounding gas to be absorbed. The gassing takes place over the entire contact surface of the fluid and goes on until the entire fluid film is saturated.

After the film in Fig. 13.8 has moved a short development distance, it takes on a fully developed state, i.e. the fluid mechanics of the film flow can be described by the following differential equation for one-dimensional, axi-symmetric flows of fluids with constant density and constant viscosity:

$$-\frac{\mathrm{d}\Pi}{\mathrm{d}z} + \mu \frac{1}{r}\frac{\mathrm{d}}{\mathrm{d}r}\left(r\frac{\mathrm{d}U_z}{\mathrm{d}r}\right) + \rho g_z = 0. \tag{13.81}$$

The externally imposed pressure gradient ($\mathrm{d}\Pi/\mathrm{d}z$) is zero for film flows, so that with $g_z = g$ it holds that:

$$\frac{\mathrm{d}}{\mathrm{d}r}\left(r\frac{\mathrm{d}U_z}{\mathrm{d}r}\right) = -\frac{\rho g}{\mu}r. \tag{13.82}$$

After a first integration, one obtains:

$$\frac{\mathrm{d}U_z}{\mathrm{d}r} = -\frac{\rho g}{2\mu}r + \frac{C_1}{r} \tag{13.83}$$

and after a second integration:
$$U_z = -\frac{\rho g}{4\mu} r^2 + C_1 \ln r + C_2. \tag{13.84}$$

With the boundary conditions indicated in Fig. 13.8b, one obtains:
$$r = R; \quad U_z = 0: \quad 0 = -\frac{\rho g}{4\mu} R^2 + C_1 \ln R + C_2, \tag{13.85}$$

$$r = (R+\delta); \quad \frac{dU_z}{dr} = 0: \quad 0 = -\frac{\rho g}{2\mu}(R+\delta) + C_1 \frac{1}{(R+\delta)}, \tag{13.86}$$

$$C_1 = +\frac{\rho g}{2\mu}(R+\delta)^2, \tag{13.87}$$

$$C_2 = +\frac{\rho g}{4\mu} R^2 - \frac{\rho g}{2\mu}(R+\delta)^2 \ln R. \tag{13.88}$$

Thus the velocity profile can be expressed as:
$$U_z = \frac{\rho g}{4\mu} R^2 \left[1 - \left(\frac{r}{R}\right)^2 + 2\left(1+\frac{\delta}{R}\right)^2 \ln\left(\frac{r}{R}\right)\right]. \tag{13.89}$$

The fluid volume flowing in the film can be computed by the following integration:
$$\dot{Q} = \int_R^{R+\delta} 2\pi r U_z \, dr = \frac{\pi \rho g}{2\mu} R^2 \int_R^{R+\delta} \left[1 - \left(\frac{r}{R}\right)^2 + 2\left(1+\frac{\delta}{R}\right)^2 \ln\left(\frac{r}{R}\right)\right] r \, dr,$$

$$\dot{Q} = \frac{\pi \rho g}{2\mu} R^2 \left[\frac{r^2}{2} - \frac{r^4}{4R^2} + 2\left(1+\frac{\delta}{R}\right)^2 \frac{r^2}{2}\left(\ln\frac{r}{R} - \frac{1}{2}\right)\right]\Bigg|_R^{R+\delta}. \tag{13.90}$$

Thus for $\dot{Q}$ the following final relation results:
$$\dot{Q} = \frac{\pi \rho g R^4}{4\mu} \left[2\left(1+\frac{\delta}{R}\right)^2 - \frac{1}{2} + \left(1+\frac{\delta}{R}\right)^4 \left(2\ln\left(1+\frac{\delta}{R}\right) - \frac{3}{2}\right)\right]. \tag{13.91}$$

For the maximum velocity of the film flow, the following relationships hold:
$$(U_z)_{\max} = \frac{\rho g R^2}{4\mu} \left[1 - \left(\frac{R+\delta}{R}\right)^2 + 2\left(1+\frac{\delta}{R}\right)^2 \ln\left(\frac{R+\delta}{R}\right)\right], \tag{13.92}$$

$$(U_z)_{\max} = \frac{\rho g R^2}{4\mu} \left[-2\left(\frac{\delta}{R}\right) - \left(\frac{\delta}{R}\right)^2 + 2\left(1+\frac{\delta}{R}\right)^2 \ln\left(1+\frac{\delta}{R}\right)\right]. \tag{13.93}$$

Finally, it should be mentioned with regard to film flows that they stay laminar for small Reynolds numbers only, i.e. they behave for small Re-numbers as indicated above. The above equations can only be applied to small film thicknesses and fluids with relatively large kinematic viscosities. In chemical engineering, a number of film flows occur which fulfil these requirements for the existence of laminar flows.

13.7 Pipe Flow (Hagen–Poiseuille Flow)

The laminar fully developed pipe flow is another important fluid flow which can be treated as stationary, one-dimensional flow, i.e. by solving the following differential equation:

$$-\frac{d\Pi}{dz} + \mu \frac{1}{r}\frac{d}{dr}\left(r\frac{dU_z}{dr}\right) + \rho g_z = 0. \tag{13.94}$$

When considering the horizontal pipe flow as indicated in Fig. 13.9, the following simplified differential equation holds, as $g_z = 0$:

$$\left(r\frac{dU_z}{dr}\right) = \frac{1}{\mu}\left(\frac{d\Pi}{dz}\right)r. \tag{13.95}$$

This equation expresses the fact that the external pressure gradient imposed on the fluid is maintained in equilibrium by viscous forces acting also on the fluid, so that a non-accelerated flow results.

The boundary conditions for this flow are:

$$r = 0; \quad \frac{dU_z}{r} = 0 \quad \text{and for} \quad r = R; \quad U_z = 0.$$

The flow occurring in the cylindrical pipe indicated in Fig. 13.9 requires a pressure gradient to be maintained in the developed state ($d\Pi/z$), i.e. this quantity has to be applied externally for a pipe flow to be established. For the resultant flow velocity one obtains the following differential equation:

$$\frac{d}{dr}\left(r\frac{dU_z}{dr}\right) = \frac{1}{\mu}\left(\frac{d\Pi}{dz}\right)r. \tag{13.96}$$

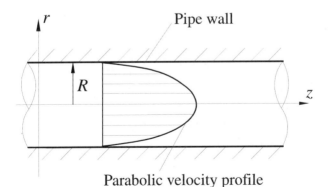

Fig. 13.9 Laminar flow in a pipe

By a first integration of (13.96), the following results:

$$\frac{dU_z}{dr} = \frac{1}{2\mu}\left(\frac{d\Pi}{dz}\right)r + \frac{C_1}{r}. \qquad (13.97)$$

By a second integration one obtains:

$$U_z = \frac{1}{4\mu}\left(\frac{d\Pi}{dz}\right)r^2 + C_1 \ln r + C_2. \qquad (13.98)$$

Applying the boundary conditions:

$$r \to 0;\quad \frac{dU_z}{dr} \to 0 \quad \text{and} \quad r = R:\ U_2 = 0. \qquad (13.99)$$

C_1 and C_2 can be determined:

$$C_1 = 0 \quad \text{and} \quad C_2 = -\frac{1}{4\mu}\left(\frac{d\pi}{dz}\right)R^2. \qquad (13.100)$$

Thus the equation for the velocity distribution $U_z(r)$ for the laminar pipe flow reads:

$$U_z = -\frac{R^2}{4\mu}\left(\frac{d\Pi}{dz}\right)\left[1 - \left(\frac{r}{R}\right)^2\right]. \qquad (13.101)$$

The velocity profile is parabolic and U_z is positive; the minus sign takes into account the presence of a negative pressure gradient in the z direction, i.e. the pressure decreases in the $+z$ direction and the fluid thus flows in that direction.

The volume flow through the pipe (volume per unit time) can be computed as follows:

$$\dot{Q} = \int_0^R 2\pi r U_z \, dr = -\frac{\pi R^4}{8\mu}\left(\frac{d\Pi}{dz}\right) \qquad (13.102)$$

or rewritten:

$$\left(\frac{d\Pi}{dz}\right) = \frac{\Delta p}{\Delta z} = -\frac{8\mu\dot{Q}}{\pi R^4}. \qquad (13.103)$$

In the case of a laminar pipe flow, the pressure drop per unit pipe length is proportional to the dynamic viscosity of the flowing fluid and the volume flow rate, as well as inversely proportional to the fourth power of the pipe radius.

The mean velocity results as:

$$\tilde{U} = \frac{\dot{Q}}{\pi R^2} = -\frac{R^2}{8\mu}\left(\frac{d\Pi}{dz}\right). \qquad (13.104)$$

The above connection between the volume flow, the inner radius R of the pipe, the viscosity of the flow medium and the resultant pressure gradient is known

13.7 Pipe Flow (Hagen–Poiseuille Flow)

as the Hagen–Poiseuille law. It was found by Hagen in 1839 and by Poiseuille in 1840/41 independently of one another in experimental investigations. The experimental confirmation of the above-derived relations stresses the validity of the assumptions made for the pipe flow and beyond that the fact that the validity of the Navier–Stokes equations for the description of fluid flows of Newtonian media hold.

The momentum loss to the wall of the pipe, due to the laminar, fully developed pipe flow, can be computed as

$$\tau_w = -\mu \left(\frac{dU_z}{dr}\right)_w = \frac{1}{2}\left(\frac{d\Pi}{dz}\right) R. \qquad (13.105)$$

The friction coefficient can thus be calculated as follows:

$$c_f = \frac{\tau_w}{\frac{\rho \tilde{U}^2}{2}} = \frac{2\tau_w(2R)}{\tilde{U}\mu\left(\frac{\tilde{U}2R}{\nu}\right)} = \frac{\left(\frac{d\Pi}{dz}\right)R(2R)}{\frac{R^2}{8}\left(\frac{d\Pi}{dz}\right)Re} = \frac{16}{Re}, \qquad (13.106)$$

i.e. we obtain the following functional relationship:

$$c_f = \frac{16}{Re} \quad \text{with} \quad Re = \frac{\tilde{U}2R}{\nu}. \qquad (13.107)$$

The representation of the friction coefficient as function of the Reynolds number yields, in a diagram with double-logarithmic axes, a straight line with the gradient (-1).

Further insight into the fluid flow and the molecule interactions, taking place in viscous mediums, can be gained by computing the energy dissipation in the pipe flow by the action of the fluid viscosity. Based on the general relationship for the energy dissipation per unit volume in a Newtonian fluid, one obtains:

$$\frac{dE_{\text{diss}}}{dV} = 2\mu\left[\left(\frac{\partial U_r}{\partial r}\right)^2 + \left(\frac{1}{r}\frac{\partial U}{\partial \varphi} + \frac{U_r}{r}\right)^2 + \left(\frac{\partial U_z}{\partial z}\right)^2\right]$$

$$+ \mu\left[r\frac{\partial}{\partial r}\left(\frac{U_\varphi}{r}\right) + \frac{1}{r}\left(\frac{dU_r}{d\varphi}\right)\right]^2 + \mu\left[\frac{1}{r}\left(\frac{\partial U_z}{\partial \varphi}\right) + \left(\frac{\partial U_\varphi}{\partial z}\right)\right]^2$$

$$+ \mu\left[\left(\frac{\partial U_r}{\partial z}\right) + \left(\frac{\partial U_z}{\partial r}\right)\right]^2. \qquad (13.108)$$

When considering all the simplifications which were introduced for the derivation of (13.94), the above general relationship for the energy dissipation of a viscous pipe flow can be described as follows:

$$\frac{dE_{\text{diss}}}{dV} = \mu\left(\frac{\partial U_z}{\partial r}\right)^2 = \mu\left(\frac{dU_z}{dr}\right)^2. \qquad (13.109)$$

By introducing $dV = 2\pi r\, dz\, dr$, one obtains

$$dE_{\text{diss}} = \mu \left(\frac{dU_z}{dr}\right)^2 2\pi r\, dz\, dr. \tag{13.110}$$

dU_z/dr can be written as:

$$\frac{dU_z}{dr} = -\frac{1}{2\mu}\left(\frac{d\Pi}{dz}\right) r. \tag{13.111}$$

Thus the dissipated energy per unit length of a pipe flow can be calculated as:

$$\frac{dE_{\text{diss}}}{dz} = \frac{\pi}{2\mu}\left(\frac{d\Pi}{dz}\right)^2 r^3\, dr. \tag{13.112}$$

On integrating this equation, one obtains the energy dissipated per unit pipe length dz:

$$\frac{dE_{\text{diss}}}{dz} = \frac{\pi}{2\mu}\left(\frac{d\Pi}{dz}\right)^2 \int_0^R r^3\, dr = \frac{\pi}{8\mu}\left(\frac{d\Pi}{dz}\right)^2 R^4, \tag{13.113}$$

i.e. the pressure gradient that has to be applied per unit length of the pipe serves for supplying the mechanical energy dissipated into heat, per unit length of the fluid motion. Considering:

$$\dot{Q} = \frac{\pi R^4}{8\mu}\left(\frac{d\Pi}{dz}\right), \tag{13.114}$$

(13.113) can be written as:

$$\frac{dE_{\text{diss}}}{dz} = \dot{Q}\left(\frac{d\Pi}{dz}\right) \quad \text{or} \quad \Delta E_{\text{diss}} = \dot{Q}\Delta P_{\text{diss}}. \tag{13.115}$$

This relationship expresses that the pressure gradient to be applied per unit length of the pipe corresponds to the energy dissipated per unit length of the pipe and per unit volume flow:

$$\frac{d\Pi}{dz} = \frac{1}{\dot{Q}}\frac{dE_{\text{diss}}}{dz}. \tag{13.116}$$

The validity of the above-derived relationships for the pipe flow is, however, limited to laminar flows, i.e. to Reynolds numbers which are smaller than Re_{crit}. This critical Reynolds number is for pipe flows in the range

$$Re_{\text{crit}} = \frac{\tilde{U}2R}{\nu} \lesssim 2.3 \text{ to } 2.5 \times 10^3. \tag{13.117}$$

When the Reynolds number of a pipe flow is larger than this critical value, and when no special precautions are taken to keep flow perturbations away

13.8 Axial Flow Between Two Cylinders

from the pipe flow, then the flow in the range of the critical Reynolds number changes abruptly from laminar to turbulent. In this case there is no longer a directed flow present as described by the above relationships. The flow in the pipe shows, superimposed on a mean flow field, stochastic velocity fluctuations which lead to an additional momentum transport transverse to the flow direction. This momentum transport is not covered by the above basic equations.

The most important properties of turbulent pipe-channel flows are indicated in Chap. 18 and some references are made to deviations from the laminar pipe flow as discussed here.

13.8 Axial Flow Between Two Cylinders

In chemical engineering, there are a large number of axially symmetric apparatus in which flows can be treated as stationary, fully developed flows. They are described by the following partial differential equation:

$$-\frac{d\Pi}{dz} + \mu \frac{1}{r}\frac{\partial}{\partial r}\left(r\frac{\partial U_z}{\partial r}\right) + \rho g_z = 0. \qquad (13.118)$$

Annular axial flows are among them, of the kind sketched in Fig. 13.10; the boundary conditions for this flow can be given as:
for $r = R_1 : U_z = 0$ and for $r = R_2 : U_z = 0$.

As an interesting example, the flow in a cylindrical annular channel, as shown in Fig. 13.10, will be discussed here. The annular channel is formed by two axially positioned pipes having radii R_1 and R_2.

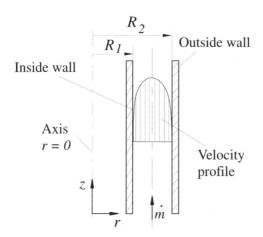

Fig. 13.10 Upwards flow for a cylindrical annular clearance

For further simplification of the derivation,
$$K = R_1/R_2, \quad \Pi^* = \Pi + \rho g z \tag{13.119}$$
are introduced. Considering the coordinate system indicated in Fig. 13.10, $g_z = -g$ holds and thus one obtains the following form of the differential equation describing the annular channel flow of Fig. 13.10

$$\mu \frac{1}{r} \frac{\partial}{\partial r}\left(r \frac{\partial U_z}{\partial r}\right) = \frac{\partial}{\partial z}(\Pi + \rho g z) = \frac{\partial \Pi^*}{\partial z}. \tag{13.120}$$

Taking into account the assumptions $\frac{\partial}{\partial z}(\cdots) = 0$ for the flow field, i.e. assuming a fully developed flow in the z direction, the above-mentioned partial differentials can be written as total differentials, as follows:

$$\frac{d}{dr}\left(r \frac{dU_z}{dr}\right) = \frac{1}{\mu} \frac{d\Pi^*}{dz} r. \tag{13.121}$$

In Sect. 13.6, it was shown that this equation has the following general solution:
$$U_z = \frac{1}{4\mu} \frac{d\Pi^*}{dz} r^2 + C_1 \ln r + C_2. \tag{13.122}$$

Based on the boundary conditions stated in Fig. 13.10, the integration constants C_1 and C_2 for the flow in a cylindrical annular clearance can be determined by (13.123) and (13.124).

From the boundary condition $r = R_1$; $U_z = 0$ results the first equation for the computation of the integration constants C_1 and C_2:

$$0 = R_1^2 + C_1 \ln R_1 + C_2. \tag{13.123}$$

On considering $r = R_2$; $U_z = 0$, one obtains:

$$0 = R_2^2 + C_1 \ln R_2 + C_2, \tag{13.124}$$

the second relationship for the computation of the integration constants C_1 and C_2. In this way, one arrives at:

$$C_1 = R_2^2 \left[1 - \left(\frac{R_1}{R_2}\right)^2\right] \frac{1}{\ln(R_1/R_2)} \tag{13.125}$$

or, considering $K = R_1/R_2$

$$C_1 = R_2^2 (1 - K^2) \frac{1}{\ln K}. \tag{13.126}$$

C_2 results as

$$C_2 = R_2^2 \left[\frac{(K^2 - 1)}{\ln K}\right] \ln R_2 - 1. \tag{13.127}$$

13.8 Axial Flow Between Two Cylinders

For the velocity distribution, the following equation results:

$$U_z = -\frac{R_2^2}{4\mu}\left(\frac{d\Pi^*}{dz}\right)\left\{\left[1-\left(\frac{r}{R_2}\right)^2\right]+\frac{K^2-1}{\ln K}\times\ln\left(\frac{r}{R_2}\right)\right\}. \quad (13.128)$$

The above equation shows that for $K \to 0$ the velocity distribution for the fully developed pipe flow is not obtained. The position of the maximum velocity is computed as

$$r = (u_z = u_{\max}) = R_z\sqrt{\frac{1-K^2}{2\ln(1/K)}}. \quad (13.129)$$

The maximum velocity is thus computed from the equation for U_z

$$(U_z)_{\max} = -\frac{1}{4\mu}\left(\frac{d\Pi^*}{dz}\right)R_2^2\left\{1-\left[\frac{1-K^2}{2\ln(1/K)}\right]\left[1-\ln\left(\frac{1-K^2}{2\ln(1/K)}\right)\right]\right\}. \quad (13.130)$$

The volume flow results as

$$\dot{Q} = -\frac{\pi^*}{8\mu}\frac{d\Pi^*}{dz}R_2^4\left[(1-K^4)-\frac{(1-K^2)^2}{\ln(1/K)}\right] \quad (13.131)$$

and for the mean velocity one obtains

$$\tilde{U}_z = \frac{\dot{Q}}{\pi(R_2^2-R_1^2)} = \frac{1}{8\mu}\left(\frac{d\Pi^*}{dt}\right)R_2^2\left[\frac{1-K^4}{1-K^2}-\frac{1-K^2}{\ln(1/K)}\right]. \quad (13.132)$$

The molecular momentum transport can be computed as

$$\tau_{r,z} = \frac{1}{2}\left(\frac{d\Pi^*}{dz}\right)R_2\left\{\left(\frac{r}{R_2}\right)-\left[\frac{1-K^2}{2\ln(1/K)}\right]\left(\frac{R_2}{r}\right)\right\}. \quad (13.133)$$

The quantity $\tau_{r,z}$ is naturally at the position $du_z/dr = 0$, i.e. at $U_z = U_{\max}$, equal to zero, so that one obtains from (13.133):

$$r(\tau_{r,z} = 0) = R_2\sqrt{\frac{1-K^2}{2\ln(1/K)}}. \quad (13.134)$$

For the annular clearance it also holds that the above relationships can only be employed for laminar flows. The additional momentum transports, occurring in turbulent flows due to the turbulent velocity fluctuations, were not taken into consideration in the above equations. Therefore, the derived equations in this section can be employed only when it has been confirmed that the flow in the considered annular channel flows in a laminar way.

13.9 Film Flows with Two Layers

The problems of steady, two-dimensional and fully developed flows of incompressible fluids, discussed in the previous chapters, can be extended to fluid flows that comprise of several non-mixable fluids. The derived basic equations for fully developed flows have to be solved, in the presence of several fluids, for each fluid flow and the boundary conditions existing in the inter-layers of the fluids have to be considered in the solutions. This is shown below for a film flow made up of two layers.

In coating technology, it is customary to insert superimposed film flows of non-mixable fluids in order to coat several films, in one process step, on to a substrate. In practice, up to 20 layers can be simultaneously applied with high accuracy. If one limits oneself to two layers, flow configurations as shown in Fig. 13.11 develop. The figure shows two superimposed film flows which are moved by gravitation on top of a plane inclined wall. Flows of this kind are described by the following differential equations:

$$0 = \mu_A \frac{d^2 U_1^A}{dx_2^2} + \rho_A g_1 \qquad (13.135)$$

and

$$0 = \mu_B \frac{d^2 U_1^B}{dx_2^2} + \rho_B g_1 \qquad (13.136)$$

with $g_1 = g\cos\beta$ and $\nu_{A,B} = \frac{\mu_{A,B}}{\rho_{A,B}}$, so that one obtains by integration:

$$U_1^A = -\left(\frac{g\cos\beta}{2\nu_A}\right) x_2^2 + C_1^A x_2 + C_2^A \qquad (13.137)$$

and

$$U_1^B = -\left(\frac{g\cos\beta}{2\nu_B}\right) x_2^2 + C_1^B x_2 + C_2^B. \qquad (13.138)$$

By this integration, four integration constants were introduced in the above relationships which have to be determined by appropriate boundary conditions:

$$x_2 = 0 \quad \rightsquigarrow \quad U_1^A = 0 \text{ (no-slip wall condition)} \quad \rightsquigarrow \quad C_2^A = 0, \qquad (13.139)$$

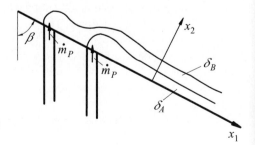

Fig. 13.11 Flow between fluid films on top of a plane, inclined wall

13.9 Film Flows with Two Layers

$$x_2 = \delta_A + \delta_B \quad \rightsquigarrow \quad \frac{dU_1^B}{dx_2} = 0 \text{ (free surface)}, \tag{13.140}$$

$$x_2 = \delta_A \quad \rightsquigarrow \quad U_1^A = U_1^B \text{ and also } \mu_A \frac{dU_1^A}{dx_2} = \mu_B \frac{dU_1^B}{dx_2}. \tag{13.141}$$

From the boundary condition for the free surface, C_1^B results as:

$$C_1^B = \frac{g \cos \beta}{\nu_B} (\delta_A + \delta_B). \tag{13.142}$$

The equality of the local film velocities in the common interface between the films yields:

$$-\frac{g \cos \beta}{2\nu_A} \delta_A^2 + C_1^A \delta_A = -\frac{g \cos \beta}{2\nu_B} \delta_A^2 + \frac{g \cos \beta}{\nu_B} (\delta_A + \delta_B) \delta_A + C_2^B. \tag{13.143}$$

The equality of the local momentum transport terms in the common interface between the films further yields:

$$-\delta_A g \cos \beta \delta_A + C_1^A = -\rho_B g \cos \beta \delta_A + \rho_B g \cos \beta (\delta_A + \delta_B). \tag{13.144}$$

From the above equation, one can deduce:

$$C_1^A = (\rho_A \delta_A + \rho_B \delta_B) g \cos \beta \tag{13.145}$$

and for C_2^B, one obtains:

$$C_2^B = -g \cos \beta \frac{\nu_A + \nu_B}{\nu_A \nu_B} \delta_A^2 + (\rho_A \delta_A + \rho_B \delta_B) g \cos \beta - \frac{g \cos \beta}{\nu_B} (\delta_A + \delta_B) \delta_A. \tag{13.146}$$

Thus, one obtains for the velocity distributions U_1^A and U_1^B:

$$U_1^A = -\left(\frac{g \cos \beta}{2\nu_A}\right) x_2^2 + [(\rho_A \delta_A + \rho_B \delta_B) \beta] x_2 \quad \text{for } 0 \leq x_2 \leq \delta_A \tag{13.147}$$

and

$$\begin{aligned} U_1^B = &-\left(\frac{g \cos \beta}{2\nu_B}\right) x_2^2 + \frac{g \cos \beta}{2\nu_B} (\delta_A + \delta_B) x_2 \\ &- g \cos \beta \frac{\nu_A + \nu_B}{2\nu_A \nu_B} \delta_A^2 + (\rho_A \delta_A + \rho_B \delta_B) g \cos \beta \\ &- \frac{g \cos \beta}{\nu_B} (\delta_A + \delta_B) \delta_A \quad \text{for } \delta_K \leq x_2 \leq \delta_B. \end{aligned} \tag{13.148}$$

For $\dot{m}_A$ and $\dot{m}_B$, we can write:

$$\dot{m}_A = \rho_A B \int_0^{\delta_A} U_1^A(x_2) \, dx_2 \quad \text{and} \quad \dot{m}_B = \rho_B B \int_{\delta_A}^{(\delta_A + \delta_B)} U_1^B(x_2) \, dx_2. \tag{13.149}$$

By integration one obtains

$$\dot{m}_A = \rho_A B \left[\left(-\frac{g\cos\beta}{2\nu_A} \right) \frac{\delta_A^3}{3} + C_1^A \frac{\delta_A^2}{2} \right]. \tag{13.150}$$

$$\dot{m}_B = \rho_B B \left[\left(-\frac{g\cos\beta}{2\nu_B} \right) \frac{(\delta_A + \delta_A)^3 - \delta_A^3}{3} + C_1^B \frac{\delta_A \delta_B + \delta_B^2}{2} + C_2^B \delta_B \right]. \tag{13.151}$$

In this way, the layer mass flows $\dot{m}_A$ and $\dot{m}_B$ can be determined, when δ_A and δ_B are given and the properties of the fluids of the coating fluids are known.

13.10 Two-Phase Plane Channel Flow

In Fig. 13.12, a plane channel flow is sketched which is composed of the flow of two superimposed non-mixable fluids, i.e. fluids A and B that flow simultaneously through a channel formed by two parallel plates. Fluid A forms a layer of thickness δ_A and has density ρ_A, viscosity μ_A and mass flow $\dot{m}_A$. The fluid that is on top of it has the density ρ_B, viscosity μ_B and mass flow $\dot{m}_B$. For both fluids, the following differential equations for the molecular momentum transport τ_{21} hold:

$$\frac{d\tau_{21}^A}{dx_2} = -\frac{d\Pi}{dx_1} \quad \text{and} \quad \frac{d\tau_{21}^B}{dx_2} = -\frac{d\Pi}{dx_1}. \tag{13.152}$$

With $\tau_{21} = -\mu\, dU_1/x_2$ the velocity field results

$$\frac{d^2 U_1^A}{dx_2^2} = \frac{1}{\mu_A} \frac{d\Pi}{dx_1} \quad \text{and} \quad \frac{d^2 U_1^B}{dx_2^2} = \frac{1}{\mu_B} \frac{d\Pi}{dx_1}. \tag{13.153}$$

Integration of (13.152) yields for both fluids

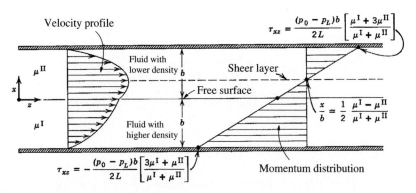

Fig. 13.12 Plane channel flow with two-layered flows; a solution is stated for $\delta = 0$

13.10 Two-Phase Plane Channel Flow

$$\tau_{21}^A = -\frac{d\Pi}{dx_1} x_2 + C_1^A \tag{13.154}$$

and

$$\tau_{21}^B = -\frac{d\Pi}{dx_1} x_2 + C_1^B. \tag{13.155}$$

Introducing the boundary conditions that the momentum transport is equal due to the common surface A and B, one obtains $\tau_{21}^A(x_2 = \delta) = \tau_{21}^B(x_2 = \delta)$:

$$-\frac{d\Pi}{dx_1}\delta + C_1^A = -\frac{d\Pi}{dx_1}\delta + C_1^B \quad \rightsquigarrow \quad C_1^A = C_1^B = C_1. \tag{13.156}$$

When carrying out the integration for the velocity fields U_1^A and U_1^B, one obtains:

$$U_1^A = -\frac{1}{2\mu_A}\frac{d\Pi}{dx_1} x_2^2 + \frac{C_1^A}{\mu_A} x_2 + C_2^A \tag{13.157}$$

and

$$U_1^B = -\frac{1}{2\mu_B}\frac{d\Pi}{dx_1} x_2^2 + \frac{C_1^B}{\mu_B} x_2 + C_2^B. \tag{13.158}$$

The coordinate system plotted in Fig. 13.12 was chosen such that the x_2 direction yields positive values for δ, i.e. the area between the two fluids lies above the plane $x_2 = 0$. In this way, one can obtain the second boundary condition that has to be imposed in the interface

$$U_1^A = (x_2 = \delta) = U_1^B(x_2 = \delta) \quad \rightsquigarrow \quad C_2^A \neq C_2^B, \tag{13.159}$$

i.e. the following relationship holds:

$$-\frac{\delta^2}{2\mu_A}\frac{d\Pi}{dx_1} + \frac{C_1\delta}{\mu_A} + C_2^A = -\frac{\delta^2}{2\mu_B}\frac{d\Pi}{dx_1} + \frac{C_1\delta}{\mu_B} + C_2^B, \tag{13.160}$$

$\delta = 0$ results in a reduction in the effort for determining C_2^A and C_2^B. This special case is discussed below. For $\delta = 0$ it results that $C_2^A = C_2^B = C_2$. The remaining integration constants can be determined with the following boundary conditions:

$$x_2 = -D \quad \rightsquigarrow \quad U_1^A = 0: \quad 0 = -\frac{d\Pi}{dx_1}\frac{1}{2\mu_A}D^2 - \frac{C_1 D}{\mu_A} + C_2, \tag{13.161}$$

$$x_2 = +D \quad \rightsquigarrow \quad U_1^B = 0: \quad 0 = -\frac{d\Pi}{dx_1}\frac{1}{2\mu_B}D^2 + \frac{C_1 D}{\mu_B} + C_2. \tag{13.162}$$

Hence one obtains for the velocity distributions in the fluids A and B:

$$U_1^A = -\frac{D^2}{2\mu_A}\frac{d\Pi}{dx_1}\left[+\frac{2\mu_A}{(\mu_A + \mu_B)} + \left(\frac{\mu_A - \mu_B}{\mu_A + \mu_B}\right)\left(\frac{x_2}{D}\right) - \left(\frac{x_2}{D}\right)^2\right] \tag{13.163}$$

and
$$U_1^B = -\frac{D^2}{2\mu_B}\frac{\mathrm{d}\Pi}{\mathrm{d}x_1}\left[+\frac{2\mu_B}{(\mu_A+\mu_B)}+\left(\frac{\mu_A-\mu_B}{\mu_A+\mu_B}\right)\left(\frac{x_2}{D}\right)-\left(\frac{x_2}{D}\right)^2\right]. \tag{13.164}$$

For the distribution of the molecular-dependent momentum transport, the following expression can be deduced:
$$\tau_{21} = -D\frac{\mathrm{d}\Pi}{\mathrm{d}x_1}\left[\left(\frac{x_2}{D}\right)-\frac{1}{2}\left(\frac{\mu_A-\mu_B}{\mu_A+\mu_B}\right)\right]. \tag{13.165}$$

When choosing in the above relations $\mu_A = \mu_B$, one obtains:
$$U_1 = \frac{-D^2}{2\mu_A}\frac{\mathrm{d}\Pi}{\mathrm{d}x_1}\left[1-\left(\frac{x_2}{D}\right)^2\right] \tag{13.166}$$

and
$$\tau_{21} = -D\frac{\mathrm{d}\Pi}{\mathrm{d}x_1}\left(\frac{x_2}{D}\right), \tag{13.167}$$

which ends up in a parabolic velocity profile with the velocity maximum in the middle of the channel and a linear τ_{21} distribution with $\tau_{21} = 0$ on the channel axis.

For $\mu_A \neq \mu_B$, the position of the velocity maximum, with $\tau_{21} = 0$, results from (13.163):
$$\frac{\delta}{D} = \frac{1}{2}\left(\frac{\mu_A-\mu_B}{\mu_A+\mu_B}\right). \tag{13.168}$$

The momentum transport to the upper wall yields:
$$\tau_W^A = -\frac{\mathrm{d}\Pi}{\mathrm{d}x_1}D\left(\frac{\mu_A+3\mu_B}{\mu_A+\mu_B}\right). \tag{13.169}$$

The momentum transport to the lower wall yields:
$$\tau_W^B = -\frac{\mathrm{d}\Pi}{\mathrm{d}x_1}D\left(\frac{3\mu_A+\mu_B}{\mu_A+\mu_B}\right). \tag{13.170}$$

The mean velocities of the partial flows A and B can be computed as:
$$\tilde{U}_1^A = -\frac{D^2}{12\mu_A}\frac{\mathrm{d}\Pi}{\mathrm{d}x_1}\left(\frac{7\mu_A+\mu_B}{\mu_A+\mu_B}\right) \tag{13.171}$$

and
$$\tilde{U}_1^B = -\frac{D^2}{12\mu_B}\frac{\mathrm{d}\Pi}{\mathrm{d}x_1}\left(\frac{\mu_A+7\mu_B}{\mu_A+\mu_B}\right). \tag{13.172}$$

The corresponding mass flows can be computed as:
$$\dot{m}_A = BD\tilde{U}_1^A \quad \text{and also} \quad \dot{m}_B = BD\tilde{U}_1^B. \tag{13.173}$$

The above treated one-dimensional flow problems are only a few of those examples available in many text books of fluid mechanics. For further examples see refs. [13.1] to [13.2].

References

13.1. Bird, R.B., Stuart, W.E., Lightfoot, E.N., "Transport Phenomena", Wiley, New York, 1960
13.2. Hutter, K., "Fluid- und Thermodynamik- eine Einfuehrung", Springer, Berlin, Heidelberg, New York 1995
13.3. Pnoeli, D., Gutfinger, Ch., "Fluid Mechanics", Cambridge University Press, Cambridge, 1992
13.4. Potter, M.C., Voss, J.F., "Fluid Mechanics", Wiley, New York, 1975
13.5. Schlichting, H., "Boundary Layer Theory", McGraw-Hill, New York, Series in Mechanical Engineering, 1979

Chapter 14
Time-Dependent, One-Dimensional Flows of Viscous Fluids

14.1 General Considerations

The flow problems, discussed in Chap. 13 for viscous fluids, were characterized by the fact, among other things, that they fulfilled the condition of stationarity, i.e. the examined flows were not dependent on time. All the derivations in Chap. 13 that led to (13.22), can be repeated, maintaining the time derivative of the velocity field in the equations. For time-dependent flows, this term cannot be set equal to zero and, hence, one obtains the basic equation for time-dependent, one-dimensional flows of viscous fluids, i.e. $U_1 = f(x_2, t)$:

$$\rho \frac{\partial U_1}{\partial t} = -\frac{\partial \Pi}{\partial x_1} + \mu \frac{\partial^2 U_1}{\partial x_2^2} + \rho g_1 \tag{14.1}$$

where $U_1(x_2, t)$ and also $\Pi(x_1, t)$ are now to be regarded as functions of both space and time. On transcribing this equation for incompressible flow, it follows that:

$$\underbrace{\frac{\partial U_1}{\partial t} = \nu \frac{\partial^2 U_1}{\partial x_2^2}}_{\text{time–dependent diffusion}} \underbrace{-\left(\frac{1}{\rho}\frac{\partial \Pi}{\partial x_1} - g_1\right)}_{\text{Source term}} \tag{14.2}$$

and one obtains an equation that is well known for dealing theoretically with transport processes. Without the source term in (14.2), it represents the fundamental equation for all transient one-dimensional, diffusion problems; e.g. for unsteady, one-dimensional, heat conduction problems it reads:

$$\frac{\partial T}{\partial t} = \alpha \frac{\partial^2 T}{\partial x_2^2} \quad \text{with} \quad \alpha = \frac{\lambda}{\rho c_p} \tag{14.3}$$

Analogous to heat-conduction problems, a number of transient, one-dimensional flow problems can be solved via analytical methods. For this purpose, it is useful to consider first the dimensionless form of (14.2) without the source term, i.e. the following equation which holds for the one-dimensional molecular momentum transport:

$$\frac{\partial U_1}{\partial t} = \nu \frac{\partial^2 U_1}{\partial x_2^2} \quad \Rightarrow \quad \frac{\partial U_1^*}{\partial t^*} = \left(\frac{\nu_c t_c}{\ell_c^2}\right) \nu^* \frac{\partial^2 U_1^*}{\partial x_2^{*2}} \qquad (14.4)$$

In this equation, the term $(\nu_c t_c)/\ell_c^2$ is the reciprocal of the product of the characteristic Reynolds and Strouhal numbers, $Re = (\ell_c U_c)/\nu_c$ and $St = \ell_c/(U_c t_c)$. It may be compared with the Fourier number of heat conduction, $Fo = (a_c t_c/\ell_c^2)$, which is normally introduced when dealing, in general, with time-dependent heat-conduction problems.

For the time-dependent, one-dimensional flow problems of viscous fluids, to be discussed in this chapter, a generalization of the considerations can be attained by setting

$$Fo = \frac{1}{ReSt} = \frac{\nu_c t_c}{\ell_c^2} = 1 \qquad (14.5)$$

such that the left- and right-hand sides of (14.4), in the dimensionless form, are of equal order of magnitude. Hence we are introducing the characteristic measures of time, length and velocity for purely diffusive flow problems as follows:

$$t_c = \frac{\ell_c^2}{\nu_c} \qquad \ell_c = \sqrt{\nu_c t_c} \qquad u_c = \frac{\nu_c}{\ell_c} \qquad (14.6)$$

If a flow is generated in a fluid, with a constant flow velocity U_0, describing a one-dimensional problem, its properties can be derived from (14.4) by the following solution ansatz:

$$\frac{U_1}{U_0} = F\left(\frac{x_2}{2\sqrt{\nu t}}\right) = F(\eta) \quad \text{with} \quad \eta = \frac{x_2}{2\sqrt{\nu t}} \qquad (14.7)$$

Introducing into (14.4) all terms of (14.7), one can carry out the follow derivations:

$$\frac{\partial U_1}{\partial t} = U_0 \frac{dF}{d\eta} \frac{\partial \eta}{\partial t} = U_0 \frac{dF}{d\eta} \frac{\partial}{\partial} \left(-\frac{\eta}{2t}\right) = -U_0 \left(\frac{\eta}{2t}\right) \frac{F}{\eta} \qquad (14.8)$$

$$\frac{\partial U_1}{\partial x_2} = U_0 \frac{dF}{d\eta} \frac{\partial \eta}{\partial x_2} = U_0 \frac{dF}{d\eta} \frac{1}{2\sqrt{\nu t}} \qquad (14.9)$$

$$\frac{\partial^2 U_1}{\partial x_2^2} = U_0 \frac{\partial}{\partial x_2}\left(\frac{dF}{d\eta}\frac{\partial \eta}{\partial x_2}\right) = U_0 \left[\frac{d^2 F}{d\eta^2}\left(\frac{\partial \eta}{\partial x_2}\right)^2 + \frac{dF}{d\eta}\frac{\partial^2 \eta}{\partial x_2^2}\right] \qquad (14.10)$$

$$\frac{\partial^2 U_1}{\partial x_2^2} = U_0 \frac{d^2 F}{d\eta^2} \frac{1}{4\nu t} \qquad (14.11)$$

When the partial derivatives in (14.8) and (14.11) are inserted into the partial differential equation (14.4), that needs to be solved, one obtains an ordinary differential equation of second order for the function $F(\eta)$:

14.1 General Considerations

$$-2\eta \frac{dF}{d\eta} = \frac{d^2 F}{d\eta^2} \quad (14.12)$$

By introducing a new function $G(\eta)$:

$$G(\eta) = \frac{dF}{d\eta} \quad (14.13)$$

one obtains from (14.12):

$$\frac{dG}{d\eta} = -2\eta G \quad \Rightarrow \quad \frac{dG}{G} = -2\eta \, d\eta \quad (14.14)$$

Through integration of (14.14), one obtains:

$$\ln G = -\eta^2 + \ln C_1' \quad \Rightarrow \quad G(\eta) = C_1' \exp(-\eta^2) \quad (14.15)$$

Hence one can express the function $G(\eta)$ as follows:

$$G(\eta) = \frac{dF}{d\eta} = C_1' \exp(-\eta^2) \quad \Rightarrow \quad F(\eta) = C_1' \int_0^\eta \exp(-\eta^2) \, d\eta + C_2. \quad (14.16)$$

Using the definition of the error function:

$$\text{erf}(\eta) = \frac{2}{\sqrt{\pi}} \int_0^\eta \exp(-\eta^2) \, d\eta \quad (14.17)$$

one obtains a general solution for one-dimensional, transient, diffusion-driven flows of an incompressible fluid:

$$F(\eta) = C_1 \text{erf}(\eta) + C_2 \quad \text{with} \quad C_1 = C_1' \frac{\sqrt{\pi}}{2} \quad (14.18)$$

From equation (14.18), one obtains the solution for U_1:

$$U_1 = U_0 \left[C_1 \text{erf}(\eta) + C_2 \right] \quad (14.19)$$

In the subsequent sections, the above general solution will be employed in order to find specific solutions for predefined initial and boundary conditions, i.e. for different flows.

A number of one-dimensional, unsteady flow problems for incompressible, viscous fluids can be dealt with more easily in cylindrical coordinates. The basic equation for such flows can now be stated for such flow problems. The derivations of the related equations start from the two-dimensional equations that were derived in Chap. 5 written in cylindrical coordinates. These equations read:

$$\frac{1}{r} \frac{\partial}{\partial r}(rU_r) + \frac{\partial U_z}{\partial z} = 0 \quad (14.20)$$

$$\rho\left(\frac{\partial U_r}{\partial t} + U_r\frac{\partial U_r}{\partial r} + U_z\frac{\partial U_r}{\partial z}\right) = -\frac{\partial P}{\partial r} + \mu\left[\frac{\partial}{\partial r}\left(\frac{1}{r}\frac{\partial (rU_r)}{\partial r}\right) + \frac{\partial^2 U_r}{\partial z^2}\right] + \rho g_r \quad (14.21)$$

$$\rho\left(\frac{\partial U_z}{\partial t} + U_r\frac{\partial U_z}{\partial r} + U_z\frac{\partial U_z}{\partial z}\right) = -\frac{\partial P}{\partial z} + \mu\left[\frac{1}{r}\frac{\partial}{\partial r}\left(r\frac{\partial U_z}{\partial r}\right) + \frac{\partial^2 U_z}{\partial z^2}\right] + \rho g_z \quad (14.22)$$

Introducing the demand for the one-dimensionality of the flow, i.e. no change of the flow field in the z direction because the flow is assumed to be fully developed:

$$\frac{\partial U_r}{\partial z} = 0 \quad \text{and} \quad \frac{\partial U_z}{\partial z} = 0 \quad (14.23)$$

With the help of these expressions, it then follows from the continuity equation (14.20):

$$\frac{1}{r}\frac{\partial}{\partial r}(rU_r) = 0 \quad \Rightarrow \quad rU_r = F(z,t) \quad (14.24)$$

Since $\frac{\partial U_r}{\partial z} = 0$ holds, the function $F(z,t) = F(t)$. Because $U_r = 0$ at the wall, the following holds: the one-dimensional, non-stationary flow of incompressible viscous mediums is unidirectional and has only a U_z component.

Hence the basic equations in cylindrical coordinates can be reduced to:

$$0 = -\frac{\partial P}{\partial r} + \rho g_r \quad (14.25)$$

$$\rho\frac{\partial U_z}{\partial t} = -\frac{\partial P}{\partial z} + \mu\left[\frac{1}{r}\frac{\partial}{\partial r}\left(r\frac{\partial U_z}{\partial r}\right)\right] + \rho g_z \quad (14.26)$$

By integration of (14.25), one obtains:

$$P = \rho g_r r + \Pi(z,t) \quad (14.27)$$

Thus the equation corresponding to the partial differential equation (14.1), but written in cylindrical coordinates, reads as follows:

$$\frac{\partial U_z}{\partial t} = \nu\left[\frac{1}{r}\frac{\partial}{\partial r}\left(r\frac{\partial U_z}{\partial r}\right)\right] - \left(\frac{1}{\rho}\frac{\partial \Pi}{\partial z} - g_z\right) \quad (14.28)$$

This generally valid equation will also be employed subsequently to deal with unsteady, one-dimensional flows of incompressible, viscous fluids that are axisymmetric.

14.2 Accelerated and Decelerated Fluid Flows

14.2.1 Stokes First Problem

Stokes (1851) was one of the first scientists who provided an analytical solution for an unsteady, one-dimensional flow problem, namely the solution for the fluid motion induced by the sudden movement of a plate and the related momentum diffusion into an infinitely extended fluid lying above the plate. In order to understand better the induced fluid movement, also observed in practice, in this and the subsequent sections flow processes are discussed which occur in fluids due to imposed wall movements. The simplest examples on the subject discussed in this chapter on wall-induced fluid motions concern the movement of plane plates. However, the general physical insights gained from these examples are not limited to the plate-induced fluid motions only, but can also be transferred to axially symmetrical flows (rotating cylinders).

Figure 14.1 shows schematically the velocity distribution which takes place in a fluid due to a wall moved at a velocity U_0. The flow setting in, due to the movement of the plate, can be expressed mathematically as follows:

$$\text{For } t < 0 : U_1(x_2, t) = 0$$
$$\text{For } t \geq 0 : U_1(x_2 = 0, t) = U_0 \quad (14.29)$$
$$U_1(x_2 \to \infty, t) = 0$$

As a consequence of the fluid viscosity (molecular momentum transport), the momentum of the fluid layer, moved in the immediate vincinity of the plate, is transferred to the layers that are further away. With progress of time, layers that are further away from the moved wall are also included in the induced fluid motion. The differential equation describing this entire process reads:

$$\frac{\partial U_1}{\partial t} = \nu \frac{\partial^2 U_1}{\partial x_2^2} \quad (14.30)$$

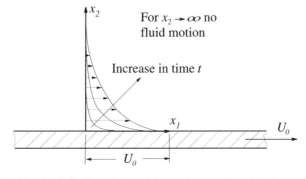

Fig. 14.1 Sketch of the flow induced by a plane wall suddenly set in motion

For this equation, the following general solution was found in Sect. 14.1 in terms of transformed variables, $\eta = \frac{x_2}{2\sqrt{\nu t}}$ and $\frac{U_1}{U_0} = F(\eta)$:

$$U_1 = U_0 \left[C_1 \operatorname{erf}(\eta) + C_2 \right] \tag{14.31}$$

For the problem of the induced plate movement, the following boundary conditions result for $0 < t < \infty$:

$$x_2 = 0, \quad \text{i.e. } \eta = 0 \quad U_1 = U_0 \tag{14.32}$$

and

$$x_2 \to \infty, \quad \text{i.e. } \eta \to \infty \quad U_1 = 0 \tag{14.33}$$

From the general solution (14.31) and for the boundary conditions (14.32) and (14.33), it follows that:

$$1 = C_1 \operatorname{erf}(0) + C_2 = C_2 \tag{14.34}$$

$$0 = C_1 \operatorname{erf}(\infty) + C_2 = C_1 + C_2 \tag{14.35}$$

i.e. the integration constants can be evaluated as:

$$C_1 = -1 \quad \text{and} \quad C_2 = +1 \tag{14.36}$$

Hence the solution reads:

$$U_1 = U_0 \left[1 - \operatorname{erf}(\eta) \right] = U_0 \left[1 - \operatorname{erf}\left(\frac{x_2}{2\sqrt{\nu t}} \right) \right] \tag{14.37}$$

This relationship shows that the movement of the plate is imposed on the fluid only with progressing time. If $\nu = 0$, then for all t and all x_2, $U_1 = 0$ holds, i.e. without the momentum transport carried out by the molecules, which is expressed by the finite viscosity of the fluid, one does not succeed in causing a fluid movement by the movement of the plate. The momentum in each fluid layer, coming about due to the movement of the plate in the x_1 direction, has to be communicated to the fluid via the molecular momentum transport. The larger the viscosity ν is, the quicker the fluid layers far away from the plate are affected, i.e. are set into motion.

For the entire physical understanding of the induced fluid motion, it is important that the force per unit area acting on the plate can be computed using:

$$\tau_w = -\mu \left(\frac{\partial U_1}{\partial x_2} \right)_{x_2=0} \tag{14.38}$$

$\frac{\partial U_1}{\partial x_2}$ can be evaluated from (14.37):

$$\frac{\partial U_1}{\partial x_2} = U_0 \frac{\mathrm{d}F(\eta)}{\mathrm{d}\eta} \frac{\partial \eta}{\partial x_2} = -\frac{U_0}{\sqrt{\pi \nu t}} \exp(-\eta^2) \tag{14.39}$$

14.2 Accelerated and Decelerated Fluid Flows

Thus for $\eta = 0$ one obtains $\left(\dfrac{\partial U_1}{\partial x_2}\right)_{x_2=0} = -\dfrac{U_0}{\sqrt{\pi \nu t}}$ and consequently:

$$\tau_w = U_0 \sqrt{\dfrac{\rho \mu}{\pi t}} \tag{14.40}$$

This relationship explains that the required force increases with increase in viscosity and density of the fluid, to be set in motion, and it decreases with progress of time. At time $t = 0$ an infinitely large force results from the derivations. However, because of the similarity relationship:

$$\eta = \dfrac{x_2}{2\sqrt{\nu t}} \tag{14.41}$$

in which the time appears in the denominator, the result for $t = 0$ is undefined. Consequently, the above statement regarding the required infinitely large shear force is not permissible. For $t \to \infty$ one can compute $\tau_w \to 0$ and $U_1 = U_0$ for all x_2, i.e. the entire fluid mass will move with the velocity of the plate if the plate motion is maintained for a long time.

14.2.2 Diffusion of a Vortex Layer

The so-called first problem of Stokes, discussed in Sect. 14.2.1, can also be dealt with by means of the vorticity equation which, for the component ω_3, can be written as follows:

$$\dfrac{\partial \omega_3}{\partial t} + U_1 \dfrac{\partial \omega_3}{\partial x_1} + U_2 \dfrac{\partial \omega_3}{\partial x_2} = \nu \left(\dfrac{\partial^2 \omega_3}{\partial x_1^2} + \dfrac{\partial^2 \omega_3}{\partial x_2^2}\right) \tag{14.42}$$

With $\omega_3 = \omega$ and $U_2 = 0$ and also $\dfrac{\partial \omega_3}{\partial x_1} = 0$, one obtains:

$$\dfrac{\partial \omega}{\partial t} = \nu \dfrac{\partial^2 \omega}{\partial x_2^2} \tag{14.43}$$

This equation describes how the vorticity, continuously produced at the plate due to its movement, is transported by molecular diffusion into the fluid above the plate. ω can be expressed in the following way:

$$\omega = -\dfrac{\partial U_1}{\partial x_2} \tag{14.44}$$

Thus, the characteristic vorticity for this flow problem can be given as $\omega_c = U_c/\ell_c = U_0 \, (\nu t)^{1/2}$, so that for ω the following similarity approach holds:

$$\omega(\eta, t) = U_0 (\nu t)^{-1/2} f(\eta) \quad \text{with} \quad \eta = \dfrac{x_2}{\sqrt{\nu t}} \tag{14.45}$$

The partial derivatives of η with respect to x_2 and t are given as:

$$\frac{\partial \eta}{\partial x_2} = (\nu t)^{-1/2} \quad \text{and} \quad \frac{\partial \eta}{\partial t} = -\frac{\eta}{2t} \qquad (14.46)$$

Therefore, one obtains

$$\begin{aligned}\frac{\partial \omega}{\partial t} &= U_0 (\nu t)^{-1/2} \left[-\frac{1}{2t} f(\eta) + f'(\eta) \frac{\partial \eta}{\partial t} \right] \\ &= -\frac{U_0}{2t} (\nu t)^{-1/2} [f(\eta) + \eta f'(\eta)]\end{aligned} \qquad (14.47)$$

$$\frac{\partial \omega}{\partial x_2} = U_0 (\nu t)^{-1/2} f'(\eta) \frac{\partial \eta}{\partial x_2} = U_0 f'(\eta) \qquad (14.48)$$

$$\frac{\partial^2 \omega}{\partial x_2^2} = U_0 f''(\eta) \frac{\partial \eta}{\partial x_2} = U_0 (\nu t)^{-1/2} f''(\eta) \qquad (14.49)$$

Introducing the equations (14.47) to (14.49) into the partial differential equation (14.43), one obtains the following ordinary differential equation for $f(\eta)$:

$$2f'' + \eta f' + f = 2f'' + (\eta f)' = 0 \qquad (14.50)$$

By integrating this equation once, one obtains:

$$2f' + \eta f = C_1 \qquad (14.51)$$

The distribution of the vorticity is symmetrical with respect to x_2, and thus $f'(\eta = 0) = 0$ holds $C_1 = 0$. With this, (14.51) can be rewritten as follows:

$$2 \frac{df}{d\eta} = -\eta f \quad \leadsto \quad \frac{df}{f} = -\frac{\eta}{2} d\eta = -d\left(\frac{\eta^2}{4}\right) \qquad (14.52)$$

Therefore, as a solution of this ordinary differential one obtains:

$$f(\eta) = C \exp\left(-\frac{\eta^2}{4}\right) \qquad (14.53)$$

With the following integration:

$$\int_0^\infty \omega \, dx_2 = -\int_0^\infty \frac{\partial U_1}{\partial x_2} dx_2 = U_0 \qquad (14.54)$$

and setting ω into the above integral, one obtains

$$C = (\pi)^{-1/2} \qquad (14.55)$$

Thus as a solution for ω one obtains

$$\omega(x_2, t) = U_0 (\pi \nu t)^{-1/2} \exp\left(-\frac{x_2^2}{\nu t}\right) \qquad (14.56)$$

14.2 Accelerated and Decelerated Fluid Flows

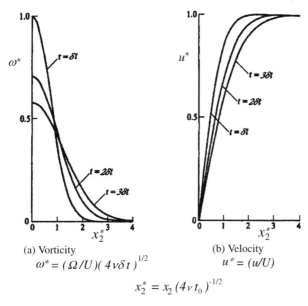

(a) Vorticity
$$\omega^* = (\Omega/U)(4\nu\delta t)^{1/2}$$

(b) Velocity
$$u^* = (u/U)$$

$$x_2^* = x_2 (4\nu t_0)^{-1/2}$$

Fig. 14.2 Diffusion of vorticity and molecular momentum transport in the fluid as a consequence of a moved plane plate

This solution corresponds to:

$$U_1(x_2, t) = U_0 \left[1 - \mathrm{erf}\left(\frac{x_2}{2\sqrt{\nu t}}\right)\right] = U_0 \mathrm{erfc}\left(\frac{x_2}{2\sqrt{\nu t}}\right) \quad (14.57)$$

The diffusion of the vorticity, expressed by (14.56) is sketched in a normalized form in Fig. 14.2 and the corresponding dimensionless velocity distribution expressed by (14.57) is plotted next to it.

For the flow shown in Fig. 14.2, the following integral parameters can be computed:

- Vorticity diffusion radius

$$\delta_\omega = \frac{2U_0}{\Omega_{max}} = (\pi\nu t)^{1/2} \quad (14.58)$$

- Displacement thickness of the flow

$$\delta_1 = \frac{2}{U_0} \int_0^\infty (U_0 - U_1)\,\mathrm{d}x_2 = \left(\frac{\nu t}{\pi}\right)^{1/2} \quad (14.59)$$

- Momentum-loss thickness of the flow

$$\delta_2 = \int_{-\infty}^{+\infty} (U_0^2 - U_1^2)\,\mathrm{d}x_2 = \frac{1}{4U_0^2}\left(\frac{\nu t}{8\pi}\right)^{1/2} \quad (14.60)$$

14.2.3 Channel Flow Induced by Movements of Plates

In this section, a one-dimensional transient flow problem of an incompressible fluid will be discussed, which cannot be solved with the help of the general solution derived in Sect. 14.1, as this solution is unable to fulfil the boundary conditions characterizing the flow problem sketched in Fig. 14.3. This fact requires the derivation of another particular solution for the differential equation characterizing the problem. It is additionally required that the new solution can satisfy the predefined boundary conditions for the problem under consideration. For this purpose, a solution path is taken which can be obtained through the well-known Fourier analysis, as employed in the theory of heat conduction. The flow problem discussed in this section will therefore serve as an example to point out the application of this known method of heat conduction in fluid mechanics.

Figure 14.3 shows schematically the flow problem to be solved. Two walls are shown that are placed in a fluid. Both walls together form a plane channel between themselves. For $t < 0$ both walls are at rest, whereas they both assume a velocity U_0 along the x_1-axis for $t \geq 0$. As a consequence of this, a fluid movement is induced, which starts at both sides of the plates and it moves inwards due to the fluid viscosity. For the problem treated in this section, the fluid flow induced between the plates and its transient progress will be discussed.

In order to obtain the solution of the general flow problem of plate-induced channel flow, the introduction of the following dimensionless quantities is recommended:

$$U^* = \frac{U_0 - U_1}{U_0} \quad \text{dimensionless velocity}$$

$$\eta = \frac{x_2}{D} \quad \text{dimensionless position coordinate}$$

$$\tau = \frac{\nu t}{D^2} \quad \text{dimensionless time}$$

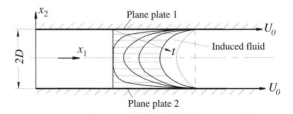

Fig. 14.3 Fluid flow induced by the movement of the walls of a plane channel

14.2 Accelerated and Decelerated Fluid Flows

The partial differential equation:

$$\frac{\partial U_1}{\partial t} = \nu \frac{\partial^2 U_1}{\partial x_2{}^2} \qquad (14.61)$$

describing the flow problem can thus be written:

$$-\frac{\nu U_0}{D^2} \frac{\partial U^*}{\partial \tau} = -\nu \frac{U_0}{D^2} \frac{\partial^2 U^*}{\partial \eta^2}$$

or

$$\frac{\partial U^*}{\partial \tau} = \frac{\partial^2 U^*}{\partial \eta^2} \qquad (14.62)$$

The initial condition expressed in dimensionless quantities:

$$\tau = 0, \quad U^*(\eta) = 1 \qquad (14.63)$$

and the boundary conditions at the walls:

$$\eta = \pm 1, \quad U^*(\eta = \pm 1) = 0 \qquad (14.64)$$

and also the demand for symmetry at the center line of the channel:

$$\eta = 0, \quad \frac{\partial U^*}{\partial \eta} = 0 \qquad (14.65)$$

define the flow problem sketched in Fig. 14.3.

For the solution of the partial differential equation (14.62), there is a classical solution path, which is based on the method of separation of the variables, i.e. the solution is sought with an ansatz of the following form:

$$U^*(\eta, \tau) = f(\eta) g(\tau) \qquad (14.66)$$

From this, it follows that the left-hand side of (14.62) can be expressed as:

$$\frac{\partial U^*}{\partial \tau} = f \frac{dg}{d\tau} \qquad (14.67)$$

and for the right-hand side:

$$\frac{\partial^2 U^*}{\partial \eta^2} = g \frac{d^2 f}{d\eta^2} \qquad (14.68)$$

The expressions in (14.67) and (14.68) are inserted into the partial differential equation (14.62) to yield:

$$\frac{1}{g} \frac{dg}{d\tau} = \frac{1}{f} \frac{d^2 f}{d\eta^2} = -\lambda^2 \qquad (14.69)$$

As the left-hand side of this ordinary differential equation depends only on the variable τ and the right-hand side only on the variable η, the equation can only be fulfilled when both sides are set equal to a constant which is introduced into (14.69) as $-\lambda^2$.

The following ordinary differential equations thus result from (14.69) for the function g:

$$\frac{dg}{d\tau} = -\lambda^2 g \qquad (14.70)$$

and the equation for f reads:

$$\frac{d^2 f}{d\eta^2} = -\lambda^2 f \qquad (14.71)$$

The general solutions of these differential equations are obtained by integrations as:

$$g = A \exp(-\lambda^2 \tau) \qquad (14.72)$$

$$f = B(\cos \lambda \eta) + C(\sin \lambda \eta) \qquad (14.73)$$

where A, B and C are integration constants. Applying the symmetry of the solution, demanded in (14.65) to the above solutions, one obtains $C = 0$, as the sine function is unable to fulfil the requirement of symmetry at $\eta = 0$. Applying the second boundary condition (14.64), one obtains:

$$B(\cos \lambda) = 0 \qquad (14.74)$$

In order to permit now a non-trivial solution of the flow problem, i.e. a solution that is different from zero, it is necessary that $B \neq 0$, i.e. the introduced quantity λ can only assume some specific values such that (14.74) fulfils the boundary conditions. Thus one obtains:

$$\lambda = \left(n + \frac{1}{2}\right)\pi \quad \text{for} \quad n = 0, \pm 1, \pm 2, \pm 3 \ldots \qquad (14.75)$$

In this way, the general solution of the problem, which fulfils the boundary conditions of the flow problem, results as:

$$U_n^* = A_n B_n \exp\left[-\left(n + \frac{1}{2}\right)^2 \pi^2 \tau\right] \cos\left[\left(n + \frac{1}{2}\right)\pi\eta\right] \qquad (14.76)$$

Since the governing differential equation is linear, one obtains the most general solution as the sum of the individual solutions stated in (14.76):

$$U^* = \sum_{n \to -\infty}^{\infty} \left\{ A_n B_n \exp\left[-\left(n + \frac{1}{2}\right)^2 \pi^2 \tau\right] \cos\left[\left(n + \frac{1}{2}\right)\pi\eta\right] \right\} \qquad (14.77)$$

Considering the symmetry of all n functions around $n = 0$ in the sum (14.77), the solution can be written as:

14.2 Accelerated and Decelerated Fluid Flows

$$U^* = \sum_{n=0}^{\infty} D_n \exp\left[-\left(n+\frac{1}{2}\right)^2 \pi^2 \tau\right] \cos\left[\left(n+\frac{1}{2}\right)\pi\eta\right] \quad (14.78)$$

In this expression, $D_n = A_n B_n + A_{-(n+1)} B_{-(n+1)}$ is an integration constant which assumes a different value for each value of n. These values can be determined from the initial condition (14.63):

$$1 = \sum_{n=0}^{\infty} D_n \cos\left[\left(n+\frac{1}{2}\right)\pi\eta\right] \quad (14.79)$$

Multiplying (14.79) by

$$\cos\left[\left(m+\frac{1}{2}\right)\pi\eta\right] d\eta$$

and integrating both sides from $\eta = -1$ to $\eta = +1$, i.e. carrying out the following integration:

$$\int_{-1}^{+1} \cos\left[\left(m+\frac{1}{2}\right)\pi\eta\right] d\eta = \sum_{n=0}^{\infty} D_n \int_{-1}^{+1} \cos\left[\left(m+\frac{1}{2}\right)\pi\eta\right] \cos\left[\left(n+\frac{1}{2}\right)\pi\eta\right] d\eta \quad (14.80)$$

one obtains on the right-hand side for all n-values being always zero, when $m \neq n$. For $m = n$ the integration on both sides yields the following conditional equation for D_m:

$$\left[\frac{\sin\left(m+\frac{1}{2}\right)\pi\eta}{\left(m+\frac{1}{2}\right)\pi}\right]_{-1}^{+1} = D_m \left[\frac{\frac{1}{2}\left(m+\frac{1}{2}\right)\pi + \frac{1}{4}\sin\left(m+\frac{1}{2}\right)2\pi\eta}{\left(m+\frac{1}{2}\right)\pi}\right]_{-1}^{+1} \quad (14.81)$$

or $\quad D_m = \dfrac{2(-1)^m}{\left(m+\frac{1}{2}\right)\pi} \Rightarrow D_n = \dfrac{2(-1)^n}{\left(n+\frac{1}{2}\right)\pi}$

With this conditional equation for D_n one obtains the final relation for the plate induced transient channel flow:

$$U^* = 2\sum_{n=0}^{+\infty} \frac{(-1)^n}{\left(n+\frac{1}{2}\right)\pi} \exp\left[-\left(n+\frac{1}{2}\right)^2 \pi^2 \tau\right] \cos\left[\left(n+\frac{1}{2}\right)\pi\eta\right] \quad (14.82)$$

or in terms of the dimensional quantities:

$$U_1 = U_0 - 2U_0 \sum_{n=0}^{\infty} \frac{(-1)^n}{\left(n+\frac{1}{2}\right)\pi} \exp\left[-\left(n+\frac{1}{2}\right)^2 \pi^2 \frac{\nu t}{D^2}\right] \cos\left[\left(n+\frac{1}{2}\right)\pi \frac{x_2}{D}\right] \quad (14.83)$$

The above infinite series has the property of converging very quickly when the dimensionless time $(\nu t/D^2)$ is large. On the other hand, the convergence is slow when $(\nu t/D^2)$ is small. Considering the derived solution (14.83) for $(\nu t/D^2) \to 0$, the result is in agreement with the solution of the plate-induced

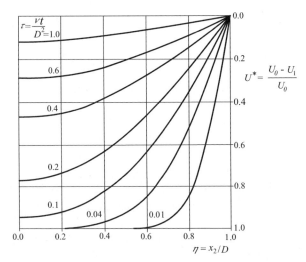

Fig. 14.4 Computed velocity distribution in the flow as a function of location and time

fluid movement treated in Sect. 4.3.2. By employing Laplace transformation for small dimensionless times, the employment of (14.37) can be recommended for the computation of the velocity distribution in the channel. This relationship has to be applied to both halves of the channel and the different positions of the coordinate systems in Figs. 14.1 and 14.3 have to be taken into consideration.

In Fig. 14.4, a graphical representation is given for the velocity distribution described by the final equation (14.83). This representation shows that, for small dimensionless times $(\nu t/D^2)$, only the fluid layers between the plane plates of Fig. 14.3 near the wall are moved. Likewise, only for a dimensionless time $(\nu t/D^2) \geq 0.04$ is a perceivable movement of the fluid in the middle of the channel obtained. For $(\nu t/D^2) \geq 1$, almost the entire fluid in the space between the plates has reached the plate velocity U_0. For $(\nu t/D^2) \to \infty$, the entire fluid moves between the plates with the velocity U_0.

On considering the final state of the plate-induced channel flow for $(\nu t/D^2) \to \infty$, one recognizes that it no longer depends on time, i.e. one should be able to compute it also by solving the partial differential equation for stationary, one-dimensional flows. The partial equation and its solution read

$$\mu \frac{\partial^2 U_1}{\partial x_2{}^2} = 0 \quad \Rightarrow \quad U_1 = C_1 x_2 + C_2 \tag{14.84}$$

Applying the boundary conditions $U_1 = U_0$ for $x_2 = \pm D$ to this solution, one obtains

$$C_1 = 0 \quad \text{and} \quad C_2 = U_0 \tag{14.85}$$

and thus $U_1 = U_0$ is obtained for plane plate-induced channel flow for $(\nu t/D^2) \to \infty$. This solution shows that for the characteristic time of this flow problem to go to infinity, all fluid moves at the constant plate velocity; the entire fluid is swept along by the plates.

14.2.4 Pipe Flow Induced by the Pipe Wall Motion

Analogous to the flow between two plates discussed in Sect. 14.2.3, which was caused by the movement of the plate walls, the pipe flow can also be treated, which is brought about by the movement of the pipe wall as sketched in Fig. 14.5. The basic equation to this problem is the partial differential equation derived from (14.28) where only the first term on the right-hand side is considered:

$$\frac{\partial U_z}{\partial t} = \nu \frac{1}{r} \frac{\partial}{\partial r}\left(r \frac{\partial U_z}{\partial r}\right) \qquad (14.86)$$

The flow problem to be studied with this equation can be defined by the following initial and boundary conditions:

$$\text{initial condition} \quad U_z(r, t=0) = 0 \quad \text{for } 0 \leq r \leq R \qquad (14.87)$$
$$\text{boundary condition} \quad U_z(R, t) = U_0 \quad \begin{array}{l}\text{moving wall} \\ \text{for all times } t \geq 0\end{array} \qquad (14.88)$$
$$\frac{\partial U_z}{\partial r}(0, t) = 0 \quad \text{symmetry} \qquad (14.89)$$

Analogous to the treatment of the channel flow, induced by the movements of the walls, the following dimensionless quantities are introduced:

$$U^* = \frac{U_0 - U_z}{U_0} \quad \text{dimensionless velocity} \qquad (14.90)$$
$$\eta = \frac{r}{R} \quad \text{dimensionless position coordinates} \qquad (14.91)$$
$$\tau = \frac{\nu t}{R^2} \quad \text{dimensionless time} \qquad (14.92)$$

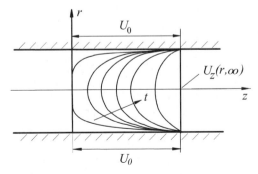

Fig. 14.5 Fluid flow in a pipe induced by the motion of the pipe walls

Thus the differential equation (14.86) can be written in dimensionless quantities as follows:

$$-\nu \frac{U_0}{R^2}\frac{\partial U_z^*}{\partial \tau} = -\nu \frac{U_0}{R^2}\left[\left(\frac{1}{\eta}\right)\frac{\partial U_z^*}{\partial \eta} + \frac{\partial^2 U_z^*}{\partial \eta^2}\right] \quad (14.93)$$

$$\frac{\partial U_z^*}{\partial \tau} = \frac{1}{\eta}\frac{\partial U_z^*}{\partial \eta} + \frac{\partial^2 U_z^*}{\partial \eta^2} \quad (14.94)$$

The following initial condition for the dimensionless velocity results:

$$\tau = 0 \qquad U^*(\eta) = 1 \quad (14.95)$$

and boundary conditions can be given for all times $t \geq 0$ as follows:

$$\eta = 1 \qquad U^* = 0 \quad (14.96)$$

and

$$\eta = 0 \qquad \frac{\partial U^*}{\partial \eta} = 0 \quad (14.97)$$

Again, the classical solution path can be chosen with the ansatz that the variables can be separated:

$$U^*(\eta, \tau) = f(\eta)g(\tau) \quad (14.98)$$

With the substitution of this ansatz into the differential equation (14.94), one obtains:

$$f\frac{\mathrm{d}g}{\mathrm{d}\tau} = \frac{g}{\eta}\frac{\mathrm{d}f}{\mathrm{d}\eta} + g\frac{\mathrm{d}^2 f}{\mathrm{d}\eta^2} \quad (14.99)$$

As g depends only on τ and f only on η, by separation of variables the following ordinary differential equations for g and f result:

$$\frac{1}{g}\frac{\mathrm{d}g}{\mathrm{d}\tau} = -\lambda^2 \quad (14.100)$$

$$\frac{1}{\eta}\frac{1}{f}\frac{\mathrm{d}f}{\mathrm{d}\eta} + \frac{1}{f}\frac{\mathrm{d}^2 f}{\mathrm{d}\eta^2} = -\lambda^2 \quad (14.101)$$

The solution for the differential equation (14.100) can be derived by integration:

$$g = C_1 \exp(-\lambda^2 \tau) \quad (14.102)$$

In order to determine the solution of the differential equation for $f(\eta)$, (14.101) can be written as follows:

$$\frac{\mathrm{d}^2 f}{\mathrm{d}\eta^2} + \frac{1}{\eta}\frac{\mathrm{d}f}{\mathrm{d}\eta} + \lambda^2 f = 0 \quad (14.103)$$

14.2 Accelerated and Decelerated Fluid Flows

From rewriting of (14.103), a Bessel differential equation results:

$$\eta^2 \frac{d^2 f}{d\eta^2} + \eta \frac{df}{d\eta} + \lambda^2 \eta^2 f = 0 \qquad (14.104)$$

and with

$$\alpha = \lambda \eta \qquad (14.105)$$

one obtains

$$\alpha^2 \frac{d^2 f}{d\alpha^2} + \alpha \frac{df}{d\alpha} + \alpha^2 f = 0 \qquad (14.106)$$

This equation has the following general solution:

$$f(\alpha) = C_2 J_0(\alpha) + C_3 Y_0(\alpha) \qquad (14.107)$$

The solution for J_0 results from the Bessel differential equation:

$$x^2 y''(x) + xy'(x) + (x^2 - p^2) y(x) = 0 \qquad (14.108)$$

which plays an essential role in many fields of theoretical physics and which has the solution

$$J_p(x) = \sum_{n=0}^{\infty} \frac{(-1)^n}{\Gamma(n+1)\,\Gamma(n+p+1)} \left(\frac{x}{2}\right)^{2n+p} \qquad (14.109)$$

where the Γ-function is defined as follows:

$$\Gamma(n) = \int_0^\infty \exp(-x)\, x^{n-1}\, dx \quad \text{for } n > 0 \qquad (14.110)$$

and can also be determined, for non-discrete values of n, by integral arguments.

The function $J_p(x)$ is defined as being a Bessel function of the first kind and of order p.

The second function required for the complete solution of the Bessel differential equation of zero order (14.106) is the Bessel function of the second kind, but also of zero order, i.e. $Y_0(\alpha)$. This function is often also called the Neumann or Weber function. Thus the solution ansatz (14.107) in terms of $J_0(\alpha)$ and $Y_0(\alpha)$ is composed of the Bessel functions of first and second kinds and of zero order.

Considering the symmetry boundary condition in (14.97) or, noting that at $\eta = 0, Y_0(\lambda \eta) \to \infty$, one finds $C_3 = 0$ in (14.107). Therefore, the solution for U^*, substituting $\alpha = \lambda \eta$, may be written as

$$U^*(\eta, \tau) = C_1 \exp(-\lambda^2 \tau) J_0(\lambda \eta) = A \exp(-\lambda^2 \tau) J_0(\lambda \eta) \qquad (14.111)$$

Application of boundary condition (14.96) yields

$$A \exp(-\lambda^2 \tau) J_0(\lambda) = 0 \tag{14.112}$$

Since setting $A = 0$ would result in a trivial solution, one must require

$$J_0(\lambda) = 0 \tag{14.113}$$

for the non-trivial solution. Therefore, one obtains multiple values of λ that satisfy the boundary conditions at the wall. The values of λ obtained from (14.113) are the 0-values of the zeroth order of Bessel functions of the first kind. From tables of Bessel functions, the solutions of (14.113) are obtained as follows:

$$\lambda_n = 2.405, 5.520, 8.654, 11.792, 14.931, 18.071, 21.212, 24.353, 27.494 \tag{14.114}$$

Each of the solutions of λ_n now constitutes an individual solution. Considering the linearity of the governing equations (14.94) and (14.97), the complete solution for $U^*(\eta, \tau)$ is obtained by linear superposition:

$$U^*(\eta, \tau) = \sum_{n=1}^{\infty} A_n \exp(-\lambda_n^2 \tau) J_0(\lambda_n \eta) = 0 \tag{14.115}$$

For readers of this book, the values for $J_n(\alpha)$ and $Y_n(\alpha)$ can be taken from Tables 14.1 and 14.2. The function $J_0(\alpha)$ and $J_1(\alpha)$ are also given in Fig. 14.6, $Y_0(\alpha)$ and $Y_1(\alpha)$ in Fig. 14.7.

In order to be able to insert the boundary conditions, it is further necessary to perform the derivatization $dU^*/d\eta$. For this it is important to know that for the following relationship for the derivative holds:

$$\frac{dJ_0(\alpha)}{dx} = -J_1(\alpha) \frac{d\alpha}{dx} \tag{14.116}$$

Thus one can write

$$\frac{dU^*}{d\eta} = -\sum_{n=1}^{\infty} A_n \exp(-\lambda_n^2 \tau) \lambda_n J_1(\lambda_n \eta) \tag{14.117}$$

Now the boundary conditions can be implemented:

$$\eta = 1 \quad \rightsquigarrow \quad U^* = 0 \quad \rightsquigarrow \quad J_0(\lambda_n) = 0 \tag{14.118}$$

and thus the following λ_n values can be determined as already explained above:

$$\lambda_n = 2.405, 5.520, 8.654, 11.792, 14.931, 18.071, 21.212, 24.353, 27.494 \tag{14.119}$$

14.2 Accelerated and Decelerated Fluid Flows

Table 14.1 Discrete values of Bessel functions of the first kind

α	$J_0(\alpha)$	$J_1(\alpha)$	α	$J_0(\alpha)$	$J_1(\alpha)$	α	$J_0(\alpha)$	$J_1(\alpha)$
0.00	+1.000	0.000	5.00	−0.178	−0.328	10.00	−0.246	+0.435
0.20	+0.990	+0.099	5.20	−0.110	−0.343	10.20	−0.250	−0.007
0.40	+0.960	+0.196	5.40	−0.041	−0.345	10.40	−0.243	−0.056
0.60	+0.912	+0.287	5.60	+0.027	−0.334	10.60	−0.228	−0.101
0.80	+0.846	+0.369	5.80	+0.092	−0.311	10.80	−0.203	−0.142
1.00	+0.765	+0.440	6.00	+0.151	−0.277	11.00	−0.171	−0.177
1.20	+0.671	+0.498	6.20	+0.202	−0.233	11.20	−0.133	−0.204
1.40	+0.567	+0.542	6.40	+0.243	−0.182	11.40	−0.090	−0.223
1.60	+0.455	+0.570	6.60	+0.274	−0.125	11.60	−0.045	−0.232
1.80	+0.340	+0.582	6.80	+0.293	−0.065	11.80	+0.002	−0.232
2.00	+0.224	+0.577	7.00	+0.300	−0.005	12.00	+0.048	−0.223
2.20	+0.110	+0.556	7.20	+0.295	+0.054	12.20	+0.091	−0.206
2.40	+0.003	+0.520	7.40	+0.279	+0.110	12.40	+0.130	−0.181
2.60	−0.097	+0.471	7.60	+0.252	+0.159	12.60	+0.163	−0.149
2.80	−0.185	+0.410	7.80	+0.215	+0.201	12.80	+0.189	−0.111
3.00	−0.260	+0.339	8.00	+0.172	+0.235	13.00	+0.207	−0.070
3.20	−0.320	+0.261	8.20	+0.122	+0.258	13.20	+0.217	−0.027
3.40	−0.364	+0.179	8.40	+0.069	+0.271	13.40	+0.218	+0.016
3.60	−0.392	+0.096	8.60	+0.015	+0.273	13.60	+0.210	+0.059
3.80	−0.403	+0.013	8.80	−0.039	+0.264	13.80	+0.194	+0.098
4.00	−0.397	−0.066	9.00	−0.090	+0.245	14.00	+0.171	+0.133
4.20	−0.377	−0.139	9.20	−0.137	+0.217	14.20	+0.141	+0.163
4.40	−0.342	−0.203	9.40	−0.177	+0.182	14.40	+0.106	+0.185
4.60	−0.296	−0.257	9.60	−0.209	+0.140	14.60	+0.068	+0.200
4.80	−0.240	−0.299	9.80	−0.232	+0.093	14.80	+0.027	+0.206
5.00	−0.178	−0.328	10.00	−0.246	+0.435	15.00	−0.014	+0.205

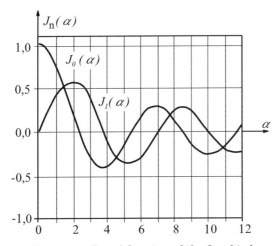

Fig. 14.6 Bessel function of the first kind

From the initial condition of the considered flow problem, one obtains

$$\tau = 0 \quad \leadsto \quad U^*(\eta, 0) = 1 = \sum_{n=1}^{\infty} A_n J_0(\lambda_n \eta) \qquad (14.120)$$

i.e. no $\lambda_{n=0}$ exists and one thus obtains

$$U^*(\eta, 0) = 1 = A_1 J_0(\lambda, \eta) + A_2 J_0(\lambda_2 \eta) + \cdots A_n J_0(\lambda_n \eta) + \cdots \qquad (14.121)$$

For determining the constants A_n one uses a special property of the Bessel function:

$$\int_0^x x J_n(ax) J_n(bx) \, \mathrm{d}x = 0 \quad \text{when } a \neq b \qquad (14.122)$$

but

$$\int_0^x x J_n^2(ax) \, \mathrm{d}x \neq 0 \quad \text{i.e. when } a = b \qquad (14.123)$$

Table 14.2 Discrete values of Bessel functions of the a second kind

α	$Y_0(\alpha)$	$Y_1(\alpha)$	α	$Y_0(\alpha)$	$Y_1(\alpha)$	α	$Y_0(\alpha)$	$Y_1(\alpha)$
0.00	$-\infty$	$-\infty$	5.00	−0.309	+0.148	10.00	+0.058	+0.249
0.20	−1.081	−3.324	5.20	−0.331	+0.079	10.20	+0.006	+0.250
0.40	−0.606	−1.781	5.40	−0.340	+0.010	10.40	−0.044	+0.242
0.60	−0.309	−1.260	5.60	−0.335	+0.057	10.60	−0.090	+0.224
0.80	−0.087	−0.978	5.80	−0.318	−0.119	10.80	−0.133	+0.197
1.00	+0.088	−0.781	6.00	−0.288	−0.175	11.00	−0.169	+0.164
1.20	+0.228	−0.621	6.20	−0.248	−0.222	11.20	−0.198	+0.124
1.40	+0.338	−0.479	6.40	−0.200	−0.260	11.40	−0.218	+0.081
1.60	+0.420	−0.348	6.60	−0.145	−0.286	11.60	−0.230	+0.035
1.80	+0.477	−0.224	6.80	−0.086	−0.300	11.80	−0.232	−0.012
2.00	+0.510	−0.107	7.00	−0.026	−0.303	12.00	−0.225	−0.057
2.20	+0.521	+0.002	7.20	+0.034	−0.293	12.20	−0.210	−0.099
2.40	+0.510	+0.101	7.40	+0.091	−0.273	12.40	−0.186	−0.137
2.60	+0.481	+0.188	7.60	+0.142	−0.243	12.60	−0.155	−0.169
2.80	+0.436	+0.264	7.80	+0.187	−0.204	12.80	−0.119	−0.194
3.00	+0.377	+0.325	8.00	+0.224	−0.158	13.00	−0.078	−0.210
3.20	+0.307	+0.371	8.20	+0.250	−0.107	13.20	−0.035	−0.218
3.40	+0.230	+0.401	8.40	+0.266	−0.054	13.40	+0.009	−0.218
3.60	+0.148	+0.415	8.60	+0.272	+0.001	13.60	+0.051	−0.208
3.80	+0.645	+0.414	8.80	+0.266	+0.054	13.80	+0.091	−0.191
4.00	−0.017	+0.380	9.00	+0.250	+0.104	14.00	+0.127	−0.167
4.20	−0.094	+0.368	9.20	+0.225	+0.149	14.20	+0.158	−0.136
4.40	−0.163	+0.326	9.40	+0.191	+0.187	14.40	+0.181	−0.100
4.60	−0.224	+0.274	9.60	+0.150	+0.217	14.60	+0.197	−0.061
4.80	−0.272	+0.214	9.80	+0.105	+0.238	14.80	+0.206	−0.020
5.00	−0.309	+0.148	10.00	+0.058	+0.249	15.00	+0.206	+0.021

14.2 Accelerated and Decelerated Fluid Flows

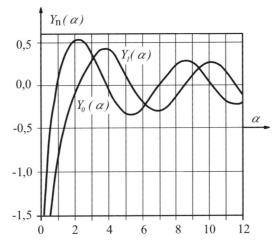

Fig. 14.7 Bessel function of the second kind

Thus the coefficients $A_1, A_2 \cdots A_n, \cdots$ in (14.117) can be determined by successive multiplication by $\eta J_0(\lambda_n \eta)$ and by the following integration:

$$\int_0^1 \eta J_0(\lambda_n \eta) \, d\eta = \int_0^1 A_n \eta J_0^2(\lambda_n \eta) d\eta \qquad (14.124)$$

Thus for each of the coefficients A_n the following relationship results:

$$A_n = \frac{\int_0^1 \eta J_0(\lambda_n \eta) d\eta}{\int_0^1 \eta J_0^2(\lambda_n \eta) d\eta} \qquad (14.125)$$

Carrying out a first step of the integration yields:

$$A_n = \frac{2}{[J_0^2(\lambda_n) + J_1^2(\lambda_n)]} \int_0^1 \eta J_0(\lambda_n \eta) d\eta \qquad (14.126)$$

By further integration one obtains:

$$A_n = \frac{2}{\lambda_n} \frac{J_1(\lambda_n)}{(J_0^2(\lambda_n) + J_1^2(\lambda_n))} \qquad (14.127)$$

Thus for the velocity distribution according to (14.111), the following expression results:

$$U^*(\eta, \tau) = \sum_{n=1}^{\infty} \frac{2}{\lambda_n} \frac{J_1(\lambda_n)}{J_0^2(\lambda_n) + J_1^2(\lambda_n)} \exp(-\lambda_n^2 \tau) J_0(\lambda_n \eta) \qquad (14.128)$$

For the gradient of the velocity profile one obtains according to (14.117):

$$\frac{\partial U^*}{\partial \eta}(\eta, \tau) = -\sum_{n=1}^{\infty} 2 \frac{J_1(\lambda_n)}{J_0^2(\lambda_n) + J_1^2(\lambda_n)} \exp(-\lambda_n^2 \tau) J_1(\lambda_n \eta) \qquad (14.129)$$

Considering that $\dfrac{\mathrm{d} J_0(\lambda_n \eta)}{\mathrm{d} \eta} = -J_1(\lambda_n \eta) \lambda_n$ is valid over the entire flow field, one can employ the above results to determine the shear-stress distribution for the considered flow problem:

$$\tau_{21} = -\mu \frac{\mathrm{d} U_z}{\mathrm{d} r} = -\mu \left(-\frac{U_0}{R} \frac{\mathrm{d} U^*}{\mathrm{d} \eta} \right) = \frac{\mu U_0}{R} \frac{\mathrm{d} U^*}{\mathrm{d} \eta} \qquad (14.130a)$$

and thus τ_{21} is given by

$$\tau_{21} = -\frac{\mu U_0}{R} \sum_{n=1}^{\infty} \frac{2 J_1(\lambda_n)}{(J_0^2(\lambda_n) + J_1^2(\lambda_n))} \exp(-\lambda_n^2 \tau) J_1(\lambda_n \eta) \qquad (14.130b)$$

For the τ_{21} value at the pipe wall, i.e. with $\eta = 1$, one obtains

$$\tau_{21}(R,t) = -\mu \frac{U_0}{R} \sum_{n=1}^{\infty} \frac{2}{1 + (J_0(\lambda_n)/J_1(\lambda_n))^2} \exp(-\lambda_n) \frac{\nu t}{R^2} \qquad (14.131)$$

i.e. a finite value, even for time $t = 0$. This is a surprising result when comparing it with $\tau_{21} \to \infty$ for $t = 0$ for the induced channel flow.

14.3 Oscillating Fluid Flows

14.3.1 Stokes Second Problem

For further deepening the physical understanding of unsteady fluid movements, induced by momentum diffusion, those fluid movements which occur due to an oscillating plate will be discussed in this section. Hence, a fluid flow problem is considered which comes about due to the oscillatory movement of a plate in a such way that the fluid movement created in the immediate vicinity of the plate is communicated to the fluid above the plate, by molecular momentum diffusion. The movement of the fluid above the plate is thus governed by the following partial differential equation:

$$\frac{\partial U_1}{\partial t} = \nu \frac{\partial^2 U_1}{\partial x_2^2} \qquad (14.132)$$

Hence the same differential equation as in the previous sections describes this flow. Its particular features are introduced by the imposed initial and boundary conditions.

14.3 Oscillating Fluid Flows

The initial and boundary conditions of the problem can be stated as follows:

$$\text{for all times } t \leq 0 : U_1(x_2, t) = 0 \quad (14.133)$$

for all times $t > 0$:

$$x_2 = 0 \quad \leadsto \quad U_1(0, t) = U_0 \cos(\omega t) \quad (14.134)$$
$$x_2 \to \infty \quad \leadsto \quad U_1(\infty, t) = 0 \quad (14.135)$$

Again, a solution is sought which can be found by the following ansatz, i.e. by separation of the variables:

$$U_1(x_2, t) = f(x_2) g(t) \quad (14.136)$$

Inserting in (14.132) results in:

$$f \frac{dg}{dt} = \nu g \frac{d^2 f}{dx_2^2} \quad (14.137)$$

By separation of the variables one can derive:

$$\frac{1}{\nu g} \frac{dg}{dt} = \frac{1}{f} \frac{d^2 f}{dx_2^2} = \pm i \lambda^2 \quad (14.138)$$

In (14.138), the constant appearing on the right-hand side was set to read $\pm i \lambda^2$ with $i = \sqrt{-1}$. This takes into consideration the fact that according to (14.134), there is a periodic stimulation of the fluid movement. Thus cosine and sine terms are expected in the solution, which can be expressed by complex terms in the exponential function. Consequently, the following differential equations have to be solved:

$$\frac{dg}{dt} - \left(\pm i \lambda^2\right) \nu g = 0 \quad (14.139)$$

$$\frac{d^2 f}{dx_2^2} - \left(\pm i \lambda^2\right) f = 0 \quad (14.140)$$

The solutions of these two differential equations yields for $U_1(x_2, t)$:

$$U_1(x_2, t) = C^* \exp\left[\pm i \lambda^2 \nu t \pm \lambda \sqrt{\pm i x_2}\right] \quad (14.141)$$

Because of the combination of positive and negative signs, four solutions result:

$$U_1^A = A \exp\left[-\frac{\lambda}{\sqrt{2}} x_2 + i \left(\lambda^2 \nu t - \frac{\lambda}{\sqrt{2}} x_2\right)\right] \quad (14.142)$$

$$U_1^B = B \exp\left[-\frac{\lambda}{\sqrt{2}} x_2 - i \left(\lambda^2 \nu t - \frac{\lambda}{\sqrt{2}} x_2\right)\right] \quad (14.143)$$

$$U_1^C = C \exp\left[+\frac{\lambda}{\sqrt{2}} x_2 + i \left(\lambda^2 \nu t - \frac{\lambda}{\sqrt{2}} x_2\right)\right] \quad (14.144)$$

$$U_1^D = D \exp\left[+\frac{\lambda}{\sqrt{2}} x_2 - i \left(\lambda^2 \nu t + \frac{\lambda}{\sqrt{2}} x_2\right)\right] \quad (14.145)$$

The last two partial solutions of the differential equations do not represent reasonable results from a physical point of view because of the requirement (14.135), as for $x_2 \to \infty$ they yield for the velocity $U_1(\infty, t) \to \infty$. Thus as a solution ansatz, that is physically meaningful, the following results:

$$U_1(x_2, t) = U_1^A(x_2, t) + U_1^B(x_2, t) \tag{14.146}$$

i.e.

$$U_1(x_2, t) = \exp\left(-\tfrac{\lambda}{\sqrt{2}} x_2\right) \left\{ A^* \exp\left[i\left(\lambda^2 \nu t - \tfrac{\lambda}{\sqrt{2}} x_2\right)\right] \right. \\ \left. + B^* \exp\left[-i\left(\lambda^2 \nu t - \tfrac{\lambda}{\sqrt{2}} x_2\right)\right] \right\} \tag{14.147}$$

The expressions in the curly brackets can be written as cosine and sine functions:

$$U_1(x_2, t) = \exp\left(-\tfrac{\lambda}{\sqrt{2}} x_2\right) \left[A \cos\left(\lambda^2 \nu t - \tfrac{\lambda}{\sqrt{2}} x_2\right) + B \sin\left(\lambda^2 \nu t - \tfrac{\lambda}{\sqrt{2}} x_2\right) \right] \tag{14.148}$$

Applying the boundary condition (14.134), one obtains

$$U_1(0, t) = U_0 \cos(\omega t) = A \cos(\lambda^2 \nu t) + B \sin(\lambda^2 \nu t) \tag{14.149}$$

and thus $B = 0$, $A = U_0$ and $\lambda = \sqrt{\omega/\nu}$ so that one obtains as a solution

$$U_1(x_2, t) = U_0 \exp\left(-\sqrt{\tfrac{\omega}{2\nu}} x_2\right) \cos\left(\omega t - \sqrt{\tfrac{\omega}{2\nu}} x_2\right) \tag{14.150}$$

This equation describes the velocity distributions stated for certain ωt values in Fig. 14.8 which are present in the fluid above the plate. In Fig. 14.8, velocities are given for $\omega t = 0, \frac{\pi}{2}, \pi, \frac{3\pi}{2}, 2\pi$.

The velocity distributions indicated in Fig. 14.8 show that the fluid movement in fluid layers, some distance away from the wall, always lags behind the movement of the plate. The amplitude of the fluid movement decreases with increasing distance from the plate. At the plate itself, the fluid movement follows exactly the movement of the plate, i.e. specifications existing due to the boundary conditions are fulfilled.

The above-mentioned phase shift is of great interest for a number of fluid motions. For practical purposes, one can state that a perceivable fluid movement can only be observed for

$$x_2 \leq 2\pi\sqrt{2}\sqrt{\tfrac{\nu}{\omega}} \tag{14.151}$$

The larger the kinematic viscosity of the fluid is, the thicker this layer becomes. Moreover, the relationship (14.151) says that high-frequency oscillations can penetrate less deep into the fluid interior than low-frequency oscillations. These kinds of results of the analytical considerations above give important insights that can be used advantageously in considerations of many externally induced fluid flows.

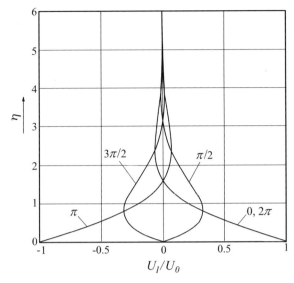

Fig. 14.8 Velocity profiles above an oscillating plane plate at fixed phases of the plate motion

14.4 Pressure Gradient-Driven Fluid Flows

14.4.1 Starting Flow in a Channel

The considerations below were carried out in order to investigate the influence of viscosity on the channel flow, setting in due to gravitational forces. It is assumed that the entire fluid in the channel in Fig. 14.9 is at rest for $t < 0$. At time $t = 0$, the fluid is set in motion, namely by the gravitational acceleration g. The setting in, non-stationary fluid flow is described by the following differential equation:

$$\rho \frac{\partial U_1}{\partial t} = \mu \frac{\partial^2 U_1}{\partial x_2^2} + \rho g \qquad (14.152)$$

This differential equation can be rewritten as

$$\frac{\partial U_1}{\partial t} = g + \nu \frac{\partial^2 U_1}{\partial x_2^2} \qquad (14.153)$$

This equation has to be solved for the following initial and boundary conditions:
Initial condition:

for $t \leq 0$ $U_1(x_2, t) = 0$ (14.154)

for $t > 0$ $\leadsto$ $U_1 \neq 0$ for $-D < x_2 < +D$ (14.155)

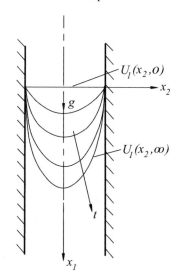

Fig. 14.9 Starting flow in a channel

The boundary conditions are:

$$U_1 = 0 \quad \text{for} \quad x_2 = \pm D \tag{14.156}$$

To solve the partial differential equation (14.153), it is recommended to introduce the following dimensionless variables:

$$U^* = \frac{U_1}{gD^2/2\nu} \quad \text{dimensionless velocity} \tag{14.157}$$

$$\eta = \frac{x_2}{D} = x_2^* \quad \text{dimensionless position coordinate} \tag{14.158}$$

$$\tau = \frac{\nu t}{D^2} = t^* \quad \text{dimensionless time} \tag{14.159}$$

On inserting these dimensionless quantities in (14.153), one obtains the partial differential equation (14.160) for the above-introduced dimensionless velocity U^*:

$$\frac{\partial U^*}{\partial \tau} = 2 + \frac{\partial^2 U^*}{\partial \eta^2} \tag{14.160}$$

with the initial and boundary conditions formulated for the dimensionless variables as follows:

$$\begin{aligned}
\text{initial conditions:} \quad & \tau \leq 0: \ U^* = 0 \text{ for } -1 < \eta < +1 \\
\text{boundary conditions:} \quad & \eta = +1: U^* = 0 \text{ for all } \tau > 0 \\
& \eta = -1: U^* = 0 \text{ for all } \tau > 0
\end{aligned} \tag{14.161}$$

When looking for a solution of the partial differential equation (14.160), the approach is to look for the stationary solution U_∞^* (occurring for $\tau \to \infty$) and for the non-stationary part U_t^*, to yield generally

14.4 Pressure Gradient-Driven Fluid Flows

$$U^* = U^*_\infty - U^*_t \tag{14.162}$$

Due to the stationarity of the flow for $\tau \to \infty$, one obtains the following partial differential equation for U^*_∞:

$$0 = 2 + \frac{\partial^2 U^*_\infty}{\partial \eta^2} \tag{14.163}$$

Taking into consideration the above boundary conditions, the following results for U^*_∞:

$$U^*_\infty = (1 - \eta^2) \tag{14.164}$$

On introducing $U^* = (1 - \eta^2) - U^*_t$ into the differential equation (14.160), one obtains a differential equation to be solved for U^*_t:

$$\frac{\partial U^*_t}{\partial \tau} = \frac{\partial^2 U^*_t}{\partial \eta^2} \tag{14.165}$$

with the following initial condition and the boundary condition:

$$\tau = 0 : U^*_t = U^*_\infty \tag{14.166}$$
$$\tau > 0 : U^*_t = 0 \text{ for } \eta = \pm 1 \tag{14.167}$$

With the ansatz for a solution by separation of variables:

$$U^*_t = f(\eta)g(\tau) \tag{14.168}$$

one obtains

$$\frac{\partial U^*_t}{\partial \tau} = f \frac{dg}{d\tau} \quad \text{and} \quad \frac{\partial^2 U^*_t}{\partial \eta^2} = g \frac{d^2 f}{d\eta^2} \tag{14.169}$$

and by insertion in (14.165):

$$\frac{1}{g} \frac{dg}{d\tau} = \frac{1}{f} \frac{d^2 f}{d\eta^2} \tag{14.170}$$

As this equation can on the left-hand side be only a function of τ and on the right-hand side only a function of η, the differential equation can be fulfilled only by setting both sides equal to a constant:

$$\frac{1}{g} \frac{dg}{d\tau} = -\lambda^2 \quad \rightsquigarrow \quad g = A \exp\left(-\lambda^2 \tau\right) \tag{14.171}$$

$$\frac{1}{f} \frac{d^2 f}{d\eta^2} = -\lambda^2 \quad \rightsquigarrow \quad f = B \cos(\lambda \eta) + C \sin(\lambda \eta) \tag{14.172}$$

where A, B and C are the constants introduced by the integration. When considering the coordinate system sketched in Fig. 14.9 and quantitatively described in Fig 14.10, yielding a solution that is symmetrical with regard to

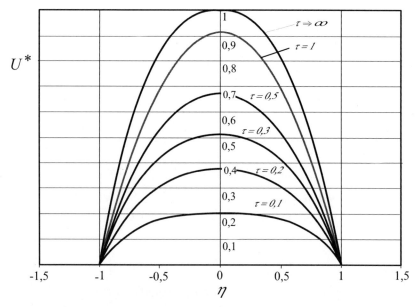

Fig. 14.10 Starting flow between two plane plates according to (14.190)

η, one has to set $C = 0$. The boundary condition (14.166) applied to (14.172), taking into consideration $C = 0$, yields:

$$0 = B\cos(\lambda_n) \tag{14.173}$$

This relationship is fulfilled for $\lambda_n = (n+1/2)\pi$ with $n = 0, \pm 1, \pm 2, \pm 3, \cdots \pm\infty$ so that one obtains as the most general term for the solution of U_t^*:

$$U_t^* = \sum_{n=-\infty}^{+\infty}\left(A_n B_n \exp\left[-\left(n+\frac{1}{2}\right)^2\pi^2\tau\right]\cos\left[\left(n+\frac{1}{2}\right)\pi\eta\right]\right). \tag{14.174}$$

Considering the symmetry of all n partial functions and setting $D_n = 2A_n B_n$, (14.174) can be written as:

$$U_t^* = \sum_{n=0}^{+\infty}D_n \exp\left[-\left(n+\frac{1}{2}\right)^2\pi^2\tau\right]\cos\left[\left(n+\frac{1}{2}\right)\pi\eta\right] \tag{14.175}$$

Taking into account the initial condition, one obtains:

$$1-\eta^2 = \sum_{n=0}^{+\infty}D_n \cos\left[\left(n+\frac{1}{2}\right)\pi\eta\right] \tag{14.176}$$

Multiplying both parts of this equation by:

$$\cos\left[\left(m+\frac{1}{2}\right)\pi\eta\right]d\eta \tag{14.177}$$

14.4 Pressure Gradient-Driven Fluid Flows

one obtains by integration from (-1) to $(+1)$:

$$\int_{-1}^{1} \left(1 - \eta^2\right) \cos\left[\left(m + \frac{1}{2}\right)\pi\eta\right] d\eta = D_m \qquad (14.178)$$

or, with the following steps of integration:

$$\left[\int_{-1}^{1} \cos\left(\left(m + \frac{1}{2}\right)\pi\eta\right) d\eta - \int_{-1}^{1} \eta^2 \cos\left(\left(m + \frac{1}{2}\right)\pi\eta\right) d\eta\right] = D_m \qquad (14.179)$$

$$\left[\int_{-1}^{1} \cos\left(\left(m + \frac{1}{2}\right)\pi\eta\right) d\eta\right] = \frac{2\sin\left(\left(m + \frac{1}{2}\right)\pi\right)}{\left(m + \frac{1}{2}\right)\pi} \qquad (14.180)$$

$$\left[\int_{-1}^{1} \eta^2 \cos\left(\left(m + \frac{1}{2}\right)\pi\eta\right) d\eta\right] = \int u \, dv = uv - \int v \, du \qquad (14.181)$$

$$u = \eta^2 \Rightarrow d = 2 * \eta \, d\eta, \ldots \quad dv = \cos\left(\left(m + \frac{1}{2}\right)\pi\eta\right) d\eta \Rightarrow v =$$

$$\frac{\sin\left(\left(m + \frac{1}{2}\right)\pi\eta\right)}{\left(m + \frac{1}{2}\right)\pi} \qquad (14.182)$$

$$\left[\int_{-1}^{1} \eta^2 \cos\left(\left(m + \frac{1}{2}\right)\pi\eta\right) d\eta\right] = \eta^2 \frac{\sin\left(\left(m + \frac{1}{2}\right)\pi\eta\right)}{\left(m + \frac{1}{2}\right)\pi}\bigg]_{-1}^{1} - \frac{2}{\left(m + \frac{1}{2}\right)\pi}$$

$$\int_{-1}^{1} \eta \sin\left(\left(m + \frac{1}{2}\right)\pi\eta\right) d\eta \qquad (14.183)$$

$$\int_{-1}^{1} \eta^2 \cos\left(\left(m + \frac{1}{2}\right)\pi\eta\right) d\eta = \frac{2\sin\left(m + \frac{1}{2}\right)\pi}{\left(m + \frac{1}{2}\right)\pi} - \frac{2}{\left(m + \frac{1}{2}\right)\pi}$$

$$\int_{-1}^{1} \eta \sin\left(m + \frac{1}{2}\right)\pi\eta \, d\eta \qquad (14.184)$$

$$\int_{-1}^{1} \eta \sin\left(m + \frac{1}{2}\right)\pi\eta \, d\eta = \int u \, dv \qquad (14.185)$$

$$u = \eta \qquad dv = \sin\left(m + \frac{1}{2}\right)\pi\eta \, d\eta$$

$$du = d\eta \qquad v = -\frac{\cos\left(\left(m + \frac{1}{2}\right)\pi\eta\right)}{\left(m + \frac{1}{2}\right)\pi}$$

$$\int_{-1}^{1} \eta \sin\left(m + \frac{1}{2}\right)\pi\eta \, d\eta = -\frac{\eta \cos\left(m + \frac{1}{2}\right)\pi\eta}{\left(m + \frac{1}{2}\right)\pi}\bigg]_{-1}^{1} + \int \frac{\cos\left(m + \frac{1}{2}\right)\pi\eta}{\left(m + \frac{1}{2}\right)\pi} d\eta$$

$$= 0 + \frac{\sin\left(m + \frac{1}{2}\right)\pi\eta}{\left[\left(m + \frac{1}{2}\right)\pi\right]^2}\bigg]_{-1}^{1} = \frac{2\sin\left(m + \frac{1}{2}\right)\pi}{\left[\left(m + \frac{1}{2}\right)\pi\right]^2} \qquad (14.186)$$

$$D_m = \frac{2\sin\left(m+\frac{1}{2}\right)\pi}{\left(m+\frac{1}{2}\right)\pi} - \frac{2\sin\left(m+\frac{1}{2}\right)\pi}{\left(m+\frac{1}{2}\right)\pi} + \frac{4\sin\left(m+\frac{1}{2}\right)\pi}{\left[\left(m+\frac{1}{2}\right)\pi\right]^3} \quad (14.187)$$

$$\sin\left(\left(m+\frac{1}{2}\right)\pi\right) = (-1)^m \quad \text{and} \quad \cos\left(\left(m+\frac{1}{2}\right)\pi\right) = 0$$

Hence, one obtains for $D_m = D_n$:

$$D_m \Rightarrow D_n = \frac{4(-1)^n}{\left(n+\frac{1}{2}\right)^3 \pi^3} \quad (14.188)$$

or for the solution of U_t^*:

$$U_t^* = 4\sum_{n=0}^{+\infty} \frac{(-1)^n}{\left(n+\frac{1}{2}\right)^3 \pi^3} \exp\left[-\left(n+\frac{1}{2}\right)^2 \pi^2 \tau\right] \cos\left[\left(n+\frac{1}{2}\right)\pi\eta\right] \quad (14.189)$$

As the complete solution for $U^* = U_\infty^* - U_t^*$, one obtains:

$$U^* = (1-\eta^2) - 4\sum_{n=0}^{+\infty} \frac{(-1)^n}{\left(n+\frac{1}{2}\right)^3 \pi^3} \exp\left[-\left(n+\frac{1}{2}\right)^2 \pi^2 \tau\right] \cos\left[\left(n+\frac{1}{2}\right)\pi\eta\right]$$
$$(14.190)$$

or in dimensional quantities:

$$U_1 = \frac{gD^2}{2\nu}\left[\left(1-\left(\frac{x_2}{D}\right)^2\right)\right] - 4\sum_{n=0}^{+\infty} \frac{(-1)^n}{\left(n+\frac{1}{2}\right)^3 \pi^3} \exp\left[-\left(n+\frac{1}{2}\right)^2 \pi^2 \frac{\nu t}{D^2}\right]$$
$$\cos\left[\left(n+\frac{1}{2}\right)\pi \frac{x_2}{D}\right]$$
$$(14.191)$$

On comparing the above derivations with those that were carried out in Sect. 14.2.3, it can easily be seen that the derivations correspond to one another. It can now easily be understood that the above derivations for the starting channel flow are equivalent to those for the fluid flow caused by a pressure gradient. On replacing the pressure gradient $-\frac{\mathrm{d}\Pi}{\mathrm{d}x_1}$ in the derivations by the gravitational force ρg, all derivations can be transferred to the pressure-driven channel flow.

14.4.2 Starting Pipe Flow

Another unsteady flow problem is the starting pipe flow, which is of certain importance in practice. It shall be discussed in this section for those conditions where for $t < 0$ a viscous fluid is at rest in an infinitely long pipe. At time $t = 0$ and for all times $t \geq 0$, a constant pressure gradient is imposed on the fluid, i.e. $-\frac{\partial \Pi}{\partial z}$ is generated along the entire pipe. The flow induced in this way is described by the partial differential equation

14.4 Pressure Gradient-Driven Fluid Flows

$$\frac{\partial U_z}{\partial t} = -\frac{1}{\rho}\frac{\partial \Pi}{\partial z} + \nu \frac{1}{r}\frac{\partial}{\partial r}\left(r \frac{\partial U_z}{\partial r}\right) \quad (14.192)$$

which was derived for one-dimensional, unsteady flows of incompressible viscous media having a constant viscosity. The initial and boundary conditions for the starting pipe flow read:

initial conditions: $t = 0 \rightsquigarrow U_z(r,t) = 0$ for $0 \leq r \leq R$ (14.193)
boundary conditions: $r = 0 \rightsquigarrow U_z =$ finally for all $t > 0$ (14.194)
$r = R \rightsquigarrow U_z = 0$ for all $t > 0$ (14.195)

One obtains the solution of the considered flow problems by introducing the following dimensionless variables:

$$U_z^* = \frac{U_z}{\left(-\frac{\partial \Pi}{\partial z}\right)\left(\frac{R^2}{4\mu}\right)} \quad \text{dimensionless velocity} \quad (14.196)$$

$$\eta = \frac{r}{R} \quad \text{dimensionless position coordinate} \quad (14.197)$$

$$\tau = \frac{\mu t}{\rho R^2} \quad \text{dimensionless time} \quad (14.198)$$

so that one has to solve the following differential equation for dimensionless quantities:

$$\frac{\partial U_z^*}{\partial \tau} = 4 + \frac{1}{\eta}\frac{\partial}{\partial \eta}\left(\eta \frac{\partial U_z^*}{\partial \eta}\right) \quad (14.199)$$

The initial and boundary conditions also need to be written for the dimensionless variables and are:

initial conditions: $\tau = 0 \rightsquigarrow U^* = 0$ for $0 \leq \eta \leq 1$ (14.200)
boundary conditions: $\eta = 0 \rightsquigarrow U^* =$ finally for $\tau > 0$ (14.201)
$\eta = 1 \rightsquigarrow U^* = 0$ for $\tau > 0$ (14.202)

To solve the above differential equation, one uses the fact that the considered flow for $\tau \to \infty$ heads for the laminar, stationary, fully developed pipe flow. The latter is introduced as a separate partial solution by the solution ansatz, where one writes $U_z^* = U_\infty^*$ for $\tau \to \infty$.

The chosen solution ansatz therefore reads for the developing velocity field:

$$U^*(\eta,\tau) = U_\infty^*(\eta) - U_t^*(\eta,\tau) \quad (14.203)$$

The stationary part of the solution, i.e $U_\infty^*(\eta)$, is obtained by solving the following differential equation:

$$0 = 4 + \frac{1}{\eta}\frac{d}{d\eta}\left(\eta \frac{dU_\infty^*}{d\eta}\right) \quad (14.204)$$

which can be derived from the above partial differential equation (14.199) by setting $\frac{\partial U^*}{\partial \tau} = 0$ for $\tau \to \infty$. By integration one obtains:

$$U^*_\infty(\eta) = -\eta^2 + C_1 \ln \eta + C_2 \qquad (14.205)$$

Employing the boundary conditions, one obtains for the integration constants $C_1 = 0$ and $C_2 = 1$ and thus for U^*_∞ the following results:

$$U^*_\infty(\eta) = 1 - \eta^2 \qquad (14.206)$$

Hence the Hagen–Poiseuille velocity distribution of the fully developed pipe flow is obtained.

On inserting now $U^*_\infty = 1 - \eta^2$ into the differential equation (14.203), one obtains:

$$U^*(\eta, \tau) = (1 - \eta^2) - U^*_t(\eta, \tau) \qquad (14.207)$$

By insertion of this relationship into the differential equation (14.199), the following differential equation can be derived:

$$-\frac{\partial U^*_t}{\partial \tau} = 4 + \frac{1}{\eta}\frac{\partial}{\partial \eta}\left[\eta \frac{\partial}{\partial \eta}(1 - \eta^2)\right] - \frac{1}{\eta}\frac{\partial}{\partial \eta}\left(\eta \frac{\partial U^*_t}{\partial \eta}\right) \qquad (14.208)$$

or, after having carried out the differentiations:

$$\frac{\partial U^*_t}{\partial \tau} = \frac{1}{\eta}\frac{\partial}{\partial \eta}\left(\eta \frac{\partial U^*_t}{\partial \eta}\right) \qquad (14.209)$$

This differential equation now has to be solved for the following initial and boundary conditions:

initial conditions: $\tau = 0 \rightsquigarrow U^*_t(\eta, 0) = U_\infty(\eta)$ for $0 \leq \eta \leq 1$ (14.210)

boundary conditions: $\eta = 0 \rightsquigarrow U^*_t =$ finite for all $\tau > 0$ (14.211)

$\eta = 1 \rightsquigarrow U^*_t = 0$ for all $\tau > 0$ (14.212)

Again, a separation ansatz for the variables η and τ is employed for solving the differential equation for U^*_t:

$$U^*_t = f(\eta) g(\tau) \qquad (14.213)$$

This ansatz leads to the following relationship:

$$\frac{1}{g}\frac{dg}{d\tau} = \frac{1}{f}\frac{1}{\eta}\frac{\partial}{\partial \eta}\left(\eta \frac{df}{d\eta}\right) = -\lambda^2 \qquad (14.214)$$

Hence it is necessary to solve the following differential equations for g and f:

$$\frac{dg}{d\tau} = -\lambda^2 g \qquad (14.215)$$

14.4 Pressure Gradient-Driven Fluid Flows

and
$$\frac{1}{\eta}\frac{d}{d\eta}\left(\eta\frac{df}{d\eta}\right)+\lambda^2 f=0 \tag{14.216}$$

These are differential equations known in the field of unsteady fluid mechanics. Their solutions are known and can be stated as follows:
$$g = A\exp(-\lambda^2\tau) \tag{14.217}$$
$$f = BJ_0(\lambda\eta)+CY_0(\lambda\eta) \tag{14.218}$$

In these general partial solutions of the differential equations (14.215) and (14.216), the quantities $J_0(\lambda\eta)$ and $Y_0(\lambda\eta)$ are Bessel functions of the first and second kind of zero order. A, B and C are integration constants which have to be determined by the initial and boundary conditions for $U_t^*(\eta,\tau)$. When employing the first boundary condition (14.211), i. e. $\eta = 0 \rightsquigarrow U_t^*$ is finite, one obtains $C = 0$, since $Y_0(0) = -\infty$. Hence, one obtains for U_t^*:
$$U_t^* = A\exp(-\lambda^2\tau)BJ_0(\lambda\eta) \tag{14.219}$$

If one demands that the second boundary condition (14.212) be fulfilled, i.e. $U_t^* = 0$ for $\eta = 1$, then $J_0(\lambda) = 0$ has to be fulfilled and only those λ values are permissible that fulfil these conditions. The following values can be computed:
$$\lambda_1 = 2,405; \quad \lambda_2 = 5,520; \quad \lambda_3 = 8,654; \quad \text{etc.} \tag{14.220}$$

i.e. there is a large number of discrete flows, one for each value of λ_n, which, summed up, result in the following general solution:
$$U_t^* = \sum_{n=1}^{\infty} A_n \exp(-\lambda_n^2\tau) J_0(\lambda_n\eta) \tag{14.221}$$

This general solution now fulfils the partial differential equation which describes the flow problem and the characteristic boundary conditions. Fulfilling the flow condition can now serve to determine the integration constant A_n, not yet defined. For $\tau \to 0$, the following holds:
$$(1-\eta^2) = \sum_{n=1}^{\infty} A_n J_0(\lambda_n\eta) \tag{14.222}$$

When one multiplies both sides of this equation by:
$$J_0(\lambda_m\eta)\eta\, d\eta \tag{14.223}$$

and integrates from 0 to 1, i.e. when one carries out the following arithmetic operations in the subsequent equation:
$$\int_0^1 J_0(\lambda_m\eta)(1-\eta^2)\eta\, d\eta = \sum_{n=1}^{\infty} A_n \int_0^1 J_0(\lambda_n\eta)J_0(\lambda_m\eta)\eta\, d\eta \tag{14.224}$$

one obtains, because of the orthogonality of the Bessel function, values of the right-hand side that are different from zero only when $m = n$. On employing known relationships for the Bessel functions, one obtains:

$$\frac{4J_1(\lambda_n)}{\lambda_n^3} = A_n \frac{1}{2}[J_1(\lambda_n)]^2 \tag{14.225}$$

or, rewritten,

$$A_m = \frac{8}{\lambda_m^3 J_1(\lambda_m)} \tag{14.226}$$

Hence one can deduce for U_t^*

$$U_t^* = 8 \sum_{n=1}^{\infty} \frac{J_0(\lambda_n \eta)}{\lambda_n^3 J_1(\lambda_n)} \exp\left(-\lambda_n^2 \tau\right) \tag{14.227}$$

and for the total velocity

$$U^* = (1 - \eta^2) - 8 \sum_{n=1}^{\infty} \frac{J_0(\lambda_n \eta)}{\lambda_n^3 J_1(\lambda_n)} \exp\left(-\lambda_n^2 \tau\right) \tag{14.228}$$

The velocity distributions computed according to (14.228) are shown in Fig. 14.11, where the development of the velocity profile can be seen. Again, for $(\nu t/R^2) = 1$ the fluid flow has reached the stationary state of fully developed laminar pipe flow with an accuracy that is sufficient in practice.

Again, similar to chapter 13, only some examples of time-dependent, one-dimensional flows are treated in this chapter. Further, treatments are found in refs. [14.1] to [14.6]

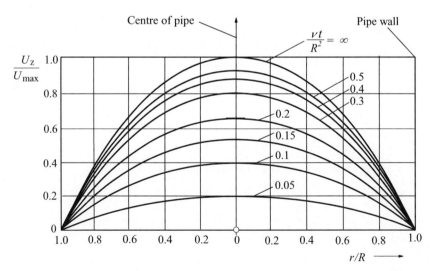

Fig. 14.11 Velocity distribution in the pipe for the starting, pressure-driven laminar pipe flow

References

14.1. Bird, R.B., Stewart, W.E., Lightfoot, E.N., "Transport Phenomena", Wiley, New York, 1960
14.2. Bosnjakovic, R., "Technische Thermodynamik", Theodor Steinkopf Verlag, Dresden, 1965
14.3. Eckert, E.R.G., Drake, R.M., Jr., "Analysis of Heat and Mass Transfer", McGraw-Hill Kogakusha, Tokyo, 1972
14.4. Sherman, F.S., "Viscous Flow", McGraw-Hill, Singapore, 1990
14.5. Stokes, G.G., "On the Effect of the Internal Friction of Fluids on the Motion of Pendulums", Cambridge Philosophical Transactions IX, 8, 1851
14.6. Stokes, G.G., "Mathematical and Physical Papers", Cambridge Philosophical Transactions III, 1–141, 1901

Chapter 15
Fluid Flows of Small Reynolds Numbers

15.1 General Considerations

As shown in the preceding chapters of this book, the integrations of the generally valid basic equations of fluid mechanics, that were derived in differential form in Chap. 5, are only successful, by means of analytical methods, when simplifications concerning the dimensionality of the considered flows are made. Furthermore, one has to choose for the treatment of transport problems in fluids very simple boundary conditions. These simple boundary conditions correspond to simple flow problems. For analytical solutions, they often have to be chosen in such a simple way, that the insights into the physics of fluid flows resulting from the solutions are of only slight practical interest. In general, this means that *practically relevant* flow problems cannot be treated by *analytical* solutions of the generally valid basic equations of fluid mechanics. The boundary conditions, which have to be imposed on the solutions, are mostly so complicated for practically interesting flows, that they can only be implemented in analytical solutions to some extent. Thus, fluid mechanics researchers, interested in analytical solutions, have only the possibility to treat such flow problems, for which the basic equations can be simplified. The solution of these equations often means that the considerations still have also to be restricted to flows with simple geometries, e.g. the flow around spheres, or the flow around cylinders. To study flows of this kind, the considerations start from the general basic equations that have been made dimensionless with the inflow velocity U_∞, a geometric dimension D, the fluid density ρ, the fluid viscosity μ, etc. This yields

$$\rho^* \left(\frac{D}{t_c U_\infty} \frac{\partial U^*}{\partial t^*} + U_i^* \frac{\partial U_j^*}{\partial x_i^*} \right) = -\frac{\Delta P_c}{\rho U_\infty^2} \frac{\partial P^*}{\partial x_j^*} + \frac{\mu}{\rho U_\infty D} \mu^* \frac{\partial^2 U_j^*}{\partial x_i^{*2}} + \frac{gD}{U_\infty^2} \rho^* g_j^* \tag{15.1}$$

or, rewritten with $St = D/(t_c U_\infty)$ (Strouhal number), $Eu = \Delta P/(\rho U_\infty^2)$ (Euler number), $Re = (U_\infty D)/\nu$ (Reynolds number) and $Fr = U_\infty^2/(gD)$ (Froude number):

$$\rho^* \left(St \frac{\partial U_j^*}{\partial t^*} + U_i^* \frac{\partial U_j^*}{\partial x_i^*} \right) = -Eu \frac{\partial P^*}{\partial x_j^*} + \frac{1}{Re} \mu^* \frac{\partial^2 U_j^*}{\partial x_i^{*2}} + \frac{1}{Fr} \rho^* g_j^* \ . \quad (15.2)$$

For stationary flows that are not influenced by gravitational forces, and where the viscosity forces are larger than the acceleration forces, i.e. for $Re < 1$, the following reduced form of the momentum equation can be employed:

$$Eu \frac{\partial^2 P^*}{\partial x_j^{*2}} + \frac{1}{Re} \mu \frac{\partial^2 U_j^*}{\partial x_i^{*2}} = 0 \quad \leadsto \quad 0 = -\frac{\partial P}{\partial x_j} + \mu \frac{\partial^2 U_j}{\partial x_i^2} \quad (j = 1,2,3). \quad (15.3)$$

This simplification of the momentum equation is valid for such flows that have the following properties:

- The geometric dimensions of the bodies are small, around which flow takes place, or the channels are small in which flows occur.
- Flows characterized by very small flow velocities (creeping flows).
- Flows of fluids with large coefficients of kinematic viscosity.

When all the above requirements for a flow are present at the same time, one arrives at conditions for the presence of smallest Reynolds numbers, i.e. at flows, where the fluid flows are characterized by *viscous* length, time and velocity scales. This will be explained at the end of the present chapter, where the properties of creeping flows are treated.

The differential equations given in Chap. 5, the continuity and the Navier–Stokes equations, read for $Re \to 0$:

$$\frac{\partial U_i}{\partial x_i} = 0 \quad \text{and} \quad 0 = -\frac{\partial P}{\partial x_j} + \mu \frac{\partial^2 U_j}{\partial x_i^2} \ . \quad (15.4)$$

These are known as the Stokes equations. For two-dimensional, fully developed flows they are identical with the equations discussed in Chap. 13, when g_j is set equal to zero. In this respect, some of the flows treated in this chapter are related to those in Chap. 13. This fact is pointed out at the appropriate place in the treatment of creeping flows.

By attempting the treatment of simplified forms of the basic equations, multidimensional flow problems can be included in the analytical treatment of flows. This will be shown with examples in Sects. 15.2–15.8. These examples have been chosen in such a way that they demonstrate, on the one hand, the multidimensionality of the possible computations achievable by simplifications and, on the other, the employment of the basic equations to practical flow problems of small Reynolds numbers. The fluid mechanics of slide bearings are considered. In addition, the rotating flow around a cylinder and the rotating flow around a sphere are discussed. For both geometries, the translatory motions and the rotating motions are considered. These considerations are carried out for viscous flows of small Reynolds numbers, employing the equations in (15.4). The considerations are carried out up to the computations of forces, to make it clear how small Reynolds number flows can be treated in a complete way. The computations of forces require derivations of the pressure distributions on the surface of bodies and also computations of the local momentum losses of flows at walls. The chosen examples are aimed to give a suitable introduction to the treatments of flows of small Reynolds numbers. Solutions of other creeping flow examples, going beyond the treatments in this chapter, are easily possible and are often described extensively in books on fluid mechanics, see refs. [15.1] to [15.7]. Hence, no further considerations of more complex flows are needed here.

15.2 Creeping Fluid Flows Between Two Plates

In this section, the flow of a viscous fluid between two parallel plates is considered, whose distance D can be regarded as being very small (Fig. 15.1). When the area-averaged mean flow velocity

$$\tilde{U} = \frac{1}{A} \int_A U_i \, \mathrm{d}A_i \qquad (15.5)$$

is also small, the conditions for the employment of the following differential equations exist. When $\rho = $ constant holds, then one can write

$$\frac{\partial U_i}{\partial x_i} = 0 \qquad (15.6)$$

$$\frac{\partial P}{\partial x_j} = \mu \frac{\partial^2 U_j}{\partial x_i^2}, \qquad (15.7)$$

i.e. the Stokes differential equations can be employed to treat the flow between two parallel plates.

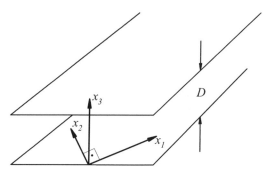

Fig. 15.1 The diagram shows the plates and the coordinate system for a flow between plates

This set of differential equations reads, written in full, for $j = 1, 2, 3$:

$$\frac{\partial U_1}{\partial x_1} + \frac{\partial U_2}{\partial x_2} + \frac{\partial U_3}{\partial U_3} = 0 \tag{15.8}$$

$$j = 1: \quad \frac{\partial P}{\partial x_1} = \mu \left(\frac{\partial^2 U_1}{\partial x_1^2} + \frac{\partial^2 U_1}{\partial x_2^2} + \frac{\partial^2 U_1}{\partial x_3^2} \right) \tag{15.9}$$

$$j = 2: \quad \frac{\partial P}{\partial x_2} = \mu \left(\frac{\partial^2 U_2}{\partial x_1^2} + \frac{\partial^2 U_2}{\partial x_2^2} + \frac{\partial^2 U_2}{\partial x_3^2} \right) \tag{15.10}$$

$$j = 3: \quad \frac{\partial P}{\partial x_3} = \mu \left(\frac{\partial^2 U_3}{\partial x_1^2} + \frac{\partial^2 U_3}{\partial x_2^2} + \frac{\partial^2 U_3}{\partial x_3^2} \right). \tag{15.11}$$

The above equations are valid within the limits $-\infty < x_1, x_2 < +\infty$ and $0 \leq x_3 \leq D$. In addition, the flow in the x_1 and x_2 directions is assumed to be fully developed, i.e. $\partial U_j / \partial x_1 = 0$ and $\partial U_j / \partial x_2 = 0$, so that one can deduce from the continuity equation:

$$\frac{\partial U_3}{\partial x_3} = 0 \quad \leadsto \quad U_3 = \text{constant}. \tag{15.12}$$

Because $U_3 = 0$ for $x_3 = 0$ and $x_3 = D$, $U_3 = 0$ holds in the entire flow region between the two channel walls.

Due to the existence of fully developed flow conditions in the x_1 and x_2 directions, the following differential equations hold:

$$\frac{\partial P}{\partial x_1} = \mu \frac{\partial^2 U_1}{\partial x_3^2}; \quad \frac{\partial P}{\partial x_2} = \mu \frac{\partial^2 U_2}{\partial x_3^2}; \quad \frac{\partial P}{\partial x_3} = 0. \tag{15.13}$$

From these equations, one obtains for U_1:

$$U_1 = \frac{1}{\mu} \left(\frac{\partial P}{\partial x_1} \right) \frac{x_3^2}{2} + C_1 x_3 + C_2 \tag{15.14}$$

and for U_2

$$U_2 = \frac{1}{\mu} \left(\frac{\partial P}{\partial x_2} \right) \frac{x_3^2}{2} + C_3 x_3 + C_4. \tag{15.15}$$

The integration constants C_1 and C_2 can be determined from the boundary conditions:

$$U_1 = 0 \text{ for } x_3 = 0 \text{ and } x_3 = D \quad \leadsto \quad C_2 = 0 \text{ and } C_1 = -\frac{D}{2\mu} \frac{\partial P}{\partial x_1}.$$

Thus one can derive for U_1:

$$U_1 = -\frac{1}{2\mu} \left(\frac{\partial P}{\partial x_1} \right) x_3 (D - x_3) \tag{15.16}$$

and likewise U_2 can be determined as:

$$U_2 = -\frac{1}{2\mu}\left(\frac{\partial P}{\partial x_2}\right)x_3(D-x_3). \tag{15.17}$$

The equations for U_1 and U_2 show that the velocities differ only due to the pressure gradients imposed in the x_1 and x_2 directions.

With the aid of the above solutions for U_1 and U_2, some further interesting considerations can be carried out. The result is that the cross-sectional mean velocities can be determined as follows, according to (15.5):

$$\tilde{U}_1(x_1,x_2) = -\frac{D^2}{12\mu}\left(\frac{\partial P}{\partial x_1}\right) \qquad \tilde{U}_2 = -\frac{D^2}{12\mu}\left(\frac{\partial P}{\partial x_2}\right), \tag{15.18}$$

where $\tilde{U}_1$ and $\tilde{U}_2$ are the area-averaged velocities in the x_1 and x_2 directions. On introducing the potential $\phi(x_1,x_2)$, driving the mean flow field to an extent that $\phi(x_1,x_2) = -\dfrac{D^2}{12\mu}P(x_1,x_2)$ holds, then the following relationship can be stated:

$$\tilde{U}_1 = \frac{\partial \phi}{\partial x_1} \quad \text{and} \quad \tilde{U}_2 = \frac{\partial \phi}{\partial x_2}. \tag{15.19}$$

These relationships between the components of the mean velocity field and the potential ϕ, express for the area-averaged fluid velocity, the driving force to be the differentials of ϕ. This essentially suggests that the flow can be regarded as having a block profile and is formally running, like the vorticity-free flow of an ideal fluid (potential flow).

15.3 Plane Lubrication Films

Already daily experience shows that a fluid film between two plates has positive properties, as far as the sliding of one plate on another is concerned. This provides insight into the film action, suggesting that the forces acting on two solid plates, that are exposed to a gliding process, can be reduced by lubrication films. One further notices that a fluid film placed between two plates is able to absorb considerable forces. All these considerations make it clear why fluid films can be used in so-called slide bearings in mechanical systems, in order to make sliding and also rotating machine elements work in practice. In order to understand the principal function of lubrication films and also their properties, the flow in a very thin film which develops between two sliding plates is investigated. Such a film which develops below a plate, having a length l and leading to film thicknesses h_1 and h_2 at the beginning and end of the plate, respectively, is plotted in Fig. 15.2. Due to $h_2 \neq h_1$, an angle of inclination α develops by which the upper plate is inclined with respect to the lower plate. Thus, the film thickness $h(x_1)$ is determined by the film and its motion. Moreover, in the treatment of the film motion, the lower plate moves at a velocity U_P relative to the upper plate.

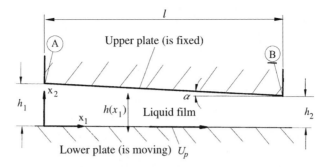

Fig. 15.2 Considerations of the fluid flow between two plates (basic flow of tripology). U_P = velocity of the moving lower plate; α = inclination angle between the plates; $R(x_1)$ = variation of plate distance

For the flow between the inclined plates, induced by the motion of the lower plate in Fig. 15.2, (15.8)–(15.11) hold the Stokes' equations in the following form, because $U_3 = 0$:

$$0 = \frac{\partial U_1}{\partial x_1} + \frac{\partial U_2}{\partial x_2} \tag{15.20}$$

$$\frac{\partial P}{\partial x_1} = \mu \left(\frac{\partial^2 U_1}{\partial x_1^2} + \frac{\partial^2 U_1}{\partial x_2^2} \right) \tag{15.21}$$

$$\frac{\partial P}{\partial x_2} = \mu \left(\frac{\partial^2 U_2}{\partial x_1^2} + \frac{\partial^2 U_2}{\partial x_2^2} \right). \tag{15.22}$$

Orders of magnitude considerations of the terms in (15.20) yield:

$$\left|\frac{\partial U_1}{\partial x_1}\right| \approx \frac{U_P}{l} \quad \text{and} \quad \left|\frac{\partial U_2}{\partial x_2}\right| = \frac{U_2}{h} = \left|\frac{\partial U_1}{\partial x_1}\right| = \frac{U_P}{\ell} \quad U_2 \approx U_P \left(\frac{h}{l}\right). \tag{15.23}$$

Because $(h/l) \ll 1$, $U_2 \ll U_P$ (thin lubrication film), the order of U_2 can be computed. Similar considerations for the terms in (15.21) and (15.22) yield:

$$\frac{\partial^2 U_1}{\partial x_1^2} = \text{term 1} \approx \frac{U_P}{l^2}; \quad \frac{\partial^2 U_1}{\partial x_2^2} = \text{term 2} \approx \frac{U_P}{h^2} \tag{15.24}$$

$$\frac{\partial^2 U_2}{\partial x_1^2} = \text{term 3} \approx \frac{U_P h}{l^3}; \quad \frac{\partial^2 U_2}{\partial x_2^2} = \text{term 4} \approx \frac{U_P}{lh}. \tag{15.25}$$

Comparisons of terms 1–4 in (15.24) and (15.25) show that term 2 dominates and that therefore the following simplified forms of (15.21) and (15.22) describe, sufficiently well, the film flow between two plates indicated in Fig. 15.2:

$$\frac{\partial P}{\partial x_1} = \mu \frac{\partial^2 U_1}{\partial x_2^2} \quad \text{and} \quad \frac{\partial P}{\partial x_2} = 0. \tag{15.26}$$

15.3 Plane Lubrication Films

For the induced film flow, the following boundary conditions hold:

$$x_2 = 0: \quad U_1(x_1, 0) = U_P \quad \text{and} \quad x_2 = h(x_1): \quad U_1(x_1, h) = 0. \quad (15.27)$$

Integrating $\partial P / \partial x_2 = 0$ yields $P = \Pi(x_1)$, so that the differential equation for the velocity field reads

$$\left(\frac{d\Pi}{dx_1}\right) = \mu \frac{\partial^2 U_1}{\partial x_2^2} \quad \rightsquigarrow \quad U_1(x_1, x_2) = \frac{1}{2\mu}\left(\frac{d\Pi}{dx_1}\right) x_2^2 + C_1 x_2 + C_2. \quad (15.28)$$

With the boundary conditions in (15.27), the two integration constants in (15.28) can be derived to read as follows:

$$C_1 = -\frac{U_P}{h(x_1)} - \frac{1}{2\mu}\left(\frac{d\Pi}{dx_1}\right) h(x_1) \quad \text{and} \quad C_2 = U_P. \quad (15.29)$$

Thus, the following equation for U_1 results:

$$U_1(x_1, x_2) = \frac{-1}{2\mu}\left(\frac{d\Pi}{dx_1}\right) x_2 [h(x_1) - x_2] + U_P \left(\frac{h(x_1) - x_2}{h(x_1)}\right). \quad (15.30)$$

The constant volume flow rate of the film flow can be computed by integration:

$$\dot{v}(x_1) = \int_0^h U_1 \, dx_2 = \frac{-1}{12\mu}\left(\frac{d\Pi}{dx_1}\right) h^3(x_1) + \frac{1}{2} U_P h(x_1). \quad (15.31)$$

Hence, the pressure gradient imposed by $\dot{v} =$ constant is computed as follows:

$$\frac{d\Pi}{dx_1} = 6\mu \left(\frac{U_P}{h^2} - \frac{2\dot{v}}{h^3}\right) = f(x_1). \quad (15.32)$$

This relationship shows that the movement of the plates and the flow with a volume flow rate $\dot{v}$ within the film lead to a pressure gradient dependent on x_1. In order to compute the pressure $P(x_1) = \Pi(x_1)$, we carry out the following integration, where $P(x_1 = 0) = P_1 = P_0$ can be set:

$$\int_{P_0}^{P(x_1)} \frac{d\Pi}{dx_1} \, dx_1 = 6\mu \int_0^{x_1} \frac{U_P}{h^2(x_1)} \, dx_1 - 12\mu \int_0^{x_1} \frac{\dot{v}}{h^3(x_1)} \, dx_1. \quad (15.33)$$

Simple geometric considerations yield:

$$h(x_1) = h_1 - \left(\frac{h_2 - h_1}{l}\right) x_1 = h_1 - x_1 \tan\alpha. \quad (15.34)$$

Hence, one can derive $dh = -\tan\alpha \, dx_1 \approx -\alpha \, dx_1$, so that the following result for the pressure distribution holds:

$$P(x_1) - P_0 = \frac{6\mu}{\alpha}\left[U_P\left(\frac{1}{h} - \frac{1}{h_1}\right) - \dot{v}\left(\frac{1}{h^2} - \frac{1}{h_1^2}\right)\right]. \tag{15.35}$$

As the ends of the slide bearing are exposed to the same fluid space, $P(x_1 = l) = P_0$ holds and one can derive from (15.35) the following:

$$\dot{v} = U_P\left(\frac{h_1 h_2}{h_1 + h_2}\right). \tag{15.36}$$

This relationship inserted in (15.32) yields for the pressure gradient in the fluid:

$$\frac{dP}{dx_1} = \frac{6\mu U_P}{h^3}(h - h_0) \quad \text{with} \quad h_0 = \frac{2h_1 h_2}{(h_1 + h_2)}. \tag{15.37}$$

From this, it readily follows that:

$$\frac{dP}{dx_1} > 0 \text{ at the location } h = h_1; \quad \frac{dP}{dx_1} < 0 \text{ at the location } h = h_2$$

$$\frac{dP}{dx_1} = 0 \text{ at the location } h = h_0,$$

i.e. the pressure distribution in the liquid film shows a maximum. For the pressure distribution itself one can write:

$$P - P_1 = \frac{\sigma\mu U_P}{\alpha}\left[\frac{(h_1 - h)(h - h_2)}{h^2(h_1 + h_2)}\right]. \tag{15.38}$$

With the above relationships (15.36)–(15.38), the volume flow rate and the pressure distribution in the film can be computed for the case that the relative velocity of the moving plates is given and the entire bearing geometry can be considered to be known. The employment of (15.38) yields $P > P_1 > P_2$, i.e. the film plotted in Fig. 15.2 produces an overpressure due to the indicated relative movement of the plates. The film is thus able to absorb forces that act on the upper plate. A pressure maximum develops within the lubrication film at which the value of the pressure can be computed to be:

$$P_{\max} - P_1 = \mu l \frac{U_P}{h_1^2}. \tag{15.39}$$

In this equation, the approximation $(h_1 - h_2)/h_1 \approx 1$ has been introduced. The resulting pressure force on the plate surfaces can be computed as follows:

$$K_P = \int_0^l P \, dx = \frac{\sigma\mu U_P}{\alpha^2}\left[\ln\frac{h_1}{h_2} - 2\frac{(h_1 - h_2)}{(h_1 + h_2)}\right]. \tag{15.40}$$

15.3 Plane Lubrication Films

The tangential force on the lower plate can be computed to yield:

$$(K_\tau)_\text{low} = \int_0^l \mu \left(\frac{\partial U}{\partial y} \right)_{y=0} \mathrm{d}x = \frac{2\mu U_P}{\alpha} \left[3\frac{(h_1 - h_2)}{(h_1 + h_2)} - 2\ln\left(\frac{h_1}{h_2}\right) \right] \quad (15.41)$$

and for the upper plate:

$$(K_\tau)_\text{up} = -\int_0^l \mu \left(\frac{\partial U}{\partial y} \right)_{y=h} = \frac{2\mu U_P}{\alpha} \left[3\frac{(h_1 - h_2)}{(h_1 + h_2)} - \ln\left(\frac{h_1}{h_2}\right) \right]. \quad (15.42)$$

The tangential forces acting on the two surfaces of the plates are not equal because the flow between the plates is partly dragged along in the film.

The flow lines developing in the film flow are shown in Fig. 15.3, together with the plotted local velocity profiles. The profiles result from (15.30), including (15.38) for the computations. The following was introduced:

$$\frac{\rho U_P^2 / l}{\mu U_P / h^2} = \frac{\rho U_P l}{\mu} \left(\frac{h^2}{l^2} \right) \ll 1. \quad (15.43)$$

The above considerations were carried out for plane flows, as the intention of the derivations was to give an introduction into the theory of tribological flows. Flows in slide bearings usually have to be treated as rotating cylinder

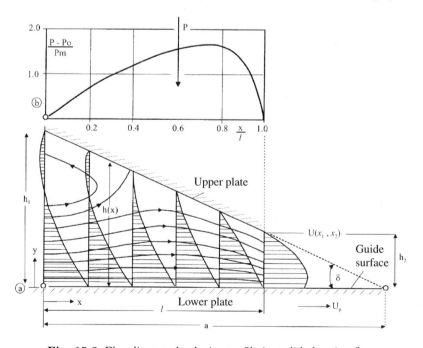

Fig. 15.3 Flow lines and velocity profile in a slide-bearing flow

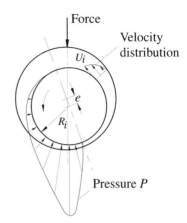

Fig. 15.4 Pressure distribution in rotating slide bearings

flows with an eccentric bearing positioning of the inner cylinder, relative to the outer one. This bearing case is plotted in Fig. 15.4, which shows the developing pressure distribution due to rotation. It is characteristic for this kind of flow that the direction of the pressure maximum is not situated in the direction of the load. Details of this flow are treated in the following chapter. They represent the basis for understanding the fluid-mechanical functioning of rotating slide bearings.

15.4 Theory of Lubrication in Roller Bearings

A roller bearing comprises a nonrotating bearing and a pivot rotating at angular velocity ω. The practically employed double-cylinder arrangement is shown in Fig. 15.5, with the following approximations being valid:

$$R_1 + h \approx R_2 + e \cos \varphi,$$
$$R_2 = R_1 + e + \delta,$$
$$h \approx \delta + e(1 + \cos \varphi).$$

The solution of the equations for the fluid motion, which is important for the lubrication flow in slide bearings, i.e. the rotating fluid motion which develops between pivot and bearing, probably represents the technically most important application of Stokes' equations. Due to the resultant fluid movement in a thin film, well-known bearing friction laws result which differ drastically to those for dry friction laws. To demonstrate this, the motion of the inner cylinder (pivot) having a radius $r = R_1$ is considered, which rotates at angular velocity ω, while the outer cylinder (bearing), having the radius R_2, is at rest. The internal rotating cylinder thus has a circumferential velocity $U = R_1\omega$. From the representation in Fig. 15.5, one can establish that the position of the bearing can be given as $r = R_1 + h$. Here $h = \delta + e(1 + \cos \varphi)$.

15.4 Theory of Lubrication in Roller Bearings

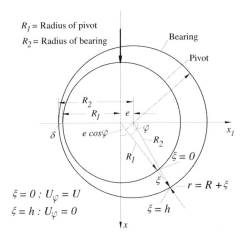

Fig. 15.5 Bearing-pivot arrangement for considered roller bearing

These relationships hold with sufficient precision for the considerations carried out. In this section, δ is the film thickness of the lubrication fluid existing at $\varphi = 180°$ and e the eccentricity of the center point of the bearing with regard to the position of the center point of the pivot.

On introducing into the considerations that the film thickness δ is very small with respect to the radius of the pivot, i.e. $\delta/R_1 \ll 1$ holds, then one can show that $\partial U_\varphi / \partial r \gg 1/r \partial U_\varphi / \partial \varphi$ holds. Order of magnitude considerations of the remaining terms in the momentum equation (5.115) demonstrate that the following differential equation can be employed for the treatment of film flows in roller bearings:

$$\frac{1}{r} \frac{\partial P}{\partial \varphi} = \mu \frac{\partial^2 U_\varphi}{\partial r^2}. \qquad (15.44)$$

As the r values appearing in the flow field of the film do not vary strongly, the following approximation holds, because $r \approx R$:

$$\frac{1}{R} \frac{\partial P}{\partial \varphi} = \mu \frac{\partial^2 U_\varphi}{\partial r^2}. \qquad (15.45)$$

Due to this simplification, it is possible to treat the problem of roller bearings in a way which comes close to that of the plane slide bearing.

Concerning the integration of the differential (15.45), one can proceed as follows. The introduction of the variable ξ yields

$$r = R_1 + \xi \quad \leadsto \quad dr = d\xi. \qquad (15.46)$$

Integration of (15.45) therefore yields:

$$U_\varphi = \frac{1}{2R_1 \mu} \left(\frac{dP}{d\varphi} \right) \xi^2 + C_1 \xi + C_2. \qquad (15.47)$$

With the boundary conditions:

$$\xi = 0 : \quad U_\varphi = U \quad \text{and} \quad \xi = h : \quad U_\varphi = 0 \tag{15.48}$$

C_1 and C_2 can be determined, so that the following relationship holds:

$$U_\varphi = \frac{1}{2R_1\mu}\left(\frac{\partial P}{\partial \varphi}\right)\xi(\xi - h) + \frac{U(h-\xi)}{h}. \tag{15.49}$$

For the volume flow rate, one can derive:

$$\dot{v} = \int_R^{R+h} U_\varphi \, dr = \int_0^h U_\varphi \, d\xi = -\frac{1}{2R_1\mu}\left(\frac{\partial P}{\partial \varphi}\right)\frac{h^3}{6} + \frac{Uh}{2}. \tag{15.50}$$

Due to the introduction of a mean film thickness h_0 with:

$$\dot{v} = -\frac{1}{2R_1\mu}\left(\frac{\partial P}{\partial \varphi}\right)\frac{h^3}{6} + \frac{Uh}{2} = \frac{Uh_0}{2} \tag{15.51}$$

and thus from (15.45), the following pressure gradient can be computed:

$$\frac{\partial P}{\partial \varphi} = \frac{6R_1\mu U(h - h_0)}{h^3} \tag{15.52}$$

and by integration the pressure distribution results:

$$P(\varphi) = P(0) + 6R_1\mu U\left[\int_0^\varphi \frac{d\varphi}{h^2} - h_0\int_0^\varphi \frac{d\varphi}{h^3}\right]. \tag{15.53}$$

With $h \approx \delta + e(1 + \cos\varphi)$ from Fig. 15.5, the following results:

$$h(\varphi) = e\left(\frac{\delta}{e} + 1 + \cos\varphi\right) = e(\alpha + \cos\varphi). \tag{15.54}$$

Because $P(2\pi) = P(0) = P_0$, according to (15.53) it must hold that:

$$\int_0^{2\pi}\frac{d\varphi}{h^2} = h_0\int_0^{2\pi}\frac{d\varphi}{h^3} \longrightarrow h_0 = \frac{\int_0^{2\pi}\frac{d\varphi}{h^2}}{\int_0^{2\pi}\frac{d\varphi}{f^3}}. \tag{15.55}$$

Considering:

$$\int_0^{2\pi}\frac{d\varphi}{(\alpha + \cos\varphi)^h} = 2\int_0^\pi \frac{d\varphi}{(\alpha + \cos\varphi)^h} \tag{15.56}$$

15.4 Theory of Lubrication in Roller Bearings

and because:

$$J_1 = \int_0^\varphi \frac{d\varphi}{(\alpha + \cos\varphi)} = \frac{2}{\sqrt{\alpha^2 - 1}} \arctan\left(\sqrt{\frac{\alpha-1}{\alpha+1}} \tan\frac{\varphi}{2}\right) \quad (15.57)$$

one obtains for $J_1(\varphi = \pi)$, $J_2(\varphi = \pi)$, etc.

$$J_1 = \frac{2\pi}{\sqrt{(\alpha^2-1)}}; \quad J_2 = -\frac{dJ_1}{d\alpha} = \frac{2\pi\alpha}{\sqrt{(\alpha^2-1)^3}}; \quad J_3 = -\frac{1}{2}\frac{dJ_2}{d\alpha} = \frac{\pi(2\alpha^2+1)}{\sqrt{(\alpha^2-1)^5}}. \quad (15.58)$$

Thus, from (15.51) one can compute

$$h_0 = e\frac{J_2}{J_3} = \frac{2e\alpha(\alpha^2-1)}{(2\alpha^2+1)} \text{ with } \alpha = \left(\frac{\delta}{e}+1\right). \quad (15.59)$$

For the pressure distribution between pivot and bearing (Fig. 15.6), one can compute

$$P(\varphi) - P_0 = 6R_1\mu U \left[\int_0^\varphi \frac{d\varphi}{(\alpha+\cos\varphi)^2} = \frac{2e\alpha(\alpha^2-1)}{(2\alpha^2+1)} \int_0^\varphi \frac{d\varphi}{(\alpha+\cos\varphi)^3}\right]. \quad (15.60)$$

From (15.52), one can see that $\frac{\partial P}{\partial \varphi} = 0$ for $h = h_0$.

On defining with the angle φ_0 the angular position, where the pressure shows an extremum, one obtains from (15.54)

$$e(\alpha + \cos\varphi_0) = \frac{2e\alpha(\alpha^2-1)}{(2\alpha^2+1)} \quad (15.61)$$

$$\cos\varphi_0 = -\frac{3\alpha}{2\alpha^2+1}. \quad (15.62)$$

With this relationship, it has been shown that the points of highest and lowest pressure are positioned on that half of the pivot circumference that comprises the narrowest film.

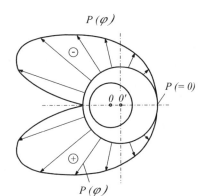

Fig. 15.6 Distribution of the normal pressure on the pivot of a roller bearing

The derived pressure distribution can be used to compute the share of the pivot force resulting from the pressure actions only.

By integration, one obtains:

$$K_P = \frac{12\pi\mu R_1^2 U}{e^2 (2\alpha^2 + 1) \sqrt{\alpha^2 - 1}}. \tag{15.63}$$

For determining the friction force, the momentum loss due to the motion of the pivot can be stated as follows:

$$\tau_{r\varphi} = -\mu \left(\frac{1}{r} \frac{\partial U_r}{\partial \varphi} + \frac{\partial U_\varphi}{\partial r} - \frac{\varphi}{r} \right) \tag{15.64}$$

and because $U_\varphi = U$ and $U_r = 0$ hold on the pivot surface $r = R_1$, one can deduce for $\tau_{r\varphi}$:

$$\tau_{r\varphi} = -\mu \left[\left(\frac{\partial U_\varphi}{\partial \xi} \right)_{\xi=0} - \frac{U}{R_1} \right] = \frac{h}{2R_1} \frac{\partial P}{\partial \varphi} + \frac{U\mu}{h} + \frac{U\mu}{R_1}. \tag{15.65}$$

With $(\partial P/\partial \varphi) = 6R\mu U(h - h_0)/h^3$ and h_0 inserted from (15.60), the following results for the pressure gradient:

$$\frac{\partial P}{\partial \varphi} = \frac{6R_1\mu U}{e^2} \left[\frac{1}{(\alpha + \cos\varphi)^2} - \frac{2\alpha(\alpha^2 - 1)}{(2\alpha^2 + 1)} \frac{1}{(\alpha + \cos\varphi)^3} \right] \tag{15.66}$$

and for $\tau_{r\varphi}$

$$\tau_{r\varphi} = \mu \frac{U(4h - 3h_0)}{h^2} = \frac{2\mu U}{e} \left[\frac{2}{(\alpha + \cos\varphi)} - \frac{3\alpha(\alpha^2 - 1)}{(2\alpha^2 + 1)} \frac{1}{(\alpha + \cos\varphi)} \right]$$

so that the torque can be computed as:

$$M = \int_0^{2\pi} \tau_{r\varphi} R_1^2 \, d\varphi = \frac{2\mu U R_1^2}{e} \frac{2\pi(\alpha^2 + 2)}{(2\alpha^2 + 1)\sqrt{\alpha^2 - 1}}. \tag{15.67}$$

In practice, the above equations can be employed as follows:

- When $K_P, \mu, U, R, e, \delta$ are known, the extremity of the pivot position and also α can be determined by means of (15.63).

With (15.67), a friction factor can be introduced as

$$f = \frac{M}{K_P R_1} = \frac{\delta}{R_1} \frac{(\alpha^2 + 2)}{3\alpha}. \tag{15.68}$$

As $\delta \approx e$ holds in practice, one can deduce

$$f_0 \approx \frac{\delta}{R_1} \quad \leadsto \quad M \approx f_0 K_P R_1. \tag{15.69}$$

15.5 The Slow Rotation of a Sphere

With this value, friction moments can be computed with sufficient accuracy for application in practice. For large values of α, the following holds for M:

$$M = \frac{2\pi\mu U R_1^2}{\delta}. \tag{15.70}$$

The above derivations have shown, in an exemplary way, that with the aid of Stokes' equations it is possible to treat technically relevant fluid flows in such a manner that not only technically interesting insights result from the derivations, but also quantitative results can be obtained.

15.5 The Slow Rotation of a Sphere

Considering the flow in a viscous fluid, which is caused by the slow rotation of a small sphere, its Reynolds number can be determined as follows:

$$Re = \frac{U_c R}{\nu} = \frac{\omega R^2}{\nu}, \tag{15.71}$$

where ω is the angular velocity of the rotation. Applying to the equations of motion in spherical coordinates the conditions that exist due to the motion of a sphere, as plotted in Fig. 15.7, then for

$$U_\phi = U_\phi(r, \theta) \tag{15.72}$$

the following differential equation holds, which represents an equation of second order. This differential equation can be derived from the general equations of motion in spherical coordinates:

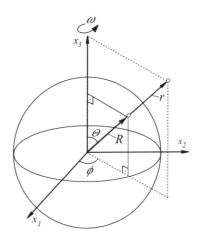

Fig. 15.7 Flow in a viscous fluid due to the rotation of a sphere around the x_3 axis

$$0 = \left[\frac{1}{r^2}\frac{\partial}{\partial r}\left(r^2\frac{\partial U_\phi}{\partial r}\right) + \frac{1}{r^2\sin\theta}\frac{\partial}{\partial\theta}\left(\sin\theta\frac{\partial U_\phi}{\partial\theta}\right) + \frac{1}{r^2\sin^2\theta}\frac{\partial^2 U_\phi}{\partial\phi^2}\right.$$
$$\left. - \frac{U_\phi}{r^2\sin^2\theta} + \frac{2}{r^2\sin\theta}\frac{\partial U_r}{\partial\phi} + \frac{2\cos\theta}{r^2\sin^2\theta}\frac{\partial U_\theta}{\partial\phi}\right]. \tag{15.73}$$

Because of the chosen rotational symmetry, the following terms are zero:

$$\frac{2}{r^2\sin\theta}\frac{\partial U_r}{\partial\phi}, \qquad \frac{2\cos\theta}{r^2\sin^2\theta}\frac{\partial U_\theta}{\partial\phi}, \qquad \text{and} \qquad \frac{1}{r^2\sin^2\theta}\frac{\partial^2 U_\phi}{\partial\phi^2}. \tag{15.74}$$

Thus, (15.73) obtains the following form:

$$0 = \frac{\partial^2 U_\phi}{\partial r^2} + \frac{2}{r}\frac{\partial U_\phi}{\partial r} + \frac{1}{r^2}\frac{\partial^2 U_\phi}{\partial\theta^2} + \frac{1}{r^2}\cot\theta\frac{\partial U_\phi}{\partial\theta} - \frac{U_\phi}{r^2\sin^2\theta}. \tag{15.75}$$

As boundary condition one can write $U_\phi(R,\theta) = R\omega\sin\theta$.

The aim of the solution of the flow problem is to find a function $U_\phi(r,\theta)$, which fulfils the differential equation (15.75), and at the same time is able to fulfil the stated boundary conditions. Such a function proves to be:

$$U_\phi = A(r)\sin\theta. \tag{15.76}$$

Insertion of (15.76) into (15.75) yields:

$$\frac{d^2 A}{dr^2}\sin\theta + \frac{2}{r}\frac{dA}{dr}\sin\theta - \frac{A}{r^2}\sin\theta + \frac{A\cos^2\theta}{r^2\sin\theta} - \frac{A}{r^2}\frac{1}{\sin\theta} = 0 \tag{15.77}$$

and as the differential equation for $A(r)$ to be solved:

$$\frac{d^2 A}{dr^2} + \frac{2}{r}\frac{dA}{dr} - \frac{2A}{r^2} = 0. \tag{15.78}$$

Integration of this differential equation yields

$$A(r) = C_1 r + \frac{C_2}{r^2}. \tag{15.79}$$

Because $U_\phi(r\to\infty,\theta) = 0$, it follows that $C_1 = 0$.

From $U_\phi(R,\theta) = R\omega\sin\theta = \frac{C_2}{R^2}\sin\theta$ it follows that $C_2 = \omega R^3$, and hence one obtains as a solution for the velocity field of the fluid of the rotating sphere:

$$U_\phi = \frac{\omega R^3}{r^2}\sin\theta. \tag{15.80}$$

In order to maintain this flow, imposed on the fluid by the rotating sphere, a moment has to be imposed continuously, which can be computed as derived

below. From
$$\tau_{r,\phi} = -\mu \left[\frac{1}{r \sin\theta} \frac{\partial U_r}{\partial \phi} + r \frac{\partial}{\partial r}\left(\frac{U_\phi}{r}\right) \right] \tag{15.81}$$

it follows for the momentum release to the fluid by the rotating sphere:
$$\tau_{R,\phi} = -\mu \left[\frac{\partial U_\phi}{\partial r} - \frac{U_\phi}{r} \right]_{r=R} = 3\mu\omega\sin\theta \tag{15.82}$$

and for this the moment can be computed to be:
$$M = -\int_0^\pi (3\mu\omega\sin\theta)(R\sin\theta)(2\pi R^2 \sin\theta)\, d\theta \tag{15.83}$$

$$M = -6\pi\mu\omega R^3 \int_0^\pi \sin^3\theta\, d\theta = 8\mu\pi R^3\omega. \tag{15.84}$$

Analogous to the above solution concerning the problem of the fluid motion around a rotating sphere, the creeping fluid motion between two concentrically positioned rotating spheres, with radii R_2 and R_1 and angular velocities ω_2 and ω_1, can be treated. The rotation is to take place again around the x_3 axis shown in Fig. 15.7. One obtains once more:

$$U_\phi = A(r)\sin\theta \quad \text{with} \quad A(r) = C_1 r + \frac{C_2}{r^2}. \tag{15.85}$$

With the boundary conditions $A(R_2) = \omega_2 R_2$ and $A(R_1) = \omega_1 R_1$, one obtains as a solution for the induced velocity field:

$$U_\phi(r,\theta) = \frac{\sin\theta}{r^2(R_2^3 - R_1^3)}\left[\omega_2 R_2^3(r^3 - R_1^3) - \omega_1 R_1^3(r^3 - R_2^3)\right] \tag{15.86}$$

and for the torque one can compute:
$$M = -8\pi\mu(\omega_2 - \omega_1)\frac{R_2^3 R_1^3}{(R_1^3 - R_2^3)}. \tag{15.87}$$

As far as the above solution is concerned, it was assumed that the eventually occurring disturbances of the flow are attenuated by viscous effects, i.e. that the computed solution is thus stable. On the basis of the treatment of creeping flows, i.e. smallest Reynolds numbers, this assumption is justified, so that the above analytically obtained results are very well obtained in experiments also.

15.6 The Slow Translatory Motion of a Sphere

The flow around a sphere is considered in this section, as induced by the straight and uniform motion of the considered sphere in a viscous fluid. As far as the equations to be solved are concerned, the problem is equivalent to the flow around a stationary sphere in a viscous fluid. The specific fluid motion is characterized by a sphere of radius R, the fluid velocity U_∞, the density ρ and the dynamic viscosity μ. From these, the Reynolds number of the flow problem is computed with $\nu = \mu/\rho$:

$$Re = \frac{RU_\infty}{\nu} < 1. \tag{15.88}$$

Stokes (1851) was the first to solve the problem of the translatory motion of a sphere in a viscous fluid by considering only the pressure and viscosity terms in the Navier–Stokes equations and neglecting all other terms in the equations of motion. The same procedure is shown below. In this context, a stationary sphere, located in the center of a Cartesian coordinate system, is assumed in the subsequent considerations. For this flow case the following boundary conditions hold:

$$U_1 = U_2 = U_3 = 0 \quad \text{for} \quad r = R$$

with $r = \sqrt{x_1^2 + x_2^2 + x_3^2}$, as shown in Fig. 15.8. This figure shows the sphere whose center is at the origin of a Cartesian coordinate system. Relative to this system, the coordinates of the spheres r, ϕ, θ are also indicated, which are subsequently employed for the treatment of the flow around the considered sphere. The surface of the sphere is located at $r = R$. Because of this assumption of the location of the flow boundary, it is advantageous to take the basic equations of fluid mechanics, for the solution of the flow around a sphere, in spherical coordinates. In this way, the boundary conditions, imposed by the presence of the sphere, can be included much more easily into the solution of the flow problem, as if the treatment of the flow in Cartesian coordinates was sought.

The following considerations are based on the diagram in Fig. 15.8. At infinity, i.e. for $r \to \pm\infty$, the following values for the velocity components will establish themselves:

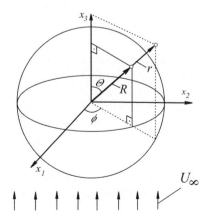

Fig. 15.8 Flow around a sphere with Cartesian and spherical coordinates

15.6 The Slow Translatory Motion of a Sphere

$$U_1 = 0, \quad U_2 = 0, \quad U_3 = U_\infty \quad \text{for} \quad r \to \infty. \tag{15.89}$$

As already mentioned, for the solution of the basic equations for flows around spheres, it seems obvious to employ the equations in spherical coordinates, in order to be able to include the boundary conditions easily into the solution. When considering the relevant terms for small Reynolds numbers, one obtains the equations below. They can be derived from the general form of the basic equations of fluid mechanics, using spherical coordinates. The resulting equations read as follows for Stokes flows:

$$\frac{\partial U_r}{\partial r} + \frac{1}{r}\frac{\partial U_\theta}{\partial \theta} + \frac{2U_r}{r} + \frac{U_\theta \cos\theta}{r} = 0 \tag{15.90}$$

$$\frac{\partial P}{\partial r} = \mu \left(\frac{\partial^2 U_r}{\partial r^2} + \frac{1}{r^2}\frac{\partial^2 U_r}{\partial \theta^2} + \frac{2}{r}\frac{\partial U_r}{\partial r} + \frac{\cot\theta}{r^2}\frac{\partial U_r}{\partial \theta} - \frac{2}{r^2}\frac{\partial U_\theta}{\partial \theta} \right.$$
$$\left. - \frac{2U_r}{r^2} - \frac{2\cot\theta}{r^2}U_\theta \right) \tag{15.91}$$

$$\frac{1}{r}\frac{\partial P}{\partial \theta} = \mu \left(\frac{\partial^2 U_\theta}{\partial r^2} + \frac{1}{r^2}\frac{\partial^2 U_\theta}{\partial \theta^2} + \frac{2}{r}\frac{\partial U_\theta}{\partial r} + \frac{\cot\theta}{r^2}\frac{\partial U_\theta}{\partial \theta} \right.$$
$$\left. + \frac{2}{r^2}\frac{\partial U_r}{\partial \theta} - \frac{U_\theta}{r^2 \sin^2\theta} \right). \tag{15.92}$$

In spherical coordinates, the boundary conditions can be stated as follows:

$$r = R: \quad U_r(R,\theta) = 0 \quad \text{and} \quad U_\theta(R,\theta) = 0 \tag{15.93}$$

$$r \to \infty: \quad U_r \to U_\infty \cos\theta \quad \text{and} \quad U_\theta \to -U_\infty \sin\theta. \tag{15.94}$$

Again, the boundary conditions suggest solving the above differential equations with the following solution ansatzes for U_r and U_θ:

$$U_r(r,\theta) = B(r)\cos\theta \qquad U_\theta(r,\theta) = -A(r)\sin\theta \tag{15.95}$$

and for the pressure:

$$P(r,\theta) = \mu C(r)\cos\theta. \tag{15.96}$$

On inserting these ansatz functions (15.95) and (15.96) into the above differential equations (15.90)–(15.92), one obtains:

$$\frac{\mathrm{d}B}{\mathrm{d}r} + \frac{2(B-A)}{r} = 0 \tag{15.97}$$

$$\frac{\mathrm{d}C}{\mathrm{d}r} = \frac{\mathrm{d}^2 B}{\mathrm{d}r^2} + \frac{2}{r}\frac{\mathrm{d}B}{\mathrm{d}r} - \frac{4(B-A)}{r^2} \tag{15.98}$$

$$\frac{C}{r} = \frac{d^2 A}{dr^2} + \frac{2}{r}\frac{dA}{dr} + \frac{2(B-A)}{r^2}. \tag{15.99}$$

From the boundary conditions for flows around spheres, the following boundary conditions result for the functions A, B and C:

$$A(R) = 0, \quad B(R) = 0, \quad A(\infty) = U_\infty, \quad \text{and} \quad B(\infty) = U_\infty. \tag{15.100}$$

The solution steps for the differential equations (15.97)–(15.99) can be stated as follows
From (15.97) it follows that:

$$A = \frac{1}{2}r\frac{dB}{dr} + B. \tag{15.101}$$

Inserting A into (15.99) results in:

$$C = \frac{1}{2}r^2\frac{d^3 B}{dr^3} + 3r\frac{d^2 B}{dr^2} + 2\frac{dB}{dr}. \tag{15.102}$$

This differential equation can be differentiated with respect to r and the result can be inserted into (15.98) to yield:

$$r^3\frac{d^4 B}{dr^4} + 8r^2\frac{d^3 B}{dr^3} + 8r\frac{d^2 B}{dr^2} - 8\frac{dB}{dr} = 0. \tag{15.103}$$

The resulting Euler differential equation can be solved by integration with particular solutions being sought of the form $B = cr^k$. If one inserts $B = cr^k$ into the above differential equations, an equation of fourth order for k results:

$$k(k-1)(k-2)(k-3) + 8k(k-1)(k-2) + 8k(k-1) - 8k = 0$$

or, written in a somewhat simplified form:

$$k(k-2)(k+3)(k+1) = 0 \tag{15.104}$$

so that the following k-values result as solutions:

$$k = 0, \quad k = 2, \quad k = -3 \quad \text{and} \quad k = -1. \tag{15.105}$$

Thus, for $B(r)$ the following general solution can be given:

$$B(r) = \frac{C_1}{r^3} + \frac{C_2}{r} + C_3 + C_4 r^2. \tag{15.106}$$

From the equations for $A(r)$ (15.101) and $C(r)$ (15.102) one obtains:

$$A(r) = -\frac{C_1}{2r^3} + \frac{C_2}{2r} + C_3 + 2C_4 r^2 \tag{15.107}$$

$$C(r) = \frac{C_2}{r^2} + 10 C_4 r. \tag{15.108}$$

15.6 The Slow Translatory Motion of a Sphere

Inserting for $A(r)$, $B(r)$ and $C(r)$ the boundary conditions (15.100), the integration constants C_1, C_2, C_3 and C_4 can be determined:

$$C_1 = \frac{1}{2}U_\infty R^3; \quad C_2 = -\frac{3}{2}U_\infty R; \quad C_3 = U_\infty; \quad C_4 = 0. \tag{15.109}$$

Thus, for $U_r(r,\theta)$, $U_\theta(r,\theta)$ and $P(r,\theta)$ the following solutions result:

$$U_r(r,\theta) = U_\infty \cos\theta \left(1 - \frac{3}{2}\frac{R}{r} + \frac{1}{2}\frac{R^3}{r^3}\right) \tag{15.110}$$

$$U_\theta(r,\theta) = -U_\infty \sin\theta \left(1 - \frac{3}{4}\frac{R}{r} - \frac{1}{4}\frac{R^3}{r^3}\right) \tag{15.111}$$

$$P(r,\theta) = -\frac{3}{2}\mu \frac{U_\infty R}{r^2} \cos\theta. \tag{15.112}$$

With these relationships, the solution for the velocity and pressure fields for the flow around a sphere are available. Although obtained by solving a set of simplified differential equations, which was obtained as a reduced set from the Navier–Stokes equations by neglecting the acceleration terms, an important set of results emerged for U_r, U_θ and P. However, the solutions obtained hold only for $Re < 1$.

When looking at the different momentum transport terms τ_{rr} and $\tau_{r\theta}$ of the flow problem discussed here, one obtains for $\rho = $ constant:

$$\tau_{rr} = -\mu\left(2\frac{\partial U_r}{\partial r}\right) \quad \text{for } r = R \text{ is} \quad \left(\frac{\partial U_r}{\partial r}\right)_{r=R} = 0 \tag{15.113}$$

and $\tau_{r,\theta} = -\mu\left(\frac{1}{r}\frac{\partial U_r}{\partial \theta} + \frac{\partial U_\theta}{\partial r} - \frac{U_\theta}{r}\right)$. From this, one obtains for $r = R$, because $U_r = U_\theta = 0$ and therefore also $\frac{\partial U_r}{\partial \theta} = 0$ and $\frac{\partial U_\theta}{\partial \theta} = 0$, at the surface of the sphere the following expression for the pressure:

$$P = \frac{3}{2}\frac{\mu U_\infty}{R}\cos\theta \quad \text{(acts on each point vertically to the surface of the sphere).} \tag{15.114}$$

In addition,

$$\tau_{r,\theta} = -\mu\frac{\partial U_\theta}{\partial r} = -\frac{3\mu U_\infty}{2R}\sin\theta \quad \text{(acts on each point tangentially to the surface of the sphere).} \tag{15.115}$$

The drag force F_D can thus be computed:

$$F_D = \iint_F (P\cos\theta - \tau_{r,\theta}\sin\theta)\,dF \tag{15.116}$$

$$F_D = -\int_0^\pi \left(\frac{3}{2}\frac{\mu U_\infty}{R}\cos^2\theta + \frac{3\mu U_\infty}{2R}\sin^2\theta\right)(2\pi R^2 \sin\theta)\,d\theta \tag{15.117}$$

or rewritten and integrated:

$$F_D = -\int_0^\pi 3\pi\mu U_\infty R(\sin\theta)d\theta = -3\pi\mu U_\infty R(-\cos\theta)\Big]_0^\pi \quad (15.118)$$

$$\boxed{F = -6\pi\mu U_\infty R}.$$

The integration over the pressure acting on the surface of the sphere, and over the momentum loss to the wall, yields the Stokes drag force W. Here it is of interest that the share of force coming from the pressure:

$$F_D = -\int_0^\pi P\cos\theta 2\pi R^2 \sin\theta\, d\theta = -3\pi\mu U_\infty R \int_0^\pi \cos^2\theta \sin\theta\, d\theta \quad (15.119)$$

with $F_D = -2\pi\mu U_\infty R$ amounts only to one-third of the total drag, i.e. two-thirds of the drag force results thus from the molecular momentum input to the surface of the sphere. This underlines the importance that must be attached to the viscosity terms in the Navier–Stokes equations for the solution of practical flow problems at small Reynolds numbers.

In order to determine now the velocity components in Cartesian coordinates, i.e. U_1, U_2 and U_3, the following equations for the transformation of coordinates are employed:

$$x_1 = r\cos\theta \rightsquigarrow U_1 = U_r\cos\theta - U_\theta\sin\theta \quad (15.120)$$
$$x_2 = r\sin\theta\cos\phi \rightsquigarrow U_2 = U_r\sin\theta\cos\phi + U_\theta\cos\theta\cos\phi - U_\phi\sin\phi \quad (15.121)$$
$$x_3 = r\sin\theta\sin\phi \rightsquigarrow U_3 = U_r\sin\theta\sin\phi + U_\theta\cos\theta\sin\phi + U_\phi\cos\phi. \quad (15.122)$$

With these equations one obtains:

$$U_1 = U_\infty\left(1 - \frac{3}{4}\frac{R}{r} - \frac{1}{4}\frac{R^3}{r^3}\right) - \frac{3}{4}\frac{U_\infty R x_1^2}{R^3}\left(1 - \frac{R^2}{r^2}\right) \quad (15.123)$$

$$U_2 = -\frac{3}{4}\frac{U_\infty R x_1 x_2}{R^3}\left(1 - \frac{R^2}{r^2}\right) \quad (15.124)$$

$$U_3 = -\frac{3}{4}\frac{U_\infty R x_1 x_3}{R^3}\left(1 - \frac{R^2}{r^2}\right). \quad (15.125)$$

For the solution of the Stokes flow around spheres, the above simplified flow equations (15.90)–(15.92) were employed. To be able to assess how large the neglected terms of the Navier–Stokes equations are in comparison with the terms considered in the solution, the acceleration:

$$\rho\left(\frac{DU_1}{Dt}\right)_{\theta=0} = \rho\left(U_r\frac{\partial U_r}{\partial r}\right)_{\theta=0} = \frac{3\rho}{2}\frac{U_\infty^2}{r^2}R\left(1 - \frac{R^2}{r^2}\right)\left(1 - \frac{3R}{2r} + \frac{R^3}{2r^3}\right) \quad (15.126)$$

15.7 The Slow Rotational Motion of a Cylinder

is compared with the pressure term:

$$\left(\frac{\partial P}{\partial r}\right)_{\theta=0} = 3\mu \frac{U_\infty R}{r^3}, \tag{15.127}$$

i.e. put in relation to one another:

$$\frac{\rho \left(\frac{DU_1}{Dt}\right)_{\theta=0}}{\left(\frac{\partial P}{\partial r}\right)_{\theta=0}} = \frac{U_\infty r}{2\nu}\left(1 - \frac{R^2}{r^2}\right)\left(1 - \frac{3R}{2r} + \frac{R^3}{2r^3}\right). \tag{15.128}$$

For large values of r, this relationship shows that the above solution should be valid only when $U_\infty r/2\nu < 1$ holds, i.e. for large values of r the requested condition for the validity of the solution is not fulfilled. As, however, for such large values for r, the terms that have been employed above, for order of magnitude considerations, become very small, it is justified to assume that the velocity and pressure fields in the immediate proximity of the sphere are not affected by influences of the introduced assumption for the validity of the obtained solution. In order to achieve the derived solution in a reliable way, it had to be assumed, however, that $(U_\infty R)/\nu \ll 1$. On the basis of such considerations, it can be assumed that the Stokes solution already does not hold any longer for the flow around a sphere for $Re \approx 1$, i.e. when $Re = (U_\infty D/\nu) \approx 1$. It is a solution for $Re < 1$.

15.7 The Slow Rotational Motion of a Cylinder

Analogous to the discussion of the slowly rotating flow around a sphere in Sect. 15.5, the rotating cylinder flow will be discussed in this section. Here, the flow which occurs in the annular clearance between two concentric rotating cylinders, with radii R_1 and R_2, will be investigated. This flow is described by the equations that are stated below and which, assuming

$$\frac{\partial U_\varphi}{\partial \varphi} = 0 \quad \text{and} \quad U_\varphi = U_\varphi(r) \tag{15.129}$$

can be derived from the general basic equations of fluid mechanics written in cylindrical coordinates:

$$\rho \frac{U_\varphi^2}{r} = \frac{\partial P}{\partial r} = \frac{dP}{dr} \tag{15.130}$$

and

$$\frac{d^2 U_\varphi}{dr^2} + \frac{d}{dr}\left(\frac{U_\varphi}{r}\right) = 0. \tag{15.131}$$

These are the differential equations for the pressure P and the flow velocity U_φ. From (15.131), one obtains by integration:

$$\frac{\mathrm{d}U_\varphi}{\mathrm{d}r} + \frac{U_\varphi}{r} = 2C_1 = \frac{1}{r}\frac{\mathrm{d}}{\mathrm{d}r}(rU_\varphi). \tag{15.132}$$

By further integration, one obtains:

$$U_\varphi = C_1 r + \frac{C_2}{r} \tag{15.133}$$

with the integration constants C_1 and C_2. These can be determined from the boundary conditions:

$$r = R_1 \quad \leadsto \quad U_\varphi = \omega_1 R_1 \quad \text{and} \quad r = R_2 \quad \leadsto \quad U_\varphi = \omega_2 R_2, \tag{15.134}$$

i.e. the following equations hold for C_1 and C_2:

$$\omega_1 R_1 = C_1 R_1 + \frac{C_2}{R_1} \quad \text{and} \quad \omega_2 R_2 = C_2 R_2 + \frac{C_2}{R_2} \tag{15.135}$$

and thus

$$C_1 = \omega_1 + \frac{R_2^2}{(R_2^2 - R_1^2)}(\omega_2 + \omega_1) \tag{15.136a}$$

$$C_2 = \frac{R_1^2 R_2^2}{(R_2^2 - R_1^2)}(\omega_2 + \omega_1). \tag{15.136b}$$

Inserting C_1 and C_2 in (15.133), one obtains:

$$U_\varphi = \frac{1}{(R_2^2 - R_1^2)}\left[\left(\omega_2 R_2^2 - \omega_1 R_1^2\right)r - \frac{R_1^2 R_2^2}{r^2}(\omega_2 - \omega_1)\right]. \tag{15.137}$$

By means of (15.137), one obtains by integration from (15.130) for the pressure distribution in the annular clearance:

$$P(r) = P_1 + \frac{\rho}{(R_2^2 - R_1^2)^2}\left[\left(\omega_2 R_2^2 - \omega_1 R_1^2\right)^2\left(\frac{r^2 - R_1^2}{2}\right) - 2R_1^2 R_2^2(\omega_2 - \omega_1)\right.$$
$$\left.\left(\omega_2 R_2^2 - \omega_1 R_1^2\right)\ln\frac{r}{R_1} - \frac{R_1^4 R_2^4}{2}(\omega_2 - \omega_1)\left(\frac{1}{r^2} - \frac{1}{R_1^2}\right)\right]. \tag{15.138}$$

The pressure at the internal cylinder wall was introduced in (15.138) with P_1, in order to determine the constant resulting from the integration of (15.130). For the pressure distribution along the periphery of the external cylinder, one can compute from (15.138)

$$P(R_2) = P_1 + \frac{\rho}{(R_2^2 - R_1^2)^2}\left[\left(\omega_2 R_2^2 - \omega_1 R_1^2\right)^2\left(\frac{R_2^2 - R_1^2}{2}\right) - 2R_1^2 R_2^2\right.$$
$$\left.(\omega_2 - \omega_1)\left(\omega_2 R_2^2 - \omega_1 R_1^2\right)\ln\frac{R_2}{R_1} - \frac{R_1^4 R_2^4}{2}(\omega_2 - \omega_1)\left(\frac{1}{R_2^2} - \frac{1}{R_1^2}\right)\right]. \tag{15.139}$$

For the molecular momentum transport, the following relationship holds:

$$\tau_{r\varphi} = -\mu \left[r \frac{d}{dr}\left(\frac{U_\varphi}{r}\right) \right]. \qquad (15.140)$$

With the aid of the solution (15.137) for U_φ, one can compute

$$\tau_{r\varphi} = \frac{-2\mu}{(R_2^2 - R_1^2)} \frac{R_1^2 R_2^2}{r^2}(\omega_2 - \omega_1). \qquad (15.141)$$

The molecular-dependent momentum input into the internal cylinder amounts to

$$\tau_{r\varphi}(r = R_1) = \frac{-2\mu}{(R_2^2 - R_1^2)} R_2^2 (\omega_2 - \omega_1) \qquad (15.142)$$

and for the external cylinder to

$$\tau_{r\varphi}(r = R_2) = \frac{-2\mu}{(R_2^2 - R_1^2)} R_1^2 (\omega_2 - \omega_1). \qquad (15.143)$$

The circumferential forces acting on the cylinder can therefore be computed as follows:

$$\begin{aligned} F r(r = R_1) &= \tau_{r\varphi}(r = R_1) 2\pi R_1 L = F_1 \\ F_\varphi(r = R_2) &= \tau_{r\varphi}(r = R_2) 2\pi R_2 L = F_2. \end{aligned} \qquad (15.144)$$

From the relationships for the forces, one can see that the resulting circumferential forces are directly proportional to the viscosity, a fact which is used in the production of viscosimeters to measure the viscosities of fluids.

15.8 The Slow Translatory Motion of a Cylinder

The considerations carried out at the end of Sect. 15.6 show that performing fluid-flow computations with induced simplifications into basic equations, can lead to solutions for which, in some subdomains of the flow field, the assumptions made for the simplifications are no longer valid. This fact has resulted in some regions for the Stokes solution of the flow around a sphere, e.g. in regions where $U_\infty r/\nu \geq 1$. In these regions far away from the sphere, the Reynolds number of the flow becomes too large. There, the acceleration terms, neglected in the Stokes solution ansatz, prove to be no longer small in comparison with the considered pressure terms. Basically, unsatisfactory argumentations had to be used to justify the validity of the obtained solution. Strictly, only experimental investigations for determining the drag force on the sphere could confirm the correctness of the argumentation.

The problematic nature shown for the flow around a sphere becomes even clearer when one looks at the corresponding cylindrical problem, i.e. the two-dimensional flow around a cylinder. It shows indeed that for the plane flow around a cylinder of a viscous fluid no solution at all can be found by

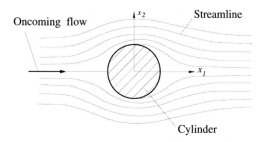

Fig. 15.9 Diagram of the flow around a cylinder

employing the differential equations (15.4) or (15.6) and (15.7) or (15.8)–(15.11). It is thus very problematic not to employ the complete set of basic equations when solving flow problems, but to use reduced sets of differential equations. The latter approach is, however, often necessary for the reason that analytical solutions for the complete set of the basic equations are not available.

For the flow around a cylinder, shown in Fig. 15.9, the following differential equations hold, when assuming a flow possessing a small Reynolds number, i.e. to postulate the validity of the Stokes equations:

$$\frac{\partial U_1}{\partial x_1} + \frac{\partial U_2}{\partial x_2} = 0 \tag{15.145}$$

$$\frac{\partial P}{\partial x_1} = \mu \left(\frac{\partial^2 U_1}{\partial x_1^2} + \frac{\partial^2 U_1}{\partial x_2^2} \right) \tag{15.146}$$

$$\frac{\partial P}{\partial x_2} = \mu \left(\frac{\partial^2 U_2}{\partial x_1^2} + \frac{\partial^2 U_2}{\partial x_2^2} \right) \tag{15.147}$$

with the boundary conditions:

$$U_1 = U_2 = 0 \quad \text{for} \quad r = R \tag{15.148}$$

$$U_1 = U_\infty, \ U_2 = 0 \quad \text{for} \quad r \to \infty. \tag{15.149}$$

The analytical solution that can now be determined for the above differential equations, proves to be of a form such that the boundary conditions introduced for $r = R$ and $r \to \infty$ lead to two solutions that contradict one another. Thus, it is consequently not possible to solve the simplified flow equations (15.145)–(15.147), in which the acceleration terms were neglected, for the boundary conditions stated in (15.148) and (15.149). This insight into the problem suggests leaving the acceleration terms in the basic equations. For the plane problem of the flow around a cylinder, for $\rho = $ constant and stationary flow conditions, one obtains the following set of differential equations:

$$\frac{\partial U_1}{\partial x_1} + \frac{\partial U_2}{\partial x_2} = 0 \tag{15.150}$$

15.8 The Slow Translatory Motion of a Cylinder

$$\rho\left(U_1 \frac{\partial U_1}{\partial x_1} + U_2 \frac{\partial U_1}{\partial x_2}\right) = -\frac{\partial P}{\partial x_1} + \mu\left(\frac{\partial^2 U_1}{\partial x_1^2} + \frac{\partial^2 U_1}{\partial x_2^2}\right) \tag{15.151}$$

$$\rho\left(U_1 \frac{\partial U_2}{\partial x_1} + U_2 \frac{\partial U_2}{\partial x_2}\right) = -\frac{\partial P}{\partial x_2} + \mu\left(\frac{\partial^2 U_2}{\partial x_1^2} + \frac{\partial^2 U_2}{\partial x_2^2}\right). \tag{15.152}$$

With $U_1 = U_\infty + u_1$ and $U_2 = u_2$, the following set of equations results, assuming $U_\infty \gg u_1$ to yield the *generalized Stokes equations*:

$$\frac{\partial u_1}{\partial x_1} + \frac{\partial u_2}{\partial x_2} = 0 \tag{15.153}$$

$$\rho U_\infty \frac{\partial u_1}{\partial x_1} = -\frac{\partial P}{\partial x_1} + \mu\left(\frac{\partial^2 u_1}{\partial x_1^2} + \frac{\partial^2 u_1}{\partial x_2^2}\right) \tag{15.154}$$

$$\rho U_\infty \frac{\partial u_2}{\partial x_2} = -\frac{\partial P}{\partial x_2} + \mu\left(\frac{\partial^2 u_2}{\partial x_1^2} + \frac{\partial^2 u_2}{\partial x_2^2}\right). \tag{15.155}$$

Introducing the potential function $\phi(x_1, x_2)$, one obtains with $\frac{\partial^2 \phi}{\partial x_i^2} = 0$, according to a solution path proposed by Lamb (1911), the following ansatz for u_1 and u_2:

$$u_1 = \frac{\partial \phi}{\partial x_1} + \frac{1}{2k}\frac{\partial \chi}{\partial x_1} - \chi \quad \text{and} \quad u_2 = \frac{\partial \phi}{\partial x_2} + \frac{1}{2k}\frac{\partial \chi}{\partial x_2}, \tag{15.156}$$

where the quantities ϕ and χ fulfil the following differential equations:

$$\frac{\partial^2 \phi}{\partial x_1^2} + \frac{\partial \phi}{\partial x_2} = 0 \quad \text{and} \quad \frac{\partial^2 \chi}{\partial x_1^2} + \frac{\partial^2 \chi}{\partial x_2^2} - 2k\frac{\partial \chi}{\partial x_1} = 0. \tag{15.157}$$

Equations (15.153)–(15.155) are all fulfilled, when one inserts for the pressure:

$$P = P_\infty - \rho U_\infty \frac{\partial \phi}{\partial x_1}. \tag{15.158}$$

For $\phi(x_1, x_2)$, the following ansatz can be found, in order to fulfil the differential equation (15.157):

$$\phi = A_0 \ln r + A_1 \frac{\partial \ln r}{\partial x_1} + A_2 \frac{\partial^2 \ln r}{\partial x_1^2} + \cdots. \tag{15.159}$$

For $\chi(x_1, x_2)$, one introduces:

$$\chi = \psi \exp(kx_1) \tag{15.160}$$

so that for the determination of ψ the following differential equation results:

$$\frac{\partial^2 \psi}{\partial x_1^2} + \frac{\partial^2 \psi}{\partial x_2^2} - k^2 \psi = 0 \tag{15.161}$$

or in cylindrical coordinates:

$$\frac{\partial^2 \psi}{\partial r^2} + \frac{1}{r}\frac{\partial \psi}{\partial r} + \frac{1}{r^2}\frac{\partial^2 \psi}{\partial \varphi^2} - k^2\psi = 0. \qquad (15.162)$$

On now looking for the solution of this equation, which depends on r, one obtains the following ordinary differential equation:

$$\frac{d^2\phi}{dr^2} + \frac{1}{r}\frac{d\phi}{dr} - k^2\phi = 0. \qquad (15.163)$$

This differential equation is determined by the Bessel function $K_0(kr)$ and its derivatives, so that the following ansatz seems reasonable:

$$\chi = -U_\infty + \exp(kx)\left[B_0 K_0(kr) + B_1 \frac{\partial K_0(kr)}{\partial x_1} + B_2 \frac{\partial^2 K_0(kr)}{\partial x_1^2} + \cdots\right]. \qquad (15.164)$$

Because $\quad \dfrac{\partial(\ln r)}{\partial x_1} = \dfrac{x_1}{r^2} = \dfrac{\cos\varphi}{r} \quad$ and $\quad \dfrac{\partial^2 \ln r}{\partial x_1^2} = -\dfrac{\cos 2\theta}{r^2} \qquad (15.165)$

it can be derived that:

$$\phi = A_0 \ln r + A_1 \frac{\cos\varphi}{r} - A_2 \frac{\cos 2\varphi}{r^2}. \qquad (15.166)$$

For the function χ one can write, on introducing the Mascheroni constant, $\gamma = 1.7811$ or $\ln\gamma = 0.57722$:

$$\chi = -U_\infty - B_0 \left[\ln\left(\frac{\gamma}{2}kr\right) + kr\cos\varphi \ln\left(\frac{\gamma}{2}kr\right)\right] - B_1 \frac{\cos\varphi}{r} \qquad (15.167)$$

so that for the velocity field in cylindrical coordinates one can write:

$$U_r(r,\varphi) = \frac{A_0}{r} - \frac{A_1\cos\varphi}{r^2} + U_\infty\cos\varphi - B_0\left[\frac{1}{2kr} + \frac{1}{2}\cos\varphi \right. \qquad (15.168)$$
$$\left. - \frac{1}{2}\cos\varphi \ln\left(\frac{\gamma}{2}kr\right)\right] + B_1\frac{\cos\varphi}{2kr^2}$$

$$U_\varphi(r,\varphi) = -\frac{A_1\sin\varphi}{r^2} - U_\infty\sin\varphi - B_0\frac{\sin\varphi}{2}\ln\left(\frac{\gamma}{2}kr\right) + \frac{B_1\sin\varphi}{2kr^2}. \qquad (15.169)$$

Including the boundary conditions for $r = R$, one obtains:

$$\frac{A_0}{R} - \frac{B_0}{2kR} = 0 \quad \rightsquigarrow \quad A_0 = \frac{B_0}{2k} \qquad (15.170)$$

$$\frac{A_1}{R^2} + U_\infty - \frac{B_0}{2}\left[1 - \ln\left(\frac{\gamma}{k}kR\right)\right] + \frac{B_1}{2kR^2} = 0 \qquad (15.171)$$

$$-\frac{A_1}{R^2} - U_\infty - \frac{B_0}{2}\left[\ln\left(\frac{\gamma}{2}kR\right)\right] + \frac{B_1}{2kR^2} = 0. \qquad (15.172)$$

Thus, one obtains for the integration constants:

15.8 The Slow Translatory Motion of a Cylinder

$$A_0 = \frac{4\nu}{1 - 2\ln\left(\frac{\gamma}{2}kR\right)} = \frac{2U_\infty}{k\left[1 - 2\ln\left(\frac{\gamma}{2}kR\right)\right]} \tag{15.173}$$

$$B_0 = \frac{4U_\infty}{1 - 2\ln\left(\frac{\gamma}{2}kR\right)} \tag{15.174}$$

$$A_1 - \frac{B_1}{2k} = \frac{-U_\infty R^2}{1 - 2\ln\left(\frac{\gamma}{2}kR\right)}. \tag{15.175}$$

This yields for the velocity components *in proximity of the cylinder*:

$$U_r(r,\varphi) = \frac{U_\infty \cos\varphi}{1 - 2\ln\left(\frac{\gamma}{2}kR\right)}\left[-1 + \frac{R^2}{r^2} + 2\ln\left(\frac{r}{R}\right)\right] \tag{15.176}$$

$$U_\varphi(r,\varphi) = \frac{-U_\infty \sin\varphi}{1 - 2\ln\left(\frac{\gamma}{2}kR\right)}\left[1 - \frac{R^2}{r^2} + 2\ln\left(\frac{r}{R}\right)\right]. \tag{15.177}$$

For *large distances* the following equations hold for the velocity components:

$$U_r(r,\varphi) = \frac{A_0}{r} + \frac{1}{2}B_0 \exp(kr\cos\varphi)\left[K_0'(kr) - \cos\varphi K_0(kr)\right] \tag{15.178}$$

$$U_\varphi(r,\varphi) = \frac{1}{2}B_0 \exp(kr\cos\varphi) K_0(kr) \sin\varphi, \tag{15.179}$$

where for large arguments (kr) the following asymptotic relationships hold:

$$K_0(kr) \approx \sqrt{\frac{\pi}{2kr}} \exp(-kr) \tag{15.180}$$

and

$$K_0'(kr) \approx -\sqrt{\frac{\pi}{2kr}} \exp(-kr). \tag{15.181}$$

The pressure can be computed as:

$$P = P_\infty - \rho U_\infty A_0 \frac{\cos\varphi}{r} + \rho U_\infty A_1 \frac{\cos 2\varphi}{r^2}. \tag{15.182}$$

For the drag force, the following equation results:

$$F_D = 2\pi\rho U_\infty A_0. \tag{15.183}$$

When A_0 is inserted, the Lamb equation for the drag force per unit length of a cylinder results:

$$F_D = \frac{8\pi\mu U_\infty}{\left[1 - 2\ln\left(\frac{\gamma}{2}kR\right)\right]}. \tag{15.184}$$

Although the acceleration terms are taken into consideration, the above relation for F_D, i.e. (15.184), can only be employed for small values for $Re = \frac{U_\infty R}{\nu} < 1$.

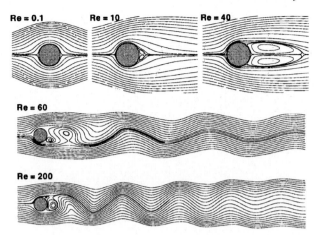

Fig. 15.10 Stream lines for flows around a cylinder at different Reynolds numbers

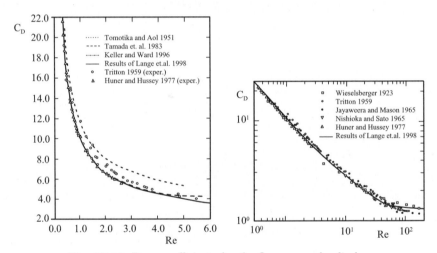

Fig. 15.11 Drag coefficients for the flows around cylinders

If one does not want to have the above limitations of the derived solution of the considered flow problem, i.e. if one seeks a solution without any restrictions for the flow around a cylinder, one has to solve the complete set of equations numerically. Such solutions are nowadays possible for $Re \leq 10.000$ by direct solutions of the continuity and Re−equations. They lead to the results shown in Fig. 15.10 for the stream lines of the flows. In Fig. 15.11, solutions for the drag coefficient of fluid flows for small Reynolds numbers are shown. Fluid flows information for small Re−number flows around spheres are provided in Fig. 15.12.

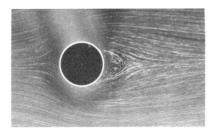 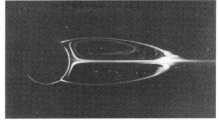

Fig. 15.12 Recirculating flow regions behind spheres, from the book of Van Dyke [15.8]

15.9 Diffusion and Convection Influences on Flow Fields

In paragraphs 3 and 5 the analogy of heat conduction and molecular momentum transport is underlined and, in order to emphasize the significance of this analogy, the general form of the momentum equations was transformed into the transport equation for vorticity (see Chap. 5):

$$\rho\left(\frac{\partial \omega_j}{\partial t} + U_i \frac{\partial \omega_j}{\partial x_i}\right) = \rho \omega_j \frac{\partial U_j}{\partial x_i} + \mu \frac{\partial^2 \omega_j}{\partial x_i^2}. \tag{15.185}$$

For the two-dimensional flow around a cylinder, $\omega_3 = \omega$ is the only component which is unequal to zero, and this fact allows one to write the vorticity equation as a scalar equation. Because $\rho \omega_i \frac{\partial U_j}{\partial x_i} = 0$, this equation reads:

$$\rho\left(\frac{\partial \omega}{\partial t} + U_i \frac{\partial \omega}{\partial x_i}\right) = \mu \frac{\partial^2 \omega}{\partial x_i^2}. \tag{15.186}$$

On comparing this equation with the heat or mass transport equations for convective and diffusive transport:

$$\rho c_v \left(\frac{\partial T}{\partial t} + U_i \frac{\partial T}{\partial x_i}\right) = \lambda \frac{\partial^2 T}{\partial x_i^2} \text{ and } \rho\left(\frac{\partial c}{\partial t} + U_i \frac{\partial c}{\partial x_i}\right) = D \frac{\partial^2 c}{\partial x_i^2}. \tag{15.187}$$

one sees that one can understand the influence of walls on flows in such a way, that at the boundary of the body the vorticity ω is produced. The vorticity is then transported from the body to the fluid, by molecular diffusion, into the moving fluid, see Fig. 15.13.

In order to understand now the interaction between convection and diffusion, it is thus possible to consider the diffusive and convective heat transport, and to transfer the insight gained in this way to the vorticity and its transport in the flow field.

On considering a heated cylinder with a small diameter when a sudden temperature increase takes place, it can be seen that in a time Δt, a heat front dissipates as follows, due to heat conduction:

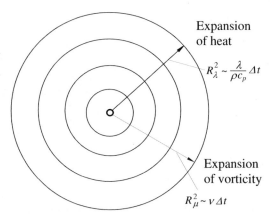

Fig. 15.13 Spreading of head and momentum by diffusion only

$$R_\lambda^2 = \text{const} \frac{\lambda}{\rho c_v} \Delta t, \tag{15.188}$$

where R_λ is a measure of the radial propagation, the heat has moved in the time Δt. For the vorticity one can write in an analogous way

$$R_\mu^2 = \text{const}\, \nu\, \Delta t. \tag{15.189}$$

For the diffusion velocity one thus obtains:

$$u_\lambda = \frac{R_\lambda}{\Delta_t} = \text{const} \frac{1}{R_\lambda}\left(\frac{\lambda}{\rho c_v}\right) \tag{15.190}$$

or

$$u_\mu = \frac{R_\mu}{\Delta_t} = \text{const} \frac{1}{R_\mu} \nu. \tag{15.191}$$

When a fluid now moves convectively at a small flow velocity $U_1 = U_\infty$, the state illustrated in Fig. 15.14 results, which is characterized by the fact that a point can be found on the x_1 axis, at which the dissipation velocity is $U_\mu = U_\infty$, so that

$$(x_1)_\lambda = R_\lambda = \text{const} \frac{1}{U_\infty}\left(\frac{\lambda}{\rho c_v}\right) \tag{15.192}$$

or

$$(x_1)_\mu = R_\mu = \text{const} \frac{\nu}{U_\infty}. \tag{15.193}$$

With this it can be understood that in the presence of an inflow the influence of the cylinder on the temperature or velocity field can have an effect in a limited area only, as Fig. 15.14 shows. To the right of point $(x_1)_\mu$, there is no information at all about the body lying in Fig. 15.14. The insights explained in Fig. 15.14 are important when one has to find in the inflow domain of a

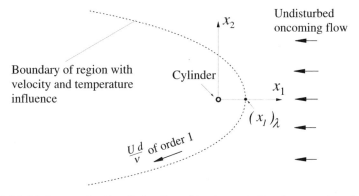

Fig. 15.14 Finite area for the dissipation of heat or rotational momentum for small Reynolds numbers

cylinder the area in which inflow conditions have to be imposed that are not disturbed by the cylinder. According to (15.193), one obtains

$$\left(\frac{x_1}{D}\right)_\mu > \mathrm{const}\frac{\nu}{U_\infty D} = \frac{\mathrm{const}}{Re} \qquad (15.194)$$

or

$$\left(\frac{x_1}{D}\right)_\lambda > \mathrm{const}\frac{\nu}{U_\infty D}\left(\frac{\lambda}{\mu c_v}\right) = \mathrm{const}\frac{1}{RePr}. \qquad (15.195)$$

Equation (15.194) shows that with decreasing Reynolds number the integration area increases, which has to be covered with a numerical grid when numerical integration procedures are employed, in order to install the boundary conditions holding at infinity. An additional extension of the computation area results for Peclet numbers, $Pe = (RePr) < 1$, i.e. for $Pr < 1$, when the temperature field of a flow around a cylinder also has to be computed.

References

15.1. F.S. Sherman: "Viscous Flow" McGraw Hill, New York, 1990
15.2. S.W. Yuan: "Foundations of Fluid Mechanics", Civil Engineering and Mechanics Series, Mei Ya, Taipei, Taiwan (1971)
15.3. H. Lamb: "On the Uniform Motion of a Sphere Through a Viscous Fluid", Phil. Mag. 21, p. 120 (1911)
15.4. H. Lamb: "Hydrodynamics", Dover, New York, 1945
15.5. H. Schlichting: "Boundary Layer Theory", 6th Edition, McGraw Hill, New York (1968)
15.6. R.B. Bird, W.E. Stewart, E.N. Lightfoot: "Transport Phenomena", Wiley, New York (1960)
15.7. C.S. Yih: "Fluid Mechanics - A Concise Introduction to the Theory", West River, Ann Arbor, MI, 1979
15.8. M. van Dyke: "The Album of Fluid Motion", The Parabolic Press, Stanford, California, 1982

Chapter 16
Flows of Large Reynolds Numbers Boundary-Layer Flows

16.1 General Considerations and Derivations

In Chap. 15, flows that were characterized by small Reynolds numbers (Re) were considered, i.e. fluid flows were treated that were diffusion-dominated and where convection played a secondary role. This can be expressed by small Re, e.g. when taking Re as a ratio of forces:

$$Re = \frac{U_c L_c}{\nu_c} = \frac{\rho_c U_c^2}{\mu_c \frac{U_c}{L_c}} = \frac{\text{acceleration forces}}{\text{viscosity forces}}, \quad (16.1)$$

where ρ_c and μ_c represent the density and viscosity characterizing a fluid, respectively, U_c represents a characteristic velocity and L_c is a length characterizing the flow domain.

Equivalent considerations on the significance of Re can, however, also be expressed by the ratio of times typical for diffusion and convection processes in the considered fluid flows:

$$Re = \frac{U_c L_c}{\nu_c} = \frac{L_c^2/\nu_c}{L_c/U_c} = \frac{\text{diffusion times}}{\text{convection times}} \quad (16.2)$$

or by the corresponding velocities that are typical for diffusion and convection processes:

$$Re = \frac{U_c L_c}{\nu_c} = \frac{U_c}{\nu_c/L_c} = \frac{\text{convection velocities}}{\text{diffusion velocities}}. \quad (16.3)$$

Considering flows of large Re, i.e. flows in which the acceleration forces are large in comparison with the viscous forces, or in which the diffusion times are large in comparison with the convection times, or the convection velocities are large in comparison with the diffusion velocities, it can be shown that, e.g., the influences of wall boundaries on flows are limited to small areas near the walls. This is sketched in Fig. 16.1, which shows the flow around a flat plate and indicates there the small region near the wall, where viscous influences can be observed. The contents of this figure results from the extended

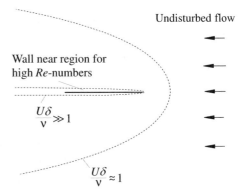

Fig. 16.1 Area limitations for diffusion processes with $Re \approx 1$ and in the case of a flow around a plate for $Re \gg 1$

considerations which were carried out at the end of Chap. 15. Applying the insights gained from Sect. 15.9 to the flow around a flat plate, a large region results for $Re \approx 1$, in which diffusion processes are present. In this region, information about the presence of the plate (around which the flow passes) is available. When, on the other hand, conditions exist that are characterized by $Re \gg 1$, the influence of the diffusion remains restricted to a small region very close to the plate. There, a so-called wall boundary layer forms. Boundary layers of this kind are thus the characteristic properties of flows of high Re. Such flows can therefore be subdivided into body-near regions, where viscous influences on flows have to be considered, and regions that are distant from the wall, which can be regarded as being free from viscous influences.

The above considerations show clearly that special treatments are necessary, in order to derive the equations that can be employed as approximations of the Navier–Stokes equations for $Re \gg 1$ to solve flow problems. Looking for derivations where the viscous terms, because $Re \gg 1$, are completely neglected in the differential equations describing the flow results in the Euler equations:

$$\rho^* \overbrace{\left[\frac{L_c}{t_c U_c} \frac{\partial U_j^*}{\partial t^*} + U_i^* \frac{\partial U_j^*}{\partial x_i^*} \right]}^{\text{standardized Euler equation}} = -\frac{\Delta P_c}{\rho_c U_c^2} \frac{\partial P^*}{\partial x_j^*} + \overbrace{\frac{\nu_c}{U_c L_c} \frac{\partial^2 U_j^*}{\partial x_i^{*2}}}^{\Rightarrow 0 \text{ because } Re \Rightarrow \infty} . \quad (16.4)$$

These equations are not applicable to solving wall boundary-layer flow problems. For the derivation of the boundary-layer equations, one has rather to apply considerations as proposed by Prandtl (1904, 1905). They are based on order of magnitude considerations of the terms in the Navier–Stokes equations, taking into consideration the differences in the times and velocities in diffusion and convection transport processes. If one neglects the viscous terms entirely, this would be equivalent to a reduction of the order of the

16.1 General Considerations and Derivations

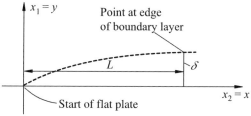

$x_1 = x$ where x = the boundary-layer coordinate in the flow direction

$x_2 = y$ where y = the boundary-layer coordinate vertical to the flow direction

Fig. 16.2 Boundary-layer thickness along a plane flat plate and its development

basic equations describing the fluid flow. Hence it would not be possible to implement all the boundary conditions characterizing a flow, and flows that result in this way from the differential equations as solutions would show considerable deficits.

The transition from the generally valid Navier–Strokes equations to the boundary-layer equations, as indicated in Fig. 16.1, is an essential step which has to be taken into consideration, in order to admit only such simplifications of the Navier–Stokes equations which result in physically still reasonable solutions. The resultant equations are called the boundary-layer equations.

When choosing L as a distance in the flow direction along the flat plate flow indicated in Fig. 16.2, δ is the resultant boundary-layer thickness, Fig. 16.1 showing that $\delta \ll L$ holds for large Re. Considering the times:

$$\Delta t_c = L/U_\infty \quad \text{convection time,}$$
$$\Delta t_D = \delta^2/\nu \quad \text{diffusion time}$$

one obtains, as $\Delta t_c = \Delta t_D$ for the boundary point $\delta(L)$:

$$\frac{L}{U_\infty} = \frac{\delta^2}{\nu} \quad \rightsquigarrow \quad \frac{\delta}{L} = \left(\frac{1}{\underbrace{\frac{U_\infty \delta}{\nu}}}\right). \tag{16.5}$$

The boundary-layer thickness δ, normalized with the development length, is proportional to the reciprocal of the Reynolds number, formed with the external flow velocity and the boundary-layer thickness: $(\delta/L) = 1/Re_\delta$. Paying attention to the context in (16.3), it results for the diffusion velocity occurring in the y-direction that $U_D = U_y \approx \nu/\delta$, so that for considerations of the orders of magnitude of the terms in the Navier–Stokes equations, the following normalization can be carried out:

$$x^* = \frac{x}{L}; \quad y^* = \frac{y}{\delta}; \quad U_x^* = \frac{U_x}{U_\infty}; \quad U_y^* = \frac{U_y}{\nu/\delta}; \quad t^* = \frac{t}{L/U_\infty}; \quad P^* = \frac{P}{P_\infty}. \tag{16.6}$$

Introducing these normalized quantities into the two-dimensional Navier–Stokes equations with constant fluid properties:

$$\frac{\partial U_1}{\partial x_1} + \frac{\partial U_2}{\partial x_2} = 0, \tag{16.7}$$

$$\rho\left[\frac{\partial U_1}{\partial t} + U_1\frac{\partial U_1}{\partial x_1} + U_2\frac{\partial U_1}{\partial x_2}\right] = -\frac{\partial P}{\partial x_1} + \mu\left[\frac{\partial^2 U_1}{\partial x_1^2} + \frac{\partial^2 U_1}{\partial x_2^2}\right], \tag{16.8}$$

$$\rho\left[\frac{\partial U_2}{\partial t} + U_1\frac{\partial U_2}{\partial x_1} + U_2\frac{\partial U_2}{\partial x_2}\right] = -\frac{\partial P}{\partial x_2} + \mu\left[\frac{\partial^2 U_2}{\partial x_1^2} + \frac{\partial^2 U_2}{\partial x_2^2}\right] \tag{16.9}$$

the below-stated derivations can be carried out and for the continuity equation one obtains:

$$\frac{U_\infty}{L}\frac{\partial U_x^*}{\partial x^*} + \frac{\nu}{\delta^2}\frac{\partial U_y^*}{\partial y^*} = 0 \rightsquigarrow \frac{\partial U_x^*}{\partial x^*} + \underbrace{\frac{\nu L}{U_\infty \delta^2}}_{\approx 1}\frac{\partial U_y^*}{\partial y^*} = 0, \tag{16.10}$$

where the term $\nu L/U_\infty\delta^2 = (L/U_\infty)/(\delta^2/\nu)$ is the relationship for the ratio of the convection and diffusion times. According to (16.5), both terms are equal, so that both gradients in the continuity equation are of the same order of magnitude and therefore have to be carried along in the boundary-layer equations. Thus the continuity equation for boundary-layer flows reads:

$$\frac{\partial U_x}{\partial x} + \frac{\partial U_y}{\partial y} = 0. \tag{16.11}$$

For the momentum equation in the x-direction, the normalization yields:

$$\rho\left[\frac{U_\infty^2}{L}\frac{\partial U_x^*}{\partial t^*} + \frac{U_\infty^2}{L}U_x^*\frac{\partial U_x^*}{\partial x^*} + \frac{\nu U_\infty}{\delta^2}U_y^*\frac{\partial U_x^*}{\partial y^*}\right]$$
$$= -\frac{P_\infty}{L}\frac{\partial P^*}{\partial x^*} + \mu\left[\frac{U_\infty}{L^2}\frac{\partial^2 U_x^*}{\partial x^{*2}} + \frac{U_\infty}{\delta^2}\frac{\partial^2 U_x^*}{\partial y^{*2}}\right]. \tag{16.12}$$

On dividing the entire equation by $\rho U_\infty^2/L$, one obtains:

$$\frac{\partial U_x^*}{\partial t^*} + U_x^*\frac{\partial U_x^*}{\partial x^*} + \underbrace{\frac{\nu L}{U_\infty\delta^2}}_{\approx 1}U_y^*\frac{\partial U_x^*}{\partial y^*} = -\frac{P_\infty}{\rho U_\infty^2}\frac{\partial P^*}{\partial x^*} + \underbrace{\frac{\nu}{U_\infty L}}_{\delta/(L Re_\delta)}\frac{\partial^2 U_x^*}{\partial x^{*2}}$$
$$+ \underbrace{\frac{\nu L}{U_\infty\delta^2}}_{\approx 1}\frac{\partial^2 U_x^*}{\partial y^{*2}}. \tag{16.13}$$

With $\delta/L \approx 1/Re_\delta$, the first of the two viscous terms in (16.13), multiplied by $1/Re_\delta^2$, can be regarded as negligible for $Re_\delta \gg 1$. Thus for the boundary

16.1 General Considerations and Derivations

layer form of the x-momentum equation the following equation holds:

$$\rho \left[\frac{\partial U_x}{\partial t} + U_x \frac{\partial U_x}{\partial x} + U_y \frac{\partial U_x}{\partial y} \right] = -\frac{\partial P}{\partial x} + \mu \frac{\partial^2 U_x}{\partial y^2}. \tag{16.14}$$

Analogous derivations yield for the two-dimensional y-momentum equation:

$$\rho \left[\frac{\nu U_\infty}{\delta L} \frac{\partial U_y^*}{\partial t^*} + \frac{U_\infty \nu}{\delta L} U_x^* \frac{\partial U_y^*}{\partial x^*} + \frac{\nu^2}{\delta^3} U_y^* \frac{\partial U_y^*}{\partial y^*} \right] = -\frac{P_\infty}{\delta} \frac{\partial P^*}{\partial y^*}$$

$$+ \mu \left[\frac{\nu}{\delta L^2} \frac{\partial^2 U_y^*}{\partial x^{*2}} + \frac{\nu}{\delta^3} \frac{\partial^2 U_y^*}{\partial y^{*2}} \right]. \tag{16.15}$$

On dividing this equation also by $\rho U_\infty^2/L$, the following equation results:

$$\frac{\nu}{\delta U_\infty} \frac{\partial U_y^*}{\partial t^*} + \frac{\nu}{\delta U_\infty} U_x^* \frac{\partial U_y^*}{\partial x^*} + \frac{\nu}{U_\infty \delta} \underbrace{\frac{\nu L}{U_\infty \delta^2}}_{\approx 1} U_y^* \frac{\partial U_y^*}{\partial y^*} = -\frac{P_\infty L}{\rho U_\infty^2 \delta} \frac{\partial P^*}{\partial y^*}$$

$$+ \frac{\nu}{U_\infty \delta} \frac{\nu}{U_\infty L} \frac{\partial^2 U_y^*}{\partial x^{*2}} + \left(\frac{\nu}{U_\infty \delta} \right)^2 \frac{L}{\delta} \frac{\partial^2 U_y^*}{\partial y^{*2}} \tag{16.16}$$

or rewritten:

$$\frac{1}{Re_\delta} \left[\frac{\partial U_y^*}{\partial t^*} + U_x^* \frac{\partial U_y^*}{\partial x^*} + U_y^* \frac{\partial U_y^*}{\partial y^*} \right] = -\frac{P_\infty}{\rho U_\infty^2} Re_\delta \frac{\partial P^*}{\partial y^*}$$

$$+ \left[\frac{1}{Re_\delta^3} \frac{\partial^2 U_y^*}{\partial x^{*2}} + \frac{1}{Re_\delta} \frac{\partial^2 U_y^*}{\partial y^{*2}} \right]. \tag{16.17}$$

From (16.17), it can be seen that all acceleration and viscous terms can be neglected when compared with terms in (16.14). Because $Re \gg 1$, they are very small in comparison with the corresponding terms in the x-momentum equation (16.14). Thus the y-momentum equation results in the following equation:

$$\frac{\partial P}{\partial y} = 0. \tag{16.18}$$

This equation expresses the fact that the pressure in a boundary layer, vertical to the flow direction, does not change. The boundary layer thus experiences, up to the wall, the pressure change imposed in the x-direction on the outer flow. This means for many problem solutions that the pressure distribution $P(x,y) = P_\infty(x)$ is known, so that, through the boundary-layer equations for the solution of flow problems, only the velocity components U_x and U_y have to be determined.

Looking at the above derivations, the two-dimensional boundary-layer equations can be stated as follows, on the basis of the above order-of-magnitude considerations:

$$\frac{\partial U_x}{\partial x} + \frac{\partial U_y}{\partial y} = 0, \tag{16.19a}$$

$$\rho \left[\frac{\partial U_x}{\partial t} + U_x \frac{\partial U_x}{\partial x} + U_y \frac{\partial U_x}{\partial y} \right] = -\frac{\partial P}{\partial x} + \mu \frac{\partial^2 U_x}{\partial y^2}, \tag{16.19b}$$

$$\frac{\partial P}{\partial y} = 0. \tag{16.19c}$$

Equations (16.19) are, as can easily be shown, a set of parabolic differential equations. They can be solved with the corresponding boundary conditions for some flow geometries and thus make it possible to compute the velocity distributions in boundary-layer flows with simpler equations than the Navier–Stokes equations. There are numerous text books that describe the above boundary-layer equations, e.g. see refs [16.4] to [16.9]

16.2 Solutions of the Boundary-Layer Equations

In the preceding section the boundary-layer equations were derived:

$$\frac{\partial U_x}{\partial x} + \frac{\partial U_y}{\partial y} = 0, \tag{16.20a}$$

$$\rho \left[\frac{\partial U_x}{\partial t} + U_x \frac{\partial U_x}{\partial x} + U_y \frac{\partial U_x}{\partial y} \right] = -\frac{\partial P}{\partial x} + \mu \frac{\partial^2 U_x}{\partial y^2}, \tag{16.20b}$$

$$\frac{\partial P}{\partial y} = 0. \tag{16.20c}$$

For the outer flow, where no viscous effects occur, the pressure distribution can be determined through the Euler form of the momentum equation:

$$\frac{\partial U_\infty}{\partial t} + U_\infty \frac{\partial U_\infty}{\partial x} = -\frac{1}{\rho} \frac{\partial P}{\partial x}. \tag{16.21}$$

Because of (16.20c), the momentum equation (16.20b) can be written as follows, taking (16.21) into account:

$$\frac{\partial U_x}{\partial t} + U_x \frac{\partial U_x}{\partial x} + U_y \frac{\partial U_x}{\partial y} = \frac{\partial U_\infty}{\partial t} + U_\infty \frac{\partial U_\infty}{\partial x} + \nu \frac{\partial^2 U_x}{\partial y^2}. \tag{16.22}$$

It is necessary to solve this equation together with (16.20a), in order to compute boundary-layer flows. However, the validity of this equation has strictly been verified only for Cartesian coordinates. It should be pointed out, however, that it holds also for curved coordinates, when the radius of

16.2 Solutions of the Boundary-Layer Equations

curvature of the flow lines is large in comparison with the boundary layer thickness δ.

In order to solve the boundary layer equation, it is recommended to introduce the stream function Ψ, so that the continuity equation is eliminated:

$$U_1 = \frac{\partial \Psi}{\partial x_2} = \frac{\partial \Psi}{\partial y} = U_x; \quad U_2 = -\frac{\partial \Psi}{\partial x_1} = -\frac{\partial \Psi}{\partial x} = U_y. \quad (16.23)$$

Thus, according to (16.22), the following partial differential equation for the stream function Ψ can be derived:

$$\frac{\partial^2 \Psi}{\partial t \partial y} + \frac{\partial \Psi}{\partial y}\frac{\partial^2 \Psi}{\partial x \partial y} - \frac{\partial \Psi}{\partial x}\frac{\partial^2 \Psi}{\partial y^2} = \frac{\partial U_\infty}{\partial t} + U_\infty \frac{\partial U_\infty}{\partial x} + \nu \frac{\partial^3 \Psi}{\partial y^3}. \quad (16.24)$$

We therefore have to deal with a partial differential equation of third order, which has to be solved for the stream function $\Psi(x, y, t)$. Hence the solution of the equation requires one to state three boundary conditions and suitable initial conditions. Attention has to be paid to the fact that the different boundary-layer flows are given by the corresponding boundary and initial conditions. The transport processes occurring in the boundary-layers are all described, however, by the differential equation for Ψ.

When stationary flow conditions exist, from (16.24) the following equation results for $\frac{\partial}{\partial t}\left(\frac{\partial \Psi}{\partial x}\right) = 0$:

$$\frac{\partial \Psi}{\partial y}\frac{\partial^2 \Psi}{\partial x \partial y} - \frac{\partial \Psi}{\partial x}\frac{\partial^2 \Psi}{\partial y^2} = U_\infty \frac{dU_\infty}{dx} + \nu \frac{\partial^3 \Psi}{\partial y^3}. \quad (16.25)$$

This equation was stated by Blasius (1908) for the case of a flow over a plane plate with $U_\infty = $ constant and was also solved by him analytically.

For the case of stationary boundary layer flows, von Mises (1927) attributed the boundary-layer equations to a non-linear partial differential equation of second order, which corresponded to the equation typical for heat conduction. The essential points of the derivation of the von Mises differential equation can be summarized as shown below.

The derivations proposed by von Mises start also from the stream function Ψ, which is, however, introduced into the derivation as an independent variable, so that the following holds:

$$U_x(x, y) = V_x(x, \Psi) \quad \text{and} \quad U_y(x, y) = V_y(x, \Psi). \quad (16.26)$$

With this, the relationships below can be stated:

$$\frac{\partial U_x}{\partial x} = \frac{\partial V_x}{\partial x} + \frac{\partial V_x}{\partial \Psi}\frac{\partial \Psi}{\partial x} = \frac{\partial V_x}{\partial x} - U_y \frac{\partial V_x}{\partial \Psi}, \quad (16.27)$$

$$\frac{\partial U_x}{\partial y} = \frac{\partial V_x}{\partial \Psi}\frac{\partial \Psi}{\partial y} = U_x \frac{\partial V_x}{\partial \Psi} = V_x \frac{\partial V_x}{\partial \Psi}. \quad (16.28)$$

For the second derivatives with respect to y, one obtains the following intermediate result:

$$\frac{\partial^2 U_x}{\partial y^2} = \frac{\partial}{\partial y}\left(\frac{\partial U_x}{\partial y}\right) = \frac{\partial}{\partial y}\left(U_x \frac{\partial V_x}{\partial \Psi}\right) = U_x^2 \frac{\partial V_x}{\partial \Psi} + U_x \left(\frac{\partial V_x}{\partial \Psi}\right)^2$$

$$= U_x \left[U_x \frac{\partial V_x}{\partial \Psi} + \left(\frac{\partial V_x}{\partial \Psi}\right)^2\right] = U_x \frac{\partial}{\partial \Psi}\left(V_x \frac{\partial V_x}{\partial \Psi}\right). \tag{16.29}$$

Thus the following second derivative can be deduced:

$$\frac{\partial^2 U_x}{\partial y^2} = U_x \frac{\partial}{\partial \Psi}\left[\frac{\partial}{\partial \Psi}\left(\frac{1}{2}V_x^2\right)\right] = V_x \frac{\partial^2}{\partial \Psi^2}\left(\frac{1}{2}V_x^2\right). \tag{16.30}$$

On inserting (16.27)–(16.30) into the boundary layer form of the stationary momentum equation, the following results:

$$V_x \frac{\partial V_x}{\partial x} = U_\infty \frac{dU_\infty}{dx} + \nu V_x \frac{\partial^2}{\partial \Psi^2}\left(\frac{V_x^2}{2}\right) \tag{16.31}$$

or rewritten:

$$\frac{\partial V_x^2}{\partial x} = \frac{dU_\infty^2}{dx} + \nu V_x \frac{\partial^2}{\partial \Psi^2}\left(V_x^2\right). \tag{16.32}$$

If one now introduces a new function:

$$\mathcal{V}(x, \Psi) = U_\infty^2 - V_x^2 \tag{16.33}$$

so that $V_x = \sqrt{U_\infty^2 - \mathcal{V}}$ holds, the differential equation (16.32) adopts the so-called von Mises form:

$$\frac{\partial \mathcal{V}}{\partial x} = \nu \sqrt{U_\infty^2 - \mathcal{V}} \frac{\partial^2 \mathcal{V}}{\partial \Psi^2}. \tag{16.34}$$

The von Mises differential equation has to satisfy the boundary conditions:

$$\begin{aligned} \Psi = 0: \quad & U_x = 0, \quad \text{i.e. } \mathcal{V} = U_\infty^2, \\ \Psi \to \infty: \quad & U_x \to U_\infty, \quad \text{i.e. } \mathcal{V} = 0. \end{aligned} \tag{16.35}$$

The above general solution ansatz for the boundary layer equations are applied to different flows in subsequent chapters.

16.3 Flat Plate Boundary Layer (Blasius Solution)

The flow over a flat plate, sketched in Fig. 16.3, represents the flow of a fluid having constant fluid properties and also a constant inflow velocity. This inflow hits, at the origin of the x–y coordinate system, an infinitely extended

16.3 Flat Plate Boundary Layer (Blasius Solution)

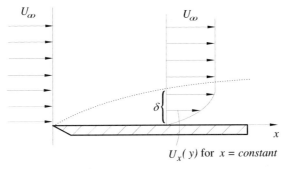

Fig. 16.3 Formation of a plate boundary layer with $\frac{\partial P}{\partial x} = 0$ and $\frac{dU_\infty}{dx} = 0$

flat plate, positioned in the x–y coordinate system, so that along the flat plate a boundary layer flow forms. For the latter the boundary layer equations (16.20a)–(16.20c) hold, with the simplifications according to (16.21):

$$\frac{\partial P}{\partial x} = 0 \tag{16.36}$$

and

$$\frac{\partial U_\infty}{\partial x} = \frac{dU_\infty}{dx} = 0 \tag{16.37}$$

so that the boundary layer equations hold as follows:

$$U_x \frac{\partial U_x}{\partial x} + U_y \frac{\partial U_x}{\partial y} = \nu \frac{\partial^2 U_x}{\partial y^2}, \tag{16.38}$$

$$\frac{\partial U_x}{\partial x} + \frac{\partial U_y}{\partial y} = 0 \tag{16.39}$$

with the boundary conditions:

$$y = 0: \quad U_x = U_y = 0 \quad \text{and} \quad y \to \infty: \quad U_x \to U_\infty. \tag{16.40}$$

On introducing the stream function Ψ for the elimination of the continuity equation:

$$U_x = \frac{\partial \Psi}{\partial y} \quad \text{and} \quad U_y = -\frac{\partial \Psi}{\partial x} \tag{16.41}$$

one obtains the following differential equation for the x-momentum transport:

$$\frac{\partial \Psi}{\partial y} \frac{\partial^2 \Psi}{\partial x \partial y} - \frac{\partial \Psi}{\partial x} \frac{\partial^2 \Psi}{\partial y^2} = \nu \frac{\partial^3 \Psi}{\partial y^3}. \tag{16.42}$$

Blasius proposed a similarity solution for (16.42), such that the solution was obtained with the ansatz

$$\frac{U_x}{U_\infty} = F(\eta) \quad \text{with} \quad \eta = y\sqrt{\frac{U_\infty}{\nu x}}. \tag{16.43}$$

This ansatz takes into consideration that $\eta \approx \dfrac{y}{\delta}$ with $\delta \approx \sqrt{\nu t} = \sqrt{\dfrac{\nu x}{U_\infty}}$ can be set. For the stream function one can write

$$\Psi = \int_0^y U_x \, dy = U_\infty \sqrt{\dfrac{\nu x}{U_\infty}} \int_0^\eta F(\eta) \, d\eta = \sqrt{U_\infty \nu x}\, f(\eta). \tag{16.44}$$

For the different terms in (16.38) and (16.39), the following relationships can thus be derived:

$$U_x = \dfrac{\partial \Psi}{\partial y} = \dfrac{d\Psi}{d\eta}\dfrac{\partial \eta}{\partial y} = U_\infty \dfrac{df}{d\eta} = U_\infty f'(\eta), \tag{16.45a}$$

$$U_y = -\dfrac{\partial \Psi}{\partial x} = -\left[\dfrac{1}{2}\sqrt{\dfrac{U_\infty \nu}{x}} f(\eta) + \dfrac{\partial \Psi}{\partial \eta}\dfrac{\partial \eta}{\partial x}\right]$$

$$= -\dfrac{1}{2}\sqrt{\dfrac{\nu U_\infty}{x}} f(\eta) - \sqrt{U_\infty \nu x}\left(-\dfrac{\eta}{2x}\right) f'(\eta) \tag{16.45b}$$

or rewritten for U_y:

$$U_y = \dfrac{1}{2}\sqrt{\dfrac{\nu U_\infty}{x}} \left(\eta f'(\eta) - f(\eta)\right). \tag{16.46}$$

For the further terms in (16.38) and (16.39), we can deduce

$$\dfrac{\partial U_x}{\partial x} = \dfrac{\partial^2 \Psi}{\partial x \partial y} = -\dfrac{U_\infty}{2}\dfrac{\eta}{x} f''(\eta), \tag{16.47}$$

$$\dfrac{\partial U_x}{\partial y} = \dfrac{\partial^2 \Psi}{\partial y^2} = U_\infty \sqrt{\dfrac{U_\infty}{\nu x}} f''(\eta), \tag{16.48}$$

$$\dfrac{\partial^2 U_x}{\partial y^2} = \dfrac{\partial^3 \Psi}{\partial y^3} = \dfrac{U_\infty^2}{\nu x} f'''(\eta). \tag{16.49}$$

On introducing all the derived terms into (16.42), one obtains

$$-\dfrac{U_\infty^2}{2x}\eta f' f'' + \dfrac{U_\infty^2}{2x}[\eta f' - f] f'' = \nu \dfrac{U_\infty^2}{x\nu} f''' \tag{16.50}$$

or, after complete rearrangement:

$$ff'' + 2f''' = 0. \tag{16.51}$$

This is the ordinary differential equation derived by Blasius in his Göttingen doctoral thesis. As he showed, it can be integrated numerically. This integration results, with the following boundary conditions for f and f':

$$\eta = 0: \quad f = 0; \quad f' = 0 \quad \text{and} \quad \eta \to \infty: \quad f' \to 1 \tag{16.52}$$

16.3 Flat Plate Boundary Layer (Blasius Solution)

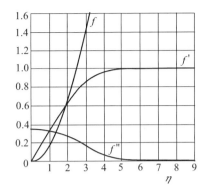

Fig. 16.4 The numerical solution of the Blasius boundary-layer equation yields the above-stated distributions of $f(\eta)$, $f'(\eta)$, $f''(\eta)$

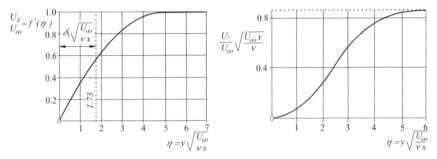

Fig. 16.5 Longitudinal and transversal velocity distributions in a boundary layer

in the distributions shown in Fig. 16.4 for f, f' and f''. The given functional values can be employed at each point η, in order to compute $U_x/U_\infty = f'(\eta)$ and $U_y/U_\infty = (1/2U_\infty)\sqrt{\nu U_\infty/x}(\eta f' - f)$. The distribution of U_x/U_∞ is presented in Fig. 16.5a and $\frac{U_y}{U_\infty}\sqrt{\frac{U_\infty x}{\nu}}$ in Fig. 16.5b. Both figures indicate velocity distributions as they are also found in experimental investigations. This is shown in Fig. 16.6. The above considerations have shown that, by introducing the boundary-layer equations, it has been possible to handle an important flow theoretically, namely the viscous flow over a flat plate.

It is interesting to see from Fig. 16.4 that the U_y velocity component at the outer edge of the boundary layer, i.e. for $\eta \to \infty$, adopts the value

$$(U_y)_\infty = 0.8604 U_\infty \sqrt{\frac{\nu}{xU_\infty}}. \tag{16.53}$$

This velocity component directed out of the boundary layer flow region comes from the fact that with increasing length along the plate of the flow and thus increasing boundary layer thickness, the fluid is being forced away from the wall.

Further values concerning the velocity profiles of the flat plate boundary layer can be taken from Table 16.1, in which $f(\eta)$, $f'(\eta)$ and $f''(\eta)$ are

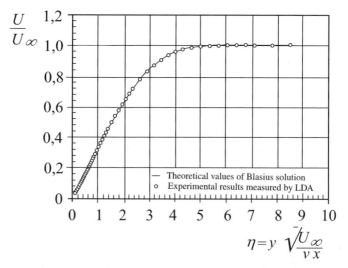

Fig. 16.6 Agreement of experimental and theoretical results for a flat plate boundary layer

indicated. This table can be employed for the determination of all properties of the flat plate boundary-layer flow. The computed values are in good agreement with the experiments, as can be seen from Fig. 16.6.

16.4 Integral Properties of Wall Boundary Layers

In the preceding considerations of boundary-layer properties, δ was used as a quantity, without being defined precisely. It was introduced into the considerations from derivations of the molecular momentum diffusion as $\delta \approx \sqrt{\nu t_D}$ with $t_D = t_c = (x/U_\infty)$, so that the following holds:

$$\delta \approx \sqrt{\frac{\nu x}{U_\infty}}. \tag{16.54}$$

It is possible to make a somewhat more precise statement, of what δ is, by means of the definition of the displacement thickness δ_1, which indicates the extent to which the flow, originally arriving with U_∞, was displaced by the plate, due to the boundary-layer development (integral theorem for mean properties):

$$\delta_1 U_\infty = \int_0^\infty (U_\infty - U_x)\,dy \tag{16.55}$$

or rewritten in terms of δ_1 and integrated:

$$\delta_1 = \int_0^\infty \left(1 - \frac{U_x}{U_\infty}\right) dy = \sqrt{\frac{\nu x}{U_\infty}} \int_0^\infty [1 - f'(\eta)]\,d\eta = 1.73\sqrt{\frac{\nu x}{U_\infty}}. \tag{16.56}$$

16.4 Integral Properties of Wall Boundary Layers

Table 16.1 Solution values of the Blasius equation according to Howarth

$\eta = y\sqrt{\dfrac{U_\infty}{\nu x}}$	f	$f' = \dfrac{U_x}{U_\infty}$	f''
0	0	0	0.33206
0.2	0.00664	0.06641	0.33199
0.4	0.02656	0.13277	0.33147
0.6	0.05974	0.19894	0.33008
0.8	0.10611	0.26471	0.32739
1.0	0.16557	0.32979	0.32301
1.2	0.23795	0.39378	0.31659
1.4	0.32298	0.45627	0.30787
1.6	0.42032	0.51676	0.29667
1.8	0.52952	0.57477	0.28293
2.0	0.65003	0.62977	0.26675
2.2	0.78120	0.68132	0.24835
2.4	0.92230	0.72899	0.22809
2.6	1.07252	0.77246	0.20646
2.8	1.23099	0.81152	0.18401
3.0	1.39682	0.84605	0.16136
3.2	1.56911	0.87609	0.13913
3.4	1.74696	0.90177	0.11788
3.6	1.92954	0.92333	0.09809
3.8	2.11605	0.94112	0.08013
4.0	2.30576	0.95552	0.06424
4.2	2.49806	0.96696	0.05052
4.4	2.69238	0.97587	0.03897
4.6	2.88826	0.98269	0.02948
4.8	3.08534	0.98779	0.02187
5.0	3.28329	0.99155	0.01591
5.2	3.48189	0.99425	0.01134
5.4	3.68094	0.99616	0.00793
5.6	3.88031	0.99748	0.00543
5.8	4.07990	0.99838	0.00365
6.0	4.27964	0.99898	0.00240
6.2	4.47948	0.99937	0.00155
6.4	4.67938	0.99961	0.00098
6.6	4.87931	0.99977	0.00061
6.8	5.07928	0.99987	0.00037
7.0	5.27926	0.99992	0.00022
7.2	5.47925	0.99996	0.00013
7.4	5.67924	0.99998	0.00007
7.6	5.87924	0.99999	0.00004
7.8	6.07923	1.00000	0.00002
8.0	6.27923	1.00000	0.00001
8.2	6.47923	1.00000	0.00001
8.4	6.67923	1.00000	0.00000
8.6	6.87923	1.00000	0.00000
8.8	7.07923	1.00000	0.00000

On now choosing $\delta = 3\delta_1$ as a more precise definition of the thickness of the boundary layer, as proposed by Prandtl, one obtains

$$\delta = 5.2\sqrt{\frac{\nu x}{U_\infty}}. \qquad (16.57)$$

According to Table 16.1, $u(\delta)$ shows a deviation of δ from the external velocity U_∞ by about 0.5%.

Analogous to the above-computed displacement thickness, which was defined and computed as "mass loss thickness of the boundary layer", the momentum-loss thickness can also be defined and computed. Here, the introduction of the momentum deficit into the integral took place with $\Delta U = (U_\infty - U_x)$:

$$\rho U_\infty^2 \delta_2 = \rho \int_0^\infty U_x \left(U_\infty - U_x\right) dy. \qquad (16.58)$$

Solved in terms of δ_2 and integrated, this yields

$$\delta_2 = \int_0^\infty \frac{U_x}{U_\infty}\left[1 - \frac{U_x}{U_\infty}\right] dy = \sqrt{\frac{\nu x}{U_\infty}} \int_0^\infty f'\left[1 - f'\right] dy = 0.664\sqrt{\frac{\nu x}{U_\infty}}. \qquad (16.59)$$

Hence, a comparison of the various thickness gives $\delta_2 = 0.384, \delta_1 = 0.128\delta$.

It is important to take into consideration that the boundary-layer equations become valid only with a certain distance x from the front edge of the plate, i.e. the boundary-layer equations hold only from a certain Re_x, with $Re_x = U_\infty x/\nu$.

For smaller values of Re_x, the complete Navier–Stokes equations have to be employed to compute the velocity field. The solutions of these equations have to consider, moreover, that the front edge of the plate represents a singularity, which requires special attention when carrying out numerical solutions, as shown by Carrier and Lin [16.3], Boley and Friedman [16.2], Shi et al. [16.7], etc.

A further quantity, which can be derived from the solutions of the Blasius boundary layer equation, is the friction coefficient c_f, defined as

$$c_f = \frac{|\tau_w|}{\frac{\rho}{2}U_\infty^2},$$

where $|\tau_w|$ = local momentum loss to the wall and $\frac{1}{2}\rho U_\infty^2$ = stagnation pressure of the outer flow.

The local momentum loss $|\tau_w|$ is computed as

$$|\tau_w| = \mu \left(\frac{\partial U_x}{\partial y}\right)_{y=0} = \frac{\mu U_\infty f''(0)}{\sqrt{\nu x/U_\infty}} = \frac{0.332}{\sqrt{Re_x}}\rho U_\infty^2 \qquad (16.60)$$

16.4 Integral Properties of Wall Boundary Layers

and for c_f for boundary-layer flows it therefore holds that

$$c_f = \frac{0.664}{\sqrt{Re_x}}. \qquad (16.61)$$

Often it is sufficient for boundary-layer flows to state their integral properties as indicated in (16.56), (16.57) and (16.59), i.e. the quantities

$$\delta = 5.2\sqrt{\frac{\nu x}{U_\infty}} \quad \text{boundary-layer thickness,}$$

$$\delta_1 = 1.73\sqrt{\frac{\nu x}{U_\infty}} \quad \text{displacement thickness,}$$

$$\delta_2 = 0.664\sqrt{\frac{\nu x}{U_\infty}} \quad \text{momentum-loss thickness}$$

serve for indicating integral properties of boundary-layer flows (Fig. 16.7).

General considerations of the integral form of the boundary-layer equations, that are based on (16.38) and (16.39), originate from von Karman. He proposed to integrate the equation

$$\frac{\partial U_x}{\partial t} + U\frac{\partial U_x}{\partial x} + U_y\frac{\partial U_x}{\partial y} = -\frac{1}{\rho}\frac{\mathrm{d}P}{\mathrm{d}x} + \nu\frac{\partial^2 U_x}{\partial y^2} \qquad (16.62)$$

from $y = 0$ to $y = \delta(x)$, where (16.62), with the aid of the continuity equation, can be rewritten as follows:

$$\frac{\partial U_x}{\partial t} + \frac{\partial(U_x^2)}{\partial x} + \frac{\partial(U_xU_y)}{\partial y} = -\frac{1}{\rho}\frac{\mathrm{d}P}{\mathrm{d}x} + \frac{\mu}{\rho}\frac{\partial^2 U_x}{\partial y^2}. \qquad (16.63)$$

Now applying integration from 0 to δ, one obtains

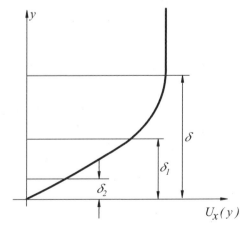

Fig. 16.7 Illustration of the displacement thickness δ_1, the momentum loss thickness δ_2, and the boundary-layer thickness δ

$$\frac{\partial}{\partial t}\int_0^\delta U_x\,dy + \int_0^\delta \frac{\partial(U_x^2)}{\partial x}\,dy + [U_xU_y]_0^\delta = -\frac{1}{\rho}\frac{dP}{dx}\delta + \frac{\mu}{\rho}\int_0^\delta \frac{\partial^2 U_x}{\partial y^2}\,dy. \quad (16.64)$$

For $\int_0^\delta \frac{\partial(U_x^2)}{\partial x}\,dy$ we can write:

$$\int_0^\delta \frac{\partial(U_x^2)}{\partial x}\,dy = \frac{d}{dx}\int_0^\delta U_x^2\,dy - U_\infty^2\frac{d\delta}{dx}. \quad (16.65)$$

Moreover, introducing $[U_xU_y]_0^\delta = -U_\infty\int_0^\delta \frac{\partial U_x}{\partial x}\,dy$ and rewriting:

$$U_\infty\int_0^\delta \frac{\partial U_x}{\partial x}\,dy = U_\infty\frac{d}{dx}\int_0^\delta U_x\,dy - U_\infty^2\frac{d\delta}{dx} \quad (16.66)$$

yields with $\dfrac{\mu}{\rho}\int_0^\delta \dfrac{\partial^2 U_x}{\partial y^2}\,dy = -\dfrac{\tau_w(x)}{\rho}$ the following equation:

$$\frac{\partial}{\partial t}\int_0^\delta U_x\,dy + \frac{d}{dx}\int_0^\delta U_x^2\,dy - U_\infty\frac{d}{dx}\int_0^\delta U_x\,dy = -\frac{1}{\rho}\frac{dP}{dx}\delta - \frac{\tau_w}{\rho}. \quad (16.67)$$

For stationary flows, with consideration of $-\dfrac{1}{\rho}\dfrac{dP}{dx} = U_\infty\dfrac{dU_\infty}{dx}$, one can therefore write:

$$\frac{d}{dx}\int_0^\delta U_x^2\,dy - U_\infty\frac{d}{dx}\int_0^\delta U_x\,dy - U_\infty\delta\frac{dU_\infty}{dx} = -\frac{\tau_w}{\rho} \quad (16.68)$$

or, in a somewhat rewritten way:

$$\frac{d}{dx}\int_0^\delta U_x^2\,dy - \frac{d}{dx}\int_0^\delta U_\infty U_x\,dy + \frac{dU_\infty}{dx}\int_0^\delta U_x\,dy - \frac{dU_\infty}{dx}\int_0^\delta U_\infty\,dy = -\frac{\tau_w}{\rho} \quad (16.69)$$

so that the following equation can be deduced:

$$\frac{d}{dx}\int_0^\delta U_x(U_\infty - U_x)\,dy + \frac{dU_\infty}{dx}\int_0^\delta (U_\infty - U_x)\,dy = +\frac{\tau_w}{\rho}. \quad (16.70)$$

16.4 Integral Properties of Wall Boundary Layers

As outside δ, i.e. for $\delta \to \infty$, there is no more change of the velocity profile, one can write

$$\frac{\mathrm{d}}{\mathrm{d}x}\underbrace{\int_0^\infty U_x(U_\infty - U_x)\mathrm{d}y}_{U_\infty^2 \delta_2} + \frac{\mathrm{d}U_\infty}{\mathrm{d}x}\underbrace{\int_0^\infty (U_\infty - U_x)\mathrm{d}y}_{U_\infty \delta_1} = +\frac{\tau_w}{\rho}. \qquad (16.71)$$

Therefore, the following integral relationship holds:

$$\frac{\mathrm{d}}{\mathrm{d}x}(U_\infty^2 \delta_2) + \frac{\mathrm{d}U_\infty}{\mathrm{d}x}(U_\infty \delta_1) = \frac{\tau_w}{\rho}. \qquad (16.72)$$

This equation can be rewritten in the following form:

$$\frac{\mathrm{d}\delta_2}{\mathrm{d}x} + (2\delta_2 + \delta_1)\frac{1}{U_\infty}\frac{\mathrm{d}U_\infty}{\mathrm{d}x} = \frac{\tau_w}{\rho U_\infty^2} = \frac{c_f}{2}. \qquad (16.73)$$

For the Blasius boundary layer, this equation reduces to

$$\frac{\mathrm{d}\delta_2}{\mathrm{d}x} = \frac{c_f}{2}. \qquad (16.74)$$

This result can be verified by inserting (16.59) and (16.61) into (16.74).

The basic idea behind the Karman integral consideration of the boundary-layer equations is the fact that, for determining integral properties of boundary-layer flows, one does not require the exact distribution of $U_x/U_\infty = f'(\eta)$. When giving an approximate function $U_x/U_\infty = g(y/\delta)$, the general character of boundary-layer flows is already captured. The Karman integral equations allow one to determine good approximations for $\delta(x)$ and $c_f(x)$:

$$\frac{U_x}{U_\infty} = A + B\eta + C\eta^2 + D\eta^3 \qquad (16.75)$$

with $\eta = y/\delta$ and the boundary conditions

$$\begin{aligned} y = 0 : U_x = 0 \quad &\text{and} \quad \frac{\partial^2 U_x}{\partial y^2} = 0, \\ y = \delta : U_x = U_\infty \quad &\text{and} \quad \frac{\partial U_x}{\partial y} = 0. \end{aligned} \qquad (16.76)$$

With these values, $f'(\eta) = \frac{3}{2}\eta - \frac{1}{2}\eta^3$ can be stated as the velocity profile and the values deduced:

$$\delta = 4.641\sqrt{\frac{\nu x}{U_\infty}}; \quad c_f = \frac{1.293}{\sqrt{Re_x}}; \quad \delta_1 = 1.74\sqrt{\frac{\nu x}{U_\infty}}. \qquad (16.77)$$

These are close to the values derived from the Blasius solution of flat-plate boundary-layer flow.

16.5 The Laminar, Plane, Two-Dimensional Free Shear Layer

On allowing two parallel flows of identical fluids, that differ only in having different fluid velocities, to interact with one another, a flow results that is defined as a laminar, plane, two-dimensional free shear layer. Such a flow is sketched in Fig. 16.8, which shows that the features of the flow are generated by the cross-flow molecular momentum transport that occurs along the flow. The velocity gradient in the shear layer decreases with increasing distance to $x = 0$. This takes place because of the momentum transport, i.e. because of the momentum transport from the region of high velocity to the region of low velocity, as indicated in Fig. 16.8.

The flow sketched in Fig. 16.8 has properties which were employed for the derivations of the boundary-layer equations for flows.

- In the flow direction, there exists a convection-dominated momentum transport. In the cross-flow direction, a diffusion-dominated momentum transport occurs.

Likewise, the flow has no pressure gradient, i.e. $\frac{\partial P}{\partial x} = \frac{\partial P}{\partial y} = 0$, so that one can state

- The form of the boundary-layer equations, which was deduced by Blasius for the plate boundary-layer, also holds for the laminar, plane, two-dimensional free shear layer.

Thus, (16.42) has to be employed to treat free shear flows of the kind sketched in Fig. 16.8, i.e. one obtains the solution by the following differential equation:

$$\frac{\partial \Psi}{\partial y} \frac{\partial^2 \Psi}{\partial x \partial y} - \frac{\partial \Psi}{\partial x} \frac{\partial^2 \Psi}{\partial y^2} = \nu \frac{\partial^3 \Psi}{\partial y^3}. \qquad (16.78)$$

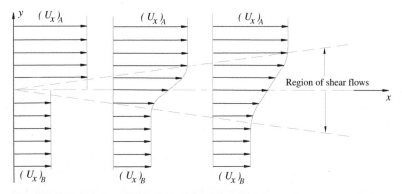

Fig. 16.8 Formation of a laminar, free shear layer by momentum diffusion

16.6 The Plane, Two-Dimensional, Laminar Free Jet

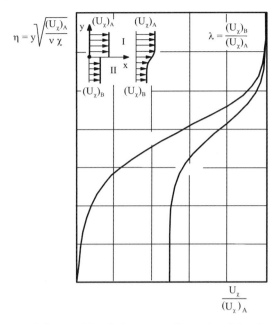

Fig. 16.9 Diagram of the considered plane, two-dimensional, laminar free shear layer

With the ansatzes $\eta = y\sqrt{(U_x)_A/\nu x}$ and $\Psi = \sqrt{\nu(U_x)_A\,x}\,f(\eta)$, the following ordinary differential equation results:

$$ff'' + 2f''' = 0 \tag{16.79}$$

which has to be solved for the following boundary conditions, describing the considered free shear-layer flow:

$$\begin{aligned} \eta = +\infty: & \quad f' = 1, \\ \eta = -\infty: & \quad f' = \lambda = \frac{(U_x)_B}{(U_x)_A}, \\ \eta = 0: & \quad f = 0. \end{aligned} \tag{16.80}$$

The solution of (16.79) has again to be carried out numerically, similarly to the Blasius solution for the flat plate, as there is no analytical solution of (16.79) available. The solution obtained by Lock [16.6] is indicated in Fig. 16.9 for $\lambda = 0$, i.e. $(U_\infty)_B = 0$, and also for $\lambda = 0.5$.

16.6 The Plane, Two-Dimensional, Laminar Free Jet

Another flow with boundary-layer character will now be investigated. This flow is usually referred to as a two-dimensional, laminar free jet, sketched in Fig. 16.10.

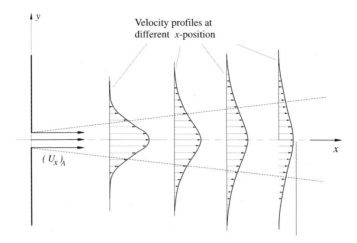

Fig. 16.10 Sketch of the considered plane, two-dimensional laminar free jet

Below a two-dimensional free jet is considered, which can be generated in the plane $x = 0$, by a flow from a narrow slit located in the x–y plane. The jet propagates in the x-direction and the propagation is such, that the x-axis is the symmetry axis. As the propagation of the jet occurring in the y-direction is small in comparison with the propagation in the flow direction and as, moreover, only stationary main flow conditions will be considered, the boundary-layer equation for $\frac{dP}{dx} = 0$ can be employed to study the flow:

$$\frac{\partial \Psi}{\partial y} \frac{\partial^2 \Psi}{\partial x \partial y} - \frac{\partial \Psi}{\partial x} \frac{\partial^2 \Psi}{\partial y^2} = \nu \frac{\partial^3 \Psi}{\partial y^3}. \tag{16.81}$$

This is again the boundary-layer equation for Ψ as it was employed in the case of the flow along a flat plate and also when the plane shear layer was considered. The difference from the plate boundary-layer flow occurs due to the boundary conditions, which for the free jet flow are as follows:

$$\begin{aligned} y &= 0: \quad U_y = 0 \text{ and } (\partial U_x/\partial y) = 0, \text{ as is the symmetry axis,} \\ y &= \infty: \quad U_x \to 0, \text{ as there is no background flow.} \end{aligned} \tag{16.82}$$

For the free jet, the total momentum can be derived as:

$$I_{\text{ges}} = \int_{-\infty}^{+\infty} \rho U_x^2 \, dy = 2 \int_0^{+\infty} \rho U_x^2 \, dy = 2\rho \left(U_x\right)_A^2 b. \tag{16.83}$$

The momentum is constant along the x-axis. This follows from (16.68), which holds for the free jet as given below:

$$\frac{d}{dx} \int_0^{+\infty} U_x^2 \, dy - U_\infty \overbrace{\frac{d}{dx} \int_0^{+\infty} U_x \, dy}^{=0} - \overbrace{U_\infty \delta \frac{dU_\infty}{dx}}^{=0} = \frac{\mu}{\rho} \overbrace{\int_0^{+\infty} \frac{\partial^2 U_x}{\partial y^2} \, dy}^{=0} \tag{16.84}$$

16.6 The Plane, Two-Dimensional, Laminar Free Jet

so that one can derive

$$\frac{d}{dx}\int_0^{+\infty} U_x^2\, dy = 0 \quad \leadsto \quad I_{ges} = 2\int_0^{+\infty} \rho U_x^2\, dy = \text{constant}. \tag{16.85}$$

For deriving the similarity solution for the free jet flow, the boundary-layer equation can be solved with the following ansatzes:

$$\eta = x^\alpha y \quad \text{and} \quad \Psi = x^\beta f(\eta). \tag{16.86}$$

From this the different terms in the boundary layer equation can be expressed as follows in terms of the introduced quantities η and Ψ:

$$U_x = \frac{\partial \Psi}{\partial y} = x^{(\alpha+\beta)} f', \tag{16.87}$$

$$U_y = -\frac{\partial \Psi}{\partial x} = -x^{(\beta-1)}(\alpha \eta f' + \beta f), \tag{16.88}$$

$$\frac{\partial U_x}{\partial y} = \frac{\partial^2 \Psi}{\partial y^2} = x^{(2\alpha+\beta)} f'', \tag{16.89}$$

$$\frac{\partial^2 \Psi}{\partial x \partial y} = x^{(\alpha+\beta-1)}(\alpha \eta f'' + \alpha f' + \beta f'), \tag{16.90}$$

$$\frac{\partial^3 \Psi}{\partial y^3} = x^{(3\alpha+\beta)} f'''. \tag{16.91}$$

The boundary-layer equation (16.81) adopts the following form when (16.87)–(16.91) are inserted into (16.81):

$$x^{(2\alpha+2\beta-1)}\left[(\alpha+\beta) f'^2 - \beta f f''\right] = \nu x^{(3\alpha+\beta)} f'''. \tag{16.92}$$

This equation becomes a physically correct equation for $f(\eta)$, when the exponents of the x-terms are equal, hence we can write:

$$2\alpha + 2\beta - 1 = 3\alpha + \beta \quad \leadsto \quad \beta = \alpha + 1. \tag{16.93}$$

Moreover, one can deduce from the total momentum equation:

$$I_{ges} = 2\int_0^{+\infty} \rho U_x^2\, dy = 2\rho x^{(\alpha+2\beta)}\int_0^{+\infty} f'^2\, d\eta = \text{constant} \tag{16.94}$$

Hence, we obtain as an additional requirement for α and β:

$$\alpha + 2\beta = 0 \tag{16.95}$$

so that from (16.93) one can deduce:

$$\alpha = -\frac{2}{3} \quad \text{and} \quad \beta = \frac{1}{3}, \tag{16.96}$$

i.e. the following similarity ansatzes hold:

$$\eta = yx^{-\frac{2}{3}} \quad \text{and} \quad \Psi = x^{\frac{1}{3}} f(\eta). \tag{16.97}$$

By means of these ansatzes the differential equation (16.81) turns into the following equation to determine $f(\eta)$:

$$(f')^2 + ff'' + 3\nu f''' = 0 \tag{16.98}$$

with the boundary conditions coming from (16.82):

$$\eta = 0: \ f = 0 \quad \text{and} \quad f'' = 0, \tag{16.99}$$
$$\eta \to \infty: \ f' \to 0. \tag{16.100}$$

Moreover, to eliminate also the factor 3ν from the differential equation (16.98) in order to obtain a generally valid equation to determine $f'(\eta)$, the following ansatzes are finally chosen:

$$\tilde{\eta} = \frac{1}{3\nu^{\frac{1}{2}}} \frac{y}{x^{\frac{2}{3}}} \quad \text{and} \quad \Psi = \nu^{\frac{1}{2}} x^{\frac{1}{3}} \tilde{f}(\tilde{\eta}). \tag{16.101}$$

With this one obtains the following ordinary differential equation:

$$\tilde{f}'^2 + \tilde{f}\tilde{f}'' + \tilde{f}''' = 0 \tag{16.102}$$

with the following boundary conditions:

$$\begin{array}{l} y = 0: \partial U_x/\partial y = 0 \text{ and } U_y = 0 \rightsquigarrow \tilde{\eta} = 0: \tilde{f}'' = 0 \text{ and } \tilde{f} = 0, \\ y \to \infty: U_x = 0 \qquad\qquad\qquad\quad \rightsquigarrow \tilde{\eta} \to \infty: \tilde{f}' = 0. \end{array} \tag{16.103}$$

On integrating the differential equation (16.102) once, one obtains:

$$\tilde{f}\tilde{f}' + \tilde{f}'' = C_1. \tag{16.104}$$

The resulting integration constant yields, due to the employed boundary conditions $C_1 = 0$, as for $\tilde{\eta} = 0$, $\tilde{f}$ and also $\tilde{f}''$ are equal to zero. This can readily be deduced from the boundary conditions, so that the following differential equation results:

$$\tilde{f}\tilde{f}' + \tilde{f}'' = 0. \tag{16.105}$$

The solution of this differential equation can be obtained through the ansatz:

$$\xi = \int_0^F \frac{\mathrm{d}F}{1-F^2} = \frac{1}{2}\ln\left(\frac{1+F}{1-F}\right) = \tanh^{-1} F. \tag{16.106}$$

16.6 The Plane, Two-Dimensional, Laminar Free Jet

From this it follows that:
$$F = \tanh \xi = \frac{1 - \exp(-2\xi)}{1 + \exp(-2\xi)} \tag{16.107}$$

and from (16.104) it follows that $\frac{dF}{d\xi} = 1 - \tanh^2 \xi$ and thus for U_x the following relationship holds:
$$U_x = \frac{2}{3} A^2 x^{-\frac{1}{3}} \left(1 - \tanh^2 \xi\right). \tag{16.108}$$

The constant A contained in this equation is determined through the constancy of the total momentum of the free jet:

$$I_{\text{ges}} = 2 \int_0^\infty U_x^2 \rho \, dy \quad \leadsto \quad I_{\text{ges}} = \frac{4}{3} A^3 \rho \nu^{\frac{1}{2}} \int_0^\infty (1 - \tanh^2) \, d\xi,$$

$$I_{\text{ges}} = \frac{16}{9} \rho A^3 \nu^{\frac{1}{2}} \tag{16.109}$$

or solved for A:
$$A = 0.826 \left(\frac{I_{\text{ges}}}{\rho \nu^{\frac{1}{2}}}\right)^{\frac{1}{3}}. \tag{16.110}$$

For the velocity components one thus obtains:
$$U_x = 0.454 \left(\frac{I_{\text{ges}}^2}{\rho^2 \nu}\right)^{\frac{1}{3}} \left(1 - \tanh^2 \xi\right) x^{\frac{1}{3}}, \tag{16.111}$$

$$U_y = 0.55 \left(\frac{I_{\text{ges}} \nu}{\rho x^2}\right)^{\frac{1}{3}} \left[2\xi \left(1 - \tanh^2 \xi\right) - \tanh \xi\right] \tag{16.112}$$

and for $\xi = 0.275 \left(\dfrac{I_{\text{ges}}}{\rho \nu^2}\right)^{\frac{1}{3}} \dfrac{y}{x^{\frac{2}{3}}}.$

The velocity profile that can be computed from the above equations is shown in Fig. 16.11. The U_y component of the velocity field is computed at the edge of the jet:
$$U_y(\xi_\infty) = -0.55 \left(\frac{I_{\text{ges}} \nu}{\rho x^2}\right)^{\frac{1}{3}}.$$

It is negative and this indicates that the free jet flow continuously sucks in fluid from the outer flow, so that the mass flow of the free jet increases in the flow direction.

The mass flow can be computed at each point x of the free jet:
$$\dot{m} = \rho \int_{-\infty}^{+\infty} U_x \, dy, \tag{16.113}$$

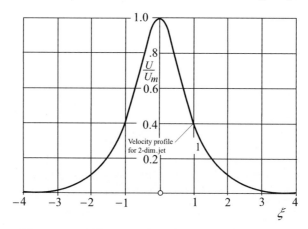

Fig. 16.11 Velocity profile of the plane free jet flow

$$\dot{m} = 3.3 \left(\frac{I_{\text{ges}}}{\rho} \nu x \right)^{\frac{1}{3}}. \qquad (16.114)$$

The fluid input into the free jet flow takes place due to the viscosity of the fluid, i.e. the entrainment is caused by the molecular momentum transport.

16.7 Plane, Two-Dimensional Wake Flow

Additional flows that are important in practice and that can be treated with the aid of the boundary-layer equations, are wake flows showing the features sketched in Fig. 16.11. This figure shows a plane, two-dimensional wake flow as occurs, e.g., behind a plane plate or a cylinder located with the main axis perpendicular to the x–y plane. Such flows are characterized by a momentum deficit, which corresponds in its integral properties to the flow resistance force of the plate or cylinder around which a flow passes. This can be computed by employing the integral form of the momentum equation:

$$K_w = 2\rho B \int_0^\infty U_1 (U_\infty - U_1) dy, \qquad (16.115)$$

where $U_1(y)$ represents the velocity profile of the wake flow existing at a certain x-position, B is the width of the plate in z-direction, and U_∞ corresponds to the main flow in the x-direction at $y \to \infty$. For the infinitely thin plane plate, one obtains the result

$$K_w = 2\rho B U_\infty^2 \delta_2. \qquad (16.116)$$

For the cylinder one obtains

16.7 Plane, Two-Dimensional Wake Flow

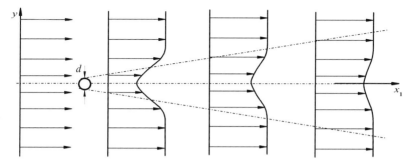

Fig. 16.12 Wake flow behind a body around which a flow takes place

$$K_w = c_w \frac{\rho}{2} U_\infty^2 \ell d. \tag{16.117}$$

Although the actual flow structure near the plate or cylinder around which the flow passes can be complicated, the flow in the downward field proves to be of the kind that becomes independent of the body which generated the wake flow. In the downstream region, the flow has a boundary-layer flow structure, as the fluid flow in the x-direction of the wake takes place convectively and the transverse distribution is established by diffusion.

For the treatment of the wake flow, sketched in Fig. 16.12, the velocity difference can be expressed as

$$u(x_1, x_2) = U_\infty - U_1(x_1, x_2) \tag{16.118}$$

and it is this difference that is introduced into the boundary-layer equations. When considering that the pressure in the entire flow region is constant and that moreover, because $u(x_1, x_2) << U_\infty$ holds, one can write

$$(U_\infty - u)\frac{\partial}{\partial x_1}(U_\infty - u) \approx U_\infty \frac{\partial u}{\partial x_1} = \nu \frac{\partial^2 u}{\partial x_2^2} \tag{16.119}$$

and, hence, one obtains the differential equation that is to be solved for the wake flow. The boundary conditions can be stated as

$$x_2 = 0 : \frac{\partial u}{\partial x_2} = 0 \quad \text{and} \quad x_2 \to \infty : u = 0. \tag{16.120}$$

Similarly to the earlier ansatzes, the following relationships are introduced.

$$\eta = x_2 \sqrt{\frac{U_\infty}{\nu x_1}} \quad \text{and} \quad u = U_\infty C \left(\frac{x_1}{\ell}\right)^{-1/2} f(\eta) \tag{16.121}$$

and (16.121) introduced into (16.115) yields

$$K_w = \rho B U_\infty^2 C \left(\frac{\nu \ell}{U_\infty}\right)^{1/2} \int_{-\infty}^{+\infty} f(\eta) \, d\eta. \tag{16.122}$$

On introducing (16.121) into (16.119), one obtains the ordinary differential equation to be solved for wake flows:

$$f'' + \frac{1}{2}\eta f' + \frac{1}{2}f = 0 \tag{16.123}$$

with the boundary conditions:

$$\eta = 0 : f'(\eta) = 0 \quad \text{and} \quad \eta \to \infty : f(\eta) \to 0. \tag{16.124}$$

Equation (16.123) can be rewritten and integrated:

$$\frac{d^2 f}{d\eta^2} + \frac{d}{d\eta}\left(\frac{1}{2}\eta f\right) = 0, \quad \frac{d}{d\eta}\left(\frac{df}{d\eta} + \frac{1}{2}\eta f\right) = 0 \tag{16.125}$$

so that following final relationship holds:

$$\frac{df}{d\eta} + \frac{1}{2}\eta f = C \quad \leadsto \quad C = 0, \quad \text{since} \quad \eta = 0 \frac{df}{d\eta} = 0. \tag{16.126}$$

In this way, the solution for $f(\eta)$ is obtained as an exponential function of η^2:

$$f(\eta) = \exp\left(-\frac{1}{4}\eta^2\right). \tag{16.127}$$

With this, the following can be computed from (16.122):

$$K_w = \rho B U_\infty^2 C \left(\frac{\nu \ell}{U_\infty}\right)^{1/2} \int_{-\infty}^{+\infty} \exp\left(-\frac{1}{4}\eta^2\right) d\eta. \tag{16.128}$$

For the flat plate the value of K_w can be computed by integration along both plate sides:

$$K_w = 1.328 B \rho U_\infty^2 \sqrt{\frac{\nu \ell}{U_\infty}} \tag{16.129}$$

so that for C in (16.128) one can deduce $C = 0.664/\sqrt{\pi}$. Hence, one obtains for the velocity profile of the wake flow behind a flat plate:

$$U_1(x_1, x_2) = U_\infty - U_\infty \frac{0.664}{\sqrt{\pi}} \left(\frac{x_1}{\ell}\right)^{-1/2} \exp\left(-\frac{1}{4}\frac{x_2^2 U_\infty}{x_1 \nu}\right). \tag{16.130}$$

This somewhat asymptotic solution is given in Fig. 16.13, namely for the difference velocity $u(x_1, x_2) \leadsto U(\eta)$, normalized with the local maximum of the "deficit velocity":

$$U_{\max} = U_\infty \frac{0.664}{\sqrt{\pi}} \left(\frac{x_1}{\ell}\right)^{-1/2}, \tag{16.131}$$

i.e. $U(\eta)/U_{\max}$ is plotted in Fig. 16.13.

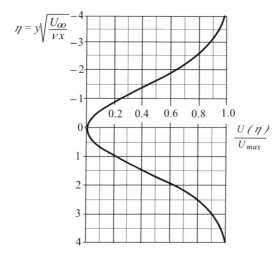

Fig. 16.13 Solution for the normalized wake flow behind a plate

16.8 Converging Channel Flow

The boundary-layer flows considered above, were all characterized by an outer flow with constant velocity. They thus represent the easiest flows which can be solved with aid of the boundary-layer equations.

Based on the solution procedure derived by Blasius, in this section the boundary layer of a flow will be considered, whose imposed flow is given by the following velocity distribution:

$$U_\infty(x) = -\frac{U_Q}{x} = U(x), \qquad (16.132)$$

where $U_Q = \dot{Q}/x$, i.e. is the velocity existing at $x = 1$. When considering that the velocity distribution corresponds to that of a sink flow in a convergent channel with an aperture angle α between the plane walls, then the volume flow rate flowing through the region α $\dot{Q} = \alpha x h U_Q$ is given for $h = 1$ by αU_Q, i.e. the following holds:

$$U_Q = \frac{\dot{Q}}{\alpha}. \qquad (16.133)$$

Because of the above-suggested velocity distribution of the flow, outside the developing boundary layers, a pressure gradient results in the stationary boundary-layer equation:

$$\frac{\partial \Psi}{\partial y}\frac{\partial^2 \Psi}{\partial x \partial y} - \frac{\partial \Psi}{\partial x}\frac{\partial^2 \Psi}{\partial y^2} = -\frac{1}{\rho}\frac{\mathrm{d}P}{\mathrm{d}x} + \nu\frac{\partial^3 \Psi}{\partial y^3}. \qquad (16.134)$$

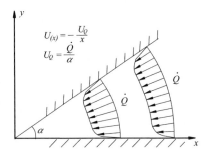

Fig. 16.14 Diagram to explain boundary-layer formation in a converging two-dimensional channel flow

The resulting pressure gradient can be computed as:

$$\frac{1}{\rho}\frac{dP}{dx} = -U_\infty(x)\frac{dU(x)}{dx} = \frac{\dot{Q}}{\alpha x}\frac{\dot{Q}}{\alpha^2} = \frac{\dot{Q}^2}{\alpha^2 x^3}. \tag{16.135}$$

In order to obtain the sought similarity solution for the boundary layer which forms at the walls of a converging channel flow (Fig. 16.14), the following similarity variable is introduced:

$$\eta = y\sqrt{\frac{-U_\infty}{x\nu}} = \frac{y}{x}\sqrt{\frac{U_Q}{\nu}} = \frac{y}{x}\sqrt{\frac{\dot{Q}}{\alpha\nu}}. \tag{16.136}$$

For the stream function of the flow we introduce

$$\Psi(x,y) = -\sqrt{\nu U_Q}f(\eta) = -\sqrt{\frac{\dot{Q}\nu}{\alpha}}f(\eta) \tag{16.137}$$

so that for the different terms in (16.38) the following relationships can be derived:

$$U_x = \frac{\partial\Psi}{\partial y} = \frac{\partial\Psi}{\partial\eta}\frac{\partial\eta}{\partial y} = -\frac{\dot{Q}}{\alpha}\frac{1}{x}f'(\eta), \tag{16.138}$$

$$U_y = -\frac{\partial\Psi}{\partial x} = -\frac{\partial\Psi}{\partial\eta}\frac{\partial\eta}{\partial x} = -\frac{\dot{Q}}{\alpha}\frac{y}{x^2}f'(\eta), \tag{16.139}$$

$$\frac{\partial^2\Psi}{\partial x\partial y} = \frac{\partial}{\partial x}\left(\frac{\partial\Psi}{\partial y}\right) = \frac{\dot{Q}}{\alpha}\frac{1}{x^2}f'(\eta) + \frac{\dot{Q}}{\alpha}\frac{\eta}{x^2}f''(\eta), \tag{16.140}$$

$$\frac{\partial^2\Psi}{\partial y^2} = \frac{\partial}{\partial y}\left(\frac{\partial\Psi}{\partial y}\right) = -\frac{\dot{Q}}{\alpha}\frac{1}{x}f''(\eta)\frac{1}{x}\sqrt{\frac{\dot{Q}}{\alpha\nu}}, \tag{16.141}$$

$$\frac{\partial^3\Psi}{\partial y^3} = -\frac{\dot{Q}}{\alpha}\frac{1}{x}f'''(\eta)\frac{1}{x^2}\frac{\dot{Q}}{\alpha\nu}. \tag{16.142}$$

On inserting (16.134) and also (16.136)–(16.140) in the boundary-layer equation (16.132), one obtains the following differential equation for $f(\eta)$; only

16.8 Converging Channel Flow

a few terms remain, since the products of the mixed derivatives $f'f''$ cancel out:
$$f''' - (f')^2 + 1 = 0 \tag{16.143}$$
with the boundary conditions at $\eta = 0$, $f' = 0$ and for $\eta \to \infty$, $f' = 1$ and $f'' = 0$.

The integration of the above differential equation becomes analytically possible with the ansatz
$$F(\eta) = f'(\eta) \tag{16.144}$$
so that (16.143) can be written as
$$F'' = F^2 - 1 \quad \text{with} \quad F(0) = 0 \quad \text{and} \quad F(\infty) = 1. \tag{16.145}$$

On multiplying both sides with the first derivative $F'(\eta)$, it is possible, by partial integration, to obtain the following solution:
$$(F')^2 - \frac{2}{3}(F-1)^2(F+2) = C, \tag{16.146}$$
where C is the integration constant, which, because $F \to 1$ and $F' \to 0$ as $\eta \to \infty$, adopts the value $C = 0$. Hence, from (16.146) it follows that
$$F' = \frac{dF}{d\eta} = \sqrt{\frac{2}{3}(F-1)^2(F+2)} \tag{16.147}$$

or rewritten:
$$d\eta = \frac{dF}{\sqrt{\frac{2}{3}(F-1)^2(F+2)}}. \tag{16.148}$$

Hence, the following equation holds:
$$\eta = \int_0^F \frac{dF}{\sqrt{\frac{2}{3}(F-1)(F+2)}}. \tag{16.149}$$

The integral can be given in a closed form:
$$\eta = \sqrt{\frac{3}{2}} \left(\tanh^{-1} \frac{\sqrt{2+F}}{\sqrt{3}} - \tanh^{-1} \sqrt{\frac{2}{3}} \right). \tag{16.150}$$

Solved in terms of $F = \frac{U_x}{U} = f'(\eta)$:
$$f'(\eta) = \frac{U_x}{U} = 3 \tanh^2 \left[\frac{\eta}{\sqrt{2}} + \ln\left(\sqrt{3}+\sqrt{2}\right) \right] - 2 \tag{16.151}$$

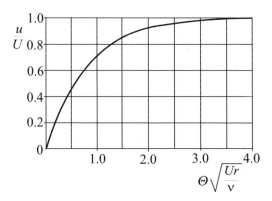

Fig. 16.15 Velocity distribution in a laminar boundary layer forming at the walls of a converging channel

or rewritten:

$$f'(\eta) = \frac{U_x}{U} = 3\tanh^2\left(\frac{\eta}{\sqrt{2}} + 1.146\right) - 2. \quad (16.152)$$

Introducing also $\Theta = y/x$ and $\dot{Q} = rU\alpha$, one can rewrite η as:

$$\eta = \Theta\sqrt{\frac{Ur}{\nu}} \quad (16.153)$$

and the velocity distribution shown in Fig. 16.15 results for the boundary layer at the walls of a converging, plane, two-dimensional channel flow.

The above-obtained relationship (16.152):

$$U_x = \frac{3\dot{Q}}{\alpha x}\tanh^2\left(\frac{\eta}{\sqrt{2}} + 1.146\right) - 2 \quad (16.154)$$

can also be written as:

$$\tanh^2[Z] = 1 - \frac{1}{\cosh^2[Z]} = 1 - \frac{4}{(\exp[Z] + \exp[-Z])^2}. \quad (16.155)$$

This can be inserted to yield the velocity distribution:

$$U_x = \frac{\dot{Q}}{\alpha x}\left[1 - \frac{12}{\left[(\sqrt{3} + \sqrt{2})\exp(\eta/\sqrt{2}) + (\sqrt{3} - \sqrt{2})\exp(-\eta/\sqrt{2})\right]^2}\right]. \quad (16.156)$$

References

16.1. G. K. Batchelor: "An Introduction to Fluid Dynamics", Cambridge University Press, Cambridge (1967)
16.2. B. A. Boley and M. B. Friedman: On the Viscous Flow Around the Leading Edge of a Flat Plate, JASS 26, 453–454 (1959)

16.3. G. F. Carrier and C. C. Lin: On the Nature of the Boundary Layer Near the Leading Edge of a Flat Plate, Q. Appl. Math. VI, 63–68 (1948)
16.4. D. Pnueli and C. Gutfinger: "Fluid Mechanics", Cambridge University Press, Cambridge (1992)
16.5. H. Schlichting: "Boundary Layer Theory", 6th edition, McGraw-Hill, New York (1968)
16.6. F. S. Sherman: "Viscous Flow", McGraw-Hill, Singapore (1990)
16.7. J. M. Shi, F. Durst and M. Breuer: Matching Asymptotic Computation of the Viscous Flow Around the Leading Edge of a Flat Plate, in preparation.
16.8. F. Durst: "Strömungslehre – Einführung in die Theorie der Strömungen", 4, Springer, Berlin Heidelberg New York (1996)
16.9. C. S. Yih: "Fluid Mechanics – A Concise Introduction to the Theory", West River Press, Ann Arbor, MI (1979)

Chapter 17
Unstable Flows and Laminar-Turbulent Transition

17.1 General Considerations

It is common practice to categorize flows as laminar or turbulent, i.e. to employ a special state of the flow to perform a subdivision: into laminar flows, i.e. in such flows in which the momentum, heat and mass transport processes are molecular dependent, and into such flows in which turbulence-dependent transport processes occur in addition. For the considerations presented in this chapter, a further subdivision is appropriate, so that grouping into four sub-groups is made:

- *Stable Laminar Flows.* A laminar flow may fulfill all requirements of the basic equations of fluid mechanics. It may also satisfy the initial and boundary conditions characteristic for the flow. Yet it must not represent a solution such as one finds in corresponding experimental investigations. Disturbances of the flow, as always occur in experiments, are often not considered in solutions of the basic equations governing fluid flows. Only such laminar flows that prove stable towards disturbances that act from the outside, i.e. attenuate the imposed disturbances, are defined as stable laminar flows.
- *Unstable Laminar Flows.* A laminar flow is considered unstable when disturbances introduced into it are amplified, but a certain "regularity" in the excited disturbance is maintained, i.e. due to the disturbance the investigated flow merges into a new laminar flow state. If this "new laminar flow state" is stable towards newly introduced disturbances, we have a bifurcation of the laminar flow. Here it is important to understand that the flow occurring after the imposed disturbance can be stationary or non-stationary.
- *Transitional Flows.* When the disturbances introduced in a laminar flow are amplified and result in flows that appear orderly in parts, but show also temporarily and/or spatially irregular fluctuations of all flow quantities, we speak of a transitional flow state. Intermittent laminar and

turbulent flow states occur, i.e. phases occur in the flow in which the flow is laminar and phases in which the flow shows turbulent characteristics. Flows that are in a transitional state still show clear characteristics that depend on the imposed disturbances.

- *Turbulent Flows.* It is now easily possible to imagine, on the basis of the considerations presented above, that disturbances are introduced into flows to such an extent that fluid motions result from them that are "out of control." Such turbulent fluctuations of all flow quantities are superimposed on corresponding mean flow quantities and are characterized by high non-stationarity and by high three-dimensionality. Turbulence-dependent transport processes of momentum, heat and mass are superimposed on the molecular-dependent transport process. A closed treatment of turbulent transport processes is at present only possible for small Reynolds numbers ($Re \leq 40{,}000$) by employing numerical computation procedures. The treatment of turbulent flows at high Reynolds numbers remains a problem of fluid mechanics that has not been solved.

In the preceding chapters, flows were investigated that are assumed to be stable, i.e. without concrete proof it was assumed that disturbances which are introduced into the flow are damped out by viscosity. In order to understand now the causes of stable laminar flows, i.e. the causes of the attenuation of disturbances by viscosity, the simplified one-dimensional momentum equation is considered:

$$\rho \frac{\partial U_1}{\partial t} = \mu \frac{\partial^2 U_1}{\partial x^2}. \tag{17.1}$$

When applying this to a flow with a constant flow velocity U_0, on which a disturbance with amplitude u_A is superimposed, the following equation applies for the total velocity field:

$$U_1 = U_0 + u_A \sin\left(2\pi \frac{x}{\lambda}\right) \quad \text{with} \quad U_0 = \text{constant.} \tag{17.2}$$

On now forming the temporal derivative ($\partial U_1/\partial t$) and the spatial derivative ($\partial^2 U_1/\partial x^2$), one obtains from (17.2) the following differential equation for the disturbance. It indicates how the temporal change of the amplitude of the imposed fluctuations behaves in time:

$$\frac{du_A}{dt} = -\nu \frac{4\pi^2}{\lambda^2} u_A. \tag{17.3}$$

This differential equation can be solved by separation of the variables. Thus, for the initial condition $u_A = (u_A)_0$, for the time $t \geq 0$ the following solution can be derived:

$$u_A(t) = (u_A)_0 \exp\left(-\nu \frac{4\pi^2}{\lambda^2} t\right). \tag{17.4}$$

This solution makes it clear that the viscosity terms in the momentum equations can be considered to lead to attenuations of imposed disturbances, i.e. disturbances which are introduced into a laminar flow field will be damped

17.1 General Considerations

due to the viscosity of the fluid. As expressed by (17.4), the attenuation of short-wave disturbances, i.e. disturbances with small λ values, turns out to be stronger, so that these receive stronger damping in the course of time. It is this attenuation effect, caused by the viscosity of a fluid, which ensures that many laminar flows possess high stability. This means that they show strong resistance against external disturbances.

As concerns the possible mechanisms of amplification of disturbances, these can be manifold and some are discussed in an introductory way in subsequent sections. Generally it can be said, however, that gradients of flow and/or fluid properties can be stated as causes of amplification. When they act on introduced disturbances such that an exponential excitation takes place, the latter can be described as follows:

$$u_A = (u_A)_0 \exp(\alpha t). \tag{17.5}$$

When a viscosity-dependent attenuation exists at the same time, the temporal development of the amplitude of a disturbance can be stated in a simplified way, and the following net result can be assumed to be valid:

$$u_A = (u_A)_0 \exp\left[(\alpha - \beta)t\right]. \tag{17.6}$$

When the viscosity-dependent attenuation term β proves to be larger than the amplification-caused term α, i.e. $\beta > \alpha$, we have a stable laminar flow. When, on the other hand, the amplification term α dominates, i.e. $\alpha > \beta$, we have an unstable flow. This means that the flow field determined from the Navier–Stokes equations for given initial and boundary conditions will not form in practice. Due to the above-postulated exponential increase of the disturbance introduced into the flow, a transition into a turbulent flow is to be expected. When the excitation takes place in another, non-exponential form, other unstable flow states, as mentioned in the above points, can form.

To make clear now what is to be understood by a stable laminar flow state, reference is made to the backward-facing, double-sided step flow, which is illustrated in Fig. 17.1. It shows a symmetrical solution for $Re \leq 200$. When imposing temporal disturbances on these flows, the temporal change of the separation lengths x_2, shown in Fig. 17.1, indicates that, after abandoning the imposed disturbances, the separation and reattachment lengths, that are characteristic for the step flow, are attained again. The flow is thus, for the investigated Reynolds number, stable towards the imposed disturbances. At higher Reynolds numbers, i.e. for $Re \geq 200$, this stability no longer exists. The flow abandons its symmetry, and two separate regions of different lengths and shapes occur.

For further explanations of the processes that take place with unstable laminar flows, reference is made to the flow through a rectangular channel. The latter is characterized by secondary flows as shown in Fig. 17.2. These so-called secondary flows represent fluid motions in a plane vertical to the main flow. Depending on the Reynolds number, a certain secondary flow pattern develops as the so-called bifurcation diagram demonstrates, which

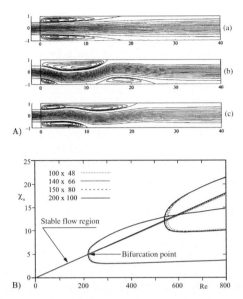

Fig. 17.1 (**A**) Stability of double-sided step flow (sudden expansion): (**a**) symmetrical solution (unstable); (**b**) asymmetric solution (stable solution of the first bifurcation); (**c**) asymmetric solution (unstable solution of the second bifurcation). (**B**) Bifurcation diagram for a flow with sudden extension

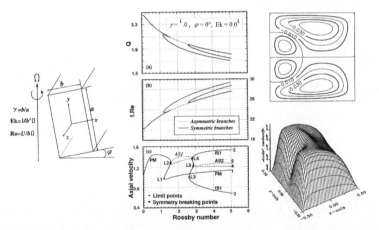

Fig. 17.2 Pattern of the secondary flow in a rectangular channel and corresponding bifurcation diagram

is also shown in Fig. 17.2. Detailed numerical investigations show, however, that certain patterns of secondary flow can only be obtained by imposed disturbances that are well directed.

17.1 General Considerations

Another possibility of bifurcation of an unstable laminar flow is given in Fig. 17.3. This figure shows that the laminar flow around a cylinder posseses a symmetry of the flow field for small Re. For $Re \geq 46$ the "symmetry" is broken. A non-stationary vortex flow (Hopf bifurcation) develops with alternating vortices relieving one another. In the wake of the cylinder, the so-called Karman vortex street results as the form of the laminar flow around a cylinder which proves to be a stable flow form for $Re > 46$.

To illustrate a transitional flow, reference is made to Fig. 17.4, in which a region of turbulent flow is shown that is embedded in a laminar plate flow. It can clearly be seen that the turbulence-dependent flow disturbances are

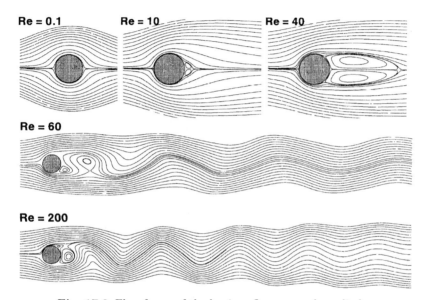

Fig. 17.3 Flow forms of the laminar flows around a cylinder

Fig. 17.4 Turbulent spot to illustrate the transitional, laminar-to-turbulent flow state of a flow; see van Dyke [17.1]

spatially and temporally limited. Likewise, a certain, still clearly distinctive structure of the flow can be recognized. All of these are clear characteristics of flows that are in a transitional, laminar-to-turbulent flow state.

As a turbulent state of a flow, the photograph illustrates a grid wake flow, which is made visible by introduced smoke. The "filaments" of smoke, introduced near the grid, are structured by the turbulent fluid motions and form a typical picture of an almost isotropic flow field. By isotropy of a flow a local property is understood, namely the independence of the mean properties of the flow of considered directions (Fig. 17.5).

That the flow states described above can occur within *one* flow region is illustrated in Fig. 17.6. This figure shows an open jet coming out of a nozzle

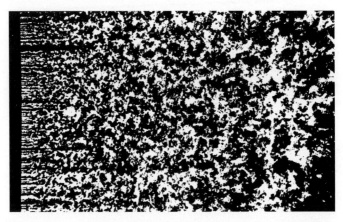

Fig. 17.5 Turbulent grid wake flow: making turbulent flow properties visible by introduced smoke; see van Dyke [17.1]

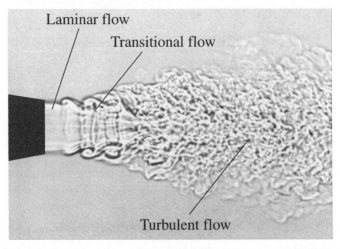

Fig. 17.6 Subsonic open jet with areas of laminar, transitional and turbulent flow; see van Dyke [17.1]

which shows laminar flow behavior in the immediate vicinity of the nozzle. The laminar flow becomes unstable and leads to a laminar-to-turbulent transition behavior, before further downstream the flow becomes completely turbulent. For the last-named flow region it is characteristic that no distinctively coherent flow structures can be recognized any more. This makes it clear that the transitional flow behavior can occur in parts of flows. In front of these transitional sub-domains the flow is laminar and behind these domains (looked at in the flow direction) the flow is turbulent.

17.2 Causes of Flow Instabilities

Flow instabilities show features that are manifold, and there is an extensive literature available on their causes and on the resultant flow structures that can be observed. It shall be the task here to discuss some of the causes that occur and the resultant flow phenomena. The chapter tries to give an introduction to the considerations that have to be carried out to investigate flow instabilities theoretically. The methods employed in the considerations are also part of the presentations, but they are limited to selected examples. They were chosen, however, such that they clearly illustrate the full attraction of fluid-mechanical investigations of unstable flows. Hence, the aim of the presented material is to ensure that students of fluid mechanics receive an early introduction to the broad (but also very specialized) field of non-stationary flow investigations, and that they are made familiar with the available solution methods. Here, it is important always to take the below-mentioned steps towards a solution of the posed instability problems:

(a) Determine the main flow field as an analytical or numerical solution of the Navier–Stokes equations and for the boundary conditions determining the flow problem.
(b) Utilize the basic equations to derive equations to treat flow disturbances. Carry out considerations for small amplitudes of the disturbances, i.e. use the linearized equations to compute the disturbance. In this way, linear, partial differential equations result and have to be treated for the disturbances of all flow quantities to be considered.
(c) Obtain solutions of the resulting linear, partial differential equation system for given disturbances, in order to investigate the increase or decrease in the amplitude of the disturbances in space and time.
(d) The linear, partial differential equation system can be solved, to some extent, for the general propagations of disturbances. Solutions do not exist for all disturbances, so that one is often forced to carry out further simplifications. The derivation of the Orr–Sommerfeld equation results from such a simplification.

(e) Finally, interpretations of the results obtained are needed for deciding the ranges of parameter values for stable or unstable, laminar flows.

Steps (a)–(e) are shown in parts of the subsequent derivations.

17.2.1 Stability of Atmospheric Temperature Layers

In Chap. 6, the pressure distribution in the atmosphere was considered for very different temperature distributions, e.g. also for the following temperature distribution:

$$T(x_2) = T_0 \left(1 - \frac{x_2}{c}\right), \tag{17.7}$$

which states a linear temperature decrease with increasing height above sea level. In this relationship $\gamma = T_0/c$ is the (existing) temperature gradient:

$$\frac{dT}{dx_2} = \frac{T_0}{c} = \text{constant}. \tag{17.8}$$

For the temperature distribution in the atmosphere, given by (17.7), the corresponding pressure distribution could be found by integration of the following equation:

$$\frac{dP}{dx_2} = -g\rho = -\frac{g}{RT}P \tag{17.9}$$

resulting in

$$P = P_0 \left(1 - \frac{x_2}{c}\right)^{\frac{gc}{RT_0}} = P_0 \left(T_0 - \gamma x_2\right)^{\frac{g}{R\gamma}}. \tag{17.10}$$

In the considerations in Chap. 6, it was indirectly assumed that the chosen temperature distribution is stable, i.e. introduced disturbances leave the imposed temperature distribution undisturbed. Here, the extent to which this holds true will be investigated, i.e. up to what temperature layer the imposed temperature gradient is stable.

To be able to carry out the required stability considerations, a fluid element is considered which is deflected upwards for a short time. Here the deflection of the fluid element is considered under adiabatic conditions. When the deflection leads to a buoyancy force on the fluid element, increasing the induced deflection, the temperature distribution is considered to be unstable. When the deflected fluid element experiences a restoring force, which is directed such that a reduction of the deflection takes place, we have a stable temperature distribution in the considered atmosphere (Fig. 17.7).

Taking the temperature $T_\Re$ as the temperature of a fluid particle which experiences an adiabatic upward movement from z to $z + dz$, the following relationship holds under adiabatic conditions:

17.2 Causes of Flow Instabilities

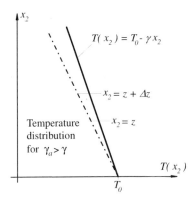

Fig. 17.7 Stability of temperature distribution in the atmosphere

$$\frac{T_{\Re}}{(T_{\Re})_z} = \left(\frac{P_{\Re}}{(P_{\Re})_z}\right)^{\frac{\kappa-1}{\kappa}} = \left(\frac{P}{P_z}\right)^{\frac{\kappa-1}{\kappa}}, \tag{17.11}$$

where $T_{\Re}$ and $(T_{\Re})_z$ are the fluid-element temperature and $P_{\Re}$ and $(P_{\Re})_z$ represent the corresponding pressures at the considered heights.

From (17.11) one obtains

$$d(\ln T_{\Re}) = \frac{\kappa - 1}{\kappa} d(\ln P). \tag{17.12}$$

From (17.9), it follows that

$$d(\ln P) = -\frac{g}{RT} dx_2 \tag{17.13}$$

so that the following differential equation holds:

$$d(\ln T_{\Re}) = -\frac{\kappa - 1}{\kappa} \frac{g}{RT} dx_2. \tag{17.14}$$

From $T = T_0 \left(1 - \frac{x_2}{c}\right)$, it follows that

$$dT = -\frac{T_0}{c} dx_2 = -\gamma\, dx_2. \tag{17.15}$$

From (17.14) and (17.15), one obtains

$$\frac{dT_{\Re}}{T_{\Re}} = -\frac{\kappa - 1}{\kappa} \frac{g}{\Re \gamma} \frac{dT}{T}. \tag{17.16}$$

On introducing the adiabatic temperature gradient, common in meteorology:

$$\gamma_A = \left(\frac{\kappa - 1}{\kappa}\right) \frac{g}{R} \tag{17.17}$$

into the above relationship, one can then write

$$\frac{dT_\Re}{T_\Re} = \frac{\gamma_A}{\gamma}\frac{dT}{T} \tag{17.18}$$

or integrated:

$$\frac{T_\Re}{(T_\Re)_z} = \left(\frac{T}{T_z}\right)^{(\gamma_A/\gamma)}. \tag{17.19}$$

Due to the deflection of the fluid particle from position z, where it was in equilibrium with its surroundings, a fluid motion results due to the presence of the buoyancy force, for which the following equation of motion holds:

$$\rho_\Re \frac{d^2 z}{dt^2} = (\rho - \rho_\Re) g \tag{17.20}$$

or, after further rewriting:

$$\frac{d^2 z}{dt^2} = g\left[\left(\frac{T_\Re}{T}\right) - 1\right]. \tag{17.21}$$

Thus, because $(T_\Re)_z = T_z$, (17.21) holds:

$$\frac{d^2 z}{dt^2} = g\left[\frac{T_\Re}{(T_\Re)_z}\left(\frac{T_z}{T}\right) - 1\right] \tag{17.22}$$

and thus one can write for the subsequent considerations

$$\frac{d^2 z}{dt^2} = g\left[\left(\frac{T(z)}{T_z}\right)^{\frac{\gamma_A - \gamma}{\gamma}} - 1\right]. \tag{17.23}$$

For these considerations, it is important to know whether the existing linear temperature gradient $\gamma = T_0/c$ is larger or smaller than the adiabatic temperature gradient of the atmosphere, i.e. $\frac{(\kappa-1)}{\kappa}\frac{g}{R} \equiv \frac{g}{c_p} \lessgtr \gamma$ decides the stability of the temperature distribution in the atmosphere.

Hence, the following considerations can be carried out:

- A positive deflection z causes $T(z) < T_z$ because $T = T_0(1 - z/c)$, so that for $\gamma_A > \gamma$ it follows that $d^2z/dt^2 < 0$, i.e. the deflected fluid element experiences a restoring force. The considered temperature distribution is stable.
- When there is a value of $\gamma = T_0/c > \gamma_A$, a fluid element deflected in the positive z-direction experiences a positive force, i.e. the induced deflection of the fluid element is increased. This temperature distribution therefore is unstable. The smallest fluctuations of temperature and/or pressure in the atmosphere will therefore "under these conditions" lead to the formation of forces disturbing the considered temperature distribution.

17.2 Causes of Flow Instabilities

Summarizing, it can therefore be stated that for $T = T_0(1 - z/c) = (T_0 - \gamma z)$ the following holds:

$\gamma < \frac{g}{c_p}$: stable temperature distribution,
$\gamma > \frac{g}{c_p}$: unstable temperature distribution.

These insights into the physics of aerostatics have to be considered when employing (17.10) for the pressure distribution in the atmosphere. It holds only for $\gamma < \frac{g}{c_p}$.

The above derivations make it clear also that there are mechanisms present in the atmosphere which often are not noticed and which are suited to reduce temperature fields with strong gradients in the atmosphere. When the local temperature gradient reaches γ-values that are larger than g/c_p, the higher temperatures lying below the considered point will rise upwards when disturbances occur. An intermixing of the air layers results, such that $\gamma \leq g/c_p$ is achieved.

17.2.2 Gravitationally Caused Instabilities

In order to investigate the gravitation-dependent instability of an interface between two fluids, two infinitely extended fluids are considered which have a common interface surface in the plane x_1–x_3 (see Fig. 17.8). Here, the density of the upper fluid is ρ_A and that of the lower fluid is ρ_B. It is moreover assumed that the surface tension in the interface layer is given by σ. Due to the assumption that the fluid A expands in $0 \leq x_2 < +\infty$ and the fluid B in $-\infty < x_2 \leq 0$, an instability problem results that is spatially dependent only on x_2.

As it is assumed that the considered fluids in the upper and lower regions are viscous media, the considerations that should be carried out, concerning possible instabilities, should be based on the Navier–Stokes equations. When, however, one starts from the assumption that the following holds:

$$\sqrt{g\ell} \gg \frac{\nu}{\ell}, \qquad (17.24)$$

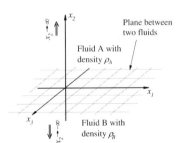

Fig. 17.8 Stratified fluids and stability of their interface

where ν/ℓ is the characteristic "viscous velocity" and $\sqrt{g\ell}$ is the characteristic "gravitation velocity", one can assume for the stability considerations to be carried out here that gravitation effects dominate when compared with viscous influences. These facts allow the employment of the "viscosity-free" form of the basic equations, in order to investigate the instability caused by gravitation, i.e. the instability of the fluids with the common interface shown in Fig. 17.8. As the considerations to be carried out start from the assumption that the fluids are at rest before the action of a disturbance sets in, the fluid motion imposed by the disturbance will be irrotational from the beginning. Therefore, it is recommended to treat the considered instability problem by introducing the potentials ϕ_A and ϕ_B for fluids A and B.

It is understandable that in the case of the influence of a disturbance on the interface surface, i.e. on $x_2 = 0$ for all x_1–x_3 values, one can expect for $x_2 \to \pm\infty$ that the velocities reach $(U_2)_A = (U_2)_B = 0$, so that one can set, without limiting the universal validity of the considerations:

$$\phi_A \to 0 \text{ for } x_2 \to +\infty \quad \text{and} \quad \phi_B \to 0 \text{ for } x_2 \to -\infty. \tag{17.25}$$

As the solutions for ϕ_A and ϕ_B must, for viscous-free flows, satisfy the Laplace equation, the following ansatzes can be chosen:

$$\phi_A(x_i, t) = C_A \exp(\alpha t - kx_2) S(x_1, x_3), \tag{17.26}$$

$$\phi_B(x_i, t) = C_B \exp(\alpha t + kx_2) S(x_1, x_3), \tag{17.27}$$

where $S(x_1, x_3)$ has to satisfy the following partial differential equation:

$$\left(\frac{\partial^2}{\partial x_1^2} + \frac{\partial^2}{\partial x_3^2} + k^2 \right) S(x_1, x_3) = 0. \tag{17.28}$$

On defining with η the deflection of the interface surface, the following kinematic relationship holds:

$$\frac{\partial \phi_A}{\partial x_2} = \frac{\partial \phi_B}{\partial x_2} = \frac{\partial \eta}{\partial t} \quad \text{for} \quad x_2 = \eta. \tag{17.29}$$

This relationship indicates that at the interface surface $(U_2)_A$ has to be equal to $(U_2)_B$ and that the velocity is given by the deflection velocity of the interface surface. Strictly there exists an equality for the normal components of the velocities. For small deflections of the interface it can generally be assumed, however, that the normal components of velocity are equal to the vertical components. By introducing this equality into the considerations to be carried out, a linearization of the problem is introduced, i.e. the subsequent considerations can be assigned to the field of the linear instability theory.

When considering the relations (17.26) and (17.27) for $x_2 = 0$, one can write for $\eta(x_1, x_3, t)$

$$\eta = C \exp(\alpha t) S(x_1, x_3). \tag{17.30}$$

17.2 Causes of Flow Instabilities

With (17.26), (17.27) and (17.30), one obtains from (17.29):

$$-kC_A = kC_B = C\alpha. \tag{17.31}$$

For the pressure difference between fluids A and B, one obtains:

$$P_A - P_B = \sigma \left(\frac{1}{R_A} + \frac{1}{R_B} \right) = \sigma \left(\frac{\partial^2 \eta}{\partial x_1^2} + \frac{\partial^2 \eta}{\partial x_2^2} \right). \tag{17.32}$$

With (17.30) one obtains, with consideration of (17.28):

$$P_A - P_B = -\sigma k^2 \eta. \tag{17.33}$$

Statements on P_A and P_B can also be obtained via the Bernoulli equation:

$$\frac{P_A}{\rho_A} = -\frac{\partial \phi_A}{\partial t} - g\eta; \quad \frac{P_B}{\rho_B} = -\frac{\partial \phi_B}{\partial t} - g\eta. \tag{17.34}$$

Hence one can derive

$$\rho_A(-\alpha C_A - gC) - \rho_B(-\alpha C_B - gC) = -\sigma k^2 C. \tag{17.35}$$

The elimination of C_A, C_B and C from (17.30) and (17.35) allows the following derivation for α:

$$\alpha^2 = \frac{g(\rho_A - \rho_B)k}{(\rho_A + \rho_B)} - \frac{\sigma k^3}{(\rho_A + \rho_B)}, \tag{17.36}$$

where α indicates the growth rate of a disturbance with time, (see (17.29)), so that

$$\frac{g(\rho_A - \rho_B)}{(\rho_A + \rho_B)} - \frac{\sigma k^2}{(\rho_a + \rho_B)} \geq 1 \tag{17.37}$$

or solved in terms of k^2:

$$k^2 < \frac{g}{\sigma}(\rho_A - \rho_B) > \frac{2\pi}{\ell}. \tag{17.38}$$

This relationship expresses that in the case of infinitely extended fluids with a common interface there always exists an ℓ which fulfills the condition for instability when $\rho_A > \rho_B$. Fluids with a common horizontal interface, where the heavy fluid is above, are inherently unstable. The fluids tend to "turn over," i.e. the heavier fluid tends to move to the lower location.

17.2.3 Instabilities in Annular Clearances Caused by Rotation

In the preceding considerations in this chapter, instabilities of static fluids were considered see also refs. [17.2]. A flow which proves unstable for certain parameter combinations was treated by Taylor [17.7]. He considered the

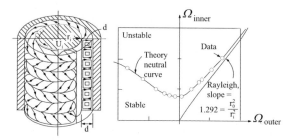

Fig. 17.9 Diagram of the vortex development for the Taylor annular-clearance flow and instability diagram

laminar flow between two rotating cylinders, as sketched in Fig. 17.9. There, the inner cylinder is assumed to rotate at a velocity $(U_\varphi)_1 = R_1\omega_1$ and the outer cylinder at $(U_\varphi)_2 = R_2\omega_2$. For $U_r = 0$ and $U_z = 0$, the following system of equations results for the flow in the annular clearance between the two cylinders:

$$\rho\frac{U_\varphi^2}{r} = \frac{dP}{dr} \quad \text{and} \quad \frac{d^2 U_\varphi}{dr^2} + \frac{1}{r}\frac{dU_\varphi}{dr} - \frac{U_\varphi}{r^2} = 0. \tag{17.39}$$

The second differential equation for U_φ is of the Euler type and thus allows particular solutions of the kind:

$$U_\varphi = C_k r^k \quad \rightsquigarrow \quad \frac{d^2 U_\varphi}{d\varphi^2} = C_k k(k-1) r^{k-2},$$

$$\frac{1}{r}\frac{dU_\varphi}{dr} = C_k k r^{k-1} \quad \text{and} \quad \frac{U_\varphi}{r^2} = C_k r^{k-2}. \tag{17.40}$$

This yields for k the general equation:

$$k(k-1) + (k-1) = (k+1)(k-1) = 0 \tag{17.41}$$

and thus $k_1 = 1$ and $k_2 = -1$ are obtained. Therefore, the general solution of the differential equation for the velocity U_φ reads:

$$U_\varphi = C_1 r + \frac{C_2}{r}. \tag{17.42}$$

The integration constants C_1 and C_2 result from the boundary conditions $U_\varphi(R_1) = R_1\omega_1$ and $U_\varphi(R_2) = R_2\omega_2$, so that one obtains:

$$C_1 = \frac{\omega_2 R_2^2 - \omega_1 R_1^2}{R_2^2 - R_1^2} \quad \text{and} \quad C_2 = \frac{(\omega_1 - \omega_2) R_1^2 R_2^2}{(R_2^2 - R_1^2)} \tag{17.43}$$

and thus for the velocity distribution $U_\varphi(r)$:

$$U_\varphi(r) = \frac{1}{r(R_2^2 - R_1^2)} \left[\left(\omega_2 R_2^2 - \omega_1 R_1^2\right) r^2 + (\omega_1 - \omega_2) R_1^2 R_2^2 \right]. \tag{17.44}$$

17.2 Causes of Flow Instabilities

The derivations carried out above show that the solution stated in (17.44) for the flow problem indicated in Fig. 17.9 fulfills the Navier–Stokes equations and the corresponding boundary conditions. The derivations carried out in order to obtain the analytical solution leave, however, the question of the stability of the solution unanswered, i.e. the extent to which disturbances introduced into the flow are attenuated or amplified still has to be resolved. Taylor demonstrated that the question can be solved through purely analytical considerations. In accordance with Taylor's considerations, we supplement the above-indicated considerations by means of the following ansatzes for the velocity components in the φ-, r- and z-directions:

$$U_r = u'_r; \quad U_\varphi = U_\varphi(r) + u'_\varphi \quad \text{and} \quad U_z = u'_z. \tag{17.45}$$

On entering these velocity ansatzes into the basic equations and neglecting the terms of second order, i.e. carrying out linear stability considerations, one obtains the following equation system for the determination of the disturbances u'_r, u'_φ and u'_z:
Continuity equation:

$$\frac{\partial}{\partial r}(ru'_r) + \frac{\partial}{\partial z}(ru'_z) = 0. \tag{17.46}$$

Momentum equations:

$$\frac{\partial u'_r}{\partial t} = -\frac{1}{\rho}\frac{\partial p'}{\partial r} + 2\left(C_1 + \frac{C_2}{r^2}\right)u'_\varphi + \nu\left[\frac{\partial^2 u'_r}{\partial z^2} + \frac{\partial^2 u'_r}{\partial r^2} + \frac{1}{r}\frac{\partial u'_r}{\partial r} - \frac{u'_r}{r^2}\right], \tag{17.47}$$

$$\frac{\partial u'_\varphi}{\partial t} = -2C_1 u'_r + \nu\left[\frac{\partial^2 u'_\varphi}{\partial z^2} + \frac{\partial^2 u'_\varphi}{\partial r^2} + \frac{1}{r}\frac{\partial u'_\varphi}{\partial r} - \frac{u'_\varphi}{r^2}\right], \tag{17.48}$$

$$\frac{\partial u'_z}{\partial t} = -\frac{1}{\rho}\frac{\partial p'}{\partial z} + \nu\left[\frac{\partial^2 u'_z}{\partial z^2} + \frac{\partial^2 u'_z}{\partial r^2} + \frac{1}{r}\frac{\partial u'_z}{\partial r}\right]. \tag{17.49}$$

Here, the boundary conditions $u'_r = u'_\varphi = u'_z = 0$ for $r = R_1$ and $r = R_2$ hold. For their solution, the following ansatzes are now chosen:

$$u'_r = u_1(r)\cos(\ell z)\exp(\beta t), \quad u'_\varphi = u_2(r)\cos(\ell z)\exp(\beta t) \text{ and}$$
$$u'_z = u_B(r)\sin(\ell z)\exp(\beta t). \tag{17.50}$$

Hence, the following equation system results for u_1, u_2 and u_3, which all depend only on the position coordinate r:

$$\nu\left[\frac{d^2 u_2}{dr^2} + \frac{1}{r}\frac{du_2}{dr} - \frac{u_2}{r^2} - \ell'^2 u_2\right] = 2C_1 u_1, \tag{17.51}$$

$$\frac{\nu}{\lambda}\frac{d}{dr}\left[\frac{d^2 u_3}{dr^2}+\frac{1}{r}\frac{du_3}{dr}-\ell'^2 u_3\right]=-2\left(C_1+\frac{C_2}{r^2}\right)u_2$$
$$-\nu\left[\frac{d^2 u_1}{dr^2}+\frac{1}{r}\frac{du_1}{dr}-\frac{u_1}{r^2}-\ell'^2 u_1\right], \qquad (17.52)$$

$$\frac{du_1}{dr}+\frac{u_1}{r}+\lambda u_3=0, \qquad (17.53)$$

where the following holds for ℓ':

$$\ell'^2=\ell^2+\frac{\beta}{\nu}. \qquad (17.54)$$

The system of equations for u_1, u_2 and u_3 can be solved by ansatzes of Fourier–Bessel series, where the development takes place in terms of the Bessel function:

$$z_1(k_\alpha r)=\alpha_1 J_1(k_\alpha r)+\alpha_2 N_1(k_\alpha r), \qquad (17.55)$$

where α_1 and α_2 are chosen such that the following holds:

$$z_1(k_\alpha R_1)=z_1(k_\alpha R_2)=0. \qquad (17.56)$$

Here

$$u_1(r)=\sum_{\alpha=1}^{\infty}A_\infty z_1(k_\alpha r), \qquad (17.57)$$

where the coefficients A_α have to be determined by the following relationships:

$$A_\alpha=\frac{1}{H_\alpha}\int_{R_1}^{R_z}ru_1(r)z_1(k_\alpha r)\,dr \text{ with } H_\alpha=\int_{R_1}^{R_z}rz_1^2(k_\alpha r)\,dr. \qquad (17.58)$$

For u_2 one obtains in accordance with (17.51):

$$\nu\left(\frac{d^2 u_2}{dr^2}+\frac{1}{r}\frac{du_2}{dt}-\frac{1}{r^2}-\lambda'^2\right)=2C_1\sum_{\alpha=1}^{\infty}A_\alpha z_1(k_\alpha r) \qquad (17.59)$$

so that one obtains:

$$u_2(r)=\sum_{\alpha=1}^{\infty}B_\alpha z_1(k_\alpha r) \qquad (17.60)$$

with the coefficients B_α being:

$$B_\alpha=-\frac{2C_1 A_\alpha}{\nu(k_d^2+\lambda'^2)}. \qquad (17.61)$$

The boundary conditions $u_2(R_1)=u_2(R_2)=0$ supply, however, $u_2(r)=0$.

17.2 Causes of Flow Instabilities

To determine $u_3(r)$, the differential equation (17.53) is employed and a known property of the Bessel function is implemented:

$$\frac{1}{k}\frac{d}{dr}z_0(kr) = -z_1(kr) \tag{17.62}$$

so that one obtains

$$\frac{d}{dr}\left(\frac{d^2 u_3}{dr^2} + \frac{1}{r}\frac{du_3}{dr} - \lambda'^2 u_3\right) = 0. \tag{17.63}$$

For this differential equation the following solution results:

$$u_3(r) = z_0(i\lambda' r) + \text{constant}. \tag{17.64}$$

By employing the above solution, Taylor was able to demonstrate that in the case of given values of ω_1, ω_2, R_1 and R_2 the quantities

$$\lambda \text{ and } \lambda' = \sqrt{\lambda^2 + \frac{\beta}{\nu}} \tag{17.65}$$

are linked with one another.

Taylor carried out the above-indicated analysis in detail, assuming $(R_2 - R_1) \ll \frac{1}{2}(R_2 + R_1)$, i.e. his results hold for narrow annular clearances. His results can be summarized as follows, in consideration of the stability diagram in Fig. 17.10:

- When both cylinders, forming the annular clearance, rotate in the same direction, the stability of the flow is always guaranteed when $R_1^2 \omega_1 < R_2^2 \omega_2$ holds (see Fig. 17.10).
- The elementary stability criterion given above is not applicable when the rotating cylinders possess opposite directions of rotation.
- For $\frac{\omega_1}{\omega_2} = \frac{R_2^2}{R_1^2} = 1.292$, the stability curve shown in Fig. 17.10 results. Here the loss of stability is characterized by the so-called Taylor vortices that form in the annular clearance. The directions of rotation of these vortices alternate.
- With $\delta = (R_2 - R_1)$ and $U_1 = R_1 \omega_1$, the critical Taylor number Ta, where the instability starts, can be stated as follows:

$$Ta = \frac{U_1 \delta}{\nu}\sqrt{\frac{\delta}{R_1}} \geq 41.3. \tag{17.66}$$

The flow forming in the annular clearance, due to the treated instability, is laminar again.
- The energy which is introduced through the drive of the inner cylinder drives the secondary vortices and, after the steady state of the flow has established itself, also the energy dissipation in the vortices.

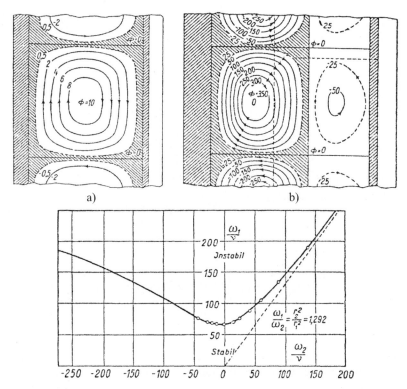

Fig. 17.10 (a) Vortex shapes between cylinders rotating in the same direction after occurrence of the Taylor instability. (b) Vortex development with cylinders rotating in the opposite direction after occurrence of the Taylor instability

It is interesting that in the cases of opposite directions of rotation of the two cylinders, i.e. for $\omega_2 = 0$, the value for ω_1 for which the Taylor instability occurs increases again when departing from $\omega_2 = 0$. It is further interesting that the annular vortices change with the opposite mode of rotation. As Fig. 17.10 shows, two vortex series form, one which is to be assigned to the rotation of the inner cylinder and another which belongs to the rotation of the outer cylinder.

17.3 Generalized Instability Considerations (Orr–Sommerfeld Equation)

Section 17.2 showed to some extent how to proceed with considerations in the framework of the linear instability theory. The approach presented there will now be generalized for flow computations for which the following disturbance ansatzes hold, see also refs. [17.6]:

$$\hat{u}_1 = U_x(y) + u'_x(x,y,t); \quad \hat{u}_2 = u'_y(x,y,t); \quad \hat{u}_3 = 0, \tag{17.67}$$

17.3 Generalized Instability Considerations

i.e. a one-dimensional stationary velocity field $U_x(y)$ is assumed as given, into which two-dimensional time-dependent disturbances are introduced, or can form in the flow. Furthermore, for the pressure we assume

$$\hat{P} = P(x) + p'(x, y, t), \tag{17.68}$$

where for all fluctuation quantities we introduce:

$$\frac{a'}{A} \ll 1; \quad \text{i.e.} \quad \frac{u'_x}{U_x} \ll 1; \quad \frac{u'_y}{U_x} \ll 1; \quad \frac{p'}{P} \ll 1. \tag{17.69}$$

On introducing the quantities defined in (17.67) and (17.68) into the Navier–Stokes equations, the following system of equations results:

$$\frac{\partial u'_x}{\partial x} + \frac{\partial u'_y}{\partial y} = 0, \tag{17.70}$$

$$\frac{\partial u'_x}{\partial t} + (U_x + u'_x)\frac{\partial u'_x}{\partial x} + u'_y\left(\frac{dU_x}{dy} + \frac{\partial u'_x}{\partial y}\right)$$
$$= -\frac{1}{\rho}\left(\frac{dP}{dx} + \frac{\partial p'}{\partial x}\right) + \nu\left(\frac{\partial^2 u'_x}{\partial x^2} + \frac{\partial^2 u'_x}{\partial y^2} + \frac{d^2 U_x}{dy^2}\right), \tag{17.71}$$

$$\frac{\partial u'_y}{\partial t} + (U_x + u'_x)\frac{\partial u'_y}{\partial x} + u'_y\frac{\partial u'_y}{\partial y} = -\frac{1}{\rho}\frac{\partial p'}{\partial y} + \nu\left(\frac{\partial^2 u'_y}{\partial x^2} + \frac{\partial^2 u'_y}{\partial y^2}\right). \tag{17.72}$$

These differential equations yield for the special case that no disturbances exist:

$$0 = -\frac{1}{\rho}\frac{dP}{dx} + \nu\frac{d^2 U_x}{dy^2}. \tag{17.73}$$

Disregarding all squared terms occurring in the disturbance quantities and after subtraction of (17.73), one obtains the following system of equations for the considered disturbance quantities u'_x, u'_y and p':

$$\frac{\partial u'_x}{\partial x} + \frac{\partial u'_y}{\partial y} = 0, \tag{17.74}$$

$$\frac{\partial u'_x}{\partial t} + U_x\frac{\partial u'_x}{\partial x} + u'_y\frac{dU_x}{dy} = -\frac{1}{\rho}\frac{\partial p'}{\partial x} + \nu\left(\frac{\partial^2 u'_x}{\partial x^2} + \frac{\partial^2 u'_y}{\partial y^2}\right), \tag{17.75}$$

$$\frac{\partial u'_y}{\partial t} + U_x\frac{\partial u'_y}{\partial x} = -\frac{1}{\rho}\frac{\partial p'}{\partial y} + \nu\left(\frac{\partial^2 u'_y}{\partial x^2} + \frac{\partial^2 u'_y}{\partial y^2}\right). \tag{17.76}$$

The introduced two-dimensionality of the disturbances allows the elimination of the differential equation (17.74), resulting from the continuity equation, by introducing a stream function for the velocity field of the disturbances:

$$u'_x = \frac{\partial \Psi'}{\partial y} \quad \text{and} \quad u'_y = -\frac{\partial \Psi'}{\partial x}. \tag{17.77}$$

If one expresses (17.75) and (17.76) in this disturbance stream function Ψ', one obtains the following differential equations for Ψ' and p':

$$\frac{\partial^2 \Psi'}{\partial y \partial t} + U_x \frac{\partial^2 \Psi'}{\partial x \partial y} - \frac{\partial \Psi'}{\partial x} \frac{dU_x}{dy} = -\frac{1}{\rho} \frac{\partial p'}{\partial x} + \nu \left(\frac{\partial^3 \Psi'}{\partial x^2 \partial y} + \frac{\partial \partial^3 \Psi'}{\partial \partial y^3} \right), \quad (17.78)$$

$$-\frac{\partial \Psi'}{\partial x \partial t} - U_x \frac{\partial^2 \Psi'}{\partial x^2} = -\frac{1}{\rho} \frac{\partial p'}{\partial y} - \nu \left(\frac{\partial^3 \Psi'}{\partial x^3} + \frac{\partial^3 \Psi'}{\partial x \partial y^2} \right). \quad (17.79)$$

By differentiation of (17.78) with respect to y and (17.79) with respect to x, the terms $\partial^2 p'/\partial x \partial y$ can be eliminated, so that one only obtains a differential equation for Ψ'. The latter contains, however, terms of fourth order and can be written as follows:

$$\left(\frac{\partial}{\partial t} + U_x \frac{\partial}{\partial x} \right) \left(\frac{\partial^2 \Psi'}{\partial x^2} + \frac{\partial^2 \Psi'}{\partial y^2} \right) - \frac{d^2 U_x}{dy^2} \frac{\partial \Psi'}{\partial x}$$
$$= \nu \left(-\frac{\partial^4 \Psi'}{\partial x^4} + 2 \frac{\partial^4 \Psi'}{\partial x^2 \partial y^x} - \frac{\partial^4 \Psi'}{\partial y^4} \right). \quad (17.80)$$

All of the above differential equations clearly express that disturbances, as expressed by (17.67) and (17.68), are governed by the conservation laws of fluid mechanics and the given undisturbed flow field $U_x(y)$. Special solutions can be obtained for the following ansatz for $\Psi'(x, y, t)$:

$$\Psi' = f(y) \exp\left[i \left(kx - \omega t \right) \right]. \quad (17.81)$$

Here k is always set as real, i.e. waves of wavelength $\lambda = 2\pi/k$ are considered in the direction of the x coordinate, whose behavior are described by the differential equation (17.80).

The quantity ω in the exponential term of the ansatz (17.81) can adopt complex values:

$$\omega = \omega_R + i\omega_I. \quad (17.82)$$

On proving that $\omega_I < 0$ holds, Ψ' decreases with time and the fluid flow $U_x(y)$ can be regarded as stable. For $\omega_I > 0$, the disturbance is excited with time, i.e. the flow $U_x(y)$, investigated for instabilities, proves to be unstable with respect to the imposed disturbances.

The differential equation (17.80) is of fourth order, so that its integration requires the implementation of four boundary conditions. They can be stated for plane channel flows and wall boundary-layer flows as follows:

- For plane channel flow it results for $y = 0$ and $y = 2H$ that $u'_x = u'_y = 0$, i.e.

$$f(0) = f'(0) = f(2H) = f'(2H) = 0. \quad (17.83)$$

- For flat plate boundary layer flow one obtains, because of the no-slip condition at the wall

$$f(0) = f'(0) = 0.$$

17.3 Generalized Instability Considerations

For the outer flow one can state, because of the lack of viscosity forces, i.e. $d^2 U_x/dy^2 = 0$, that

$$f'' - kf = 0 \quad \leadsto \quad f = \omega \exp(-ky). \tag{17.84}$$

On setting $\Psi' = f(y)\exp[ik(x - ct)]$, the well-known Orr–Sommerfeld differential equation results:

$$(kU_x - \omega)(f'' - k^2 f) - kU_x'' f = \frac{\nu}{ik}(f'''' - 2k^2 f'' + k^4 f). \tag{17.85}$$

This usually needs to be solved numerically for investigating the stability of a certain flow, using the undisturbed velocity distribution $U_x(y)$ and the assumed wavelength k in the equation and employing the above-indicated boundary conditions, e.g. for:

$$\begin{aligned} y = 0: &\quad f(0) = f'(0) = 0, \\ y \to \infty: &\quad f(y) = f'(y) \leadsto 0. \end{aligned} \tag{17.86}$$

For physical considerations it is also appropriate to introduce $c = \omega/k = c_R + ic_I$. For stability considerations one therefore has to look for the solution of an eigenvalue problem and to determine it for each wavelength of a disturbance, i.e. for each $\lambda = 2\pi/k$. The wavelength range which leads to negative values of the imaginary part of c is defined as stable, i.e. the investigated flow is stable with respect to these disturbances. Thus, it is determined by successive computations, for which wavelength the imaginary part of c is positive. This then leads to an insight into whether for a solution $U_x(y)$, that we have for a flow, the flow field $U_x(y)$ changes abruptly into a flow state differing from its undisturbed state.

When two-dimensional disturbances are imposed on considered flows, the behavior of flows with two-dimensional velocity profiles can nowadays be investigated numerically, e.g. plane channel flow with sudden cross-section widening indicated in Fig. 17.11. This figure illustrates an inner flow, which is given by a fully symmetric inflow in plane A, and shows symmetrical boundary conditions between planes A and B, and in B a symmetrical profile of the outflow exists. In spite of this, flow investigations show that the flow profiles between planes A and B are asymmetric from a certain Re_s value onwards. This is stated in Fig. 17.11b, which shows that from $\Re_s \approx 150$ onwards separate regions with differing lengths and locations form.

The results in Fig. 17.11 for two-dimensional plane channel flow were obtained numerically. For $Re < Re_s$ the numerical computations yielded symmetrical velocity profiles, i.e. for this Re range the viscosity influences were strong enough to attenuate disturbances in the flow. Thus it was possible at all locations on the symmetry axis to obtain and maintain $U_2 = 0$ for all times. For $Re > Re_s$ this important condition for the symmetry of the flow could not be fulfilled any longer, and the asymmetry of the flow indicated in the lower half of Fig. 17.11 developed. This kind of investigation also

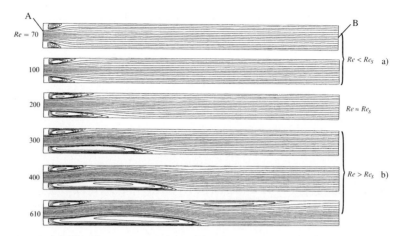

Fig. 17.11 (a) Two-dimensional plane channel flows with sudden expansion of cross-sectional area; (b) computation results with an asymmetry of the flow

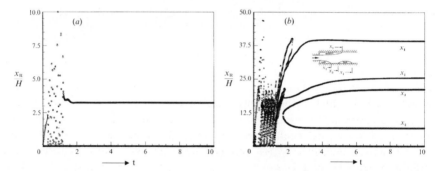

Fig. 17.12 Computations of the separation regions for imposed disturbances for flows of (a) $Re = 70$ and (b) $Re = 610$

yielded that the shorter separation region, characteristic for the asymmetry, can occur either above or below, depending on the imposed disturbance.

When drawing the separation lengths as a function of the Reynolds number, one obtains a so-called bifurcation diagram with $x_2/x_1 = 1$ for smaller Reynolds numbers, i.e. for $Re < Re_s$. This diagram would show that a bifurcation point S occurs for $Re = Re_s$. Two branches of the characteristic function x_2/x_1, typical of the asymmetry of the flow, start at $Re = Re_s$. The symmetrical solution turns out to become unstable for $Re > Re_s$, i.e. the smallest disturbances that are introduced into the flow lead to a break of the symmetry of the flow.

In order to demonstrate that the asymmetric flow fields that develop for $Re < Re_s$ are stable, the numerical solutions were exposed to considerable disturbances in time. This is shown in Fig. 17.12, which shows (a) temporal changes of the x_2 positions characterizing the separation region behind the step. When the temporal disturbances are removed, the dimensions of the

17.4 Classifications of Instabilities

The complexity of the behavior of flow fluctuations, resulting from induced disturbances due to flow instabilities, becomes clear when considering the ansatz for the stream function for velocity fluctuations:

$$\Psi'(x_1, x_2, t) = f(x_2) \exp\left[i(kx_1 - \omega t)\right]. \tag{17.87}$$

On introducing $k = k_R + ik_I$ and $\omega = \omega_R + i\omega_I$, one obtains with $f(x_2) = F(x_2)\exp[-i\theta(x_2)]$

$$\Psi'(x_1, x_2, t) = \exp\left(\omega_I t - k_I x_1\right)\left\{F(x_2)\cos\left[k_R x_1 - \omega_R t - \theta(x_2)\right]\right\}. \tag{17.88}$$

If all values k_R, k_I, ω_R and ω_I are unequal to zero, a disturbance is described by $\Psi'(x_1, x_2, t)$, which is extremely complex and can only be classified as difficult to describe and even more difficult to "mentally digest." When keeping the space point constant for considerations within a certain range, i.e. x_1, x_2 constant, (17.88) describes a disturbance (unstable flow) increasing with time, or a disturbance (stable flow) decreasing with time. Here, it has to be taken into consideration that an insight into the physics of stability or instability of a flow is only gained for one point of the flow field. It cannot be transferred to other points. On adding other disturbances to make the complexity of stability considerations still clearer, and if one looks at flow disturbances for constant values of x_2 and t, increases or decreases of disturbances in the x_1-direction look different. One can see that direct considerations of (17.88) lead only to a very limited extent to a deeper understanding of the stability or instability of flows.

When the information on an existing instability for one point of the flow field holds for the entire flow field, one talks about an *absolute instability*. This occurs, e.g., when in a rotating annular flow a disturbance is introduced which then transits from a flow free of Taylor vortices to a flow in which Taylor vortices occur in the entire flow field.

On introducing into a flow a disturbance, which is "carried away" like, e.g., the surface wave caused by a stone thrown into water, and when the disturbance then increases, one talks of a *convective instability*. The difference to the absolute instability is indicated in Fig. 17.13. In 17.13a, the increase of the disturbance in the entire part of the flow field with time is

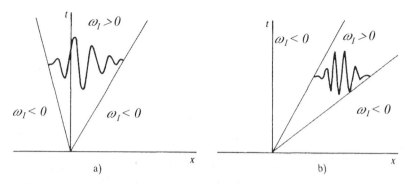

Fig. 17.13 Schematic diagrams of (**a**) absolute flow instability and (**b**) convective instability

indicated. This is characteristic of the presence of an absolute instability of the flow. In Fig. 17.13b, the increase of an induced disturbance with simultaneous movement in space is illustrated; this is characteristic of the presence of a convective instability.

In the presence of a convective instability, "feed-back" mechanisms can occur, such that the disturbance returns back, due to reflections, to the place from where the disturbance started. This causes a new disturbance there, which again is transported in a convective way and is again reflected, so that an instability only occurs in an embedded part of the system. In this case, one talks of a *global instability*. When there are global instabilities, subdivisions with regard to the feed-back mechanisms and the interactions of the reflected disturbances with the initial flows can be made. This is a research-active field of modern fluid mechanics.

In the literature, further classifications of possible instabilities have been made. We start from the following ansatz for the stream function of a disturbance that is allowed to increase with space and time:

$$\Psi'(x_1, x_2, t) = f(x_2) \exp\left[i\left(kx_1 - \omega t\right)\right] \quad (17.89)$$

and considering the corresponding Orr–Sommerfeld equation:

$$(kU_1 - \omega)\left(f'' - k^2 f\right) - kU_1'' f + iRe^{-1}\left[f'''' - 2k^2 f'' + k^4 f\right] = 0. \quad (17.90)$$

Simple stability considerations now start with the assumption that $k_I = 0$, so that as the basic disturbance a sine wave serves in the x_1-direction. For this case, for given values of Re and k_R, the eigenvalues of ω_R and ω_I are determined. When ω_I is positive, the amplitude of the disturbing wave grows with time; we have a *time instability* of the flow. On comparing this approach with the diagrams on the absolute flow instability in Fig. 17.13, it becomes clear that the time analysis of flow instabilities is appropriate when one wants to analyze a flow with regard to the presence of an absolute instability. Often time analysis is also chosen for reasons of simplicity, even when it is evident

17.5 Transitional Boundary-Layer Flows

that a convective instability is involved. The reason for this can be taken from the Orr–Sommerfeld equation (17.90). In this equation, k appears as a factor in several terms, so that a certain simplification occurs in the solution procedure when k is real. Strictly, one would have to set $\beta_I = 0$ (see the text above (17.88)), when investigating the spatial instabilities of a flow and thus carrying out investigations for given values for Re and ω_R to determine the eigenvalues of k_R and k_I. When k_I proves negative, we have an amplification of the amplitude of the disturbance with x_1, i.e. a spatial instability of the flow is present. Recent investigations of flow instabilities mostly carried out temporal and spatial analyses of the stability or instability of flows.

17.5 Transitional Boundary-Layer Flows

By relatively simple experimental investigations, one can detect that flows can be in a laminar or in a turbulent flow state. Velocity sensors introduced into a certain stationary flow lead to signals as shown in Fig. 17.14. In a flow defined as laminar, the velocity signal shows a temporally constant velocity. In a flow defined as being turbulent, there exists, however, a time variation of the local velocity which shows velocity fluctuations around a mean value. Both velocity variations with time are shown in Fig. 17.14. This makes clear the difference in the time variations of the velocity signals. The figure also shows the differences in the profiles of the velocity distributions.

The field of fluid mechanics that treats the transition of the laminar state of a flow into a turbulent one, is defined as "fluid mechanics of transitional flows." Usually the laminar-to-turbulent transition of a flow takes place as an intermittent process, such that a flow state is initially laminar. Introduced flow disturbances are then excited for a short time, subsequently experiencing an attenuation again. The flow transits initially only in an intermittent way from the laminar into the turbulent flow state, where over the duration of the turbulent phase, a comparison concerning the entire observation time, a so-called "intermittent factor" can be introduced, i.e.

$$I_F = \frac{\sum \Delta t_\mathrm{turb}}{T}. \tag{17.91}$$

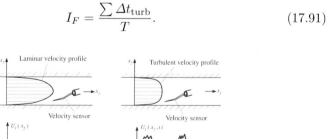

Fig. 17.14 Laminar and turbulent flow states at plane channel flows

For $I_F = 0$ the considered flow is, at the place of measurement, in its laminar state and for $I_F = 1$ it has reached its turbulent state. For the entire region $0 < I_F < 1$, there exists the so-called laminar-to-turbulent transitional range, with relatively abrupt changes from the laminar to the turbulent state of the flow. Investigations of these abrupt changes belong to the important investigations which constitute modern fluid mechanics research. An essential sub-domain of the current investigations in this field of research involves the laminar-to-turbulent transition *in boundary layers*. The latter will be treated here in an introductory way, with emphasis on wall boundary layers.

The basic idea of the present laminar-to-turbulent transition research starts from the assumption that the transition from the laminar to the turbulent state of a flow is concerned with the increase in disturbances. Thus, transition research is a sub-domain of stability research carried out in fluid mechanics. Its theory is thus based, especially where boundary-layer flows are concerned, on the Orr–Sommerfeld equation:

$$(kU_1 - \omega)(f'' - k^2 f) - kU_1'' f + iRe^{-1} \left[f'''' - 2k^2 f'' + k^4 f \right] = 0. \quad (17.92)$$

For $\nu = 0$ this differential equation can be stated as follows:

$$(kU_1 - \omega)(f'' - k^2 f) - kU_1'' f = 0. \quad (17.93)$$

This equation (Rayleigh equation) is only of second order, and therefore only two of the boundary conditions formulated in the preceding section can be introduced into the solution. They are usually employed as follows:

$$y = 0 \quad \leadsto \quad f(0) = 0 \quad \text{and} \quad y \to \infty \quad \leadsto \quad f(\to \infty) = 0. \quad (17.94)$$

Neglecting the viscosity term in (17.92) leads to a drastic simplification of the differential equation to be solved, because of the above-mentioned reduction of the order. This was probably the reason why all initial studies on the stability of flows were based on the Rayleigh equation (17.93). From this results the following insight:

> All laminar velocity profiles which show an inflection point, i.e. for which at one location of the velocity profile $d^2 U_1/dy^2 = 0$ holds, are unstable.

With this criterion, we have a necessary (Rayleigh) and a sufficient (Tollmien) condition for the occurrence of flow instabilities. This fact alone makes it clear that the curvature of velocity profiles has an important influence on the stability of a flow.

Intuitively one would assume that the inclusion of a viscosity term in (17.92), i.e. applying the Orr–Sommerfeld equation rather than the Rayleigh equation, leads to an attenuation of introduced disturbances. This, however, cannot be confirmed. It rather turns out that the solution of the Orr–Sommerfeld equation always has to be employed to obtain the approximately correct disturbance behavior of a flow in its initial phase. The insights gained from such solutions can be plotted in so-called instability diagrams

17.5 Transitional Boundary-Layer Flows

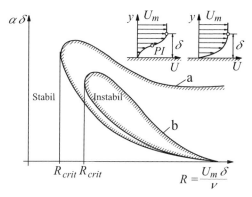

Fig. 17.15 Instability diagram for boundary layers with and without viscosity term in the Orr–Sommerfeld equation

as indicated in Fig. 17.15 for a flat plate boundary layer. This figure shows that only a relatively narrow range of wavelengths and frequencies of disturbances have to be classified as "dangerous" for the stability of the boundary layer. There always exists, for each investigated disturbance, a lower limit for each Reynolds number of the boundary layer and an upper limit. Outside the so-called critical Re range that is characteristic of each disturbance, boundary-layer flows prove to be stable.

When carrying out numerical computations, it results for $C = \omega/k$ that

$$\frac{C_R}{U_\infty} = 0.39, \quad k\delta_1 = 0.36 \quad \text{and} \quad \frac{\omega_R \delta_1}{U_\infty} = 0.15.$$

It is interesting that the smallest wavelength of the disturbances which can act in an unstable way on boundary layers is fairly long:

$$\lambda_{\min} = \frac{2\pi}{0.36}\delta_1 = 17.5\delta_1 = 6\delta.$$

For smaller wavelengths of disturbances, boundary layers prove to be stable.

As the critical Reynolds number for the laminar-to-turbulent transition, numerical computations yield

$$Re_{\text{crit}} = 520$$

a value which is lower than the corresponding experimentally obtained value:

$$(Re_{\text{crit}})_{\exp} \approx 950.$$

The difference is probably due to the fact that the numerically determined value $Re_{\text{crit}} = 520$, ascertained from stability considerations, represents the "point of neutral instability," whereas the experimentally determined value probably represents the "point of the laminar-to-turbulent transition that occurs abruptly." The two are, as can easily be understood, not necessarily identical.

References

17.1. van Dyke, M.: "An Album of Fluid Motion", Parabolic Press, Stanford, CA (1982)
17.2. Currie, I.G.: "Fundamental Mechanics of Fluids", McGraw-Hill, New York (1974)
17.3. Oertel Jr., H.: "Prandtl-Führer durch die Strömungslehre", Vieweg, Braunschweig/Wiesbaden (2001)
17.4. Potter, M.C., Foss, J.F.: "Fluid Mechanics", Wiley, New York (1975)
17.5. Schlichting, H.: "Boundary Layer Theory", McGraw-Hill, New York (1979)
17.6. Sherman, F.S.: "Viscous Flow", McGraw-Hill, Singapore (1990)
17.7. Taylor, G.: Stability of a Viscous Liquid Contained Between Two Rotating Cylinders, Proc. R. Soc. A, 223, 289 (1923); Proc. Int. Congr. Appl. Mech., Delft, 1924; Z. Angew. Math. Mech., 5, 250 (1925)

Chapter 18
Turbulent Flows

18.1 General Considerations

In Chap. 17, we pointed out that special flow properties exist to justify the classification of fluid motions into laminar, transitional and turbulent flows. As laminar were designated all flows which proved stable towards disturbances introduced from outside, resulting in flows with a high degree of order and in which diffusion phenomena are characterized only by molecular diffusion. Laminar flows can be dependent on time, as the Karman vortex street shows, which is depicted in Fig. 18.1. As long as the flow shows the high degree of order which is characteristic of it, it is laminar. This means that the viscosity of the fluid is able, in stable laminar flows, to attenuate sufficiently fluctuations of the flow properties that would otherwise disturb the orderliness of the flow. Perturbation attenuations of this kind usually occur at low Reynolds numbers of all flows.

When considering flows at high Reynolds numbers, one finds that flow phenomena such as the Karman vortex street, visually perceivable thanks to flow visualization techniques, lose their "regularity," i.e. stochastic fluctuations of all flow properties are observed, as indicated in Fig. 18.2. These fluctuations occur superimposed on the mean flow characteristics.

The fluid motions, known to be of high regularity for laminar flows, do not exist any longer as orderly in turbulent flows, i.e. in flows of high Reynolds numbers. At high Reynolds numbers, a flow state exists which stands out for its strong irregularity, in connection with an extremely high diffusivity which can exceed the molecular-dependent transport processes by several orders of magnitude. Connected with that is an increased intermixing of the fluid and an increased transport rate of the momentum, and also increased heat and mass transport. All these characteristics led to the introduction of the term "turbulence" for the state of flows with this strongly irregular flow behavior, in order to give a clear expression of the differing character compared with laminar flows. These differences have to be considered also when treating

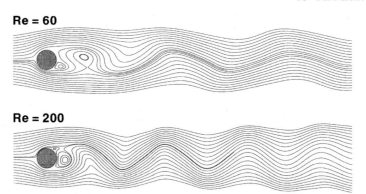

Fig. 18.1 The Karman vortex street, at low Reynolds numbers a time-varying laminar flow

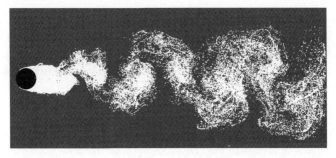

Fig. 18.2 The Karman vortex street (at high Reynolds numbers) in a turbulent flow

turbulent flows theoretically, i.e. turbulent flows require a specific treatment which differs from that of laminar flows.

On introducing into a turbulent flow field a velocity sensor which is capable of measuring the local instantaneous velocity, such a measurement results in a velocity dependence on time, as indicated in Fig. 18.3. At a point in space, the signal is characterized by strong fluctuations of the flow velocity in time, which can be stated as deviations $u'_j(x_i, t)$ from a mean value $\overline{U}_j(x_i)$, the latter being a constant with respect to time. Here, the time mean value $\overline{U}_j(x_i)$ is defined as follows:

$$\overline{U}_j(x_i) = \lim_{T \to \infty} \frac{1}{T} \int_0^T \hat{U}_j(x_i, t)\, dt, \tag{18.1}$$

where $\hat{U}_j(x_i, t)$ indicates the instantaneous value of the velocity (see Fig. 18.3) and T is the integration time over which the indicated time-averaging takes place.

Thus, $\hat{U}_j(x_i, t)$ can be taken as a quantity which allows one to consider the local flow velocity, varying over time, as the sum of a quantity that is constant with respect to time and a quantity that is fluctuating in time. This

18.1 General Considerations

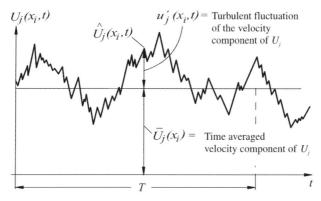

Fig. 18.3 Time velocity path at a point x_i within a turbulent flow field

decomposition of the instantaneous velocity $\hat{U}_j(x_i, t)$ into a time-averaged part $\overline{U}_j(x_i)$ and a fluctuating part $u'_j(x_i, t)$ has advantages, as will be shown later. It was introduced by Reynolds to treat turbulent flows.

The above definition of the mean velocity states, for $T \to \infty$, an equality in area:

$$\underbrace{T\overline{U}_i}_{\text{rectangular area}} = \underbrace{\lim_{T \to \infty} \int_0^T \hat{U}_i(t)\mathrm{d}t}_{\text{integral over time-dependent signal}}. \tag{18.2}$$

When considering this definition of the time mean value of the velocity, then for the quantity $u'_j(x_i, t)$, designated as turbulent velocity fluctuation, the following equation holds:

$$u'_j(x_i, t) = \hat{U}_j(x_i, t) - \overline{U}_j(x_i). \tag{18.3}$$

When applying to this relationship the operator $\lim_{T \to \infty} \frac{1}{T} \int_0^T ()\mathrm{d}t$, the following can be carried out:

$$\lim_{T \to \infty} \frac{1}{T} \int_0^T u'_j(x_i, t)\, \mathrm{d}t = \lim_{T \to \infty} \frac{1}{T} \int_0^T \left[\hat{U}_j(x_i, t) - \overline{U}_0(x_i)\right] \mathrm{d}t$$

$$= \underbrace{\lim_{T \to \infty} \frac{1}{T} \int_0^T \hat{U}_j(x_i, t)\, \mathrm{d}t}_{= \overline{U}_j} - \underbrace{\lim_{T \to \infty} \frac{1}{T} \int_0^T \overline{U}_j(x_i)\, \mathrm{d}t}_{= \overline{U}_j}. \tag{18.4}$$

It can be concluded that the two integrals shown on the right-hand side of (18.4) are equal and their difference yields 0, i.e. the following holds for the time average of turbulent velocity fluctuations:

$$\lim_{T \to \infty} \frac{1}{T} \int_0^T u'_j(x_i, t)\, \mathrm{d}t = \overline{u'_j(x_i, t)} = 0, \tag{18.5}$$

where the overbar on $u'_j(x_i,t)$ represents a simplified way of writing the carried out time averaging. On designating the turbulent velocity fluctuations with $u'_j(x_i,t)$ (or simplifying this to u'_j), then the following can be said:

- The time average of the turbulent velocity fluctuations u'_j is equal to zero per definition. Hence, there is a way to present turbulence in local, time-varying quantities, in a form such that the turbulent fluctuations of all flow quantities, that are introduced into the considerations, show a time mean value that is zero.

For the fluctuating velocity quantity u'_j, moments of higher order can also be defined:
$$\overline{u'^n_j} = \lim_{T\to\infty} \frac{1}{T} \int_0^T u'^n_j \, dt, \tag{18.6}$$

which in general show values different from zero. Especially for the rms value of the turbulent velocity fluctuations, the following holds:
$$\sigma_i = \sqrt{\overline{u'^2_i}} = \sqrt{\lim_{T\to\infty} \frac{1}{T} \int_0^T \overline{u'^2_i} \, dt}. \tag{18.7}$$

This can be employed for the definition of the turbulence intensity:
$$Tu = \frac{\sqrt{\frac{1}{2}\overline{u_i u_i}}}{\overline{U}_{\text{tot}}} = \frac{\sqrt{\frac{1}{2}\left(\overline{u^2_1} + \overline{u^2_2} + \overline{u^2_3}\right)}}{\overline{U}_{\text{tot}}}. \tag{18.8}$$

This quantity represents a measure of the intensity of the turbulent fluctuations of the velocity components with respect to the local mean value $\overline{U}_{\text{tot}}$. As shown in (18.8), it is often usual to take as a relative value the mean value of the total velocity vector, i.e. $\overline{U}_{\text{tot}} = \sqrt{\overline{U_j U_j}}$. The Tu-value is often only around a few percent for some flows and, because of this, one speaks of a turbulent flow of low intensity. When the value is around 10% and more, the flow is defined as highly turbulent. Highly turbulent flows occur mostly in industrial flow systems. It is the task of fluid mechanics to develop and bring to application measurement techniques and numerical solution methods which allow investigations of turbulent flows with low and high turbulence intensities. In practice, it is often sufficient to only have information on time mean values of turbulent quantities of an investigated flow field.

The introductory explanations above indicate clearly that turbulent flow fields show a complex behavior, providing strong property variations in space and time, so that detailed considerations are only worth the effort if special insights into the physics of turbulence are needed. For practical flow considerations, it is sufficient to treat turbulent flow processes by means of their statistical mean properties, i.e. to describe the most important characteristics of turbulent flows by statistical mean values. In this chapter, the most important methods of statistical flow considerations are summarized and explained briefly, in order to employ them subsequently in the treatment of turbulent flows. More details are found in refs. [18.4] to [18.9].

18.2 Statistical Description of Turbulent Flows

As emphasized in Sect. 18.1, turbulent velocity fields are characterized by strong irregularities of all their properties, e.g. strong changes of their velocity and pressure in space and time. To register them, at all times and at all locations, is not only a task which is difficult to solve and that exceeds our present measuring and representation capacities of fluid mechanical processes, but moreover constitutes a task whose solution is not worth striving for. The solution would result in such a large amount of information that it could not be possible to process them further, or to exploit them, in order to gain new insights into fluid mechanical processes. The large amount of information which turbulent flows possess due to their time and space behavior, therefore does not serve to deepen our fluid mechanical knowledge, nor does it help to improve fluid-flow equipment and/or its installation. As fluid-flow information is only useful to the extent to which it can be mentally grasped and exploited further, it is necessary to reduce appropriately the large amount of information available in turbulent flow fields. In today's turbulent flow research, this is done by mainly limiting investigations to two types of questions relating to turbulent flows:

- How do the *local* turbulent fluctuations of the velocity components and pressure vary around the corresponding mean values? What correlations exist between the fluctuating quantities, and what physical significance do these correlations have?
- How are neighboring turbulent fluctuations of the velocity components and pressure correlated with one another, and what physical significance do these correlations have?

To be able to give answers to these questions, one uses in turbulence research methods of statistics and nearly all the terminology related to it. The distribution of the local turbulent flow fluctuations and the turbulent pressure fluctuations are recorded by the probability density distribution $\wp(u'_j)$ or $\wp(p')$, or by their Fourier transforms, the so-called characteristic function $\varphi(k)$. In order to describe the existing correlation between neighboring points in terms of space and/or time, one uses appropriate correlation functions or their Fourier transforms. To describe the locally occurring fluctuations in time, the auto-correlation function of the fluctuations is used and its Fourier transforms, or their corresponding energy spectrum. All these quantities (probability density distribution, characteristic function, auto-correlation function, energy spectrum, etc.) result from the instantaneous values of the velocity components and pressure describing the turbulent flow field, by applying mathematical operators which are explained this chapter in a somewhat summarizing way. It is very important for the further comprehension of the description of the characteristics of turbulent flows to understand the employment of these operators and to realize their physical significance. Further details are provided in refs. [18.1], [18.3].

18.3 Basics of Statistical Considerations of Turbulent Flows

18.3.1 Fundamental Rules of Time Averaging

For the treatment of turbulent flows, a method of consideration was introduced by Reynolds (1895), where the instantaneous values of the velocity components, pressure, density, temperature, etc., are replaced by mean values (which are defined as constant in terms of time) to which the corresponding time-varying, turbulent, fluctuating quantities, deviating from the mean values, are additively superimposed. Consequently, the instantaneous values can be written as follows:

Velocity components	$\hat{U}_j(x_i, t) = \overline{U_j}(x_i) + u'_j(x_i, t),$	(18.9)
Pressure	$\hat{P}(x_i, t) = \overline{P}(x_i) + p'(x_i, t),$	(18.10)
Temperature	$\hat{T}(x_i, t) = \overline{T}(x_i) + t'(x_i, t),$	(18.11)
Density	$\hat{\rho}(x_i, t) = \overline{\rho}(x_i) + \rho'(x_i, t).$	(18.12)

The above quantities with overbars on them are the time-averaged values and the quantities with a "hat," ^, are the corresponding instantaneous values. The values with primes, ', represent the turbulent fluctuations.

When applying the time-averaging operator:

$$\lim_{T \to \infty} \frac{1}{T} \int_0^T (\cdots) \, dt \qquad (18.13)$$

to the instantaneous values of the above quantities, one obtains the time mean values, and this makes clear that the averaging in time over the turbulent fluctuations of the quantities has the value 0. This means that the following holds:

$$\lim_{T \to \infty} \frac{1}{T} \int_0^T u'_j(x_i, t) \, dt = 0 \quad \text{and} \quad \lim_{T \to \infty} \frac{1}{T} \int_0^T p'(x_i, t) \, dt = 0 \qquad (18.14)$$

18.3 Statistical Considerations

and further for the temperature fluctuations t' and the density fluctuations ρ':

$$\lim_{T\to\infty} \frac{1}{T} \int_0^T t'(x_i, t)\, \mathrm{d}t = 0 \quad \text{and} \quad \lim_{T\to\infty} \frac{1}{T} \int_0^T \rho'(x_i, t)\, \mathrm{d}t = 0. \quad (18.15)$$

Generally, we can therefore write:

$$\lim_{T\to\infty} \frac{1}{T} \int_0^T \alpha'(x_i, t)\, \mathrm{d}t = 0, \quad (18.16)$$

where $\alpha'(x_i, t)$ stands for any randomly varying turbulent flow property.

When applying time averaging to derivatives of the quantity $\hat{\alpha} = \bar{\alpha} + \alpha'$, the following can be shown to be valid:

$$\overline{\frac{\partial \hat{\alpha}}{\partial x_i}} = \lim_{T\to\infty} \frac{1}{T} \int_0^T \frac{\partial \hat{\alpha}}{\partial x_i}\, \mathrm{d}t = \frac{\partial}{\partial x_i}\left[\lim_{T\to\infty} \frac{1}{T} \int_0^T (\bar{\alpha} + \alpha')\, \mathrm{d}t\right] = \frac{\partial \bar{\alpha}}{\partial x_i}. \quad (18.17)$$

Furthermore, the following fundamental rules of time averaging can be stated:

$$\overline{(\hat{\alpha} + \hat{\beta})} = \bar{\alpha} + \bar{\beta}, \quad (18.18)$$

$$\overline{\bar{\alpha}\alpha'} = 0 \quad \text{and} \quad \overline{\hat{\alpha}\alpha'} = \overline{\alpha'^2}, \quad (18.19)$$

$$\overline{\bar{\alpha}\bar{\beta}} = \bar{\alpha}\bar{\beta} \quad \text{and} \quad \overline{\hat{\alpha}\hat{\beta}} = \bar{\alpha}\bar{\beta} + \overline{\alpha'\beta'}. \quad (18.20)$$

With the help and the consequent application of the integration rules, stated above for the time averaging procedure, further relationships can be derived for combinations of the functions $\hat{\alpha}(t)$ and $\hat{\beta}(t)$.

When applying to the product of the instantaneous functions $\hat{\alpha}(t)$, $\hat{\beta}(t)$ and $\hat{\gamma}(t)$ the time averaging rules above, the following relationship results:

$$\overline{\hat{\alpha}\hat{\beta}\hat{\gamma}} = \overline{(\bar{\alpha}+\alpha')(\bar{\beta}+\beta')(\bar{\gamma}+\gamma')} = \overline{(\bar{\alpha}\bar{\beta} + \alpha'\bar{\beta} + \beta'\bar{\alpha} + \alpha'\beta')(\bar{\gamma}+\gamma')}$$

$$= \bar{\alpha}\bar{\beta}\bar{\gamma} + \overline{\alpha'}\bar{\beta}\bar{\gamma} + \overline{\beta'}\bar{\alpha}\bar{\gamma} + \overline{\alpha'\beta'}\bar{\gamma} + \bar{\alpha}\bar{\beta}\overline{\gamma'} + \overline{\alpha'\bar{\beta}\gamma'} + \overline{\beta'\bar{\alpha}\gamma'} + \overline{\alpha'\beta'\gamma'}$$

$$= \bar{\alpha}\bar{\beta}\bar{\gamma} + \overline{\alpha'\beta'}\bar{\gamma} + \overline{\alpha'\gamma'}\bar{\beta} + \overline{\beta'\gamma'}\bar{\alpha} + \overline{\alpha'\beta'\gamma'}. \quad (18.21)$$

In this way, triple products of the mean values of $\hat{\alpha}$, $\hat{\beta}$ and $\hat{\gamma}$ are obtained, products of correlations of two quantities multiplied by the mean value of the third quantity, and triple correlations of the turbulent fluctuation quantities α', β' and γ' result.

The above time averaging rules are employed in the subsequent sections, in order to derive equations for the mean values from the basic equations of fluid mechanics. The latter are usually formulated for the instantaneous values of the velocity, pressure, etc. In all the derivations in this chapter, the fluid properties are assumed to be constant, and especially $\rho = $ constant. The resultant equations derived in this way indicate the mean volume change by the time-averaged continuity equation, the mean momentum transport by the time-averaged momentum equation and the mean energy transport by the time-averaged energy equation. On subtracting these equations from the equations for the corresponding instantaneous values, one obtains transport equations for the fluctuating quantities. The latter can be employed for gaining information on these properties of turbulent flows.

18.3.2 Fundamental Rules for Probability Density

For the introduction of the probability density function for the velocity components $\hat{U}_j(x_i, t)$, the velocity axis in Fig. 18.4 is subdivided into equal sections $\Delta \hat{U}_j$. The velocity distribution is plotted along the time axis, as obtained at a fixed x_i measuring location in a turbulent flow field. For the considerations carried out here, the velocity distribution over time can be assumed to be given for each velocity component $\hat{U}_j (j = 1, 2, 3)$. The horizontal lines of the subdivision of the $\hat{U}_j$ axis now lead for each velocity interval ΔU_j to a corresponding period of time $(\Delta t)_\alpha = f[(\hat{U}_j)_\alpha, \Delta U_j]$. This time interval indicates how long the velocity trace stays in the corresponding time interval. On summarizing all time intervals which are assigned to the same velocity interval, the probability density function can be defined as follows:

$$\lim_{\Delta \hat{U}_j \to 0} \wp \left[\left(\hat{U}_j \right)_\alpha \right] \left(\Delta \hat{U}_j \right)_\alpha = \lim_{T \to \infty} \frac{1}{T} \lim_{N \to \infty} \sum_{\alpha=1}^{N} (\Delta t)_\alpha \qquad (18.22)$$

or rewritten:

$$\wp \left[\left(\hat{U}_j \right)_\alpha \right] = \lim_{T \to \infty} \frac{1}{T} \lim_{\Delta U_j \to 0} \frac{1}{(\Delta \hat{U}_j)_\alpha} \sum_{\alpha=1}^{N} (\Delta t)_\alpha. \qquad (18.23)$$

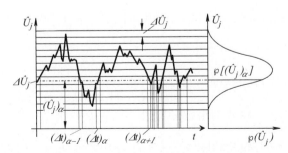

Fig. 18.4 Time path of the velocity and resultant probability density distribution

18.3 Statistical Considerations

The probability density distribution defined in this way contains all the required information that indicates in which amplitude range the velocity moves in time at a given measuring location. It indicates, moreover, with which probability the amplitude of the fluctuating velocity components occur in the course of time. The resultant function is also shown in Fig. 18.4, as a distribution function, which is plotted along the $\hat{U}_j$ axis. It describes, in a time-averaged way, the amplitude of the velocity fluctuations occurring at a point in the turbulent flow. The computation of time-averaged values can thus also take place through the corresponding probability density distribution, as the entire amplitude distribution for the turbulent velocity fluctuations is recorded in it.

For the mean value of the velocity, discussed in Sect. 18.1:

$$\overline{U}_{j,(x_i)} = \lim_{T \to \infty} \frac{1}{T} \int_0^T \hat{U}_j(x_i, t)\, dt \tag{18.24}$$

the time-related increment dt can be written as

$$\frac{dt}{T} = \wp\left(\hat{U}_j\right) d\hat{U}_j. \tag{18.25}$$

For the integral $\int_{-\infty}^{+\infty} \wp\left(\hat{U}_j\right) d\hat{U}_j$, one can therefore derive from (18.25):

$$\int_{-\infty}^{+\infty} \wp\left(\hat{U}_j\right) d\hat{U}_j = \frac{1}{T} \int_0^T dt = 1, \tag{18.26}$$

i.e. the "area" below the probability density distribution has the value 1.

The computation of the mean value of the local instantaneous flow velocity $\hat{U}_j(x_i, t)$, as given in (18.25), can be computed with the help of the probability density function. This means that two possibilities exist for computing the mean value, computation in the time domain or in the probability density domain:

$$\overline{U}_j(x_i) = \lim_{T \to \infty} \frac{1}{T} \int_0^T \hat{U}_j(x_i)(t)\, dt = \int_{-\infty}^{+\infty} \wp(\hat{U}_j)\, \hat{U}_j\, d\hat{U}_j. \tag{18.27}$$

Here, $\wp(\hat{U}_j)$ is to be considered as being given for a fixed location x_i. $\hat{U}_j = \hat{U}_j(x_i, t)$ represents the instantaneous value of the jth velocity component.

It is usual in turbulence research to state the probability density distribution only for the turbulent fluctuations, i.e. $\wp(u'_j)$. This probability density distribution arises from the distribution shown in Fig. 18.4 by a parallel displacement of the ordinate axis by the amount of the mean velocity $\overline{U}_j$. With this parallel displacement the form of the probability density function does not change and it thus provides, in this new coordinate system, the amplitude values of the turbulent fluctuations only.

Analogous to the time mean value computation with (18.27), one can also compute the following moments of the velocity fluctuations. On the one hand, computations can be carried out in the time domain of the velocity and, on the other hand, in the probability density domain. Both methods yield the time-averaged properties for the nth moment of the turbulent velocity fluctuations:

$$\overline{(u'_j)^n} = \lim_{T \to \infty} \frac{1}{T} \int_0^T [u'_j(t)]^n \, dt = \int_{-\infty}^{+\infty} \wp(u'_j)(u'_j)^n \, du'_j. \quad (18.28)$$

Of special importance in turbulence research is the second moment $\overline{u'^2_j}$, which is employed for the definition of the turbulence intensity, $\alpha = 1, 2, 3$:

$$Tu_\alpha = \frac{\sigma_\alpha}{\overline{U}_\alpha} = \frac{\sqrt{\overline{u'^2_\alpha}}}{\overline{U}_\alpha} \quad \text{or} \quad Tu_j = \frac{\sigma_j}{\overline{U}_j} = \frac{\sqrt{\overline{u'^2_j}}}{\overline{U}_j}. \quad (18.29)$$

Moreover, the "standardized third moment" of the turbulent velocity fluctuations is included in several considerations, which allows statements about the "skewness" of the probability density distribution of the turbulent velocity fluctuations. Here, the "skewness" is defined in the following way as the standardized value of the turbulent velocity fluctuations:

$$S_j = \frac{\overline{u'^3_j}}{\sigma_j^3} \quad \text{with} \quad \sigma_j^3 = \left(\sqrt{\overline{u'^2_j}}\right)^3. \quad (18.30)$$

For the corresponding standardized fourth moment, one often finds the term "flatness" used in the literature. It represents another important property of the probability density distribution of turbulent velocity fluctuations. As for the skewness above, the flatness is again defined as a standardized quantity (kurtosis):

$$F_j = \frac{\overline{u'^4_j}}{\sigma_j^4} \quad \text{with} \quad \sigma_j^4 = \left(\sqrt{\overline{u'^2_j}}\right)^4. \quad (18.31)$$

The above moments of higher order for the velocity fluctuations are defined as central moments of the probability density distributions of the corresponding components of the turbulent velocity field.

In turbulence research, it is often necessary to define correlations between the different velocity fluctuations. They are computed from the different time-varying velocity fluctuations u'_i and u'_j as time integration over the products of the fluctuations, according to the equation below. They can also be computed, in a corresponding way, using the two-dimensional probability density distributions.

18.3 Statistical Considerations

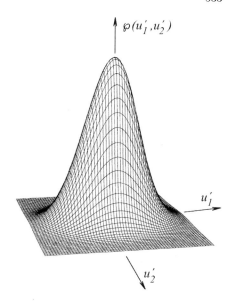

Fig. 18.5 Diagram of a two-dimensional probability density distribution

$$\overline{u_i'^n u_j'^m} = \lim_{T \to \infty} \frac{1}{T} \int_0^T u_i'^n u_j'^m \, dt = \int_{-\infty}^{+\infty} u_i'^n u_j'^m \, \wp\left(u_i', u_j'\right) du_i' \, du_j'. \quad (18.32)$$

It is again possible to compute these correlations in the time domain of the velocity field or in the probability density domain. It is important to emphasize again that the probability density distributions in (18.25), for the components of the turbulent velocity fluctuation, are probability density distributions for the turbulent fluctuations at one point in the flow field.

Two-dimensional probability density distributions $\wp(u_1', u_2')$, as shown in Fig. 18.5, are of special importance in the subsequent treatment of turbulent flows. The above considerations about probability can now be carried out for two-dimensional functions $\wp(u_1', u_2')$, and this results in the following relationship:

$$\int_{-\infty}^{+\infty}\int_{-\infty}^{+\infty} \wp(u_1', u_2') \, du_1' \, du_2' = 1 \quad \text{and} \quad 0 \le \wp(u_1', u_2') \le 1. \quad (18.33)$$

When deriving from this two-dimensional distribution the one-dimensional distribution that applies to the u_1' component of the turbulent velocity field, i.e. deriving $\wp(u_1)$, the following relationship holds:

$$\wp(u_1') = \int_{-\infty}^{+\infty} \wp(u_1', u_2') \, du_2'. \quad (18.34)$$

By means of the two-dimensional distribution $\wp(u'_1, u'_2)$, combined moments of the turbulent velocity components u'_1 and u'_2 can be computed:

$$\overline{u'_1{}^n u'_2{}^m} = \int\!\!\!\int_{-\infty}^{+\infty} u'_1{}^n u'_2{}^m \wp(u'_1, u'_2)\, du'_1\, du'_2. \tag{18.35}$$

For $n = m = 1$, the covariance of the velocity components u_1 and u_2 results:

$$\overline{u'_1 u'_2} = \int\!\!\!\int_{-\infty}^{+\infty} u'_1 u'_2 \wp(u'_1, u'_2)\, du'_1\, du'_2. \tag{18.36}$$

This integration results in an expression for a correlation existing between u'_1 and u'_2, i.e. it shows to what extent the velocity fluctuations u'_1 and u'_2 experience correlated changes. Since information of this kind is needed again and again, for special considerations in the derivations in subsequent sections, the significance of correlations between turbulent fluctuations will be explained here briefly. It is important to point out that two turbulent velocity fluctuations that show no correlation with one another are, in the statistical sense, not necessarily independent. This becomes clear when one considers the definitions stated below of independence of two turbulent velocity fluctuations and compares this definition with the condition for the variables to be uncorrelated:

- Two velocity fluctuations $u'_1(t)$ and $u'_2(t)$ are considered to be statistically independent when the following relationship for their probability density distributions holds:

$$\wp(u'_1, u'_2) = \wp(u'_1)\wp(u'_2), \tag{18.37}$$

i.e. the probability density of one component of the velocity fluctuations is not influenced by the distribution of the second.
- Two velocity fluctuations $u'_1(t)$ and $u'_2(t)$ are considered to be uncorrelated when their covariance is zero, i.e. when the following relationship holds:

$$\overline{u'_1 u'_2} = \int_{-\infty}^{+\infty}\!\!\int_{-\infty}^{+\infty} u'_1 u'_2 \wp(u'_1, u'_2)\, du'_1\, du'_2 = 0. \tag{18.38}$$

Two variables are always uncorrelated when the probability density distribution $\wp(u'_1, u'_2)$ is fully symmetrical, i.e. when it fulfills the following condition:

$$\wp(+u'_1, +u'_2) = \wp(+u'_1, -u'_2) = \wp(-u'_1, +u'_2) = \wp(-u'_1, -u'_2). \tag{18.39}$$

Such a symmetrical probability density distribution is shown in Fig. 18.6, where the isolines $\wp(u'_1, u'_2) = $ constant are shown and lines of equal probability density. The latter indicate the probability density distribution vertical to the $u'_1 - u'_2$ plane.

18.3 Statistical Considerations

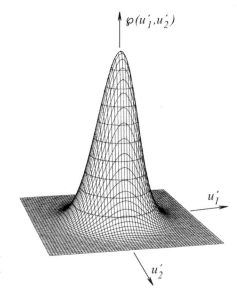

Fig. 18.6 Two-dimensional symmetrical probability density distribution with isolines

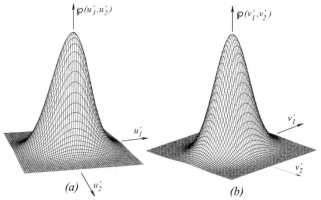

Fig. 18.7 Probability density distribution for isotropic turbulent flows

The probability density distributions shown in Fig. 18.7a, b are identical with that in Fig. 18.5, but are different in the directions of the coordinate axes. The latter leads to the finite covariances $\overline{v_1' v_2'} \neq 0$ indicated in the figures, i.e. to correlations between v_1' and v_2'. Thus it becomes evident that a correlation, existing between two turbulent velocity fluctuations is dependent on the choice of the coordinate system.

For the two coordinate systems indicated in Fig. 18.7 the following holds, where α is the angle of rotation between the two coordinate systems:

$$\begin{aligned} v_1' &= u_1' \cos\alpha + u_2' \sin\alpha, \\ v_2' &= -u_1' \sin\alpha + u_2' \cos\alpha. \end{aligned} \quad (18.40)$$

Using these relationships, one can compute by multiplication and time averaging:
$$\overline{v'_1 v'_2} = \overline{u'_1 u'_2} \cos 2\alpha - (\overline{u'^2_1} - \overline{u'^2_2}) \cos\alpha \sin\alpha. \qquad (18.41)$$
This relationship makes it clear that $\overline{v'_1 v'_2}$ is only equal to $\overline{u'_1 u'_2}$ and only equal to zero when the condition $\overline{u'^2_1} = \overline{u'^2_2}$ is fulfilled (see Fig. 18.6), i.e. when the flow field is isotropic. By isotropy one understands here a property of the flow field that shows:

- No directionality of all time-averaged local flow quantities which describe a turbulent flow field.

In addition, a flow field can have properties which are designated as spatially homogeneous. For the homogeneity of a flow field, the following holds:

- The time-averaged parameters describing the turbulent flow field are independent of the position of the measuring location.

In Fig. 18.8, the two-dimensional probability density distribution of a turbulent isotropic flow field is shown. When the same probability density distribution exists in each space point and this satisfies the isotropy requirements, the turbulence is defined as being homogeneous *and* isotropic (see Fig. 18.8).

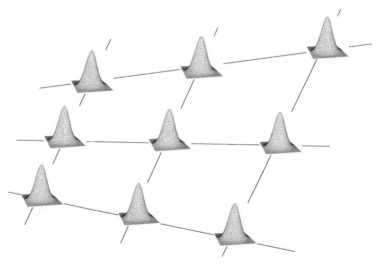

Fig. 18.8 Spatial distributions of the probability density distributions for isotropic and homogeneous turbulence

18.3.3 Characteristic Function

For a number of considerations, the Fourier transform of the probability density distribution is employed, which is usually defined as a characteristic function of the flow field:

$$\varphi(k) = \int_{-\infty}^{+\infty} \wp(u'_j) \exp(iku'_j(t)) \, du'_j, \tag{18.42}$$

where $i = \sqrt{-1}$ represents the imaginary unit of a complex number $z = x+iy$. Considering the identity of the operators:

$$\lim_{T \to \infty} \frac{1}{T} \int_0^T (\cdots) \, dt = \int_{-\infty}^{+\infty} (\cdots) p(u'_j) \, du'_j \tag{18.43}$$

the characteristic function of the velocity fluctuations $u'_j(t)$ can be computed as follows:

$$\varphi(k) = \lim_{T \to \infty} \frac{1}{T} \int_0^T \exp(iku'_j(t)) \, dt. \tag{18.44}$$

The significance of this function lies, on the one hand, in the experimental field of turbulence research, where one finds that the convergence of $\wp(u'_i)$ is bad and that this, with decreasing $\Delta u'_j$, leads to very long measuring times. The measurement of $\varphi(k)$, on the other hand, is connected to a fairly rapid convergence and $\wp(u'_j)$ can thus be computed from $\varphi(k)$ as follows (inverse Fourier transformation):

$$\wp(u'_j) = \int_{-\infty}^{+\infty} \varphi(k) \exp(-iku'_j) \, dt. \tag{18.45}$$

The multidimensional characteristic function can also be stated as follows:

$$\varphi(k_j) = \lim_{T \to \infty} \frac{1}{T} \int_0^T \exp(i\{k_j\}\{u_j\}) \, dt \tag{18.46}$$

or by $u'_j = \{u_1, u_2, u_3\}$ and $k_j = \{k, l, m\}$:

$$\varphi(k, l, m) = \lim_{T \to \infty} \frac{1}{T} \int_0^T \exp[i(ku_1(t) + lu_2(t) + mu_3(t)] \, dt. \tag{18.47}$$

When considering the one-dimensional characteristic function $\varphi(k)$ and its definition, the following holds:

$$\left.\frac{d\varphi}{dk}\right|_{k=0} = \int_{-\infty}^{+\infty} \wp(u'_j) i u'_j \, du'_j = 0 \qquad (18.48)$$

or quite generally:

$$\left.\frac{d^n\varphi}{dk^n}\right|_{k=0} = i^n \overline{u'^n_j}. \qquad (18.49)$$

The characteristic function enters as a horizontal line into the axis $k = 0$. At the point $k = 0$, the derivations of the functions $\varphi(k)$ are linked to the central moments of the velocity components. Because of this, the characteristic function can be written as a Taylor series of the corresponding central moments of the turbulent velocity fluctuations:

$$\varphi(k) = \sum_{n=0}^{\infty} \frac{(ik)^n}{n!} \overline{u'^n_i}. \qquad (18.50)$$

These positive properties of the characteristic function can also be put to very good use in analytical considerations, but in this chapter only its use in experimental studies is explained.

18.4 Correlations, Spectra and Time-Scales of Turbulence

In order to obtain information on the structure of turbulence, two different ways of consideration have gained acceptance, characterized as follows:

- The turbulent velocity fluctuations $u'_j(x_i, t)$, which in general are functions of space and time, are usually measured for a fixed location and thus can be regarded as time series. Information on $u'_j(t)$, for a preset value of x_i, is therefore recorded for fixed points in space. Its time-averaged properties can also be provided in the form of probability density distributions, characteristic functions, etc.
- The turbulent velocity fluctuations $u'_j(x_i, t)$ can also be considered for a fixed time, yielding information on the spatial distributions of the turbulence of the flow field. Information on $u'_j(x_i)$ is in this way recorded for fixed points x_i at the same time t. The entire information on turbulence can also be provided in the form of two-point probability density distributions, or multi-point probability density distributions, depending on the information sought.

18.4 Correlations, Spectra and Time-Scales of Turbulence

For considerations of signals varying over time at a fixed point in space, the question concerning the time interval over which the turbulent velocity fluctuations are correlated with one another can be answered. This question can be answered using the autocorrelation function $\mathbf{R}(\tau)$, which is defined as follows:

$$\mathbf{R}(\tau) = \lim_{T \to \infty} \frac{1}{T} \int_0^T u_i'(t) u_i'(t+\tau) \, dt. \tag{18.51}$$

With $t' = t + \tau$, the following holds for processes that are stationary in a time-averaged manner:

$$\overline{u_j'^2(t)} = \overline{u_j'^2(t')} = \text{constant for } \tau = 0. \tag{18.52}$$

This constant "effective value" of the turbulent velocity fluctuations can be employed for the standardization of the autocorrelation function and thus for the introduction of the autocorrelation coefficient $\rho(\tau)$:

$$\rho(\tau) = \frac{1}{\overline{u_j'^2}} \mathbf{R}(\tau) = \frac{1}{\overline{u_j'^2}} \lim_{T \to \infty} \frac{1}{T} \int_0^T u_j'(t) u_j'(t+\tau) \, dt. \tag{18.53}$$

For the autocorrelation coefficient, the following general properties hold:

$$\rho(\tau) = \rho(-\tau) \quad \text{symmetric with the } \tau = 0 \text{ axis}, \tag{18.54}$$

$$\rho(0) = 1 \quad \text{and} \quad \rho(\tau) \leq 1. \tag{18.55}$$

A typical result for $\rho(\tau)$ is shown in Fig. 18.9. By means of the autocorrelation coefficient of a turbulent flow field, through $\rho(\tau)$, typical time-scales of turbulence can be introduced. As the integral time-scale the following quantity is defined:

$$I_t = \int_0^\infty \rho(\tau) \, d\tau = \frac{1}{\overline{u_j'^2}} \int_0^\infty \mathbf{R}(\tau) \, d\tau, \tag{18.56}$$

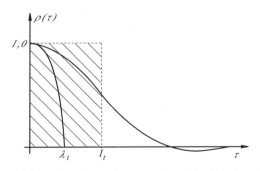

Fig. 18.9 Autocorrelation function and time-scales of turbulent velocity fluctuations

I_t corresponds, therefore, to the surface below the $\rho(\tau)$ distribution, and this means that the following identity of the surfaces in Fig. 18.9 holds:

$$\overline{u'^2_j} I_t = \int_0^\infty \mathbf{R}(\tau)\, \mathrm{d}\tau. \tag{18.57}$$

It is a characteristic property of turbulent flows that they show velocity fluctuations having finite integral time-scales. The integral time-scale I_t is a quantity which shows the order of magnitude of the period of time over which the velocity fluctuations $u'_j(t)$ are correlated with one another. $I_t = 0$ means that there is no correlation. Such a "degenerated" turbulent flow field cannot exist in reality; it lacks essential elements for maintaining turbulent flow fluctuations. Hence, turbulence contains structures of finite time durations. In fact, turbulent flows contain an entire spectrum of vortex-like structures.

In addition to the integral time-scale of turbulence, a micro time-scale λ_t can also be introduced, which is defined through the curvature of the autocorrelation coefficient function at the point $\tau = 0$:

$$\frac{\mathrm{d}^2 \rho(\tau)}{\mathrm{d}\tau^2} = -\frac{2}{\lambda_t^2}. \tag{18.58}$$

On expanding $\rho(\tau)$ in a Taylor series around $\tau = 0$ and considering the symmetry of the $\rho(\tau)$ distribution, then for the small τ values the following parabolic function holds:

$$\rho(\tau) = 1 - \frac{\tau^2}{\lambda_t^2} \pm \cdots \tag{18.59}$$

so that by repeated derivatives the above definition equation (18.58) for λ_t can be derived. Moreover, for the parabola arising from the Taylor series expansion, it can be derived that this parabola cuts the $\rho(\tau) = 0$ axis (abscissa) at $\tau = \lambda_t$ (see Fig. 18.9).

Based on the relationship

$$\frac{\mathrm{d}^2}{\mathrm{d}t^2}\left(u'^2_j\right) = 2 u'_j \frac{\mathrm{d}^2 u'_j}{\mathrm{d}t^2} + 2\left(\frac{\mathrm{d}u'_j}{\mathrm{d}t}\right)^2 \tag{18.60}$$

valid for all time-averaged turbulent flow processes which are stationary, we can derive

$$\frac{\mathrm{d}^2}{\mathrm{d}t^2}\left(\overline{u'^2_j}\right) = 0 = 2\overline{u'_j \frac{\mathrm{d}^2 u'_j}{\mathrm{d}t^2}} + 2\overline{\left(\frac{\mathrm{d}u'_j}{\mathrm{d}t}\right)^2}, \tag{18.61}$$

i.e. the following relationship holds:

$$\overline{\left(\frac{\mathrm{d}u'_j}{\mathrm{d}t}\right)^2} = -\overline{u'_j \frac{\mathrm{d}^2 u'_j}{\mathrm{d}t^2}}. \tag{18.62}$$

18.4 Correlations, Spectra and Time-Scales of Turbulence

In consideration of the properties of the autocorrelation function, one can write

$$\overline{\left(\frac{du'_j}{dt}\right)^2} = \frac{2}{\lambda_t^2}\overline{u'^2_j} \tag{18.63}$$

or, expressed in terms of λ_t^2:

$$\lambda_t^2 = \frac{2\overline{u'^2_i}}{\overline{\left(\dfrac{du'_i}{dt}\right)^2}}. \tag{18.64}$$

This shows that the micro time-scale of the turbulence can also be determined from the double of the rms value of the turbulent velocity fluctuations divided by the rms value of the time derivative of the turbulent velocity fluctuations.

In conclusion, one should mention, with regard to the above considerations, that turbulence comprises an entire spectrum of time-scales or corresponding frequencies, all of which one can imagine to lie in the range between the integral time-scale I_t and the micro time-scale λ_t. This distribution of scales is determined by the total distribution of the $\rho(\tau)$ function which for $\tau = 0$ has the value 1, and for all finite τ values the $\rho(\tau)$ values satisfy the requirement

$$\rho(\tau) < \frac{1}{\tau}. \tag{18.65}$$

As for $\tau \longrightarrow \infty$ (18.65) also holds: the integral time-scale can be computed from $\rho(\tau)$ always to have a finite value.

The considerations presented above can also be carried out in the spectral range. The spectral energy density distribution $S(\omega)$ is given as follows:

$$S(\omega) = \int_{-\infty}^{+\infty} \frac{1}{2\pi}\rho(\tau)\exp\{-i\omega\tau\}\,d\tau. \tag{18.66}$$

Thus in reverse, the autocorrelation coefficient $S(\omega)$ can be computed from the spectral energy density distribution by Fourier transformation:

$$\rho(\tau) = \int_{-\infty}^{+\infty} S(\omega)\exp\{i\omega\tau\}\,d\omega. \tag{18.67}$$

For $S(0)$, the following relationship results from the above equation:

$$S(0) = \int_{-\infty}^{+\infty} \frac{1}{2\pi}\rho(\tau)\,d\tau = 2\int_0^{\infty} \frac{1}{2\pi}\rho(\tau)\,d\tau = \frac{I_t}{\pi}. \tag{18.68}$$

With this, the value of the energy spectrum for $\omega = 0$ is determined by the integral time-scale of the turbulence in the following way:

$$S(0) = \frac{I_t}{\pi}. \tag{18.69}$$

The significance of the spectral energy-density distribution also becomes clear when one considers the Fourier coefficients of the turbulent velocity fluctuations $u'_j(t)$:

$$a_T(\omega, t) = \frac{1}{T} \int_t^{t+T} u'_j(t') \exp\{i\omega t'\} \, dt' \tag{18.70}$$

and the time average of the square of this value:

$$\lim_{T \to \infty} \overline{|a_T(\omega, t)|^2} = \overline{u'^2_i} S(\omega). \tag{18.71}$$

The spectral energy density distribution thus represents the energy of $u'_j(t)$ at the frequency ω, i.e. the following relationship holds:

$$\frac{dE(\omega)}{d\omega} = \overline{u'^2_i} S(\omega). \tag{18.72}$$

The total energy can thus be computed as:

$$E_{\text{ges}} = \overline{u'^2_j} \int_0^\infty S(\omega) \, d\omega. \tag{18.73}$$

18.5 Time-Averaged Basic Equations of Turbulent Flows

It has been stressed in the previous sections that turbulent flows possess complex properties, and one therefore limits oneself to the determination of the time-averaged properties of turbulent flows, i.e. one does not try to recover the time-varying properties of the flow field. Because of this, turbulent flows can therefore be theoretically better treated by the Reynolds equations instead of the Navier–Stokes equations. In order to derive the Reynolds equations, the instantaneous velocity $\hat{U}_j(t)$ is replaced by the sum of the mean velocity $\overline{U}_j$ and the fluctuation velocity $u'_j(t)$, i.e. $\hat{U}_j = \overline{U}_j + u'_j$ and analogously $\hat{\rho} = \bar{\rho} + \rho'$, $\hat{P} = \bar{P} + p'$, etc. Introducing these decomposed quantities into the Navier–Stokes equations and, by time averaging the equations, a new set of equations results for the mean values of the flow properties, the so-called Reynolds equations. The corresponding derivations are shown below.

18.5.1 The Continuity Equation

In Chap. 5, the continuity equation was derived as a mass-conservation equation which holds in the following form for the instantaneous values of the density and the velocity components:

$$\frac{\partial \hat{\rho}}{\partial t} + \frac{\partial}{\partial x_i}\left(\hat{\rho}\hat{U}_i\right) = 0. \tag{18.74}$$

By introducing for the instantaneous values the time mean values and the corresponding turbulent fluctuations, the continuity equation can also be written as follows:

$$\frac{\partial}{\partial t}\left(\bar{\rho} + \rho'\right) + \frac{\partial}{\partial x_i}\left[\left(\bar{\rho} + \rho'\right)\left(\overline{U}_i + u'_i\right)\right] = 0. \tag{18.75}$$

By time-averaging this equation, one obtains, applying the time-averaging rules indicated in Sect. 18.3.1:

$$\frac{\partial \bar{\rho}}{\partial t} + \frac{\partial}{\partial x_i}\left(\bar{\rho}\overline{U}_i + \overline{\rho' u'_i}\right) = 0. \tag{18.76}$$

This time-averaged equation can now be subtracted from (18.75), so that one obtains for the instantaneous density changes:

$$\frac{\partial \rho'}{\partial t} + \frac{\partial}{\partial x_i}\left(\bar{\rho}u'_i + \overline{U}_i\rho'\right) = 0. \tag{18.77}$$

For fluids with constant density, the above equations can be written in a simplified way:

$$\bar{\rho} = \hat{\rho} = \text{constant and thus } \rho' = 0 \quad \Rightarrow \quad \frac{\partial \overline{U}_i}{\partial x_i} = 0 \text{ and } \frac{\partial u'_i}{\partial x_i} = 0. \tag{18.78}$$

These two final equations can now be employed for dealing with turbulent fluid flows. In connection with this, it has to be taken into consideration that the covariance between ρ' and u'_j, i.e. $\overline{\rho' u'_j}$ appearing in the continuity equation for the mean values, represents three unknowns, $\overline{\rho' u'_1}$, $\overline{\rho' u'_2}$ and $\overline{\rho' u'_3}$, which have to be considered when solving fluid flow problems with variable density. These quantities represent turbulence-dependent mean mass flows in the x_1-, x_2- and x_3-directions, which appear superimposed on the mass flows due to the mean flow field. It is interesting to see from equations for fluids with constant density that time averaging of the continuity equation does not result in an additional unknown, i.e.

- For fluids of constant density, the time-averaging of the continuity equation does not result in additional turbulent transport terms. Through the continuity equation no "additional unknowns" are introduced for fluids with constant density, when dealing theoretically with turbulent flows, using the time-averaged continuity equation.

18.5.2 The Reynolds Equation

In Chap. 5, the equation of momentum was stated in the following form:

$$\hat{\rho}\left[\frac{\partial \hat{U}_j}{\partial t} + \hat{U}_i \frac{\partial \hat{U}_j}{\partial x_i}\right] = -\frac{\partial \hat{P}}{\partial x_j} - \frac{\partial \hat{\tau}_{ij}}{\partial x_i} + \hat{\rho} g_j \quad (18.79)$$

and for Newtonian fluids with $\hat{\rho} = $ constant the $\hat{\tau}_{ij}$ term can be expressed as:

$$\hat{\tau}_{ij} = -\mu \left(\frac{\partial \hat{U}_j}{\partial x_i} + \frac{\partial \hat{U}_i}{\partial x_j}\right) \quad (18.80)$$

so that (18.79) can be written as:

$$\hat{\rho}\left[\frac{\partial \hat{U}_j}{\partial t} + \hat{U}_i \frac{\partial \hat{U}_j}{\partial x_i}\right] = -\frac{\partial \hat{P}}{\partial x_j} + \frac{\partial}{\partial x_i}\left[\mu\left(\frac{\partial \hat{U}_j}{\partial x_i} + \frac{\partial \hat{U}_i}{\partial x_j}\right)\right] + \hat{\rho} g_j. \quad (18.81)$$

With the continuity equation for fluids of constant density, i.e. for $\hat{\rho} = $ constant:

$$\frac{\partial \hat{U}_i}{\partial x_i} = 0 \quad \text{and thus} \quad \frac{\partial^2 \hat{U}_i}{\partial x_i \partial x_j} = \frac{\partial}{\partial x_j}\left(\frac{\partial \hat{U}_i}{\partial x_i}\right) = 0. \quad (18.82)$$

Hence the Navier–Stokes equations, for $j = 1, 2, 3$, can be written as follows:

$$\hat{\rho}\left[\frac{\partial \hat{U}_j}{\partial t} + \hat{U}_i \frac{\partial \hat{U}_j}{\partial x_i}\right] = -\frac{\partial \hat{P}}{\partial x_j} + \mu \frac{\partial^2 \hat{U}_j}{\partial x_i^2} + \hat{\rho} g_j \quad (18.83)$$

or by including the continuity equation:

$$\hat{\rho}\left[\frac{\partial \hat{U}_j}{\partial t} + \frac{\partial}{\partial x_i}\left(\hat{U}_i \hat{U}_j\right)\right] = -\frac{\partial \hat{P}}{\partial x_j} + \mu \frac{\partial^2 \hat{U}_j}{\partial x_i^2} + \hat{\rho} g_j. \quad (18.84)$$

If one introduces:

$$\hat{\rho} = \bar{\rho} \quad \text{and} \quad \rho' = 0; \quad \hat{U}_j = \bar{U}_j + u'_j \quad \text{also} \quad \hat{P} = \bar{P} + p', \quad (18.85)$$

18.5 Time-Averaged Basic Equations of Turbulent Flows

(18.84) results in

$$\bar{\rho}\left[\frac{\partial}{\partial t}(\overline{U}_j + u'_j) + \frac{\partial}{\partial x_i}\left[(\overline{U}_i + u'_i)(\overline{U}_j + u'_j)\right]\right]$$
$$= -\frac{\partial}{\partial x_j}(\bar{P} + p') + \mu \frac{\partial^2}{\partial x_i^2}(\overline{U}_j + u'_j) + \bar{\rho} g_j$$
(18.86)

and after completion of time-averaging of all terms of the equation

$$\bar{\rho}\left[\underbrace{\frac{\partial \overline{U}_j}{\partial t}}_{=0} + \frac{\partial}{\partial x_i}\left(\overline{U}_i \overline{U}_j + \overline{u'_i u'_j}\right)\right] = -\frac{\partial \bar{P}}{\partial x_j} + \mu \frac{\partial^2 \overline{U}_j}{\partial x_i^2} + g_j.$$
(18.87)

Rearranging the terms, one obtains

$$\bar{\rho}\frac{\partial}{\partial x_i}\left(\overline{U}_i \overline{U}_j\right) = -\frac{\partial \bar{P}}{\partial x_j} + \frac{\partial}{\partial x_i}\left[\mu \frac{\partial \overline{U}_j}{\partial x_i} - \bar{\rho}\overline{u'_i u'_j}\right] + \bar{\rho} g_j.$$
(18.88)

Considering the continuity equation for fluids of constant density, one obtains

$$\bar{\rho}\overline{U}_i \frac{\partial \overline{U}_j}{\partial x_i} = -\frac{\partial \bar{P}}{\partial x_j} + \underbrace{\frac{\partial}{\partial x_i}\left[\mu \frac{\partial \overline{U}_j}{\partial x_i} - \bar{\rho}\overline{u'_i u'_j}\right]}_{-(\tau_{ij})_{\text{tot}}} + \bar{\rho} g_j.$$
(18.89)

This equation shows that, due to the time-averaging of the non-linear terms on the left-hand side of the Navier–Stokes equations, additional terms are introduced into the Reynolds equations, which can be interpreted as additional momentum transport terms, so that for a turbulent fluid flow the following holds:

$$(\tau_{ij})_{\text{tot}} = -\mu \frac{\partial \overline{U}_j}{\partial x_i} + \bar{\rho}\overline{u'_i u'_j}.$$
(18.90)

The additional terms represent a tensor which can be stated as follows:

$$\overline{u'_i u'_j} = \begin{pmatrix} \overline{u'^2_1} & \overline{u'_1 u'_2} & \overline{u'_1 u'_3} \\ \overline{u'_2 u'_1} & \overline{u'^2_2} & \overline{u'_2 u'_3} \\ \overline{u'_3 u'_1} & \overline{u'_3 u'_2} & \overline{u'^2_3} \end{pmatrix}.$$
(18.91)

This tensor is called in the literature the Reynolds "stress tensor." It is diagonally symmetrical, i.e. the following holds:

$$\left|\overline{u'_i u'_j}\right| = \left|\overline{u'_j u'_i}\right|.$$
(18.92)

Furthermore, because $\hat{\rho} = \bar{\rho} =$ constant, the following relationship holds:

$$\frac{\partial}{\partial x_i}\overline{u'_i u'_j} = \overline{u'_i \frac{\partial u'_j}{\partial x_i}}$$
(18.93)

as the $u'_j \frac{\partial u'_i}{\partial x_i}$ term is equal to zero, because of the continuity equation, written for the fluctuating velocity components.

The diagonal terms appearing in the Reynolds stress tensor can be interpreted as "normal stresses", whose significance for the time-averaged transport of momentum is negligible in many fluid flows. The non-diagonal terms, i.e. the terms $\overline{u'_i u'_j}$ for $i \neq j$, represent, in many fluid flows, the main transport terms of the momentum which are due to the turbulent velocity fluctuations. This makes it clear that the splitting of the instantaneous velocity components into a mean component and a turbulent fluctuation leads to a division of the total momentum transport in a mean part and in a turbulent part. The total momentum transport develops on the one hand due to the mean flow field, $\rho \overline{U}_i \overline{U}_j$, and on the other due to correlations of the turbulent velocity fluctuations, $\rho \overline{u'_i u'_j}$. This subdivision is very useful for many considerations in fluid mechanics of turbulent flows. It means, however, as far as the momentum transport equations are concerned, that six additional unknown quantities appear, namely all terms of the Reynolds stress tensor. The introduction of the instantaneous values of the velocity fluctuations, and the employed time averaging, thus results in a system of equations which is not closed: it contains more unknowns than equations that are at disposal for the solution of fluid-flow problems. It is nowadays one of the main tasks of turbulence research to link the additional unknowns of the Reynolds stress tensor to the components of the mean flow field in such a way that independent additional equations result, which can be employed to solve fluid flow problems. Deriving such additional relationships is called turbulence modeling. The main elements of turbulence modeling are summarized in Sect. 18.7.

18.5.3 Mechanical Energy Equation for the Mean Flow Field

In the preceding section, the momentum equations for Newtonian fluids of constant density were derived as follows:

$$\bar{\rho} \overline{U}_i \frac{\partial \overline{U}_j}{\partial x_i} = -\frac{\partial \bar{P}}{\partial x_j} + \frac{\partial}{\partial x_i}\left[\mu \frac{\partial \overline{U}_j}{\partial x_i} - \bar{\rho}\overline{u'_i u'_j}\right] + \bar{\rho} g_j. \tag{18.94}$$

On multiplying this equation by $\overline{U}_j$, one obtains

$$\bar{\rho} \overline{U}_i \overline{U}_j \frac{\partial \overline{U}_j}{\partial x_i} = -\overline{U}_j \frac{\partial \bar{P}}{\partial x_j} + \overline{U}_j \frac{\partial}{\partial x_i}\left[\mu \frac{\partial \overline{U}_j}{\partial x_i} - \bar{\rho}\overline{u'_i u'_j}\right] + \bar{\rho} g_j \overline{U}_j \tag{18.95}$$

18.5 Time-Averaged Basic Equations of Turbulent Flows

or rearranging this equation for $g_j = 0$:

$$\bar{\rho} \bar{U}_i \frac{\partial}{\partial x_i}\left(\frac{1}{2}\bar{U}_j^2\right) = -\frac{\partial}{\partial x_j}(\bar{P}\bar{U}_j) + \frac{\partial}{\partial x_i}\left[\mu \bar{U}_j \frac{\partial \bar{U}_j}{\partial x_i} - \overline{\bar{\rho} u_i' u_j'} \bar{U}_j\right] \qquad (18.96)$$

$$-\mu \frac{\partial \bar{U}_j}{\partial x_i} \frac{\partial \bar{U}_j}{\partial x_i} + \overline{\bar{\rho} u_i' u_j'} \frac{\partial \bar{U}_j}{\partial x_i}.$$

The different terms of the mechanical energy equation, derived for a unit volume and a unit time and by time averaging, can now be interpreted as follows, if the considerations are carried out for a fixed control volume:

$$\iiint_{\delta V} \bar{\rho} \bar{U}_i \frac{\partial}{\partial x_i}\left(\frac{1}{2}\bar{U}_j^2\right) dV = \iint_{\delta O} \left(\frac{1}{2}\bar{U}_j^2\right) \bar{\rho} \bar{U}_i \, dF_i.$$

The above term represents the mean kinetic energy of the flow field flowing in and out through the surfaces of a control volume. (As the energy equation was derived from the momentum equation, it comprises only terms which can be designated mechanical energy terms.)

$$-\iiint_{\delta V} \frac{\partial}{\partial x_j}(\bar{P}\bar{U}_j) \, dV = -\iint_{\delta O} \bar{P}\bar{U}_j \, dF_j.$$

The above term indicates what pressure energy per unit time flows in and out of the control volume.

$$\iiint_{\delta V} \frac{\partial}{\partial x_i}\left[\mu \bar{U}_j \frac{\partial \bar{U}_j}{\partial x_i}\right] dV = \iint_{\delta O} \mu \bar{U}_j \frac{\partial \bar{U}_j}{\partial x_i} \, dF_i$$

$$= \iint_{\delta O} \mu \frac{\partial \frac{1}{2}\bar{U}_j^2}{\partial x_i} \, dF_i.$$

This above term indicates the molecule-caused inflow and outflow of the kinetic energy of the fluid flow.

$$-\iiint_{\delta V} \frac{\partial}{\partial x_i}\left[\overline{\bar{\rho} u_i' u_j'} \bar{U}_j\right] dV = -\iint_{\delta O} \overline{\bar{\rho} u_i' u_j'} \bar{U}_j \, dF_i.$$

The term above describes the turbulence-dependent transport of the energy resulting from u_j' and U_j interactions.

$$-\iiint_{\delta V} \mu \frac{\partial \bar{U}_j}{\partial x_i} \frac{\partial \bar{U}_j}{\partial x_i} \, dV.$$

This term indicates how the dissipation of the mean energy by viscosity takes place.

$$\iiint_{\delta V} \overline{\rho u'_i u'_j} \frac{\partial \overline{U}_j}{\partial x_i} \, dV.$$

The last term of (18.96) describes how the energy of the mean flow field is turned into turbulence, i.e. how turbulence is produced by the interaction of turbulence with the mean flow field. (The energy withdrawn from the mean flow field leads to the production of energy of turbulent fluid-flow fluctuations.)

The product of the negative correlation of the turbulent velocity fluctuations u'_i and u'_j and the gradient of the mean flow field yields the following term, which is called production term of turbulence:

$$-\overline{\rho u'_i u'_j} \frac{\partial \overline{U}_j}{\partial x_i}. \tag{18.97}$$

This term occurs in the differential transport equation of the turbulent kinetic energy. Because of the symmetry of the Reynolds stress tensor, it can also be written as follows:

$$-\overline{\rho u'_i u'_j} \frac{\partial \overline{U}_j}{\partial x_i} = -\overline{\rho u'_i u'_j} \bar{D}_{ij} \tag{18.98}$$

with $\bar{D}_{ij} = \frac{1}{2} \left(\frac{\partial \overline{U}_j}{\partial x_i} + \frac{\partial \overline{U}_i}{\partial x_j} \right)$ = tensor of the time-averaged deformation rate. It appears with a positive sign on the right-hand side of the above energy equation for the mean flow field, and this means that the energy serving for the production of turbulent velocity fluctuations is withdrawn from the mean flow field. The actual dissipation term in the mean flow energy equation appears:

$$\mu \frac{\partial \overline{U}_j}{\partial x_i} \frac{\partial \overline{U}_j}{\partial x_i} \tag{18.99}$$

the significance of which, relative to the dissipation or production term of turbulence, is negligible for many turbulent fluid flows. This can be assessed by the subsequent order of magnitude considerations:

$$\overline{\rho u'_i u'_j} \frac{\partial \overline{U}_j}{\partial x_i} \rightsquigarrow \overline{\rho u_c^2} \frac{\overline{U}_c}{L_c} \rightsquigarrow \rho (Tu)^2 \frac{\overline{U}_c^3}{L_c}, \tag{18.100}$$

$$\mu \frac{\partial \overline{U}_j}{\partial x_i} \frac{\partial \overline{U}_j}{\partial x_i} \rightsquigarrow \mu \frac{\overline{U}_c^2}{L_c^2}, \tag{18.101}$$

where $\overline{u'^2_c}$ represents the effective value of a characteristic velocity fluctuation present in the flow field, $(Tu)^2 \approx \overline{u_c^2}/\overline{U_c^2}$ represents the square of the corresponding turbulence intensity, U_c being a characteristic velocity and L_c a characteristic dimension of the flow geometry. With this, the following relationship holds:

18.5 Time-Averaged Basic Equations of Turbulent Flows

$$\frac{\overline{\rho u_i' u_j'} \frac{\partial \overline{U}_j}{\partial x_i}}{\mu \frac{\partial \overline{U}_j}{\partial x_i} \frac{\partial \overline{U}_j}{\partial x_i}} \rightsquigarrow \frac{\bar\rho (Tu)^2 \overline{U}_c^3 L_c^2}{L_c \mu \overline{U}_c^2} = \left(\frac{\overline{U}_c L_c}{\nu}\right)(Tu)^2 = Re(Tu)^2. \quad (18.102)$$

As turbulence always occurs for large Reynolds numbers, e.g. $Re = 10^4$, (18.102) shows that even for $Tu = 20\%$, a comparatively large degree of turbulence, the viscous dissipation is negligible compared with the turbulence production. Similar considerations also hold for the terms

$$\frac{\overline{\rho u_i' u_j'} \overline{U}_j}{\mu \overline{U}_j \frac{\partial \overline{U}_j}{\partial x_i}} \rightsquigarrow \frac{\rho \overline{U}_c^2 \overline{U}_c^3 L_c^2}{\mu \overline{U}_c \overline{U}_c L_c} \rightsquigarrow \left(\frac{\overline{U}_c L_c}{\nu}\right)(Tu)^2 = Re(Tu)^2 \quad (18.103)$$

so that for many practical computations the transport equation for the "mechanical energy" of the mean velocity field of a turbulent flow can be written in a simplified way, as the molecule-dependent energy transport and molecular dissipations terms can be neglected compared with the turbulence-dependent production term:

$$\bar\rho \overline{U}_i \frac{\partial}{\partial x_i}\left(\frac{1}{2}\overline{U}_j^2\right) = -\frac{\partial}{\partial x_j}\left(\bar{P}\overline{U}_j\right) - \bar\rho \frac{\partial}{\partial x_i}\left(\overline{u_i' u_j'} \overline{U}_j\right) + \overline{\rho u_i' u_j'} \frac{\partial \overline{U}_j}{\partial x_i}. \quad (18.104)$$

This somewhat simplified energy equation contains all terms that are important for the mechanical energy transport occurring for the mean flow field in the more practical, i.e. non-academic, applications of fluid mechanics.

In order to underline the significance of the energy transport equation for the mean flow field, for the comprehension of the turbulence production, we shall discuss the fully developed turbulent flow between two plane plates for which (18.104) reads as follows:

$$0 = -\frac{\partial}{\partial x_1}\left(\bar{P}\overline{U}_1\right) - \bar\rho \frac{\partial}{\partial x_2}\left[(\tau_{ges})_{21} \overline{U}_1\right] + \overline{\rho u_1' u_2'} \frac{\partial \overline{U}_1}{\partial x_2}. \quad (18.105)$$

Integration over the control volume, indicated in Fig. 18.10, results in the following relationship:

$$0 = (P_A - P_B)\dot{Q} - \iiint_{\delta V}\left(-\overline{\rho u_1' u_2'} \frac{\partial \overline{U}_1}{\partial x_2}\right) dV. \quad (18.106)$$

The entire "pressure energy" generated per unit time by a ventilator for air or a pump for water, i.e. $(P_A - P_B)\dot{Q}$, serves for the production of the turbulent kinetic energy in the flow.

When carrying out similar considerations for the turbulent Couette flow indicated in Fig. 18.11, the energy transport equation for this flow reads

$$0 = -\iint_{\delta O} (\tau_{ges})_{21} \overline{U}_1 \, dF_1 - \iiint_{\delta V}\left(-\overline{\rho u_1' u_2'} \frac{\partial \overline{U}_1}{\partial x_2}\right) dV. \quad (18.107)$$

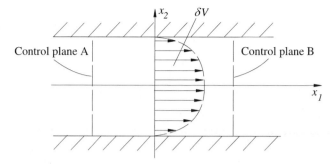

Fig. 18.10 Plane channel flow (the employed pump energy $(P_A - P_B)\dot{Q}$ serves for the production of turbulent, kinetic energy)

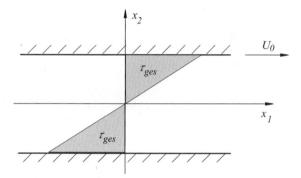

Fig. 18.11 Couette flow (the employed movement energy serves for the production of turbulent, kinetic energy)

This equation expresses that the entire "driving energy" put into the plate movement, indicated by U_0 in Fig. 18.11, is used for the production of turbulent kinetic energy in the flow.

The above examples of turbulent flows make the significance of fluid mechanical efforts clear to reduce the production of turbulence wherever possible, e.g. by the addition of additives to flowing fluids. By using additives such as high molecular weight polymers or surfactants, the production of the turbulent energy can be considerably suppressed. The achieved reduction of turbulence production represents a considerable saving of the pump energy which has to be introduced for the propulsion of turbulent flows.

18.5.4 Equation for the Kinetic Energy of Turbulence

In addition to the considerations of the energy balance for the mean flow field, it is instructive for the understanding of some of the physics of turbulent flows to consider the energy balance of the turbulent part of the flow. For this the energy equation for the kinetic energy of turbulence is employed, which again

18.5 Time-Averaged Basic Equations of Turbulent Flows

is derived from the general momentum equation:

$$\hat{\rho}\left[\frac{\partial \hat{U}_j}{\partial t} + \hat{U}_i \frac{\partial \hat{U}_j}{\partial x_i}\right] = -\frac{\partial \hat{P}}{\partial x_j} + \mu \frac{\partial^2 \hat{U}_j}{\partial x_i^2} + \hat{\rho} g_j. \qquad (18.108)$$

On introducing in this equation $\hat{\rho} = \bar{\rho} = \text{constant}$ and $\hat{U}_j = \overline{U}_j + u'_j$, $\hat{P} = \bar{P} + p'$ and neglecting g_j, one obtains:

$$\hat{\rho}\left[\frac{\partial}{\partial t}(\overline{U}_j + u'_j) + \frac{\partial}{\partial x_i}(\overline{U}_i + u'_i)(\overline{U}_j + u'_j)\right]$$
$$= -\frac{\partial}{\partial x_j}(\bar{P} + p') + \mu \frac{\partial^2}{\partial x_i^2}(\overline{U}_j + u'_j). \qquad (18.109)$$

Multiplying this equation by $(U_j + u'_j)$, the following relationship results if, additionally, the continuity equation is taken into account:

$$\hat{\rho}\left[\frac{\partial}{\partial t}\frac{1}{2}(\overline{U}_j + u'_j)^2 + (\overline{U}_i + u'_i)\frac{\partial}{\partial x_i}\frac{1}{2}(\overline{U}_j + u'_j)^2\right]$$
$$= -\frac{\partial}{\partial x_j}[(\bar{P} + p')(\overline{U}_j + u'_j)] + \mu(\overline{U}_j + u'_j)\frac{\partial^2}{\partial x_i^2}(\overline{U}_j + u'_j) \qquad (18.110)$$

and after time averaging the entire equation:

$$\bar{\rho}\overline{U}_i \frac{\partial}{\partial x_i}\left[\frac{1}{2}\left(\overline{U}_j^2 + \overline{u'^2_j}\right)\right] + \bar{\rho}\frac{\partial}{\partial x_i}\left[\overline{U}_j \overline{u'_i u'_j}\right] + \bar{\rho}\frac{\partial}{\partial x_i}\left(\overline{u'_i u'^2_j}\right)$$
$$= -\frac{\partial}{\partial x_j}\left(\bar{P}\overline{U}_j + \overline{p'u'_j}\right) + \overline{U}_j \mu \frac{\partial^2 \overline{U}_j}{\partial x_i^2} + \overline{u'_j \frac{\partial^2 u'_j}{\partial x_i^2}}. \qquad (18.111)$$

Subtracting the energy equation for the mean flow field:

$$\bar{\rho}\overline{U}_i \frac{\partial}{\partial x_i}\left(\frac{1}{2}\overline{U}_j^2\right) = -\frac{\partial}{\partial x_j}(\bar{P}\overline{U}_j) + \overline{U}_j \mu \frac{\partial^2 \overline{U}_j}{\partial x_i^2} - \overline{U}_j \bar{\rho}\frac{\partial}{\partial x_i}\overline{u'_i u'_j} \qquad (18.112)$$

one obtains as the transport equation for the turbulent kinetic energy:

$$\bar{\rho}\overline{U}_i \frac{\partial}{\partial x_i}\left(\frac{1}{2}\overline{u'^2_j}\right) = -\frac{\partial}{\partial x_j}\left(\overline{p'u'_j}\right) + \frac{\partial}{\partial x_j}\left(\overline{\mu u'_j \frac{\partial u'_j}{\partial x_i}}\right) - \frac{\bar{\rho}}{2}\frac{\partial}{\partial x_i}\left(\overline{u'_i u'^2_j}\right)$$
$$-\bar{\rho}\overline{u'_i u'_j}\frac{\partial \overline{U}_j}{\partial x_i} - \mu \overline{\frac{\partial u'_j}{\partial x_i}\frac{\partial u'_j}{\partial x_i}}. \qquad (18.113)$$

The different terms of this equation can now be interpreted as follows, showing their significance for the kinetic energy balance for a control volume:

$$\iiint_{\delta V} \bar{\rho}\overline{U}_i \frac{\partial}{\partial x_i}\left(\frac{1}{2}\overline{u'^2_j}\right) dV = \iiint_{\delta V} \bar{\rho}\frac{\partial}{\partial x_i}\left[\overline{U}_i\left(\frac{1}{2}\overline{u'^2_j}\right)\right] dV = \iint_{\delta O} \frac{\bar{\rho}}{2}\overline{u'^2_j} U_i \, dF_i.$$
$$(18.114)$$

The left-hand side term in this equation represents the *turbulent* kinetic energy, transported in and out of a control volume by convection.

$$-\iiint_{\delta V} \frac{\partial}{\partial x_j} \left(\overline{p'u'_j}\right) \mathrm{d}V = -\iint_{\delta 0} \overline{p'u'_j}\, \mathrm{d}F_j.$$

This is the pressure-velocity correlation responsible for the redistribution of the energy of the turbulence from the j component to the other components of the turbulent velocity fluctuations.

$$\iiint_{\delta V} \frac{\partial}{\partial x_i}\left[\overline{\mu u'_j \frac{\partial u'_j}{\partial x_i}}\right] \mathrm{d}V = \iint_{\delta 0} \overline{\mu u'_j \frac{\partial u'_j}{\partial x_i}}\, \mathrm{d}F_i$$

$$= \iint_{\delta 0} \mu \frac{\partial \overline{\tfrac{1}{2} u'^2_j}}{\partial x_i}\, \mathrm{d}F_i.$$

The above term describes the molecular transport of the turbulent kinetic energy into and out of the control volume.

$$-\frac{\bar{\rho}}{2}\iint_{\delta V} \frac{\partial}{\partial x_i}\left(\overline{u'_i u'^2_j}\right) \mathrm{d}V = -\frac{\bar{\rho}}{2}\iint_{\delta 0} \overline{u'_i u'^2_j}\, \mathrm{d}F_i.$$

The above term is the diffusive transport of the turbulent kinetic energy by velocity fluctuations into and out of the considered control volume.

$$\bar{\rho}\iiint_{\delta V} \left(-\overline{u'_i u'_j}\right) \frac{\partial \overline{U}_j}{\partial x_i}\, \mathrm{d}V.$$

This term represents the production term of the turbulent kinetic energy which appeared in the energy equation for the mean flow field, but with inverse sign.

$$-\iiint_{\delta V} \mu \overline{\frac{\partial u'_j}{\partial x_i} \frac{\partial u'_j}{\partial x_i}}\, \mathrm{d}V.$$

By the above term the viscous dissipation of turbulent kinetic energy is described, caused by the turbulent velocity fluctuations.

When defining the following total term as the "turbulent diffusion" of the turbulent kinetic energy:

$$D_j = \overline{p'u'_j} - \overline{\mu u'_i \frac{\partial u'_i}{\partial x_j}} + \frac{\bar{\rho}}{2}\overline{u'_j u'^2_i} \tag{18.115}$$

the production of turbulent kinetic energy as:

$$P_k = -\rho \overline{u'_i u'_j} \frac{\partial \overline{U}_j}{\partial x_i} \tag{18.116}$$

18.6 Scales of Turbulent Flows

and the dissipation of turbulent kinetic energy as

$$\epsilon_k = \mu \overline{\frac{\partial u'_j}{\partial x_i} \frac{\partial u'_j}{\partial x_i}} \tag{18.117}$$

the equation for the turbulent kinetic energy can be written in the following way:

$$\bar{\rho}\overline{U}_i \frac{\partial k}{\partial x_i} = -\frac{\partial D_j}{\partial x_j} + P_k - \epsilon_k, \tag{18.118}$$

where $k = (1/2\overline{u'^2_j})$ was introduced, i.e. $k = 1/2(\overline{u'^2_1} + \overline{u'^2_2} + \overline{u'^2_3})$. The mean transport of turbulent kinetic energy is kept "in balance" by the diffusion, production and dissipation of turbulent kinetic energy at a point in the flow.

Earlier, it was shown that in the equation for the energy of the mean flow the viscous dissipation, caused by the mean flow field, when compared with the production term of the turbulent kinetic energy, can be neglected. This is not the case for the turbulent dissipation in the above equation for the *turbulent kinetic energy*, i.e. ϵ_k cannot be neglected with respect to P_k. This can, on the other hand, be shown through the following order of magnitude considerations:

$$\frac{P_k}{\epsilon_k} \sim \frac{\bar{\rho} u_c^2 \overline{U}_c l_c^2}{L_c \mu u_c^2} = \left(\frac{\overline{U}_c L_c}{\nu}\right)\left(\frac{l_c}{L_c}\right)^2 = Re \left(\frac{l_c}{L_c}\right)^2, \tag{18.119}$$

where l_c/L_c is the ratio of a characteristic length scale of the considered turbulence to a length scale characterizing the mean flow. For the ratio l_c/L_c, the following relationship holds in general:

$$\frac{l_c}{L_c} = Re^{-m}, \tag{18.120}$$

where $m = 1/2$ can be set, i.e. in the equation for the turbulent kinetic energy the viscous dissipation cannot be neglected. It is an essential part of (18.118), describing the transport of turbulent kinetic energy.

18.6 Characteristic Scales of Length, Velocity and Time of Turbulent Flows

In the preceding sections, order of magnitude considerations were made in which scales of length, velocity and time of turbulent flows were employed. So, for the characterization of the mean flow field, a characteristic mean velocity $\overline{U}_c$, a characterizing length L_c and a time scale t_c were introduced. Here, L_c is of the order of the dimensions of the total flow extension, i.e. for internal flows L_c has the linear cross-section dimension of the flow channel or pipe where the fluid flows. The area-averaged or the time-averaged velocity of the flow field can be introduced as the characteristic mean velocity. For the characteristic time, t_c, the following ratio holds:

$$t_c = \frac{L_c}{\overline{U}_c}. \tag{18.121}$$

If one considers the turbulent velocity fluctuations that occur superimposed on the mean flow field, it is easy to see that the integral time-scale of the turbulence, introduced in Sect. 18.3, always has to be of the order of magnitude of the characteristic time scale of the mean flow field, i.e. the largest vortices that the turbulent part of a flow field possesses have time scales which correspond to those of the mean flow field. Generally, the following relationship holds:

$$I_t \leq t_c = \frac{L_c}{\overline{U}_c}. \tag{18.122}$$

Following a suggestion of Kolmogorov, so-called micro-scales can be introduced to characterize the turbulent flow field:

$$l_K = \text{Kolmogorov's length scale},$$
$$u_k = \text{Kolmogorov's velocity scale},$$
$$\tau_k = (l_K/u_K) = \text{Kolmogorov's time scale}.$$

The length, velocity and time scales introduced by Kolmogorov are determined in such a way that they characterize that part of the spectrum of the turbulent velocity fluctuations in which the energy production of the turbulent vortices is equal to the dissipation. Thus, assuming isotropic turbulence, one can introduce:

$$\epsilon \approx P \approx \frac{u_K^3}{l_K} \quad \text{for the relationship for production}, \tag{18.123}$$

$$\epsilon \approx \nu \frac{u_K^2}{l_K^2} \quad \text{for the relationship for dissipation}. \tag{18.124}$$

From these results, we can deduce:

$$u_K^6 = \epsilon^2 l_K^2 = \epsilon^3 \frac{l_K^6}{\nu^3} \quad \text{or} \tag{18.125}$$

$$1 = \epsilon \frac{l_K^4}{\nu^3} \rightsquigarrow l_K = \left(\frac{\nu^3}{\epsilon}\right)^{\frac{1}{4}}. \tag{18.126}$$

From the equality of the terms for production and dissipation, it follows that:

$$u_K = \frac{\nu}{l_K} \quad \text{and thus} \quad u_K = (\nu\epsilon)^{\frac{1}{4}}. \tag{18.127}$$

For the Kolmogorov time scale, the following expression results:

$$\tau_K = \left(\frac{\nu}{\epsilon}\right)^{\frac{1}{2}}. \tag{18.128}$$

18.6 Scales of Turbulent Flows

The Reynolds number resulting on the basis of the above-introduced micro-length scale and micro-velocity scale is:

$$Re_K = \frac{l_K u_K}{\nu} = 1. \tag{18.129}$$

The characteristic turbulent eddy quantities, determined by Kolmogorov's scales of turbulence, are those that represent the viscous effects which damp the turbulent velocity fluctuations. These smallest eddies are assumed to convert the kinetic energy of turbulence into heat. Because of these characteristic properties, the following definitions are available in the literature for the smallest scales of turbulence:

Kolmogrov's scales = micro-scales = viscous eddy scales:

$$l_K = \left(\frac{\nu^3}{\epsilon}\right)^{\frac{1}{4}}; \quad u_K = (\nu\epsilon)^{\frac{1}{4}}; \quad \tau_K = \left(\frac{\nu}{\epsilon}\right)^{\frac{1}{2}}. \tag{18.130}$$

From these scales, characterizing the smallest vortices of a turbulent flow, the Taylor micro-scale has to be distinguished, which is defined as follows:

$$\tau_T = \frac{l_T}{u_T} = \frac{l_T}{\overline{U}_c}. \tag{18.131}$$

The extensions of the different scales can perhaps best be illustrated in schematic form; see Fig. 18.12, which shows that the Taylor micro-scale defines an eddy size which is located between the smallest viscous eddies and the large eddies having quantities of the dimension of the geometric extension of the mean flow. Taking this into account, it can be shown that the following hold:

$$\frac{\overline{U}_c^3}{L_c} = \nu \frac{\overline{U}_c^2}{l_T^2}; \quad \left(\frac{l_T}{L_c}\right) = \frac{1}{Re^{\frac{1}{2}}} = Re^{-\frac{1}{2}}, \tag{18.132}$$

where $Re = (U_c L_c)/\nu$:

Considering that the following holds:

$$\lambda_T = l_K \frac{\overline{U}_c}{u_c} \tag{18.133}$$

then inserting this relationship into the equation for $\frac{l_T}{L_C}$ and taking into account $\frac{l_K u_K}{\nu} = 1$, a further important relationship follows from the above derivations:

$$\frac{l_T}{L_c} = \left(\frac{l_K}{l_T}\right)^2; \quad l_T = L_c l_k^2. \tag{18.134}$$

The different length scales of turbulence have proven to be very useful in formulations of turbulence models, which are summarized in the subsequent sections. The presentations to follow were chosen such that they are suitable for introduction into turbulence modeling. More detailed descriptions can be

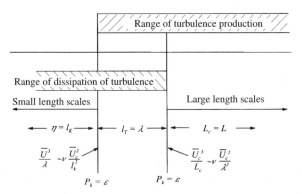

Fig. 18.12 Length scales of turbulence and contributions to production and dissipation

taken from the available specialized literature on turbulence modeling, given in the list of references at the end of this chapter.

The above derivations indicate the differences in the structure of turbulent flows at small and high Reynolds numbers. For flows with the same integral dimensions (see Fig. 18.12), the flow at a large Reynolds number proves to be "micro-structured," i.e. the smallest eddies have small dimensions, whereas for the small Reynolds number the flow appears "macro-structured." The Taylor length scale proves always to be larger than the Kolmogorov micro length, and the difference between the two becomes larger with increasing Reynolds number. From the above relationships, one can compute $l_T/l_K = (Re)^{1/4}$. For a Reynolds number of approximately $Re = 10^4$, l_T is approximately 10 times larger than l_K (Fig. 18.13).

In order to characterize the complex nature of turbulent flows, the following Reynolds numbers are often employed:

$$Re_K = \frac{l_K U_K}{\nu} = 1, \quad Re_{K,c} = \frac{l_K U_c}{\nu}, \quad Re_\lambda = \frac{l_T U_c}{\nu}, \tag{18.135}$$

$$Re_L = \frac{L_c U_c}{\nu} = Re. \tag{18.136}$$

Moreover, for the relationships of the characteristic length scales, the following expression holds:

$$\frac{l_K}{L_c} = Re_L^{-\frac{3}{4}} = Re_\lambda^{-\frac{3}{2}}, \quad \frac{\lambda_T}{L_c} = Re_L^{-\frac{1}{2}} = Re_\lambda^{-1} = \left(\frac{l_K}{l_T}\right)^2,$$

$$\frac{l_K}{l_T} = Re_L^{-\frac{1}{4}} = Re_\eta^{-1} = Re_\lambda^{-\frac{1}{2}}. \tag{18.137}$$

These relationships are often employed when considering turbulent flows, in order to carry out order of magnitude considerations regarding the characteristic properties of turbulence.

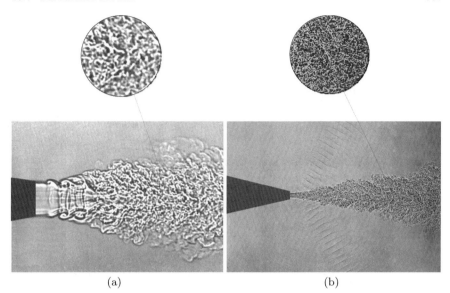

Fig. 18.13 (a) Micro-structure of the turbulence of small Reynolds numbers and (b) micro-structure of the turbulence of higher Reynolds numbers

18.7 Turbulence Models

18.7.1 General Considerations

When limiting oneself to considerations of the components of *mean* velocity and the moments of turbulent fluctuation quantities, as far as the solutions of fluid mechanical problems are concerned, there is the possibility of solving flow problems theoretically: *instead of the Navier–Stokes equations, the Reynolds equations are solved*. In order to derive the latter set of partial differential equations, the instantaneous velocity $\hat{U}_j(x_i, t)$ was replaced in Sect. 18.2 by the sum of the local mean velocity $\overline{U}_j(x_i)$ and the local fluctuation velocity $u'_i(x_i, t)$, i.e. in Sect. 18.2, it was shown that the following can be set:

$$\hat{U}_j = \overline{U}_j + u'_j. \tag{18.138}$$

In Sect. 18.3, it was shown moreover that the introduction of this relationships into the continuity equation and the Navier–Stokes equations results, after time-averaging all terms in the equations, in a new system of equations which has to be solved numerically for the flow problems to be investigated. The derivations again yield, for $\rho =$ constant, four differential equations:

Continuity equation for $j = 1, 2, 3$:
$$\frac{\partial U_i}{\partial x_i} = 0. \tag{18.139}$$

Reynolds equation:
$$\bar{\rho}\bar{U}_i \frac{\partial \bar{U}_j}{\partial x_i} = -\frac{\partial \bar{P}}{\partial x_j} + \frac{\partial}{\partial x_i}\left(\mu \frac{\partial \bar{U}_j}{\partial x_i} - \bar{\rho}\overline{u'_i u'_j}\right) + \bar{\rho}g_j. \tag{18.140}$$

The derivations have led, however, to the introduction of new unknowns given by the following fourth-order tensor:
$$\overline{u'_i u'_j} = \begin{pmatrix} \overline{u'^2_1} & \overline{u'_1 u'_2} & \overline{u'_1 u'_3} \\ \overline{u'_2 u'_1} & \overline{u'^2_2} & \overline{u'_2 u'_3} \\ \overline{u'_3 u'_1} & \overline{u'_3 u'_2} & \overline{u'^2_3} \end{pmatrix}, \tag{18.141}$$

which can be referred to as "Reynolds momentum transport terms," but are often also called Reynolds stress terms.

The introduction of additional unknowns into the basic equations of fluid mechanics lead to a non-closed system of equations and this requires additional information in order to obtain solutions from (18.139) and (18.140). This required information, often formulated as additional differential equations, represents general statements on the interrelation between the Reynolds momentum transport terms and the mean velocity field. The derivations of these additional differential equations require additional physical insights into the turbulent velocity correlations (18.141) and their dependence on the mean flow field. The development of model considerations for these unknown terms is an important field of research in fluid mechanics. Considerations of this kind are carried out in the form of turbulence models by various research groups. The basics of such models are described in the following sections only in an introductory manner. Further insight into this important sub-domain of fluid mechanics shows that the introduction of turbulence models serves to derive "closing assumptions" for the Reynolds equations. Succeeding in formulating additional equations, i.e. producing physically appropriate turbulence models, a new system of equations results, based on the Reynolds equations, which can be employed to solve turbulent flow problems.

The formulation of physically appropriate additional equations, i.e. to provide physically appropriate information for the correlations $\overline{u'_i u'_j}$ for the treatment of the Reynolds equations, can partly be realized by hypothetical assumptions, as was often done in the past when setting up turbulence models. There is, however, the possibility to obtain the information required for turbulence models by means of detailed experimental investigations in different turbulent flows. For this purpose, local measurements of the instantaneous velocity of turbulent flows are necessary. Such measurements can be achieved with the help of hot wire and hot film anemometry and also laser Doppler anemometry. Those measuring techniques provide the necessary resolution

18.7 Turbulence Models

with respect to time and space variations of the flow, to carry out all the measurements required for detailed turbulence modeling. The measuring methods can be considered to be fully developed, so that the required measurements can be carried out without serious application problems. Such measurements contribute considerably to deepening the comprehension of the physics of turbulence and make it possible to introduce additional information in the form of new equations to yield numerical solution procedures for turbulent flows.

For measurements in wall boundary layers, hot wire and hot film anemometers have been employed with great success for determining the mean velocity $\overline{U}_i$ and the fluctuation quantities $\overline{u'_i u'_j}$ (see Sect. 18.8). Flows of this kind can be investigated reliably with hot wire anemometers because of their characteristic properties. However, in the case of very thin boundary layers, inherent disturbances may occur which are caused by the introduced measuring sensors. By special shaping of the employed measuring sensors, these disturbances and the resulting measurement errors can be kept low.

Most measuring methods, requiring the introduction of measuring sensors into flows, measure the flow velocity only indirectly, i.e. with most measuring instruments physical quantities are recorded which are functions of the flow velocity. Unfortunately, the measuring quantities are often also functions of the properties of the state of the flow medium. The latter have to be known and have to be adapted already when calibrating the measuring method, in order to make the interpretation of the measured data possible in terms of velocity. When fluctuations of the fluid properties occur during the attempted velocity measurements, e.g. in two-phase flows, flows with chemical reactions etc., they have to be known to be able to determine reliably the required velocity values.

The above-mentioned difficulties in the employment of indirect measuring techniques for flow velocities, such as hot wire and hot film anemometry, have led to the development of the laser Doppler technique, which measures flow velocities directly. By measuring the time which a particle needs to flow through an interference pattern with a well-defined fringe distance, the velocity of light-scattering particles can be determined. Such measurements can be carried out locally and do not depend on the unknown thermodynamic properties of the flow fluid. Measurements are possible in one- and two-phase flows, and also in combustion systems and in the atmosphere. The measuring technique can moreover be employed in particle-loaded flows, i.e. in media as they often occur in practice. Its application requires, however, optical access to the measuring point and transparency of the flow medium. In this respect the employment of laser Doppler anemometry is limited, but its application makes the determination of flow velocities possible in a large number of flows that are not accessible to other measuring methods.

Due to considerable developments in applied mathematics over the last few decades, new methods to solve numerically systems of partial differential equations, such as those describing fluid flows, have appeared. These

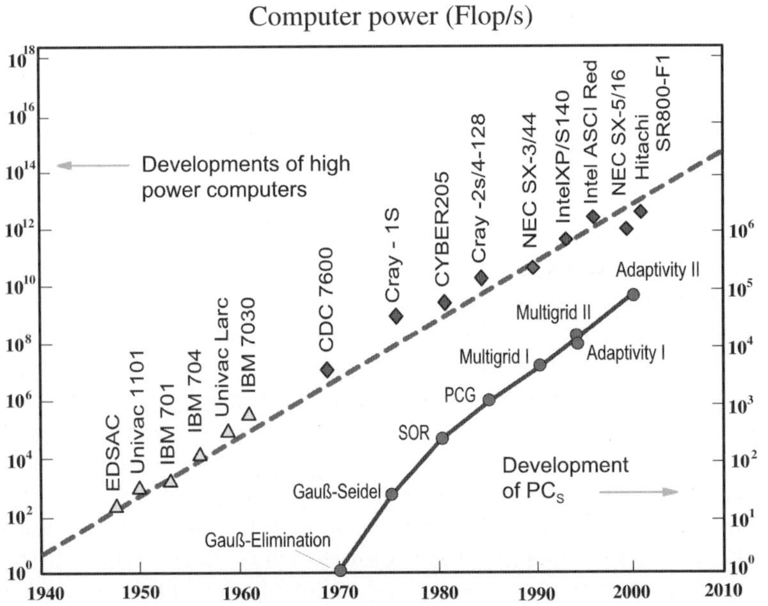

Fig. 18.14 Increase in computing and computer power in the employment of mathematical methods and of high-speed computers

developments were supported by an increase in computing power, as shown in Fig. 18.14.

Considering also the increased performance of high-performance computer systems, as also indicated in Fig. 18.14, it becomes understandable that it is possible nowadays to obtain direct numerical solutions of turbulent flows at least at small Reynolds numbers. Such solutions lead to important insights that can be employed for the development of refined turbulence models. In the following section, it is shown how the knowledge gained by experimental and numerical investigations of turbulent flows can be used to formulate turbulence models for the unknown Reynolds momentum transport terms. Whereas in the past turbulence model developments had to be carried out without such knowledge, numerically obtained information on the behavior of turbulent flows is nowadays available.

18.7.2 General Considerations Concerning Eddy Viscosity Models

In Sect. 18.5, the basic equations of fluid mechanics were employed, i.e. the continuity equation, the Navier–Stokes equation and the mechanical energy equation, in order to introduce into them time-averaged quantities of

18.7 Turbulence Models

turbulent property fluctuations. After time averaging of the resulting equations and taking into account $\partial/\partial t(\overline{\cdots}) = 0$, the following relationships could be derived:

- Continuity equations:
  - Mean flow field:
  $$\frac{\partial \overline{U}_i}{\partial x_i} = 0. \qquad (18.142)$$
  - Fluctuating flow field:
  $$\frac{\partial u'_i}{\partial x_i} = 0. \qquad (18.143)$$

- Navier–Stokes equations (for $g_j = 0$)
  - Mean flow field:
  $$\overline{\rho} \overline{U}_i \frac{\partial \overline{U}_j}{\partial x_i} = -\frac{\partial \overline{P}}{\partial x_j} + \frac{\partial}{\partial x_i}\left[\mu \frac{\partial \overline{U}_j}{\partial x_i} - \overline{\rho u'_i u'_j}\right]. \qquad (18.144)$$
  - Fluctuating flow field:
  $$\overline{\rho} \overline{U}_i \frac{\partial \overline{\frac{1}{2} u'_j u'_j}}{\partial x_i} = -\overline{\rho u'_i u'_j} \frac{\partial \overline{U}_j}{\partial x_i} - \frac{\partial}{\partial x_i}\overline{\left[u'_i\left(\frac{\overline{\rho} u'_j u'_j}{2} + p'\right)\right]}$$
  $$-\mu \overline{\frac{\partial u'_j}{\partial x_i} \frac{\partial u'_j}{\partial x_i}} + \mu \frac{\partial \overline{\frac{1}{2} u'_j u'_j}}{\partial x_i \partial x_i}. \qquad (18.145)$$

- Thermal energy equation:
  - Mean temperature field:
  $$\rho c_p \overline{U}_i \frac{\partial \overline{T}}{\partial x_i} = \frac{\partial}{\partial x_i}\left(\lambda \frac{\partial \overline{T}}{\partial x_i} - \rho c_p \overline{u'_i T'}\right). \qquad (18.146)$$
  - Fluctuating temperature field:
  $$U_i \frac{\partial}{\partial x_i}\left(\frac{1}{2}\overline{T'^2}\right) = -\frac{\partial}{\partial x_i}\left[\frac{1}{2}\overline{T'^2 u'_i} - a\frac{\partial}{\partial x_i}\left(\frac{1}{2}\overline{T'^2}\right)\right] - \overline{u'_i T'}\frac{\partial \overline{T}}{\partial x_i} - a\overline{\left(\frac{\partial T'}{\partial x_i}\right)^2} \qquad (18.147)$$

  with $a = \lambda/\rho c_p$.

Regarding the solutions of laminar flow problems, it had been possible, by employing the continuity and the Navier–Stokes equations, to solve a number of problems analytically, i.e. the set of differential equations that was available constituted a closed system of partial differential equations for these cases.[1] When considering the corresponding system of equations for turbulent flows, e.g. the continuity equation and the momentum equation for the mean flow

---

[1] By a closed differential system, one understands a system in which the number of unknown variables and the number of equations available are equal.

field, it can be seen from the above statements that the system of equations is not closed. There appear six additional unknown quantities, namely the correlations of the velocity fluctuations u'_i and u'_j, i.e. the elements of the following tensor:

$$\overline{u'_i u'_j} = \begin{pmatrix} \overline{u'^2_1} & \overline{u'_1 u'_2} & \overline{u'_1 u'_3} \\ \overline{u'_2 u'_1} & \overline{u'^2_2} & \overline{u'_2 u'_3} \\ \overline{u'_3 u'_1} & \overline{u'_3 u'_2} & \overline{u'^2_3} \end{pmatrix}. \tag{18.148}$$

This shows that the derivations of the time-averaged equations, generally holding for turbulent flows, have led to a closing problem which has to be solved before solutions of the above equations for turbulent flows can be sought. The development of suitable closing assumptions is tackled in flow research by turbulence model developments:

- By turbulence modeling, one understands the development of closing assumptions, which are formulated in the form of additional equations and which are employed in addition to the averaged continuity and Navier–Stokes equations, i.e. the Reynolds equations, for the solution of flow problems.

Such closing assumptions should make use of experimentally or numerically obtained information, so that soundly based assumptions for the properties of turbulent flow quantities can be employed as additional equations, needed for numerical solutions of turbulent flow problems.

In the literature, a large number of different turbulence models have been proposed, developed and employed for flow problem solutions, based on the Reynolds equation for turbulent flows. They can be classified as follows. All the following models have one thing in common, according to a suggestion of Boussinesq: they set the transport mechanism of turbulent velocity fluctuations equal to the transport of molecules in an isotropic Newtonian fluid, i.e. it is assumed that the following relationship holds:

$$\overline{\bar{\rho} u'_i u'_j} = -\bar{\rho} \nu_T \left(\frac{\partial \overline{U}_j}{\partial x_i} + \frac{\partial \overline{U}_i}{\partial x_j} \right), \tag{18.149}$$

ν_T is defined as the eddy viscosity, which has to be regarded as an unknown quantity, and it represents a property of the turbulent flow and not the fluid. It is the task of the above eddy viscosity models, often also called "first-order closure models," to provide good model assumptions for the eddy viscosity ν_T. To reach this goal, the considerations below based on characteristic velocity and length scales need to be considered:

$$\rho u_c^2 \sim \rho \nu_T \frac{u_c}{l_c} \quad \rightsquigarrow \quad \nu_T \sim u_c l_c, \tag{18.150}$$

i.e. defining equation of the eddy viscosity:

$$\nu_T = \frac{-\overline{u'_i u'_j}}{\left(\dfrac{\partial \overline{U}_j}{\partial x_i} + \dfrac{\partial \overline{U}_i}{\partial x_j} \right)}. \tag{18.151}$$

18.7 Turbulence Models

Order of magnitude considerations can be carried out, employing characteristic units of velocity and length scales of the considered turbulence. For simple turbulence model considerations, ν_T is often treated as a scalar quantity, although, by definition, it constitutes a fourth-order tensorial quantity. Strictly, by the introduction of ν_T the assumption of isotropy of a turbulent flow field is introduced into turbulence.

The characteristic scales of length and velocity, i.e. l_c and u_c, used in the different models are $k = \frac{1}{2} u'_k u'_k$

$$\text{Analytical models} \quad l_c: \quad \left|\frac{\partial \overline{U}_j}{\partial x_i}\right| l_c \quad \leadsto \quad \nu_T \sim l^2 \left|\frac{\partial \overline{U}_j}{\partial x_i}\right|, \qquad (18.152)$$

$$\text{One-equation models} \quad l_c: \quad k^{\frac{1}{2}} \quad \leadsto \quad \nu_T \sim l_c k^{\frac{1}{2}}, \qquad (18.153)$$

$$\text{Two-equation models} \quad \frac{k^{\frac{3}{2}}}{\epsilon} : \quad k^{\frac{1}{2}} \quad \leadsto \quad \nu_T \sim \frac{k^2}{\epsilon}. \qquad (18.154)$$

"Analytical turbulence models" are characterized by describing the characteristic length scale of the considered turbulence, l_c, as a local property of the turbulent flow field by means of an analytical relationship. This states the distribution of the length scale as a function of the location in the flow. The characteristic velocity is often given by the local gradient of the mean velocity field, multiplied by the characteristic length scale of turbulence. With this the turbulent eddy viscosity can be stated to be a product of the square of the characteristic length scale of turbulence and the local gradient of the mean velocity field. With the eddy viscosity introduced in this way, an additional equation results, expressing ν_T, which can be employed for the solution of turbulent flow problems. However, as a new unknown the characteristic length scale of turbulence is added, which has to be given as a function of the location, i.e. it has to be determined from experiments and introduced into the analytical equations for ν_T.

The one-equation turbulence models, indicated in Fig. 18.15, maintain the analytical description of the turbulent length scale, but solve a transport equation for the turbulent kinetic energy k. The eddy viscosity, appearing in the Boussinesq assumption (18.149) for the correlation $-\overline{u'_i u'_j}$, can thus be defined as a quantity which is computed from the product of the analytically described length scale of turbulence, l_c, and the $(k)^{\frac{1}{2}}$ value calculated with the help of a transport equation. In the transport equation for k, the occurring correlations of turbulent property fluctuations of higher order (see (18.145)), which were introduced by averaging the equations, are replaced by suitable modeling assumptions.

Two-equation turbulence models represent extensions with respect to lower equation models, where the locally existing length scale of turbulence, i.e. l_c, is defined with the help of the turbulent dissipation, ϵ, and this latter property of turbulence is computed from a transport equation. When this

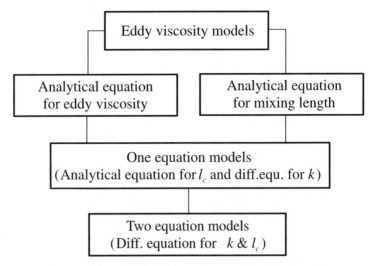

Fig. 18.15 Classification of eddy viscosity models for turbulent flows

quantity is computed locally by solving a separate transport equation, it can be combined with the locally determined turbulent kinetic energy, k, to a length scale of turbulence, so that the following holds:

$$l_c = \frac{k^{\frac{3}{2}}}{\epsilon}. \tag{18.155}$$

Experience with the employment of turbulence models in practice has shown that two-equation models are at least necessary to obtain mathematical formulations of turbulence model equations that have a certain general validity. As shown in Sect. 18.7.5, the model constants appearing in the transport equations for k and ϵ, can be determined from direct numerical computations, or from experimental studies of basic flows, and can later be applied to a wide range of practically relevant flows. In this way, solutions for the mean flow field and the most important turbulence quantities result and the computed quantities agree with experimental data with a precision sufficient for practical application. Employing analytical models of turbulence, or one-equation turbulence models, does not lead to the same generally valid applicability of once determined constants appearing in the transport equations. Similarly, in lower order models the analytical expressions far the turbulent velocity and length scale are far less generally applicable. For the analytical turbulence models and the one-equation turbulence models, it is important to find, for each flow geometry, new turbulence information for the model quantities and to formulate them in appropriate equations. Hence specific turbulence models result.

A further generalization of the applicability of turbulence models for the Reynolds differential equations was reached by the development of the so-called Reynolds stress turbulence models. These models do not use eddy viscosity formulations for the turbulent transport quantities, but solve a transport equation for each of the cited $\overline{u'_i u'_j}$ terms. These transport equations for the $\overline{u'_i u'_j}$ terms can be derived from the Navier–Stokes equations; however, they lead to correlation terms of higher order, for which also modeling assumptions have to be introduced. Experience seems to show that these model assumptions for higher-order terms can be found more easily in a generally valid form than transport assumptions for correlations of lower order. This is one of the main reasons for the generally wide employment of the Reynolds stress turbulence models. In comparison with two-equation turbulence models, they show an increased number of partial differential equations which have to be solved. The availability of increased computing capacity and increasing computing performance, already makes this kind of turbulence modeling appear interesting for practical computations.

Although the above-cited turbulence models use essentially local quantities for determining turbulent transport quantities, the solutions of the differential equations describing the turbulence properties include also global information on the entire flow field. This takes into account that the computed turbulence quantities comprise the effect of an entire spectrum of quantities, and thus also those from eddy sizes of the order of magnitude of the entire flow field. The transport equations solved, e.g for k and ϵ, take into account also the values of their corresponding quantities existing at the boundaries of the flows. Local transport processes enter into the computations also. Nevertheless, the cited turbulence models give grounds for discussion as to whether they include sufficiently all essential properties of turbulence and their effects on the unknown $\overline{u'_i u'_j}$ correlations. It can be hoped, however, that open questions will be solved by detailed experimental investigations, such as are possible nowadays with modern methods of flow measurements, as laser Doppler anemometry and hot wire anemometry.

18.7.3 Zero-Equation Eddy Viscosity Models

In the preceding section, the introduction of a turbulent eddy viscosity, as suggested by Boussinesq, was explained briefly as a quantity only characterizing sufficiently isotropic turbulence. The postulated analogy between the molecule-dependent momentum transport and turbulence-dependent momentum transport led to the following ansatz:

$$(\tau_{\text{turb}})_{ij} = \rho \overline{u'_i u'_j} = -\rho \nu_T \left(\frac{\partial \overline{U}_j}{\partial x_i} + \frac{\partial \overline{U}_i}{\partial x_j} \right) \qquad (18.156)$$

which, for plane-parallel flows, is reduced to a single term (e.g. x_1 = flow direction; x_2 = vertical to the flow direction and to the wall):

$$\overline{\rho u_2' u_1'} = \rho \nu_T \frac{\partial \overline{U}_1}{\partial x_2}. \tag{18.157}$$

For the determination of ν_T, Prandtl made use of simple considerations taken from the kinetic gas theory. This theory was developed by Boltzmann and used for the derivation of the molecule-dependent momentum transport, i.e. for the introduction of fluid viscosity. By introducing so-called turbulent eddies, it is possible, following Prandtl's suggestion, to postulate "a uniform size for the moment transporting eddies" of a turbulent flow field. These turbulent eddies perform stochastic motions in space. They are assumed to cover a path length l_m, the so-called turbulent mixing length, before they interact with other turbulent eddies, exchanging their momentum. This means that the actually occurring continuous interaction between turbulent flow sub-domains is modeled by a process, where the interaction of turbulence eddies is postulated to take place only after passing a finite distance. The extent to which this simplified model process can derive properly the actually occurring turbulent transport processes, has to be demonstrated by comparisons of theoretically derived insights into turbulent transport processes with corresponding experimental results.

Taking into account the above model explanations, a turbulent field can be subdivided into turbulent eddies in such a way that there are n_T eddies per unit volume. They carry out stochastic motions, so that $(1/6n_T)$ of the turbulent eddies move, on average, in each of the positive and negative directions of a Cartesian coordinate system. During a time Δt the below-cited number of turbulence eddies is moving through the area a^2 of a control volume, where $|u_2'|$ was assumed to be the velocity of the eddies in the x_2-direction, i.e. in the consideration carried out here, $u_c = |u_2'|$ is set.

$$\frac{1}{6} n_T a^2 |u_2'| \Delta t = \frac{1}{6} n_T a^2 u_c \Delta t. \tag{18.158}$$

On attributing to each turbulent eddy the mass Δm_T, then for the mass transported through an area a^2 in a time Δt in the positive and negative x_2-direction, the following relationship holds:

$$\frac{1}{6} n_T \Delta m_T a^2 u_c \Delta t = \frac{\rho}{6} a^2 u_c \Delta t. \tag{18.159}$$

Connected with this mass transport is the momentum transport. The transport occurring in the positive direction can be given as

$$J^+ = \frac{\rho}{6} a^2 u_c \Delta t \overline{U}_1 (x_2 - l_m) \tag{18.160}$$

and in the negative direction

$$J^- = -\frac{\rho}{6} a^2 u_c \Delta t \overline{U}_1 (x_2 + l_m). \tag{18.161}$$

18.7 Turbulence Models

The momentum transport difference which occurs over an area a^2 in a time Δt therefore is:

$$\Delta J = J^+ + J^- = \frac{\rho}{6}a^2 u_c \Delta t \left[\overline{U}_1\left(x_2 - l_m\right) - \overline{U}_1\left(x_2 + l_m\right)\right]. \tag{18.162}$$

Per unit time and unit area this results in the following relationships:

$$\frac{\Delta J}{a^2 \Delta t} = (\tau_{\text{turb}})_{21} = \frac{\rho}{6}u_c\left[\overline{U}_1\left(x_2 - l_m\right) - \overline{U}_1\left(x_2 + l_m\right)\right]. \tag{18.163}$$

After carrying out Taylor series expansion for the velocity difference $U_1(x_2 - l_m)$ and $U_2(x_2 + l_m)$ and subtraction, one obtains

$$(\tau_{\text{turb}})_{21} = \frac{\rho}{6}u_c\left[-2\left(\frac{\partial \overline{U}_1}{\partial x_2}\right)l_m\right] \tag{18.164}$$

or rewritten:

$$(\tau_{\text{turb}})_{21} = -\frac{\rho}{3}u_c l_m \left(\frac{\partial \overline{U}_1}{\partial x_2}\right). \tag{18.165}$$

Taking into account that the contribution occurring due to $|u'_2|$ is given by the term:

$$|u'_2| = \left|\frac{\partial \overline{U}_1}{\partial x_2}\right| l_m = u_c \tag{18.166}$$

the following expression holds:

$$(\tau_{\text{turb}})_{21} = -\rho \overline{u'_2 u'_1} = -\frac{\rho}{3}l_m^2 \left|\frac{\partial \overline{U}_1}{\partial x_2}\right|\left(\frac{\partial \overline{U}_1}{\partial x_2}\right). \tag{18.167}$$

On introducing the Prandtl mixing length as:

$$l_p^2 = \frac{1}{3}l_m^2 \tag{18.168}$$

one obtains the final relationship put forward by Prandtl for the turbulent momentum transport:

$$(\tau_{\text{turb}})_{21} = -\rho l_p^2 \left|\frac{\partial \overline{U}_1}{\partial x_2}\right|\left(\frac{\partial \overline{U}_1}{\partial x_2}\right) \tag{18.169}$$

and thus the turbulent viscosity can be expressed as:

$$\nu_T = l_p^2 \left|\frac{\partial \overline{U}_1}{\partial x_2}\right|. \tag{18.170}$$

Corresponding to the representations in Sect. 18.7.2, for the Prandtl mixing length model, the following time and velocity scales of turbulence result:

$$l_c = l_p, \quad u_c = l_p \left| \frac{\partial \overline{U}_1}{\partial x_2} \right|. \tag{18.171}$$

In the treatment of turbulence, l_p has to be considered as an unknown quantity in the above derivations. It has to be given analytically as a local quantity for each turbulent flow field to be investigated, before solutions of the basic equations for turbulent flows can be sought. To make this clearer, the turbulent channel flow, for which the following basic equations hold, will be treated below as an example.

Continuity equation:

$$\left(\frac{\partial \overline{U}_2}{\partial x_2} \right) = 0 \quad \rightsquigarrow \quad \overline{U}_2 = \text{constant} = 0. \tag{18.172}$$

Reynolds equations:

$$0 = -\frac{\partial \bar{P}}{\partial x_1} + \frac{\partial}{\partial x_2} \left(\rho \nu \frac{\mathrm{d}\overline{U}_1}{\mathrm{d}x_2} - \rho \overline{u'_2 u'_1} \right), \tag{18.173}$$

$$0 = -\frac{\partial \bar{P}}{\partial x_2} + \frac{\mathrm{d}}{\mathrm{d}x_2} \left(-\rho \overline{u'^2_2} \right) \tag{18.174}$$

and the boundary conditions:

$$x_2 = \pm D \quad : \quad \overline{U}_1 = 0; \ \overline{u'^2_2} = 0; \ \rho \overline{u'_2 u'_1} = 0$$
$$\text{and } P(x_1, x_2) = P_w(x_1) \tag{18.175}$$

$$x_2 = 0 \quad \text{symmetry conditions} \quad \frac{\partial}{\partial x_2}(\cdots) = 0. \tag{18.176}$$

The integration of the second Reynolds equation yields:

$$P(x_1, x_2) = P_w(x_1) - \rho \overline{u'^2}_2. \tag{18.177}$$

From the second Reynolds equation, the following results, therefore, for the pressure gradient in the flow direction:

$$\frac{\partial P}{\partial x_1} = \frac{\mathrm{d}P_w}{\mathrm{d}x_1}. \tag{18.178}$$

Because of the assumption of a fully developed flow field, also with regard to the turbulent quantities, the following relationship holds:

$$\frac{\partial}{\partial x_1} \left(-\rho \overline{u'^2_2} \right) = 0. \tag{18.179}$$

This relationship expresses that the change in pressure in the flow direction, that occurs in the entire flow field, is equal to the pressure gradient along the wall.

18.7 Turbulence Models

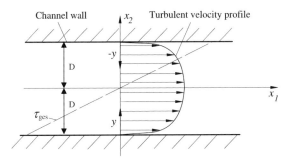

Fig. 18.16 Turbulent fully developed, plane channel flow

As an integral relationship for the channel flow plotted in Fig. 18.16, one can write:

$$2\tau_w BL = L\left(\frac{dP_w}{dx_1}\right) 2DB \tag{18.180}$$

or

$$\tau_w = D\frac{\partial P}{\partial x_1} \quad \rightsquigarrow \quad u_\tau^2 = \frac{\tau_w}{\rho} = \frac{D}{\rho}\frac{dP_w}{dx_1}. \tag{18.181}$$

The first of the Reynolds equations can thus be written as follows:

$$\frac{\partial P}{\partial x_1} = \frac{dP_w}{dx_1} = \frac{\rho}{D}u_\tau^2 = \rho\nu\frac{d}{dx_2}\left[\left(1+\frac{\nu_T}{\nu}\right)\frac{d\overline{U}_1}{dx_2}\right] \tag{18.182}$$

or expressed by the wall coordinate y:

$$y = D - x_2 \quad \text{and} \quad dy = -dx_2, \tag{18.183}$$

$$\frac{u_\tau^2}{D} = \nu\frac{d}{dy}\left[\left(1+\frac{\nu_T}{\nu}\right)\frac{d\overline{U}_1}{dy}\right]. \tag{18.184}$$

With $\nu_T = l_P^2\left|\dfrac{d\overline{U}_1}{dy}\right|$ and the linear ansatz suggested by Prandtl:

$$l_P = \kappa y \tag{18.185}$$

the following differential equation results:

$$\frac{u_\tau^2}{\nu D} = \frac{d}{dy}\left[\left(1+\frac{\kappa^2 y^2}{\nu}\left|\frac{d\overline{U}_1}{dy}\right|\right)\frac{d\overline{U}_1}{dy}\right]. \tag{18.186}$$

To solve the above equations, it is recommended to carry out the first integration for the momentum equation in the following form:

$$\frac{\partial P}{\partial x_1} = \frac{dP_w}{dx_1} = -\frac{\tau_w}{D} = -\frac{d}{dy}(\tau_{tot})_{21} \tag{18.187}$$

and
$$(\tau_{tot})_{21} = \frac{\tau_w}{D}y + C = -\tau_w\left(1 - \frac{y}{D}\right), \quad (18.188)$$

where $y = D - x_2$ was considered. Thus the following equation holds:

$$-\rho\nu\frac{d\overline{U}_1}{dy} + \rho\overline{u'_2 u'_1} = -\tau_w\left(1 - \frac{y}{D}\right), \quad (18.189)$$

$$\rho\nu\frac{d\overline{U}_1}{dy} - \rho\nu_T\frac{d\overline{U}_1}{dy} = \tau_w\left(1 - \frac{y}{D}\right). \quad (18.190)$$

On setting $U_1^+ = \frac{\overline{U}_1}{u_\tau}$ and $y^+ = \frac{y u_\tau}{\nu}$, one obtains:

$$\left(1 - \frac{\nu_T}{\nu}\right) dracU_1^+ y^+ = \left(1 - \frac{y^+}{D^+}\right) \quad (18.191)$$

and with

$$\frac{\nu_T}{\nu} = \frac{l_P^2}{\nu}\left|\frac{dU_1}{dy}\right| = (l_P^+)^2 \left|\frac{dU_1^+}{dy^+}\right|, \quad (18.192)$$

$$\left[1 + (l_P^+)^2 \left|\frac{dU_1^+}{dy^+}\right|\right]\left(\frac{dU_1^+}{dy^+}\right) = \left(1 - \frac{y^+}{D^+}\right). \quad (18.193)$$

On introducing the linear relation suggested by Prandtl:

$$l_P^+ = \kappa y^+ \quad (18.194)$$

one obtains:

$$\left[1 + \kappa^2 y^{+2}\left|\frac{dU_1^+}{dy^+}\right|\right]\left(\frac{dU_1^+}{dy^+}\right) = \left(1 - \frac{y^+}{D^+}\right). \quad (18.195)$$

For the upper half-plane of the channel flow $\left|\frac{dU_1^+}{dy^+}\right| = \left(\frac{dU_1^+}{dy^+}\right)$. The following equation thus holds:

$$\left(\frac{U_1^+}{y^+}\right)^2 + \frac{1}{\kappa^2 y^{+2}}\left(\frac{U_1^+}{y^+}\right) - \frac{1}{\kappa^2 y^{+2}}\left(1 - \frac{y^+}{D^+}\right) = 0 \quad (18.196)$$

or solved for $\frac{dU_1^+}{dy^+}$ it results (for $\frac{dU_1^+}{dy^+} > 0$):

$$\frac{dU_1^+}{dy^+} = -\frac{1}{2\kappa^2 y^{+2}} + \sqrt{\underbrace{\frac{1 + 4\kappa^2 y^{+2}\left(1 - \frac{y^+}{D^+}\right)}{4\kappa^4 y^{+4}}}_{A}} \quad (18.197)$$

18.7 Turbulence Models

or rewritten:
$$\frac{U_1^+}{y^+} = \frac{\left[-\frac{1}{2\kappa^2 y^2} + \sqrt{A}\right]\left[\frac{1}{2\kappa^2 y^2} + \sqrt{A}\right]}{\frac{1}{2\kappa^2 y^2} + \sqrt{A}} \tag{18.198}$$

to yield the following result:
$$\frac{U_1^+}{y^+} = \frac{2\left(1 - \frac{y^+}{D^+}\right)}{1 + \sqrt{1 + 4\kappa^2 y^{+2}\left(1 - \frac{y^+}{D^+}\right)}} \tag{18.199}$$

so that U_1^+ can be computed as follows:
$$U_1^+ = \int_0^{y^+} \frac{2\left(1 - \frac{y^+}{D^+}\right)}{1 + \sqrt{1 + 4\kappa^2 y^{+2}\left(1 - \frac{y^+}{D^+}\right)}} \, dy^+. \tag{18.200}$$

Detailed considerations of the turbulence behavior near walls indicated that the attenuation of the turbulence taking place near walls is not fully taken into account by the linear assumptions for l_p proposed by Prandtl. Through considerations of the Stokes problem of a viscous flow oscillating parallel to a fixed wall, van Driest derived an attenuation factor which can be introduced into the assumptions for the Prandl mixing length. Considering this leads to the equation

$$l_{vD} = l_P \left[1 - \exp\left(-\frac{y\frac{U_\tau}{\nu}}{A^+}\right)\right], \tag{18.201}$$

where $A^+ = 26$ was determined by van Driest from experimental results. For large values of y^+, the above attenuation factor approaches the value 1 and the van Driest mixing length theory and the Prandtl assumptions merge, i.e. become identical. Near the wall, owing to the viscous attenuation, a reduction of the mixing length takes place, which is taken into account in the exponential term of the van Driest assumption.

Finally, some considerations on the Prandtl mixing length theory and the derived final relationships for the turbulent momentum transport are recommended:

$$-\overline{u_1' u_2'} = -l_P u_c \left(\frac{\partial U_1}{\partial x_2}\right) = -u_c^2 \left(\frac{l_P}{u_c}\right)\left(\frac{\partial \overline{U}_1}{\partial x_2}\right). \tag{18.202}$$

On introducing two time scales:

Characteristic time scale of the turbulence: $\tau_c = \dfrac{l_P}{u_c}$,

Characteristic time scale of the mean flow field: $\tau_M = \dfrac{1}{\left(\dfrac{\partial \bar{U}_1}{\partial x_2}\right)}$ the following holds:

$$-\overline{u'_1 u'_2} = -u_c^2 \frac{\tau_c}{\tau_M}. \tag{18.203}$$

For turbulent flows for which the turbulence is exposed for a long time to a mean flow field with constant deformation, a constant relationship of the above time scale develops, so that, at least in some regions of the flow, the following can be assumed to hold:

$$\overline{u'_1 u'_2} = \text{constant} \cdot u_c^2. \tag{18.204}$$

For this reason, and for a wide range of such turbulent flows, one can write

$$R_{12} = \frac{\overline{u'_1 u'_2}}{\sqrt{\overline{u'^2_1}}\sqrt{\overline{u'^2_2}}} = \text{constant}. \tag{18.205}$$

In spite of the fact that a constant deformation rate in turbulent wall boundary layers is not guaranteed and that therefore turbulence elements experience differing deformations on their way through the flow field, the above relationship also holds over wide ranges of turbulent boundary layers. This is indicted in Fig. 18.17.

When turbulent modeling is desired only for one class of flows, e.g. turbulent wall boundary layers, R_{12} can also be stated from experiments as a function of the location. For the experimentally obtained distribution in Fig. 18.17, one can derive

$$R_{12} = f\left(\frac{r}{R}\right) = \frac{\overline{u'_1 u'_2}}{\sqrt{\overline{u'^2_1}}\sqrt{\overline{u'^2_2}}} \approx \frac{\overline{u'_1 u'_2}}{k}. \tag{18.206}$$

Knowing k, via R_{12} the local value for $\overline{u'_1 u'_2}$ can be computed and employed in the momentum equation for computing the mean flow field.

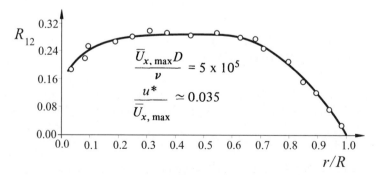

Fig. 18.17 Correlation coefficient R_{12} for turbulent pipe flows

18.7.4 One-Equation Eddy Viscosity Models

The class of turbulence models, discussed in Sect. 18.7.3, tries to describe the momentum transport properties of turbulent flows with the help of a single parameter, namely the Prandtl mixing length. The latter is introduced into the considerations in a geometry-specific way and is stated in the form of an algebraic equation, where the velocity characteristic of the turbulence is introduced via the mixing length and the gradient of the mean velocity field. In order to achieve a model expansion, the characteristic velocity typical of the turbulent momentum transport is set as follows:

$$u_c = k^{\frac{1}{2}} \quad \text{with} \quad k = \frac{1}{2}\overline{u'_i u'_i}. \tag{18.207}$$

With the characteristic length l_c of one-equation models, thus $\nu_T = (l_c)_1 k^{\frac{1}{2}}$ results. Here k is the local turbulent kinetic energy, which is described by the following transport equation:

$$U_i \frac{\partial k}{\partial x_i} = \frac{\partial}{\partial x_i}\left(-\overline{\frac{p'u'_i}{\rho}} + 2\nu\overline{u'_j \frac{\partial u'_j}{\partial x_i}} - \frac{1}{2}\overline{u'_j u'_j u'_i}\right) - \overline{u'_i u'_j}\frac{\partial U_j}{\partial x_i} - 2\nu\overline{\left(\frac{\partial u'_j}{\partial x_i}\right)^2} \tag{18.208}$$

so that the turbulent transport properties in (18.208) change with the modifications of turbulence in a flow field. It is customary in turbulence modeling to rewrite the above equation as follows:

$$U_i \frac{\partial k}{\partial x_i} = -\underbrace{\frac{\partial D_i}{\partial x_i}}_{\text{diffusion}} + \underbrace{P}_{\text{production}} - \underbrace{\epsilon}_{\text{dissipation}}, \tag{18.209}$$

where the following relationships hold:

$$D_i = -\left(-\overline{\frac{p'u'_i}{\rho}} + 2\nu\overline{u'_j \frac{\partial u'_j}{\partial x_i}} - \frac{1}{2}\overline{u'_j u'_j u'_i}\right) \tag{18.210}$$

and

$$P = -\overline{u'_i u'_j}\frac{\partial U_j}{\partial x_i}; \quad \epsilon = -\nu\overline{\frac{\partial u'_j}{\partial x_i}\frac{\partial u'_j}{\partial x_i}}. \tag{18.211}$$

With the modeling assumptions for D_i, P and ϵ, customary in present-day turbulence modeling research, the following modeled k equation results:

$$U_i \frac{\partial k}{\partial x_i} = \underbrace{\frac{\partial}{\partial x_i}\left[\left(\nu + \frac{\nu_T}{\sigma_k}\right)\frac{\partial k}{\partial x_i}\right]}_{\text{diffusion}} + \underbrace{\nu_T\left(\frac{\partial U_j}{\partial x_i} + \frac{\partial U_i}{\partial x_j}\right)\frac{\partial U_j}{\partial x_i}}_{\text{production}} - \underbrace{C_D \frac{k^{\frac{3}{2}}}{(l_c)^1}}_{\text{dissipation}}. \tag{18.212}$$

The second term on the right-hand side of (18.212) represents the production of turbulence by the mean flow field, which was stated as follows:

$$-\overline{u'_i u'_j}\frac{\partial U_j}{\partial x_i} = \nu_T\left(\frac{\partial U_j}{\partial x_i}+\frac{\partial U_i}{\partial x_j}\right)\frac{\partial U_j}{\partial x_i}. \quad (18.213)$$

The first term on the right-hand side of (18.212) also contains σ_k, a quantity which indicates how the turbulent diffusion of k is related to the turbulent momentum diffusion ν_T, i.e. σ_k states the relationship of turbulent momentum dissipation to energy dissipation. The third term on the right-hand side of (18.212) states the turbulent dissipation. For this term, order of magnitude considerations suggest, for equilibrium flows:

$$P \sim \frac{u_c^3}{l_c} \sim \epsilon \quad \rightsquigarrow \quad \epsilon = C_D \frac{k^{\frac{3}{2}}}{(l_c)_1}. \quad (18.214)$$

The constants introduced into this one-equation turbulence model have to be determined from experiments. Here, one makes use of the two-dimensional form of the k equation for a boundary-layer flow:

$$U_x\frac{\partial k}{\partial x}+U_y\frac{\partial k}{\partial y} = \frac{\partial}{\partial y}\left[\left(\nu+\frac{\nu_T}{\sigma_T}\right)\frac{\partial k}{\partial y}\right]+\nu_T\left(\frac{\partial U_x}{\partial y}\right)^2 - C_D\frac{k^{\frac{3}{2}}}{(l_c)_1}. \quad (18.215)$$

In the turbulent equilibrium range of the flow (inertial sub-range), it holds that $P = \epsilon$, i.e. it can be stated that:

$$\nu_T\left(\frac{dU_x}{dy}\right)^2 = +C_D\frac{k^{\frac{3}{2}}}{(l_c)_1}. \quad (18.216)$$

On taking into account that the following relationships are valid:

$$\tau_{xy} = -\rho\nu_T\left(\frac{dU_x}{dy}\right) \approx -\tau_w \quad (18.217)$$

and considering that $\tau_w/\rho = u_\tau^2$, one obtains $\tau_w/\rho\left(\dfrac{\partial U_x}{\partial y}\right) = c_D\dfrac{k^{3/2}}{(l_c)_1}$, and with $\dfrac{\partial U_x}{\partial y} \approx \dfrac{k^{1/2}}{(l_c)_1}$ it can be stated that

$$k^+ = \frac{k}{u_\tau^2} = C_D^{-\frac{1}{2}}. \quad (18.218)$$

As in Sect. 18.7.3, for the one-equation k-l model, assumptions were chosen in turbulence considerations which "in wall boundary layers" take into account the closeness of the wall and which lead to differing assumptions for $(l_c)_1$ values, depending on whether one considers the diffusion term or the dissipation term:

18.7 Turbulence Models

$$(l_c)_{1,\nu} = C_D^{\frac{1}{4}} \kappa y \left[1 - \exp\left(-A_\nu R_T\right)\right] \quad \text{with} \quad A_\nu = 0.016 \quad (18.219)$$

and

$$(l_c)_{1,D} = C_D^{\frac{1}{4}} \kappa y \left[1 - \exp\left(-A_D R_T\right)\right] \quad \text{with} \quad A_D = 0.26, \quad (18.220)$$

where $R_T = \dfrac{k^{\frac{1}{2}} y}{\nu}$ represents the Reynolds number formed by wall disturbance:

$$(l_c)_1 = l_P C_D^{\frac{1}{4}}. \quad (18.221)$$

From boundary layer data, the value $C_D \approx 0.09$ results. With this value, it is now possible to integrate the Reynolds equations employing the *k-l* one-equation turbulence model.

In order to determine the C_D value, the equilibrium region of a boundary layer was employed. Figure 18.18 shows where this region is located.

It is also customary to employ other characteristic ranges of turbulent flows to determine the "free constants" in turbulence models. Altogether the following ranges are available, introduced here by terms customary in the English literature on turbulence.

Equilibrium range:
$$0 = P - \epsilon. \quad (18.222)$$

Decay range:
$$\overline{U}_i \frac{\partial k}{\partial x_i} = -\epsilon. \quad (18.223)$$

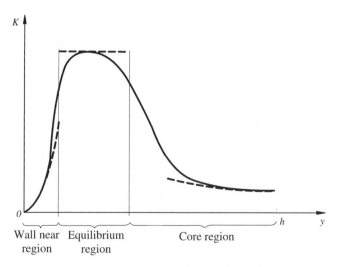

Fig. 18.18 Ranges of turbulent channel flow

Rapid distortion range:
$$\overline{U}_i \frac{\partial k}{\partial x_i} = P. \tag{18.224}$$

The above equations are useful in discussions of different turbulent flow regions.

18.7.5 Two-Equation Eddy Viscosity Models

Practical experience with turbulence models shows that, as far as their general applicability is concerned, the k-l one-equation model represents a considerable improvement over the zero-equation models. However, very good computations of turbulent flows can only be carried out by zero- and one-equation models when small flow accelerations or decelerations occur, i.e. flows with strong pressure gradients can only be computed in an unsatisfactory way with k-l one-equation models. Experience shows that the limitations of the applicability of k-l one-equation models are due to the algebraic form of the characteristic length scale l, which often means that the applicability remains restricted to such flows which were applied for deriving the k-l one-equation models. This is the reason for the introduction of two-equation eddy viscosity models. One of these models is the k-ϵ model, which is based on the solution of the following two differential equations:

k equation:

$$\underbrace{\frac{\partial k}{\partial t} + \overline{u}_i \frac{\partial k}{\partial x_i}}_{\frac{Dk}{Dt}} = \underbrace{-\overline{u_i u_j} \frac{\partial \overline{U}_j}{\partial x_i}}_{P} - \underbrace{\frac{\partial}{\partial x_i} \overline{u_i \left(\frac{u_j u_j}{2} + \frac{P}{\rho} \right)}}_{T} - \underbrace{\nu \overline{\frac{\partial u_j}{\partial x_i} \frac{\partial u_j}{\partial x_i}}}_{\epsilon} + \underbrace{\nu \frac{\partial^2 k}{\partial x_i \partial x_i}}_{D},$$
$$\tag{18.225}$$

where P = production term, T = transport term and ϵ = dissipation term and D = diffusion term.

ϵ equation:

$$\frac{\partial \epsilon}{\partial t} + \overline{u}_i \frac{\partial \epsilon}{\partial x_i} = \underbrace{-2\nu \overline{\frac{\partial u_j}{\partial x_i} \frac{\partial u_k}{\partial x_i}} \frac{\partial \overline{U}_j}{\partial x_k} - 2\nu \overline{\frac{\partial u_j}{\partial x_k} \frac{\partial u_j}{\partial x_i}} \frac{\partial \overline{U}_k}{\partial x_i}}_{P_\epsilon^1 \qquad\qquad P_\epsilon^2}$$

$$\underbrace{-2\nu \overline{u_k \frac{\partial u_j}{\partial x_i} \frac{\partial^2 U_j}{\partial x_k \delta x_i}} - 2\nu \overline{\frac{\partial u_j}{\partial x_i} \frac{\partial u_k}{\partial x_i} \frac{\partial u_j}{\partial x_k}}}_{P_\epsilon^3 \qquad\qquad P_\epsilon^4}$$

$$\underbrace{-\frac{\partial}{\partial x_k} \overline{\nu u_k \frac{\partial u_j}{\partial x_i} \frac{\partial u_j}{\partial x_i}}}_{T_\epsilon} - \underbrace{\frac{\partial}{\partial x_k} \frac{2\nu}{\rho} \overline{\frac{\partial u_k}{\partial x_i} \frac{\partial p}{\partial x_i}}}_{M_\epsilon}$$

18.7 Turbulence Models

$$-2\nu^2 \overline{\frac{\partial^2 u_j}{\partial x_i \delta x_k} \frac{\partial^2 u_j}{\partial x_i \delta x_k}}_{\gamma>0} + \underbrace{\nu \frac{\partial^2 \epsilon}{\partial x_k \delta x_k}}_{D_\epsilon}. \qquad (18.226)$$

The eddy viscosity can be defined with k and ϵ, i.e. with $u_c = k^{1/2}$ and $l_c = \dfrac{k^{3/2}}{\epsilon}$, as follows:

$$\nu_T = C_\mu \frac{k^2}{\epsilon}.$$

Here, k and ϵ are determined from the above-modeled differential equations. In this context, it is useful to know that in turbulence modeling the above-cited equation for k is considered to be sufficiently well modeled as concerns practical computations, whereas similarly satisfactory modeling assumptions do not exist for the ϵ equation. The model equations often used in present-day flow computations are the following:

k equation:

$$\frac{\partial k}{\partial t} + \overline{U}_i \frac{\partial k}{\partial x_i} = \nu_t \left(\frac{\partial \overline{U}_j}{\partial x_i} + \frac{\partial \overline{U}_i}{\partial x_j} \right) \frac{\partial \overline{U}_j}{\partial x_i} + \frac{\partial}{\partial x_i} \frac{\nu_t}{\sigma_K} \frac{\partial k}{\partial x_i} - \epsilon + \nu \frac{\partial^2 k}{\partial x_i \partial x_i}. \qquad (18.227)$$

With ν_t as

$$\nu_t \cong 0.09 \frac{k^2}{\epsilon}.$$

ϵ equation:

$$\frac{\partial \epsilon}{\partial t} + \overline{U}_i \frac{\partial \epsilon}{\partial x_i} = c_{\epsilon 1} \frac{\epsilon}{R} \nu_t \left(\frac{\partial \overline{U}_j}{\partial x_i} + \frac{\partial \overline{U}_i}{\partial x_j} \right) \frac{\partial \overline{U}_j}{\partial x_i} - c_{\epsilon 2} f_\epsilon \frac{\epsilon^2}{k}$$

$$+ \frac{\partial}{\partial x_i} \frac{\nu_t}{\sigma_\epsilon} \frac{\partial \epsilon}{\partial x_i} + \nu \frac{\partial^2 \epsilon}{\partial x_i \partial x_i}. \qquad (18.228)$$

Here, the modeling of the last term in the ϵ equation is based on the assumption:

$$P \sim k \frac{k^{\frac{1}{2}}}{l_c} \epsilon \quad \leadsto \quad \nu \left(\frac{U}{\lambda^2} \right)^2 = \frac{\epsilon^2}{k}. \qquad (18.229)$$

As concerns the boundary layer formulation of the Reynolds and k-ϵ turbulence model equations, valid for high Reynolds numbers, they can be given as follows:

$$\frac{\partial U}{\partial x} + \frac{\partial V}{\partial y} = 0, \qquad (18.230)$$

$$U \frac{\partial U}{\partial x} + V \frac{\partial U}{\partial y} = F_x - \frac{1}{\rho} \frac{\partial P}{\partial x} + \frac{\partial}{\partial y} \left[(\nu + \nu_T) \frac{\partial U}{\partial y} \right], \qquad (18.231)$$

$$U \frac{\partial K}{\partial x} + V \frac{\partial K}{\partial y} = \frac{\partial}{\partial y} \left(\frac{\nu_T}{\sigma_K} \frac{\partial K}{\partial y} \right) + \nu_T \left(\frac{\partial U}{\partial y} \right)^2 - \epsilon, \qquad (18.232)$$

$$U\frac{\partial \epsilon}{\partial x} + V\frac{\partial \epsilon}{\partial y} = \frac{\partial}{\partial y}\left(\frac{\nu_T}{\sigma_\epsilon}\frac{\partial \epsilon}{\partial y}\right) + C_{\epsilon 1}\frac{\nu_T \epsilon}{K}\left(\frac{\partial U}{\partial y}\right)^2 - C_{\epsilon 2}\frac{\epsilon^2}{K}, \quad (18.233)$$

$$\nu_T = C_\mu \frac{K^2}{\epsilon},$$

where $C_\mu = 0.09$, $\sigma_K = 1.0$, $\sigma_\epsilon = 1.3$, $C_{\epsilon 1} = 1.45$, $C_{\epsilon 2} = 2.0$.

In order to determine c_μ, the ansatz $\nu_T \left(\frac{\partial U}{\partial y}\right)^2 = \epsilon \rightsquigarrow c_\mu = c_D = \left(\frac{k}{u_\tau^2}\right)^2 = 0.09$ again holds. This can also be determined from measurements of wall boundary layers.

18.8 Turbulent Wall Boundary Layers

Turbulent boundary-layer flows, whose essential properties are determined by the presence of a wall, are called wall boundary layers. As classic examples one can cite

- Internal flows: plane channel flows and pipe flows
- External flows: plane plate flows and film flows

These flows are sketched in Fig. 18.19. Their essential feature is the momentum loss to a wall which is characteristic of all flows in Fig. 18.19, i.e. wall momentum loss exists in all the indicated flow cases and, in addition, the properties of the fluid, the density ρ and the dynamic viscosity μ characterize the fluid.

In order to discuss the properties of turbulent boundary layers in an introductory way, the fully developed, two-dimensional, plane, turbulent

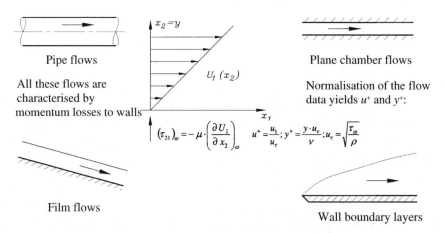

Fig. 18.19 Examples of internal and external wall boundary layers

18.8 Turbulent Wall Boundary Layers

channel flow is subjected to more detailed considerations below. From the Reynolds equations (see Sect. 18.5.2), the following reduced set of equations can be deduced for fully developed, plane channel flows with the above-cited properties:

$$x_1 - \text{momentum equation:} \quad 0 = -\frac{\partial P}{\partial x_1} + \frac{\mathrm{d}}{\mathrm{d}x_2}\left(\mu \frac{\mathrm{d}\overline{U}_1}{\mathrm{d}x_2} - \rho\overline{u'_1 u'_2}\right). \quad (18.234)$$

$$x_2 - \text{momentum equation:} \quad 0 = -\frac{\partial P}{\partial x_2} - \rho\overline{u'_2{}^2}. \quad (18.235)$$

$$x_3 - \text{momentum equation:} \quad 0 = -\frac{\mathrm{d}}{\mathrm{d}x_2}\left(\overline{u'_2 u'_3}\right). \quad (18.236)$$

The last partial differential equation (18.236) can be integrated and yields, because of the wall boundary condition $\overline{u'_2 u'_3} = 0$, that the correlation $\overline{u'_2 u'_3}$ in the entire plane perpendicular to the plates of the channel has the value $\overline{u'_2 u'_3} = 0$.

The integration of the second differential equation (18.235) yields

$$P(x_1, x_2) = P_w(x_1) - \rho\overline{u'_2{}^2}, \quad (18.237)$$

where $\rho\overline{u'_2{}^2} = f(x_2)$, since in the x_1-direction the flow was assumed to be fully developed. Nevertheless, the above relationship expresses that the pressure in a turbulent channel flow changes slightly over the cross-section. The change proves to be so small, however, that it can be neglected for practical considerations of the properties of fully developed, two-dimensional, plane, turbulent channel flows. Thus, from (18.237) the following results for the pressure gradient:

$$\frac{\partial P}{\partial x_1} \approx \frac{\mathrm{d}P_w}{\mathrm{d}x_1}. \quad (18.238)$$

Inserted in (18.234), one obtains:

$$\frac{\mathrm{d}P_w}{\mathrm{d}x_1} = \underbrace{\frac{\mathrm{d}}{\mathrm{d}x_2}\left(\mu \frac{\mathrm{d}\overline{U}_1}{\mathrm{d}x_2} - \rho\overline{u'_1 u'_2}\right)}_{\tau_{\text{ges}}} = \frac{\mathrm{d}\tau_{\text{ges}}}{\mathrm{d}x_2}. \quad (18.239)$$

If one introduces, for scaling the above equations, the velocity and length scales:

$$u_\tau = \sqrt{\frac{\tau_w}{\rho}} \quad \text{and} \quad \ell_e = \frac{\nu}{u_\tau} \quad (18.240)$$

the momentum equation (18.239) can be written in general form as follows:

$$\frac{\mathrm{d}U_1^+}{\mathrm{d}y^+} = 1 - \frac{y^+}{Re_\tau} + \left(\overline{u'_1 u'_2}\right)^+. \quad (18.241)$$

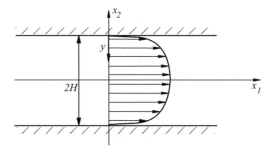

Fig. 18.20 Fully developed turbulent, plane channel flow

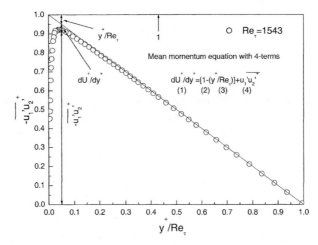

Fig. 18.21 Terms of the momentum equation for two-dimensional channel flow

Here the wall coordinate y is introduced: $y = H - x_2$.

$$U_1^+ = \frac{\overline{U_1}}{u_\tau}; \quad y_c^+ = \frac{y u_\tau}{\nu}; \quad Re_\tau = \frac{H u_\tau}{\nu} \quad \text{and} \quad \left(\overline{u_1' u_2'}\right)^+ = \frac{\overline{u_1' u_2'}}{u_\tau^2}. \quad (18.242)$$

With these standardized quantities introduced into (18.242), (18.241) can be derived. The latter relation comprises four terms which can all be seen in Fig. 18.21.

In Fig. 18.21, the horizontal line represents the value 1. The quantity $-y^+/Re_\tau$ and the quantity $-\overline{u_1' u_2'}$ in the equation and also dU^+/dy^+ are also shown in Fig. 18.21.

From (18.239), we obtain

$$\tau_{\text{ges}} = \frac{\tau_w}{H} x_2 = \frac{\tau_w}{H} (h - y)$$

so that the different terms stated in (18.241) can be shown as in Fig. 18.21. It is evident that the term dU_1^+/dy^+ represents, over wide ranges of the flow, the smallest value in the standardized momentum equation (18.241).

18.8 Turbulent Wall Boundary Layers

In order to obtain information on dU_1^+/dy^+ for plane channel flows, laser Doppler and hot wire measurements were carried at the Institute for Fluid Mechanics (LSTM) of Friedrich-Alexander-University, Erlangen-Nürnberg to achieve $U_1(y)$ distributions experimentally. In connection with detailed shear stress measurements, the presentation of the normalized measured values of the velocity gradient was achieved in the form

$$\ln\left(\frac{dU_1^+}{dy^+}\right) = f\left(\ln y^+\right) \qquad (18.243)$$

and from this it was determined that, for high Reynolds numbers, the measured and normalized mean velocity gradients plotted in Fig. 18.21 can be described as follows:

$$\ln\left(\frac{dU_1^+}{dy^+}\right) = -\ln y^+ + 1 \equiv \ln\frac{e}{y^+}. \qquad (18.244)$$

From this, $U_1^+ = e \ln y^+ + B$ can be obtained, i.e. the standardized velocity distribution in a plane channel flow can, over a wide range of the channel cross-section, be described by a logarithmic velocity distribution:

$$U_1^+ = \frac{1}{\kappa}\ln y^+ + B \quad \text{with} \quad \begin{array}{l}\kappa = 1/e \\ B = 10/e\end{array}. \qquad (18.245)$$

These values were found through the experimental investigations at LSTM Erlangen.

Figure 18.22 shows that the double-logarithmic plotting yields a line with gradient -1. This is due to the fact that the logarithmic law is valid for the normalized velocity distribution.

In the literature there has been a number of investigations to determine the value of κ and the additive constant B to represent with these the logarithmic boundary velocity law. The following represents a summary of these investigations and the resultant values. The large variation in the values is mainly due to the use of measurement techniques which do not permit sufficiently local measurements of the mean flow velocities. Furthermore, effects which arise from flows of low Reynolds numbers were included in the evaluations of κ and B in some literature data. If one considers all the possible influences, i.e. permits only reliable hot wire and laser Doppler measurements to enter the evaluations, then one obtains the values indicated in (18.245) for κ and B (Fig. 18.23).

If one employs only reliable measurements for dU_1^+/dy^+, the relationship stated in (18.241) allows the determination of $(\overline{u_1' u_2'})^+$ values for turbulent channel flows. Distributions of these turbulent transport terms, plotted in a normalized form, are shown in Fig. 18.24.

By means of a plane channel measuring test section with glass side walls and the use of an LDA velocity-measuring system, information on the

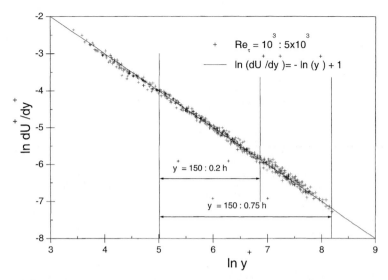

Fig. 18.22 Representations of experimental investigations for determining the logarithmic wall law

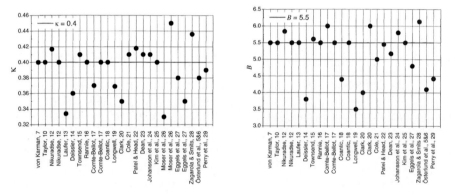

Fig. 18.23 Scatter of the κ and B values in the experimental determination of $U_1^+ = f(y^+)$ for wall boundary layers

turbulent velocity fluctuations existing in flow direction, could also be obtained at LSTM Erlangen. This information is shown in a summarized way in Fig. 18.25. In this figure they are compared with corresponding results of numerical flow computations.

Detailed measurements carried out at LSTM Erlangen have confirmed interesting new results, e.g. that the local turbulence intensity $\sqrt{\overline{u_1'^2}}/U_1$ does not adopt a constant wall value. The result shows that the constant wall value of this velocity ratio depends on the Reynolds number of the flow (see Fig. 18.26). Although small discrepancies can be seen as compared with the

18.8 Turbulent Wall Boundary Layers

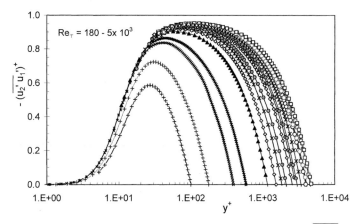

Fig. 18.24 Standardized turbulent momentum transport terms $(\overline{u_1' u_2'})$ for plane channel flows

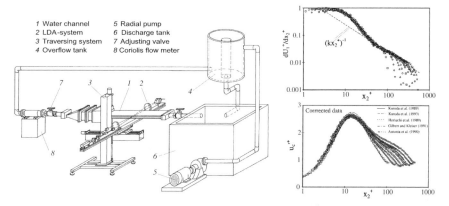

Fig. 18.25 Plane channel flow and LDA system. Measurement results for standardized turbulent velocity fluctuations in the flow direction

values obtained by numerical investigations, a general trend exists. The remaining discrepancies between the experimental and numerical data can, in all probability, be attributed to mistakes in the numerical computations, as the computed values were not subjected to corrections concerning the finite numerical grid spacings employed.

The investigations described above are limited to such wall roughnesses for which it holds that

$$\frac{\delta_s u_\tau}{\nu} \leq \epsilon = 2.72, \qquad (18.246)$$

i.e. the values $\kappa = 1/e$ and $B = 10/e$ are valid only for flows with high Reynolds numbers and for walls which can be considered to be hydraulically

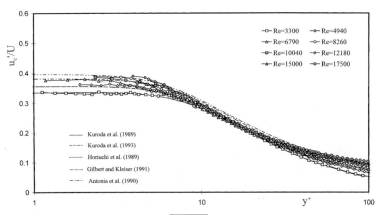

Fig. 18.26 Turbulence intensity $\sqrt{u_1'^2}/U_1$ near the wall as a function of the Reynolds number

smooth. For rough walls, an amendment proves necessary for which the following considerations hold:

$$U_1 = f(y, \rho, \mu, \tau_w, \delta_s) \quad \leadsto \quad U_1^+ = f\left(\frac{y}{\delta_s}\right), \tag{18.247}$$

where δ_s represents the "sand roughness" of the wall. Similarity considerations show that the following logarithmic wall law for rough channel walls can be derived:

$$U_1^+ = \frac{1}{\kappa}\ln y^+ + B - \Delta B\left(\delta_s^+\right). \tag{18.248}$$

Of particular interest is that point in the viscosity-controlled sub-layer of the flow where the sub-layer $U_1^+ = y^+$ and the layer $U_1^+ = 1/\kappa(\ln y^+)$ have the same values and the same gradients:

$$U_1^+ = y^+ = \frac{1}{\kappa}\ln y^+ \quad \text{and} \quad \frac{dU_1^+}{dy^+} = 1 = \frac{1}{y^+}. \tag{18.249}$$

From this it follows that at $y^+ = e$ this consideration is fulfilled for $\kappa = 1/e$, a value that also resulted from the measurements at LSTM, Erlangen. For the logarithmic velocity profile with maximum roughness δ_s^+ for which a viscosity dominated sub-layer still exists, it results that $\Delta B(\delta_s^+) = B = 10/e$, see Fig. 18.27. With this, we can see that a constant representation of the normalized velocity distributions for hydromechanically smooth and rough channel walls can be presented. Nevertheless, many questions concerning detailed problems of turbulent wall boundary layers have still to be answered and need to be investigated with the help of modern measuring and computation techniques. It is especially necessary to extend the results obtained here for fully developed, two-dimensional, plane, turbulent channel flows to pipe flows, and also to flat plate flows and turbulent film flows.

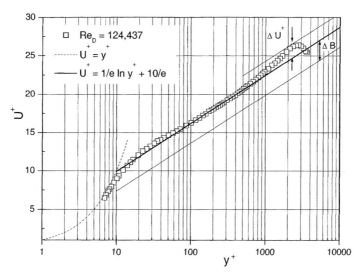

Fig. 18.27 Modification of the additive constants of the logarithmic wall law due to roughness

References

18.1. Arpaci, V. S., Larsen, P. S.: "Convection Heat Transfer", Central Book Company, Taipei, Taiwan (1984)
18.2. Biswas, G., Eswaran, V.: "Turbulent Flows, Fundamentals, Experiments and Modeling", IIT Kanpur Series of Advanced Texts, Narosa Publishing House, New Delhi (2002)
18.3. Hinze, J. O.: "Turbulence", 2nd edition, McGraw-Hill, New York (1975)
18.4. Pope, S. B.: "Turbulent Flows", Cambridge University Press, 2002
18.5. Schlichting, H.: "Boundary Layer Theory", McGraw-Hill, New York (1979)
18.6. Tennekes, H., Lumley, J. L.: "A first Course in Turbulence", MIT Press, Cambridge, Mass., USA, (1972)
18.7. Townsend A. A.: "The Structure of Turbulent Shear Flows", 2nd edition, Cambridge University Press, Cambridge, UK (1976)
18.8. White, F. M.: "Viscous Fluid Flow", Kingsport Press, Kingsport (1982)
18.9. Wilcox, D. C.: "Turbulence Modeling for CFD", DCW Industries, La Canada (1993)

Chapter 19
Numerical Solutions of the Basic Equations*

19.1 General Considerations

The considerations in Chaps. 13–16 showed that analytical solutions of the basic equations of fluid mechanics can often only be obtained when simplified equations or fully developed flows and small or large Reynolds numbers, respectively, are considered and if, in addition, one limits oneself to flow problems which are characterized by simple boundary conditions. Even with these simplifications, the derivations carried out did not result in analytical solutions for all flow problems to be solved, but merely reduced the flow describing *partial* differential equations to *ordinary* differential equations. As shown in the previous chapters, the latter could be solved by means of current analytical methods. Moreover, the boundary conditions characterizing the flow problems could also be implemented into these solutions. Thus it was demonstrated that the methods known in applied mathematics for solving ordinary differential equations represent an important tool for the theoretically working fluid mechanics researcher. Although analytical techniques for the solution of flow problems no longer have the significance they had in the past, it is part of a good education in fluid mechanics to teach these methods to students and for the latter to learn them.

When considering the partial differential equations discussed in the preceding chapters, they can all be brought into the subsequent general form

$$A \frac{\partial^2 \Phi}{\partial x^2} + 2B \frac{\partial^2 \Phi}{\partial x \partial y} + C \frac{\partial^2 \Phi}{\partial y^2} + D \frac{\partial \Phi}{\partial x} + E \frac{\partial \Phi}{\partial y} + F\Phi = g(x, y). \quad (19.1)$$

When designating as the discriminant of the differential equation (19.1)

$$d := AC - B^2 \quad (19.2)$$

---
*Important contributions to this chapter were made by my son Dr. -Ing. Bodo Durst.

one defines the differential equation as parabolic, hyperbolic or elliptic when for d the following holds:

Parabolic differential equation: $d = 0$ (one-parameter characteristics),
Hyperbolic differential equation: $d < 0$ (two-parameter characteristics),
Elliptic differential equation: $d > 0$ (no real characteristics).

This classification of the differential equation orients itself by the equations for parabolas, hyperbolas and ellipses of the field of plane geometry, in which the equation:
$$ax^2 + 2bxy + cy^2 + dx + ey + f = 0 \tag{19.3}$$
describes parabolas ($ac - b^2 = 0$), hyperbolas ($ac - b^2 < 0$) and ellipses ($ac - b^2 > 0$). Accordingly, for the differential equations discussed in the preceding chapters, one can characterize:

Diffusion equation $\quad \frac{\partial U}{\partial t} = \nu \frac{\partial^2 U}{\partial x^2}$, i.e. it holds $A = \nu$, $B = C = 0$ and thus $d = 0$. The diffusion equation has "parabolic properties."

Wave equation $\quad \frac{\partial^2 U}{\partial t^2} = c^2 \frac{\partial^2 U}{\partial x^2}$, i.e. it holds $A = c^2$, $B = 0$, $C = -1$ and thus $d = -c^2 < 0$. The wave equation is hyperbolic.

Potential equation $\quad \frac{\partial^2 \Phi}{\partial x^2} + \frac{\partial^2 \Phi}{\partial y^2} = 0$, i.e. $A = C = 1$, $B = 0$ and thus $d > 0$. The potential equation shows "elliptical behavior."

The stationary boundary-layer equations are, as can be demonstrated when analyzing them according to the above representations, parabolic differential equations. Such equations have a property which is important for the numerical solution to be treated in this section. The solution of the differential equation determined at a certain point of a flow field, does not depend on the boundary conditions that lie downstream. This makes it possible to find a solution for the entire flow field via "forward integration," i.e. the solution can be computed in a certain level of the flow field alone from the values of the preceding level. This is characteristic for parabolic differential equations which thus, in the case of numerical integration, possess advantages which the elliptic differential equations do not have. This becomes clear when looking at Fig. 19.1, which shows which subdomain of a flow field acts on the properties at a point P, i.e. determines its flow properties.

In order to be able to solve differential equations numerically, it is necessary to cover the flow region with a "numerical grid," as e.g. indicated in Fig. 19.2 for a special flow problem, where a structured grid is shown. This is installed over a plane plate which carries out the following motion:

$U_1(x_2 = 0, t < 0) = 0 \quad$ The plate rests for all times $t < 0$,
$U_1(x_2 = 0, t \geq 0) = U_0 \quad$ The plate moves at constant velocity for $t \geq 0$.

By the motion of the plate, the fluid above the plate is, due to the molecular momentum transport, set in motion.

19.1 General Considerations

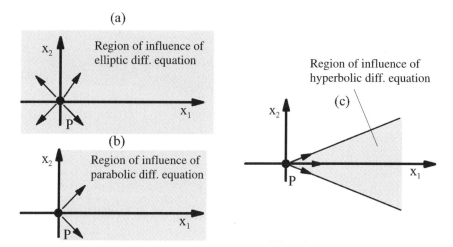

Fig. 19.1 Regions of influence for properties in point P for elliptic (**a**), parabolic (**b**) and hyperbolic (**c**) differential equations

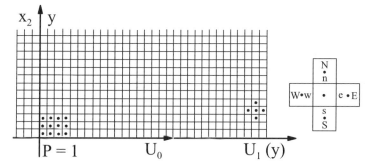

Fig. 19.2 Numerical grid for computing the fluid motion for sudden motion of the plate

The momentum input into the fluid is, assuming an infinitely long plate in the x_1 direction, described by the following differential equation:

$$\frac{\partial U_1}{\partial t} = \nu \frac{\partial^2 U_1}{\partial x_2{}^2} = \nu \frac{\partial^2 U}{\partial y^2} . \tag{19.4}$$

To be able to describe the solution of this differential equation on the numerical grid of Fig. 19.2, discretizations of the first derivative in time and second derivative in space in the differential equation (19.4) are necessary. An analysis of the differential equation shows that the discriminant is $d = 0$, i.e. the differential equation is parabolic. The solution concerning time t can thus be computed from the solution concerning time $t - \Delta t$. Inversely it holds, however, that the solution concerning time $(t + \Delta t)$ cannot be used to compute the solution concerning time t. When using the "finite-difference method" for discretization (see Sect. 19.3), for the time a forward-difference formulation and for y a central-difference formulation can be employed, from which the following finite-difference equation for the differential equation (19.4) results:

$$\frac{U_\beta^{\alpha+1} - U_\beta^\alpha}{\Delta t} = \nu \left[\frac{U_{\beta+1}^\alpha - 2U_\beta^\alpha + U_{\beta-1}^\alpha}{(\Delta y)^2} \right]. \tag{19.5}$$

With this the velocity U in the time interval $(\alpha + 1)$ can be computed explicitly:

$$U_\beta^{\alpha+1} = U_\beta^\alpha - \frac{\nu \Delta t}{(\Delta y)^2} \left(U_{\beta+1}^\alpha - 2U_\beta^\alpha + U_{\beta-1}^\alpha \right), \tag{19.6}$$

i.e. a discrete solution of the differential equation is possible. The discretized equation is defined as *consistent* when for the transition $\Delta y \to 0$ and $\Delta t \to 0$ (19.6) turns into the original differential equation (19.4). Here, checking of the consistency can be done through an analysis treating the truncation error. When for Δy and $\Delta t \to 0$ the truncation error heads towards zero, the employed differentiation method is consistent.

For the reasonable application of numerical computation methods, it is moreover necessary that the "discretization method used is stable," i.e. yields stable solutions. The required *stability* is generally guaranteed when perturbations introduced into the solution process are attenuated by the discretization method. However, when an excitation of occurring perturbations takes place, the chosen discretization method is defined as unstable. With respect to fluid-mechanical problems, for stability considerations of the solution methods employed, the diffusion number Di and the Courant number Co are of importance:

$$Di = \frac{\nu \Delta t}{(\Delta y)^2} \quad \text{and} \quad Co = \frac{U \Delta t}{\Delta y}. \tag{19.7}$$

Here, the diffusion number indicates the ratio of the time interval Δt chosen in the discretization to the diffusion time $(\Delta y^2)/\nu$, while the Courant number states the ratio of the chosen time interval Δt to the convection time $(\Delta y/U)$. It is understandable that the numerical computation method can provide stable solutions only when the chosen time intervals Δt can resolve the physically occurring diffusion and convection times in the chosen numerical grid. Details are explained in the subsequent paragraphs.

It is also important for a chosen discretization method that the *convergence* of the solution is guaranteed. This means that the numerical solution, with continuous improvements of the numerical grid, agrees more and more with the exact solution of the differential equation. Yet the "Lax equivalence theorem" says that stable and consistently formulated discretizations of linear initial-value problems lead to convergent solutions, i.e. at least for linear differential equations, it can be shown that the consistency, stability and convergence of discretization methods are closely linked properties of a discretization method employed for the solution of differential equations. In the case of existing consistency, it is sufficient for a discretization method to prove its stability, in order to be able to predict reliably its convergence also.

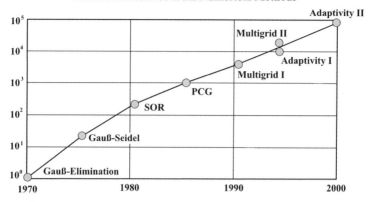

Fig. 19.3 Increase in performance of mathematical methods for solving basic equations in fluid mechanics

As far as numerical solutions of the basic equations of fluid mechanics are concerned, their solution regarding engineering problems is connected to high computational efforts. By efficient numerical computational methods and the employment of the currently available high computer power, this can nowadays be managed. Developments in applied mathematics have contributed to this (see Fig. 19.3) and have led to a continuous increase in the performance of computational methods for numerical solutions of the basic equations of fluid mechanics. Figure 19.3 shows the increase in performance of numerical computational methods, which has led, on average, to a tenfold increase every 8 years. Combining this with the increase in computational power, which can be said to have had a tenfold increase every 5 years (see Fig. 19.4), it becomes understandable why numerical fluid mechanics has been gaining increased significance in recent years. It is the field of numerical fluid mechanics to which the greatest importance has to be attached in the near future. The above increases in computing and computer power have significance with regard to solutions of engineering fluid flow problems. It is therefore imperative that modern fluid-mechanical education has an emphasis on numerical fluid mechanics. In this chapter, only an introduction to this important field can be given. These are detailed treatments of numerical fluid mechanics given in refs. [19.1] to [19.6].

19.2 General Transport Equation and Discretization of the Solution Region

In Chap. 5, the basic equations of fluid mechanics were derived and stated in the form indicated below:

- Continuity equation:

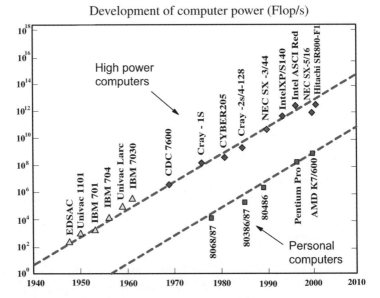

Fig. 19.4 Increase in the performance of high-speed computers and of personal computers

$$\frac{\partial \rho}{\partial t} + \frac{\partial (\rho U_i)}{\partial x_i} = 0. \tag{19.8}$$

- Navier–Stokes equation:

$$\frac{\partial (\rho U_j)}{\partial t} + \frac{\partial (\rho U_i U_j)}{\partial x_i} = -\frac{\partial P}{\partial x_j} - \frac{\partial \tau_{ij}}{\partial x_i} + \rho g_j \tag{19.9}$$

$$\tau_{ij} = -\mu \left(\frac{\partial U_j}{\partial x_i} - \frac{\partial U_i}{\partial x_j} \right) + \frac{2}{3} \mu \delta_{ij} \frac{\partial U_k}{\partial x_k}.$$

- Energy equation:

$$\frac{\partial (\rho c_v T)}{\partial t} + \frac{\partial (\rho c_v U_i T)}{\partial x_i} = -\frac{\partial q_i}{\partial x_i} - \tau_{ij} \frac{\partial U_j}{\partial x_i} - P \frac{\partial U_i}{\partial x_i}, \tag{19.10}$$

$$q_i = -\lambda \frac{\partial T}{\partial x_i},$$

where $\tau_{ij} \frac{\partial U_j}{\partial x_i}$, in the last equation, represents the dissipation and $P \frac{\partial U_i}{\partial x_i}$ the work done during expansion. These equations can be transferred into a general transport equation in such a way that the following equation holds:

$$\frac{\partial (\rho \Phi)}{\partial t} + \frac{\partial}{\partial x_i} \left(\rho U_i \Phi - \Gamma \frac{\partial \Phi}{\partial x_i} \right) = S_\Phi, \tag{19.11}$$

where Φ and S_Φ for the different equations are indicated in Table 19.1

Considerations on the numerical solution of the basic equations of fluid mechanics can thus be restricted to equations of the form indicated in (19.11).

19.2 General Transport Equation and Discretization

Table 19.1 Φ, Γ and S_Φ values for the general transport equation

Equation	Φ	Γ	S	Remarks
Continuity	1	0	0	–
Momentum	u_j	μ	$\dfrac{\partial}{\partial x_i}\left[\mu\left(\dfrac{\partial U_i}{\partial x_j}\right)\right] - \dfrac{2}{3}\dfrac{\partial}{\partial x_j}\left(\mu\dfrac{\partial U_k}{\partial x_k}\right) + g_j\rho - \dfrac{\partial P}{\partial x_j}$	Newtonian fluid and compressible flow
Energy	T	$\dfrac{\lambda}{c_p}$	$\dfrac{\rho T}{c_p}\left(\dfrac{\partial \nu}{\partial T}\right)_p \dfrac{DP}{Dt} - \dfrac{\tau_{ij}}{c_p}\dfrac{\partial U_j}{\partial x_i}$	–
	T	$\dfrac{\lambda}{c_p}$	$\dfrac{1}{c_p}\dfrac{DP}{Dt} - \dfrac{\tau_{ij}}{c_p}\dfrac{\partial U_j}{\partial x_i}$	Ideal gas
	T	$\dfrac{\lambda}{c_p}$	$-\dfrac{\tau_{ij}}{c_p}\dfrac{\partial U_j}{\partial x_i}$	Incompressible or isobaric

The latter comprises a relationship which indicates the variations in terms of time of $(\rho\Phi)$ and changes in space in the form of a convection term $(\rho U_i \Phi)$ and also a diffusion term $(-\Gamma\frac{\partial \Phi}{\partial x_i})$. In the source term S, all those terms of the considered basic equations are contained that cannot be placed in the general convection and diffusion terms.

With the general transport equation (19.11), a constant description in terms of time and space is available of all the physical laws to which fluid motions are subjected. However, the numerical solution of the transport equation requires a discretization of the equation with respect to time and space. Thus, through the numerical solutions of flow problems, solutions are sought only of the flow quantities at determined points in space, which are arranged at distances of Δx_i. The determination of these points, before starting a numerical solution of a flow problem, is defined as grid generation, i.e. the entire flow region is subdivided into discrete subdomains. In general, it is usual to do the subdivision already in such a way that certain advantages result for the sought numerical solution method. Regular (structured) and irregular (unstructured) grids can, in principle, be employed; experience shows, however, that the efficiency of numerical solution algorithms is influenced particularly disadvantageously by the irregularity of the numerical grid. On the other hand, it holds that in the presence of complex flow boundaries the grid generation is considerably facilitated by unstructured grids. Geometrically complex boundaries of flow regions can be introduced more easily into the numerical computations to be carried out via unstructured grids. With this, in the case of strongly irregular grids, triangles, quadrangles, tetrahedrons, hexahedrons, prisms, etc., can be employed in combination.

Structured grids are characterized by the general property that the neighboring points surrounding a considered grid point correspond to a firm

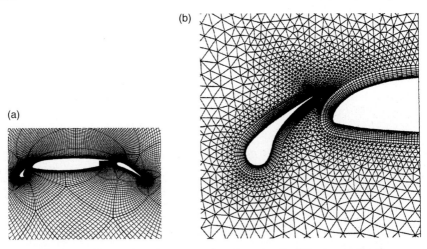

Fig. 19.5 Examples of structured and unstructured numerical grids

pattern in the entire solution region. Thus, for structured grids, it is generally only necessary to store the coordinates of the grid points, as the information on the relationship of a grid point to its neighbors, which is required for the discretization method, is determined by the structure of the grid. In the case of unstructured grids, such firm relationships of neighboring grid points do not exist but have to be stored for each grid point individually.

Figure 19.5 makes clear the difference between structured, (a), and unstructured, (b), numerical grids. It can easily be seen that for unstructured grids there is no regularity in the order of a grid point relative to its neighboring points. Without detailed explanations, it becomes clear that the missing structure in the grid order provides high flexibility for arranging the grid points over the entire solution regions such, that in areas with a high demand for grid points many points can be placed. In particular, it is easily possible with unstructured grids to capture corners and edges of flow geometry such, that they are sufficiently resolved for supplying a good numerical solution of a flow problem. However, a considerable disadvantage is that unstructured grids require high computer storage core spaces. Besides the grid points themselves, i.e. their position in the flow region, neighborhood relationships between the grid points have to be stored via index fields.

With the example of the one-dimensional stationary convection-diffusion equation without sources, it will be demonstrated which advantage finite-volume methods have in this respect, e.g. as against finite-difference methods:

$$\frac{d}{dx}\left(\rho U \Phi - \Gamma \frac{d\Phi}{dx}\right) = 0. \tag{19.12}$$

The term $\rho U \Phi$ represents the convective share of the flux density of Φ and $-\Gamma \frac{d\Phi}{dx}$ the diffusive share. This equation will now be integrated via a random volume in terms of space which extends in the x direction from W to E For the considered one-dimensional case, integration over a finite volume thus yields

$$\int_W^E \frac{\mathrm{d}}{\mathrm{d}x}\left(\rho U \Phi - \Gamma \frac{\mathrm{d}\Phi}{\mathrm{d}x}\right) \mathrm{d}x \cdot 1 \cdot 1 = 0 \tag{19.13}$$

or, when carrying out the integration:

$$\left(\rho U \Phi - \Gamma \frac{\mathrm{d}\Phi}{\mathrm{d}x}\right)_W = \left(\rho U \Phi - \Gamma \frac{\mathrm{d}\Phi}{\mathrm{d}x}\right)_E. \tag{19.14}$$

In words, this means that the entire flux of the quantity Φ, which at point W flows *into* the volume, has to flow *out* of the volume at point E, as no sources are present in (19.12). This shows that finite-volume methods allow conservative, discrete formulations of the integrations of the differential equations.

19.3 Discretization by Finite Differences

After having explained briefly the discretization of the flow region in Sect. 19.2, the discretization of the general transport equation (19.11) has to be explained. One possibility of such a discretization is given by the finite-difference method, which, for the time being, will be explained for the case of the one-dimensional, stationary convection-diffusion equation without source terms, i.e. for the equation

$$\frac{\mathrm{d}}{\mathrm{d}x}\left(\rho U \Phi - \Gamma \frac{\mathrm{d}\Phi}{\mathrm{d}x}\right) = 0. \tag{19.15}$$

Starting from a grid point β, $\Phi(x + \Delta x)$ can be represented via Taylor series expansion as follows:

$$\Phi_{\beta+1} = \Phi_\beta + \left(\frac{\mathrm{d}\Phi}{\mathrm{d}x}\right)_\beta \Delta x_{\beta+1} + \frac{1}{2}\left(\frac{\mathrm{d}^2\Phi}{\mathrm{d}x^2}\right)_\beta \Delta x_{\beta+1}^2 \pm 0\left(\Delta x_{\beta+1}^3\right). \tag{19.16}$$

In the same way, $\Phi_{\beta-1}$ can be stated:

$$\Phi_{\beta-1} = \Phi_\beta - \left(\frac{\mathrm{d}\Phi}{\mathrm{d}x}\right)_\beta \Delta x_{\beta-1} + \frac{1}{2}\left(\frac{\mathrm{d}^2\Phi}{\mathrm{d}x^2}\right)_\beta \Delta x_{\beta-1}^2 \pm 0\left(\Delta x_{\beta-1}^3\right). \tag{19.17}$$

By subtraction of (19.17) from (19.16), one obtains

$$\Phi_{\beta+1} - \Phi_{\beta-1} = \left(\frac{\mathrm{d}\Phi}{\mathrm{d}x}\right)(\Delta x_{\beta+1} + \Delta x_{\beta-1})$$
$$- \frac{1}{2}\left(\frac{\mathrm{d}^2\Phi}{\mathrm{d}x^2}\right)\left(\Delta x_{\beta+1}^2 - \Delta x_{\beta-1}^2\right) + \cdots$$

For grids having equal distances between the grid points, the second term on the right-hand side is equal to zero, so that for $\Delta x_{\beta+1} = \Delta x_\beta = \Delta x$ the following expression holds:

$$\frac{d\Phi}{dx} = \frac{\Phi_{\beta+1} - \Phi_{\beta-1}}{2\Delta x} + O\left(\Delta x^2\right). \tag{19.18}$$

For the terms with second derivatives one obtains by addition of (19.17) and (19.16)

$$\Phi_{\beta+1} + \Phi_{\beta-1} = 2\Phi_\beta + \left(\frac{d^2\Phi}{dx^2}\right)\frac{1}{2}\left(\Delta x_{\beta+1}^2 + \Delta x_{\beta-1}^2\right) + O\left(\Delta x^3\right) \tag{19.19}$$

so that for $\Delta x_{\beta+1} = \Delta x_{\beta-1} = \Delta x$ the following expression holds again:

$$\frac{d^2\Phi}{dx^2} = \frac{\Phi_{\beta+1} - 2\Phi_\beta + \Phi_{\beta-1}}{\Delta x^2} + O\left(\Delta x^3\right). \tag{19.20}$$

Thus, the differential equation (19.15) for Γ = constant can be stated as a finite difference equation as follows:

$$\frac{(\rho U)_{\beta+1}\Phi_{\beta+1} - (\rho U)_{\beta-1}\Phi_{\beta-1}}{\Delta x} - \Gamma\frac{\Phi_{\beta+1} - 2\Phi_\beta + \Phi_{\beta-1}}{\Delta x^2} = 0. \tag{19.21}$$

Ordered according to the unknowns $\Phi_{\beta+1}$, Φ_β and $\Phi_{\beta-1}$, one obtains:

$$\left[(\rho U)_{\beta+1}\frac{1}{\Delta x} - \frac{\Gamma}{\Delta x^2}\right]\Phi_{\beta+1} + \frac{2\Gamma}{\Delta x^2}\Phi_\beta + \left[(\rho U)_{\beta-1}\frac{1}{\Delta x} - \frac{\Gamma}{\Delta x^2}\right]\Phi_{\beta-1} = 0. \tag{19.22}$$

Hence, a linear system of equations results for the unknown Φ_β, which has to be solved to obtain, at each point of the numerical grid, a solution for all variables Φ of the considered flow field. In this way, a solution path has been found for the quantities Φ describing a flow. On considering now the solutions for Φ_β in the solution area from border W to border E, indicated in Fig. 19.6, the integral U, where $u = U$ is introduced:

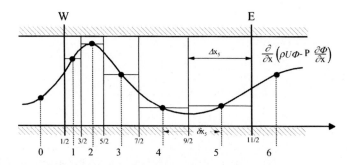

Fig. 19.6 One-dimensional computational area

19.3 Discretization by Finite Differences

$$\int_W^E \frac{d}{dx}\left(\rho u \Phi - \Gamma \frac{d\Phi}{dx}\right) dx \cdot 1 \cdot 1 \approx \sum_{\beta=1}^{5} \left[\frac{\left((\rho u)_{\beta+1} \Phi_{\beta+1} - (\rho u)_{\beta-1} \Phi_{\beta-1}\right)}{x_{\beta+1} - x_{\beta-1}} \right.$$
$$\left. - \frac{\left(\Gamma_{\beta+1}\left(\frac{d\Phi}{dx}\right)_{\beta+1} - \Gamma_{i-1}\left(\frac{d\Phi}{dx}\right)_{\beta-1}\right)}{x_{\beta+1} - x_{\beta-1}} \right] = 0$$
(19.23)

can only be determined by discrete integration, which has also been included in (19.23). On writing this equation in full for the six supporting points in Fig. 19.6, one obtains:

$$\left[\frac{((\rho u)_2 \Phi_2 - (\rho u)_0 \Phi_0) - \left(\Gamma_2 \left(\frac{d\Phi}{dx}\right)_2 - \Gamma_0 \left(\frac{d\Phi}{dx}\right)_0\right)}{x_2 - x_0} \right] (x_{\frac{3}{2}} - x_{\frac{1}{2}}) +$$

$$\left[\frac{((\rho u)_3 \Phi_3 - (\rho u)_1 \Phi_1) - \left(\Gamma_3 \left(\frac{d\Phi}{dx}\right)_3 - \Gamma_1 \left(\frac{d\Phi}{dx}\right)_1\right)}{x_3 - x_1} \right] (x_{\frac{5}{2}} - x_{\frac{3}{2}}) +$$

$$\left[\frac{((\rho u)_4 \Phi_4 - (\rho u)_2 \Phi_2) - \left(\Gamma_4 \left(\frac{d\Phi}{dx}\right)_4 - \Gamma_2 \left(\frac{d\Phi}{dx}\right)_2\right)}{x_4 - x_2} \right] (x_{\frac{7}{2}} - x_{\frac{5}{2}}) +$$

$$\left[\frac{((\rho u)_5 \Phi_5 - (\rho u)_3 \Phi_3) - \left(\Gamma_5 \left(\frac{d\Phi}{dx}\right)_5 - \Gamma_3 \left(\frac{d\Phi}{dx}\right)_3\right)}{x_5 - x_3} \right] (x_{\frac{9}{2}} - x_{\frac{7}{2}}) +$$

$$\left[\frac{((\rho u)_6 \Phi_6 - (\rho u)_4 \Phi_4) - \left(\Gamma_6 \left(\frac{d\Phi}{dx}\right)_6 - \Gamma_4 \left(\frac{d\Phi}{dx}\right)_4\right)}{x_6 - x_4} \right] (x_{\frac{11}{2}} - x_{\frac{9}{2}}) = 0.$$
(19.24)

When one compares (19.23) with (19.24), one recognizes that the finite-difference method employed for the discretization does not furnish the same result as the integration yielding (19.23). This leads to the fact that the chosen discretization method turns out to be non-conservative. Generally, it can be said that discretizations by means of finite-difference methods require special measures to produce conservative discrete formulations of the basic equations of fluid mechanics. When choosing in (19.24) $\Delta x_p = \Delta x_i = \Delta x_{i+1}$, one obtains

$$\frac{1}{2}\left[(\rho u)_5 \Phi_5 - \Gamma_5 \left(\frac{d\Phi}{dx}\right)_5 + (\rho u)_6 \Phi_6 - \Gamma_6 \left(\frac{d\Phi}{dx}\right)_6\right] =$$
$$\frac{1}{2}\left[(\rho u)_0 \Phi_0 - \Gamma_0 \left(\frac{d\Phi}{dx}\right)_0 + (\rho u)_1 \Phi_1 - \Gamma_1 \left(\frac{d\Phi}{dx}\right)_1\right], \quad (19.25)$$

i.e. all internal fluxes drop out and consequently a conservative form of the conservation equation for the unknown discrete variables Φ_β results. In this equation, both sides represent admissible approximations of the flows into and out of the solution area. The discretized equation is thus a conservative approximation of the general transport equation and the discretization scheme for this case is consequently conservative. However, one recognizes that for finite-difference methods special measures have to be taken to force the conservativeness. Not least, it is assumed in the derivations that the numerical grid employed does not show a strong non-equidistance. When the latter assumption is not fulfilled, the method is not conservative and, moreover, the order of the discretization method is reduced by an order of magnitude. It is therefore emphasized once again that finite-difference methods are not necessarily conservative. In addition, connected with this, non-conservative formulations yield disadvantageous reductions of the order of the accuracy of the solution methods.

19.4 Finite-Volume Discretization

19.4.1 General Considerations

The notation used in this section is represented in Fig. 19.7. Considered is a point P and its neighbors located in direction of the coordinate axes, e.g. of a Cartesian coordinate system. The neighboring points in the x–y plane are named West, South, East and North, corresponding to their position relative to P, and the two points in the z direction are referred to as Top and Bottom points.

For the considerations to follow, around point P a control volume is formally installed, so that P is the center of this control volume. The boundary surfaces of the control volumes are marked according to the respective neighboring points, but in lower-case letters. Terms which generally would read the same for all neighboring points, are stated with an index Nb to abbreviate the notation. Accordingly, terms for the boundary surface of the control volume are given the index cf.

As the present considerations always start from the assumption that *all* points W, S, E, N, T and B are grid points and that the grid points are located exactly in the centers of neighboring control volumes, it is sufficient to store only the coordinates of the control-volume boundary surfaces

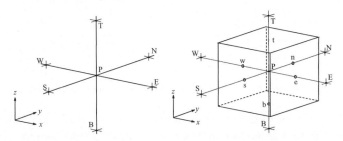

Fig. 19.7 Cartesian grid and control volume with characteristic point

19.4 Finite-Volume Discretization

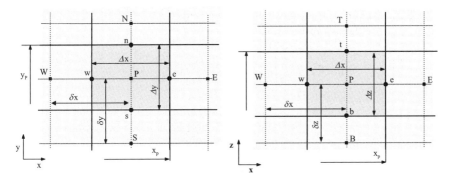

Fig. 19.8 Cross planes for explaining the basic ideas of interpolation

for numerical flow computations via finite-volume methods. All other information with regard to the grid can be computed back from these data. The distance between P and E, for example, which we denote δx_e, can be computed as follows (see Fig. 19.8):

$$\delta x_e = 1/2(x_E + x_P) - 1/2(x_P + x_W) = 1/2(x_E - x_W). \tag{19.26}$$

In the case of non-stationary processes, a discrete representation of the coordinate time is necessary. Discrete time planes result for the process, which also have to be marked. Here, the style of marking is t^α, or $t^{\alpha+1}$ for each new time level and $t^{\alpha-1}$ for each old time level.

When, for examining the conservativeness of the formulation of discretization, the integration of (19.15) around a point P_β results in the following relationship:

$$\iiint_{\Delta V_\beta} \frac{\mathrm{d}}{\mathrm{d}x}\left(\rho u \Phi - \Gamma \frac{\mathrm{d}\Phi}{\mathrm{d}x}\right) \mathrm{d}V = \int_{w_\beta}^{e_\beta} \frac{\mathrm{d}}{\mathrm{d}x}\left(\rho u \Phi - \Gamma \frac{\mathrm{d}\Phi}{\mathrm{d}x}\right) \mathrm{d}x \cdot 1 \cdot 1 = 0 \tag{19.27}$$

and thus the integral yields

$$\left(\rho u \Phi - \Gamma \frac{\partial \Phi}{\partial x}\right)_{e_\beta} - \left(\rho u \Phi - \Gamma \frac{\partial \Phi}{\partial x}\right)_{w_\beta} = 0. \tag{19.28}$$

This equation says for the individual control volume that the convective and diffusive flows at the East surface and the West surface are exactly the same. The integration can thus be carried out by summing over all five control volumes of the computation region indicated in Fig. 19.6:

$$\sum_{\beta=1}^{5}\left[\left(\rho u \Phi - \Gamma \frac{\partial \Phi}{\partial x}\right)_{e_\beta} - \left(\rho u \Phi - \Gamma \frac{\partial \Phi}{\partial x}\right)_{w_\beta}\right] = 0. \tag{19.29}$$

When taking into consideration that the common surface of two neighboring control volumes are identical (boundary surface e_β is at the same time boundary surface $w_{\beta+1}$), most of the fluxes in the last equation cancel each other and the following remains:

$$\left(\rho u \Phi - \Gamma \frac{\partial \Phi}{\partial x}\right)_{e_5} - \left(\rho u \Phi - \Gamma \frac{\partial \Phi}{\partial x}\right)_{w_1} = 0. \quad (19.30)$$

The two surfaces which remain in the consideration are those which limit the computational region in Fig. 19.6, i.e. W and E. Thus it holds that

$$\left(\rho u \Phi - \Gamma \frac{\partial \Phi}{\partial x}\right)_E - \left(\rho u \Phi - \Gamma \frac{\partial \Phi}{\partial x}\right)_W = 0. \quad (19.31)$$

This is *exactly* the same equation as one would obtain by integration over the total region; see (19.6). One recognizes that a discretization method based on finite volumes is *inherently conservative*.

19.4.2 Discretization in Space

The model differential equation for a general scalar quantity Φ was stated in Sect. 19.2. In the same section, it was shown how, by inserting different expressions for the individual terms of this differential equation, the basic equations of fluid mechanics can be rederived. In the present section, we shall only consider the model equation and carry out the discretization by means of it. The discrete forms of individual fluid equations can then be derived by inserting the expressions for Φ, Γ and S, respectively.

For the representations to be carried out here, the following transport equation is thus considered:

$$\frac{\partial}{\partial t}(\rho \Phi) + \frac{\partial}{\partial x_i}\left(\rho u_i \Phi - \Gamma \frac{\partial \Phi}{\partial x_i}\right) = S_\Phi. \quad (19.32)$$

Here, the total flux of Φ consists of the partial fluxes:

$$\rho u_i \Phi = \text{convective flux} \quad \text{and} \quad -\Gamma \frac{\partial \Phi}{\partial x_i} = \text{diffuse flux}$$

and can be stated as follows:

$$f_i = \rho u_i \Phi - \Gamma \frac{\partial \Phi}{\partial x_i} \quad (19.33)$$

so that the transport equation consequently reads:

$$\frac{\partial}{\partial t}(\rho \Phi) + \frac{\partial f_i}{\partial x_i} = S_\Phi. \quad (19.34)$$

19.4 Finite-Volume Discretization

This equation is now nominally integrated over all control volumes of the computational domain. We consider as a substitute the control volume around a considered point P:

$$\iiint_{\Delta V} \frac{\partial}{\partial t}(\rho \Phi) \mathrm{d}V + \iiint_{\Delta V} \frac{\partial f_i}{\partial x_i} \mathrm{d}V = \iiint_{\Delta V} S_\Phi \mathrm{d}V = \iiint_{\Delta V} S \mathrm{d}V \qquad (19.35)$$

and treat the individual expressions of this equation separately.

Applying the Gauss integral theorem, for the second term on the left-hand side of (19.35), the following holds:

$$\iiint_{\Delta V} \frac{\partial f_i}{\partial x_i} \mathrm{d}V = \iint_{\Delta A} f_i \, \mathrm{d}A_i \qquad (19.36)$$

with ΔA being the surface of the control volume.

The integration over the entire control-volume surface can also be represented as the sum of the integrations over the individual boundary surfaces:

$$\iint_{\Delta A} f_i \mathrm{d}A_i = \iint_{\Delta A_w} f_i \mathrm{d}A_i + \iint_{\Delta A_e} f_i \mathrm{d}A_i + \iint_{\Delta A_s} f_i \mathrm{d}A_i + \iint_{\Delta A_n} f_i \mathrm{d}A_i +$$

$$\iint_{\Delta A_b} f_i \mathrm{d}A_i + \iint_{\Delta A_t} f_i \mathrm{d}A_i. \qquad (19.37)$$

The individual external surface normals of the boundary surfaces can be stated plainly, e.g. always are in these considerations:

$$\mathbf{dA}_w = \begin{pmatrix} -\mathrm{d}x_2 \mathrm{d}x_3 \\ 0 \\ 0 \end{pmatrix} \quad \text{and} \quad \mathbf{dA}_t = \begin{pmatrix} 0 \\ 0 \\ \mathrm{d}x_1 \mathrm{d}x_2 \end{pmatrix}. \qquad (19.38)$$

On introducing the scalar products $f_i \, \mathrm{d}A_i$ into (19.37), one obtains:

$$\iint_{\Delta A} f_i \mathrm{d}A_i = \iint_{\Delta A_e} f_1 \mathrm{d}x_2 \mathrm{d}x_3 - \iint_{\Delta A_w} f_1 \mathrm{d}x_2 \mathrm{d}x_3 +$$

$$\iint_{\Delta A_n} f_2 \mathrm{d}x_1 \mathrm{d}x_3 - \iint_{\Delta A_s} f_2 \mathrm{d}x_1 \mathrm{d}x_3 +$$

$$\iint_{\Delta A_t} f_3 \mathrm{d}x_1 \mathrm{d}x_2 - \iint_{\Delta A_b} f_3 \mathrm{d}x_1 \mathrm{d}x_2. \qquad (19.39)$$

For the discretization, certain approximations are necessary, and they can be taken as approximations on three different planes. The first approximation is

$$\iint_{\Delta A_{cf}} f_i \mathrm{d}x_j \mathrm{d}x_k \approx F_{icf}. \tag{19.40}$$

The mean-value theorem of the integral calculus says that a value $\overline{f}_{icf}$ can always be found on the surface A_{cf} so that the above relation is exactly fulfilled, i.e. that:

$$\iint_{\Delta A_{cf}} f_i \mathrm{d}x_j \mathrm{d}x_k = \overline{f}_{icf} A_{cf}$$

holds. To state this value a priori is not possible, however, as exactly f_{icf} quantities are supposed to be calculated. In order to state a computational method now, the value in the center of the control-volume surface f_{icf} is used as an approximate value of $\overline{f}_{icf}$.

For Cartesian geometries, it holds that $\Delta A_{cf} = \Delta x_j \Delta x_k$ $(j \neq k)$ and, although $A_e = A_w$, both expressions will be used further. With analogous approximations also for the other two directions, the following relationship holds:

$$\iiint_{\Delta V} \frac{\partial f_i}{\partial x_i} \mathrm{d}V = F_e - F_w + F_n - F_s + F_t - F_b. \tag{19.41}$$

The first expression on the left-hand side of (19.35), and also the source term of this equation, are approximated in the same way, namely by using the value in the control-volume center as an approximation for the mean value over the control volume:

$$\iiint_{\Delta V} \left(\frac{\partial}{\partial t}(\rho \Phi) \right) \mathrm{d}V \approx \frac{\partial}{\partial t}(\rho \Phi)_P \Delta V = \frac{\partial}{\partial t}(\rho \Phi)_P \Delta x_1 \Delta x_2 \Delta x_3 \tag{19.42}$$

and

$$\iiint_{\Delta V} S \, \mathrm{d}V \approx S_P \Delta x_1 \Delta x_2 \Delta x_3. \tag{19.43}$$

On inserting the last three equations into (19.35), one obtains:

$$\frac{\partial}{\partial t}(\rho \Phi)_P \Delta V + F_e - F_w + F_n - F_s + F_t - F_b = S_P \Delta V \tag{19.44}$$

and with the expressions for the total flow it results that:

$$\frac{\partial}{\partial t}(\rho \Phi)_P \Delta V + \left(\rho u_1 \Phi - \Gamma \frac{\partial \Phi}{\partial x_1} \right)_e \Delta A_e - \left(\rho u_1 \Phi - \Gamma \frac{\partial \Phi}{\partial x_1} \right)_w \Delta A_w +$$
$$\left(\rho u_2 \Phi - \Gamma \frac{\partial \Phi}{\partial x_2} \right)_n \Delta A_n - \left(\rho u_2 \Phi - \Gamma \frac{\partial \Phi}{\partial x_2} \right)_s \Delta A_s +$$
$$\left(\rho u_3 \Phi - \Gamma \frac{\partial \Phi}{\partial x_3} \right)_t \Delta A_t - \left(\rho u_3 \Phi - \Gamma \frac{\partial \Phi}{\partial x_3} \right)_b \Delta A_b = S_P \Delta V. \tag{19.45}$$

19.4 Finite-Volume Discretization

In order to increase the clarity of indexing in what follows, the variables x_1, x_2 and x_3 are replaced by x, y and z and the velocities u_1, u_2 and u_3 by u, v and w.

The next approximation step, for the derivation of a finite-volume computational method, is the linearization of the expressions in question. If one considers, e.g., the term $\rho u \Phi$, one recognizes that for $\Phi = u$ the unknown u occurs as u^2, i.e. the term is nonlinear. To be able to treat such terms, in the computational method to be derived, in a simple way, one linearizes them by considering the mass flow density (ρu) and the diffusion coefficient Γ independently of the unknown quantity Φ. For computing Φ one falls back on the values $(\rho u)^*$ and Γ^*, worked out in preceding computational steps. Thus the values $(\rho u)^*$ and Γ^* are known for the considered computational step, and one looks for the final solution in several steps. The values with an asterisk are each taken from the preceding iteration of the computations:

$$\left(\rho u \Phi - \Gamma \frac{\partial \Phi}{\partial x}\right)_{cf} \approx (\rho u)^*_{cf} \Phi_{cf} - \Gamma^*_{cf} \left(\frac{\partial \Phi}{\partial x}\right)_{cf} \quad (19.46)$$

and thus it follows from (19.44):

$$\frac{\partial}{\partial t}(\rho^*_P \Phi_P)\Delta V + \left[(\rho u)^*_e \Phi_e - \Gamma^*_e \left(\frac{\partial \Phi}{\partial x}\right)_e\right]\Delta A_e -$$
$$\left[(\rho u)^*_w \Phi_w - \Gamma^*_w \left(\frac{\partial \Phi}{\partial x}\right)_w\right]\Delta A_w +$$
$$\left[(\rho v)^*_n \Phi_n - \Gamma^*_n \left(\frac{\partial \Phi}{\partial y}\right)_n\right]\Delta A_n - \left[(\rho v)^*_s \Phi_s - \Gamma^*_s \left(\frac{\partial \Phi}{\partial y}\right)_s\right]\Delta A_s +$$
$$\left[(\rho w)^*_t \Phi_t - \Gamma^*_t \left(\frac{\partial \Phi}{\partial z}\right)_t\right]\Delta A_t - \left[(\rho w)^*_b \Phi_b - \Gamma^*_b \left(\frac{\partial \Phi}{\partial z}\right)_b\right]\Delta A_b = S_P \Delta V. \quad (19.47)$$

In the subsequent considerations, it will be necessary again and again to have the discrete form of the continuity equation at disposal. The continuity equation in its general form is obtained from the equation for the general scalar quantity by setting $\Phi = 1$, $\Gamma = 0$ and $S = 0$. With the same values, the discrete form results from the above equation as

$$\frac{\partial \rho^*_P}{\partial t}\Delta V + (\rho u)^*_e \Delta A_e - (\rho u)^*_w \Delta A_w + (\rho v)^*_n \Delta A_n - (\rho v)^*_s \Delta A_s$$
$$+ (\rho w)^*_t \Delta A_t - (\rho w)^*_b \Delta A_b = 0 \quad (19.48)$$

or, by abbreviating the mass fluxes by m_{cf}:

$$\frac{\partial \rho^*_P}{\partial t}\Delta V + m^*_e - m^*_w + m^*_n - m^*_s + m^*_t - m^*_b = 0. \quad (19.49)$$

In (19.47) there are values of Φ and $\frac{\partial \Phi}{\partial x}$, which have to be computed on the surfaces of the control volume. However, in the usually employed computational methods only values of Φ at the grid points are stored, as only they are of interest for the solution and later computations. As a consequence, the values Φ_{cf} have to be expressed as functions of the values Φ_{Nb} of the grid points neighboring the considered control volume surface. Computing the values Φ_{cf} from Φ_{Nb}, required for the computational methods, is the actual difficulty when discretizing the basic fluid mechanics equations by means of finite-volume methods. There is the problem, as already stated above, when applying the mean integral value theorem, that the behavior of Φ as a function of space should be known between two grid points, in order to be able to derive the exact values of Φ_{cf} or $\frac{\partial \Phi}{\partial x}|_{cf}$ at each of the considered control volume surfaces. However, it is exactly this behavior which has to be computed, and it is therefore necessary for the derivations of the computational scheme to assume a behavior. This assumption represents the third approximation step of the finite-volume method considered here.

A certain understanding of the introduced approximation can be found by considering a simplified problem. The equation for a one-dimensional, stationary flow problem, without sources, is

$$\frac{\mathrm{d}}{\mathrm{d}x}\left(\rho u \Phi - \Gamma \frac{\mathrm{d}\Phi}{\mathrm{d}x}\right) = 0. \quad (19.50)$$

For this equation, an analytical solution can be found:

$$\Phi \sim \exp\left[\frac{(\rho u)x}{\Gamma}\right]. \quad (19.51)$$

In Fig. 19.9, the behavior of this solution of the above equation is represented for a boundary value problem between two points P_l and P_u with the corresponding functional values Φ_l and Φ_u. The functional relationship is given as a function of the Peclet number. The Peclet number represents the ratio between the convective and diffuse transport of Φ.

On considering Fig. 19.9 and also the solution of the transport equation (19.50), it is actually obvious to assume an exponential behavior of Φ between the grid points also in the multi-dimensional flow case. However, the computation of exponential functions on a computer is very costly compared with

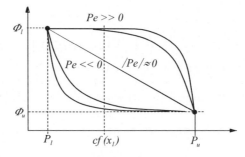

Fig. 19.9 Dependence of the behavior of Φ on the mass current density (Peclet number)

19.4 Finite-Volume Discretization

other operations, and one is inclined to approximate the actual functional behavior by a polynomial. The various approximation ansatzes employed for this differs in the order of the polynomial used. In the present section, those polynomials of zero and first order will be considered, which lead to the so-called *Upwind* or central differential methods.

Owing to the assumption of a linear distribution of Φ between the two grid points, the derivation for the diffusive transport term can approximately be replaced by:

$$\left(\frac{\partial \Phi}{\partial x}\right)_{cf} = \frac{\Phi_u - \Phi_l}{\delta x_{cf}}. \tag{19.52}$$

The flow through the control-volume surface cf can then be approximated as follows:

$$-\Gamma^*_{cf}\left(\frac{\partial \Phi}{\partial x}\right)_{cf}\Delta A_{cf} = \frac{\Gamma^*_{cf}\Delta A_{cf}}{\delta x_{cf}}(\Phi_l - \Phi_u) = D^*_{cf}(\Phi_l - \Phi_u) \tag{19.53}$$

and one obtains for the different control-volume sides:

$$cf = w, s, b: \quad -\Gamma^*_{cf}\left(\frac{\partial \Phi}{\partial x}\right)_{cf}\Delta A_{cf} = D^*_{cf}(\Phi_{Nb} - \Phi_P), \tag{19.54}$$

$$cf = e, n, t: \quad -\Gamma^*_{cf}\left(\frac{\partial \Phi}{\partial x}\right)_{cf}\Delta A_{cf} = D^*_{cf}(\Phi_P - \Phi_{Nb}), \tag{19.55}$$

where Nb is the neighboring point of cf with the direction being given by P and the location of the considered surface.

For the approximation of the convective fluxes, differing approximate considerations can now be used, yielding different computational methods; they are explained below.

19.4.2.1 Upwind Method

The upwind method approximates the behavior of Φ between two grid points by a polynomial of zero order, i.e. a constant. From Fig. 19.9, one recognizes quickly that this approximation is good for large Peclet numbers, i.e. for situations in which the convective transport is predominant. For this case, the value of Φ at the P-surface differs only slightly from the value at the grid point located upwind of cf. It is thus also clear by which value of Φ the approximation should be introduced, i.e. always by the one located upwind of cf. For a flow in the positive coordinate direction, this is Φ_l, and for the negative direction, it is Φ_u. It is therefore necessary to be able to determine the flow direction at the control-volume surfaces. For this purpose, we shall use the mass flux:

$$m^*_{cf} = (\rho u)^*_{cf}\Delta A_{cf}. \tag{19.56}$$

For each control-volume surface, the direction of the mass flux has to be determined by means of its plus/minus sign. Therefore, the following needs to be considered:

$$cf = w, s, b: \quad \Phi_{cf} = \begin{cases} \Phi_{Nb} & \text{for } m^*_{cf} > 0 \\ \Phi_P & \text{for } m^*_{cf} < 0, \end{cases} \quad (19.57)$$

$$cf = e, n, t: \quad \Phi_{cf} = \begin{cases} \Phi_P & \text{for } m^*_{cf} > 0 \\ \Phi_{Nb} & \text{for } m^*_{cf} < 0. \end{cases} \quad (19.58)$$

To be able to represent all possible combinations of the above expressions by a uniform notation, we introduce a unit "step function," which is defined as follows:

$$e(x) = \begin{cases} 1 & \text{for } x \geq 0 \\ 0 & \text{for } x < 0. \end{cases} \quad (19.59)$$

This function is sketched in Fig. 19.10. The function for which the step is carried out for negative x is given as follows:

$$e(-x) = \begin{cases} 0 & \text{for } x > 0 \\ 1 & \text{for } x \leq 0 \end{cases} \quad (19.60)$$

and it is represented by Fig. 19.10b. With this, the functions $e(x)$ given in (19.57) can be written as:

$$cf = w, s, b: \quad \Phi_{cf} = e\left(m^*_{cf}\right)\Phi_{Nb} + e\left(-m^*_{cf}\right)\Phi_P, \quad (19.61)$$
$$cf = e, n, t: \quad \Phi_{cf} = e\left(-m^*_{cf}\right)\Phi_{Nb} + e\left(m^*_{cf}\right)\Phi_P. \quad (19.62)$$

On inserting these expressions for the convective and also the diffusive terms into (19.47), one obtains:

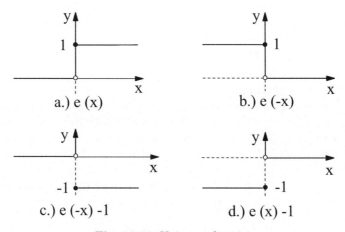

Fig. 19.10 Unit step function

19.4 Finite-Volume Discretization

$$\frac{\partial}{\partial t}(\rho_P^* \Phi_P)\Delta V + (m_e^*(e(-m_e^*)\Phi_E + e(m_e^*)\Phi_P) + D_e^*(\Phi_P - \Phi_E)) -$$
$$(m_w^*(e(m_w^*)\Phi_W + e(-m_w^*)\Phi_P) + D_w^*(\Phi_W - \Phi_P)) +$$
$$(m_n^*(e(-m_n^*)\Phi_N + e(m_n^*)\Phi_P) + D_n^*(\Phi_P - \Phi_N)) -$$
$$(m_s^*(e(m_s^*)\Phi_S + e(-m_s^*)\Phi_P) + D_s^*(\Phi_S - \Phi_P)) +$$
$$(m_t^*(e(-m_t^*)\Phi_T + e(m_t^*)\Phi_P) + D_t^*(\Phi_P - \Phi_T)) -$$
$$(m_b^*(e(m_b^*)\Phi_B + e(-m_b^*)\Phi_P) + D_b^*(\Phi_B - \Phi_P)) = S_P \Delta V$$
(19.63)

and with a little rearrangement of the terms of this equation, the total expression then reads:

$$\rho_P^* \frac{\partial \Phi_P}{\partial t}\Delta V + (m_e^*(e(-m_e^*)\Phi_E + (e(m_e^*) - 1)\Phi_P) + D_e^*(\Phi_P - \Phi_E)) -$$
$$(m_w^*(e(m_w^*)\Phi_W + (e(-m_w^*) - 1)\Phi_P) - D_w^*(\Phi_P - \Phi_W)) +$$
$$(m_n^*(e(-m_n^*)\Phi_N + (e(m_n^*) - 1)\Phi_P) + D_n^*(\Phi_P - \Phi_N)) -$$
$$(m_s^*(e(m_s^*)\Phi_S + (e(-m_s^*) - 1)\Phi_P) - D_s^*(\Phi_P - \Phi_S)) +$$
$$(m_t^*(e(-m_t^*)\Phi_T + (e(m_t^*) - 1)\Phi_P) + D_t^*(\Phi_P - \Phi_T)) -$$
$$(m_b^*(e(m_b^*)\Phi_B + (e(-m_b^*) - 1)\Phi_P) - D_b^*(\Phi_P - \Phi_B)) +$$
$$\underbrace{\left(\frac{\partial \rho_P^*}{\partial t}\Delta V + m_e^* - m_w^* + m_n^* - m_s^* + m_t^* - m_b^*\right)}_{= 0}\Phi_P = S_P \Delta V.$$
(19.64)

As indicated before, the term covered by the brace represents the continuity equation in its discrete form and therefore disappears because of (19.49). A necessary requirement for this is that the iteration process secures the mass conservation after each iteration step.

The above equation contains the expressions $e(-m_{cf}^*) - 1$ and $e(m_{cf}^*) - 1$. These functions are also represented in Fig. 19.10. One realizes from their behavior that the following holds:

$$e(x) - 1 = -e(-x) \quad \text{and} \quad e(-x) - 1 = -e(x). \tag{19.65}$$

On considering this $e(x)$ and $e(-x)$ behavior, equation (19.64) can be simplified further:

$$\rho_P^* \frac{\partial \Phi_P}{\partial t}\Delta V +$$
$$(-m_e^* e(-m_e^*) + D_e^*)(\Phi_P - \Phi_E) - (-m_w^* e(m_w^*) - D_w^*)(\Phi_P - \Phi_W) +$$
$$(-m_n^* e(-m_n^*) + D_n^*)(\Phi_P - \Phi_N) - (-m_s^* e(m_s^*) - D_s^*)(\Phi_P - \Phi_S) +$$
$$(-m_t^* e(-m_t^*) + D_t^*)(\Phi_P - \Phi_T) - (-m_b^* e(m_b^*) - D_b^*)(\Phi_P - \Phi_B) = S_P \Delta V$$
(19.66)

and one can introduce the abbreviation a_{Nb} for the coefficients of the terms $(\Phi_P - \Phi_{Nb})$:

$$Nb = W, S, B : a_{Nb} = D_{cf}^* + m_{cf}^* e(m_{cf}^*),$$
$$Nb = E, N, T : a_{Nb} = D_{cf}^* - m_{cf}^* e(-m_{cf}^*). \quad (19.67)$$

The expressions, which now also contain the unit step function, can also be expressed by means of the function $\max[a, b]$, which exists in most programming languages and which expresses the larger of the two values:

$$Nb = W, S, B : a_{Nb} = D_{cf}^* + \max[m_{cf}^*, 0],$$
$$Nb = E, N, T : a_{Nb} = D_{cf}^* + \max[0, -m_{cf}^*]. \quad (19.68)$$

When using the coefficients a_{Nb}, one obtains from (19.65):

$$\rho_P^* \frac{\partial \Phi_P}{\partial t} \Delta V + a_E(\Phi_P - \Phi_E) + a_W(\Phi_P - \Phi_W) +$$
$$a_N(\Phi_P - \Phi_N) + a_S(\Phi_P - \Phi_S) + \quad (19.69)$$
$$a_T(\Phi_P - \Phi_T) + a_B(\Phi_P - \Phi_B) = S_P \Delta V.$$

After inserting the following abbreviation:

$$\hat{a}_P = a_E + a_W + a_N + a_S + a_T + a_B = \sum_{Nb} a_{Nb} \quad (19.70)$$

one finally obtains:

$$\rho_P^* \frac{\partial \Phi_P}{\partial t} \Delta V + \hat{a}_P \Phi_P - \sum_{Nb} a_{Nb} \Phi_{Nb} = S_P \Delta V. \quad (19.71)$$

This equation is the discrete analogue of (19.32) after discretization of the differentials with respect to space by means of the upwind method. The coefficients can be computed according to (19.68).

19.4.2.2 Central Difference Method

The central difference method approximates the exponential behavior of Φ, between two grid points, by a polynomial of first order. This corresponds to the assumption of a linear variation of Φ, which represents a good approximation for small Peclet numbers. The behavior of Φ is approximated all the better by this method, the more the diffusive transport prevails in the flow and the more diffusive transports of properties are present.

19.4 Finite-Volume Discretization

A linear behavior of Φ between two grid points P_l and P_u can be expressed by

$$\Phi(x) = \Phi_l + \frac{\Phi_u - \Phi_l}{\delta x_{cf}} [x - 1/2(x_l + x_{ll})] \quad \text{for} \quad x_l \leq x \leq x_u. \quad (19.72)$$

On representing δx_{cf} by the stored coordinates at the grid points, the following results:

$$\delta x_{cf} = 1/2(x_u - x_{ll}). \quad (19.73)$$

The control surface is just located at point $x = x_l$, so that one can write

$$\begin{aligned}\Phi_{cf} &= \Phi_l + \frac{\Phi_u - \Phi_l}{1/2(x_u - x_{ll})}(x_l - 1/2 x_l - 1/2 x_{ll}) \\ &= \Phi_l + (\Phi_u - \Phi_l)\frac{(x_l - x_{ll})}{(x_u - x_{ll})}.\end{aligned} \quad (19.74)$$

By the definition of the interpolation coefficient as

$$\eta_{cf} = \frac{x_l - x_{ll}}{x_u - x_{ll}} \quad (19.75)$$

an equation can be derived for the interpolation of the control surface values, employing the values of the neighboring grid points:

$$\Phi_{cf} = \eta_{cf} \Phi_u + (1 - \eta_{cf}) \Phi_l. \quad (19.76)$$

For all occurring control-volume sides, one thus obtains

$$cf = w, s, b: \quad \Phi_{cf} = \eta_{cf} \Phi_P + (1 - \eta_{cf}) \Phi_{Nb}, \quad (19.77)$$
$$cf = e, n, t: \quad \Phi_{cf} = \eta_{cf} \Phi_{Nb} + (1 - \eta_{cf}) \Phi_P. \quad (19.78)$$

On inserting these expressions and again the approximations for the diffusive flows (19.54) and (19.47) in (19.47), one obtains

$$\begin{aligned}\frac{\partial}{\partial t}(\rho_P^* \Phi_P)\Delta V &+ (m_e^*(\eta_e \Phi_E + (1-\eta_e)\Phi_P) + D_e^*(\Phi_P - \Phi_E)) - \\ &(m_w^*(\eta_w \Phi_P + (1-\eta_w)\Phi_W) + D_w^*(\Phi_W - \Phi_P)) + \\ &(m_n^*(\eta_n \Phi_N + (1-\eta_n)\Phi_P) + D_n^*(\Phi_P - \Phi_N)) - \\ &(m_s^*(\eta_s \Phi_P + (1-\eta_s)\Phi_S) + D_s^*(\Phi_S - \Phi_P)) + \\ &(m_t^*(\eta_t \Phi_T + (1-\eta_t)\Phi_P) + D_t^*(\Phi_P - \Phi_T)) - \\ &(m_b^*(\eta_b \Phi_P + (1-\eta_b)\Phi_B) + D_b^*(\Phi_B - \Phi_P)) = S_P \Delta V\end{aligned} \quad (19.79)$$

from which the finite difference form of the continuity equation can also be separated:

$$\rho_P^* \frac{\partial \Phi_P}{\partial t} \Delta V + (m_e^*(\eta_e \Phi_E - \eta_e \Phi_P) + D_e^*(\Phi_P - \Phi_E)) -$$
$$(m_w^*((\eta_w - 1)\Phi_P + (1 - \eta_w)\Phi_W) - D_w^*(\Phi_P - \Phi_W)) +$$
$$(m_n^*(\eta_n \Phi_N - \eta_n \Phi_P) + D_n^*(\Phi_P - \Phi_N)) -$$
$$(m_s^*((\eta_s - 1)\Phi_P + (1 - \eta_s)\Phi_S) - D_s^*(\Phi_P - \Phi_S)) +$$
$$(m_t^*(\eta_t \Phi_T - \eta_t \Phi_P) + D_t^*(\Phi_P - \Phi_T)) -$$
$$(m_b^*((\eta_b - 1)\Phi_P + (1 - \eta_b)\Phi_B) - D_b^*(\Phi_P - \Phi_B)) +$$
$$\underbrace{\left(\frac{\partial \rho_P^*}{\partial t}\Delta V + m_e^* - m_w^* + m_n^* - m_s^* + m_t^* - m_b^*\right)\Phi_P}_{= 0} = S_P \Delta V.$$
(19.80)

Appropriate rearrangement of the terms leads to the form:

$$\rho_P^* \frac{\partial \Phi_P}{\partial t} \Delta V +$$
$$(-m_e^* \eta_e + D_e^*)(\Phi_P - \Phi_E) - (m_w^*(\eta_w - 1) - D_w^*)(\Phi_P - \Phi_W) +$$
$$(-m_n^* \eta_n + D_n^*)(\Phi_P - \Phi_N) - (m_s^*(\eta_s - 1) - D_s^*)(\Phi_P - \Phi_S) +$$
$$(-m_t^* \eta_t + D_t^*)(\Phi_P - \Phi_T) - (m_b^*(\eta_b - 1) - D_b^*)(\Phi_P - \Phi_B) = S_P \Delta V$$
(19.81)

and by introducing the following coefficients:

$$Nb = W, S, B: \quad a_{Nb} = D_{cf}^* + C_{cf}^*(1 - f_{cf}), \tag{19.82}$$

$$Nb = E, N, T: \quad a_{Nb} = D_{cf}^* - C_{cf}^* f_{cf} \tag{19.83}$$

this equation can be simplified to read:

$$\rho_P^* \frac{\partial \Phi_P}{\partial t} \Delta V + a_E(\Phi_P - \Phi_E) + a_W(\Phi_P - \Phi_W) +$$
$$a_N(\Phi_P - \Phi_N) + a_S(\Phi_P - \Phi_S) + \tag{19.84}$$
$$a_T(\Phi_P - \Phi_T) + a_B(\Phi_P - \Phi_B) = S_P \Delta V$$

or, abbreviated:

$$\rho_P^* \frac{\partial \Phi_P}{\partial t} \Delta V + \hat{a}_P \Phi_P - \sum_{Nb} a_{Nb} \Phi_{Nb} = S_P \Delta V. \tag{19.85}$$

The last relationship is the discrete analogue of (19.32), when using the central difference method for discretization. The coefficients can now be computed according to (19.82) and (19.83).

At first glance, (19.85) is identical with (19.71), as for the description of the coefficients the same notation is used. The above derivations show, however, that different coefficients appear in the equations. The similarity of the two equations will prove in the next section to be advantageous, as it allows, with

19.4 Finite-Volume Discretization

the two equations to be handled at the same time, the discretization of the differentiation with respect to time. However, there are also approaches which permit the two methods to be mixed with a weighting factor (hybrid method or *deferred correction schemes*). In such cases, a distinction has to be made concerning the notation between the coefficients of the different discretization schemes.

19.4.3 Discretization with Respect to Time

In order to simplify the subsequent considerations of discretizing the derivatives with respect to time, we restrict our considerations to incompressible fluids in the parts to follow. The deduction of the analogue equations for compressible fluids can be realized, however, according to the same procedure.

For an incompressible fluid ($\rho = $ constant), for (19.71) and (19.85) it holds that:

$$\rho \frac{\partial \Phi_P}{\partial t} \Delta V + \hat{a}_P \Phi_P - \sum_{Nb} a_{Nb} \Phi_{Nb} = S_P \Delta V. \tag{19.86}$$

When this equation is integrated over a time interval, the following results:

$$\rho \Delta V \int_{\Delta t} \frac{\partial \Phi_P}{\partial t} \mathrm{d}t + \int_{\Delta t} \hat{a}_P \Phi_P \mathrm{d}t - \int_{\Delta t} \sum_{Nb} a_{Nb} \Phi_{Nb} \mathrm{d}t = \Delta V \int_{\Delta t} S_P \mathrm{d}t. \tag{19.87}$$

The first integral of this equation can be computed to give:

$$\int_{t^{\alpha-1}}^{t^{\alpha}} \frac{\partial \Phi_P}{\partial t} \mathrm{d}t = \Phi_P^{\alpha} - \Phi_P^{\alpha-1}. \tag{19.88}$$

The remaining integrals are approximated by means of the mean-value theorem of integration:

$$\int_{t^{\alpha-1}}^{t^{\alpha}} \hat{a}_P \Phi_P \mathrm{d}t = \overline{\hat{a}_P \Phi_P} \Delta t \approx \hat{a}_P^{\tau} \Phi_P^{\tau} \Delta t, \tag{19.89}$$

$$\int_{t^{\alpha-1}}^{t^{\alpha}} S_P \mathrm{d}t = \bar{S}_P \Delta t \approx S_P^{\tau} \Delta t, \tag{19.90}$$

where Φ_P^{τ} defines the value Φ_P at a point in the interval $[t^{\alpha-1}, t^{\alpha}]$. With these approximations, (19.87) can be written as:

$$\frac{\rho \Delta V}{\Delta t} (\Phi_P^{\alpha} - \Phi_P^{\alpha-1}) + \hat{a}_P^{\tau} \Phi_P^{\tau} - \sum_{Nb} a_{Nb}^{\tau} \Phi_{Nb}^{\tau} = S_P^{\tau} \Delta V. \tag{19.91}$$

In general, for numerical computations, so-called two-time-level methods are employed, where the value Φ_P^{α} of the new time level is computed from the values Φ_{Nb} and Φ_P of the new and/or the old time level. More complex methods, which use three or even more time levels, offer higher precision. This is correct, but they require greater numerical effort. The requirement for storage of data increases and methods of lower order have to be employed to be able to begin computing at the first time intervals to avoid divergence of the solution.

The different methods for discretizing variables with respect to time, differ only in the choice of τ. Following the type of equations which result from different values of τ, the corresponding methods are called explicit or implicit methods.

In the explicit case, $t^\tau = t^{\alpha-1}$ is chosen and thus the sought value Φ_P^α is computed only from the values Φ_{Nb} and Φ_P of the old time level. Equation (19.91) therefore reads

$$\frac{\rho_0 \Delta V}{\Delta t}(\Phi_P^\alpha - \Phi_P^{\alpha-1}) + \hat{a}_P^{\alpha-1}\Phi_P^{\alpha-1} - \sum_{Nb} a_{Nb}^{\alpha-1}\Phi_{Nb}^{\alpha-1} = S_P^{\alpha-1}\Delta V \qquad (19.92)$$

or, rearranged with $n = \alpha$ and $0 = \alpha - 1$:

$$\Phi_P^n = \Phi_P^o - \frac{\Delta t}{\rho_0 \Delta V}(\hat{a}_P^o \Phi_P^o - \sum_{Nb} a_{Nb}^o \Phi_{Nb}^o - S_P^o \Delta V). \qquad (19.93)$$

This is an explicit equation for Φ_P^α as, except for the sought value Φ_P^n, all other values are known from the preceding time interval. Generally, explicit methods have the disadvantage that the size of the time interval is limited. This can be understood and explained by considerations of the numerical stability of the method. Another disadvantage is, that explicit methods do not describe the time behavior of the diffusive transport processes in the same way that the initial differential equation does. When an explicit method is used in numerical computations, the information on a modification of the boundary conditions per time interval is only carried by one grid point. This is different from the actual physical behavior, as such information, due to diffusion, is immediately transferred to the entire computational area.

In this respect, implicit methods are often better suited to reflect the actual physical process, which also explains their higher numerical stability. Implicit methods use, among other things, $t^\tau = t^\alpha$, with which, as a consequence results from (19.91), the simplest implicit method of first order results:

$$\frac{\rho \Delta V}{\Delta t}(\Phi_P^\alpha - \Phi_P^{\alpha-1}) + \hat{a}_P^\alpha \Phi_P^\alpha - \sum_{Nb} a_{Nb}^\alpha \Phi_{Nb}^\alpha = S_P^\alpha \Delta V. \qquad (19.94)$$

As also values from the new time interval are used, influences caused by modifications to the boundary conditions can spread within one time interval over the entire computational area. The above relationship represents an implicit equation for Φ_P^α as unknown values of the neighboring grid points appear in the equation also. Here, it has to be taken into account that in all considerations up to now, one grid point has been considered to represent all

19.4 Finite-Volume Discretization

others. The inclusion of all grid points results in as many equations as there are unknowns. Subsequently, the resulting system of equations is closed and can be solved.

When the coefficients of Φ_P^α are appropriately factored out and combined into a new coefficient $\check{a}_P^n$ with $\check{a}_P^\alpha = \hat{a}_P^\alpha + \frac{\rho_\alpha \Delta V}{\Delta t}$, the following relationship results:

$$\check{a}_P^\alpha \Phi_P^\alpha - \sum_{Nb} a_{Nb}^\alpha \Phi_{Nb}^\alpha = S_P^\alpha \Delta V + \frac{\rho \Delta V}{\Delta t} \Phi_P^{\alpha-1}, \qquad (19.95)$$

which represents the finite form of (19.32) after discretization of the derivatives with respect to time.

19.4.4 Treatments of the Source Terms

In the above section, it was mentioned that (19.95) represents an implicit equation to compute Φ_P^α and that, for solving it, an entire system of equations has to be solved with the help of a corresponding algorithm. Here, mostly iterative solution algorithms are used, which have to have a large coefficient a_P of the central point as a convergence condition (diagonal dominance of the resulting coefficient matrix). For each point of the solution area,

$$a_P \geq \sum_{Nb} a_{Nb} \qquad (19.96)$$

should therefore be fulfilled. Without the source term, this is automatically fulfilled by the previous discretization, as $\hat{a}_P = \sum_{Nb} a_{Nb}$ and $\check{a}_P = \hat{a}_P + \frac{\rho_0 \Delta V}{\Delta t}$. With the discretization of the source term, steps that lead to a reduction of a_P should therefore be avoided.

It is possible that the source term S is not a linear function of Φ, however. Nevertheless, this term can be linearized by splitting it into an independent and a dependent part:

$$S_P^\alpha = S_P^{\alpha\prime} \Phi_P^\alpha + S_c^\alpha, \qquad (19.97)$$

where S_c^α designates the part of S_P^α, which does not explicitly depend on Φ_P^α. $S_P^{\alpha\prime}$ can be replaced by $S_P^{\alpha\prime*}$, which is computed with known values of Φ_P from previous iterations; thus, a linearization of the source term is obtained. By insertion of (19.97) into (19.95), the following results:

$$\check{a}_P^\alpha \Phi_P^\alpha - S_P^{\alpha\prime*} \Phi_P^\alpha \Delta V - \sum_{Nb} a_{Nb}^\alpha \Phi_{Nb}^\alpha = S_c^\alpha \Delta V + \frac{\rho \Delta V}{\Delta t} \Phi_P^{\alpha-1}. \qquad (19.98)$$

In order to not endanger the diagonal dominance of the coefficient matrix, $S_P^{\alpha\prime*}$ has to be negative. When this condition cannot be observed, it is better for the stability of the iterative solution method to compute the entire source term from known values and leave it in the right-hand side of the equation.

When subsequently $a_P^\alpha = \breve{a}_P^\alpha - S_P^{\alpha'*}\Delta V$ and $b = S_c^\alpha \Delta V + \frac{\rho \Delta V}{\Delta t}\Phi_P^{\alpha-1}$ is set, the completely discretized form of (19.32) results:

$$a_P^\alpha \Phi_P^\alpha - \sum_{Nb} a_{Nb}^\alpha \Phi_{Nb}^\alpha = b. \tag{19.99}$$

19.5 Computation of Laminar Flows

When one analyzes the general Navier–Stokes equations and the energy equation for an incompressible fluid, i.e. the general transport equation, it can be demonstrated that we have a set of partial differential equations which shows a parabolic time response and an elliptic space behavior. Because of this time response and space behavior, initial conditions at the point of time $t = 0$ have to be given, and the boundary conditions have to be specified along the entire area of the borders of the flow area. It is usual to include the following boundary conditions into the numerical computations:

- The *Dirichlet* boundary conditions: Descriptions of the values of all variables along the boundaries of the computational area
- The *Neumann* boundary conditions: Descriptions of the gradients (or of the diffusive fluxes) of the variables along the boundaries of the computational area
- A combination of the Dirichlet and Neumann boundary conditions
- Periodic boundary conditions

Thus, for practical computations, specific boundary conditions result for:

- Solid walls
- Symmetry planes
- Inflow planes
- Outflow planes

which can be considered separately; for this purpose, the $s - n$ coordinate system of Fig. 19.11 is used.

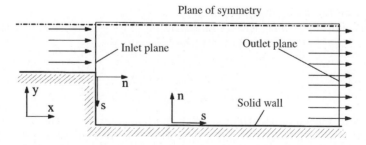

Fig. 19.11 Diagram explaining possible boundary conditions

19.5.1 Wall Boundary Conditions

On walls the no-slip boundary condition is employed. For impermeable surfaces, both velocities are set at zero, i.e.

$$U_s = U_n = 0. \tag{19.100}$$

For the temperature either the wall temperature or the wall heat flux values can be specified:

$$T = T_w \quad \text{or} \quad \frac{\partial T}{\partial n} = -\frac{Pr}{\mu c_p} Q_w, \tag{19.101}$$

where T_w designates the wall temperature and Q_w the wall heat flux per unit area (heat flux density).

19.5.2 Symmetry Planes

At symmetry planes, the normal gradients (or fluxes) of the tangential velocity and all scalar variables are zero. In addition, the velocity normal to the symmetry plane disappears:

$$\frac{\partial U_s}{\partial n} = U_n = \frac{\partial T}{\partial n} = 0. \tag{19.102}$$

19.5.3 Inflow Planes

At inflow planes the profiles of U_s, U_n and T are normally prescribed by tabulated data or by analytical functions.

19.5.4 Outflow Planes

If the outflow planes are of the type where the flow shows parabolic behavior and the plane is sufficiently far away from the flow region of interest, a fully developed flow can be assumed, i.e. the gradients in the flow direction can be neglected:

$$\frac{\partial U_s}{\partial n} = \frac{\partial U_n}{\partial n} = \frac{\partial T}{\partial n} = 0. \tag{19.103}$$

The specification of profiles for U_s, U_n and T is also possible.

With boundary conditions of this kind, laminar flows as represented in Figs. 19.12–19.14 can now be computed.

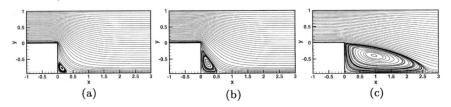

Fig. 19.12 Computations of flows with different Reynolds numbers in a two-dimensional flow channel with a backward facing step; $Re = 10^{-4}$ (**a**), $Re = 10$ (**b**), $Re = 100$ (**c**)

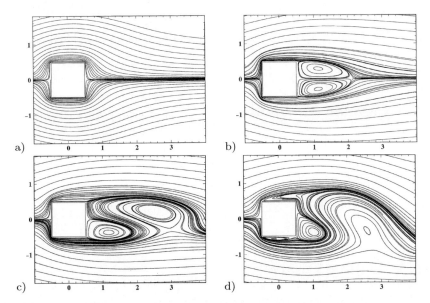

Fig. 19.13 Computations of flows for the flow around a two-dimensional cylinder with square cross-sectional area; $Re = 1$ (**a**), $Re = 30$ (**b**), $Re = 60$ (**c**), $Re = 200$ (**d**)

19.6 Computations of Turbulent Flows

19.6.1 Flow Equations to be Solved

Computations of turbulent flows, in the presence of high Reynolds numbers, require the solution of the Reynolds transport equations that were derived in Chap. 17 and for which, for *two-dimensional* flows, areas can be stated as follows:

Mass Conservation (Continuity Equation):

$$\frac{\partial \rho}{\partial t} + \frac{\partial (\rho U)}{\partial x} + \frac{\partial (\rho V)}{\partial y} = 0. \tag{19.104}$$

19.6 Computations of Turbulent Flows

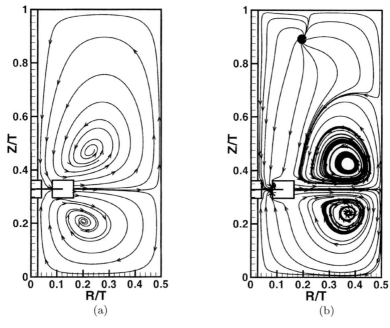

Fig. 19.14 Results of laminar flow computations in a stirred vessel with inserted Rushton turbine; (a) $Re = 1$, (b) $Re = 100$; Breuer (2002)

Momentum Conservation in the x-Direction:

$$\frac{\partial(\rho U)}{\partial t} + \frac{\partial}{\partial x}(\rho U^2 + \rho\overline{u^2}) + \frac{\partial}{\partial y}(\rho UV + \rho\overline{uv}) = -\frac{\partial P}{\partial x}. \tag{19.105}$$

Momentum Conservation in the y-Direction:

$$\frac{\partial(\rho V)}{\partial t} + \frac{\partial}{\partial x}(\rho UV + \rho\overline{uv}) + \frac{\partial}{\partial y}(\rho V^2 + \rho\overline{v^2}) = -\frac{\partial P}{\partial y}. \tag{19.106}$$

For turbulent flows, all variables designated by capital letters and the thermodynamic fluid properties have to be interpreted as time-averaged quantities. Fluctuating flow quantities are designated by lower-case letters. Mean values of turbulent velocity fluctuation, resulting in correlations of the fluctuations, are indicated by overbars. The molecule-dependent momentum transports in the momentum equations are neglected in the following, which is justified by the assumption that high Reynolds numbers are involved. This assumption is a customary approximation, when computing elliptical turbulent flows.

The correlations $-\rho\overline{u^2}$, $-\rho\overline{uv}$ and $-\rho\overline{v^2}$ represent the momentum transport by the turbulent fluctuating velocity components. They act like "stresses" on the considered fluid elements and are therefore defined as *Reynolds stresses*. These stresses are additional unknowns in the equation system (19.104)–(19.106) and have to be related via a *turbulence model* to "known quantities."

To make a complete solution of the flow equations possible, the set of equations needs to be closed.

The $k-\epsilon$ turbulence model (Launder and Spalding, 1972) makes use of the eddy-viscosity hypothesis, which relates the Reynolds stresses to the mean flow deformation rates in the following way:

$$-\rho\overline{u^2} = 2\mu_t \frac{\partial U}{\partial x} - \frac{2}{3}\rho k, \tag{19.107}$$

$$-\rho\overline{v^2} = 2\mu_t \frac{\partial V}{\partial y} - \frac{2}{3}\rho k, \tag{19.108}$$

$$-\rho\overline{uv} = \mu_t \left(\frac{\partial U}{\partial y} + \frac{\partial V}{\partial x} \right). \tag{19.109}$$

The proportionality factor μ_t is the eddy viscosity. The quantity k, appearing in (19.107) and (19.108), is the turbulent kinetic energy, which is equal to half the sum of the normal Reynolds stresses (divided by the density):

$$k = \frac{1}{2}\left(\overline{u^2} + \overline{v^2} + \overline{w^2}\right) \tag{19.110}$$

and w is the fluctuation of the velocity in the z direction. The eddy viscosity μ_t is not a fluid property, but depends on the local turbulent flow conditions.

When one writes the above x–y momentum equations in the form which corresponds to the general transport equation, one obtains:

$$\frac{\partial(\rho U)}{\partial t} + \frac{\partial}{\partial x}\left(\rho U^2 - \mu_t \frac{\partial U}{\partial x}\right) + \frac{\partial}{\partial y}\left(\rho UV - \mu_t \frac{\partial U}{\partial y}\right)$$
$$= -\frac{\partial P^*}{\partial x} + S_U,$$

$$\frac{\partial(\rho V)}{\partial t} + \frac{\partial}{\partial x}\left(\rho UV - \mu_t \frac{\partial V}{\partial x}\right) + \frac{\partial}{\partial y}\left(\rho V^2 - \mu_t \frac{\partial V}{\partial y}\right)$$
$$= -\frac{\partial P^*}{\partial y} + S_V \tag{19.111}$$

with the source terms:

$$S_U = \frac{\partial}{\partial x}\left(\mu_t \frac{\partial U}{\partial x}\right) + \frac{\partial}{\partial y}\left(\mu_t \frac{\partial V}{\partial x}\right), \tag{19.112}$$

$$S_V = \frac{\partial}{\partial x}\left(\mu_t \frac{\partial U}{\partial y}\right) + \frac{\partial}{\partial y}\left(\mu_t \frac{\partial V}{\partial y}\right). \tag{19.113}$$

The modified pressure term P^* is equal to:

$$P^* = P + \frac{2}{3}\rho k. \tag{19.114}$$

In most cases relevant in practice we have $P \gg \frac{2}{3}\rho k$, so that $P = P^*$ can be set without introducing a serious error.

19.6 Computations of Turbulent Flows

The conservation of the time-averaged or ensemble-averaged energy results in a transport equation for the temperature described by

$$\frac{\partial(\rho c_p T)}{\partial t} + \frac{\partial}{\partial x}\left(\rho c_p UT + \rho c_p \overline{ut}\right) + \frac{\partial}{\partial y}\left(\rho c_p VT + \rho c_p \overline{vt}\right) = S_T, \quad (19.115)$$

where t is the fluctuating value of T, and $-\rho c_p \overline{ut}$ and $-\rho c_p \overline{vt}$ represent turbulent energy-flux values. The terms for the molecular transport were already neglected in (19.115). For Prandtl numbers around 1, this agrees with the assumption that was made when deriving the conservation equation for the momentum of turbulent flows.

For turbulent flows, the turbulent energy transport terms are related to the mean temperature gradients by an eddy diffusivity concept:

$$-\rho c_p \overline{ut} = \frac{\mu_t c_p}{Pr_t}\frac{\partial T}{\partial x}, \quad (19.116)$$

$$-\rho c_p \overline{vt} = \frac{\mu_t c_p}{Pr_t}\frac{\partial T}{\partial y}. \quad (19.117)$$

The eddy viscosity μ_t is employed in the expression from the $k-\epsilon$ turbulence model. The turbulent Prandtl number Pr_t is an empirical constant which is specified in the next section. The introduction of (19.116) into (19.115) yields the temperature equation to be solved for turbulent flow predictions:

$$\frac{\partial(\rho c_p T)}{\partial t} + \frac{\partial}{\partial x}\left(\rho c_p UT - \frac{\mu_t c_p}{Pr_t}\frac{\partial T}{\partial x}\right) + \frac{\partial}{\partial y}\left(\rho c_p VT - \frac{\mu_t c_p}{Pr_t}\frac{\partial T}{\partial y}\right) = S_T. \quad (19.118)$$

The definition of μ_t, in many practical cases, takes place through the $k-\epsilon$ turbulence model, where dimensional considerations lead to the following relationship concerning the eddy viscosity:

$$\mu_t = \rho c_\mu k^2/\epsilon,$$
$$k = \text{turbulent kinetic energy}, \quad (19.119)$$
$$\epsilon = \text{turbulent rate of dissipation},$$

where c_μ is an empirical constant which is given below. Local values of k and ϵ are obtained by solving the semi-empirical transport equations for k and ϵ, which read as follows:

$$\frac{\partial(\rho k)}{\partial t} + \frac{\partial}{\partial x}\left(\rho U k - \frac{\mu_t}{\sigma_k}\frac{\partial k}{\partial x}\right) + \frac{\partial}{\partial y}\left(\rho V k - \frac{\mu_t}{\sigma_k}\frac{\partial k}{\partial y}\right)$$
$$= P_k - \rho\epsilon, \quad (19.120)$$

$$\frac{\partial(\rho\epsilon)}{\partial t} + \frac{\partial}{\partial x}\left(\rho U\epsilon - \frac{\mu_t}{\sigma_\epsilon}\frac{\partial \epsilon}{\partial x}\right) + \frac{\partial}{\partial y}\left(\rho V\epsilon - \frac{\mu_t}{\sigma_\epsilon}\frac{\partial \epsilon}{\partial y}\right)$$
$$= \frac{\epsilon}{k}(c_{\epsilon 1}P_k - c_{\epsilon 2}\rho\epsilon). \quad (19.121)$$

The terms on the left-hand sides of these equations represent the changes of k and ϵ with time and their "transport" through the time-averaged and fluctuating turbulent motions in the flow. The right-hand sides contain the production and "destruction" rates. In the k equation, the "k destruction rate" is set equal to the dissipation rate ϵ multiplied by the density. The production rate P_k is defined as

$$P_k = \mu_t \left\{ 2\left(\frac{\partial U}{\partial x}\right)^2 + 2\left(\frac{\partial V}{\partial y}\right)^2 + \left[\left(\frac{\partial U}{\partial y}\right) + \left(\frac{\partial V}{\partial x}\right)\right]^2 \right\}. \quad (19.122)$$

For the empirical constants $c_{\epsilon 1}$, $c_{\epsilon 2}$, σ_k, σ_ϵ and c_μ, usually the standard values suggested by Launder and Spalding (1974) are assumed. For flows limited by walls and for free flows, values of 0.6 and 0.86 are recommended for the turbulent Prandtl number. The $k - \epsilon$ model constants are summarized in Table 19.2.

The transport equations presented above for turbulent flows can be written, in a general form, as follows:

$$\frac{\partial(\rho\Phi)}{\partial t} + \frac{\partial\partial}{\partial\partial x}\left(\rho U \Phi - \Gamma_\Phi \frac{\partial \Phi}{\partial x}\right) + \frac{\partial}{\partial y}\left(\rho V \Phi - \Gamma_\Phi \frac{\partial \Phi}{\partial y}\right) = S_\Phi, \quad (19.123)$$

where Φ represents U, V, T, k or ϵ. The diffusion coefficients Γ_Φ and the source terms S_Φ for turbulent flows are compiled in Table 19.3. For laminar flows, the eddy viscosity and the turbulent Prandtl number are replaced by corresponding molecular values and the k and ϵ transport equations do not have to be solved.

19.6.2 Boundary Conditions for Turbulent Flows

19.6.2.1 Wall Boundary Conditions

For specifying the wall boundary conditions, usually the wall function method (Launder and Spalding, 1972) is used, which bridges the viscous wall-near

Table 19.2 Empirical constants of the $k - \epsilon$ turbulence models

c_μ	$c_{\epsilon 1}$	$c_{\epsilon 2}$	σ_k	σ_ϵ	Pr_t
0.09	1.44	1.92	1.0	1.3	0.6–0.86

Table 19.3 Diffusivity and source terms for general transport equation (19.123)

Φ	Γ_Φ	S_Φ
U	μ_t	$-\partial P/\partial x + \partial/\partial x(\mu_t \partial U/\partial x) + \partial/\partial y(\mu_t \partial V/\partial x)$
V	μ_t	$-\partial P/\partial y + \partial/\partial x(\mu_t \partial U/\partial y) + \partial/\partial y(\mu_t \partial V/\partial y)$
T	$\mu_t c_p / Pr_t$	S_T
k	μ_t/σ_k	$P_k - \rho\epsilon$
ϵ	μ_t/σ_ϵ	$\epsilon/k(c_{\epsilon 1} P_k - \rho c_{\epsilon 2} \epsilon)$

19.6 Computations of Turbulent Flows

regions with empirical assumptions. Using wall functions in the near-wall region, instead of resolving the viscous sublayers, offers two advantages:

- The computational time and the required computer storage of data are both reduced because the high gradients of all the dependent variables near the wall need not be resolved.
- Some of the assumptions made in the derivation of the $k - \epsilon$ models lose their validity in the viscosity-dominant near-wall zone, hence this fact does not need to be considered.

For the wall functional method, the first grid node away from the wall, for the numerical computations, has to be localized in the fully turbulent area, according to the diagram in Fig. 19.15. Typical dimensionless wall distances which characterize this region are stated below.

In the case of the wall-parallel velocity, the wall functional method suggests an equation which relates the wall shear stress τ_w (i.e. the momentum flow through the control volume close to the wall) to the velocity at the first grid node $U_{s,c}$ away from the wall and to the distance n_c of this point to the wall. The basis for this equation is the logarithmic velocity law:

$$\frac{U_{s,c}}{\sqrt{\tau_w/\rho}} = \frac{1}{\kappa} \ln\left(\frac{n_c\sqrt{\rho\tau_w}}{\mu}\right) + C = \frac{1}{\kappa} \ln\left(E\frac{n_c\sqrt{\rho\tau_w}}{\mu}\right). \quad (19.124)$$

Values for the *von Karman constant* κ and the roughness parameter C are given in Table 19.4. It should be noted that the value given for C in Table 19.4, holds for hydrodynamically smooth surfaces only. For rough walls, other values of C are needed.

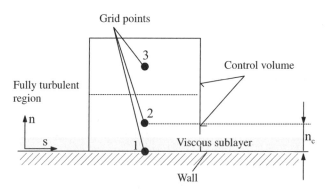

Fig. 19.15 Control volume near the wall

Table 19.4 Constants in the logarithmic wall law

κ	C	$E = e^{\kappa C}$
0.41	5.2	8.43171

Equation (19.124) is transcendental in τ_w and becomes singular at separation points (where $\tau_w \to 0$). Because of these problems, a modified form of (19.124) is often used. The extension is built on the following three assumptions for the flow in the near-wall control volumes:

- Directly near the wall, a Couette flow exists, with $\delta/\delta s = 0$, and $U_n = 0$
- The production and dissipation rates are in equilibrium, i.e. local equilibrium of the turbulence
- There is a layer of constant stress, with $-\rho \overline{u_s u_n} = \tau_w$

The logarithmic law for the mean velocity and the above three assumptions does not hold in the proximity of or within separation areas.

When assuming a Couette flow, the eddy-viscosity relationship (19.109) becomes

$$-\rho \overline{u_s u_n} = \mu_t \frac{\partial U_s}{\partial n} = \rho c_\mu \frac{k^2}{\epsilon} \frac{\partial U_s}{\partial n} \qquad (19.125)$$

and the local equilibrium condition (production rate = dissipation rate) is expressed as follows:

$$-\rho \overline{u_s u_n} \frac{\partial U_s}{\partial n} = \rho \epsilon. \qquad (19.126)$$

On inserting (19.126) into (19.125), the following results:

$$k = \frac{-\overline{u_s u_n}}{\sqrt{c_\mu}} \qquad (19.127)$$

and the assumption of a layer of constant stress leads to

$$\tau_w/\rho = \sqrt{c_\mu} k. \qquad (19.128)$$

From (19.128) and the logarithmic velocity law (19.124), one obtains, by simple algebraic derivations, an explicit relationship for the wall shear stress:

$$\tau_w = \frac{\rho \kappa c_\mu^{1/4} k_c^{1/2}}{\ln(E n_c^*)} U_{s,c} \qquad (19.129)$$

with the dimensionless wall distance

$$n_c^* = \frac{\rho c_\mu^{1/4} k_c^{1/2} n_c}{\mu}. \qquad (19.130)$$

Equations (19.129) and (19.130) are used in most computer programs. These equations hold in the range

$$30 < n_c^* < 500. \qquad (19.131)$$

One should therefore place the grid nodes of the wall-nearest control volumes carefully in this range.

19.6.2.2 Standard Velocity

As in the case of laminar flows, the standard velocity at the wall, and also its gradient, are equal to zero. The wall function takes care of the difference from the correct velocity gradient.

19.6.2.3 Temperature

The near-wall distribution for the temperature is based on similarity arguments for the inner wall layer, from which (for low Mach numbers) a linear relationship results between the temperature and the velocity. The often used temperature law reads

$$\frac{(T_c - T_w)\rho c_p c_\mu^{1/4} k_c^{1/2}}{Q_w} = \frac{Pr_t}{\kappa} \ln n_c^* + C_Q(Pr). \qquad (19.132)$$

The additive constant C_Q is a function of the molecular Prandtl number Pr. To determine it, an empirical relationship is employed:

$$C_Q = 12.5 Pr^{2/3} + 2.12 \ln Pr - 5.3 \quad \text{for } Pr > 0.5, \qquad (19.133)$$
$$C_Q = 12.5 Pr^{2/3} + 2.12 \ln Pr - 1.5 \quad \text{for } Pr \leq 0.5. \qquad (19.134)$$

Equation (19.132) can easily be resolved according to the wall heat flux Q_w, which is the quantity of interest for the implementation of the wall temperature into the finite-volume methods computational procedure.

19.6.2.4 Turbulent Kinetic Energy

Immediately near the wall, the turbulent kinetic energy k changes quadratically with the wall distance, i.e. $k \propto n_c^2$. At the same time, the diffusive wall flow of k has the value zero:

$$\frac{\mu_t}{\sigma_k} \frac{\partial k}{\partial n} = 0. \qquad (19.135)$$

In addition to the application of (19.135), the production and dissipation rates in the wall-nearest control volumes are determined approximately, logically assuming a Couette flow, a local equilibrium and a constant stress layer, as discussed previously. This is described in the following. With (19.122), the production rate of the kinetic energy in the wall-nearest control volumes is computed with

$$P_k = \frac{\tau_w^2}{\rho \kappa c_\mu^{1/4} k_c^{1/2} n_c}, \qquad (19.136)$$

where the velocity gradient is derived from the logarithmic law (19.129) and the local shear tension is set equal to the wall shear stress.

The dissipation rate, which appears in the k equation (19.121), is approximated by assuming a linear dependence of the near-wall length scale:

$$L = \frac{k^{3/2}}{\epsilon} = \frac{\kappa}{c_\mu^{3/4}} n. \tag{19.137}$$

Equation (19.137) holds under local equilibrium conditions and for a logarithmic velocity law. From this follows an equation for ϵ in the near-wall control volume in the form

$$\epsilon_c = c_\mu^{3/4} \frac{k_c^{3/2}}{\kappa n_c}. \tag{19.138}$$

19.6.2.5 Dissipation Rate

According to the treatment of boundary conditions in practice, for the dissipation rate ϵ, its value is determined at the first computational node away from the wall by employing (19.138).

19.6.2.6 Symmetry Planes

Along the symmetry planes (or symmetry lines) of flow, the standard gradients (diffusive flows) of the dependent variables and the normal velocity U_n are set equal to zero:

$$\frac{\partial U_s}{\partial n} = U_n = \frac{\partial T}{\partial n} = \frac{\partial k}{\partial n} = \frac{\partial \epsilon}{\partial n} = 0. \tag{19.139}$$

19.6.2.7 Inflow Planes

Usually inlet flow profiles are derived from experimental data or from other empirical information and are employed as inflow conditions. The turbulence quantities often refer to the inflow velocity U_{in} by specifying a relative turbulence intensity Tu, defined as

$$Tu = \frac{\sqrt{\overline{u^2}}}{U_{in}}. \tag{19.140}$$

Typical values for Tu lie between 1 and 20%. For an isotropic turbulent flow, Tu is related to k through

$$k = \frac{3}{2}(Tu\, U_{in})^2. \tag{19.141}$$

19.6 Computations of Turbulent Flows

At the inflow plane dissipation rates can be specified, assuming that the quantity of the larger turbulence vortices is proportional to a typical scale of length H of the cross flow area, i.e.

$$\epsilon = \frac{k^{3/2}}{aH}. \qquad (19.142)$$

Characteristic values for the proportional factor a are of the order of magnitude of 0.01–1.

19.6.2.8 Outflow Planes

The same approach as described in Sect. 19.5.4, is usually employed for outflow planes. In addition to the equations stated there, the gradients of k and ϵ assigned to the flow direction are set to zero.

The above equations can now be solved numerically for two-dimensional flows, including the indicated boundary conditions. For the illustration of typical computational results, different flow problems are shown in Figs. 19.16–19.18.

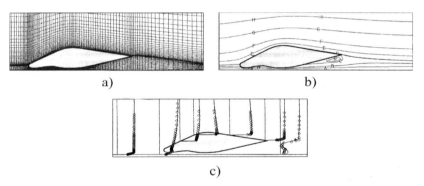

Fig. 19.16 Flows around a near-ground model vehicle – computational result – LDA measurements

Fig. 19.17 Subcritical flow around a circular cylinder at $Re = 3{,}900$; Breuer (2002)

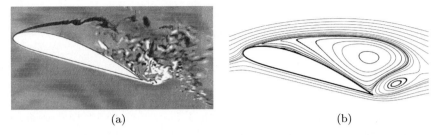

Fig. 19.18 Separated flow around a wing at $Re = 20{,}000$; (**a**) non-stationary rotational-power distribution; (**b**) time-averaged streamlines; Breuer (2002)

References

19.1. Schäfer, M.: Numerik im Maschinenbau, Springer, Berlin Heidelberg New York, 1999
19.2. Cebeci, T. and Bradshaw, P.: Physical and Computational Aspects of Convective Heat Transfer, Mei Ya Publications, Springer, Berlin Heidelberg New York, Tokyo, 1984
19.3. Ferziger, J.L. and Peric, M.: Computational Methods for Fluid Dynamics, 2nd revised edition, Springer, Berlin Heidelberg New York, 1999
19.4. Peyret, R. and Taylor, T.D.: Computational Methods for Fluid Flow, Springer Series in Computational Physics, Springer, Berlin Heidelberg New York, 1983
19.5. Wilcox, D.C.: Turbulence Modeling, DCW Industries, Griffin Printing, Glendale, CA, 1993
19.6. Launder, B.E. and Spalding, D.B.: Mathematical Models of Turbulence, Acadmic Press, London (1972)

Chapter 20
Fluid Flows with Heat Transfer

20.1 General Considerations

The derivations carried out in Chap. 5 to yield the basic equations of fluid mechanics also include considerations of the energy transport and, based on these considerations, different forms of the energy equation were derived. There it could be shown that the local mechanical energy equation, stated as a differential equation for a fluid, does not represent an independent equation, as it can be derived from the generally formulated momentum equation. This was the reason to subtract this equation from the total energy equation, in order to obtain the thermal energy equation that can be written as follows:

$$\rho\left(\frac{\partial e}{\partial t} + U_i \frac{\partial e}{\partial x_i}\right) = -\frac{\partial \dot{q}_i}{\partial x_i} - P\frac{\partial U_i}{\partial x_i} - \tau_{ij}\frac{\partial U_j}{\partial x_i}, \qquad (20.1)$$

where ρ is the density of the fluid, e its inner energy, t the time, U_i the fluid velocity, $\dot{q}_i$ the heat flux, P the pressure and τ_{ij} the molecular-dependent momentum transport. This equation can now be employed for a thermodynamically ideal fluid, i.e. for $\rho = $ constant and thus for $(\partial U_i / \partial x_i) = 0$, and also for

$$\dot{q}_i = -\lambda \frac{\partial T}{\partial x_i} \quad \text{and} \quad \tau_{ij} = -\mu\left(\frac{\partial U_j}{\partial x_i} + \frac{\partial U_i}{\partial x_j}\right) \qquad (20.2)$$

in the following form for heat transfer computations, where $c_v = c_p = c = $ was taken into consideration because of $\rho = $ constant:

$$\rho c \left(\frac{\partial T}{\partial t} + U_i \frac{\partial T}{\partial x_i}\right) = \lambda \frac{\partial^2 T}{\partial x_i^2} + \mu \left(\frac{\partial U_j}{\partial x_i}\right)^2. \qquad (20.3)$$

Together with the continuity equation:

$$\frac{\partial U_i}{\partial x_i} = 0 \qquad (20.4)$$

and the momentum equations for $j = 1, 2, 3$:

$$\rho \left(\frac{\partial U_j}{\partial t} + U_i \frac{\partial U_j}{\partial x_i} \right) = -\frac{\partial P}{\partial x_j} + \mu \frac{\partial^2 U_j}{\partial x_i^2} + \rho g_j \qquad (20.5)$$

one obtains a system of five differential equations for the five unknowns U_1, U_2, U_3, P and T, which can be solved with suitable boundary conditions. Thus, flow problems can be solved that occur coupled with heat-transfer problems, but do not lead to considerable modifications of the fluid properties such as ρ, μ and λ. This is often the case in technical applications of fluid mechanics, where the actual fluid flow processes are sometimes of secondary importance, the main importance being given to the heat transfer and heat dissipation of a system. This significance of heat transfer is the actual reason for including the present chapter in a book on fluid mechanics. It serves as an introduction for students of fluid mechanics into an important field of applications of fluid-mechanical know-how.

In order to present, in an easily understandable way, some general properties of heat transfer, as compared with momentum transfer, equations (20.3) and (20.5) are rearranged:

$$\frac{\partial T}{\partial t} + U_i \frac{\partial T}{\partial x_i} = \frac{\lambda}{\rho c} \frac{\partial^2 T}{\partial x_i^2} + \nu \left(\frac{\partial U_j}{\partial x_i} \right)^2 \quad \text{with} \quad \frac{\nu}{Pr} = \frac{\lambda}{\rho c} \qquad (20.6)$$

and

$$\frac{\partial U_j}{\partial t} + U_i \frac{\partial U_j}{\partial x_i} = \nu \frac{\partial^2 U_j}{\partial x_i^2} + \rho g_j - \frac{\partial P}{\partial x_j}, \qquad (20.7)$$

where ν is the viscous diffusion coefficient and $a = \lambda/\rho c$ is the thermal diffusion coefficient. From their interrelationship, the Prandtl number results:

$$Pr = \frac{\nu}{a} = \frac{\mu}{\rho} \frac{\rho c}{\lambda} = \frac{\mu c}{\lambda}$$

which expresses how the molecular-dependent momentum transport relates to the molecular-dependent heat transport. The Prandtl number can thus be employed to demonstrate in which way momentum transport and heat transport relate relatively to one another.

For small Pr, e.g. for metallic melts for equal development lengths, the thermal boundary layer of a plate flow has developed more intensely than the "velocity boundary layer" (Fig. 20.1).

The facts illustrated in Fig. 20.1 can also be expressed in development lengths for momentum and temperature boundary layers, such that for $\delta_u = \delta_T$ the following relationships hold:

$$\frac{L_T}{\sqrt{a}} = \frac{L_U}{\sqrt{\nu}} \quad \leadsto \quad \frac{L_U}{L_T} = \sqrt{\frac{\nu}{a}} = \sqrt{Pr}. \qquad (20.8)$$

For equal development length, the following results for the boundary layer thicknesses can be derived:

20.1 General Considerations

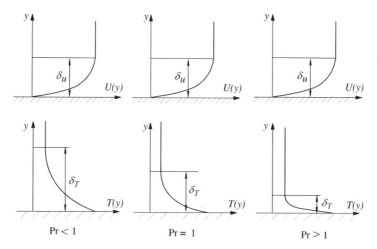

Fig. 20.1 Thicknesses of temperature-boundary layers for different Prandtl numbers

$$\frac{\delta_U}{\delta_T} = \sqrt{Pr}. \tag{20.9}$$

This means that the molecular-dependent momentum transport is larger than the molecular-dependent heat transport when $Pr > 1$ (see Fig. 20.1).

When extending the similarity considerations in Chap. 7, where from the dimensionless momentum and τ_{ij} transport equations:

$$\rho^* \left(\frac{\ell_c}{U_c t_c} \frac{\partial U_j^*}{\partial t^*} + U_i^* \frac{\partial U_j^*}{\partial x_j^*} \right) = -\frac{\Delta P}{\rho_c U_c^2} \frac{\partial P^*}{\partial x_j^*} - \frac{\tau_w}{\rho_c U_c^2} \frac{\partial \tau_{ij}^*}{\partial x_i^*}, \tag{20.10}$$

$$\tau_{ij}^* = \frac{\mu_c U_c}{\tau_w \ell_c} \left[-\mu^* \left(\frac{\partial U_j^*}{\partial x_i^*} + \frac{\partial U_i^*}{\partial x_j^*} \right) + \frac{2}{3} \mu^* \delta_{ij} \frac{\partial U_k^*}{\partial x_k^*} \right] \tag{20.11}$$

the following characteristic quantities of a flow can be derived:

$$u_c = u_\tau = \sqrt{\frac{\tau_w}{\rho}}, \quad \ell_c = \frac{\nu}{u_\tau} \quad \text{and} \quad t_c = \frac{\nu}{u_\tau^2}. \tag{20.12}$$

When carrying out corresponding considerations for the energy equation, one obtains the following derivations:

- Energy equation:

$$\rho c_p \left(\frac{\partial T}{\partial t} + U_i \frac{T}{x_i} \right) = -\frac{\partial q_i}{\partial x_i} + \left(\frac{\partial P}{\partial t} + U_i \frac{\partial P}{\partial x_i} \right) - \tau_{ij} \frac{\partial U_j}{\partial x_i}. \tag{20.13}$$

- Fourier law of heat transfer

$$q_i = -\lambda \frac{\partial T}{\partial x_i}. \tag{20.14}$$

- Dimensionless form of energy equation:

$$\rho^* c_p^* \left(\frac{\ell_c'}{t_c' U_c} \frac{\partial T^*}{\partial t^*} + U_i^* \frac{\partial T^*}{\partial x_i^*} \right) = - \boxed{\frac{\dot{q}_c}{\rho_c (c_p)_c \Delta T_c U_c}} \frac{\partial \dot{q}_i^*}{\partial x_i^*} - \frac{\Delta P_c}{\rho_c (c_p)_c \Delta T_c}$$

$$\left(\frac{\ell_c'}{t_c' U_c} \frac{\partial P^*}{\partial t^*} + U_i^* \frac{\partial P^*}{\partial x_i^*} \right) - \boxed{\frac{\tau_w}{\rho_c c_{p,c} \Delta T_c}} T_{ij}^* \frac{\partial U_j^*}{\partial x_i^*}. \quad (20.15)$$

- Characteristic temperature difference and characteristic heat transport:

$$\boxed{\Delta T_c = \frac{\tau_w}{\rho_c (c_p)_c}} \quad \text{and} \quad \boxed{\dot{q}_c = \tau_w U_c}. \quad (20.16)$$

- Characteristic units of length and time:

$$\dot{q}_c q_i^* = -\frac{\lambda_c \Delta T_c}{\ell_c} \lambda^* \frac{\partial T^*}{\partial x_i^*} \rightsquigarrow \ell_c' = \frac{\lambda_c \Delta T_c}{\dot{q}_c} = \frac{\lambda_c \nu_c}{\mu_c (c_p)_c U_c} = \frac{1}{Pr} \left(\frac{\nu_c}{u_\tau} \right) \quad (20.17)$$

so that $\ell_T = \ell_U/Pr$ holds and $t_T = t_U/Pr$. These derivations represent the facts stated in Fig. 20.1 and in (20.8). These facts will turn up again in the examples cited in the subsequent representations of flows with heat transfer.

20.2 Stationary, Fully Developed Flow in Channels

A simple flow with heat transfer is the stationary fully developed flow in channels with a wall in motion. In Fig. 20.2, the basic geometry of the two-dimensional version of this flow is sketched, and the boundary conditions are also stated. For this flow, the following equations for two-dimensional flows hold, given below for $\rho = $ constant and for constant viscosity and constant heat conductivity:

$$\frac{\partial U_1}{\partial x_2} + \frac{\partial U_2}{\partial x_2} = 0, \quad (20.18)$$

$$\rho \left(\frac{\partial U_1}{\partial t} + U_1 \frac{\partial U_1}{\partial x_1} + U_2 \frac{\partial U_1}{\partial x_2} \right) = -\frac{\partial P}{\partial x_1} + \mu \left(\frac{\partial^2 U_1}{\partial x_1^2} + \frac{\partial^2 U_1}{\partial x_2^2} \right) + \rho g_1, \quad (20.19)$$

$$\rho \left(\frac{\partial U_2}{\partial t} + U_1 \frac{\partial U_2}{\partial x_1} + U_2 \frac{\partial U_2}{\partial x_2} \right) = -\frac{\partial P}{\partial x_2} + \mu \left(\frac{\partial^2 U_2}{\partial x_1^2} + \frac{\partial^2 U_2}{\partial x_2^2} \right) + \rho g_2, \quad (20.20)$$

$$\rho c \left(\frac{\partial T}{\partial t} + U_1 \frac{\partial T}{\partial x_1} + U_2 \frac{\partial T}{\partial x_2} \right) = \lambda \left(\frac{\partial^2 T}{\partial x_1^2} + \frac{\partial^2 T}{\partial x_2^2} \right) + \mu \left[\left(\frac{\partial U_1}{\partial x_1} \right)^2 + \left(\frac{\partial U_2}{\partial x_2} \right)^2 + \left(\frac{\partial U_1}{\partial x_2} \right)^2 + \left(\frac{\partial U_2}{\partial x_1} \right)^2 \right]. \quad (20.21)$$

20.2 Stationary, Fully Developed Flow in Channels

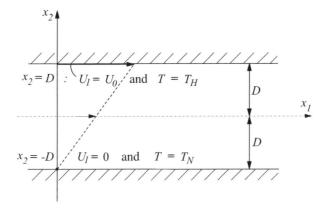

Fig. 20.2 Plane flow in a channel with one wall in motion and with wall temperatures T_H and T_N

Because of the assumed stationarity of the overall problem, the following holds:
$$\frac{\partial(\cdots)}{\partial t} = 0 \text{ for all quantities}$$
and moreover also $(\partial U_i/\partial x_1) = 0$, due to the fully developed flow in the x_2 direction and $(\partial T/\partial x_1) = 0$, as the temperature field also is assumed to be fully developed.

With the above assumptions, one obtains from the continuity equation:
$$\frac{\partial U_1}{\partial x_1} = 0 \rightsquigarrow \frac{\partial U_2}{\partial x_2} = 0 \rightsquigarrow U_2 = 0. \qquad (20.22)$$

Equations (20.19) and (20.20) reduce for the flow sketched in Fig. 20.2 to the following forms:
$$0 = -\frac{\partial P}{\partial x_1} + \mu \frac{\partial^2 U_1}{\partial x_2^2}; \quad 0 = -\frac{\partial P}{\partial x_2} - \rho g. \qquad (20.23)$$

From (20.23) $P(x_1, x_2) = -\rho g x_2 + \Pi(x_1)$, and thus the equation for the velocity field can be written as:
$$0 = -\frac{d\Pi}{dx_1} + \mu \frac{d^2 U_1}{dx_2^2}. \qquad (20.24)$$

Standardization of this equation with $x_2^* = x_2/D$ and $U_1^* = U_1/U_0$ yields:
$$\frac{d^2 U_1^*}{dx_2^{*2}} = \frac{D^2}{\mu U_0} \frac{d\Pi}{dx_1} = -A. \qquad (20.25)$$

For the energy equation (20.21), assuming stationarity and a fully developed temperature field, one obtains:

$$0 = \lambda \frac{\mathrm{d}^2 T}{\mathrm{d}x_2{}^2} + \mu \left(\frac{\mathrm{d}U_1}{\mathrm{d}x_2} \right)^2. \tag{20.26}$$

This equation yields with $T^* = (T - T_N)/(T_H - T_N)$, with $T_H =$ the high wall temperature and $T_N =$ the low wall temperature,

$$\frac{\mathrm{d}^2 T^*}{\mathrm{d}x_2^{*2}} = -\frac{\mu U_0^2}{\lambda (T_H - T_N)} \left(\frac{\mathrm{d}U_1^*}{\mathrm{d}x_2^*} \right)^2, \tag{20.27}$$

where the normalization factor on the right-hand side represents the Brinkmann number:

$$Br = \frac{\mu U_0^2}{\lambda (T_H - T_N)} = \left(\frac{\mu c}{\lambda} \right) \left[\frac{U_0^2}{c(T_H - T_N)} \right] = PrEc \tag{20.28}$$

with $Pr = \frac{\mu c}{\lambda}$ (Prandtl number) and $Ec = \frac{U_0^2}{c(T_H - T_N)}$ (Eckert number).

Integration of (20.25) yields

$$\frac{\mathrm{d}U_1^*}{\mathrm{d}x_2^*} = -Ax_2^* + C_1 \quad \rightsquigarrow \quad U_1^* = -\frac{A}{2}x_2^{*2} + C_1 x_2^* + C_2. \tag{20.29}$$

With the boundary conditions $x_2^* = -1$, $U_1^* = 0$ and $x_2^* = 1$, $U_1^* = 1$, one obtains for the integration constants $C_1 = \frac{1}{2}$ and $C_2 = \frac{1}{2}(A+1)$. Thus the equation for the normalized velocity distribution reads

$$U_1^* = \frac{1}{2}(1 + x_2^*) + \frac{1}{2}A(1 - x_2^{*2}). \tag{20.30}$$

This equation expresses that the resulting velocity distribution is composed of the Couette flow $\frac{1}{2}(1+x_2^*)$ moved by U_0 and the pressure-driven Poiseuille flow $\frac{1}{2}A(1-x_2^{*2})$. The linear differential equation (20.25) leads to the superposition of the Couette and Poiseuille flows.

In order to obtain the solution of the normalized temperature equation, one first computes the velocity gradient:

$$\frac{\mathrm{d}U_1^*}{\mathrm{d}x_2^*} = \frac{1}{2} - Ax_2^*. \tag{20.31}$$

With (20.27), one obtains the following differential equation for the temperature distribution:

$$\frac{\mathrm{d}^2 T^*}{\mathrm{d}x_2^{*2}} = -Br \left(\frac{1}{4} - Ax_2^* + A^2 x_2^{*2} \right). \tag{20.32}$$

On integrating this equation, one can derive

$$T^* = -Br \left(\frac{1}{8}x_2^{*2} - \frac{A}{6}x_2^{*3} + \frac{A^2}{12}x_2^{*4} \right) + C_1 x_2^* + C_2. \tag{20.33}$$

20.3 Natural Convection Flow

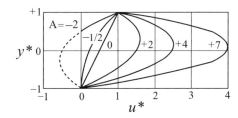

Fig. 20.3 Velocity profiles for different values of A

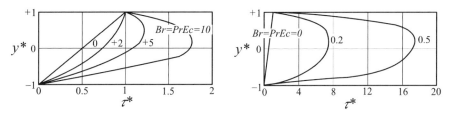

Fig. 20.4 Temperature profiles for different Brinkmann numbers

With the boundary conditions $x_2^* = -1$, $T^* = 0$ and $x_2^* = 1$, $T^* = 1$, one can derive for the integration constants C_1 and C_2:

$$C_1 = \frac{1}{2} - Br\frac{A}{6} \quad \text{and} \quad C_2 = \frac{1}{2} + Br\left(\frac{1}{8} + \frac{A^2}{12}\right). \tag{20.34}$$

Equations (20.34) inserted in (20.33) yield, after rearranging the terms,

$$T^* = \frac{1}{2}(1 + x_2^*) + \frac{Br}{8}\left(1 - x_2^{*2}\right) - \frac{BrA}{6}\left(x_2^* - x_2^{*3}\right) - \frac{BrA^2}{12}\left(1 - x_2^{*4}\right). \tag{20.35}$$

The resulting temperature distribution shows a first term corresponding to the pure heat conduction, i.e. a linear change of the temperature from T_N at the lower non-moving wall to T_H at the upper moving wall. The second term results from the dissipative heat due to the linear velocity profile of the Couette flow and the remaining terms from the dissipation shear of the flow, which results from the parabolic velocity profile of the Poiseuille flow.

The velocity and temperature profiles that followed from the above derivations are shown in Figs. 20.3 and 20.4 for different parameters A and Br.

20.3 Natural Convection Flow Between Vertical Plane Plates

In the preceding sections, flows were considered for which it was assumed that the fluid properties, such as the density ρ, the dynamic viscosity μ and the heat conduction λ, are constant. They could therefore be considered as predefined and did not enter into the fluid-mechanical considerations of the quantities of the flow problem as unknowns that were to be computed. Thus the complexity of flow-problem solutions was considerably reduced, as with constant values for ρ, μ and λ the strong coupling between the momentum equations and the energy equation was broken. For the solution of flow problems, it was therefore sufficient to solve the continuity and the momentum equations, i.e. the energy equation had only to be employed when, in addition to the knowledge of the flow field, information on the temperature field of the fluid was needed.

In this section, a flow problem will be considered for which it is no longer permissible to neglect the density modifications that occur. Restrictively, it will be assumed, however, that only small density modifications arise, so that the following holds:

$$\rho = \rho_0 + \Delta\rho \approx \rho_0 \left[1 - \beta_0 \left(T - T_0\right)\right] \quad \text{with } \beta_0 = -\frac{1}{\rho_0}\left(\frac{\partial \rho}{\partial T}\right)_p. \quad (20.36)$$

With this, the equations of fluid mechanics can be stated as follows:

$$\frac{\partial U_i}{\partial x_i} = 0, \quad (20.37)$$

$$\rho_0 \left(\frac{\partial U_j}{\partial t} + U_i \frac{\partial U_j}{\partial x_i}\right) = -\frac{\partial}{\partial x_j}\left(P - \rho_0 g_j x_j\right) + \mu \frac{\partial^2 U_j}{\partial x_i^2} + (\rho - \rho_0) g_j, \quad (20.38)$$

$$(\rho - \rho_0) = -\rho_0 \beta_0 (T - T_0), \quad (20.39)$$

$$\frac{\partial T}{\partial t} + U_i \frac{\partial T}{\partial x_i} = \left(\frac{\lambda_0}{\rho_0 c_{p0}}\right) \frac{\partial^2 T}{\partial x_i^2}. \quad (20.40)$$

These equations can be employed for examining flows driven by density differences, i.e. with the above set of partial differential equations natural convection flows can be described mathematically.

From these equations, one obtains for two-dimensional flow conditions $\partial/\partial x_3(\ldots) = 0$, and for fully developed flows $\partial/\partial x_2(U_j) = 0$, the following simplified equations:

- Momentum equation:

$$\rho \frac{\partial U_2}{\partial t} = -\frac{\partial \Pi}{\partial x_2} + \rho g \beta_0 (T - T_0) + \mu \frac{\partial^2 U_2}{\partial x_1^2}. \quad (20.41)$$

- Energy equation:

$$\rho c_p \frac{\partial T}{\partial t} = \lambda \frac{\partial^2 T}{\partial x_1^2} + \mu \left(\frac{\mathrm{d}U_2}{\mathrm{d}x_1}\right)^2 \quad (20.42)$$

20.3 Natural Convection Flow

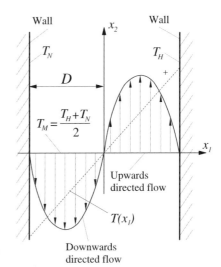

Fig. 20.5 Free convective flow between vertical plates

where, concerning the axial directions, the coordinate system stated in Fig. 20.5 was chosen.

Equations (20.41) and (20.42) can now be simplified for stationary flows, that run without an external pressure gradient, i.e. $\partial \Pi / \partial x_2 = 0$. For such flows, the basic equations describing natural convection hold as follows:

- Momentum equation:

$$0 = \rho g \beta_0 (T - T_0) + \mu \frac{\mathrm{d}^2 U_2}{\mathrm{d} x_1{}^2} \,. \tag{20.43}$$

- Energy equation:

$$0 = \frac{\lambda}{\rho c_p} \frac{\mathrm{d}^2 T}{\mathrm{d} x_1{}^2} + \frac{\nu}{c_p} \left(\frac{\mathrm{d} U_2}{\mathrm{d} x_1} \right)^2 . \tag{20.44}$$

For the flow driven by natural convection between two plates, the momentum equation results in the following form:

$$0 = \mu \frac{\mathrm{d}^2 U_2}{\mathrm{d} x_1{}^2} + \rho_M g \beta_M (T - T_M) \tag{20.45}$$

and the energy equation, taking into consideration the above assumption, can be written as:

$$0 = \lambda \frac{\mathrm{d}^2 T}{\mathrm{d} x_1{}^2} + \mu \left(\frac{\mathrm{d} U_2}{\mathrm{d} x_1} \right)^2 , \tag{20.46}$$

where $T_M = 1/2(T_H + T_N)$. The subsequent boundary conditions describe the natural convection flow problem sketched in Fig. 20.5:

$$U_2(D) = U_2(-D) = 0, \qquad (20.47)$$

$$T(D) = T_H; \quad T(-D) = T_N. \qquad (20.48)$$

Introducing the so-called buoyancy-viscosity parameter A:

$$A = \frac{\beta_M g \mu D}{\lambda} \qquad (20.49)$$

the basic equations can be normalized as stated below, introducing the following dimensionless quantities:

$$x_1^* = x_1/D; \qquad U_2^* = \frac{\rho D U_2}{\mu}; \qquad T^* = \frac{(T - T_N)}{(T_H - T_N)}. \qquad (20.50)$$

One obtains in this way the following dimensionless equations for the resultant velocity and temperature distribution:

$$\frac{\mathrm{d}^2 U_2^*}{\mathrm{d} x_1^{*2}} = -Gr T^* \quad \text{and} \quad \frac{\mathrm{d}^2 T^*}{\mathrm{d} x_1^{*2}} = -\frac{A}{Gr}, \qquad (20.51)$$

where the Grashof number Gr results from the derivations as follows:

$$Gr = \frac{g \rho^2 D^3 \beta (T_H - T_N)}{\mu^2}. \qquad (20.52)$$

When considering that the buoyancy-viscosity parameter A assumes very small values for most fluids and that moreover Gr assumes large values for buoyancy-driven flows, relevant in practice, then for the dimensionless temperature distribution, the following is obtained to a good approximation:

$$\frac{\mathrm{d}^2 T^*}{\mathrm{d} x_1^{*2}} = 0 \quad \leadsto \quad T^* = C_1 x_1^* + C_2. \qquad (20.53)$$

With $T^* = -1$ for $x_1^* = -1$ and $T^* = 1$ for $x_1^* = 1$ one obtains $C_1 = 1$ and $C_2 = 0$ and thus

$$T^* = x_1^* \qquad (20.54)$$

T^* inserted in (20.51) yields

$$\frac{\mathrm{d}^2 U_2^*}{\mathrm{d} x_1^{*2}} = -Gr x_1^* \quad \text{and hence} \quad U_2^* = -\frac{Gr}{6} x_1^{*3} + C_1 x_1^* + C_2. \qquad (20.55)$$

With the boundary conditions at $x_1^* = 1$: $U_2^* = 0$ and at $x_1^* = -1$: $U_2^* = 0$, one obtains $C_1 = Gr/6$ and $C_2 = 0$ and thus

$$U_2^* = \frac{Gr}{6} \left(x_1^* - x_1^{*3} \right). \qquad (20.56)$$

The resulting temperature distribution emerges from this analysis as linear. It thus represents the distribution typical for pure heat conduction. On the other hand, the velocity distribution is described by a point-symmetrical cubic function as sketched in Fig. 20.5. Along the wall with the higher temperature, an upward directed flow forms, and on the side of the cool wall a flow forms that is directed downwards. Flows of this kind can occur between the planes of insulating-glass windows when these have been dimensioned incorrectly.

20.4 Non-Stationary Free Convection Flow Near a Plane Vertical Plate

The combined flow and heat-transfer problem discussed in this section, deals with the diffusion of heat from a vertical wall, heated suddenly and brought to a temperature T_W at time $t = 0$. The diffusion of heat takes place into an infinitely extended field extending into a half-plane. The density modifications in the fluid, caused by the heat diffusion, result in buoyancy forces, and these in turn lead to a fluid movement that can be treated analytically as a free convection flow. With it the basic equations, expanded in the momentum equations by the Oberbeck/Bussinesq terms as stated in Sect. 20.3, can be given as indicated below in the form of a system of one-dimensional equations. Basically equations result for an unsteady flow providing a basis for the sought solution. In this context the following was taken into consideration:

- Because of $\partial U_2/\partial x_2 = 0$, due to the fully developed flow in the x_2 direction, one obtains from the two-dimensional continuity equation $U_1 =$ constant. As $U_1 = 0$ at the wall is given, one obtains $U_1 = 0$ in the entire flow area.
- With the above insights, the left-hand side of the x_2 momentum equation reduces to the term $\rho_0(\partial U_2/\partial t)$, so that the following system of equations holds for the considered natural convective flow problem:

  – Momentum equation:
  $$\rho_0 \frac{\partial U_2}{\partial t} = \mu_0 \frac{\partial^2 U_2}{\partial x_1^2} + (\rho - \rho_0)g. \tag{20.57}$$

  – Energy equation:
  $$\rho_0 c_{p,0} \frac{\partial T}{\partial t} = \lambda_0 \frac{\partial^2 T}{\partial x_1^2}. \tag{20.58}$$

The dissipation term in the energy equation $\mu \, (dU_2/dx_1)^2$ was neglected here for reasons stated in Sect. 20.3. For further details, see also ref. [20.4].

For the further explanation of the problem to be examined here, it should be said that for all times $t < 0$ the following holds: $U_2(x_1, t) = 0$ and $T(x_1, t) = T_0$ for $x_1 \geq 0$, i.e. in the entire area filled with fluid there is initially no flow, and the fluid has the same temperature everywhere.

For all times $t \geq 0$, the following boundary conditions will hold: $U_2(0, t) = 0$ (no-slip condition at the wall) and $T(0, t) = T_W$ (sudden increase of the wall temperature). Moreover, the flow problem to be examined is described for $x_1 \to \infty$ by $U_2(\infty, t) = 0$ and $T(\infty, t) = T_0$.

The velocity and temperature fields sketched in Fig. 20.6 indicate the diffusion processes that take place and how they contribute to the initiation of the described buoyancy flow. The molecular diffusion of the temperature field is evident, together with the induced fluid movement and the momentum loss to the wall. Important for the quantitative information to be derived here is the presence of an analytical solution of the buoyancy problem indicated in Fig. 20.6.

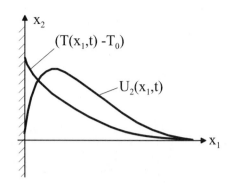

Fig. 20.6 Unsteady natural convection flow at a flat vertical plate

The above flow problem has to be solved as a one-dimensional, unsteady natural convection flow problem, namely as a similarity solution of the equation system (20.57) and (20.58). To derive the solution, we introduce the similarity variable:

$$\eta = \frac{x_1}{2\sqrt{\nu_0 t}} \tag{20.59}$$

and for the dependent variables $U_2(x_1, t)$ and $T(x_1, t)$ for $x_1 \geq 0$ the similarity ansatz:

$$U_2(x_1, t) = [\beta_0 (T_W - T_0) gt] F(\eta) \tag{20.60}$$

and

$$T(x_1, t) = (T_W - T_0) G(\eta). \tag{20.61}$$

These ansatzes are introduced in this particular form with the aim of conserving the dimensionless forms of the differential equations describing the problem and, moreover, to transfer the partial differential equations into ordinary differential equations.

The above ansatzes (20.60) and (20.61) hold for $Pr = 1$ and are solved below for this special case. More general solutions for $Pr \neq 0$ were given by Illingworth [20.1] and can be looked up there. The special case discussed here suffices to introduce students of fluid mechanics to the field of natural convection flows.

With the above similarity ansatz, one obtains for the derivative in the differential equation for U_2:

$$\rho_0 \frac{\partial U_2}{\partial t} = \rho_0 \beta_0 (T_W - T_0) g \frac{\partial}{\partial t} [tF(\eta)] \tag{20.62}$$

or the derivative executed with respect to t:

$$\rho_0 \frac{\partial U_2}{\partial t} = \rho_0 \beta_0 (T_W - T_0) g \left(F - \frac{1}{2} \eta F' \right). \tag{20.63}$$

Similarly, for the first derivative with respect to x_1:

20.4 Non-Stationary Free Convection Flow

$$\mu_0 \frac{\partial U_2}{\partial x_1} = \mu_0 \beta_0 \left(T_W - T_0\right) gt F' \frac{1}{2\sqrt{\nu_0 t}} \tag{20.64}$$

and thus for the second derivative:

$$\mu_0 \frac{\partial^2 U_2}{\partial x_1^2} = \mu_0 \beta_0 \left(T_W - T_0\right) gt F'' \frac{1}{4\nu_0 t} \tag{20.65}$$

or in consideration of $\nu_0 = (\mu_0/\rho_0)$ transcribed as:

$$\mu_0 \frac{\partial^2 U_2}{\partial x_1^2} = \rho_0 \beta_0 \left(T_W - T_0\right) g \left(\frac{1}{4} F''\right). \tag{20.66}$$

For the gravitation term in the U_2 differential equation, one obtains:

$$(\rho - \rho_0)g = \rho_0 \beta_0 (T - Z_0)g = \rho_0 \beta_0 (T_W - T_0)g G(\eta). \tag{20.67}$$

Insertion of (20.63), (20.66) and (20.67) into the momentum equation to be solved, yields the following ordinary differential equation for $F(\eta)$:

$$F'' + 2\eta F' - 4F + 4G = 0. \tag{20.68}$$

For the derivatives in the energy equation in terms of time, one obtains:

$$\rho_0 c_0 \frac{\partial T}{\partial t} = \rho_0 c_0 \left(T_w - T_0\right) \frac{\partial}{\partial t} \left[G(\eta)\right] \tag{20.69}$$

and after carrying out the differentiation:

$$\rho_0 c_0 \frac{\partial T}{\partial t} = \rho_0 c_0 \left(T_W - T_0\right) \frac{\eta}{2t} G'. \tag{20.70}$$

Deriving the second derivative with respect to x_1 yields:

$$\lambda_0 \frac{\partial^2 T}{\partial x_1^2} = G'' \lambda_0 \frac{\rho}{4\mu_0 t} (T_W - T_0) \tag{20.71}$$

and for $Pr = 1$

$$\lambda_0 \frac{\partial^2 T}{\partial x_1^2} = G'' \frac{1}{4t} \rho_0 c_0 (T_W - T_0). \tag{20.72}$$

Insertion of (20.70) and (20.71) into the energy equation yields:

$$G'' - 2\eta G' = 0. \tag{20.73}$$

The boundary conditions for the solution of the above ordinary differential equations (20.68) and (20.73) read:

$$\begin{aligned}
x_1 = 0 &: U_2(0, t) = 0 &&\rightsquigarrow&& \eta = 0 : F(0) = 0 \\
&\; T(0, t) = T_W &&\rightsquigarrow&& G(0) = 1 \\
x_1 \to \infty &: U_2(\infty, t) = 0 &&\rightsquigarrow&& \eta = 1 : F(1) = 0 \\
&\; T(\infty, t) = T_0 &&\rightsquigarrow&& G(1) = 0
\end{aligned}$$

As a solution of the differential equation (20.73), one obtains:

$$G(\eta) = 1 - \mathrm{erf}(\eta). \tag{20.74}$$

The solution of the differential equation (20.68), with $G(\eta)$ inserted, can be obtained as a solution of the homogeneous differential equation for $F(\eta)$:

$$F'' + 2\eta F' - 4F = 0 \tag{20.75}$$

and with the particular solution $F(\eta) = \mathrm{erf}(\eta)$ and adding the homogeneous solution results in

$$F(\eta) = \frac{2}{\sqrt{\pi}} \eta \exp\left(-\eta^2\right) - 2\eta^2 \mathrm{erf}(\eta). \tag{20.76}$$

With this, the solutions for F and G, as shown in Fig. 20.7, can be determined from (20.74) and (20.76). With a decrease in G with increase in η, a decrease in temperature with increasing distance from the wall is indicated. The $F(\eta)$ distribution relates to the velocity distributions for the natural convection. This convective flow forms due to the buoyancy forces induced by density differences near the wall.

The parts of the similarity solutions of the equation system (20.57) and (20.58) represented in Fig. 20.7 show, on the one hand, the normalized temperature profile $G(\eta)$ that develops due to the temperature diffusion from

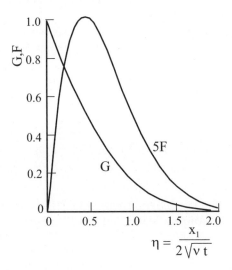

Fig. 20.7 Solutions $F(\eta)$ and $G(\eta)$ for the free convection flow along a plane vertical plate

the heated wall into the fluid. The figure shows, moreover, the standardized velocity profile, which is caused by buoyancy and which is strongly influenced by the molecule-dependent momentum loss to the wall. Because of the assumed fully developed flow in the x_2 direction, for the temperature field (20.61) and the velocity field (20.60) physically convincing solutions result from the differential equations. Altogether the flow and temperature distributions are understood as examples of many buoyancy flows that exist in nature in a large variety. For a number of these flows, driven by temperature fields, analytical solutions exist.

20.5 Plane-Plate Boundary Layer with Plate Heating at Small Prandtl Numbers

In Chap. 16, the two-dimensional boundary-layer equations were derived from the general Navier–Stokes equations according to a procedure suggested by Prandtl. On extending these derivations to boundary-layer flows with heat transfer, one obtains on the following assumptions:

$x_1 = $ flow direction; $x_3 = $ direction with $\partial/\partial x_3 (\cdots) = 0$

the following equations for $x_1 = x$, $x_2 = y$, $U_1 = U$, $U_2 = V$:

Stationary Compressible Flows (Boundary-Layer Equations)

$$\frac{\partial}{\partial x}(\rho U) + \frac{\partial}{\partial x}(\rho V) = 0, \qquad (20.77)$$

$$\rho\left(U\frac{\partial U}{\partial x} + V\frac{\partial U}{\partial y}\right) = -\frac{dP}{dx} + \frac{\partial}{\partial y}\left(\mu\frac{\partial U}{\partial y}\right) + \rho g_x \beta(T - T_\infty), \qquad (20.78)$$

$$\rho c_p\left(U\frac{\partial T}{\partial x} + V\frac{\partial T}{\partial y}\right) = U\frac{dP}{dx} + \lambda\frac{\partial^2 T}{\partial y^2} + \mu\left(\frac{\partial U}{\partial y}\right)^2, \qquad (20.79)$$

$\rho = $ constant   or $\dfrac{P}{\rho} = RT$ and $\mu = \mu(T),\ \lambda(T),\ c_p(T)$.   (20.80)

Hence, there are four differential equations for U, V, P, ρ and T, which can be solved with the boundary conditions defining the respective problem. For incompressible flows, we obtain the following:

Stationary Incompressible Flows (Boundary-Layer Equations)

$$\frac{\partial U}{\partial x} + \frac{\partial V}{\partial y} = 0, \qquad (20.81)$$

$$\rho_\infty \left(U \frac{\partial U}{\partial x} + V \frac{\partial U}{\partial y} \right) = -\frac{\mathrm{d}P}{\mathrm{d}x} + \mu_\infty \frac{\partial^2 U}{\partial y^2} - \rho_\infty g_x \beta_\infty (T - T_\infty), \quad (20.82)$$

$$\rho_\infty c_{p\infty} \left(U \frac{\partial T}{\partial x} + V \frac{\partial T}{\partial y} \right) = \lambda \frac{\partial^2 T}{\partial y^2} + \mu_\infty \left(\frac{\partial U}{\partial y} \right)^2 + \rho_\infty g_x \beta_\infty (T - T_\infty). \qquad (20.83)$$

This system of partial differential equations can be solved for $\rho = \rho_\infty =$ constant, $\lambda = \lambda_\infty =$ constant, $c_p = c_{p\infty} =$ constant and $\mu = \mu_\infty =$ constant for plane-plate boundary-layer flows by including the boundary conditions that characterize the flow and heat-transfer problem, in order to compute U, V and T. The externally imposed pressure gradient $(\mathrm{d}P/\mathrm{d}x)$ can often be assumed to be given for this kind of flow.

In order to integrate the equations, it is recommended also to include in the considerations the influence of the Prandtl number on the solution. Here, it has to be taken into consideration that the Prandtl numbers of the fluids considered in this book, are able to cover the wide range that is indicated in Fig. 20.8.

For boundary-layer flows with very small Prandtl number, i.e. boundary layers of melted metals, thermal boundary layers result, which are many times thicker than the fluid boundary layers (see Fig. 20.9). It is therefore understandable that it is recommended, for small Prandtl numbers, to treat boundary-layer flows with heat transfer, such that the fluid boundary layer is entirely neglected. From the continuity equation (20.81), it follows that the gradients of V in y direction and U in x direction are connected in the following way:

$$\frac{\partial V}{\partial y} = -\frac{\mathrm{d}U}{\mathrm{d}x} \qquad (20.84)$$

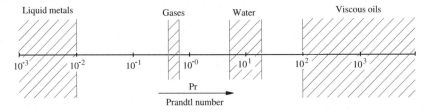

Fig. 20.8 Domains of Prandtl numbers for different fluids (fluids and gases)

20.5 Plane-Plate Boundary Layer

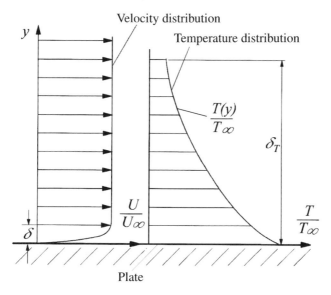

Fig. 20.9 Thermal boundary layer and thermal boundary layer for small Prandtl numbers

and thus
$$V = -\frac{dU}{dx} y. \tag{20.85}$$

For the analytical considerations to be carried out, the similarity variable:
$$\eta = \frac{1}{2} y \sqrt{\frac{U_\infty}{ax}} \tag{20.86}$$

is introduced. From the energy equation, one obtains:
$$U_x \frac{\partial T}{\partial x} - y \frac{dU}{dx} \frac{\partial T}{\partial y} = a \frac{\partial^2 T}{\partial y^2} \quad (Pr \ll 1). \tag{20.87}$$

For $T = T_W$ for $y = 0$ and $T = T_\infty$ for $y \to \infty$ (and this for all x positions), one obtains for a constant external flow, i.e. $U(x) = U_\infty = $ constant,
$$U_\infty \frac{\partial T}{\partial x} = a \frac{\partial^2 T}{\partial y^2} \quad \text{with } a = \frac{\lambda_\infty}{\rho_\infty c_{p,\infty}}. \tag{20.88}$$

For the standardized temperature $T^* = (T - T_\infty)/\Delta T_w$ with $\Delta T_w = (T_w - T_\infty)$, one obtains:
$$U_\infty \frac{\partial T^*}{\partial x} = a \frac{\partial T^*}{\partial y} \tag{20.89}$$

and with the similarity ansatz:
$$T^* = f(\eta) \tag{20.90}$$

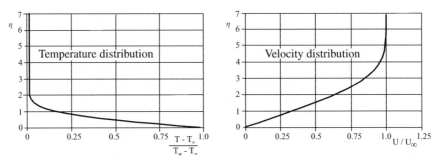

Fig. 20.10 Temperature distribution and velocity distribution for a plate boundary-layer flow with constant wall temperature at small Prandtl numbers

one obtains for the derivative in the above differential equation (20.89)

$$\frac{\partial T^*}{\partial x} = f'\left(-\frac{1}{4x}y\sqrt{\frac{U_\infty}{ax}}\right) = \eta f'\frac{1}{2x}, \tag{20.91}$$

$$\frac{T^*}{y} = f'\left(\frac{1}{2}\sqrt{\frac{U_\infty}{ax}}\right) \quad \text{and} \quad \frac{\partial^2 T^*}{\partial y^2} = f''\left(\frac{1}{4}\frac{U_\infty}{ax}\right). \tag{20.92}$$

From (20.91) and (20.92), the following ordinary differential equation for $f(\eta)$ can be derived:

$$f'' + 2\eta f' = 0. \tag{20.93}$$

With the following boundary conditions:
For all $x \geq 0$:

$$T^*(y = 0) = 1 \quad \rightsquigarrow \quad \eta = 0: \quad f(\eta) = 1,$$
$$T^*(y \to \infty) = 0 \quad \rightsquigarrow \quad \eta = 1: \quad f(\eta) = 0$$

one obtains for the temperature distribution

$$\left(\frac{T - T_\infty}{T_W - T_\infty}\right) = 1 - \operatorname{erf}(\eta) = 1 - \frac{2}{\sqrt{\pi}}\int_0^\eta \exp\left(-\eta^2\right)\,\mathrm{d}\eta. \tag{20.94}$$

This temperature distribution is given as function of η in Fig. 20.10 together with the corresponding temperature distribution. For more details, see ref. [20.5]

20.6 Similarity Solution for a Plate Boundary Layer with Wall Heating and Dissipative Warming

In Chap. 15, boundary-layer flows were discussed and an introduction was given to the solution of the boundary-layer equations by means of similarity ansatzes. The flat plate flow with heat transfer discussed here, is likewise

20.6 Similarity Solution for a Plate Boundary Layer

based on the solution of the boundary-layer equations for the stationary flow around plates suggested by Blasius, i.e. on the solution of the following equations:

$$\frac{\partial U}{\partial x} + \frac{\partial V}{\partial y} = 0, \tag{20.95}$$

$$\rho\left(U\frac{\partial U}{\partial x} + V\frac{\partial U}{\partial y}\right) = \mu\frac{\partial^2 U}{\partial y^2}. \tag{20.96}$$

For the discussion of the heat transfer, the boundary-layer form of the energy equation is included. For the solution to be sought, the temperature dependences of the material values ρ, μ, λ, and thus also the buoyancy forces, are neglected in the energy equation:

$$\rho c_p\left(U\frac{\partial T}{\partial x} + V\frac{\partial T}{\partial y}\right) = \lambda\frac{\partial^2 T}{\partial y^2} + \mu\left(\frac{\partial U}{\partial y}\right)^2. \tag{20.97}$$

For flat plate flow with wall heating, the following boundary conditions result:

$$y = 0 : U = V = 0 \quad \text{and} \quad T = T_W, \tag{20.98}$$

$$y = 0 : U \to U_\infty \quad \text{and} \quad T \to T_\infty. \tag{20.99}$$

For the solution of the flat plate boundary layer problem, it is important to realize that (20.95) and (20.96), for the determination of the velocity field, are decoupled from the energy equation (20.97) if the material properties are assumed to be independent of the temperature. This assumption is made here. Thus, for the velocity field the solution suggested by Blasius can be taken (see Chap. 16).

With $\eta = y\sqrt{U_\infty/\nu x}$ and $\psi = \sqrt{\nu x U_\infty} f(\eta)$, one obtains $U = U_\infty f'(\eta)$ and $V = \frac{1}{2}\sqrt{\nu U_\infty/x}\,(\eta f' - f)$.

From the momentum equation, the following differential equation for the quantity f can be derived:

$$ff'' + 2f''' = 0. \tag{20.100}$$

As shown in Chap. 16, $f(\eta)$ and $f'(\eta)$ can be determined numerically from this, and thus U and V can be determined for the boundary conditions reading as follows:

$$\eta = 0: \quad f = f' = 0 \quad \text{and} \quad \eta \to \infty: \quad f' \to 1.$$

The energy equation can be treated as follows:

$$\left[U_\infty f'\frac{dT}{d\eta}\frac{\partial \eta}{\partial x} + \frac{1}{2}\sqrt{\frac{\nu U_\infty}{x}}(\eta f' - f)\frac{dT}{d\eta}\frac{\partial \eta}{\partial y}\right]$$

$$\frac{\lambda}{\rho c_p}\frac{\partial^2 T}{\partial \eta^2}\left(\frac{\partial \eta}{\partial y}\right)^2 + \frac{\mu}{\rho c_p}U_\infty^2 f''^2\frac{U_\infty}{\nu x} \tag{20.101}$$

or rewritten for further considerations:

$$-\frac{1}{2}f\left(\frac{dT}{d\eta}\right) = \underbrace{\frac{\lambda}{\mu c_p}}_{1/Pr}\frac{d^2 T}{d\eta^2} + \frac{U_\infty^2}{c_p}(f'')^2 \qquad (20.102)$$

in order to obtain the final form:

$$0 = \frac{d^2 T}{d\eta^2} + \frac{Pr}{2}f\frac{dT}{d\eta} + 2Pr\frac{U_\infty^2}{2c_p}(f'')^2. \qquad (20.103)$$

On now introducing

$$\Theta(\eta) = \frac{T - T_\infty}{T_W - T_\infty} \qquad (20.104)$$

one obtains the following ordinary differential equation of Θ:

$$0 = \Theta'' + \frac{Pr}{2}f\Theta' + 2Pr\frac{U_\infty^2}{2c_p}(f'')^2. \qquad (20.105)$$

Without dissipative heating of the boundary layer, given by the last term in (20.105), one obtains the differential equation

$$\Theta'' + \frac{Pr}{2}f\Theta' = 0 \qquad (20.106)$$

which has to be solved for the boundary conditions:

$$\eta = 0 \rightsquigarrow \theta = 1 \quad \text{and} \quad \eta \to \infty \rightsquigarrow \theta \to 0.$$

A solution of this equation is possible, as $f(\eta)$ is known. It was derived as a solution of the continuity and momentum equations. It was suggested by Pohlhansen [20.3] and is given in Fig. 20.11 for different Prandtl numbers:

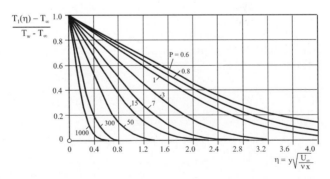

Fig. 20.11 Temperature distribution at a plane heated plate with temperature difference $(T_W - T_\infty)$

20.7 Vertical Plate Boundary-Layer Flows

For heat-transfer computations, the following equations hold:

$$\dot{q}(x) = -\lambda \left(\frac{\partial T}{\partial y}\right)_0 = -\lambda \left(\frac{\partial T}{\partial \eta}\right)_0 \left(\frac{\partial \eta}{\partial y}\right), \quad (20.107)$$

$$\dot{q}(x) = -\lambda \left(\frac{\partial \Theta}{\partial \eta}\right)_0 (T_W - T_\infty) \sqrt{\frac{U_\infty}{\nu x}}. \quad (20.108)$$

For the technically interesting fields the following holds:

$$\left(\frac{\partial \Theta}{\partial \eta}\right)_0 \approx \frac{1}{3}(Pr)^{1/3} \quad (20.109)$$

so that for the local heat flow we obtain

$$\dot{q}(x) = -\frac{1}{3}\lambda Pr^{1/3}(T_W - T_\infty)\sqrt{\frac{U_\infty}{\nu x}}. \quad (20.110)$$

The amount of heat which is released from the tip of the plate up to the plate-length L can be obtained by integration:

$$\dot{Q}(L) = +\frac{1}{6}\lambda Pr^{1/3}(T_W - T_\infty)\sqrt{\frac{U_\infty}{\nu}} x^{-3/2}. \quad (20.111)$$

In this way, important information for technical engineering can be gained from the above derivations.

20.7 Vertical Plate Boundary-Layer Flows Caused by Natural Convection

Near a vertical plane, which was heated up to temperature T_W, a natural convection boundary-layer flow develops, which is caused by natural convection and is directed upwards. It is described by the boundary-layer equations by $dP/\partial x = 0$:

$$\frac{\partial U}{\partial x} + \frac{\partial V}{\partial y} = 0, \quad (20.112)$$

$$\rho\left(U\frac{\partial U}{\partial x} + V\frac{\partial U}{\partial y}\right) = \mu\frac{\partial^2 U}{\partial y^2} + \rho g \beta (T - T_\infty), \quad (20.113)$$

$$\rho c_p \left(U\frac{\partial T}{\partial x} + V\frac{\partial T}{\partial y}\right) = \lambda \frac{\partial^2 T}{\partial y^2}. \quad (20.114)$$

For $\beta = \frac{1}{T_\infty}$ and $\theta = \frac{T-T_\infty}{T_W-T_\infty}$, these equations can be transcribed as follows and can be employed for the solution of the velocity and temperature fields:

$$\frac{\partial U}{\partial x} + \frac{\partial V}{\partial y} = 0, \tag{20.115}$$

$$\rho \left(U \frac{\partial U}{\partial x} + V \frac{\partial U}{\partial y} \right) = \mu \frac{\partial^2 U}{\partial y^2} + \rho g \frac{T_W - T_\infty}{T_\infty} \Theta, \tag{20.116}$$

$$U \frac{\partial \Theta}{\partial x} + V \frac{\partial \Theta}{\partial y} = a \frac{\partial^2 \Theta}{\partial y^2}. \tag{20.117}$$

The introduction of the stream function $U = \frac{\partial \psi}{\partial y}$ and $V = -\frac{\partial \psi}{\partial x}$ eliminates the continuity equation and makes possible the similarity ansatz given below:

$$\Psi = 4\nu A x^{3/4} f(\eta) \quad \text{with} \quad \eta = A \frac{y}{\sqrt[4]{x}} \tag{20.118}$$

with A being

$$A = \sqrt[4]{\frac{g(T_W - T_\infty)}{4\nu^2 T_\infty}}. \tag{20.119}$$

The velocity components U and V can be computed as follows:

$$U = 4\nu x^{1/2} A^2 f' \quad \text{and} \quad V = \nu A x^{-1/4} (\eta f' - 3f)$$

and the derivatives yield:

$$\frac{\partial U}{\partial x} = \frac{\nu A^2}{\sqrt{x}} (2f' - \eta f''),$$

$$\frac{\partial V}{\partial y} = \frac{\nu A^2}{\sqrt{x}} (\eta f'' - 2f').$$

With this results the following set of differential equations, which can be employed to determine the similarity functions $f(\eta)$ and $\theta(\eta)$:

$$f''' + 3ff'' - 2f'^2 + \Theta = 0 \quad \text{and} \quad \Theta'' + 3 Pr f \Theta' = 0, \tag{20.120}$$

where the following boundary conditions determine the problem:

$$\eta = 0: \quad f = f' = 0 \quad \text{and} \quad \Theta = 1, \tag{20.121}$$

$$\eta \to \infty: \quad f' = 0 \quad \text{and} \quad \Theta = 0. \tag{20.122}$$

The numerical integration of the differential equation system was carried out by Pohlhansen [20.3] and Ostrach [20.2] and led to the solutions stated in Fig. 20.12 for different Prandtl numbers.

On computing now the heat flow existing locally per unit time and unit area, one obtains:

$$\dot{q}(x) = -\lambda \left(\frac{\partial T}{\partial y} \right)_0 = -\lambda \left(\frac{\partial \Theta}{\partial \eta} \right)_0 \left(\frac{\partial \eta}{\partial y} \right) (T_W - T_\infty) \tag{20.123}$$

20.8 Similarity Considerations for Flows with Heat Transfer

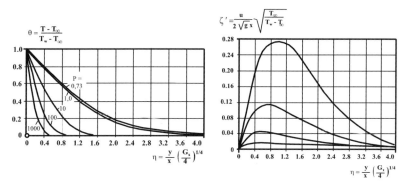

Fig. 20.12 Temperature and velocity distribution at a heated vertical plane plate caused by natural convection

or, after carrying out the differentiation $\left(\frac{\partial \eta}{\partial y}\right)$, one obtains:

$$\dot{q}(x) = -\lambda A \frac{1}{\sqrt[4]{x}} (T_W - T_\infty) \left(\frac{\partial \Theta}{\partial \eta}\right). \tag{20.124}$$

For $Pr = 0.73$ one obtains ($\frac{\partial \Theta}{\partial \eta}$)$_0 \approx \frac{1}{2}$, so that the following expression holds for $\dot{q}(x)$:

$$\dot{q}(x) \approx -\frac{\lambda A}{2\sqrt[4]{x}} (T_W - T_\infty). \tag{20.125}$$

Integration over the plate length L results in the heat transfer per unit width:

$$\dot{Q} = \frac{2}{3} L^{3/4} \lambda A (T_w - T_\infty). \tag{20.126}$$

On computing the Nusselt number, averaged over L, one obtains $\dot{Q} = \lambda (Nu)_L (T_W - T_\infty)$:

$$(Nu)_L = 0.667 A L^{3/4} \tag{20.127}$$

a relationship which is well confirmed by experimental results.

20.8 Similarity Considerations for Flows with Heat Transfer

In Chap. 6, general considerations on the similarity of fluid flows were carried out. They can be extended to flows with heat transfer as shown below.

The momentum equations of fluid mechanics can be written as follows:

$$\rho \left(\frac{\partial U_j}{\partial t} + U_j \frac{\partial U_j}{\partial x_i}\right) = -\frac{\partial P}{\partial x_j} - \frac{\partial \tau_{ij}}{\partial x_i} + \rho g_j. \tag{20.128}$$

As far as Newtonian fluids are concerned, for the molecular-dependent momentum transport one can write

$$\tau_{ij} = -\mu \left(\frac{\partial U_j}{\partial x_i} + \frac{\partial U_i}{\partial x_j} \right) + \frac{2}{3}\mu \delta_{ij} \frac{\partial U_k}{\partial x_k}. \qquad (20.129)$$

In order to write (20.128) in dimensionless form, the molecular-dependent momentum transport to the wall, τ_W, is introduced. All other quantities are made dimensionless with characteristic quantities of the flow and heat-transfer system. Therefore, the following quantities can be introduced:

$$\rho = \rho_c \rho^*; \quad U_j = U_c U_j^*; \quad t = t_c t^*; \quad x_i = l_c x_i^*; P = \Delta P_c P^*; \quad \tau_{ij} = \tau_w \tau_{ij}^*$$

and $g_j = 0$, so that one obtains the following equation:

$$\rho^* \left(\frac{l_c}{U_c t_c} \frac{\partial U_j^*}{\partial t^*} + U_i^* \frac{\partial U_j^*}{\partial x_j^*} \right) = -\frac{\Delta P_c}{\rho_c U_c^2} \frac{\partial P^*}{\partial x_j^*} - \frac{\tau_w}{\rho_c U_c^2} \frac{\partial \tau_{ij}^*}{\partial x_i^*}. \qquad (20.130)$$

On introducing $U_c^2 = \tau_w/\rho = u_\tau^2$ and $\Delta P_c = \tau_w$, the right-hand side of the equation reduces to a dimensionless form, where the quantities are written with an asterisk. The dimensionless groups of variables before the asterisked quantities are now all made equal to 1.

On applying the dimensionless quantities stated in (20.130) also to (20.129), one obtains

$$\tau_{ij}^* = -\frac{\mu_c U_c}{\tau_w l_c} \left[\mu^* \left(\frac{\partial U_j^*}{\partial x_i^*} + \frac{\partial U_i^*}{\partial x_j^*} \right) + \frac{2}{3}\mu^* \delta_{ij} \frac{\partial U_k^*}{\partial x_k^*} \right]. \qquad (20.131)$$

This equation becomes dimensionless on introducing as characteristic length $l_c = \nu_c/u_\tau$.

On introducing all these characteristic quantities into (20.128) and (20.129) to make them also dimensionless, one obtains the following forms of these two equations:

$$\rho^* \left(\frac{\partial U_j^*}{\partial t^*} + U_i^* \frac{\partial U_j^*}{\partial x_i^*} \right) = -\frac{\partial P^*}{\partial x_j^*} - \frac{\tau_{ij}^*}{x_i^*} \qquad (20.132)$$

and

$$\tau_{ij}^* = -\mu^* \left(\frac{\partial U_j^*}{\partial x_i^*} + \frac{\partial U_i^*}{\partial x_j^*} \right) + \frac{2}{3}\mu^* \delta_{ij} \frac{\partial U_k^*}{\partial x_k^*}. \qquad (20.133)$$

For similarity considerations, the following quantities were introduced as dimensionless velocity, dimensionless pressure difference and dimensionless length and time scales:

$$U_c = u_\tau = \sqrt{\tau_w/\rho}; \quad \Delta P_c = \tau_w; \quad l_c = \nu_c/u_\tau; \quad t_c = \nu_c/u_\tau^2. \qquad (20.134)$$

On extending the above-mentioned dimensionless considerations to the general form of the energy equation:

$$\rho c_P \left(\frac{\partial T}{\partial t} + U_i \frac{\partial T}{\partial x_i} \right) = -\frac{\partial q_i}{\partial x_i} + \left(\frac{\partial P}{\partial t} + U_i \frac{\partial P}{\partial x_i} \right) - \tau_{ij} \frac{\partial U_j}{\partial x_i} \quad (20.135)$$

and the Fourier law for heat conductivity:

$$\dot{q}_i = -\lambda \frac{\partial T}{\partial x_i} \quad (20.136)$$

the dimensionless form of the energy equation can be derived as follows:

$$\rho^* c_P^* \left(\frac{l_c}{t_c U_c} \frac{\partial T^*}{\partial t^*} + U_i^* \frac{\partial T^*}{\partial x_i^*} \right) = -\boxed{\frac{\dot{q}_c}{\rho_c c_{P,c} \Delta T_c U_c}} \frac{\partial \dot{q}_i^*}{\partial x_i^*} - \frac{\Delta P_c}{\rho_c c_{P,c} \Delta T_c}$$

$$\left(\frac{l_c'}{t_c' U_c} \frac{\partial P^*}{\partial t^*} + U_i^* \frac{\partial P^*}{\partial x_i^*} \right) - \boxed{\frac{\tau_w}{\rho_c c_{P,c} \Delta T_c}} \tau_{ij}^* \frac{\partial U_j^*}{\partial x_i^*}. \quad (20.137)$$

Looking at (20.137), one sees that the following quantities have to be introduced as characteristic quantities, in order to conserve the dimensionless form of the energy equation equivalent to (20.132):

$$\boxed{\Delta T_c = \frac{\tau_w}{\rho_c c_{P,c}}} \quad \text{and} \quad \boxed{\dot{q}_c = \tau_w U_c}. \quad (20.138)$$

Standardization of the Fourier law leads to the following result:

$$\dot{q}_c \dot{q}_i^* = -\frac{\lambda_c \Delta T_c}{l_c} \lambda^* \frac{\partial T^*}{\partial x_i^*} \quad \rightarrow \quad l_c' = \frac{\lambda_c \Delta T_c}{\dot{q}_c} = \frac{\lambda_c \nu_c}{\mu_c c_{P,c} U_c} = \frac{1}{Pr}\left(\frac{\nu_c}{u_\tau}\right), \quad (20.139)$$

where $Pr l_c = l_c$ and $Pr t_c = t_c$ can be stated. This represents the connection between the characteristic length and time scales for the heat *and* momentum transport.

References

20.1. Illingworth, C. R.: Some solutions of the equations of flow of a viscous compressible fluid, Proc. Cambridge Phil. Soc., 46, pp. 469–478, 1950
20.2. Ostrach, S.: An analysis of laminar free-convection flow and heat transfer about a flat plate parallel to ghe direction of the generating bridge force, NACA-Report 1111, 1953
20.3. Pohlhansen, E.: Der Wärmeaustausch zwischen festen Körpern und Flüssigkeiten mit kleiner Reibung und kleiner Wärmeleitung, ZAMM 1, pp. 115–121, 1921
20.4. Müller, U. and Ehrhard, P.: Freie Konvektion und Wärmeübertragung, C. F. Müller, Springer, Berlin Heidelberg New York, 1999
20.5. Schlichting, H.: Boundary Layer Theory, McGraw-Hill, New York, 1979

Chapter 21
Introduction to Fluid-Flow Measurement

21.1 Introductory Considerations

The derivation of the Reynolds equations, as a basis for numerical flow investigations, led to a system of differential equations which, in addition to the mean values of the components of the flow velocity and the static pressure, contain also turbulent transport terms. These terms represent, for the turbulent momentum transport, time mean values of the products of velocity fluctuations. These transport terms were derived from the Navier–Stokes equations by introducing into the equation mean velocity components and turbulent velocity fluctuation, and by averaging them afterwards with respect to time. Although the turbulent transport terms were derived formally, as new unknowns of the flow field, physical importance can be attached to them. They represent, in the averaged momentum equations, additional diffusive momentum-transport terms, which occur in flows due to turbulent velocity fluctuations that occur superimposed on the mean velocity field.

When one wants to solve the Reynolds equations, it is important to find additional relationships for these correlations of the turbulent velocity fluctuations $\overline{u_i u_j}$. These relationships can be formulated by hypothetical assumptions, and this approach played an important role in the past when setting up turbulence models. Today, however, it is considered for certain that reliable information on the time-averaged properties of turbulent flows can only be obtained by detailed experimental investigations of different flows. To gain the necessary information requires local measurements of the instantaneous velocity of turbulent flows. Such measurements can be made by means of hot-wire or hot-film anemometry and laser Doppler anemometry that possess the necessary resolution in terms of time and space for local velocity measurements in flows. These methods also permit to carry out the necessary measurements in a relatively short time. Such measurements contribute to the understanding of the physics of turbulent flows and make it possible to introduce additional information into the computations of turbulent flows for the above-mentioned correlations of turbulent velocity fluctuations.

For measurements in wall-boundary layers, hot-wire and hot-film anemometers have been applied with great success for determining the mean velocities U_i and the fluctuation velocity correlations $\overline{u_i u_j}$. Depending on their characteristic properties, many turbulent flows can be investigated by means of hot-wire and hot-film anemometers. However, in the case of thin boundary layers, inherent perturbations can occur which are caused by the measuring sensors introduced for velocity measurements. Special designs of measuring sensors are required to keep these measuring errors at a low level in wall-bound boundary layers. Turbulent flows also occur in regions with back flow, and these regions possess a number of properties which prevent the employment of hot-wire anemometers for precise measurements, or limit their application to only some flows. Of these properties of hot wires and hot films, that negatively influence the measurements, only the perturbing influence of the hot-wire support on the actual measurement is mentioned. In addition, high turbulence intensity in regions of flow separation should be mentioned, which leads to insurmountable difficulties concerning the interpretation of the hot-wire signals.

Hot-wire and hot-film measuring devices are based on measurements of the convective heat transfer that occurs due to the fluid flowing over heated elements and providing in this way a measure for the local flow velocity. These measurements, however, require the sensor to possess a higher temperature than the fluid, which can lead in liquid flows to decomposition of the fluid medium. This is indicated in Fig. 21.1 by means of a photographic recording published by Eckelmann [21.9]. Difficulties with the employment of hot-wire and hot-film anemometers, such as one encounters in measurements in industrial conditions, are likewise indicated in Fig. 21.1. When giving up the strict control of the particles carried along in the fluid medium, natural

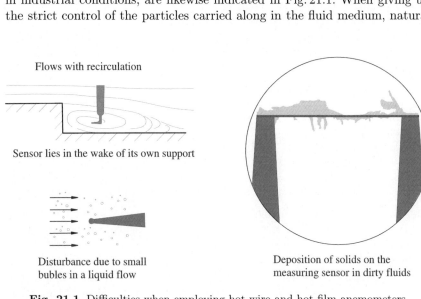

Flows with recirculation

Sensor lies in the wake of its own support

Disturbance due to small bubles in a liquid flow

Deposition of solids on the measuring sensor in dirty fluids

Fig. 21.1 Difficulties when employing hot-wire and hot-film anemometers

21.1 Introductory Considerations

contamination occurs inherent for all components exposed to the fluid. Reliable measurements with hot-wire and hot-film anemometers are therefore often possible only under laboratory conditions. In practical flow studies, one accepts that dirt is deposited on the measuring sensors continuously with time. The dirt layer developing on measuring sensors forms a heat insulation that was not taken into account when the measuring sensor was calibrated. In order to avoid measuring errors, caused in this way, recalibration of the measuring probe has to be carried out at short time intervals, which can be very time consuming.

Most measuring methods that require the insertion of measuring probes in flows measure the flow velocity of interest only indirectly, i.e. with most flow-measuring instruments physical quantities are measured which are functions of the flow velocity. Direct measurements of the local flow velocity are often not carried out. Unfortunately, the measured quantities, through which the flow velocity is determined, are mostly also functions of the thermodynamic state properties of the fluid medium. These fluid-property influences have to be known and have to be taken into account in the calibration of the sensor, in order to make the interpretation of the final measured data possible, i.e. to yield the flow velocity through measurements. When fluctuations of the state parameters of the fluid medium occur during measurements, e.g. in two-phase flows, flows with chemical reactions, etc., these have to be known in order to determine accurately the local velocity with hot-wire anemometers. However, in practical measurements it is often not possible to know the fluid properties at all measuring times, and they can therefore not be employed in the interpretation of the measured velocity signals.

The above-mentioned difficulties in the employment of indirect measuring techniques for flow velocities have led to the development of the laser Doppler anemometry, which allows, almost directly, the local flow velocity to be measured. By measuring the time which a particle needs for passing through a well-defined interference pattern, the flow velocity of the particle is determined. Such measurements do not depend on the often unknown properties of the flow fluid. Measurements are possible in one- and two-phase flows, and also in combustion systems and in the atmosphere. The measuring technique can moreover be employed in fluid media filled with particles, as they often occur in practice. However, its employment requires optical access to the measuring point and thus a sufficient transparency of the flow medium. In this respect, the employment of laser Doppler anemometry is also limited. Its application allows, nevertheless, the determination of flow velocities in a number of flows that are important in practice, but cannot be investigated by other measuring methods.

In addition to the above-mentioned measuring techniques for determining local, time-resolved fluid velocities, the determination of pressure distributions is also very important in experimental fluid mechanics. This is given consideration in the next section which treats the measurement of static pressures. Introductory presentations of the measurement of dynamic pressures follow, in order to show that stagnation-pressure probes can be employed for the determination of local velocities.

In addition to measurements of local velocities and pressures, measurements of wall-shear stresses are often required in fluid-mechanical investigations. Although reference to measurements of these quantities occasionally is made, they are not at the center of the considerations in this chapter. The following considerations have rather to be understood as an introduction to flow-measuring techniques as a subfield of fluid measurements. The introduced practice-oriented basics of flow and pressure measurements should help to round off or complete the education of students in fluid mechanics.

21.2 Measurements of Static Pressures

In the discussion of flows with boundary-layer behavior (see Chap. 16), it was shown that the momentum equation in the cross flow direction reduces to $\partial P/\partial y = 0$. This important property of the pressure field of boundary-layer flows is often employed for the measurement of wall pressures, as it permits one to determine easily the total pressure distribution in boundary-layer flows without probes having to be inserted into the flow. In order to determine now the pressure distribution $P(x)$, with x being the flow direction, it is sufficient to drill holes into the walls of the test rig, such as channels and pipes, as shown in Fig. 21.2. However, for precise wall pressure measurements, it is very important that the bore diameter is kept small, about the order of

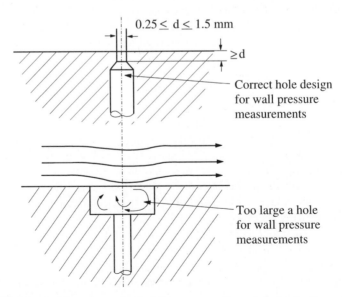

Fig. 21.2 Measurements of wall pressures through holes in the walls

21.2 Measurements of Static Pressures

magnitude of 0.5 mm diameter. Too large bores lead to recirculation flows as indicated in Fig. 21.2. These lead to measuring errors that yield wall pressures that do not correspond to the true values. Similar errors can be caused by disturbances introduced by burrs, edges, etc., that remain after drilling of the pressure-measurement holes. Further information on making out pressure holes can be found in [21.1] and [21.2].

A somewhat difficult task is the measurement of static pressures in flows, where the pressure in the flow is not obtainable through wall-pressure measurements as described above. In this case, probes have to be inserted into the flow of the kind shown in Fig. 21.3. These static pressure-measuring probes have a streamline design and are provided with one or more holes in their walls. These holes are connected through pressure tubes of small diameter, located in the interior of the probe, to pressure-measuring sensors through which the desired pressure information is obtained. As indicated in Fig. 21.3, it is very important that the appropriate location of the holes is determined experimentally in such a way, that actually the local static pressure is measured by the probe. Often, with sufficient precision, potential-theory computations are suited to determine the best location for the holes in the flow direction. The probe design shown in Fig. 21.4 proved valid for the employment of static pressure measurements in flows.

In order to measure static pressures, it is necessary to employ corresponding pressure-measuring instruments. These are briefly described below.

The simplest of these measuring instruments is the U-tube manometer shown in Fig. 21.5. If a pressure difference develops between points A and B, the liquid levels in the two tubes will adjust themselves in such a way, that the existing pressure difference is compensated by $g\rho_F h$, with g being the

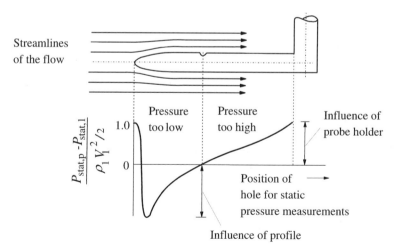

Fig. 21.3 Measurements of wall pressures with pressure probes

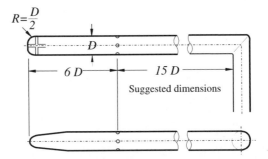

Fig. 21.4 Pressure-measuring probe and location of static pressure-measuring hole

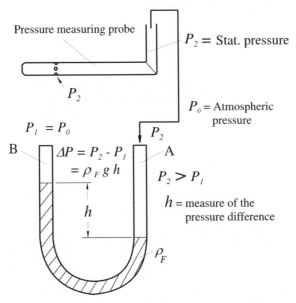

Fig. 21.5 U-tube manometer for the measurement of pressure differences

gravitation constant, ρ_F the density of the fluid in the U-tube and h the difference in the height of the surface layers.

The distance between the layers of the two fluid levels h is therefore in direct relation to the desired pressure difference and can be read by means of an installed scale. It can easily be seen that the measurable pressure range of the instrument can be adapted to each measuring problem by the selection of different fluids (ρ_F). The fluids most commonly used in practice are alcohol, water and mercury, allowing h to be adapted to the measuring problems due to the chosen fluid density.

When placing a large fluid reservoir at a location in the U-tube manometer and inclining the actual measuring tube at an angle of (ρ_F) towards the horizontal line, as shown in Fig. 21.6, one obtains a considerable improvement in the sensitivity of the manometer. A water column h develops, with

21.2 Measurements of Static Pressures

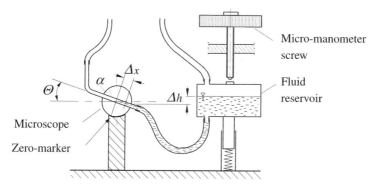

Fig. 21.6 Inclined-tube manometer with zero adjustment

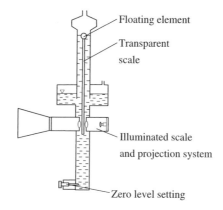

Fig. 21.7 Betz manometer with optical reading system for low pressure difference measurements

this set-up, i.e. a displacement of the fluid over a distance of $h/\sin\theta$ in the inclined arm of the instrument. The zero position, i.e. "zero adjustment" of the instrument, is obtained by adjusting the height of the fluid reservoir.

In the Betz manometer, as shown in Fig. 21.7, a floating element with an attached measuring scale is used to measure the change in the fluid level. The scale, which is made of glass, is projected by an optical enlarging system on to a screen, where another interpolation scale is installed. By this means the reading precision of the measuring instrument is considerably improved.

In the employment of all manometers, a stable and vibration-free base is of importance, in order to avoid vibrational influences on displays. Mobile pipes between the pressure probe and the manometer employed have to be avoided also, in order to avoid fluctuations of the fluid levels caused by pipe movements, or by volume changes of the pipes. In particular, however, the room temperature during a measurement has to be kept constant, as the density of the fluid (ρ_F) is temperature-dependent, so that temperature changes during the measurements are not measured as pressure changes in the flow.

In today's experimental fluid mechanics, the employment of U-tubes as manometers is less common. Developments of pressure sensors have led to measuring instruments that nowadays can be employed successfully to measure pressures in fluid-flow experiments.

21.3 Measurements of Dynamic Pressures

In fluid mechanics, mechanical probes are employed for measuring the total pressure $P_\infty + (\rho/2)U_\infty^2$. For these probes, different designs have been developed over the years. Particularly simple designs of total-pressure probes, which are named Pitot probes, after the French physicist Henri Pitot (1695–1771), are sketched in Fig. 21.8. Probes of this kind are very suited for total-pressure measurements; however, they require precise adjustment, orientating the probe in the flow direction, so that the angle influence of the pressure distribution is eliminated. Only if this precaution is taken, one can measure the desired total pressure correctly.

In some measurements, Pitot tubes of the kind sketched in Fig. 21.9 are used that possess a nozzle around the actual Pitot tube. These probes are less direction-sensitive. The nozzle serves a correcting device, guiding the fluid flow, so that it passes the entrance of the Pitot tube in a more parallel way.

On combining a Pitot probe and a probe for measuring the static pressure, one obtains the so-called Prandtl probe, that permits the following quantity

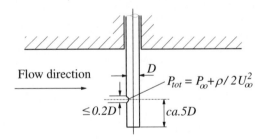

Fig. 21.8 Designs of Pitot probes to measure the total pressure P_{tot}

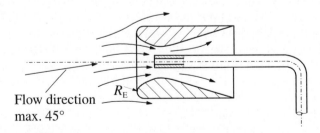

Fig. 21.9 Pitot probes mounted in a Venturi nozzle

21.3 Measurements of Dynamic Pressures

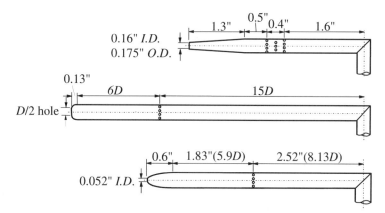

Fig. 21.10 Designs of Prandtl probes, shown with original dimensions, according to data from the N.P.L. in England

to be found:
$$\Delta P = P_{ges} - P = \frac{\rho}{2}U_\infty^2 \qquad (21.1)$$

In Fig. 21.10, practical probe designs for Prandtl probes are shown. Prandtl probes measure the total pressure at a corresponding probe tip and, through the pressure holes placed downstream in the probe walls, the static pressure is measured. Using pressure-measuring instruments to obtain the total pressure and the static pressure at the same time, results in the displayed pressure difference $\Delta P = P_{tot} - P_{stat}$. This leads to the determination of the flow velocity:
$$U_\infty = \sqrt{2\frac{\Delta P}{\rho}} \qquad (21.2)$$

Disregarding the perturbations introduced into a flow by the probe, flow-velocity measurements with Prandtl probes can be categorized as unproblematic when they are employed in laminar flows. Employment in turbulent flows, on the other hand, is not free of problems. The reason for this can be found in the continuous change of the flow direction and the continuous change in velocity magnitude which occur in turbulent flows. This leads to integral pressure measurements which can only be interpreted with difficulty, as far as the local mean velocity of the flows is concerned, as it exits at the measuring location. It is important to emphasize that this problem cannot be removed, even when using pressure-measuring instruments with large time constants, despite the often held opinion in this respect. Due to the hose connections from the probe to the measuring instrument, large time constants already exist for most pressure measurements in fluid mechanics, so that even the employment of fast pressure-measuring devices is not suited to resolve the fast time-dependent pressure.

Hence, pressure probes are not suited to study turbulent flows. In this respect, it is important to point out that the Prandtl tube measures the dynamic pressure, but it is not suited to investigate the dynamics of turbulent flow fields. Such investigations are better carried out with hot-wire and laser Doppler anemometers.

When special measuring probes are employed, in spite of the above mentioned limitations, for investigations of turbulent velocity fluctuations, the frequency characteristic of the probe has to be determined before the employment in turbulent flows fields. The probe has to be calibrated dynamically. Such calibrations are difficult. Yet, efforts of this kind are undertaken again and again in various research efforts in fluid mechanics. This indicates that in spite of major problems with pressure-measuring probes, the measurement of fast varying pressures is of importance in fluid mechanics. Because of this, newly developed pressure sensors are employed, utilizing different mechanisms to measure pressure and to convert the pressure into suitable electronic signals. A pressure probe consists, in general, of the actual sensor element having low mechanical inertia, a surface as large as possible and very short transmission elements which connect the sensor to the location where the electrical signal is recorded. Spring-plunger devices and also capacitive and inductive transducers are employed in this respect. All these devices have reached an advanced state of development and nowadays can be adapted to support pressure measurements in fluid mechanics research.

21.4 Applications of Stagnation-Pressure Probes

In the preceding section, it was shown that stagnation pressure probes, in connection with pressure-measuring instruments, can be employed in order to measure velocities of flows. Stagnation probes are in principle open tubes, one side of which is exposed to the arriving flow, while the other side is connected to a pressure element. The pressure measured in this way is, in principle, except for measurements at very low static pressures, proportional to the mean force per unit surface of a fluid particle located in the stagnation point of the flow. Measurement of the stagnation pressure and the static pressure allow one to obtain the sought velocity information, if it is possible to derive simple relationships between the velocity and the measured pressures. This is in general possible only for non-viscous media. In viscous media, one has to take the influence of the viscosity of the flowing fluid into account. This was not taken into account when employing the simple evaluation equations (see the equations in Sect. 21.2). In the following, it is shown how great the influences of viscosity on measurement with pressure probes can be.

By appropriate analytical derivations, the total pressure $(P_\infty + (\rho/2)U_\infty^2)$ can be set in relation to certain properties of fluid flows, e.g. to the local flow velocity and the locally existing static pressure. Corresponding analytical

21.4 Applications of Stagnation-Pressure Probes

considerations start, however, from assumptions that often do not exist in experimental investigations. As an example, the measured pressure P_{ges} is not in agreement with the pressure that would be measured if one permitted the locally existing flow velocity to stagnate in an isotropic manner, i.e. $P_{\text{ges,i}} = P_{\text{ges}}$. Considerations taking geometric and dynamic influences into account, allow one to express the pressure relation $P_{\text{ges,i}}/P_{\text{stat}}$, existing for real probes, by multiplying the measured pressure by a correcting function. The latter depends on the following parameters:

1. The Reynolds number of the flow, formed with the local flow velocity, the kinematic viscosity of the fluid and the probe diameter, $Re = (U_\infty D)/\nu$
2. The Mach number of the flow, Ma
3. The Prandtl number of the flow, Pr
4. The ratio of the heat capacitances, $\kappa = c_p/c_v$
5. The intensity of the flow turbulence, k^2/U^2
6. The Knudsen number, i.e. the ratio of the mean free path length of the molecules to the probe radius as a characteristic quantity for a slip velocity, λ/R
7. The ratio of the relaxation time of the molecules as an assembly to a characteristic macroscopic time interval, $t_m/(R/U)$
8. The angle of the inflow, i.e. the angle between inflow direction and probe axis, α

In the subsequent paragraphs, some experimentally found dependencies of the measured stagnation pressure on some of the above-mentioned influencing quantities are given and discussed.

Provided that the Reynolds number $Re = (UD)/\nu$ is larger than approximately 100, the above treated simple description of the flow around the total-pressure measurements is sufficiently accurate, i.e. when treated by the theory of ideal fluid flows. At smaller values of Re, however, the stagnation pressure increases as a function of the Reynolds number and also depends on the probe geometry. The measured total pressure can be expressed as

$$P_{\text{ges}} = P + \frac{\rho}{2}U_\infty^2 + 2\mu\beta \tag{21.3}$$

From this pressure coefficient, C_p results:

$$C_p = \frac{P_{\text{ges}} - P}{\frac{\rho}{2}U_\infty^2} = \frac{2\mu\beta}{\frac{\rho}{2}U_\infty^2} \tag{21.4}$$

The influence of viscosity on the pressure coefficient (C_p) is shown in Fig. 21.11, indicating the expected deviation at low Re-number.

When compressibility influences occur, the deviations between the measured total pressure and the ideal value can be stated as follows:

$$\frac{P_{\text{ges}}}{P_{\text{ges,i}}} = 1 + \frac{M_\infty^2}{Re}\left[\frac{2\kappa C_1\left(1 - C_3 M_\infty^2\right)}{1 + \frac{C_2}{\sqrt{Re}}}\right]\left(1 + \frac{\kappa - 1}{2}M_\infty^2\right)^{\frac{-\kappa}{\kappa-1}} \tag{21.5}$$

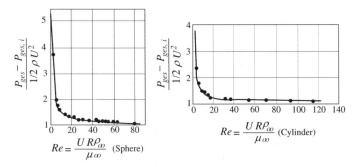

Fig. 21.11 Viscosity influences on total-pressure measurements

All of the above explanations, relating to influences that exist for the practical application of pressure probes, indicate that the application of stagnation-pressure probes for flow investigations is linked to a multitude of practical problems. Nevertheless, owing to the simplicity of application of pressure probes in practical measurements and owing to their robustness, they are still employed today in practical fluid mechanics.

21.5 Basics of Hot-Wire Anemometry

21.5.1 Measuring Principle and Physical Principles

Hot-wire anemometry is a measuring technique that permits electrical measurement of local flow velocities, and it has been employed successfully for a long time in experimental fluid mechanics and also in other fields. With hot-wire and hot-film probes, local, rapidly changing fluid flows, requiring high resolutions of measurements in terms of space and time, can be investigated.

Hot-wire and hot-film measurements are based on the measurement of the release of heat from hot sensors to the medium flowing around them. Hot-wire anemometry is therefore an indirect measurement technique, as it is not the flow velocity which is measured, but the velocity-dependent heat release from a thin, heated wire to the fluid medium surrounding it. In this respect, the technique takes advantage of this velocity-dependent heat transfer of hot wires and hot films for measuring the local flow velocity.

The heat release of the probe depends not only on the flow velocity, however, but also on other quantities:

(a) The temperature difference between the sensor and the fluid medium
(b) The physical properties of the fluid medium and the sensors
(c) The dimensions of the sensors and the design of the hot-wire prongs

When keeping the influences (a), (b) and (c) constant, the sensor only reacts to the flow velocity, and the heat, withdrawn from the sensor, is then

21.5 Basics of Hot-Wire Anemometry

a direct measure of the flow velocity existing at the sensor at the moment of measurement, i.e. a hot-wire anemometer can be built with a high time resolution.

The basic element of hot-wire anemometry is a cylindrical sensor that can be heated in a controlled way and whose electrical resistivity depends on the temperature. The temperature and related resistivity change of the wire, due to velocity changes, can be appropriately recorded electronically in a bridge circuit. The probe itself represents a bridge arm of this bridge circuit. Of the different bridge circuits in practice, the Wheatstone bridge has proved to be especially suited for hot-wire measurements.

From the heat loss of the sensor, which in the state of thermal equilibrium has to be equal to the heat produced electrically over the wire:

$$\dot{Q} = IE = I^2 R = \frac{E^2}{R},$$

where I is the electric current, E is the applied voltage and R is the resistance of the wire, the flow velocity can be determined using two circuit variants. From the need for measurements to have one dependent quantity to measure the heat loss, one can either keep constant the electric current I, or the temperature of the sensor through the wire resistance R. In the first case one talks of the constant-current anemometer (CCA) and in the second case of the constant-temperature anemometer (CTA).

In the constant-current operation of a hot-wire anemometry, the Wheatstone bridge is operated with a constant electric current. For this kind of operation, the resistance of the energy source has to be large in comparison with the total resistance of the bridge, in order to keep the current operating the bridge constant at all measuring times. The temperature and resistance changes of the hot wire, due to velocity changes of the fluid flow, induce in the circuit in Fig. 21.12, an imbalance of the voltage at the vertical bridge diagonal, e.g. the voltage between ports A and B. The resulting bridge output signal is amplified and is then displayed as a measure of the flow velocity of the fluid.

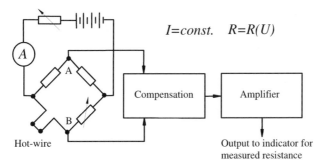

Fig. 21.12 Principal circuit of a constant-current anemometer for hot-wire measurements

One disadvantage of the velocity measurements by constant-current hot-wire anemometry is the small bandwidth of the system. This disadvantage can be attributed to the thermal inertia of the hot wire. The upper frequency of a hot wire of 5 µm diameter is approximately 100 Hz in constant-current operation. By using very thin wires, the time constant can be reduced, but thin wires are very sensitive to mechanical influences and can therefore easily be destroyed by the flow as a result of mechanical stresses.

Moreover, for the constant-current anemometry, operational difficulties exist. Due to the increasing heat losses to the flowing fluid at high velocities, the supply current has to be increased at high velocities. On the one hand, this leads to an increased sensitivity of the wire, i.e. the wire reacts more strongly to occurring velocity changes. However, when carrying out measurements the risk increases that the hot wire will burn, e.g. when the velocity decreases suddenly in a flow, the current cannot be removed fast enough from the wire and, hence, the wire burns. Finally, the dependence of the time constant of the hot wire on the mean flow velocity makes an adjustment of the compensation network to the corresponding flow velocity necessary. When using hot-film probes in their constant current mode, which possess high time constants, a compensation amplifier with a complicated control circuit is required. This special amplifier has to have very precisely the opposite frequency response to the hot-wire probe and therefore is hard to design and build in practice, let alone its employment in measurements.

Nowadays, the constant-current operation of hot wires is employed almost only for measurements of fluid temperatures. For this application, the constant heating power is reduced, in order to decrease the response of the probe to the flow velocity, so that almost exclusively temperature changes in the fluid lead to imbalances of the bridge. The bridge voltage at the horizontal diagonals in Fig. 21.13 is then a measure of the instantaneous flow temperature.

The basic idea of constant-temperature anemometry (CTA) results in an electronically achieved compensation of the thermal inertia of the probe by a fast electronic voltage feedback, guaranteeing the operation of the sensor at a constant temperature, i.e. at a constant wire resistance.

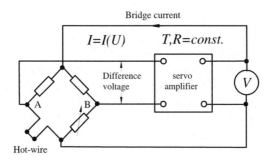

Fig. 21.13 Principle electrical circuit of a constant-temperature anemometer

21.5 Basics of Hot-Wire Anemometry

In the case of the balanced bridge, no voltage difference exists between the entrance ports of the servo-amplifier. Velocity changes in the flow, however, result in temperature and corresponding resistance changes of the hot-wire sensor, which cause voltage differences at the servo-amplifier input ports. The exit of the servo-amplifier is back-coupled to the vertical parts of the bridge, as shown in Fig. 21.13, with a polarity such that the bridge adjusts itself automatically to the new heat transfer situation. Through this back-coupling, a signal is generated which is not influenced considerably by the thermal inertia of the sensor, i.e. the upper frequency of the hot-wire anemometer response is raised by several orders of magnitude compared with constant-current operation of a hot wire. The upper frequency can reach up to approximately 1.2 MHz at high flow velocities. This upper frequency of the constant-temperature anemometer is essentially determined by the frequency response of the feedback amplifier and not by the time constant of the wire.

Advantages of constant-temperature operation are the above-mentioned large bandwidth and the possibility of choosing high operating temperatures of the sensor, to obtain a very high sensitivity to velocity changes. A disadvantage is the unstable behavior of the servo-amplifier in some extreme operational cases.

21.5.2 Properties of Hot-Wires and Problems of Application

As measuring sensors for hot-wire measurements, usually hot wires with a cylindrical form, with typical diameters of a few µm and a length larger than 200 times the wire diameter, are used. For hot-wire sensors, the wire is mounted between the tips of special supports to which the wire is soldered or welded. Because of this, certain mechanical demands are required to permit the wire to be placed between the tips of the two supports (prongs). The stated diameters and lengths of hot wires employed are a compromise between the required mechanical strength and the upper frequency of the measuring system. Typically, measuring wires with a diameter of 5 µm and a length of 1–2 mm are employed for flow measurements.

Special measuring requirements, in different velocity fields, place different requirements on hot-wire probes and make the employment of appropriate sensors necessary, usually possessing special probe geometries. Thicker wire sensors are employed when a higher mechanical stability is required; thinner wires are employed when higher frequencies are required.

The dominating and for hot-wire probes decisive property of the sensor material is the dependence of the electrical resistance on temperature. This dependence can be stated as follows:

$$R = R_0[1 + \alpha_1(T - T_0) + \alpha_2(T - T_0)^2 + \cdots], \tag{21.6}$$

where R is the wire electrical resistance at the operating temperature T, R_0 is the corresponding value at the reference temperature T_0, and α_1 and α_2 are the thermal resistance coefficients. Preferred are hot-wire materials with α_1 values as high as possible and extremely small α_2 values. In such cases the square term in (21.6) can be neglected, so that the electrical resistance changes practically linearly with temperature. For platinum as an example, the values for the resistance coefficients are

$$\alpha_1 = 3.5 \times 10^{-3} \, \text{K}^{-1}; \quad \alpha_2 = -5.5 \times 10^{-7} \, \text{K}^{-2} \tag{21.7}$$

and for tungsten

$$\alpha_1 = 5.2 \times 10^{-3} \, \text{K}^{-1}; \quad \alpha_2 = 7.0 \times 10^{-7} \, \text{K}^{-2}. \tag{21.8}$$

In Fig. 21.14, the variations of the electrical resistance with temperature are given for some pure metals.

During hot-wire measurements, the temperature along the hot wire usually differs which is explained in detail later; the measured resistivity has a mean value $R = \int_0^1 [R(z)/A(z)] \mathrm{d}z$. $A(z)$ is the hot-wire cross-section and z the coordinate along the hot wire, i.e. in direction of the flow of the electric current. $R(z)$ can therefore be considered the local resistance of the hot wire.

Another important parameter for the sensitivity of the anemometer is the overheating relation $\beta_T = (T - T_0)/T_0$, where again T is the hot-wire temperature and T_0 the reference temperature in K. Of more practical importance is the relation $\beta_R = (R - R_0)/R_0$, where R is the resistance of the sensor at the operating temperature T and R_0 the resistance at the reference temperature T_0. From the above, the following holds: $\beta_R = \alpha_1 T_0 \beta_T$ (with $\alpha_2 = 0$). In practice, one chooses the operating temperature of the hot wire as high as possible, in order to obtain a high sensitivity for the velocity changes and also a reduction of the influence of the flow-medium temperature. As a rule, tungsten

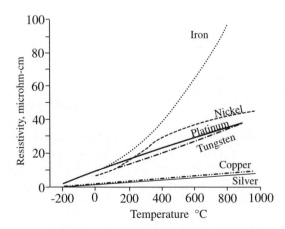

Fig. 21.14 Temperature-resistivity behavior of hot-wire materials

21.5 Basics of Hot-Wire Anemometry

wires coated with platinum tolerate temperatures up to 200–300°C. At high flow temperatures, sensors made of platinum and a 10% platinum-rhodium alloy are employed, permitting operating temperatures up to 750°C.

Measurements of flow velocities of fluids impose requirements which make the employment of special sensors necessary. These sensors consist of differently shaped elements of quartz glass, on to which thin-film layers (e.g. nickel) have been coated. Film probes can be shaped conically, like a wedge, or can have other shapes, so that they fulfill the requirements enforced by differing measurement problems. Film sensors are moreover coated with a quartz layer for protection, so as to be less sensitive towards environmental influences. In addition, the quartz layer provides electrical insulation for the film sensor and thus makes it applicable to electrically conductive fluids.

In fluid flow measurements, use can be made of different types of probes with hot-wire or hot-film elements, in order to measure wall-shear stress information in addition to carrying out local velocity measurements. Shear stress sensors are formed as flat heating elements, as shown, e.g. in Fig. 21.15, among other sensor shapes. For measurements in liquids, sensors of thin metal films are provided with a protective layer (insulation), in order to avoid electrolytic interactions between the sensor and measuring fluid. All of this makes

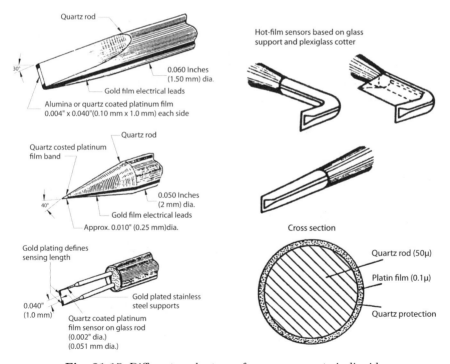

Fig. 21.15 Different probe types for measurements in liquids

it clear that flow-measurement technology nowadays employs hot elements extensively, in order to measure fluid mechanically relevant quantities, when carrying out experimental flow investigations.

The employment of hot-film technology for flow measurements in fluids requires special skills and much care from the experimentalist, in order to obtain reliable velocity measurements. The above-mentioned special designs of film sensors are required because of the special properties of liquids. The most important of these properties disturbing the execution of hot-film measurements are as follows:

1. The boiling temperature of fluids is low.
2. Organic fluids can decompose.
3. Fluids generally possess electrical conductivity.
4. Fluids dissolve gases and these can be set free.
5. Fluids are usually more contaminated than gases.
6. In water and other fluids salts are dissolved.
7. Tap water contains algae, bacteria and microorganisms.

In order to be able to obtain reproducible results when doing measurements in fluid flows, the above-mentioned special properties of fluids have to be taken into account.

Because of point 1 above, when carrying out hot-film measurements, the operating temperature of the sensor has to stay below the boiling temperature of the flow medium, as otherwise boiling of the fluid at the heated sensor occurs. For practical reasons, it is important to consider that lower operating temperatures, compared with the boiling temperature, have to be chosen for the mean temperature of the hot film. As the temperature distribution of commercially available cylindrical hot-film probes, having a length of only 20–30 times their wire diameter, show a steep temperature maximum in the wire center, as can be deduced from measurements with an infrared detector (Fig. 21.16). It is this maximum temperature that must not exceed the boiling temperature when carrying out hot-film measurements in liquids. When the temperature distribution along the wire is not taken into consideration when setting the overheating temperature, the boiling temperature of the fluid can easily be exceeded locally in the probe center and evaporation of the fluid can occur. This leads to local modifications of the heat transfer between sensor and fluid and thus to erroneous measurements.

Organic fluids decompose (point 2) after exceeding a critical temperature, lower than the boiling temperature. This can lead to depositions on the probe surface, which usually result in decreases in the anemometry output voltages.

Electrical conductivity of a fluid (point 3) leads to electrolysis at the sensor surface of uncoated and, hence, unprotected films. Due to unprotected exposure of the sensor to the liquid, gas bubbles (H_2 or O_2 bubbles in water) are generated and the sensor material is worn away from the wire surface as a result of electrolysis, which is manifested by an increase in the cold

21.5 Basics of Hot-Wire Anemometry

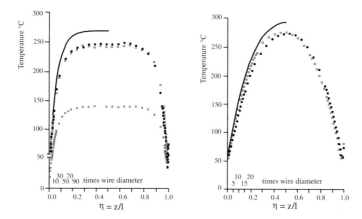

Fig. 21.16 Temperature distributions along the hot wires (Champagne et al., 1967). *Left*: length of a hot wire having a diameter of 400 μm in two overheating relations. *Right*: length of a hot wire of 99 μm diameter

resistance of the sensor. The increased electrical resistance due to the wearing away of sensor material, caused by the locally weakened cross-section of the wire, leads to an increase in the local probe temperature, which in turn intensifies the electrolysis at this point, until finally the wire or film sensor breaks. Therefore, sensors working in electrolytes always require a thin quartz layer for protection, in order to separate the wire or film from the electrically conductive flow medium.

There is also a decrease in the heat transfer between the sensor surface and the fluid, caused by degassing of the gases dissolved in the liquid (point 4). This inherently leads to non-reproducible velocity measurements. Once a gas bubble has deposited at the sensor surface, usually the formation of further bubbles takes place very quickly. They modify the heat release to the flowing medium, and thus the evaluation of the measurements can no longer be based on the carried out calibration. This can be remedied by degassing the flow fluid before the measurement. In the simplest case it is already sufficient to leave the fluid to stand quietly for some time, so that small air bubbles are discharged by rising in and leaving the fluid. However, it is better to bring about degassing by heating or by creating an under-pressure above the fluid surface before starting the measurements.

Dirt particles also deposit on the sensor used in fluids (point 5) and modify thus the heat transfer between sensors and the fluid flow. Therefore, the fluid should be kept as clean as possible during one series of measurements. This can be effected by continuous filtration with sufficiently small filter pores. Covering the flow channel with a protective cover is also recommended, in order to avoid the continuous entry of dirt. Contaminated sensors have to be cleaned. Dust can be removed mechanically, e.g. by brushing it off. It is also usual to rinse the probes in methanol to remove depositions of dirt in this way, or to clean the sensor in an ultrasound bath.

The most commonly used liquid in fluid mechanics is water. The salts dissolved in water generally lead to depositions on the sensor surface (point 6). Calcium carbonate is an essential part of the layers deposited on hot-film sensors. Calcification of the sensor is substantially stopped if the operating temperature of the sensor is below 60°C.

Finally, tap water contains algae, bacteria and microorganisms (point 7). In measurements with hot wires and hot films, slimy depositions can form on the sensor surface and thus lead to deterioration of the heat transition. In order to minimize these depositions, the flow channel should be set up in a dark room and not be exposed to solar or light radiation. Moreover, adding small amounts of borax is also recommended to stop the development of algae and microorganisms.

In general, the disadvantageous influences outlined in points 1–7 result in non-reproducible velocity measuring results and require regular recalibrations of the sensors employed for flow measurements in fluids. The same holds for corrosive changes, structural changes and other uncontrollable influences to which the wire or film material is exposed in the experiments, or during storage between experiments. Only by continuous surveillance of the calibration of the probes can negative influences on the measurements be excluded.

21.5.3 Hot-Wire Probes and Supports

As already mentioned, the actual sensor, in the case of hot-wire probes, is mounted between the two tips of two prongs acting as holders, and the wire ends are, as a rule, soldered on these wire holders. The prongs of the probe are inserted in a ceramic body acting as probe holder. Normally the hot wire is a platinum-coated tungsten wire.

In order to reduce the heat conduction from the hot wire to the cold holder tips and to be able do define the active sensor length more precisely, copper-plated or gold-plated probes have been developed, in which the sensor ends welded on to the prongs are copper-plated or gold-plated (Fig. 21.17). Thus

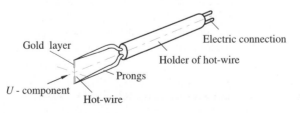

Fig. 21.17 Single-wire gold-plated sensor with prongs and probe filter

21.5 Basics of Hot-Wire Anemometry

the temperature distribution along the active sensor length is more uniform than with non-plated probes. Because of the larger distance of the prongs from the measuring point, the flow field in the region of the active sensor part is less disturbed. The described single-wire probes can show different configurations, according to the application purpose. Probes with equally long, straight prongs, where the hot wire forms an angle of 90° with the axes of both prongs, are employed for measurements of mean flow velocities and of the velocity fluctuation in the main flow direction. Figure 21.18 shows such a hot-wire sensor, which is oriented in the flow such that it is measuring the $\bar{U}$ velocity component, indicated in Fig. 21.18.

Probes with unequally long straight prongs, where the hot wire forms an angle of 45° with the axes of the two prongs, serve for measuring Reynolds shear stresses. They are used sequentially with straight probes, as shown in Figs. 21.18 and 21.19. Additional measurements with ±45° then yield $\overline{u'^2}$, $\overline{v'^2}$ and $\overline{u'v'}$, i.e. all elements of the Reynolds stress tensor.

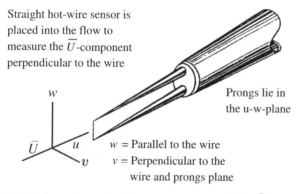

Fig. 21.18 Straight hot-wire probe for measurements of the $\bar{U}$ component and u' fluctuations in a turbulent flow

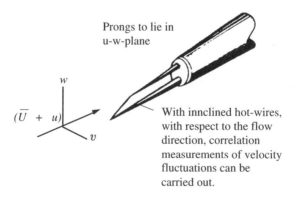

Fig. 21.19 Inclined probe for measurement of combined $\overline{u'w'}$ and $\overline{u'v'}$ term

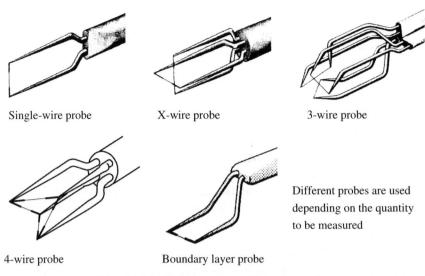

Fig. 21.20 Different types of hot-wire probes

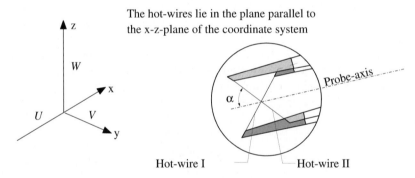

Fig. 21.21 X-probe for simultaneous measurement of the second velocity component

In Fig. 21.20, different probe holders, also for multiwire probes, are shown, giving a good overview over those wires used today for fluid velocity measurements.

For determining the flow direction in a plane, where two velocity components are located in this plane, and carrying out measurements in one measuring operation, so-called X-probes are used with two wires or films standing perpendicular to one another, as shown in Fig. 21.21. The wires are mounted parallel to the x–z plane and thus consist of two combined "inclined probes" with a probe inclination of $\pm 45°$, as shown in Fig. 21.19.

The following velocity relationships result for wires A and B:

$$U^A_{\text{gem}} = U \cos \alpha + W \sin \alpha$$
$$U^B_{\text{gem}} = U \cos \alpha - W \sin \alpha$$

21.5 Basics of Hot-Wire Anemometry

By addition and subtraction it is possible, as the above equations show, to determine the instantaneous U and W components of the velocity field.

In practice, it is also usual to employ three-wire probes, in order to measure all three velocity components simultaneously. At this point, we only want to mention this fact. It is the object of this section of the book, to give an introduction to flow-measurement technology and for this purpose the above references to a few hot-wire probe geometries suffice.

For boundary-layer investigations, it is extremely important to carry out measurements close to walls. For such investigations probes with wire holders are employed, which are formed in such a way that they permit measurements near to the walls. Such a probe with specially formed prongs is shown in Fig. 21.22. It is oriented such that the u component of the fluid is measured. Traversing takes place in the y direction, to obtain the U velocity profile.

Different demands are placed on the geometry of the hot-wire probes. In order to keep the inevitable introduction of disturbances into the flow by hot-wire probes low and to obtain a good spatial resolution and a high vibration resistivity of the probe, the probes should, on the one hand, have probe lengths as short as possible. On the other hand, in order to reduce the disturbing influence of the prongs, a large distance between the prongs would be required. Small wire diameters are required for high resolution in terms of time and space. Large diameters, on the other hand, ensure high mechanical strength and smaller wire strains when mechanically stressed. By compromising, nowadays optimized probes are available which permit reliable measurements by means of hot-wire anemometers.

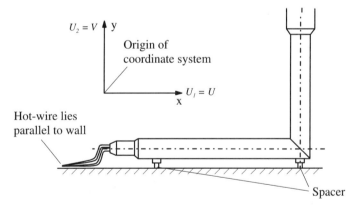

Fig. 21.22 Probe for boundary-layer investigations with special prong arrangement

21.5.4 Cooling Laws for Hot-Wire Probes

The basis for determining the flow velocity by means of hot-wire probes, is the heat transfer from the heated sensor to the medium flowing around the sensor. The heat can be transferred from the sensor by radiation $\dot{Q}_R$, conduction $\dot{Q}_C$, free convection $\dot{Q}_{FC}$, and especially by forced convection $\dot{Q}_{con}$ (Fig. 21.23).

In the thermal equilibrium state, the supplied electric power is

$$\dot{Q}_{el} = IE = I^2 R = E^2/R \qquad (21.9)$$

equal to the heat output carried off by the sensor:

$$I^2 R = \dot{Q}_R + \dot{Q}_C + \dot{Q}_{FC} + \dot{Q}_{con} \qquad (21.10)$$

The radiant heat $\dot{Q}_{RSt}$ can be computed according to the equation:

$$\dot{Q}_R = k\sigma A \left(T^4 - T_m^4\right), \qquad (21.11)$$

where σ is the Stefan–Boltzmann constant, A the heat-radiating surface of the sensor, T is the operating temperature and T_m the temperature of the flow medium. The factor k is at around 0.1 and takes into account that the radiation of hot wires amounts to about 10%, at the very most, of the radiation that a black body of equal dimensions would have. Except for extreme cases, the heat loss of hot wires due to radiation can be neglected, as it is only a small percentage of the heat which is transferred from the sensor by forced convection.

The heat conduction $\dot{Q}_C$ from the hot sensor into the cold prongs is, according to Fourier:

$$\dot{Q}_C = -2\lambda_D \left(\frac{\mathrm{d}T}{\mathrm{d}x}\right) \underbrace{\frac{\pi d^2}{4}}_{\text{end of sensor}}, \qquad (21.12)$$

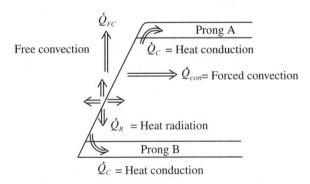

Fig. 21.23 Heat balance at the sensor in general form

21.5 Basics of Hot-Wire Anemometry

where λ_D is the heat conductivity of the wire, d the wire diameter and $\mathrm{d}T/\mathrm{d}x$ the temperature gradient. The factor 2 before λ_D, is present because of the two prongs needed to hold the wire. For computing $\dot{Q}_C$, it is necessary to know the temperature gradient at the wire ends. The temperature variation along the sensor depends implicitly on the dimensionless heat-transfer coefficient expressed by the Nusselt number (Nu). With hot-wire probes, the heat loss $\dot{Q}_C$, the so-called wire end loss to the prong, amounts to about 10–20% of the total heat loss from the sensor. Seen relatively, this proportion is the larger the smaller is the ratio of wire length to wire diameter.

The heat carried off from the sensor, due to free convection $\dot{Q}_{FC}$, gains in importance when the buoyancy forces acting on the fluid flow considerably influence the flow field around the wire. The characteristic dimensionless quantity, which allows one to describe this influence, is the Grashof number:

$$Gr = \frac{g\beta\Delta T L^3}{\nu}, \tag{21.13}$$

where g is the gravitation acceleration, β is the compressibility coefficient, ν is the kinetic viscosity and ΔT is the wire overheating temperature.

According to Collis and Williams (1959), free convection can be neglected in the case when

$$Re > Gr^{1/3} \tag{21.14}$$

The Grashof number, e.g. for a hot wire of $2.5\,\mu m$ diameter in an air stream at 300 K is about 6×10^{-7}; therefore, for Reynolds numbers larger than 0.01, no considerable free convection effects on the heat transfer of a hot wire are to be expected. This means that for the usually employed hot wires in air, free convection can be neglected at flow velocities larger than $0.1\,\mathrm{m\,s^{-1}}$.

In the case of velocity measurements with hot wires, the dominating heat-transfer component from the wire to the flow medium surrounding it, takes place by forced convection $\dot{Q}_{con}$. The latter can be calculated as follows:

$$\dot{Q}_{con} = \alpha \pi l d(T - T_m), \tag{21.15}$$

where T is the wire temperature, $d = 2r$ is the wire diameter, l is the wire length, T_m is the fluid temperature and α is the heat-transfer coefficient. It can be computed with the help of the Fourier law:

$$\dot{Q}_{Zk} = -\lambda l \int_0^{2\pi} \left(\frac{\partial T}{\partial r}\right)_{r=R} R\,\mathrm{d}\varphi, \tag{21.16}$$

where λ is the heat conduction of the fluid.

The dimensionless heat-transfer coefficient at the sensor is defined as

$$Nu = \frac{\alpha d}{\lambda} \quad \text{(Nusselt number)} \tag{21.17}$$

From the above two equations, the heat transfer by convection $\dot{Q}_{con}$ can be computed as

$$\dot{Q}_{con} = Nu\pi l\lambda(T - T_m) \tag{21.18}$$

Thus a simplified energy balance at the hot-wire sensor reads

$$\frac{E^2}{R} = 2\lambda A \left.\frac{dT}{dx}\right|_{\text{wire end}} + Nu\pi l\lambda(T - T_m) \tag{21.19}$$

For handling this equation further, a general heat-transfer law has to be formulated for hot-wire probes. The similarity theory of heat transfer states that for geometrically similar flow and heat transfer problems, the temperature and velocity fields are similar, when the dimensionless characteristic quantities are equal. In general, the heat-transfer laws are described by relationships between the Reynolds, Prandtl, Mach, Grashof and Knudsen numbers, of the length-to-diameter ratio of the sensor elements, the overheating ratio, the orientation of the probe in the flow field and other parameters.

$Nu = Nu($	$Re,$	$Pr,$	$Gr,$	$Ma,$	$Kn,$	$l/d,$	$\Delta T \ldots)$
	flow influence	fluid characteristics	buoyancy influence	compress-ability influence	influence of the molecule structure	geometry of the sensor	overheating of the hot wire

For general considerations, the Nusselt number would have to be determined individually for every flow field examined and the probe employed, in order to formulate generally a law that takes into account the above complexity of the dependencies.

For practical applications of hot-wire anemometry in gas flows, the flow velocities are usually higher than $0.1\,\text{m s}^{-1}$, and the influence of the Grashof number on the heat transfer must therefore not be taken into account. The same holds for the Mach number influence of the flow. When this characteristic number does not exceed a certain limit, e.g. $Ma \approx 0.3$, the compressibility effects on the heat transfer can be neglected. Only in special cases, such as in strongly diluted gases, e.g. in measurements of wind speeds at high atmospheric altitudes, the diameter of the sensor can be equal to or even smaller than the free pathlength of the molecules. In the normal case, ℓ (mean free path of the molecules) $\ll d$ (wire diameter), i.e. the heat transfer from the hot wires is not influenced by the Knudsen number, i.e. for all measurements continuum mechanics is applicable.

Moreover, assuming a large length-to-diameter relation of the considered hot wire $[l/d > 400]$, the heat transfer is two-dimensional. With these assumptions, the "Nusselt number dependence" reads:

$$Nu = Nu(Re, Pr, \Delta T, \ldots) \tag{21.20}$$

In spite of these introduced simplifications, it is very difficult to formulate a general law for the heat transfer by theoretical means. The heat transfer from

21.5 Basics of Hot-Wire Anemometry

Table 21.1 Heat-transfer laws

Reference	Validity range	A	B	n	s
Collis and Williams (1959)	$0.02 < Re < 44$	0.24	0.56	0.45	0.17 influence
	$0.02 < Re < 140$	0	0.48	0.51	0.17 of the temperature
Hilpert (1933)	$1 < Re < 4$	0	0.89	0.33	0
	$4 < Re < 40$	0	0.82	0.38	0
	$40 < Re < 4,000$	0	0.61	0.46	0
	$1 < Re < 4$	0	0.872	0.330	0.0825 influence
	$4 < Re < 40$	0	0.802	0.385	0.09625 of
	$40 < Re < 4,000$	0	0.600	0.466	0.1165 the temperature
King (1914)	$Pe = RePr \gg 1$	$\frac{1}{\pi}$	$\sqrt{\frac{2}{\pi}}\sqrt{Pr}$	0.5	0 for $Pr \gg 1$
Koch and Gartshore (1972)	$Re < 4.2$	0.72	0.80	0.45	-0.67
Kramers (1946)	$0.01 < Re < 1,000$	$0.42\, Pr^{0.2}$	$0.5\, Pr^{0.33}$	0.5	0
McAdams	$0.1 < Re < 1,000$	0.32	0.43	0.52	0

the hot-wire sensor is determined by the complex flow field which is developed near the wires. Some of the heat-transfer laws, formulated and available in the literature, are stated in Table 21.1. They are stated considering the following form of a fitted relationship:

$$Nu = [A(Pr, \Delta T) + b(Pr, \Delta T) Re^n] \left(\frac{T - T_m}{T_m}\right)^s \qquad (21.21)$$

The constants used in (21.21) are given in Table 21.1.

Already in 1914 King formulated in his research work, which was fundamental for hot-wire technology, a theoretical solution for the heat transfer from an evenly heated infinitely long cylinder, assuming a two-dimensional incompressible and friction-free potential flow:

$$Nu = \frac{1}{\pi} + \sqrt{\frac{2}{\pi}}\sqrt{RePr} \quad \text{valid for} \quad RePr = Pe > 0.08 \qquad (21.22)$$

This relationship obtained by King (1914) for the Nusselt number is still employed today in experimental hot-wire anemometry, not in the above original form, but in a modified form which is better suited for flow measurements. In practical applications it computes successfully, with empirically found coefficients, the heat transfer laws for hot wires.

If one has decided on an independent representation of the experimental data by a known heat-transfer law, or having found laws of one's own in an investigated flow medium for a particular hot-wire probe, one can easily obtain the anemometer output voltage (measurement value) from a simplified energy balance. The heat-transfer law formulated by McAdams, for example becomes, in this manner, the fundamental relationship for flow velocity measurements:

$$E^2/R = \lambda \frac{\pi d^2}{2} \left. \frac{dT}{dx} \right|_{\text{End of the wire}} + 0.32\pi l \lambda \left(T - T_m\right)$$
$$+ 0.43\pi R l \lambda \left(T - T_m\right) \left(\frac{d}{\nu}\right)^{0.52} U^{0.52} \quad (21.23)$$

The fundamental procedure, when determining the flow velocity from a hot-wire measurement, would then be the following. For a certain hot-wire probe, with known geometric parameters d, l and operating values R (wire resistivity) or T, one obtains the voltage-velocity function dependent on the temperature, the pressure and the thermodynamic properties of the flow medium, in addition the excess temperature $T - T_m$ and the temperature gradients at the sensor end. Knowing these parameters, the desired velocity behavior can be determined from the measured voltage behavior. After all these explanations, it is worth mentioning that in practical hot-wire anemometry direct calibration of the hot-wire sensor is preferred.

21.5.5 Static Calibration of Hot-Wire Probes

The approach described above for determining the heat loss of hot wires permits the velocity behavior to be determined, for velocity measurements, without calibration. However, for this purpose the geometric dimensions of the measuring sensors and the operating values of the entire anemometer have to be known precisely. Experience has shown, however, that a precise knowledge of all the influencing quantities cannot be obtained with sufficient precision for the commercially available hot-wire probes. Because of the complicated processes, when drawing thin wires, the diameter of the active sensor element, to give only one reason, cannot be obtained with high accuracy. Uncertainties also occur when determining the sensor length, due to the welding of the wire to the prongs. There are also other influences acting on the validity of analytical heat-transfer laws, such as aging of the wire material, homogeneity of the wire alloy and corrosion of the sensor material. For all these reasons, in measurement practice, preference is given to the experimental determination of the voltage-velocity function in suitable calibration channels, i.e. the hot wire is directly calibrated and then employed for measurements. The probe considered for flow investigations is placed in a low-turbulence airstream of known and adjustable velocities and the anemometer output voltage E, as a function of the flow velocity U, is determined in the range considered for the planned measurements, employing the calibrated sensor. The static calibration curve, determined in this way, is obtained by plotting the anemometer output voltage as a function of the known calibration

21.5 Basics of Hot-Wire Anemometry

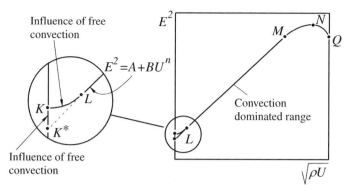

Fig. 21.24 Fundamental diagram of the calibration curve of a hot-wire probe

velocity. There is a non-linear dependence of the anemometer output voltage on the flow velocity.

In order to study the heat transfer from hot-wire sensors over a wide velocity range, i.e. from very low velocities up to high velocities, but $Ma < 0,3$, the supplied electric energy, which is proportional to the square of the voltage, is plotted as a function of $\sqrt{\rho U}$ (Norman, 1967). The reasons for this type of plotting will be discussed later, but equation (21.24) already makes clear the necessity for this type of functional behavior.

One can divide the calibration curve into sub-ranges which physically obey different laws. The sub-range of the calibration curve between L and M in Fig. 21.24 is important for air flows in practical flow cases. It can be approximated analytically as follows:

$$E^2 = A + BU^n \tag{21.24}$$

This relationship is just a modification of King's law for the heat loss from a heated cylinder. One has thus taken over, for explaining Fig. 21.24, the fundamentally existing analytical function between the energy loss and the velocity of King's equation. The parameters A, B and n are determined by calibration, as all the assumptions made by King with regard to the properties of sensors are not known in practice, or do not apply exactly. In the area L to M in Fig. 21.24, A, B and n are almost constant, as results from measurements. In the sub-range of the calibration curve between K and L free convection dominates. With increasing flow velocity or, more precisely, with increasing Mach number, the probe reaches its maximum cooling, and then decreases with further increase of the Mach number. In the sub-range M–N–Q, it is not possible to attribute only one velocity value to each measured value E, i.e. the function in this area is not unique. In measuring practice the hot wire is often employed only in the range L–M.

As the calibration of an employed hot-wire anemometer is every-day routine work for a flow-measurement technician, it is necessary to explain step

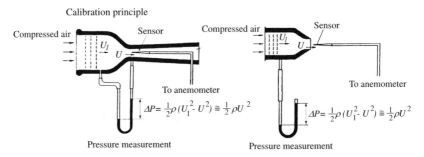

Fig. 21.25 Calibration channels with mounted hot-wire probe and pressure measuring device

by step how to proceed in calibrating a commercially available hot-wire anemometer in an air jet. The probe is mounted, for the calibration, directly in or shortly after the nozzle outlet of a calibration channel (Fig. 21.25). The hot-wire sensor is oriented towards the outcoming flow. It is thus ensured that the probe is located in an area of uniform velocity and low turbulence intensity. In this region, the geometry of the nozzle also defines the flow direction.

The calibration of a hot wire is carried out for many velocity points over the entire velocity range that is of interest for a particular set of measurements. For flow velocities that are not too low and not too high, as mentioned above the calibration follows a law as given by Fig. 21.24. Practical application of a calibrated hot wire is often limited to the range where this simple analytical expression for the $E = f(U)$ dependence can be found, i.e. to the L–M range in Fig. 21.24.

The following data serve as an example of a typical velocity calibration. The room temperature for these measurements was $t_{\text{Atm}} = 19.5°C$ and the atmospheric pressure was $p_{\text{Atm}} = 756\,\text{mmHg}$.

The measured anemometer output voltages E and the pressure difference read from the manometer and the atmospheric pressure, Δp, are given in the first two lines of Table 21.2.

To evaluate the $E^2(U)$ relationship from the data given in Table 21.2, the calibration velocity U has to be computed from the measured pressure differences Δp. Assuming an incompressible friction-free flow, the Bernoulli theorem between the cross-sections in front of and directly behind the calibration nozzle reads

$$\frac{\rho}{2}U_1^2 + p_1 = \frac{\rho}{2}U^2 + p \tag{21.25}$$

With $p = p_{\text{Atm}}$ and $p_1 - p_{\text{Atm}} = \Delta p$, one obtains:

$$\Delta p = \frac{\rho}{2}\left(U^2 - U_1^2\right) \tag{21.26}$$

21.5 Basics of Hot-Wire Anemometry

Table 21.2 Typical data of a hot-wire calibration

Anemometer output voltage (V)	Nozzle pressure Δp at manometer (mm W^{-1} s^{-1})	Calibration velocity (m s^{-1}) $U = 1.2821 \sqrt{\Delta p \,(\text{mm W}^{-1}\text{s}^{-1})} \times 9.8066$	E^2	$U^{0.4356}$
2.88	1.04	4.09	8.29	1.85
3.01	2.11	5.83	9.06	2.15
3.16	4.24	8.27	9.98	2.51
3.28	7.20	10.77	10.76	2.82
3.36	10.25	12.85	11.29	3.04
3.46	15.20	15.65	11.97	3.31
3.54	20.17	18.03	12.53	3.52
3.62	27.42	21.02	13.10	3.77
3.68	33.47	23.23	13.54	3.93
3.73	39.88	25.35	13.91	4.09
3.79	43.34	27.62	14.36	4.24
3.83	54.53	29.65	14.67	4.38
3.90	67.26	32.93	15.21	4.58

When the area ratio of the nozzle inlet to the nozzle outlet is larger than 1:16, as in the present calibration, the velocity U_1 in the above equation can be neglected without great loss of accuracy. Hence, one obtains for the calibration velocity from the Δp measurements:

$$U = \sqrt{\frac{2}{\rho}\Delta p} \qquad (21.27)$$

The density of the air, which is not known yet, can be computed from the law for ideal gases:

$$p = \rho R T \qquad (21.28)$$

Under the calibration conditions mentioned here, ρ is given by:

$$\rho = \frac{P_{\text{Atm}}}{RT_{\text{Atm}}} = \frac{756 \times 133.3}{283\,(273 + 19.5)} = 1.2167\,\text{N s}^2\,\text{m}^{-4} \qquad (21.29)$$

$\left(\text{with } 133.3\,\text{N m}^{-2}/\text{mmHg} = 1;\quad R_{\text{air}} = 283\,\text{m}^2\,\text{s}^{-2}\,\text{K}^{-1}\right)$

With this result, the calibration velocity can be computed:

$$U = 1.2821\sqrt{\Delta p}\,, \qquad (21.30)$$

where $\sqrt{\Delta p}$ represents the pressure difference read from the manometer. In the above equation, Δp has to be multiplied by 9.8066 (1 mm W s $\hat{=}$ 9.8066 n m^{-2}) and then only the root has to be extracted and finally to be multiplied by 1.2821. In this way, one obtains the calibration velocities stated in the third column of Table 21.2. Figure 21.26 shows a typical calibration curve of a hot-wire probe. It was obtained by plotting the voltage measured at the anemometer outlet as a function of the computed calibration velocities.

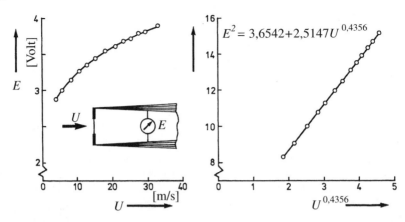

Fig. 21.26 Calibration curve taken during calibration, in two manners of representation

The application of hot-wire anemometers to determine the local velocity of a fluid flow from voltage measurements, is occasionally helped by applying an analytical expression for the voltage-velocity laws. As already mentioned, in the velocity range investigated here, this law can be represented by the modified King law:

$$E^2 = A + BU^n \tag{21.31}$$

The calibration task lies in the determination of the constants A, B and n from the measured data. This can be done graphically, by plotting the square of the anemometer output voltage against U^n. However, the exponent n is not known a priori, so that a variation of n is required, until the correct exponent n yields the measurement points lying on a straight line. The exponent n depends somewhat on the flow velocity; in a limited velocity range a constant exponent n can be defined, however. The gradient of the straight line in the $E^2 - U^n$ diagram corresponds to the constant B. The voltage value in the point of intersection of the extrapolated straight line and the E axis provides the constant A.

The values for the constants A, B, n can also be determined numerically. In an iteration procedure, the exponent n is changed systematically, and for each n value the other remaining constants A and B are evaluated from the calibration data by applying the method of least-squares fit. When a minimum of the "square of errors" between the analytical expression and the calibration data is obtained, A, B and n are taken as best fits.

Finally it should be emphasized that the constant A does not agree with the output voltage of the anemometer at zero velocity. This is understandable, as two differing mechanisms of heat transfer define this quantity. In the case of the E^2 measurement at $U = 0$, the heat release by free convection dominates and in the case of extrapolation of the data to $U = 0$, the heat release is due to forced convection.

21.6 Turbulence Measurements with Hot-Wire Anemometers

Velocity measurements by means of hot-wire anemometers require a detailed knowledge of the directional sensitivity of the hot wire. The latter is determined by direct calibration of the hot wire in a flow with known direction. Rotation of the wire leads to the velocity-angle representation of the outlet signal of a hot-wire anemometer shown in Fig. 21.27. However, this information is insufficient for using a hot-wire anemometer in turbulent flows, where the angle of the local velocity changes continuously. For the evaluation of the resulting output signal, it is necessary to know the velocity-angle dependence of the hot-wire signal analytically. In this way, it is possible to record the complex connections between turbulent velocity fluctuations and the angle dependence on the HDA output signal quantitatively. Here, it is usual to introduce an effective cooling velocity which for the velocity components vertical and parallel to the hot wire can be expressed as follows:

$$\hat{U}^2_{\text{eff}} = \hat{U}^2_{\text{per}} + k^2 \hat{U}^2_{\text{par}} , \qquad (21.32)$$

where $\hat{U}_{\text{per}}$ is the momentary velocity component vertical to the hot wire and U_{par} the momentary parallel component; see, e.g. Hinze [21.1]. Equation (21.32) is characterized by the fact that the velocity components U_{per} and U_{par} are chosen as components, i.e. the components are expressed relative to the wire. For the flow measurements it is important, however, that the velocity components are obtained for the measurements relative to a space-fixed coordinate system x_i. This makes it necessary to express $\hat{U}_{\text{per}}$ and $\hat{U}_{\text{par}}$

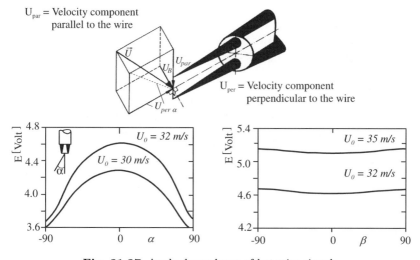

Fig. 21.27 Angle dependence of hot-wire signals

by the components $\hat{U}_i$. In this respect, for the velocity vector and the position vector of the hot wire the following holds:

$$\hat{U}_i = \{\hat{U}_1, \hat{U}_2, \hat{U}_3\} \quad \text{and} \quad \ell_i = \{\cos\alpha_1, \cos\alpha_2, \cos\alpha_3\} \tag{21.33}$$

Hence, $\hat{U}^2_{\text{per}}$ and $\hat{U}^2_{\text{par}}$ can be given as follows:

$$U^2_{\text{per}} = \Big[\hat{U}_1^2(\cos^2\alpha_2 + \cos^2\alpha_3) + \hat{U}_2^2(\cos^2\alpha_1 + \cos^2\alpha_3) + \hat{U}_3^2(\cos^2\alpha_1 + \cos^2\alpha_2)$$
$$- 2\hat{U}_1\hat{U}_2 \cos\alpha_1 \cos\alpha_2 - 2\hat{U}_1\hat{U}_3 \cos\alpha_1 \cos\alpha_3 - 2\hat{U}_2\hat{U}_3 \cos\alpha_2 \cos\alpha_3\Big] \tag{21.34}$$

and

$$U^2_{\text{par}} = \Big[\hat{U}_1^2 \cos^2\alpha_1 + \hat{U}_2^2 \cos^2\alpha_2 + \hat{U}_3^2 \cos^2\alpha_3 + 2\hat{U}_1\hat{U}_2 \cos\alpha_1 \cos\alpha_2$$
$$+ 2\hat{U}_1\hat{U}_3 \cos\alpha_1 \cos\alpha_2 + 2\hat{U}_2\hat{U}_3 \cos\alpha_2 \cos\alpha_3\Big] \tag{21.35}$$

From these equations, the effective cooling velocity indicated in (21.32) can be given as follows:

$$\hat{U}^2_{\text{eff}} = \Big\{\Big[\hat{U}_1^2\left(k^2 \cos^2\alpha_1 + \cos^2\alpha_2 + \cos^2\alpha_3\right)$$
$$+ \hat{U}_2^2\left(\cos^2\alpha_1 + k^2\cos^2\alpha_2 + \cos^2\alpha_3\right) + \hat{U}_3^2\left(\cos^2\alpha_1 + \cos^2\alpha_2 + k^2\cos^2\alpha_3\right)\Big]$$
$$- 2\left(1-k^2\right)\Big[\hat{U}_1\hat{U}_2 \cos\alpha_1 \cos\alpha_2 + \hat{U}_1\hat{U}_3 \cos\alpha_1 \cos\alpha_3 + \hat{U}_2\hat{U}_3 \cos\alpha_1 \cos\alpha_3\Big]\Big\} \tag{21.36}$$

When expressing the momentary value of $\hat{U}_i = U_i + u_i$, i.e. when introducing the mean flow velocity U_i and the turbulent fluctuation velocity u_i, i.e.

$$\hat{U}_1 = U_1 + u_1, \quad \hat{U}_2 = U_2 + u_2, \quad \hat{U}_3 = U_3 + u_3 \tag{21.37}$$

(21.36) can be written as follows:

$$\hat{U}^2_{\text{eff}} = \Big\{\big[(U_1^2 + 2U_1u_1 + u_1^2)\left(k^2 \cos^2\alpha_1 + \cos^2\alpha_2 + \cos^2\alpha_3\right)$$
$$+ (U_2^2 + 2U_2u_2 + u_2^2)\left(\cos^2\alpha_1 + k^2\cos^2\alpha_2 + \cos^2\alpha_3\right)$$
$$+ (U_3^2 + 2U_3u_3 + u_3^2)\left(\cos^2\alpha_1 + \cos^2\alpha_2 + k^2\cos^2\alpha_3\right)\big] - 2\left(1-k^2\right)$$
$$\times \big[(U_1U_2 + U_1u_2 + U_2u_1 + u_1u_2)\cos\alpha_1 \cos\alpha_2$$
$$+ (U_1U_3 + U_1u_3 + U_3u_1 + u_1u_3)\cos\alpha_1 \cos\alpha_3$$
$$+ (U_2U_3 + U_2u_3 + U_3u_2 + u_2u_3)\cos\alpha_2 \cos\alpha_3\big]\Big\} \tag{21.38}$$

When one now considers the output signal of a hot-wire anemometer, $\hat{E}$, this is connected to the effective cooling velocity of the wire as follows:

$$\hat{E} = \left(A + B\hat{U}^n_{\text{eff}}\right)^{\frac{1}{2}} \tag{21.39}$$

21.6 Turbulence Measurements with Hot-Wire Anemometers

In order to explain the application of hot-wire anemometry for measurements in turbulent flows, the following sequence of measurements needs to be considered, for which in Fig. 21.28 the selected hot-wire positions are shown. For the equation to be given below, it is assumed that:

$$\hat{U}_1 = Q_1 + q_1, \qquad \hat{U}_2 = q_2, \qquad \hat{U}_3 = q_3 \tag{21.40}$$

hold. The position of the hot wire is described by the following directional vector:

$$n_i = \{\sin \alpha, \cos \alpha, 0\}$$

With this, the effective cooling velocity is computed as:

$$\begin{aligned}\hat{U}_{\text{eff}}^2 &= \left(Q_1^2 + 2Q_1 q_1 + q_1^2\right)\left(k^2 \sin^2 \alpha + \cos^2 \alpha\right) + q_2^2 \\ &+ \left(\sin^2 \alpha + k^2 \cos^2 \alpha\right) + q_3^2 + 2\left(1 - k^2\right) q_1 q_2 \sin \alpha \cos \alpha \\ &+ 2\left(1 - k^2\right) Q_1 q_2 \sin \alpha \cos \alpha \end{aligned} \tag{21.41}$$

Rearrangement of the above equation yields:

$$\begin{aligned}\hat{U}_{\text{eff}} = Q_1 \cos \alpha \Bigg\{ &1 + \left(k^2 \tan^2 \alpha\right) + 2\left(1 + k^2 \tan^2 \alpha\right) \frac{q_1}{Q_1} \\ &+ 2\left[(1 - k^2) \tan \alpha\right] \frac{q_2}{Q_1} + \left(1 + k^2 \tan^2 \alpha\right) \frac{q_1^2}{Q_1^2} + \left(k^2 + \tan^2 \alpha\right) \frac{q_2^2}{Q_1^2} \\ &+ \left(1 + \tan^2 \alpha\right) \frac{q_3^2}{Q_1^2} + 2\left[(1 - k^2) \tan \alpha\right] \frac{q_1 q_2}{Q_1^2} \Bigg\}^{\frac{1}{2}} \end{aligned} \tag{21.42}$$

By series expansion and after neglecting terms of higher order the following relationships result:

$$\begin{aligned}\hat{U}_{\text{eff}}(\alpha) = Q_1 \cos \alpha \Bigg\{ &1 + k^2 \frac{1}{2} \tan^2 \alpha - k^4 \frac{1}{8} \tan^4 \alpha + \left(1 + k^2 \frac{1}{2} \tan^2 \alpha \right.\\ &\left. - k^4 \frac{1}{8} \tan^4 \alpha\right) \frac{q_1}{Q_1} + \left[\tan \alpha - k^2 \tan \alpha \left(1 + \frac{1}{2} \tan^2 \alpha\right)\right. \\ &\left. + k^4 \frac{1}{2} \tan^3 \alpha \left(1 + \frac{3}{4} \tan^2 \alpha\right)\right] \frac{q_2}{Q_1} \\ &+ \left(\frac{1}{2 \cos^2 \alpha} - k^2 \frac{\tan^2 \alpha}{4 \cos^2 \alpha} + \frac{3}{16} k^4 \frac{\tan^4 \alpha}{\cos^2 \alpha}\right) \frac{q_3^2}{Q_1^2} \\ &- \frac{3}{8} k^4 \tan^4 \alpha \frac{q_1^2}{Q_1^2} \left[k^2 \left(\frac{1}{2} + \tan^2 \alpha 1 + \frac{1}{2} \tan^2 \alpha\right)\right] \\ &- \frac{3}{2} k^4 \tan^2 \alpha \left(\frac{1}{2} + \tan^2 \alpha + \frac{5}{8} \tan^4 \alpha\right) \frac{q_2^2}{Q_1^2} \\ &- k^4 \tan^4 \alpha \frac{q_1^3}{Q_1^3} + \left[-k^2 \left(\frac{1}{2} + \tan^2 \alpha + \frac{1}{2} \tan^4 \alpha\right)\right. \\ &\left. + k^4 \left(\frac{3}{4} \tan^2 \alpha + \frac{3}{4} \tan^4 \alpha + \frac{9}{2} \tan^6 \alpha\right)\right] \frac{q_1 q_2^2}{Q_1^3} \end{aligned}$$

$$\begin{aligned}
&+ k^4 \left(6\tan^3\alpha - \frac{9}{4}\tan^5\alpha\right) q_2 \frac{q_1^2}{Q_1^3} \\
&+ \left[k^2 \frac{\tan^2\alpha}{4\cos^2\alpha} + k^4 \tan^4\alpha \left(\frac{3}{4}\tan^2\alpha + \frac{3}{\cos^2\alpha}\right)\right] q_1 \frac{q_3^2}{Q_1^3} \\
&+ \left[k^2 \frac{\tan^2\alpha}{2\cos^2\alpha}\left(1 + \frac{3}{2}\tan^2\alpha\right) + k^4 \left(\frac{3}{4}\tan^3\alpha - \frac{3\tan^3\alpha}{2\cos^2\alpha}\right.\right.\\
&\left.\left. + 3\frac{\tan^5\alpha}{\cos^2\alpha}\right)\right]\frac{q_2 q_3^2}{Q_1^3} + \left[-k^2\tan\alpha\left(\frac{1}{2} + \tan^2\alpha + \frac{1}{2}\tan^4\alpha\right)\right. \\
&\left. + k^4 \left(\frac{1}{2}\tan\alpha + \frac{15}{4}\tan^3\alpha + \frac{15}{4}\tan^5\alpha - \frac{1}{16}\tan^7\alpha\right)\right]\frac{q_2^3}{Q_1^3}\Bigg\}
\end{aligned}$$
(21.43)

When one introduces (21.43) in (21.39), one obtains for the time-averaged voltage of a hot-wire anemometer

$$E^2 - A \cong BQ_1^n \cos^n\alpha \left(1 + \frac{n}{2}k^2\tan^2\alpha\right) \tag{21.44}$$

For the momentary value, considering only terms of first order, the following results:

$$\begin{aligned}
E^2 + 2Ee - A &\cong BQ_1^n \cos_\alpha^n \left[1 + \frac{n}{2}k^2\tan^2\alpha + n\left(1 + k^2\tan^2\alpha\right)\right. \\
&\left. \times \frac{q_1}{Q_1} + n\left(1 - k^2\right)\tan\alpha \frac{q_2}{Q_2}\right]
\end{aligned} \tag{21.45}$$

By subtraction of (21.44) from (21.45) and squaring the difference, one obtains

$$[2E]^2 e^2 \cong n^2 B^2 Q_1^{2n} \cos^{2n}\alpha \left[\left(1 + k^2\tan^2\alpha\right)\frac{q_1}{Q_1}\right. \\
\left. + \left(1 - k^2\right)\tan\alpha\frac{q_2}{Q_1}\right]^2 \tag{21.46}$$

and, hence, the following final equations can be employed for evaluation of hot-wire anemometer signals:

$$\left(E^2 - A\right)^2 = B^2 Q_1^{2n} \cos^{2n}\alpha \left(1 + k^2\tan^2\alpha\right) \tag{21.47}$$

or rewritten

$$\left(\frac{2E}{E^2 - A}\right)^2 e^2 \cong n^2 \left[\left(\frac{q_1}{Q_1}\right) + \frac{(1-k^2)\tan\alpha}{(1+k^2\tan^2\alpha)}\left(\frac{q_2}{Q_1}\right)\right]^2 \tag{21.48}$$

By time-averaging, one obtains

$$\begin{aligned}
\left(\frac{2E}{E^2 - A}\right)^2 \overline{e^2} = n^2 &\left\{\overline{\left(\frac{q_1}{Q_1}\right)^2} + \left[\frac{(1-k^2)\tan\alpha}{(1+k^2\tan^2\alpha)}\right]^2 \overline{\left(\frac{q_2}{Q_1}\right)^2}\right. \\
&\left. + \frac{2(1-k^2)\tan\alpha}{(1+k^2\tan^2\alpha)}\frac{\overline{q_1 q_2}}{Q_1^2}\right\}
\end{aligned} \tag{21.49}$$

21.6 Turbulence Measurements with Hot-Wire Anemometers

Three measurements with the angular positions $\alpha_1 = 0$, $\alpha_2 = \pi/4$ and $\alpha_3 = -\pi/4$ yield

$$\overline{\left(\frac{q_1^2}{Q_1^2}\right)} = \frac{1}{n^2}\left[\frac{2E_{\alpha_1}}{(E_{\alpha_1}^2 - A)}\right]^2 \overline{e_{\alpha_1}^2} \tag{21.50}$$

$$\overline{\left(\frac{q_2^2}{Q_1^2}\right)} = \frac{1}{2n^2}\left[\frac{(1+k^2)}{(1-k^2)}\right]^2 \left[\left(\frac{2E_{\alpha_2}}{E_{\alpha_2}^2 - A}\right)^2 \overline{e_{\alpha_2}^2} + \left(\frac{2E_{\alpha_3}}{E_{\alpha_3}^2 - A}\right)^2 \overline{e_{\alpha_3}^2} - \left(\frac{2E_{\alpha_1}}{E_{\alpha_1}^2 - A}\right)^2 \overline{e_{\alpha_1}^2}\right] \tag{21.51}$$

$$\overline{\left(\frac{q_2 q_2}{Q_1^2}\right)} = \frac{1}{n^2}\frac{(1+k^2)}{4(1-k^2)}\left[\left(\frac{2E_{\alpha_2}}{E_{\alpha_2}^2 - A}\right)^2 \overline{e_{\alpha_2}^2} - \left(\frac{2E_{\alpha_3}}{E_{\alpha_3}^2 - A}\right)^2 \overline{e_{\alpha_3}^2}\right] \tag{21.52}$$

In order to obtain also the components in the x_1–x_3 plane, i.e. in order to measure $\overline{q_3^2}$ and $\overline{q_1 q_3}$, one chooses the wire positions in Fig. 21.28. For these positions, additional information results which can be used for measuring the subsequent quantity.

$$\overline{\left(\frac{q_3^2}{Q_1^2}\right)} = \frac{1}{2n^2}\left[\frac{(1+k^2)}{(1-k^2)}\right]^2 \left[\left(\frac{2E_{\alpha_4}}{E_{\alpha_4}^2 - A}\right)^2 \overline{e_{\alpha_4}^2} + \left(\frac{2E_{\alpha_5}}{E_{\alpha_5}^2 - A}\right)^2 \overline{e_{\alpha_5}^2} - \left(\frac{2E_{\alpha_1}}{E_{\alpha_1}^2 - A}\right)^2 \overline{e_{\alpha_1}^2}\right] \tag{21.53}$$

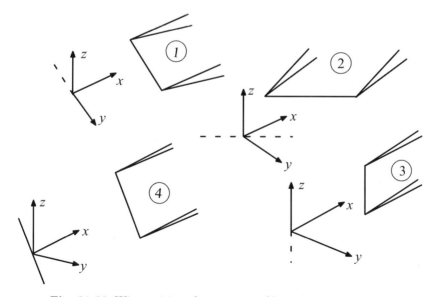

Fig. 21.28 Wire positions for sequence of hot-wire measurements

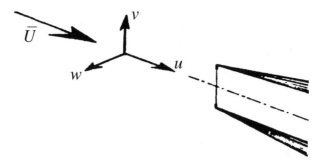

Fig. 21.29 Straight probe in flow field

$$\overline{\left(\frac{q_1 q_3}{Q_1^2}\right)} = \frac{1}{n^2}\left[\frac{(1+k^2)}{4(1-k^2)}\right]\left[\left(\frac{2E_{\alpha_4}}{E_{\alpha_4}^2 - A}\right)^2 \overline{e_{\alpha_4}^2} - \left(\frac{2E_{\alpha_5}}{E_{\alpha_5}^2 - A}\right)^2 \overline{e_{\alpha_5}^2}\right] \tag{21.54}$$

The above evaluation equations can thus be used to measure the mean flow component $Q_1 = U_1$ and the turbulence quantities $\overline{q_1^2} = \overline{u_1^2}$, $\overline{q_2^2} = \overline{u_2^2}$, $\overline{q_3^2} = \overline{u_3^2}$, $\overline{q_2 q_1} = \overline{u_2 u_1}$ and $\overline{q_1 q_3} = \overline{u_1 u_3}$. Measurements of other correlations can be carried out on the basis of correspondingly derived equations.

A usually employed quantity for describing the turbulence intensity is the degree of turbulence, Tu. The mean fluctuation velocity $\sqrt{\overline{u^2}}$ contained in it is determined from the RMS value of the anemometer output voltage (RMS value) $\sqrt{\overline{e^2}}$ of a straight hot-wire probe (Fig. 21.29), divided by the gradient of the static calibration curve of the same probe, i.e.

$$Tu = \frac{\sqrt{\overline{u^2}}}{\overline{U}} 100\,(\%) = \frac{\sqrt{\overline{e^2}}}{\frac{d\overline{E}}{d\overline{U}}\,\overline{U}} 100\,(\%) \tag{21.55}$$

When basing the computation on the modified King's law:

$$\overline{E^2} = A + B\overline{U}^n \tag{21.56}$$

differentiation yields

$$\frac{d\overline{E}}{d\overline{U}} = \frac{nB\overline{U}^{n-1}}{2\sqrt{A + B\overline{U}^n}} \tag{21.57}$$

On inserting in this differential equation once again the above King's equation one obtains, together with the theoretical exponents n, the working equation for determining the turbulence degree

$$Tu = \frac{\sqrt{\overline{u^2}}}{\overline{U}} = \frac{4 \times \overline{E}\sqrt{\overline{e^2}} \times 100}{\overline{E}^2 - A}\,(\%) \tag{21.58}$$

This equation indicates that for small fluctuations of the flow velocity, with a mean velocity value $\overline{U}$, the effective value of the velocity fluctuations is proportional to the RMS value of the voltage fluctuation of the anemometer.

21.6 Turbulence Measurements with Hot-Wire Anemometers

Fig. 21.30 Notations of the velocity component

As already emphasized, for the determination of the Reynolds momentum transport terms $\overline{u_i u_j}$, one has to employ inclined probes (probe inclined towards $\bar{U}$ by an angle α mostly $\alpha = \pm 45°$). In this case, the anemometer output voltage is made up of velocity components of contributions of the longitudinal and lateral velocity components of a space-fixed coordinate system (Fig. 21.30). Thus it was shown in previous sections above that the determination of the Reynolds momentum-transport terms becomes possible.

Simplified considerations, yet well suited for an introduction, are possible on the assumption that the hot wire is sensitive only to the vertical velocity component $U \cos \alpha$. The velocity components parallel to the hot wire and vertical on the plane formed of the hot wire and its prongs are neglected, so that the modified King's law reads:

$$\hat{E}^2 = A + B \left(\hat{U} \cos \hat{\alpha} \right)^n \tag{21.59}$$

Differentiation of this equation yields

$$2\hat{E}\, d\hat{E} = B \cos^n \hat{\alpha}\, n \hat{U}^{n-1}\, d\hat{U} - Bn \sin \hat{\alpha} \cos_\alpha^{n-1} \hat{\alpha} \hat{U}^n\, d\alpha \tag{21.60}$$

When the velocity fluctuations of a turbulent flow are small in comparison with the mean flow velocity U, one can set

$$d\hat{U} = u, \qquad \hat{U}\, d\alpha = v$$

With this simplifying assumptions, the following results:

$$2\hat{E}\, d\hat{E} = nB \left(\hat{U} \cos \hat{\alpha} \right)^n \frac{1}{\hat{U}} (u - v \tan \hat{\alpha}) \tag{21.61}$$

From the modified King's law, one obtains

$$B \left(U \cos \hat{\alpha} \right)^n = \hat{E}^2 - A, \tag{21.62}$$

where $\hat{E}$ is once again the anemometer output voltage at the flow velocity $\hat{U}$. From the latter two equations, it can be derived that

$$\frac{2\hat{E}}{\hat{E}^2 - A} d\hat{E} = \frac{n}{\bar{U}} (u - v \tan \hat{\alpha}) \tag{21.63}$$

with

$$\frac{2\hat{E}}{\hat{E}^2 - A} \approx \frac{2\bar{E}}{\bar{E}^2 - A} = D \quad \text{and} \quad d\hat{E} = e \tag{21.64}$$

Combining the equations, one obtains

$$eD = \frac{n}{\bar{U}} (u - v \tan \alpha) \tag{21.65}$$

Squared and time-averaged, one obtains the basic equation for the RMS value of the anemometer output voltage, when positioning the probe in the u, v plane:

$$D^2 \overline{e^2} = \frac{n^2}{\bar{U}^2} \left(\overline{u^2} + \overline{v^2} \tan^2 \alpha - 2\overline{uv} \tan^2 \alpha \right), \tag{21.66}$$

where $\sqrt{\overline{e^2}}$=RMS value of the anemometer output voltage.

For a measurement with the $+\alpha$ inclined and the $-\alpha$ inclined probe in the u–v plane (Fig. 21.31), one obtains the following:
$+\alpha$ inclined probe:

$$D^2 \overline{e_1^2} = \frac{n^2}{\bar{U}^2} \left(\overline{u^2} + \overline{v^2} \tan^2 \alpha - 2\overline{uv} \tan^2 \alpha \right) \tag{21.67}$$

$-\alpha$ inclined probe:

$$D^2 \overline{e_2^2} = \frac{n^2}{\bar{U}^2} \left(\overline{u^2} + \overline{v^2} \tan^2 \alpha - 2\overline{uv} \tan^2 \alpha \right) \tag{21.68}$$

The difference of (21.67) and (21.68) produces the turbulent shear stresses (up to density ρ):

$$\frac{\overline{uv}}{\bar{U}^2} \tan_\alpha^2 = \frac{1}{4n^2} D^2 \left(\overline{e_1^2} - \overline{e_2^2} \right) \tag{21.69}$$

The sum yields

$$\frac{\overline{u^2}}{\bar{U}^2} + \frac{\overline{v^2}}{\bar{U}^2} \tan^2 \alpha = \frac{1}{2n^2} D^2 \left(\overline{e_1^2} + \overline{e_2^2} \right) \tag{21.70}$$

With known probe angle α (mostly $\alpha = 45°$) and previously measured $\overline{u^2}/\bar{U}^2$ from the above equation, $\overline{v^2}/\bar{U}^2$ can be computed.

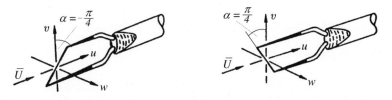

Fig. 21.31 Measurements with inclined probe in u, v plane

21.6 Turbulence Measurements with Hot-Wire Anemometers

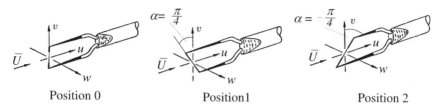

Position 0 Position 1 Position 2

Fig. 21.32 The measured signals of a linearized anemometer and a sensor sensitive only against the vertical velocity components

For the normal probe, $\alpha = 0°$, the following results from the basic equation, an already known value:

$$D^2 \overline{e_0^2} = \frac{n^2}{\bar{U}^2} \overline{u^2} \tag{21.71}$$

The still remaining turbulence intensity $\overline{w^2}/\bar{U}^2$ is similar to what is described above, with the only difference that the probe has to be positioned in the u, w plane, so that in the above equations v only has to be replaced by w. In the case of a linearized anemometer, with a voltage output proportional to U, i.e.

$$E = SU \tag{21.72}$$

the evaluation of hot-wire signals is simplified considerably. Assuming further, on the other hand, that the hot wire is only sensitive to the vertical velocity components (Fig. 21.32), one arrives at the following connections:

Position 0 : $\alpha = 0$ $e_0 = a_0 u;$ $\bar{E} = a_0 \bar{U}$ (21.73)

Position 1 : $\alpha_1 = 45°$ $e_1 = au + bv$ (21.74)

Position 2 : $\alpha_2 = -45°$ $e_2 = au - bv$ (21.75)

From these three equations, one obtains by squaring and time averaging:

$$\overline{e_0^2} = a_0^2 \overline{u^2} \tag{21.76}$$

$$\overline{e_1^2} = a^2 \overline{u^2} + b^2 \overline{v^2} + 2ab\overline{uv} \tag{21.77}$$

$$\overline{e_2^2} = a^2 \overline{u^2} + b^2 \overline{v^2} - 2ab\overline{uv} \tag{21.78}$$

From the RMS values, the flow parameters can be determined:

$$\overline{u^2} = \frac{1}{a_0^2} \overline{e_0^2} \tag{21.79}$$

$$\overline{v^2} = \frac{\overline{e_1^2} + \overline{e_2^2}}{2b^2} - \frac{a^2}{b^2} \overline{u^2} \tag{21.80}$$

$$\overline{uv} = \frac{\overline{e_1^2} - \overline{e_2^2}}{4ab} \tag{21.81}$$

When an X-probe is used in the measurements, e_1 and e_2 are measured simultaneously with two separate electric systems (Fig. 21.33). For the flow

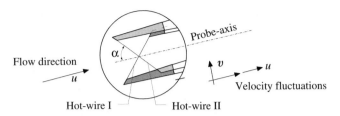

Fig. 21.33 Measurement signals of an X-probe

parameters, the following evaluation would also be possible:

$$u = \frac{1}{2a}(e_1 + e_2) \tag{21.82}$$

$$v = \frac{1}{2a}(e_1 - e_2) \tag{21.83}$$

The output signals are then processed such that the sought flow parameters can be determined, i.e. the following quantities:

$$\overline{u^2} = \frac{1}{4a^2}\overline{(e_1 + e_2)^2} \tag{21.84}$$

$$\overline{v^2} = \frac{1}{4b^2}\overline{(e_1 - e_2)^2} \tag{21.85}$$

$$\overline{uv} = \frac{1}{4ab}\left(\overline{e_1^2} - \overline{e_2^2}\right) \tag{21.86}$$

Quantities that depend on squares of differences of small voltages can only be determined very inaccurately.

21.7 Laser Doppler Anemometry

21.7.1 Theory of Laser Doppler Anemometry

The physical background of optical velocity measurements by means of laser light beams, discussed in this section, is the Doppler effect, which leads to measurable frequency changes of the laser light, that are generated by the movements of light scattering particles. The prerequisite for the applicability of optical velocity measurement procedures, therefore, is the existence of appropriate light-scattering particles, which either exist naturally in the flowing fluid, or need to be added by particle generators. These particles serve as receiver and transmitter of the incident laser light and bring about, by their motion, the desired frequency changes of the laser radiation. These frequency changes are measured and from the measurements one induces the velocities

21.7 Laser Doppler Anemometry

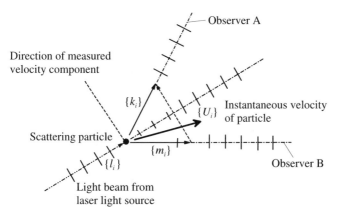

Fig. 21.34 Considerations of the Doppler shifted frequency of the scattered light

of the particles. This is the basis of optical velocity measurement by means of laser Doppler anemometry.

In Fig. 21.34, an arrangement of laser beams is shown, which is suited for explaining the LDA measurement principle. A beam, emanating from a laser-light source, has the beam direction $\{\ell_i\} = \ell_i$ and hits a scattering particle which scatters light in all directions in space, and consequently also in the directions which are given by the vectors $\{k_i\} = k_i$ and $\{m_i\} = m_i$. The moving scattered particle has the velocity $\{U_i\} = U_i$ and therefore scatters light having a Doppler shift of the frequency. In the two directions of the receivers A and B, given by the direction vectors k_i and m_i, the frequencies can be derived as

$$\nu_A = \nu \left(\frac{c - U_i \ell_i}{c - U_i k_i} \right) \quad \text{and} \quad \nu_B = \nu \left(\frac{c - U_i \ell_i}{c - U_i m_i} \right) \tag{21.87}$$

Here, the particle generating the scattering beams once acts as a moving receiver of the arriving laser radiation and at the same time also as a moving transmitter of the scattered radiation, i.e. the Doppler effect has to be applied twice, in order to deduce the relationships in (21.87). The simultaneous detection of the two light signals by observers A and B allows one to detect the frequency difference:

$$\Delta \nu = \nu_A - \nu_B = \frac{1}{\lambda} U_i (k_i - m_i) \tag{21.88}$$

by superposition of the scattered waves. This superposition is carried out in order to determine the velocity of the scattering particles by a frequency detectable by available photo detectors.

This superposition of the two waves is carried out since direct measurement of the frequencies ν_A and ν_B, which according to (21.87) already show the desired velocity dependence, is not possible, as the frequencies to be detected

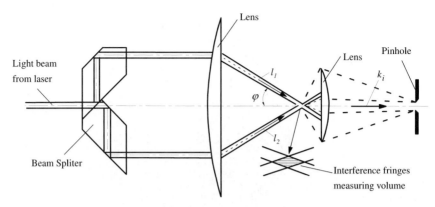

Fig. 21.35 Diagram of optical set-up of a two-beam LDA system

are in the range of 10^{15} Hz. Moreover, there is no detection system having the frequency resolution to measure relative to 10^{15} Hz, the frequency of the laser radiation, the Doppler shift, which is around 2×10^5 Hz m^{-1} s. It is therefore necessary to determine the velocity-dependent Doppler shift of the laser light from the frequencies of two scattered light signals that can be superimposed. This leads to the introduction of the two-beam anemometer, as shown in Fig. 21.35. A first summary of the early work in Laser Doppler anemometry is given by Buchhave et al. [21.3].

When applying the above derivations to the optical set-up in Fig. 21.35, one obtains in each direction given by the vector k_i the following two frequencies:

$$\nu_1 = \nu \left[\frac{c - U_i (\ell_1)_i}{c - U_i k_i} \right] \quad (21.89)$$

and

$$\nu_2 = \nu \left[\frac{c - U_i (\ell_2)_i}{c - U_i k_i} \right] \quad (21.90)$$

From this, the difference frequency can be computed, for $c \ll |U_i|$:

$$\Delta \nu = \frac{1}{\lambda} U_i \left[(\ell_2)_i - (\ell_1)_i \right] \quad (21.91)$$

This yields for the frequency difference:

$$\Delta \nu = \frac{1}{\lambda} U_i \left[(\ell_2)_i - (\ell_1)_i \right] = \frac{U_\perp 2 \sin \phi}{\lambda}, \quad (21.92)$$

where λ is the wavelength of the laser light $U_\perp$ the velocity component vertical to the axis of the optical system and φ half the angle between the scattering beams. This relationship shows that the suspended frequency is independent of the direction of observation. This is a big advantage when employing this optical system, since a large collecting aperture of the receiving optical system can be employed to detect the scattered light signal.

21.7 Laser Doppler Anemometry

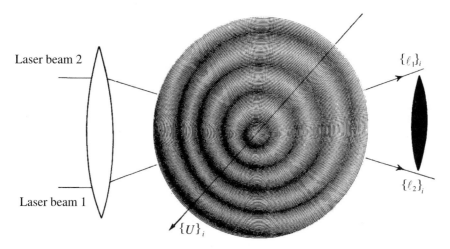

Fig. 21.36 Visual representation of LDA signals by transparents

In Fig. 21.36, an attempt is made to represent visually, by means of a method developed by Durst and Stevenson [21.4], the above light superposition processes which lead to the desired frequency difference, in order to measure $f_D = \Delta\nu$. Figure 21.36 shows the two laser beams of the optical velocity-measuring instrument and also a moving scattering particle which was assumed to be present in the measurement volume. This figure also shows scattering waves and the lens of the receiving optical system, which is needed for detecting the scattering light. When superimposing on the scattered wave of the first laser beam the second scattered wave of the second laser beam, then, due to the different frequencies of the two scattering beams, the frequency difference becomes detectable by means of a photodetector. This detector yields an electrical signal proportional to the light intensity variations whose frequency is a measure of the velocity of the particle generating the two scattered waves. The velocity component vertical to the axis of the optical system, i.e. perpendicular to the bisector of the two laser beams, is measured. The measured velocity component is located in the plane set up by the two laser beams.

The above-explained physical processes of optical velocity measurements by means of laser beams can also be explained by means of the so-called interference model; see Fig. 21.37. This model starts from the assumption that in the intersection volume of the two laser beams, interference fringes are produced, whose fringe distance is given by the geometry of the optical set-up and the wavelength of the laser light:

$$\Delta x = \frac{\lambda}{2\sin\varphi} \qquad (21.93)$$

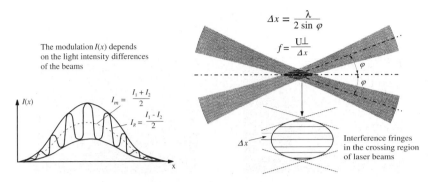

Fig. 21.37 Visual representation of the interference fringes in the measuring volume of a laser Doppler anemometer

When a scattered particle is moving through this interference pattern, it will scatter light whose intensity, for a sufficiently small particle diameter, is proportional to the local light intensity in the measuring volume. Because of this, the intensity of the scattered light shows sinusoidal fluctuations that are caused by the motion of the particle through the interference pattern. When the particle has a velocity component $U_\perp$ perpendicular to the plane representing the interference fringes, the moving particle needs the following time to cross a single fringe:

$$\Delta t = \frac{\Delta x}{U_\perp} \tag{21.94}$$

This brings about a signal frequency which can be expressed as follows:

$$f = \frac{1}{\Delta t} = \frac{U_\perp 2 \sin \varphi}{\lambda} \tag{21.95}$$

These considerations lead to the same final equation as was derived above with the help of a Doppler model. In Fig. 21.37, in turn, the modeling of Durst and Stevenson [21.4] is included, in order to explain the interference pattern in the measuring volume. Parallel interference fringes are present in the crossing region of the two incident beams, as explained above, and they are made visible.

To understand fully the signals occurring in optical velocity measurements, attention has to be given also to the fact that the incident beams have a finite expansion perpendicular to their direction of propagation and that their intensity distribution over the cross-section shows a Gaussian distribution. When taking this into account, it is understandable that a particle, that passes through the center of the measuring volume of an LDA system, brings about a signal as shown in Fig. 21.37.

When a particle is moving through the measuring volume of an LDA system, a little away from the center, due to the locally existing intensity differences of the two crossing beams, signals are to be expected as indicated in Fig. 21.38. A signal which originates from a particle that moved through

21.7 Laser Doppler Anemometry

Analytical presentation of LDA-Signals

$(A)_k$ = Amplitude of k-th signal η_k = Modulation depth of k-th signal

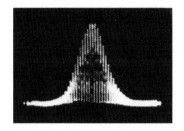
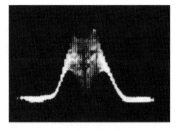

$$s(t) = \sum_{k=1}^{N}(A)_k \exp[-(t-t_k)/\tau_k^2]\{1+\eta_k \cos[2\pi v_k(t-t_k)+\phi]\}$$

Fig. 21.38 Analytical representation of LDA signals

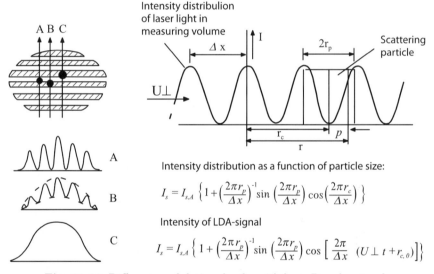

Intensity distribution as a function of particle size:

$$I_s = I_{s,A}\left\{1+\left(\frac{2\pi r_p}{\Delta x}\right)^{-1}\sin\left(\frac{2\pi r_p}{\Delta x}\right)\cos\left(\frac{2\pi r_c}{\Delta x}\right)\right\}$$

Intensity of LDA-signal

$$I_s = I_{s,A}\left\{1+\left(\frac{2\pi r_p}{\Delta x}\right)^{-1}\sin\left(\frac{2\pi r_p}{\Delta x}\right)\cos\left[\frac{2\pi}{\Delta x}(U\perp t + r_{c,0})\right]\right\}$$

Fig. 21.39 Different modulation depths with laser Doppler signals

the center of the volume, showing a good signal modulation, serves for comparison. Here, the two beams are located in the same plane, and they can therefore be considered as overlapping well. Away from the center the particle crosses first one beam, then the overlapping area of the two beams and finally the area of the second beam. The LDA signal in Fig. 21.39 reflects this final form of the signal.

The influence of the particle size is shown in Fig. 21.39. When particles are very small, they do not integrate the interference pattern in the measurement of volume, i.e. the modulation depth of the intensity distribution of the interference pattern is fully reflected in the scattering signal of a particle. In this way a signal forms as is shown by the LDA-signal in (A). When the diameter of a particle is increased, a signal shape develops as sketched in (B). When the particle corresponds to the size of the distance of the interference fringes, it is possible that the modulation disappears completely (C). In Fig. 21.39, equations are given that show these relationships, for a particle assumed to be square for the considerations carried out here. This form of the particles is sketched and it is assumed to be moving through the interference pattern.

As far as the determination of the particle velocity direction is concerned, some additional explanations are necessary. The explanations of optical velocity measurements given so far, do not permit different signals to be generated for particles with the same velocity, but different velocity directions. Therefore, a particle that moves in one direction through the interference pattern in Fig. 21.37, will yield the same signal as a particle moving in the opposite direction, having the same velocity. If, however, one changes the frequency of a laser beam by a certain amount, this leads to a moving interference pattern in the measuring volume. This can be explained in the simplest way, when looking at the optical system in Fig. 21.40. It consists of a diffraction grating, which is employed for splitting one laser beam. The two beams of first order are made, with the help of a lens, to cross one another in the measuring volume. There, in turn, one can imagine the interference-fringe pattern required for the laser Doppler measurements. This means that the coherence of the two crossing laser beams is maintained.

The interference fringes forming in the measuring volume can be viewed as an image of the grid lines present on the diffraction grating, and with this it can be understood that a rotation of the diffraction grating brings about a motion of the interference fringes in the measuring volume. Here, emphasis has to be laid on the fact that it is not the entire measuring volume that moves, but only the intensity changes continuously in the measuring volume with time. This means that only the interference pattern moves and not the measuring volume. When a particle is now moving in the same direction as the interference pattern, a smaller frequency is measured than with a motion directed against the motion of the frequency pattern. When one knows the

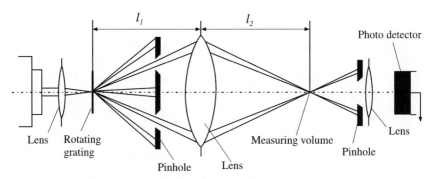

Fig. 21.40 Diagram of a laser Doppler system based on a rotating diffraction grating

motion velocity of the interference fringes, which is given by the rotation of the diffraction grating, the relative motion can be determined and, hence, also the sign of the measured velocity component. Simple and easy to understand introductions into the LDA-measuring technique are provided in references [2.5] and [2.7].

21.7.2 Optical Systems for Laser Doppler Measurements

The insights gained in the first decade of developments in laser Doppler anemometry, have led to different optical setups for functioning laser Doppler systems. All of them can be employed for contact-free velocity measurements in flowing fluids. Many available optical systems can be subdivided into three main groups which nowadays are designated as reference-beam anemometer, two-beam anemometer and two-scattering-beam anemometer; see Fig. 21.41. Practical employment of these instruments has shown that the first LDA system shown above, is suitable especially for measurements in very soiled fluids, where high particle concentrations are present and multiple particles are present in the measuring volume. The second system, the two-beam anemometer, is particularly suited for measurements, where the particle concentration is such that, on average, there is less than one particle in the measuring volume. This can be expected in all liquid flows and gas flows that appear completely transparent to the human eye.

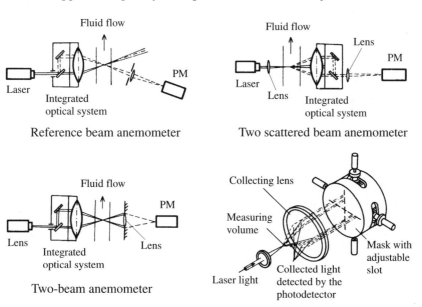

Fig. 21.41 Reference-beam, two-beam and two-scattering-beam optical systems for laser Doppler measurements

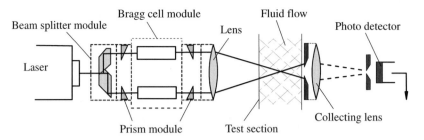

Fig. 21.42 Optical elements of a two-beam laser Doppler anemometer

When a laser beam passes through a fluid, the laser beam generally lights up intensely and this indicates that there are particles in a range that cannot be observed by the human eye and which are about a few μm in size. These particles are suited for velocity measurements with the help of a laser Doppler system and the two-beam method is particularly suited for reliable measurements. The two-scattering laser beam method needs to be employed only in special cases, when measurements of two velocity components have to be carried out with simple means.

In Fig. 21.42, the essential elements of a two-beam LDA system are shown. From this diagram, it can be deduced that the optical transmitter system is essentially composed of the laser light source and a beam-splitter unit, followed by a lens. A frequency-shifting unit, consisting of Bragg cells, is also included for measuring the direction of the flow. For the collection of the light, another lens and a photomultiplier, with an appropriate pinhole, are required. In the intersection region of the two beams of this optical set-up, one can imagine that the interference fringes, explained in Sect. 21.7.1, form for the actual measurements. To obtain good LDA signals, it is essential that these interference fringes are fully modulated, i.e. that intensity becomes zero in the dark part of the interference fringe pattern. This can be achieved by matching the intensities of the two beams, and in addition by employing optimal optical systems, i.e. optical systems with the same optical pathlengths of the two beams. This latter requirement yields in the measuring volume to zero phase differences of the superimposed laser-light waves.

For precise laser Doppler measurements, it is necessary to choose the correct laser. As most lasers have an outlet showing several axial modes, it is necessary to keep the optical pathlengths of the two laser beams of the system to be almost the same. This is guaranteed by employing an optical beam-splitting prism, as indicated in Fig. 21.42. This prism has further advantages that recommend it for employment for practical measurements. It is insensitive to adjustment and, hence, rotation of the prism in the plane, as shown in Fig. 21.42, does not lead to a directional change of the parallel beams leaving the prism. These beams are always parallel to the incident beam and always have the same distance in relation to the optical axis of the

21.7 Laser Doppler Anemometry

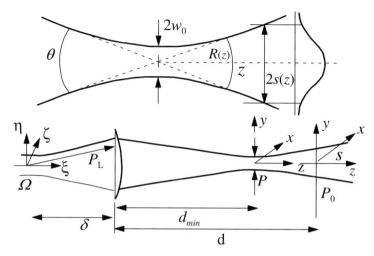

Fig. 21.43 Sketch of laser beams to explain the Gaussian beam behavior in optical systems

system. By choosing the prism shown in Fig. 21.42, LDA instruments become insensitive to adjustment and are suited for practical flow investigations.

Laser beams show a behavior that is usually referred to as that of a Gaussian beam (Fig. 21.43). For $f \to \infty$ and $\rho_L \to \infty$, one can derive from relationships, as indicated by Durst and Stevenson [21.4], analytical expressions for the variation of the wave-front curvature $R(z)$ and the beam radius $s(z)$:

$$R(z) = z\left[1 + \left(\frac{\pi \omega_0^2}{\lambda z}\right)^2\right] \quad \text{and} \quad s(z) = \omega_0 \left[1 + \left(\frac{\lambda z}{\pi \omega_0^2}\right)^2\right]^{\frac{1}{2}} \quad (21.96)$$

With the help of these equations, the radius of the two beams and the wavefront curvature at the lens of an LDA system can be computed:

$$\sigma_L = \omega_0 \left[1 + \left(\frac{\lambda z}{\pi \omega_0^2}\right)^2\right]^{\frac{1}{2}} \quad (21.97)$$

and

$$\rho_L = \delta \left[1 + \left(\frac{\pi \omega_0^2}{\lambda \sigma}\right)^2\right] \quad (21.98)$$

The expression for the light intensity is

$$I_p(r) = \frac{2}{\pi}\left(\frac{\pi \omega_0}{f \lambda}\right)^2 P_L \exp\left(-\frac{2r^2}{s^2}\right) \quad (21.99)$$

The influence of the properties of focused Gaussian beams on the mode of operation of laser Doppler anemometers was investigated by Durst and Stevenson [21.4] and Hanson [21.2]. When using a single lens to focus the two incident beams, as is usually the case in two-beam systems, the axes of the beams cross one another in the focal region of the lens. When the beam waists are not adjusted to this region, this has an increased measuring volume as a consequence. In addition, it could be shown by Durst and Stevenson [21.4] that the curvature of the wave fronts, which occurs inside the crossing region of the two beams, can lead to significant changes of the Doppler frequency, when the particles traverse different parts of the measuring volume. If properly set up LDA systems are employed, the mentioned effects are often very small and they were usually ignored in fluid flow measurements with laser Doppler anemometers, either because they were not known, or because the advantages of a set-up with a single lens outweighed these small discrepancies. For some applications, e.g. in laser Doppler measurements over large distances, investigations in extended flow fields, etc., the indicated errors can be of significance, so that appropriate steps have to be taken to ensure an optimal beam intersection region. This can be achieved in two ways:

- The waists of the two laser beams are laid into the back focal region of the transmitter lens of the LDA optical system.
- An additional optical component, consisting of a convex and a concave lens, is put between the laser and the lens of the LDA optical system. This system is used for choosing freely the position of the beam contraction with regard to the main lens of the optical system.

The above means that, to carry out now reliable LDA measurements, one has to ensure that the beams in the measuring volume of a laser Doppler anemometer are focused in such a way that we obtain parallel interference fringes, with a high modulation depth, i.e. the minimum of the light intensity should almost be zero. As far as the optical set-up of an LDA system is concerned, the following should be mentioned:

- The lens in front of the photodetector has to be chosen such that the equation:

$$\mathrm{d}_{ph} = \frac{N_{ph} M \lambda}{2 \sin \varphi} \qquad (21.100)$$

holds, where d_{ph} is the number of interference fringes which the photo multiplier sees, M is the optical enlarging relation of the detection system ($M = b/a$), λ the wavelength of the laser light and φ is half the angle between the two incident laser beams.

The diameter of the effective measuring volume can then be calculated as:

$$\mathrm{d}_m = \frac{d_{ph}}{M} = \Delta x N_{ph} = \frac{\lambda N_{ph}}{2 \sin \varphi} \qquad (21.101)$$

21.7 Laser Doppler Anemometry

This effective diameter of the measuring volume has to be smaller than the diameter of the cross-sectional region of the two laser beams of the incident optical system:

$$d_{in} = \frac{d_s}{\cos\varphi} = \frac{5}{\pi}\lambda\left(\frac{f_1}{D_1}\right)\frac{1}{\cos\varphi} \qquad (21.102)$$

The above equations are equivalent to the requirement that the number of interference fringes in the intersection area (N_{fr}) has to be larger than the number of the interference fringes N_{ph} which can be seen by the photodetector. A guideline for this relation is:

$$N_{fr} \approx \frac{5}{4}N_{ph} \qquad (21.103)$$

With this, one ensures that the outer area of the interference-fringe pattern, with its bad signal quality, does not have an influence on the carried out LDA measurements. A summary of the above equations yields for the number of fringes within the intersection region of an LDA optical system the following equation:

$$N_{fr} \approx \frac{d_{in}}{\Delta x} = \frac{10}{\pi}\frac{f_1}{D_1}\tan\varphi \qquad (21.104)$$

Details of the above considerations can be found in the book by Durst et al. [21.8]. Multidimensional laser Doppler measurements are possible, but one pair of beams is required per velocity component, i.e. per measured flow direction. An optical system which allows one to carry out such measurements with a two-beam configuration is shown in Fig. 21.44.

In the optical systems represented in Fig. 21.44, the Bragg cells are not included. When these are introduced into the two pairs of beams, measurements with frequency shifts in both pairs of the beams, for two-component measurements, are possible, i.e. the directional recognition of the flow velocities can be measured at the same time for the two velocity components.

21.7.3 Electronic Systems for Laser Doppler Measurements

When a scattered particle is traversing the measuring volume of a laser Doppler optical system, and when the scattered light resulting from this particle is detected by a photomultiplier, a signal results at the outlet of the photomultiplier as shown in Fig. 21.45. This signal comprises a low-frequency component resulting from the Gauss intensity distribution of the laser light. Furthermore, there is a high-frequency portion that is an inherent part of the laser Doppler signal. The entire signal thus has a frequency spectrum as is also shown in Fig. 21.45. It is the high-frequency component of the signal which is of interest for velocity measurements by LDA.

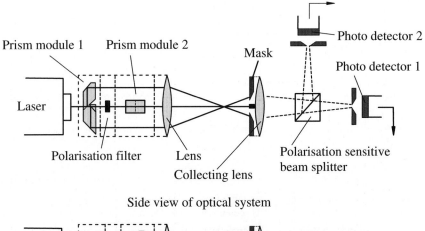

Fig. 21.44 Two-beam anemometer for two-component LDA measurements of particle velocity

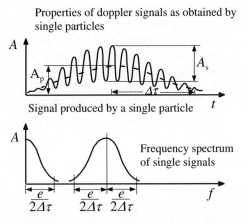

Fig. 21.45 Typical Doppler signal of a single scattered particle

In principle, it would be sufficient, for determining the Doppler frequency of a signal originating from a scattered particle, to measure the frequency by means of a spectrum analyser. The most important components of such an instrument are indicated in Fig. 21.46. This figure shows a block diagram of

21.7 Laser Doppler Anemometry

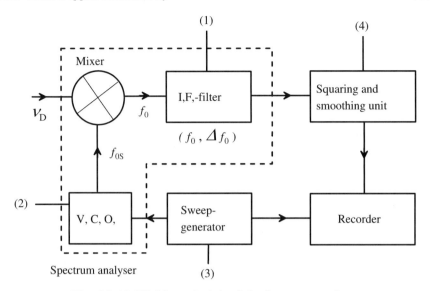

Fig. 21.46 Working principle of the frequency analyser

the most essential components for scanning the frequency range covered by the LDA photodetector signal.

In practice, scanning of the expected Doppler frequency range is done by a filter whose middle frequency is adjusted to a fixed frequency f_0. For detection of the Doppler frequency, the Doppler signal $s_k(t)$ is mixed with the signal $\cos 2\pi f_{0S} t$ of a voltage-controlled oscillator, and in this way the Doppler frequency range is scanned by varying the oscillator frequency. When the signal of the kth particle is represented by:

$$s_k(t) = a_k(t) \cos(2\pi \nu_k t + \phi_k) \tag{21.105}$$

for the momentary value of the mixer outlet (an analogue multiplier), the following signal holds:

$$s_M(t) = a_k(t) \cos(2\pi \nu_k t + \phi_k) \cos 2\pi f_{0s} t \tag{21.106}$$

The amplitude of the mixed signal is proportional to the amplitude of the photodetector signal and has frequency components $f_{0S} \pm \nu_k$.

Most of the spectrum analysers choose for displacement of the detected frequency, for their unambiguous operation, the lower of the two frequencies. When the Doppler frequency to be detected satisfies the conditions

$$f_0 - \frac{\Delta f_0}{2} \leq f_{0s} - \nu_k \leq f_0 + \frac{\Delta f_0}{2} \tag{21.107}$$

a signal passes the filter and reaches the squaring and integrating part of the electronics. After squaring and smoothing the frequency-analyser output signal, the result is recorded.

The voltage-controlled oscillator is driven by a sawtooth voltage, so that the frequency of the signal, transmitted by the mixer, increases linearly with time. As a consequence, the mixture of the Doppler frequencies ν_k and oscillator frequency, which contribute to the output signal of the analyser, also increases. When the same sawtooth voltage is used for triggering the x-basis of a plotter, the Doppler spectrum detected by the frequency analyser can be plotted. The calibration of the x-axis with respect to frequency is carried out with the voltage output of a suitable oscillator.

In order to be able also to carry out time-resolved laser Doppler measurements, so-called frequency tracking modulators can be employed. In contrast to the frequency analyser described above, they permit real-time detection of the Doppler signal. Frequency tracking demodulation yields an analogue signal whose voltage is always proportional to the component of the local fluid velocity which the optical system detects.

Another part of the signal processing with analogous instruments provides the statistical description of the flow velocity, e.g. via the mean velocity and the components of the fluctuation velocity, and also quantities which cannot be obtained by a frequency analysis, such as the turbulence spectrum and the autocorrelation function of the velocities. Difficulties result, however, from the non-ideal mode of operation of a tracker. They are caused, e.g. by the often discontinuous signal of the photodetector, which comes from the fact that only single scattered particles are traversing the measuring volume. With this demodulated output signal, one does not receive, at every point in time, information on the momentary fluid motion. Velocity measurements can only be carried out when a scattering particle is in the measuring volume. The intermittent measurements can lead to erroneous velocity statistics. Fluctuations of the recorded Doppler frequencies can also be caused by other reasons than those of velocity fluctuations in the fluid (e.g. by broadening of the frequency spectrum due to the presence of the particles in the measuring volume for finite time). The statistical evaluation of the measurements, by the combination of an anemometer and a tracker, is made more difficult.

The essential components of a frequency tracking demodulator are represented in Fig. 21.47. This figure shows that the frequency analyzer, discussed before, and also the tracker, contain three equal components [frequency mixer, band-pass filter and voltage-controlled oscillator (VCO)]. Thus the above explanations, which deal with the properties of the bandpass filter and the mode of operation of the mixer, are also of significance for the present section on frequency trackers. The integrator of the tracker corresponds to the time averaging unit of the frequency analyser. For the discriminator, there is no comparable component in the frequency analyser. The tracker has, moreover, in contrast to the frequency analyser, a closed control circuit which drives the oscillator.

The output signal of the photodetector of an LDA-optical system, similarly to the processing in a frequency analyser, is mixed with the output signal of the VCO. Here a signal $s_M(t)$, results which is led through a narrow bandpass

21.7 Laser Doppler Anemometry

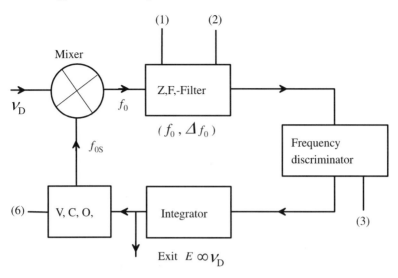

Fig. 21.47 Functional principle for offset-heterodyne tracker

filter. Thus, only those signal frequencies of the LDA signal, plus the VCO, are detected, which are located near the center frequency f_0 of the filter. Because of the small bandwidth of the bandpass filter (ZF-filter), the signal-to-noise ratio (SNR) of the LDA signal to be detected, improves considerably due to the narrowness of the band pass filter. A certain improvement of the signal-to-noise ratio can also be achieved by a filter in front of the mixer. However, one has to do this filtering with a filter bandwidth which is broad in relation to the Doppler frequency, in order to avoid attenuation of the amplitude of the Doppler signal in turbulent flows. The frequency discriminator generates a voltage which triggers the VCO in such a way that the modifications of the Doppler frequency by the VCO are compensated. This is explained below. The integrator controls the transient behavior due to individual LDA signals and the stability of the control circuit.

Trackers which operate with a narrow bandpass filter around the center frequency f_0 (see Fig. 21.47) function according to the offset-heterodyne principle. In laser Doppler anemometry, autodyne trackers have also been developed, of the kind shown in Fig. 21.48.

Whatever the actual working principles of the LDA frequency tracking demodulators are, they bring about LDA signals as sketched in Fig. 21.49. In this figure the individual Doppler bursts are shown as high-pass filtered signals, which serve as an input signal into the tracker. Every time the signal amplitude of an individual LDA signal exceeds a certain threshold value, the signal is fed to the tracker which then measures the frequency and thus changes the output voltage of the preceding signal. The output signal of the tracker therefore has the form of steps, each step being achieved by a new

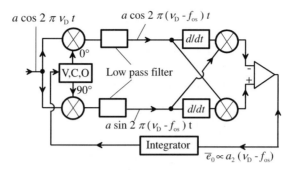

Fig. 21.48 Functional principle of an autodyne tracker for determining the LDA signal frequency

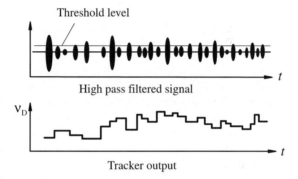

Fig. 21.49 Typical input and output signals of trackers for LDA measurements

measurement of the laser Doppler frequency. This tracker output signal, in the form of steps along the time axis, must not be considered as being disadvantageous for measuring the statistically averaged properties of a flow. The stepwise changes of velocity information take place, with sufficient concentration of particles, in time intervals which are much smaller than the characteristic times of the flow, i.e. the time scales at which velocity changes occur in the flow. It is therefore possible to integrate over several of the stepwise frequency changes by appropriate electronic devices, to achieve an averaged output signal for the actual velocity measurements. It is important to take this pre-averaging into account for precise Doppler measurements.

Finally, a signal-processing system that is extensively employed in laser Doppler anemometry, the so-called time period measurement system, has to be explained. Measurement systems of this kind are known as laser Doppler counters and are extensively applied for LDA measurements. They make use of signals which, after the photodetector, are fed, often after suitable amplification, to a bandpass filter. It is here, where the actual Doppler frequency separation from the low-frequency part of the signal takes place. The resultant signal is sketched at the top of Fig. 21.50. It is processed further in a special

21.7 Laser Doppler Anemometry

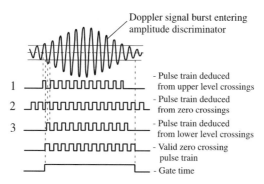

Fig. 21.50 Doppler frequency measurement with period measurement system

electronic system, in order to obtain a sequence of pulses which are also illustrated in Fig. 21.50. The entire signal processing of the counter-electronics functions as follows:

- The symmetrical signal produced by the input stage of the counter-system is processed by an amplitude discriminator and a zero-position detector to generate pulse sequences, which are fed to the several logic modules to check the validity of the frequency information. One possibility for generating the required pulse sequences that allow precise frequency measurements, is shown in Fig. 21.50. In practice, this pulse generation has proven to be a successful method for the production of the precise information on Doppler frequency.
- The signals of the two-level detectors (Schmitt triggers) and one zero-position detector are fed to an appropriate logic circuit to generate the pulse chains 1, 2 and 3. When the Doppler signal crosses the upper trigger level, it generates a Schmitt trigger output signal, which is used to provide a switch signal for a logic circuit whose output is then set to one. The zero-passage detector signal puts this output back to zero. The changing influence of the signals from the upper level detector and the zero-passage detector supplies pulse sequence 1, if the appropriate pulse passes, i.e. the LDA signal amplitude is satisfactory.
- The pulse sequence 2 is generated by the zero-passage detector only, by switching the output of an appropriate logic module in such a way that it switches to and fro between one and zero. A combination of the signals of the zero-passage detector with those of the Schmitt trigger of the lower level recognition, yields pulse sequence 3. The output of an appropriate logic module is set by the lower trigger level and put back by the zero-passage detector.
- If all three signal chains are present, one will obtain valid information and use those zero passages for which either output 1 or output 3 is set. In this way, the influence of multiple zero passages is suppressed for the largest part, i.e. good LDA measurements are obtained.

- The gate for the measurement of the duration of an LDA signal opens with the first valid zero passage and closes one zero passage after the pulse for which output 1 or output 3 was set. This corresponds to three zero passages at the end of a Doppler signal, which no longer traverse the upper or lower trigger level.

The above points show that it is possible, with the aid of counter-systems, to determine the number of zero passages of an LDA signal, and also the length of time during which the measured number of pulses is available. With this, a period-time measurement is possible, which leads to the desired Doppler frequency of the LDA signals. For each individual LDA signal arriving at the entrance of the counter-electronics with sufficiently high amplitude, a Doppler frequency can therefore be determined. The latter is now processed further, to obtain mean frequencies and standard deviations.

The measured individual frequencies of laser Doppler signals correspond to one velocity component, i.e. to $U_j(x_i, t)$. These individual measurements have now to be processed further, in order to determine the mean frequency and the RMS values of the existing deviations from the mean frequency:

$$< f_D > = \lim_{N \to \infty} \frac{1}{N} \sum_{k=1}^{N} (f_D)_k \qquad (21.108)$$

and

$$< \Delta f_D^2 > = \lim_{N \to \infty} \frac{1}{N} \sum [(f_D)_k - < f_D >]^2 \qquad (21.109)$$

These averaged quantities can be determined easily, as the knowledge of the individual frequencies, required for averaging, is known from the Doppler measurements. However, there is a difference between the *time-averaged quantities* which are of importance in turbulence, and the *particle-averaged flow quantities* which can be determined from (21.108) and (21.109). This can be explained in a simple way by means of a sketch of a temporal hypothetical flow, as shown in Fig. 21.51.

This flow shows a mean motion which is generated by the horizontally operating piston. Additional flows occur, which once show positively, once negatively imposed step changes by the motion of the vertically operating piston. This leads, at the measurement point, to a constant mean velocity with superimposed step-like flow changes.

Assuming an equal distribution of the scattered particles, it is apparent that the number of particles which pass the measuring volume depends on the actual flow velocity, and this can be expressed as follows:

$$N = c_v \mid U_\perp \mid \mid A_v \mid \qquad (21.110)$$

The number of the measured particles is proportional to the concentration of the particles in the fluid, and proportional to the flow velocity vertical to the surface of the control volume, and of course also proportional to the

21.7 Laser Doppler Anemometry

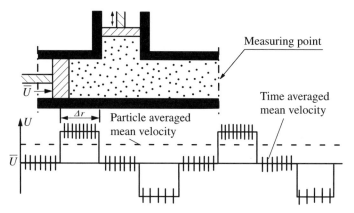

Fig. 21.51 Dependence of the particle arrival in the measuring volume on the flow velocity

surface itself. This makes it understandable why in Fig. 21.51 more particles appear at higher velocities than at lower velocities. This leads to a value of the particle-averaged velocity that is higher than the fluid time mean velocity, as indicated in Fig. 21.51. This fact was often called a "biasing error" in Laser Doppler anemometry and was presented as a principal problem of the LDA measurement technique. The above explanations make it clear that this only has to do with the fact that one usually determines ensemble-mean values, due to their easy determinability from the LDA signals. However, in most fluid flow studies mean values in terms of time, which are of importance in turbulence research, need to be measured. This difference between ensemble and time averages represents the biasing. This leads to the differences between the mean values with respect to time and with respect to particles.

In fluid mechanics, it is usual to determine time averages, e.g. for determining mean quantities of turbulent flows, which are computed as follows:

$$\overline{f_D} = \lim_{T \to \infty} \frac{1}{T} \int_0^T f_D \, dt \quad \text{and} \quad \overline{\Delta f_D^2} = \lim_{T \to \infty} \frac{1}{T} \int_0^T \Delta f_D^2 \, dt, \qquad (21.111)$$

where $\Delta f_D = (f_D(t) - \overline{f_D})$. The above integration can be carried out digitally for irregular scanning intervals Δt_k, as shown in Fig. 21.52. This leads to the following general equations:

For the time average for $\Delta t_k \neq$ constant:

$$\overline{f_D} = \lim_{N \to \infty} \frac{\sum_{k=1}^{N} (f_D)_k \Delta t_k}{\sum_{k=1}^{N} \Delta t_k} \quad \text{with} \quad T = \sum_{k=1}^{N} \Delta t_k \qquad (21.112)$$

Signal processing with time interval $\Delta t_k \neq$ constant

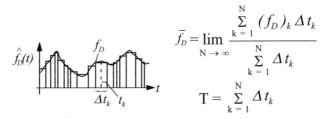

$$\bar{f}_D = \lim_{N \to \infty} \frac{\sum\limits_{k=1}^{N} (f_D)_k \Delta t_k}{\sum\limits_{k=1}^{N} \Delta t_k}$$

$$T = \sum_{k=1}^{N} \Delta t_k$$

Signal processing with time interval $\Delta t_k =$ constant

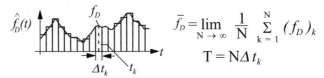

$$\bar{f}_D = \lim_{N \to \infty} \frac{1}{N} \sum_{k=1}^{N} (f_D)_k$$

$$T = N \Delta t_k$$

Fig. 21.52 Determination of the time average of the Doppler frequency for $\Delta T_k =$ constant and $\Delta t_k \neq$ constant

$$\overline{f_D} = \lim_{N \to \infty} \frac{1}{N} \sum_{k=1}^{N} (f_D)_k \quad \text{with} \quad T = N\Delta t \tag{21.113}$$

and for the moments which describe the divergence from the mean value the integration holds that

$$\overline{\Delta f_D^n} = \lim_{N \to \infty} \frac{\sum\limits_{k=1}^{N} \left[(f_D)_k - \bar{f}_D\right]^n \Delta t_k}{\sum\limits_{k=1}^{N} \Delta t_k} \tag{21.114}$$

For a stationary random process, the above equations for mean-value determination conserve their validity for all scanning intervals Δt_k, provided that the scanning process and the scanned quantity (Doppler frequency) are not correlated with one another. The dissolution of a certain frequency in a flow requires that the Δt_k values are small compared with the time measure of the flow to be registered.

When the scanning intervals Δt_k are chosen to be constant, i.e. $\Delta t_k = \Delta t =$ constant, the above equations simplify to

$$\overline{f_D} = \lim_{N \to \infty} \frac{\sum\limits_{k=1}^{N} (f_D)_k \Delta t_k}{\sum\limits_{k=1}^{N} \Delta t_k} = \lim_{N \to \infty} \frac{1}{N} \sum_{k=1}^{N} f_D = <f_D> \tag{21.115}$$

21.7 Laser Doppler Anemometry

$$\overline{\Delta f_D^n} = \lim_{N \to \infty} \frac{\sum_{k=1}^{N} \left[(f_D)_k - \overline{f_D}\right]^n \Delta t_k}{\sum_{k=1}^{N} \Delta t_k} = \lim_{N \to \infty} \frac{1}{N} \sum_{k=1}^{N} \left[(f_D)_k - \overline{f_D}\right]^n = <f_D^n> \quad (21.116)$$

In accordance with this, the time-averaged and the ensemble-averaged properties of a flow agree with one another in the special case that constant averaging time intervals are chosen. It is therefore necessary to process the obtained Doppler signals accordingly, so that one corresponding Doppler signal is attributed to a particular constant time interval Δt.

The above explanations of the behavior of LDA signals have shown that the laser Doppler signals occur at irregular time intervals, so that irregular scanning intervals are given by the Doppler signals. When this is not taken into account, the time-averaged Doppler frequency and the corresponding ensemble-averaged value can differ from one another. This is not an error of the measurement technique, but a characteristic of the chosen averaging process. The differences that occur are clear and follow from the definitions of the mean values of ensemble and time. Hence, there are no principle problems to measure biasing free averaged velocities in fluid flows.

21.7.4 Execution of LDA-Measurements: One-Dimensional LDA Systems

In the preceding sections on LDA signal processing, LDA measurements were presented without entering into the data obtained by the evaluation, in order to obtain from the measured Doppler frequencies the desired information on the flow field. It was shown that laser Doppler anemometers are linear velocity measurement value sensors with a frequency response that is given by the following equation:

$$\hat{f}_D = \frac{1}{\lambda} \hat{U}_i n_i, \quad (21.117)$$

where λ corresponds to the wavelength of the chosen laser radiation, $\hat{U}_i$ to the momentary velocity vector and $(n)_i$ to the transformation vector of the anemometer. For a stable coordinate system x_i, the two vectors of the above equation can be expressed in the following form:

$$\left(\hat{U}\right)_i = (U_1, U_2, U_3) \quad \text{and} \quad (n)_i = 2\sin\varphi \left(\cos\alpha_1, \cos\alpha_2, \cos\alpha_3\right), \quad (21.118)$$

where φ is half the angle between the light beams and α_1, α_2 and α_3 are the angles which the vector $(n)_i$ forms with the coordinate axes (Fig. 21.53). The momentary velocity components and the momentary signal frequency can be expressed as follows:

$$\hat{U}_i = U_i + u_i, \qquad \hat{f}_D = f_D + \Delta f_D \quad (21.119)$$

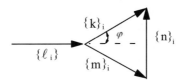

Fig. 21.53 Important parameters for the evaluation of LDA signals in measurements with one-dimensional LDA measurement systems

By combining the above equations, the following relation results:

$$(\bar{f}_D + \Delta f_D) = \frac{2\sin\varphi}{\lambda}(U_1\cos\alpha_1 + U_2\cos\alpha_2 + U_3\cos\alpha_3 \\ + u_1\cos\alpha_1 + u_2\cos\alpha_2 + u_3\cos\alpha_3) \tag{21.120}$$

This equation represents the basic relation for the evaluation of fluid mechanical quantities from frequency measurements of laser Doppler signals.

On carrying out a time average for (21.120), one obtains the following basic equation for the determination of the three velocity components U_1, U_2 and U_3. To be able to measure all three components with a one-dimensional LDA system, measurements in three different directions, α_1, α_2, and α_3 are required:

$$\bar{f}_D = \frac{2\sin\varphi}{\lambda}(U_1\cos a_1 + U_2\cos a_2 + U_3\cos a_3). \tag{21.121}$$

On deducing (21.120) from (21.121), one obtains the equation for the divergence of the frequency from the averaged frequency. With this an equation can be derived for the standard divergence from the Doppler frequency Δf_D^2:

$$\overline{\Delta f_D^2} = \frac{4\sin^2\varphi}{\lambda^2}\left[\overline{u_1^2}\cos^2 a_1 + \overline{u_2^2}\cos^2 a_2 + \overline{U_3^2}\cos^2 a_3 \\ + 2(\overline{u_1 u_2}\cos a_1 \cos a_2 + \overline{u_1 u_3}\cos a_1 \cos a_3 + \overline{u_2 u_3}\cos a_2 \cos a_3)\right] \tag{21.122}$$

The measurement of the standard divergences of the signal frequencies for six different directions of the sensitivity vector $(n_j)_i$ ($j = 1, 2, \ldots, 6$) makes possible the evaluation of all correlations $\overline{u_i u_j}$ of second order.

Similar evaluating equations can be derived for the correlations of higher order:

$$\overline{\Delta f_D^n} = \left(\frac{2\sin\varphi}{\lambda}\right)^n \overline{[u_1\cos a_1 + u_2\cos a_2 + u_3\cos a_3]^n}$$

The complexity of the evaluating equations reduces, when preferential directions are given for the transformation vectors $(n)_i$, e.g. parallel to the x_1-axis $(n)_i = (1, 0, 0)$:

$$\overline{U_1} = \frac{\lambda \bar{f}_D}{2\sin\varphi}; \quad \overline{u_1^2} = \frac{\overline{\Delta f_D^2}\lambda^2}{4\sin^2\varphi}; \quad \overline{u_1^n} = \frac{\overline{\Delta f_D^n}\lambda^n}{2^n \sin^n\varphi} \tag{21.123}$$

With this, the evaluation of the measurements which are carried out with a one-dimensional optic, is similar to the measurements with a single hot wire. However, laser Doppler anemometers are characterized by a precise Cosine-law response behavior and therefore the evaluating equations are easier. More details about Laser Doppler anemometry can be found in a recent published book by Albrecht et al. [21.10] with details about the physics of the measuring technique and good descriptions about LDA-applications.

References

21.1. Hinze JO (1975) "Turbulence", 2nd edition, McGraw-Hill, New York
21.2. Hanson S (1976) "Visualization of alignment errors and heterodyning constraints in laser Doppler velocimeters". Proceedings of the LDA Symposium, Copenhagen, 1975, pp. 176–182
21.3. Buchhave P, Delhaye JM, Durst F, George WK, Refslund K, Whitelaw JH (eds.) (1976) "The accuracy of flow measurements by laser Doppler methods". Proceedings of the LDA Symposium, Copenhagen, 1975.
21.4. Durst F, Stevenson WH (1979) "Influence of Gaussian beam properties on laser-Doppler signals", Applied Optics, 18, pp. 516–524
21.5. Goldstein RJ, (ed.) (1983) "Fluid Mechanics Measurements", Hemisphere, Washington, DC
21.6. Wiedemann J (1984) "Laser-Doppler-Anemometrie", Springer, Berlin Heidelberg New York
21.7. Ruck B (1987) "Laser-Doppler-Anemometrie", AT Fachverlag, Stuttgart
21.8. Durst F, Melling A, Whitelaw JH (1987) "Theorie und Praxis der Laser-Doppler-Anemometrie", G. Braun, Karlsruhe
21.9. Eckelmann H (1997) "Einfuehrung in die Stroemungsmesstechnik", BG Teubner, Stuttgart
21.10. Albrecht HE, Borys M, Damaschke N, Tropea C (2003) "Laser Doppler and Phase Doppler Measurement Techniques", Springer, Berlin Heidelberg New York

Index

Aerostatics, 183–187, 189
  buoyancy, 188–190
  stability, 191, 192
Axi-symmetric film flow, 376–378

Balance considerations, 73–75
Bernoulli equation, 130, 133, 134, 144, 145, 231, 232, 234, 241, 243, 254–258, 302, 304, 307, 321, 326, 341–344, 507
Blasius solution, 470–474, 479, 481
Boundary layer, 481
  equations, 12, 463–493, 588, 641–651
  flows, 209, 210, 463–493, 520, 521, 578, 641, 647
  with heating, 641–651
Bubble formation on nozzles, 175–183

Channel flow, 207, 369–371, 388–390, 407, 417–426, 489–492, 514–516, 519, 550, 568–572, 575, 578–585
Characteristic function, 516, 527, 537, 538
Classifications of instabilities, 44, 282, 517–519
Complex functions, 42–47
Complex numbers, 11, 36–43, 45
Computation
  laminar flows, 614, 615
  turbulent flows, 616–626
Connected containers, 163–168
Conservation laws, 73, 75
  for chemical species, 74, 113, 114
  for energy, 74, 113, 114
  for mass, 74, 113, 114
  for momentum, 74, 113, 114

Continuity equation, 74, 115–119, 135, 136, 148, 150, 153, 197, 200, 209, 221–224, 231, 232, 234, 251–253, 257, 258, 260, 261, 276–279, 313–316, 332, 353, 356, 362–366, 432, 466, 469, 471, 509, 513, 543–546, 558, 560, 561, 568, 592, 603, 607–609, 616, 630, 631, 641, 642, 648
Continuum mechanics, 51–55, 57–59, 61, 62, 64, 116, 678
Convection influence, 459–461, 463–465
Converging channel flow, 489–492
Coordinate systems, 17–20, 32–36, 120–124, 135–142, 184, 361–364, 443, 446, 471, 598, 635, 685, 691
Corner and sector flows, 284–288
Correlations of turbulence fluctuations, 538–542
Creeping fluid flows, 431–433
Crocco equation, 142, 146, 147

Deformation operation, 104–107, 110–112
Different wave motions, 311, 312
Differential equations, 75, 87, 89, 90, 92, 93, 96, 113–151, 275–286, 313, 316–318, 321–324, 332, 351, 365, 367, 370, 373–377, 379, 383, 384, 386, 388, 394–397, 400, 402–404, 406–409, 414–419, 422–425, 430–432, 447–449, 451, 454–456, 464, 465, 468–472, 480–484, 487, 488, 490, 491, 496, 501–515, 557–559, 576, 577, 579, 587–590, 595, 596, 632, 634, 638–642, 644–646, 648, 653, 690

719

boundary layers, 210, 464, 465, 468, 470, 490, 491, 520, 588, 642, 645, 646
chemical species, 150
dimensions forms, 638
fluid flows, 8, 123, 199, 208–211, 332, 386, 431–433, 642–648, 653
heat transfer, 197–204, 627, 628, 638, 642
molecular transport, 197, 198
numerical solutions, 587–590, 595, 596, 600, 612, 614
potential flows, 275, 281, 293
wave motions, 313, 316–318, 321, 323, 324
Differential operators, 32–36
Diffusion
influence, 459–461
of vortex layer, 399–402
Dimensional analysis, 212–218
Dipole generated flow, 291–294, 296
Discretization
in space, 600–611
of derivatives in space, 589
of derivatives in time, 589, 611–613
Divergence operator, 28
Dynamic pressures, 259, 343, 655, 660, 662

Eddy viscosity models, 560, 562, 564, 565, 573, 576
Electronic systems, 705
Elementary complex functions, 42, 43
Euler's turbine equations, 242–244
Exit velocity of nozzles, 231, 232

Fanno relation, 351
Field variables, 11, 23, 24, 27, 54, 75, 78, 83, 84, 115, 118, 146, 221
Film flow on an inclined plate, 372–376
Film flows with two layers, 386–388
Finite difference discretization, 589, 595–598
Finite volume discretization, 598
Flat plate boundary layer, 209, 470, 473, 474, 479, 514, 521, 645
Flow
forces on bodies, 302–307
with heat transfer, 344–350, 627–633, 635, 644
between vertical plates, 633
Flow around a cylinder
with circulation, 296, 306, 307
without circulation, 297

Flow between two cylinders, 383–385
Fluid flow between plates, 369–371
Fluid flow measurements, 210, 653, 669, 704
Fluid Mechanics
analytical, 4, 7, 8, 11, 73
development of, 1–14
experimental, 5, 7, 8, 655, 660, 664
historical developments of, 9–14
numerical, 5, 8, 9, 13, 14, 83, 591
significance, 1–4
sub-domains, 4–8, 12, 85
Force
on bodies, 302–307
on periodical grid of plates, 240–242
on turbine plate, 239, 240
Formation of shock waves, 338
Free fluid surface, 160, 168, 169
Free shear layer, 480, 481
Fully developed flow in channels, 630

Gas dynamics, 8, 70, 331–360
Gauss complex number plane, 38, 40
Gauss law of integration, 31, 32
General transport equation, 55, 591–593, 595, 597, 614, 618, 620
Geometric similarity, 193
continuity equation, 197–200, 209
mechanical energy equation, 198
momentum equation, 197–199, 205, 209
thermal energy equation, 194, 198
Gradient operator, 27, 28
Gravitationally caused instabilities, 505

Hagen-Poiseuille flow, 379–383
Heat transfer, 196–198, 201–204, 211, 344, 347, 627–630, 637, 641, 642, 645, 647, 649, 654, 664, 667, 670, 671, 676–681, 684
Hot wire anemometry, 664, 665, 678, 679, 687
cooling laws, 676
probes, 672
static calibration, 680
supports, 672
Hydrostatics, 10, 153–156, 158, 161, 183, 188, 190

Ideal gases, 23, 50, 58, 59, 61, 62, 64–66, 76, 79–81, 116, 123–125, 127, 129, 133, 134, 183, 184, 198, 262, 265, 310, 313, 314, 331, 332, 334, 345, 347, 349, 354, 355, 359–361

Important potential flows, 299
Incompressible flows, 257–273
  complex function, 280–282
  potential flows, 277, 280
  potential function, 276–282
  stream function, 98, 275–280
Integral laws, 29–32
Integral properties, 474–480

Jet deflection at an edge, 236
Jet flow, 481–486

Kinematic similarity, 194, 198
Kinetic energy of turbulence, 550, 555

Laminar free jet, 481–486
Laminar-turbulent transition, 12, 495, 519–522
Laplace operator, 26–28, 34
Large Reynolds number flows, 463–492
Laser Doppler anemometry, 7, 13, 558, 559, 565, 653, 655, 694–696, 701, 709, 710, 713, 717
Length scales, 207, 208, 210, 553–556, 562–564, 576, 579, 624
Longitudinal waves, 310, 313–318
Lubrication films, 433, 434, 436

Mach cone, 335, 337, 338
Mach lines, 335
Mass conservation, 73–75, 99, 100, 114–117, 221–224, 243, 256–258, 344, 345, 607, 616
Mean flow field, 125, 383, 433, 543–546, 548–554, 558, 564, 572–574
Mechanical energy equation, 128, 130, 144, 145, 221, 225, 226, 228, 245, 255, 256, 309, 546, 547, 560, 627
Mixing process in pipe, 237
Molecular heat transport, 83
Molecular mass transport, 83
Molecular momentum transport, 69, 83, 123, 125, 140, 141, 148, 200, 209, 225, 253, 354, 371, 385, 388, 394, 397, 398, 401, 453, 459, 480, 486, 588
Molecular properties, 51, 52, 54, 64, 67
Momentum
  equations, 85, 119, 120, 123, 127, 128, 130, 137–139, 144, 197–199, 209, 253, 275, 277, 303, 313, 338, 356, 362, 363, 365, 366, 459, 496, 509, 546, 617, 618, 628, 634, 637, 646, 649, 653

  on inclined plane plate, 234–236
  on plane vertical plate, 232–234
Motion of fluid elements, 85–89
  deformation, 104–107, 110–112
  divergence, 101–103
  path lines, 86–89
  relative motions, 108–112
  rotation, 104–107, 110, 111
  streak lines, 90–94
  stream function, 98–101
  stream lines, 94–101
  translation, 104–107, 110–112

Natural convection, 191, 633–636, 638, 640, 647–649
Navier-Stokes equations, 12, 123–128, 136–144, 146, 183, 277, 312, 361, 381, 430, 446, 449, 450, 464–466, 468, 476, 497, 501, 505, 509, 513, 542, 544, 545, 557, 560–562, 565, 592, 614, 641, 653
Newton's second law, 119–123
Newtonian fluids, 4, 8, 55, 69, 124, 127, 137, 197, 201, 206, 221, 364, 369, 381, 544, 546, 562, 593, 650
Non-linear propagation, 338–341
Numerical solutions, 5, 6, 8, 13, 473, 476, 501, 516, 526, 559, 560, 562, 587, 588, 590–594

One-dimensional flows and viscous fluids, 393, 394, 396, 397, 406, 426
One-equation models, 563, 573, 576
Optical systems, 696, 697, 700–706, 708
Orthogonal coordinates, 32–36
Oscillating fluid flows, 414–416
Outflow from containers, 230, 231

Partial derivative, 85, 280, 394, 400
Pipe flow, 200, 252, 258, 344–352, 355, 379–383, 385, 407, 424, 426, 572, 578, 584
Plane Couette flow, 366–369
Plane plate boundary layer, 641–644
Plane progressive waves, 325–329
Plane standing waves, 323–325
Potential flows, 275–308
Potential vortex flow, 288–291
Power of flow machines, 245–247
Pressure gradient driven fluid flows, 417–426
Pressure measuring instruments, 163–168, 657, 661, 662

Probability density, 527, 530–538
Properties of hot wires, 654, 667–672

Rankine–Hugoniot equation, 355–360
Rayleigh relation, 351–355
Reynolds equations, 542, 544–546, 557, 558, 562, 568, 569, 575, 579, 653
Roller bearing, 438–443
Rotating containers, 187, 188, 432
Rotational motion of cylinder, 451–453
Rotation operator, 27–29
Rotation of sphere, 443–445
Rules of time averaging, 120, 528–530

Scalars, 16, 17, 24
Similarity solution, 644–647
Similarity theory, 193, 197, 198, 212, 678
Small Reynolds number flows, 429–431, 447, 450, 454, 458, 461, 463, 496, 556, 557, 560
Solids and Fluids, 49–51
Sound waves in gases, 313–318
Source or sink flows, 288–291
Spectra of turbulent flows, 538–542
Stability of Flows, 517, 519, 520
Stagnation pressure, 662–664
Starting flow in channel, 417, 418, 420–422
Starting pipe flow, 422–426
Static pressures, 655–660
Statistical considerations, 528
Stereographic projection, 41, 42
Stokes' first problem, 397, 399
Stokes' law of integration, 31
Stokes' second problem, 414
Stream line, 94–98
Stream tube theory
    Bernoulli equation, 254–258
    continuity equation, 251–253, 257, 258, 260, 272
    momentum equation, 253, 254, 258, 259
    total energy equation, 256
Substantial derivative, 26, 27, 76, 84, 85, 117
Substantial variables, 27
Surface tension, 168–172, 196, 310, 326, 329, 505

Surface waves, 318, 320–322, 327, 330, 517

Taylor instability Orr-Sommerfeld equation, 512
Tensors
    first order, 15, 17–21, 24
    higher order, 23, 630
    second order, 15, 21–23
    zero order, 15–17, 24
π-Theorem, 212–214
Theory of Laser Doppler anemometry, 694
Thermal energy equation, 128, 130, 132, 133, 147, 228, 229, 257, 313, 561, 627
Thermodynamic considerations, 76–81, 146
Time dependent flows, 393, 394, 426
Time scales, 333, 538–542, 553, 554, 571, 572, 650, 651, 710
Time scales of turbulence, 538, 539
Total derivative, 85
Transitional boundary layer flows, 519
Translatory motion
    of cylinder, 453–459
    of sphere, 445–451
Transport equation for vorticity, 143, 459, 593, 614
Transport processes, 1, 12, 15, 55–58, 62, 65, 194–196, 202, 393, 464, 469, 495, 496, 523, 565, 566, 612
Transversal waves, 310, 318
Treatment of source terms, 613
Turbulence
    flows, 12, 13, 85, 385, 496, 497, 499, 500, 523–585, 616–620, 624, 653, 654, 661, 662, 673, 685, 687, 691, 713
    measurements, 685–694
    models, 14, 555, 557, 558, 560, 562–565, 573–578, 617–620, 653
    wall boundary layers, 572, 578, 582, 584
Two-equation models, 563, 564
Two-phase plane channel flow, 388–390

Unstable Flows, 495, 497, 501, 517

Vectors, 15, 17–26, 28, 30–35, 39, 57, 86, 94, 95, 101–105, 109, 110, 143, 144, 147–149, 154, 158, 168, 222, 223, 276, 277, 336, 355, 526, 686, 687, 695, 696, 715, 716

Velocity scales, 430, 554, 555, 567
Viscosity of Fluids, 69–73

Wake flow, 486–489, 500
Wall boundary
conditions, 579, 615, 616, 620

layers, 210, 464, 474–479, 514, 520, 559, 574, 578–585, 654
Wave motions, 309–330
Wave propagation, 312, 318, 337, 338

Zero-equation models, 576

Printing: Krips bv, Meppel, The Netherlands
Binding: Stürtz, Würzburg, Germany